NATO ASI Series

Advanced Science Institutes Series

A series presenting the results of activities sponsored by the NATO Science Committee, which aims at the dissemination of advanced scientific and technological knowledge, with a view to strengthening links between scientific communities.

The Series is published by an international board of publishers in conjunction with the NATO Scientific Affairs Division.

A Life Sciences B Physics	Plenum Publishing Corporation London and New York
C Mathematical and Physical Sciences D Behavioural and Social Sciences E Applied Sciences	Kluwer Academic Publishers Dordrecht, Boston and London
F Computer and Systems Sciences G Ecological Sciences H Cell Biology I Global Environmental Change	Springer-Verlag Berlin Heidelberg New York Barcelona Budapest Hong Kong London Milan Paris Santa Clara Singapore Tokyo

Partnership Sub-Series

1. Disarmament Technologies	Kluwer Academic Publishers
2. Environment	Springer-Verlag
3. High Technology	Kluwer Academic Publishers
4. Science and Technology Policy	Kluwer Academic Publishers
5. Computer Networking	Kluwer Academic Publishers

The Partnership Sub-Series incorporates activities undertaken in collaboration with NATO's Cooperation Partners, the countries of the CIS and Central and Eastern Europe, in Priority Areas of concern to those countries.

NATO-PCO Database

The electronic index to the NATO ASI Series provides full bibliographical references (with keywords and/or abstracts) to about 50 000 contributions from international scientists published in all sections of the NATO ASI Series. Access to the NATO-PCO Database compiled by the NATO Publication Coordination Office is possible in two ways:

- via online FILE 128 (NATO-PCO DATABASE) hosted by ESRIN, Via Galileo Galilei, I-00044 Frascati, Italy.

- via CD-ROM "NATO Science & Technology Disk" with user-friendly retrieval software in English, French and German (© WTV GmbH and DATAWARE Technologies Inc. 1992).

The CD-ROM can be ordered through any member of the Board of Publishers or through NATO-PCO, B-3090 Overijse, Belgium.

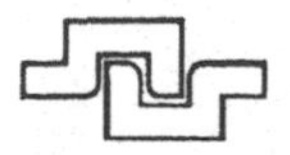

Series F: Computer and Systems Sciences, Vol. 153

Springer
Berlin
Heidelberg
New York
Barcelona
Budapest
Hong Kong
London
Milan
Paris
Santa Clara
Singapore
Tokyo

Identification, Adaptation, Learning

The Science of Learning Models from Data

Edited by

Sergio Bittanti

Politecnico di Milano
Piazza Leonardo da Vinci 32
I-20133 Milano, Italy

Giorgio Picci

Università di Padova
Via Gradenigo 6/A
I-35131 Padova, Italy

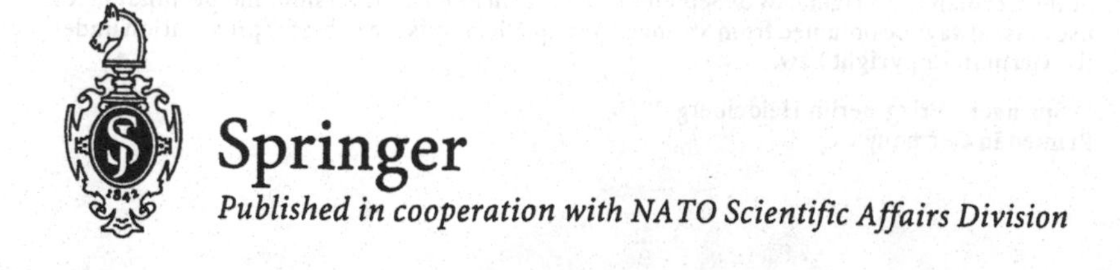

Springer

Published in cooperation with NATO Scientific Affairs Division

Proceedings of the NATO Advanced Study Institute "From Identification to Learning", held in Como, Italy, August 22 – September 2, 1994

Library of Congress Cataloging-in-Publication Data

Identification, adaptation, learning : the science of learning models from data / edited by Sergio Bittanti, Giorgio Picci.
p. cm. -- (NATO ASI series. Series F, Computer and systems sciences ; vol. 153.)
"Published in cooperation with NATO Scientific Affairs Division.
"Proceedings of the NATO Advanced Study Institute 'From Identification to Learning', held in Como, Italy, August 22-September 2, 1994"--T.p. verso.
Includes bibliographical references.

1. System identification--Congresses. 2. Linear models (Statistics)--Congresses. 3. Learning models (Stochastic processes)--Congresses. I. Bittanti, Sergio. II. Picci, Giorgio, 1942- . III. North Atlantic Treaty Organization. Scientific Affairs Division. IV. NATO Advanced Study Institute 'From Identification to Learning' (1994 : Como, Italy) V. Series: NATO ASI series. Series F, Computer and systems sciences ; no. 153.
QA402.I327 1996
003'.74'0111--dc20 96-14783
CIP

CR Subject Classification (1991): I.5–6, G.1, G.3, I.2, J.2

ISBN 978-3-642-08248-1

Printed in Germany

Dedicated to the memory of
E.J. Hannan (1921–1994)

Photograph by Peter Hall
taken at the farewell to Geof Watson
Princeton University, 1992

Preface

This book collects the lectures given at the NATO Advanced Study Institute *From Identification to Learning* held in Villa Olmo, Como, Italy, from August 22 to September 2, 1994.

The school was devoted to the themes of *Identification, Adaptation* and *Learning*, as they are currently understood in the Information and Control engineering community, their development in the last few decades, their interconnections and their applications. These titles describe challenging, exciting and rapidly growing research areas which are of interest both to control and communication engineers and to statisticians and computer scientists.

In accordance with the general goals of the Institute, and notwithstanding the rather advanced level of the topics discussed, the presentations have been generally kept at a fairly tutorial level. For this reason this book should be valuable to a variety of rearchers and to graduate students interested in the general area of Control, Signals and Information Processing. As the goal of the school was to explore a common methodological line of reading the issues, the flavor is quite interdisciplinary. We regard this as an original and valuable feature of this book.

During the two weeks of the school at Villa Olmo we have experienced a unique atmosphere and a most remarkable climate of interaction and communication between the outstanding experts gathered in Como for this occasion. It is remarkable that some of them hardly ever meet at conferences or at scientific meetings as their different fields have traditionally evolved along separate lines. The openness and active participation in discussions by both students and speakers, was a major point for the success of this Advanced Study Institute. The editors of this volume would like to thank all lecturers, and the remaining members of the Organizing Committee, S.K. Mitter and Jan C. Willems, for their helpful advice.

The superb local organization provided by the *Centro di Cultura Scientifica Alessandro Volta* deserves a primary acknowledgement, for it was a major factor for the smooth development of the school. Special thanks go to Manuela Troglio for her care in the general organizational aspects, and to Emanuela Salati for her kindness and patience in dealing with daily problems of students and teachers.

Last but not least, we would like to thank NATO for believing in this project and for the generous support of the Institute, and the Consiglio Nazionale delle Ricerche (CNR) of Italy which also provided financial funding. The general support of the *Dipartimento di Elettronica e Informazione* of the *Politecnico di Milano* is also gratefully acknowledged.

The final software layout of the book is due to Stefano Bertoncello, with the assistance of Marco Lovera.

March 1996 Sergio Bittanti and Giorgio Picci

Table of Contents

Introduction

The importance of identification, that is, automatic construction of mathematical models of dynamical systems from observed data, has grown tremendously in the last two decades. Identification techniques, in particular recursive identification techniques, have found application in diverse fields like real-time automatic control and communication systems, econometrics, geophysics, hydrology, civil engineering, bioengineering, to name just the principal areas. Naturally, the pervasive use of mathematical models in modern science and technology has been greatly stimulated by the massive diffusion of computers. One could safely say that the progress of microelectronics and computer hardware and the dramatic increase of real-time computing power available in the 1990s are leading to a shift of paradigms in the design of engineering systems. To cope with the growing complexity and the rising demand for sophistication and performance, the design of modern communication and control systems has to be based more than ever before on *quantitative models* of the signals and systems involved. For example, on-line identification algorithms have become a key ingredient in signal processing, where there is a growing demand for procedures which are adapted to the specific dynamic structure of the signal. Early examples of successful application of this principle have been model-based coding and recognition of audio and video signals. Devices based on these ideas are now part of commercially available communication systems.

The majority of physical systems, however, tend to change their dynamic characteristics in time. This change takes usually place quite slowly as compared to the dynamics which is being monitored by the identifier, but in a rather unpredictable way, either because of an unpredictable "aging" process of the system or because of erratic changes of the environment and of the operating conditions. Consequently one has to face the situation where the identified model has to be changed (in real time) so as to track closely the variations of the dynamic characteristics of the system. In modern fighter airplanes, for example, these variations can be quite dramatic and may correspond to differences of several orders of magnitude in some parameters or even in the appearance of additional dynamic terms in the model.

Many control and communication systems are required to operate in a condition of uncertain time-varying dynamic behavior of the plant or of the channel. The design of such systems calls then for controllers or filters which are able to *adapt* to the changing process and/or environment and hence in one way or another must include a logical unit (a subsystem) capable of changing the parameters or even the design philosophy of the controller (or of the filter) and adapt it to the current best mathematical model available. The analysis and design of such systems involves challenging theoretical problems, some of which are still unsolved. This is currently a very active area of systems and control theory.

Although identification and adaptation may resemble two basic steps of the biological process of "learning", the actual *theory of learning* has emerged recently quite independently, as a branch of nonparametric statistics. The important concepts here are a new notion of complexity and structural conditions on an inference problem which permit identification of the "true system" independently of the *a priori* model class one may choose to describe the data. The fundamental work of Vapnik and his collaborators on this new notion of complexity has opened the way to intensive research and much excitement. In this vein, learning and the recent emphasis on model-free (or non-parametric) statistics have provided a further impetus to the study of *neural networks* and more generally of *wavelets* and *wavelet-networks*. The in-depth study of these "universal" nonlinear model classes and of the special identification algorithms which serve for modeling observed data in this setting, has become an important part of modern identification theory. This is reflected in some of the lectures presented in the school.

The popularity of the new distribution-free ("neural-network") procedures for identification and adaptation depends on the relative ease of implementation and on the conceptual simplicity of the approach. Neural networks look like manageable nonlinear model classes which can be adapted to describe almost anything. For this reason there has been a substantial impact and a remarkable widening of the scopes in identification and signal processing. It is now possible to attack nonlinear problems which could not have been approached by the classical linear-model theory available in the past. It should be said, however, that the meaning of the numerical results (which have become exceedingly easy to produce) needs to be much better understood. The meaning and the robustness of the solutions is often not clear. Many issues related to the system-theoretical and statistical significance of the new nonlinear identification methods are still open. Some of them were the object of much discussion in the school.

In conclusion, Identification, Adaptation and Learning form together a rich area of interplay with statistics, numerical analysis, control, communication and information processing. Besides being mathematically well-founded and intellectually fascinating, this emerging area has also tremendous application capabilities. In particular it seems to be the key that will finally provide the body of *rational* methodologies necessary for understanding and designing "autonomous" and "intelligent" engineering systems.

It is unfortunate that the theory of learning and some parts of "neural" modeling and identification have been evolving quite separately from the classical disciplines of identification and adaptive control. Nowadays it is urgent to bring researchers together and teach young people a unified view of the field. In this respect we consider the initiative of this school as one filling a real need. The original title of the school suggested the existence of a cultural and methodological unity of the area, and in fact suggested the existence of a conceptual *path* leading from identification to adaptation and to the theory of learning. This idea has led to a "logically sequential" organization of the lectures which is reproduced in this book. We think that this may have been an additional stimulus, both for the audience and for the lecturers, to actually recognize the points of contacts and to unify the language.

The book is organized in thirteen chapters as follows.

In Chapter 1, Giorgio Picci and Anders Linquist introduce the essential features of the identification problem approached from the "classical" statistical point of view. A thorough review of stochastic modeling, especially state-space modeling from covariance data, is provided. Geometric methods for state-space identification (probably better known as subspace methods) are discussed in some length at a tutorial level.

Chapters 2 and 3, by Jan Maciejowski and Raimund J. Ober, respectively, are devoted to algorithms for identification of multivariable state-space models using balanced canonical forms. These forms have recently turned out quite useful in state-space identification.

The classical approach to system identification, as described in Chapter 1, is mainly based on a statistical approach. In recent years, a number of researchers have started to explore alternative approaches and come out with new ideas and new theories regarding both the modeling issue and even the very formulation of the identification problem. In this book some of these new avenues are also documented.

The so-called *behavioural* approach to modeling dynamic systems is thoroughly discussed in Chapter 4, by Paolo Rapisarda and Jan C. Willems. This approach avoids the distinction of the external variables of a system in inputs and outputs and so doing many structural issues in system theory are dealt with in a more general and natural manner.

The *Errors-in-Variable* (EIV) problem is the most general linear estimation problem in statistics and a very old one, the first formulations going probably back to the beginning of the century. It has turned out to be surprisingly difficult and for this reason EIV is sometimes described as "non-mainstream" statistics. Indeed it occurs naturally also in dynamic system modeling and identification. A state-of-the-art description of this subject is given in Chapter 5, by Manfred Deistler and Wolfgang Scherrer.

A flourishing area of recent research is identification based on worst case deterministic criteria, among which the so-called H^∞ identification theory.

This formulation of identification has been stimulated by the recent developments of robust control, where worst-case criteria play a main role in the very formulation of the problem.

One approach to the H^∞ identification problem is described in Chapter 6, by Pramod P. Khargonekar, Guoxang Gu and Jonathan Friedman. In this contribution, one goes back to a classical paradigm of the electrical engineering culture, namely input-output experiments in the frequency domain. The measured samples of the frequency response are then used to compute a robust estimate of the transfer function of the system.

A desire to reconcile probabilistic and deterministic H^∞ identification may be seen as the motivating idea of Chapter 7, written by A.A. Stoorvogel and J.H. van Schuppen. In this intriguing and informative article, several possible approximation criteria for system identification of stationary processes are introduced and compared.

Identification has always played a major role in control. One of the areas of control to which is most deeply interwined is *adaptive control.* Adaptive control is a multifaceted subject within which the so-called *self-tuning control* represents one of the keystones. The fundamental questions of convergence, stability and self-optimality in self-tuning control are discussed in Chapter 8, by Sergio Bittanti and Marco Campi. In the same paper, related important issues are discussed. In particular, the question of convergence of the least squares estimate under poor excitation conditions is clarified.

Chapters 1 to 8 deal with identification of *linear models.* However, the subject of *nonlinear systems identification* has been growing tremendously in the last few years. Particular attention has been devoted to "universal" black-box tools for identification, such as neural networks. The use of *artificial neural networks* is discussed in Chapter 9, by L. Ljung, J. Sjoberg and H. Hjalmarsson.

This naturally leads to the issues of *training* and *learning.* An overview of computational learning theory and its applications to neural network training is presented in Chapter 10, by M. Vidyasagar.

Much early and fundamental work related to this area was made in the probability and statistics literature. Thanks to the recent work, it is possible to make such a vague statements as "neural network can generalize" to precise statements such as "after adequate training, neural networks can correctly classify previously untrained inputs with high accuracy, most of the time".

As a matter of fact, system identification has most of the time been based on the idea that there is just "one" model out of a selected model class that best fits the data. However, data may change in time in a non-stationary fashion. In Chapter 11 George Cybenko describes a much more general philosophy by which, based on certain features of the current region in the data space one builds a model which is specific for that operating point and is changed when different data features are observed.

Besides neural networks, there are other important classes of functions for nonlinear black box modelization of nonparametric type, namely *wavelets*. These are discussed by A. Benveniste, A. Juditsky, B. Delyon, Q. Zhang and O.Y. Glorennec in Chapter 12.

In cases when the available information consists not only of numerical data, but also contains qualitative statements about the variables, *fuzzy logic* supplies a way of handling the data. Fuzzy modeling is discussed in Chapter 13 by Pedro Albertos.

As an underlying *leitmotiv* to all problems dealt with in the book, there is the *optimization* issue, namely the problem of finding the maximum of a function when one knows only "noisy" (uncertain) measurements of its values. Stochastic optimization methods ("searching for the best") are overviewed and surveyed in the final chapter of the book, Chapter 14, written by Georg Pflug.

Geometric Methods for State Space Identification

Anders Lindquist[1] and Giorgio Picci[2]

[1] Optimization and System Theory
Dept. of Mathematics
Royal Institute of Technology
S-10084
Sweden.

[2] Dipartimento di Elettronica e Informatica
Universita' di Padova
via Gradenigo 6/A
35131 Padova
Italy.

1. Introduction

The scope of identification theory is to construct algorithms for automatic model building from observed data. In these lectures we shall only discuss the case where the data are collected in one unrepeatable experiment and no preparation of the experiment is possible (i.e. we cannot choose the experimental conditions or the input function to the system at our will).

The observed variables, usually classified as "inputs" (u) and "outputs" (y), are measured at discrete instants of time t and collected in a string of data of finite duration T. These data are called a "time series" in the statistical literature. There is a preselected model class, say the class of finite-dimensional linear time-invariant systems of a given order and the problem is generally formulated as that of inferring a "best" mathematical model in the model class on the basis of the observed data. There may be a variety of different reasons to build models. Here we shall be chiefly interested in model building for the purpose of prediction and control. This means that the identified model should be useful for prediction or control of *future* i.e. not yet observed, data.

Essential features of the Identification Problem.

1. There are always many other variables besides the preselected "inputs" and "outputs" which influence the time evolution of the system and hence the joint dynamics of y and u during the experiment. These variables represent the unavoidable interaction of the system with its environment. For this reason, even in the presence of a true causal relation between inputs and outputs there always are some *unpredictable* fluctuations of the values taken by the measured output $y(t)$ which are not explainable in terms of past input (and/or output) history.

We cannot (and do not want to) take into account too many of these variables explicitely in the model as some of them may be inaccessible to measurement and in any case this would lead to complicated models with too many variables. We need to work with models of small complexity and treat the unpredictable fluctuations in some simple "aggregate" manner.

2. Models (however accurate) are of course always mathematical idealizations of nature. No physical phenomenon, even if the experiments were conducted in an ideal interactions-free environment can be described *exactly* by a set of differential or difference equations and even more so if the equations are a priori restricted to be linear, finite-dimensional and time-invariant. So the observables, even in an ideal "disturbance-free" situation cannot be expected to obey *exactly* any linear time-invariant model.

If we accept the arguments above it is clear that one essential issue to be addressed for a realistic formulation of the problem is a satisfactory notion of non-rigid, i.e. *approximate*, mathematical modeling of the observed data. The meaning of the word "approximate" should here be understood in the sense that a model should be able to accept as legitimate, data sets (time series) which may possibly differ slightly from each another. Imposing rigid "exact" descriptions of the type $F(u,y) = 0$ to experimental data has been criticized since the early beginnings of experimental science. Particularly illuminating is Gauss' general philosophical discussion in [27] sect. III, p. 236.

More to the point, there has been a widespread belief in the early years of control theory that identification was merely a matter of describing (exactly) the measured data by linear convolution equations of the type

$$y(t) = \sum_{t_0}^{t} h(t-\tau)u(\tau) \tag{1.1}$$

or, equivalently, by matching exactly pointwise harmonic response data with linear transfer function models. Results have always been poor and extremely sensitive to the data. New incoming data tend to change the model drastically, which means that a model determined in this way has in fact very poor predictive capabilities. The reason is that data obey exactly rigid relations of this kind "with probability zero". If in addition the model class is restricted to be finite-dimensional, which of course is what is really necessary for control applications, imposing the integral equation model (1.1) on real data normally leads to disastrous results. This is by now very well-known and documented in the literature, see e.g. [65, 70, 35, 20]. The fact, expressed in the language of numerical analysis, is that fitting rigid models to data invariably leads to ill-conditioned problems.

Gauss idea of describing data by a *distribution function* is a prime example of thinking in terms of (non-rigid) approximate models[1]. Other alternatives are possible, say using model classes consisting of a rigid "exact" model as a "nominal" object, plus an uncertainty ball around it. In this case, besides a "nominal" model, the identification procedure is required to provide at least bounds on the magnitude of the relative "uncertainty region" around the nominal model. This type of modeling philosophy is motivated by use in H^∞ control applications. Here one should provide a mathematical description of how the "dynamic" uncertainty ball is distributed in the frequency domain, rather than, as more traditionally done, in the parameter space, about the "nominal identified model".

In addition to the above we need also to introduce a mathematical description of *the data*. The data at our disposal at some fixed time instant represent only partial evidence about the behaviour of the system; we do not know the future continuation of the input and output time series, yet all possible continuations of our data must carry information about the same physical phenomenon we are about to model, and hence the possible continuations of the data cannot be "totally random" and must be related to what we have observed so far. So, in order to discover models of systems, we have to work with models of uncertain signals.

Mathematical descriptions of *uncertain signals* can be quite diverse. Possible choices are *stochastic processes*, deterministic signals with uncertainty bounds, etc. The crucial difference between theories of model building relates to the *quantitave* method for modeling uncertain signals they use[2].

In these lectures we shall eventually take the "classical" route and model uncertainty with the apparatus of probability theory. In this framework identification is phrased as a problem of mathematical Statistics.

One could argue that the basic problem of identification is, much more than designing algorithms which fit models to observed data (the easy part), the quantification of *dynamic uncertainly bounds* or the description of the *dynamic errors* incurred when using the model with future data. Any sensible identification method should provide some mathematical description of how uncertainty is distributed in time or frequency about the nominal identified model. In this respect the stochastic approach offers a very nice solution. In this setup (at least in the linear wide-sense setting) model uncertainty turns out to be equivalent to *additive* random disturbances i.e. identifying model uncertainty is equivalent to identifying models for "partially observed" stochastic processes. We shall discuss this point further in the following.

[1] A vulgar belief attributes to Gauss the invention of least squares, which is historically wrong. In Gauss' work least squares come out as a solution method for optimally fitting a certain class of *density functions* to the observed data.

[2] For this reason we would probably not classify as identification "exact modeling" where the data are "certain" signals assumed to fit exactly some finite set of (linear) relations.

1.1 Stationary Signals and the Statistical Theory of Model Building

Since identification for the purpose of prediction and control makes sense only if you can use the identified model to describe future data, i.e. different data than those employed for its calibration, at the roots of any data-based model building procedure there must be a formalization of the belief that

> *future data will continue to be generated by the same "underlying mechanism" that has produced the actual data.*

This is a vague but basic assumption on the nature of the data, which are postulated to keep being "statistically the same" in the future. Besides being inherent in the very *purpose* of collecting data for model building this assumption offers the logical background for assessing the *quality* of the identified model, by *asymptotic analysis*, i.e. by comparing finite-sample results with the "best achievable" model which could theoretically be identified with data of infinite lenght. One could probably say that *Statistics* as a discipline, is founded on asymptotic analysis, and that the wide use of Statistics and of probabilistic methods in identification is mainly motivated by the large body of effective asymptotic tools which can be applied to assess some basic "quality" features of the estimated model.

Classical Statistics traditionally starts by postulating some "urn model" whereby the data are imagined as being "drawn" at random from some universe of possible values in a "random trial" where "nature" chooses according to some probability law the current "state" of the interactions and of the experimental conditions.

It has been argued that the abstract "urn model" of probability theory looks inadequate to deal with situations like the one we have envisaged, where there is just one irrepetible experiment and there is really no sample space around from which the results of the experiment could possibly have been drawn. This critique comes from a tendendency to confuse physical reality with mathematical modelling. In effect the "urn model" is just a mathematical device which is not required to have any physical meaning or interpretation and can be used to model anything.

The critique has at least the merit of bringing up an important issue. It should be admitted that in large sectors of the literature the stochastic framework is often imposed dogmatically to practical problems (the user is normally left alone wondering if his problem is "stochastic" enough to be authorized to apply algorithm A, or his data are instead "determinstic" and he should apply algorithm B instead) and often statistical procedures are pushed to extremes where there really seems to be no physical ground for their applicability.

Yet there is a vast number of situations where a formal justification for the adoption of the probabilistic description of uncertain data can be given. Formal arguments leading to a probabilistic description of certain types of data could for example be based on the notion of "stationarity", a mathematical condition meant to capture the idea that future data should be "statistically the same" as past data. One possible line of reasoning is briefly elaborated upon below.

Let $z := \{z(t)\}_{t\in\mathbb{Z}}$ be a discrete-time signal (i.e. a sequence of real numbers). A *function of* z is just a real-valued function $f(z) := f(z(t); t \in I)$, $f : \mathbb{R}^I \to \mathbb{R}$ where I is a subinterval of $\mathbb{Z}$, possibly infinite. The *shift* σ is the map defined on real sequences as $[\sigma z](t) := z(t+1), \quad t \in \mathbb{Z}$ so that the iterated application of σ, say

$$[\sigma^t z](s) := z(t+s), \quad t, s \in \mathbb{Z}$$

transforms a signal z into its "translation by t units of time" $z_t := \{z(t+s)\}_{s\in\mathbb{Z}}$. Let us denote by $f_t(z)$ the result of applying f to the shifted sequence $\sigma^t z$, i.e. let $f_t(z) := f(z(t+s); s \in I) = f(z_t), t \in\in \mathbb{Z}$.

Definition 1.1. *A signal z will be called*

- Strict-sense *stationary if the Cesaro limit*

$$\lim_{T\to\infty} \frac{1}{T+1} \sum_{t=0}^{T} f_t(z)$$

exists for all bounded measurable functions f*;*

- Wide-sense *stationary if the limit exists for* $f(z) = z(0)$ *(so that* $f_t(z) = z(t)$*) and for all quadratic forms*[3] *in* z*.*

The two conditions for wide-sense stationarity are normally found in the literature under a variety of different names. They describe the minimum amount of structure on the data which is necessary to do a (rudimental) asymptotic analysis of an identification algorithm for linear time-invariant models. The strict-sense notion is introduced mostly for conceptual reasons. Both notions generalize in a natural way to vector-valued sequences.

The purpose of the following paragraphs is to show that (strict-sense) stationary signals admit *stationary stochastic processes* as a natural mathematical description.

First take $f(z) := I_A(z(0))$ where I_A is the indicator function of a Borel set $A \subset \mathbb{R}$ ($I_A(x) = 1$ if $x \in A$ and 0 otherwise). Then the nonnegative number

$$\nu_T(A) := \frac{1}{T+1} \sum_{t=0}^{T} I_A(z(t))$$

is just the relative frequency of visits of the signal z to the set A. In fact, for each fixed T the function $A \to \nu_T(A)$ is a *probability measure*, i.e. a countably additive set function on the Borel sets of the real line. This follows simply

[3] i.e. for all real functions f such that $f(\alpha z) = \alpha^2 f(z)$.

from the relation $I_{\cup A_k} = \sum I_{A_k}$ which is valid for any sequence of disjoint sets A_k. For a stationary sequence we have $\nu_T(A) \to \nu_0(A)$ as $T \to \infty$. It then follows readily that

Lemma 1.1. *The set function* $A \to \nu_0(A)$ *is a probability measure on* $\mathbb{R}$.

More generally, take

$$f(z) := I_A(z(0)) I_{A_1}(z(\tau_1)) \dots I_{A_n}(z(\tau_n))$$

where $\tau_1 \dots \tau_n$ are arbitrary time instants and $A, A_1 \dots A_n$ arbitrary Borel sets of the real line and consider the relative frequency

$$\nu_T(A, A_1, \tau_1, \dots A_n, \tau_n)) := \frac{1}{T+1} \sum_{t=0}^{T} I_A(z(t)) I_{A_1}(z(t+\tau_1)) \dots I_{A_n}(z(t+\tau_n))$$

of a visit to the set A followed by a visit, τ_1 instants later, to the set A_1, τ_2 instants later to the set A_2 etc.. and τ_n instants later to the set A_n. By stationarity $\nu_T(A, A_1, \tau_1, \dots A_n, \tau_n) \to \nu_n(A, A_1, \tau_1, \dots A_n, \tau_n)$ as $T \to \infty$. An easy generalization of Lemma 1.1 leads to the following statement.

Lemma 1.2. *The set function* $(A \times A_1 \dots \times A_n) \to \nu_n(A, A_1, \tau_1, \dots A_n, \tau_n)$ *is a probability measure on* $\mathbb{R}^{n+1}$ *for all time lags* $\tau_1 \dots \tau_n$. *In fact the family* $\{\nu_k\}_{k \in \mathbb{Z}_+}$ *is a* consistent family of probability distributions *in the sense of Kolmogorov, i.e.*

$$\nu_n(A, A_1, \tau_1, \dots, \mathbb{R}, \tau_n) = \nu_{n-1}(A, A_1, \tau_1, \dots A_{n-1}, \tau_{n-1})$$

for all Borel sets $A, A_1 \dots, A_{n-1}$ *and time lags* $\tau_1 \dots, \tau_n$.

It follows by a famous theorem of Kolmogorov that there is a bona - fide probability measure ν on the "sample space" $\mathbb{R}^{\mathbb{Z}}$ of all real sequences, which is the (unique) extension of the family of finite dimensional distributions $\{\nu_k\}_{k \in \mathbb{Z}_+}$ associated to a stationary signal z by the construction illustrated above. This measure is invariant with respect to the shift σ acting on the sequences of $\mathbb{R}^{\mathbb{Z}}$. In other words, tha pair $(\mathbb{R}^{\mathbb{Z}}, \nu)$ (with the natural family of measurable sets) defines a *stationary stochastic process* $\mathbf{z}$.

The moral of the story is that every stationary signal can be interpreted in a canonical way as a "representative" trajectory of a stationary process[4]. In other words,

[4] It is well known that *almost all* trajectories of a stationary process $\mathbf{z}$ are stationary signals in the sense of Definition 1.1. This is essentially the famous D.G. Birkhoff's *ergodic theorem*, see e.g. Doob [19], p. 465. A "representative" trajectory is just a trajectory belonging to the set of trajectories of ν-probability one where the Cesaro sums converge. Note that the process $\mathbf{z}$ need not be ergodic (i.e. "metrically transitive" according to the old terminology).

Proposition 1.1. *For a stationary signal* z *there always exists an "urn model" i.e. a probability space* $\{\Omega, \mathcal{A}, \mu\}$ *and a stationary process* $\mathbf{z} := \{ z(t,\omega) \,|\, t \in \mathbb{Z}, \omega \in \Omega \}$ *defined on it such that* z *is a representative trajectory of* $\mathbf{z}$*, i.e.*

$$z(t) = \mathbf{z}(t, \bar{\omega}) \quad t \in \mathbb{Z}$$

for some elementary event $\bar{\omega}$ *in the "good" set of probability one guaranteed by Birkhoff's theorem.*

So we are authorized if we wish, to think legitimately of a stationary sequence of data as being "drawn" from a population according to a stationary probability law. We shall call this probability measure the *true (probability) law* of the data.

All of the above is of course mostly of "theoretical interest" and only serves the purpose of justifying the introduction of probabilistic and statistical language in identification. Very often in practice one can make verifiable statements only about the first and second order moments of the observed data and so in the following we shall normally work under the assumption of *wide sense stationarity* only. Moreover we shall assume throughout that the time averages of all signals have been subtracted off so all data will be assumed to be *zero mean* hereafter. Hence a wide-sense stationary signal (which we shall now assume m-dimensional) is just a sequence z for which the limit

$$\lim_{T\to\infty} \frac{1}{T+1} \sum_{t=0}^{T} z(t+\tau) z(t)' := \Lambda_0(\tau) \tag{1.2}$$

exists for all $\tau \in \mathbb{Z}$.

Proposition 1.2 (Wiener). *The function* $\Lambda_0 := \tau \to \Lambda_0(\tau)$ *is a bona-fide covariance function (i.e. a symmetric positive definite matrix function)*

Proof. The function Λ_0 is the discrete-time version of $\phi(x)$ in Wiener's Generalized Harmonic Analysis [73]. □

From this result, much in the same spirit of the strict-sense Proposition 1.1 stated above, one may draw the conclusion that a wide-sense stationary signal admits as a atural probabilistic model a *stationary wide-sense stochastic process*. Here, following [19] "wide-sense process" means the equivalence class of stochastic processes (defined say on the probability space $(\mathbb{R}^m)^{\mathbb{Z}}$) with zero mean and all having the same covariance function. In certain cases it may be appropriate to take as a representative of the equivalence class the unique *Gaussian* process with (zero mean and) given covariance function. Of course the additional strict-sense probabilistic structure provides only illusory extra information (besides second-order) unless the data provide actual evidence for the choice of Gaussian distributions.

A blanket assumption during the rest of these notes will be that the input-output data extend in the future do form a stationary[5] signal z; we shall call Λ_0 the *true covariance* of this signal.

Remarks. Note that for (wide-sense) stationary signals which decay to zero as $T \to \infty$ the true covariance function is identically zero. This is not paradoxical, as a signal of this kind may intuitively be regarded as a "transient" phenomenon settling eventually to a zero steady state.

The *spectral distribution function* of the signal is a monotonic Hermitian matrix function F_0 defined on the unit circle of the complex plane $\{\zeta = e^{j\omega}\}$ by the "Fourier-like" representation formula valid for any covariance function

$$\Lambda_0(\tau) = \int_{-\pi}^{\pi} e^{j\omega\tau} dF_0(e^{j\omega}) \tag{1.3}$$

(Herglotz Theorem). If the $\Lambda_0(\tau)$ form a summable sequence (so that $\Lambda_0(\tau) \to 0$ as $\tau \to \infty$) the spectral distribution function admits a density Φ_0,

$$F_0(e^{j\omega_2}) - F_0(e^{j\omega_1}) = \int_{\omega_1}^{\omega_2} \Phi_0(e^{j\lambda}) \frac{d\lambda}{2\pi}$$

In general when the covariance function does not decay to zero, for example when there are periodic components in z, the distribution function has jumps and the density function describes only the absolutely continuous part of F_0. *Persistently exciting signals* of order n are classical examples of periodic stationary signals whose distribution function is a staircase function with exactly n jumps.

The statistical approach to identification. As we have argued in this section a reasonable mathematical description of the measured data is to model it as a finite tract of a trajectory of a stationary (wide-sense) stochastic process. The identification problem is then naturally formulated as the problem of recovering the "true" law of the process i.e. its true covariance or spectral distribution function[6] from the measured data. This of course is just the prototypical problem of Statistics.

Naturally the family of all possible "true descriptions" is an exceedingly general infinite-dimensional object and to make the problem solvable one has to choose, perhaps on the basis of some available a priori information, a manageable subclass which should be describable in terms of a finite number of real parameters. In fact, we shall ask that the model class should be compatible with *finite dimensional* prediction and control schemes. Although we keep the meaning of the term rather vague at this stage, it is very well-known that "finite-dimensional" wide-sense stationary processes can only be linear combinations of quasi-periodic (i.e. sums of sinusoids with random amplitudes)

[5] "Stationary" will mean wide-sense stationary hereafter.

[6] Of course, more generally, in a stric-sense formulation one tries to "infer" the true probability law of the underlying process.

and purely-non-deterministic processes with a *rational spectral density.* There is then very little choice for the model class. If we are interested in finite-complexity models of "truly random" (purely-non-deterministic) signals, then we must restrict to *rational spectral densities.*

1.2 Input-Output Models

Very often in "input-output" experiments one is not interested in modeling the input signals and would like to concentrate just on recovering a (causal) relation between inputs and outputs.

In the present wide-sense stochastic setup the input-output relation has in general a linear structure of the type

$$y(t) = E[\, y(t) \mid u(s); s \leq t\,] + v(t) \tag{1.4}$$

where $E[\, y(t) \mid u(s); s \leq t\,]$ is the best (in the sense of minimum variance of the error) estimate of the output $y(t)$ based on the past of u up to time t. By Wiener filtering theory it known that this estimate is described by a causal and stable linear convolution operator with a rational transfer function $F(\zeta)$. The additive term $v(t)$ is the relative "estimation error", a stationary process with rational spectrum, uncorrelated with the past of u, which models precisely the uncertainty due to disturbances etc. superimposed to the input-based prediction, $F(\zeta)u(t)$, of $y(t)$.

The structure of the "input-output" model class which results

from the assumptions of joint wide-sense stationarity and rational joint spectrum for the input and output processes is quite explicit indeed. Note that it comes out, as a formal consequence of the probabilistic setting used to describe our data. There is no arbitrariness or "user choice" at this stage, except of course for the choice of the order or the structure parameters of the transfer function. Note incidentally that identifying the model uncertainty in (1.4) means identifying a dynamic model for the additive noise process v.

A typical route which is commonly taken is to estimate the transfer function F and the noise model for v as if u was a deterministic sequence. Sometimes in the literature it is even "assumed" that u is a "deterministic" signal. This of course cannot be the real intention since it would lead to the rather absurd consequences that

$$\mathrm{E} \sum_{t,s} y(t)u(s) = \sum_{t,s} [\mathrm{E}y(t)]\, u(s) = 0$$

i.e. the input and output signals would be *completely uncorrelated.*

Estimation of the nominal input-output transfer function is generally to be understood as being "conditional on the past observed history of u". Although this may at a first sight look like a reasonable thing to do, it may lead to serious errors whenever hidden feedback links are present influencing the way in which the input variable is manufactured (i.e. introducing in u "stochastic components" correlated with the past of y).

In fact, if there is feedback from y to u the very notion of "input" looses its meaning since, as shown e.g. in [29], the input variable $u(t)$ is then also determined by a dynamical relation of the form (1.4), involving now the "output" process y playing in turn the role of an exogenous variable ("input") to determine u.

The appropriate setup for discussing these matters is within the theory of *feedback* and *causality* between stationary processess [33]. We shall not adventure into this subject in these introductory notes. We shall just content ourselves of recalling, as it has been argued in several places in the literature, that identification in the presence of feedback (and of course in the absence of any other specific information on the feedback loop) is essentially equivalent to identification of the *joint process* $[y', u']'$, in the sense of modeling the joint dynamics of the signals on the basis of the observed time-series $\{[y(t)', u(t)']'\}$. It is also for these reasons that we shall choose to restrict the scope of our discussion only to time-series identification in the rest of the paper.

2. State Space Models of Stationary Processes

From the previous section it has been seen that wide-sense stationary processes with a *rational spectral density matrix* provide a natural class of finitely-parametrized stochastic models for the identification of a wide class of observed data.

It is very well known that these processes are precisely those admitting finite-dimensional *state-space descriptions* (or *realizations*) with constant parameters. It is then natural to pose the identification problem directly in terms of recovering state-space models of y. There are different approaches to identify models of this kind and we shall discuss some recent methods (the so-called "subspace methods") in some detail later in sections 6 and 7.

In any case it is well-known that even if we restrict to *minimal models* i.e. models of the smallest possible dimension of the state space, there are in general many non-equivalent (minimal) state-space representations of the same process y. This is a significant departure from the usual deterministic linear modeling setup and brings up model choice or *identifiability* questions which should be understood well before discussing the choice of a particular statistical methodology for model building. Therefore in this and in the following two sections we shall have to review the basic facts about finite-dimensional state-space models of stationary random processes.

Consider a stationary stochastic system

$$(\Sigma) \quad \begin{cases} x(t+1) &= Ax(t) + Bw(t) \\ y(t) &= Cx(t) + Dw(t) \end{cases} \tag{2.1}$$

where $\{w(t)\}$ is p-dimensional normalized white noise , i.e.

$$\mathrm{E}\{w(t)w(s)'\} = I\delta_{ts} \quad \mathrm{E}\{w(t)\} = 0.$$

In this paper we shall think of (2.1) exclusively as a *representation* of the output process y. This representation involves *auxiliary variables* such as the *state process* x and the *generating white noise* w which are introduced for the purpose of giving the model a particular structure. These auxiliary variables play the role of "parameters" which may be eliminated producing a different model structure. For example, by eliminating x from the equations (2.1) one obtains an "input-output" representation whereby y appears as the result of processing the white noise signal w through a linear time-invariant filter

$$\xrightarrow{w} \boxed{\mathrm{W}} \xrightarrow{y} \tag{2.2}$$

of transfer function

$$W(z) = C(zI - A)^{-1}B + D. \tag{2.3}$$

We shall for the moment make the assumption that the matrix A is *stable*, i.e. the eigenvalues of A all lie inside the unit circle ($|\lambda(A)| < 1$) and that the input noise has been applied to the system for an infinitely long time, i.e. starting at $t = -\infty$. In these conditions the effect of initial conditions has died off and the system is in statistical steady state. Then

$$x(t) = \sum_{j=-\infty}^{t-1} A^{t-1-j}Bw(j)$$

and

$$y(t) = \sum_{j=-\infty}^{t-1} CA^{t-1-j}Bw(j) + Dw(t)$$

In particular, x and y are *jointly stationary*[7].

The system (2.1) can be regarded as a linear map defining x and y as linear functionals of the input noise w. In fact, since the matrix A has been assumed stable, this map will be a *causal* map. In order to capture these properties in a precise way it is convenient to think of the (components of) x and y as elements of the infinite dimensional Hilbert space of second order random variables

$$H(w) = \overline{\mathrm{span}}\{w_i(t) \mid t \in \mathbb{Z}; i = 1, 2, \dots, p\} \tag{2.4}$$

with inner product $(\xi, \eta) = \mathrm{E}\{\xi\eta\}$. Here $\overline{\mathrm{span}}$ denotes the closure of the vector space generated by linear combinations of the elements listed inside

[7] Stationarity here is always meant in the "wide sense" of second order statistics. In particular x and y being jointly stationary means that the covariance matrix $\mathrm{E}\{[x(t)'y(t)']'[x(s)'y(s)']\}$ depends only on $t - s$.

the brackets. The Hilbert space $H(w)$ is called the *ambient space* of the stochastic system (Σ). It comes equipped with a unitary *Shift operator* U which is the extension of temporal translation i.e. $Ua'w(t) = a'w(t+1)$ of the generating random variables $a'w(t)$ of the space. More generally, the symbol $H(y)$ is used to denote the Hilbert space generated by a wide-sense zero mean process y. If the process is stationary then $H(y)$ is equipped with the unitary shift of the process, U. The pair $(H(y), U)$ is called a *stationary Hilbert space.* By definition a stationary Hilbert space contains all translates $U^t\xi$ of any random variable ξ which belongs to it.

The *past subspaces* of x and y

$$H_t^-(x) = \overline{\text{span}}\{x_i(s) \mid s < t; i = 1, 2, \dots, n\} \tag{2.5}$$

$$H_t^-(y) = \overline{\text{span}}\{y_i(s) \mid s < t; i = 1, 2, \dots, m\} \tag{2.6}$$

are both contained in $H_t^-(w)$ (causality) and hence the *future space* of w

$$H_t^+(w) = \overline{\text{span}}\{w_i(s) \mid s \geq t; i = 1, 2, \dots, m\}$$

will be orthogonal to (i.e. uncorrelated with) both $H_t^-(x)$ and $H_t^-(y)$.

The finite dimensional subspace of $H(w)$

$$X_t = \text{span}\{x_1(t), x_2(t), \dots, x_n(t)\} \quad t \in \mathbb{Z},$$

is called the *state space* of the system (2.1) at the instant t.

In the following we shall always suppose that (A, B, C, D) in (2.3) is a minimal realization of W. In other words we shall assume that (A, B) is reachable and (C, A) is observable. Then, setting

$$P = \mathrm{E}\{x(0)x(0)'\},$$

it follows from stationarity that $P = \mathrm{E}\{x(t)x(t)'\}$ for all t, and hence the first equation in (2.1) yields

$$P = APA' + BB', \tag{2.7}$$

which is a Lyapunov equation. Since $|\lambda(A)| < 1$ the sum $P = \sum_{j=0}^{\infty} A^j BB'(A')^j$, converges and converges to the reachability grammian of Σ. But (A, B) is reachable, and hence $P > 0$. This implies that $\{x_1(t), x_2(t), \dots, x_n(t)\}$ is a *basis* in X_t.

Notations. We shall use the symbol $\vee$ to denote vector sum of subspaces, $+$ to denote *direct sum* and $\oplus$ to denote *orthogonal* vector sum. The orthogonal complement of a subspace A in the ambient space under consideration will be denoted by $A^\perp$. The future spaces always contain the present while the past does not [this convention will be followed generally with the only exception of Markov processes where both past and future must contain the present].

Several subspace constructions in the following are defined at some fixed reference time; by stationarity however they carry over to arbitrary time instants and we shall always implicitly mean that the relevant definition is extended by stationarity to the whole time axis.

Normally the reference time will be taken to be $t = 0$. To simplify notations the subscript $t = 0$ will normally be dropped. The symbols H^+ and H^- will denote the future and past spaces at time 0 of the process y. The orthogonal projection onto a subspace S will be denoted E^S or $\mathrm{E}[.|S]$. For Gaussian random variables this coincides with the conditional expectation given the σ-algebra generated by S. Operators like E^S or U applied to vectors will act componentwise in an obvious way.

The Coordinate-free viewpoint. The coordinate-free or *geometric* viewpoint lies at the grounds of the identification methods which will be discussed in the last sections.

The main idea here is that building state-space models of a random process (i.e. stochastic realization) is essentially a matter of constructing a space X with properties which make it the stochastic analog of a deterministic state space. Once this first basic step is done, the rest is just a matter of choosing coordinates in X and the causality structure of the model. The basic notion in this respect is the following.

Definition 2.1. *Let X be a subspace of some large stationary Hilbert space H of wide-sense random variables containing $H(y)$. Define*

$$X_t := U^t X, \quad X_t^- := \vee_{s \le t} X_s, \quad X_t^+ := \vee_{s \ge t} X_s.$$

A Markovian Splitting Subspace *X for the process y is a subspace of H making the vector sums $H^- \vee X^-$ and $H^+ \vee X^+$ conditionally orthogonal (i.e. uncorrelated) given X, denoted,*

$$H^- \vee X^- \perp H^+ \vee X^+ \mid X. \tag{2.8}$$

The subspace X is called proper, *or* purely-non-deterministic *if there are vector white noise processes w and $\bar{w}$ such that* [8]

$$H^- \vee X^- = H^-(w), \quad H^+ \vee X^+ = H^+(\bar{w})$$

Any basis vector $x(0) := [x_1(0), x_2(0), \ldots, x_n(0)]'$ in a Markovian splitting subspace X generates a stationary Markov process $x(t) := U^t x(0), t \in \mathbb{Z}$ which serves as a *state* of the the process y. If X is proper the Markov process is purely non determinstic and can be represented by a linear equation of the type $x(t+1) = Ax(t) + Bw(t)$ where A has all its eigenvalues strictly inside of the unit circle.

The fundamental characterization in this setting is the following.

---

[8] This is equivalent to requiring that

$$\cap_t H_t^- \vee X_t^- = \{0\}, \text{ and } \cap_t H_t^+ \vee X_t^+ = \{0\}.$$

Obviously in this case y must also be purely non-deterministic [68].

Theorem 2.1. *[66, 47, 49] The state space X of any stochastic ralization (2.1) is a* Markovian Splitting Subspace *for the process y.*

Conversely, given any proper Markovian splitting subspace X, to any choice of basis $x(0) = [x_1(0), x_2(0), \dots, x_n(0)]'$ in X there corresponds a stochastic ralization of y of the type (2.1) with generating input noise w.

There are formulas expressing the coefficient matrices A, B, C, D in terms of x and y. They will be given in Theorem 2.2 below.

A Markovian splitting subspace is *minimal* if it doesn't contain (properly) other Markovian splitting subspaces. Countrary to the deterministic situation minimal Markovian splitting subspaces are *non unique.* Two very important examples are the *forward and backward predictor spaces* (at time zero):

$$X_- := \mathrm{E}^{H^-} H^+ \quad X_+ := \mathrm{E}^{H^+} H^- \tag{2.9}$$

for which we have the following characterization [49].

Proposition 2.1. *The subspaces X_- and X_+ are the unique minimal splitting subspaces contained in the past H^-, and, respectively, in the future H^+, of the process y.*

The causality of the representation (2.1) can be expressed geometrically as the orthogonality relation

$$H_t^+(w) \perp X_t^- \vee H_t^-(y) \tag{2.10}$$

for all $t \in \mathbb{Z}$. One also says that Σ is a *forward* model or that it evolves *forward* in time. Note in particular, that $\mathrm{E}\{x(t)w(t)'\} = 0$.

Backward or Anticausal realizations are models where instead the past of the driving white noise is orthogonal to the future of the state and output processes. These models are useful in several instances and are as legitimate representations of y as the forward models studied so far. As a matter of fact, a random signal has no "preferred direction of time" or causality built in and admits many different sorts of causality structures, see [67].

Theorem 2.2. *[48, 47] Any choice of basis vector $x(0)$ in a (finite dimensional) proper Markovian splitting subspace X generates a stationary vector Markov process $x(t) = U^t x(0), t \in \mathbb{Z}$ such that the joint process $\begin{bmatrix} x(t) \\ y(t-1) \end{bmatrix}$ is also Markov. The joint process admits a* forward representation

$$\begin{bmatrix} x(t+1) \\ y(t) \end{bmatrix} = \begin{bmatrix} A & 0 \\ C & 0 \end{bmatrix} \begin{bmatrix} x(t) \\ y(t-1) \end{bmatrix} + \begin{bmatrix} B \\ D \end{bmatrix} w(t) \tag{2.11}$$

where $w(t)$ is the generating white noise process of $H^- \vee X^-$ i.e. $H^- \vee X^- = H^-(w)$ and

$$A = Ex(t+1)x(t)'P^{-1} \qquad B = Ex(t+1)w(t)' \tag{2.12}$$

$$C = Ey(t)x(t)'P^{-1} \qquad D = Ey(t)w(t)' \tag{2.13}$$

Dually, let $\bar{x}(0)$ *be another basis in* X *and let* $\bar{x}(t) = U^t\bar{x}(0)\, t \in \mathbb{Z}$ *be the corresponding stationary vector Markov process. The joint process* $\begin{bmatrix} \bar{x}(t) \\ y(t) \end{bmatrix}$ *is also Markov and admits a* backward representation

$$\begin{bmatrix} \bar{x}(t-1) \\ y(t-1) \end{bmatrix} = \begin{bmatrix} \bar{A} & 0 \\ \bar{C} & 0 \end{bmatrix} \begin{bmatrix} \bar{x}(t) \\ y(t) \end{bmatrix} + \begin{bmatrix} \bar{B} \\ \bar{D} \end{bmatrix} \bar{w}(t-1) \tag{2.14}$$

where $\bar{w}(t)$ *is the generating white noise process of* $H^+ \vee X^+$, *i.e.* $H^+ \vee X^+ = H^+(\bar{w})$ *and*

$$\bar{A} = E\bar{x}(t-1)\bar{x}(t)'\bar{P}^{-1} \qquad \bar{B} = E\bar{x}(t)\bar{w}(t)' \tag{2.15}$$

$$\bar{C} = Ey(t-1)\bar{x}(t)'\bar{P}^{-1} \qquad \bar{D} = Ey(t)\bar{w}(t)' \tag{2.16}$$

where $\bar{P} = E\bar{x}(t)\bar{x}(t)'$.

Taking $\bar{x}(t)$ *as the* dual basis *of* $x(t)$, *i.e.*

$$E\bar{x}(t)x(t)' = I$$

which implies

$$\bar{x}(t) = P^{-1}x(t), \quad \bar{P} = P^{-1},$$

the matrices of the backward representation $(\bar{A}, \bar{B}, \bar{C}, \bar{D})$ *are related to* (A, B, C, D) *by a one-to-one transformation. In particular,*

$$\bar{A} = A' \quad \bar{C}' = APC' + BD' \tag{2.17}$$

The formulas are asymmetric because of the asymmetry in the definition of past and future of y. This asymmetry is needed in order to avoid unnecessarily high state space dimension due to the overlap of past and future spaces of the process. For example, with the symmetric choice of including the present both in the future and in the past, a p-dimensional white noise process w (which is Markov) would admit its present space (spanned by $w(0)$) as a minimal Markovian splitting subspace and hence admit a minimal realization with a state space of dimension p. The choice here is to have the present only in $H^+(y)$.

3. Spectral Factorization

The covariance sequence of the output process y of the system (2.1), i.e.

$$\Lambda(t) := \mathrm{E}\{y(t+k)y(k)'\} = \mathrm{E}\{y(t)y(0)'\}$$

is readily computed. We see that

$$\Lambda(t) = CA^{t-1}\bar{C}' \quad \text{for } t > 0, \quad \Lambda(0) = CPC' + DD' \tag{3.1}$$

where,

$$\bar{C}' = APC' + BD'. \tag{3.2}$$

is exactly the same "backward" C matrix of (2.17) and

$$\Lambda(-t) = \Lambda(t)' = \bar{C}(A')^{t-1}C' \quad \text{for } t > 0.$$

Therefore it follows that the infinite block Hankel matrix

$$\mathbb{H} := \begin{bmatrix} \Lambda(1) & \Lambda(2) & \Lambda(3) & \cdots \\ \Lambda(2) & \Lambda(3) & \Lambda(4) & \cdots \\ \Lambda(3) & \Lambda(4) & \Lambda(5) & \cdots \\ \vdots & \vdots & \vdots & \end{bmatrix}$$

admits a factorization

$$\mathbb{H} = \begin{bmatrix} C \\ CA \\ CA^2 \\ CA^3 \\ \vdots \end{bmatrix} \begin{bmatrix} \bar{C} \\ \bar{C}A' \\ \bar{C}(A')^2 \\ \bar{C}(A')^3 \\ \vdots \end{bmatrix}' \tag{3.3}$$

and hence has finite rank bounded above by the dimension n of the state space X_t of the system Σ. Whether or not $\operatorname{rank} \mathbb{H} = n$ depends on the reachability of the pair $(A, \bar{C}')$, which is equivalent (given that (A, C) is observable by assumption) to *stochastic minimality* of the system (2.1) viewed as a representation of the output process y [47, 49, 50]. Note that

Proposition 3.1. *The backward state-output matrix $\bar{C}$ is uniquely determined by the forward parameters (A, C) and is invariant for all stochastically minimal realization (2.1) of y having the same (observable) (A, C) pair.*

We noted in the previous section that the output process y of (2.1) is a purely non-deterministic process. It is well known that this property is equivalent to

$$a'y(t) \notin H_t^-(y) \quad a \in \mathbb{R}^n.$$

i.e. for no $a \in \mathbb{R}^n$ $a'y(t)$ can be exactly equal to a linear combination of components of past variabes $y(t-1), y(t-2), \ldots$ of the process. From this it can be easily shown that the block Toeplitz matrix

$$T_k := \begin{bmatrix} \Lambda(0) & \Lambda(1) & \Lambda(2) & \cdots & \Lambda(k) \\ \Lambda(1)' & \Lambda(0) & \Lambda(1) & \cdots & \Lambda(k-1) \\ \Lambda(2)' & \Lambda(1)' & \Lambda(0) & \cdots & \Lambda(k-2) \\ \vdots & \vdots & \vdots & & \\ \Lambda(k)' & \Lambda(k-1)' & \Lambda(k-2)' & \cdots & \Lambda(0) \end{bmatrix} \tag{3.4}$$

is *(strictly) positive definite for all k.*

For a purely non-deterministic process the spectral distribution is absolutely continuous [68] and admits a density. In our case the $m \times m$ *spectral density* of y can even be computed as an ordinary Fourier transform i.e.

$$\Phi(z) = \sum_{t=-\infty}^{\infty} \Lambda(t) z^{-t}.$$

Since A is stable the series is absolutely convergent in a neighborhood of the unit circle $\{|z| = 1\}$ of the complex plane and clearly has the property

$$\Phi(1/z) = \Phi(z)'$$

which sometimes is called *para-Hermitian symmetry.* We may write

$$\Phi(z) = \Phi_+(z) + \Phi_+(1/z)' \tag{3.5}$$

where $\Phi_+(z)$ is the "causal" (i.e. analytic outside of the unit circle) component of $\Phi(z)$, given by

$$\Phi_+(z) = \frac{1}{2}\Lambda(0) + \Lambda(1)z^{-1} + \Lambda(2)z^{-2} + \cdots \tag{3.6}$$

$$= C(zI - A)^{-1}\bar{C}' + \frac{1}{2}\Lambda(0). \tag{3.7}$$

The positivity condition of the sequence of Toplitz matrices (3.4) is equivalent to positive semidefiniteness of $\Phi(z)$ on the unit circle i.e.

$$\Phi_+(e^{j\theta}) + \Phi_+(e^{-j\theta})' \geq 0 \quad \theta \in [-\pi, \pi] \tag{3.8}$$

which can be rewritten as $\Re e \Phi_+(e^{j\theta}) \geq 0$. From this, since Φ_+ has by construction all of its poles strictly inside the unit circle it is seen that it is a *positive real* function. We shall call Φ_+ the *positive real part* of Φ.

Proposition 3.2. *The transfer function W of any state space representation of the process y of the type (2.1) is a* spectral factor *of Φ, i.e.*

$$W(z)W(1/z)' = \Phi(z). \tag{3.9}$$

There is a very straightforward proof of this result in case of a stable A matrix, based on the well-known formula for computing the output spectrum of a linear time-invariant filter with stationary input (this formula is sometimes called the Wiener-Kintchine theorem).

There is however also a purely algebraic proof based on an astute decomposition of the product $W(z)W(1/z)'$ which works in general for proper rational transfer functions and does not require stability of A and stationarity of the signals involved (of course in this case the "spectrum" $\Phi(z)$ is *defined* by the formulas (3.5) plus (3.7) and need not have a probabilistic meaning). The decomposition is based on a famous trick apparently invented by Kalman and Yakubovich, namely the identity

$$P - APA' = (zI - A)P(z^{-1}I - A') + (zI - A)PA' + AP(z^{-1}I - A'). \tag{3.10}$$

Proof. A straightforward calculation shows that

$$\begin{aligned} W(z)W(1/z)' &= [C(zI-A)^{-1}B+D][B'(z^{-1}I-A')^{-1}C'+D'] \\ &= C(zI-A)^{-1}BB'(z^{-1}I-A')^{-1}C' \\ &+ C(zI-A)^{-1}BD'+DB'(z^{-1}I-A)^{-1}C'+DD' \end{aligned}$$

so, in view of (2.7), (3.10),

$$\begin{aligned} W(z)W(1/z)' &= CPC'+DD'+C(zI-A)^{-1}(APC'+BD') \\ &+ (CPA'+DB')(z^{-1}I-A')^{-1}C' \\ &= \Phi_+(z)+\Phi_+(1/z)'. \end{aligned} \tag{3.11}$$

where the last equality follows from (3.2). □

Note that (3.11) only requires existence of a solution to the Lyapunov equation $P = APA' + BB'$. In case A is stable this is of course guaranteed. In addition, W has all its poles inside the unit circle. Such a W is called a *stable* or, better, *analytic spectral factor*.

We shall need to consider also *antistable* or *coanalytic* (i.e. analytic in $\{|z| < 1\}$) spectral factors i.e. (rational) solutions of $\bar{W}(z)\bar{W}(1/z)' = \Phi(z)$, having all poles ouside of the unit circle. These spectral factors are in one-to-one correspondence with the stable factors $G(z)$ of the transpose spectrum $\Phi(z)'$ by the formula

$$\bar{W}(z) = G(1/z)$$

so that $\bar{W}(z)\bar{W}(1/z)' = G(1/z)G(z)' = \Phi(z)$.

By the same reasoning as done for stable spectral factors, antistable spectral factors turn out to be exactly the transfer functions of backward realizations of y, i.e. state-space representations of the form (2.14). For the transfer function of a backward model (2.14) can be written

$$\bar{W}(z) = \bar{C}(z^{-1}I-\bar{A})^{-1}\bar{B}+\bar{D}$$

where the $\bar{A}$ matrix is stable, i.e. has all eigenvalues inside of the unit circle. Since the realization (3.7) of Φ_+ induces a natural transpose realization for the transpose $\Phi_+(z)'$, namely

$$\Phi_+(z)' = \bar{C}(zI-A')^{-1}C' + \frac{1}{2}\Lambda(0), \tag{3.12}$$

we see that the dual choice of basis of Theorem 2.2 for the backward models is a natural one. Hence by just switching symbols according to the correspondence

$$A \leftrightarrow A' \quad C \leftrightarrow \bar{C},$$

one obtains characterizations of the family of antistable spectral factors and the corresponding backward models which are completely analogous to those for stable spectral factors and forward realizations.

An important observation to keep in mind is that even though we assumed minimality of the realization (A, B, C) in 2.1, the pair $(A, \bar{C}')$ *may not be reachable* and hence

$$\Phi_+(z) = C(zI - A)^{-1}\bar{C}' + \frac{1}{2}\Lambda(0)$$

may not be a minimal realization. This would imply that the McMillan degree of the spectrum is smaller than what appears from the spectral factorization equation (namely $2n$). In fact, we have the following proposition for the McMillan degrees of rational functions, whose proof can be found in Anderson's paper [5].

Proposition 3.3. *Let $\delta\{\cdot\}$ denote McMillan degree. Then:*

(i) If the rational matrices Z_1 and Z_2 have no poles in common, then

$$\delta(Z_1 + Z_2) = \delta(Z_1) + \delta(Z_2).$$

(ii) If W_1 and W_2 are rational matrix functions of compatible dimensions, then

$$\delta(W_1 W_2) \leq \delta(W_1) + \delta(W_2).$$

Applying this to

$$W(z)W(1/z)' = \Phi(z) = \Phi_+(z) + \Phi_+(1/z)',$$

we have

$$\delta(W) \geq \frac{1}{2}\delta(\Phi) = \delta(\Phi_+). \tag{3.13}$$

If we have equality, we say that W is a *minimal spectral factor* .

Well-known examples of minimal stable spectral factor are the *minimum phase* , sometimes also called the *outer*, and the *maximum phase* spectral factors, denoted $W_-(z)$ and $W_+(z)$ respectively. Both $W_-(z)$ and $W_+(z)$ are stable (i.e. analytic in $\{|z| \geq 1\}$) but the first has no zeros outside of the closed unit disk while the second has instead no zeros inside the open unit disk.

Dually, there are unique minimal antistable (or co-analytic) spectral factors with all the zeros outside or, respectively, inside of the unit circle, denoted[9] $\bar{W}_+$ and $\bar{W}_-$ respectively. The factor $\bar{W}_+$ is commonly called *conjugate minimum-phase* or *co-outer*.

Theorem 3.1. *All stable rational spectral factors can be constructed by postmultiplying the minimum phase factor by a stable rational matrix function $Q(z)$ such that*

$$Q(z)Q(z^{-1})' = I.$$

[9] The rationale for the subscripts will become clear from the *partial order* of realizations which we shall see in a moment.

Dually, all antistable rational spectral factors can be constructed by postmultiplying the minimum phase factor by an antistable rational matrix function $\bar{Q}(z)$ *such that*

$$\bar{Q}(z)\bar{Q}(z^{-1})' = I$$

Transfer function like Q or $\bar{Q}$ are called *all-pass* . Stable all-pass functions are called *inner*. The result above goes back to Youla's classical 1961 paper [77].

4. Spectral Factorization and the LMI

Let us now consider the following inverse problem: Given a proper rational spectral density Φ i.e. an $m \times m$ parahermitian matrix of (generic) full rank m, positive semidefinite on the unit circle, consider the problem of finding all *minimal stable* spectral factors W and the corresponding (minimal) realizations $W(z) = D + H(zI - F)^{-1}B$. (The condition that Φ is proper implies that all rational spectral factors are proper so that they have representations of this form). To solve this problem, first make the decomposition

$$\Phi(z) = \Phi_{+}(z) + \Phi_{+}(1/z)'$$

where $\Phi_{+}(z)$ has all its poles strictly inside the unit disk (so it is the positive real part of $\Phi(z)$) and compute a minimal realization

$$\Phi_{+}(z) = C(zI - A)^{-1}\bar{C}' + J,$$

where clearly

$$J + J' = \Lambda(0).$$

Note that A is a stable matrix. We shall solve the spectral factorization equation (3.9), giving a procedure to compute (F, H, B, D) from the "data" $(A, C, \bar{C}, \Lambda(0))$.

The problem can actually be reduced to finding just the B's and D's since F and H can be chosen equal for all factors.

Theorem 4.1. *Let* $(A, C, \bar{C}')$ *be a minimal realization. There is a one-to-one correpondence between minimal stable spectral factors of* $\Phi(z)$*, and symmetric* $n \times n$ *matrices* P *solving the* Linear Matrix Inequality

$$M(P) := \begin{bmatrix} P - APA' & \bar{C}' - APC' \\ \bar{C} - CPA' & \Lambda(0) - CPC' \end{bmatrix} \geq 0 \tag{4.1}$$

in the following sense:

Corresponding to each solution $P = P'$ *of (4.1), necessarily positive definite, consider the unique (modulo orthogonal transformations) full column rank matrix factor* $\begin{bmatrix} B \\ D \end{bmatrix}$ *of* $M(P)$,

$$M(P) = \begin{bmatrix} B \\ D \end{bmatrix} [B'D'] \tag{4.2}$$

and the rational matrix W parametrized in the form

$$W(z) = C(zI - A)^{-1}B + D. \tag{4.3}$$

Then (4.3) is a minimal realization of a stable minimal spectral factor of $\Phi(z)$. Conversely, for each stable minimal spectral factor W, with minimal realization $D + H(zI - F)^{-1}B$ we can choose a basis such that $F = A$ and $H = C$, and the corresponding pair $\begin{bmatrix} B \\ D \end{bmatrix}$ together with the solution $P = P'$ of the Lyapunov equation (2.7) satisfy the matrix equation (4.2) and hence the Linear Matrix Inequality (4.1).

Proof. Let $P = P'$ be a solution of (4.1) and B, D be computed as in (4.2). Then P solves the Lyapunov equation (2.7) and hence $P > 0$. Then forming the product $W(z)W(1/z)'$ it follows from the equation (3.11) above that $W = D + (zI - A)^{-1}B$ is a stable spectral factor. Note that (A, B) must be reachable for otherwise the McMillan degree of W, would be $\delta(W) < n = \frac{1}{2}\delta(\Phi)$ which contradicts (3.13). Therefore $W = D+(zI-A)^{-1}B$ is a minimal spectral factor.

To show the converse, assume $W = D+H(zI-F)^{-1}B$ is a minimal stable spectral factor. Then a $P = P' > 0$ exists solving the Lyapunov equation $BB' = P - FPF'$ and hence from the spectral factorization equation and the Kalman-Yakubovich identity we get

$$\begin{aligned} &\Phi_+(z) + \Phi_+(1/z)' = W(z)W(1/z)' = \\ &\begin{bmatrix} H(zI - F)^{-1} & I \end{bmatrix} \begin{bmatrix} BB' & BD' \\ DB' & DD' \end{bmatrix} \begin{bmatrix} (z^{-1}I - F')^{-1}H' \\ I \end{bmatrix} = \\ &HPH' + DD' + H(zI - F)^{-1}(FPH' + BD') + \\ &+(HPF' + DB')(z^{-1}I - F')^{-1}H'. \end{aligned}$$

We first argue that without loss of generality we can take $F = A$ and $H = C$. This follows readily since the equality above implies that Φ_+ is also realized by a matrix triple of the form (F, G, H). Now since we are considering only spectral factors for which $\delta(W) = \delta(\Phi_+)$, this realization must also be minimal and then the pairs (F, H) and (A, C) must be similar. In fact we may take $F = A$ and $H = C$ for all minimal spectral factors.

It is then obvious that (P, B, D) satisfy (4.2) and hence (4.1) has a positive definite solution (namely P). □

The equations (4.2) are sometimes called the *positive real equations*, for reasons to be explained below, and can be written

$$\underbrace{\begin{bmatrix} P - APA' & \bar{C}' - APC' \\ \bar{C} - CPA' & \Lambda(0) - CPC' \end{bmatrix}}_{M(P)} = \begin{bmatrix} B \\ D \end{bmatrix} \begin{bmatrix} B' & D' \end{bmatrix} \geq 0$$

The linear function $M : \mathbb{R}^{n\times n} \to \mathbb{R}^{2n\times 2n}$ depends only on $(A, \bar{C}, C, \Lambda(0))$ which are given.

One immediate consequence of Theorem 4.1 is that the *dimension* of the minimal spectral factors can be computed from the rank of the corresponding matrix $M(P)$. In fact if we agree to keep

$$\operatorname{rank}\begin{bmatrix} B \\ D \end{bmatrix}$$

full, it follows immediately from the the factorization above that the corresponding $W(z)$ is $m \times p$ where $p = \operatorname{rank} M(P)$.

It can be shown [25] that the set of solutions to the LMI (4.1)

$$\mathcal{P} := \{PJ \mid P' = P,\ M(P) \geq 0\}$$

is closed, bounded and convex. Later we shall show that there are two special elements $P_-, P_+ \in \mathcal{P}$ so that

$$P_- \leq P \leq P_+ \quad \text{for all } P \in \mathcal{P}$$

where $P_1 \leq P_2$ means that $P_2 - P_1 \geq 0$ is positive semidefinite.

For completeness, we also state the following well-known result. We have made it appear as a corollary to Theorem 4.1 although historically things went quite the other way.

Positive Real Lemma (Kalman-Yakubovich-Popov). *The family $\mathcal{P}$ is nonempty if and only if Φ_+ is positive real, i.e. (3.8) holds.*

Therefore, in our case, $\mathcal{P} \neq \emptyset$.

The Dual Positive-Real Equations. A dual of Theorem 4.1 providing a one-to-one and onto pararmetrization of minimal antistable factors in terms of the solutions $\bar{P}$ of the *dual Linear Matrix Inequality*

$$\bar{M}(\bar{P}) := \begin{bmatrix} \bar{P} - A'\bar{P}A & C' - A'\bar{P}\bar{C}' \\ C - \bar{C}\bar{P}A & \Lambda(0) - \bar{C}\bar{P}\bar{C}' \end{bmatrix} \geq 0 \tag{4.4}$$

can readily be obtained by replacing the realization $\Phi_+(z) = C(zI - A)^{-1}\bar{C}' + J$, by the transpose realization representing $\Phi_+(z)'$ and repeating verbatim the proof above, see also [48].

Then to each $\bar{P} \in \bar{\mathcal{P}}$, solution set of the dual Linear Matrix Inequality (4.4) there corresponds an antistable minimal spectral factor

$$\bar{W}(z) = \bar{C}(z^{-1}I - A')^{-1}\bar{B} + \bar{D},$$

where $\bar{B}, \bar{D}$ are determined by the analog of the matrix factorization (4.2).

In the following we shall assume that

$$R(P) := \Lambda(0) - CPC' > 0 \tag{4.5}$$

for all $P \in \mathcal{P}$. This means that all minimal state space models of y have a full-rank additive noise term in the output equation ($DD' > 0$). This condition serves here only the purpose of avoiding the use of pseudo-inverses and of simplifying the exposition. We shall see in a moment that (4.5) implies that $\Phi(z)$ is (generically) of full rank m. It is curious that a natural characterization of the spectra for which this condition holds seems still to be an open question in the literature (see however the forthcoming paper [52]). Under this assumption, if $T := -(\bar{C}' - APC')R^{-1}$, a straight-forward calculation yields

$$\begin{bmatrix} I & T \\ 0 & I \end{bmatrix} M(P) \begin{bmatrix} I & 0 \\ T' & I \end{bmatrix} = \begin{bmatrix} -\Lambda(P) & 0 \\ 0 & R \end{bmatrix},$$

where

$$\Lambda(P) = APA' - P + (\bar{C}' - APC')R(P)^{-1}(\bar{C} - CPA), \tag{4.6}$$

Hence, $M(P) \geq 0$ if and only if P satisfies the *Riccati inequality*

$$\Lambda(P) \leq 0, \tag{4.7}$$

and

$$p = \operatorname{rank} M(P) = m + \operatorname{rank} \Lambda(P).$$

If P satisfies the algebraic Riccati equation

$$\Lambda(P) = 0, \tag{4.8}$$

$\operatorname{rank} M(P) = m$, the corresponding spectral factor W is *square* $m \times m$. These P form a subfamily $\mathcal{P}_0$ in $\mathcal{P}$. If $P \notin \mathcal{P}_0$, W is rectangular.

From spectral factors to stochastic realizations. We now examine the converse of Proposition 3.2. Let W and $\bar{W}$ be two minimal square stable and antistable spectral factors. It is easy to see that such factors play the role of transfer functions of "shaping filters" of the type (2.2) for the process y. To see this we just need to manifacture two white noise processes w and $\bar{w}$ serving as input white noise processes in the two filters, the filter with transfer function W being causal (stable) and the other anticausal and hence represented in the time domain by a convolution operator integrating the input "backwards in time". Since W and $\bar{W}$ are square and invertible transfer functions, the white noise processes can be generated by passing y through the "whitening filters" W^{-1} and $\bar{W}^{-1}$. (The general idea is just the same as the classical "whitening-shaping" filter dicotomy of Bode and Shannon.) It can be shown[10] that the whitened processes $w, \bar{w}$ are in fact well-defined functionals of the history of y.

In particular, since W_- is outer, the corresponding white noise process w_- is a causal functional of y, i.e. $w_-(t-1) \in H_t^-(y)$ for all t, so that we actually have $H_t^-(w_-) = H_t^-(y)$. For this reason, w_- is called the (normalized)

[10] A precise statement of this requires spectral representation theory and can not be given here. For a similar argument see Rozanov's book [68], chapter 7.

forward innovation process of y [74]. Similarly the white noise process $\bar{w}_+$ is an anticausal functional of y, i.e. $\bar{w}_+(t) \in H_t^+(y)$, so that $H_t^+(\bar{w}_+) = H_t^+(y)$ for all t; $\bar{w}_+$ is called the (normalized) *backward innovation* process of y.

Once the white noise inputs are determined, it is rather obvious that a minimal realization (A, B, C, D) of W and (from the dual LMI) a dual minimal ralization $(A', \bar{B}, \bar{C}, \bar{D})$ of $\bar{W}$ will provide two minimal state-space stochastic realizations of the process y, the first one being causal and the second anticausal. The state processes of the two realizations will have as covariance matrices the unique solutions of the Lyapunov equations (2.7) and, respectively,

$$\bar{P} = A'\bar{P}A + \bar{B}\bar{B}'$$

In force of Theorem 4.1, each solution P of the Positive-real equations (4.1) in $\mathcal{P}_0$ will then automatically be interpretable as the state covariance matrix of the state-space realization of y corresponding to the deterministic realization (A, B, C, D) of W. Dually any solution $\bar{P}$ of the dual Positive-real equations (4.4) will be the state covariance of a backward state-space realiztion. In other words, the P (and $\bar{P}$) matrices solutions of he LMI (for the time being belonging to the subsets $\mathcal{P}_0$ and $\bar{\mathcal{P}}_0$) have the meaning of *state covariances* of minimal forward and backward realizations of y.

This picture however generalizes also to all minimal (nonsquare) spectral factors. The only difficulty in the generalization is the nonuniqueness of the white generating noises w and $\bar{w}$ associated to the spectral factors. The difficulty can be overcomed by selecting the input noises in a fixed ambient space, which is small enough to make the w's unique but also big enough to allow a solution w of the convolution equation $y = Ww$ for each minimal spectral factor W (and $\bar{W}$).

Proposition 4.1. *[50] There exists a fixed "universal" stationary Hilbert space $H \supset H(y)$ such that for each minimal spectral factor W the convolution equation $y = Ww$ has a unique (modulo orthogonal transformations) solution w with the property $H(w) \subset H$.*

Assume that the dimension of a minimal stochastic realization of y is n. Then the "universal" stationary Hilbert space H can be chosen equal to the orthogonal sum

$$H = H(y) \oplus H(z)$$

where z is a fixed n-dimensional normalized white noise process uncorrelated with y.

Hence, in order to construct all possible minimal shaping filter representations of y we need, besides the process y itself, n additional "exogenous" independent white noise generators. The filters for the actual generation of the noise processes are discussed in [50], sect. 5.2.

Once we know how to construct the w or $\bar{w}$ processes we can associate to each minimal spectral factor W ($\bar{W}$) a minimal stochastic realization of y by just picking a (minimal) realization (A, B, C, D) of W (resp. a minimal realization $(\bar{A}, \bar{B}, \bar{C}, \bar{D})$ of $\bar{W}$). The state vector of the realization will be a causal or anticausal functional in the past or future space of the white noise processes w or $\bar{w}$. Clearly if we choose all w 's in a fixed universal Hilbert space H, the state spaces X of these realizations will all be subspaces of H.

Theorem 4.2. *Let $(A, C, \bar{C}, \Lambda(0))$ be a minimal realization of $\Phi_+(z)$. There are bijective maps between the following sets*

1. $\mathcal{W}$ *: The minimal stable spectral factors W (defined modulo right multiplication by a constant orthogonal matrix)*
2. $\mathcal{P}$ *:The solutions P of the Linear Matrix inequality (4.1)*
3. $\mathcal{X}$ *: The minimal Markovian splitting subspaces for y in a fixed universal Hilbert space H as described in Proposition 4.1.*

Proof. The one-to-one and onto correspondence between $\mathcal{W}$ and $\mathcal{P}$ has been described already in Theorem 4.1. The correspondence between $\mathcal{X}$ and $\mathcal{P}$ is established through a particular choice of basis in the state space which will be described in the next section. □

Of course we have a dual version of this result in which the antistable spectral factors $\bar{\mathcal{W}}$ and the solutions of the dual Linear Matrix Inequality $\bar{\mathcal{P}}$ replace $\mathcal{W}$ and $\mathcal{P}$.

4.1 Ordering, (A,C) Pairs and Uniform Choice of Basis in $\mathcal{X}$

In this section we shall make explicit the correspondence between state covariance matrices P solutions of the LMI (4.1) and Markovian splitting subspaces in $\mathcal{X}$. In fact we shall establish a correspondence between P's and stochastic state-space realizations of y. This correspondence is intimately related to the notion of a *uniform choice of basis* in the family of minimal splitting subspaces $\mathcal{X}$, which will be defined below. This notion will be useful for understanding the geometric approach to identification and the idea of stochastic balancing.

Recall [50, sect. 6] that in the family of minimal Markovian splitting subspaces $\mathcal{X}$ one can introduce a natural partial order (denoted $\prec$), defined in terms of the *cosine of the angle* that each X makes with the future space[11] H^+, a subspace X_2 being "greater" than X_1 if it is "closer" (i.e. it makes a smaller angle) with the future than X_1.

According to this definition the forward and backward predictor spaces, X_- and X_+, defined in (2.9) are naturally the smallest and largest element in the family $\mathcal{X}$ with respect to the partial order.

[11] The (cosine of the) "angle" between subspaces is defined e.g. in [4, vol. I] p. 69. It is just the smallest *canonical correlation coefficient* of the two subspaces of random variables X and H^+. The "angle" is the largest *principal angle* between the two subspaces. For these notions in a finite-dimensional setting one may consult e.g. [32].

Consider a minimal causal realization (2.1). By minimality the components of the n–vector $x(0)$ must form a basis in the relative splitting subspace X. Now, we recall from [49], [50], that two families of bases, say $\{x(0)\}$ and $\{\bar{x}(0)\}$, for the family $\mathfrak{X}$ (i.e. each vector $x(0)$ is a basis in one X and similarly for each $\bar{x}(0)$) are called *uniformly ordered*, (or, for short, *uniform*) respectively in the *forward* or in the *backward sense*, if whenever $X_1 \prec X_2$ and $x_i(0)$ are bases in X_i, $(i = 1, 2)$, there holds

$$E^{X_1} x_2(0) = x_1(0) \tag{4.9}$$

or, respectively,

$$E^{X_2} \bar{x}_1(0) = \bar{x}_2(0) \tag{4.10}$$

the vectors $\bar{x}_i(0)$ being bases for the subspaces X_i, $(i = 1, 2)$.

Uniformly ordered bases can be constructed easily. For example pick a basis $x_+(0)$ in the "largest" state space X_+, then it can be easily seen by using the splitting property of X that the family

$$x(0) := \mathrm{E}^X x_+(0), \quad X \in \mathfrak{X}$$

is a forward-uniform basis [50]. It also follows from the definition that for all bases $x(0)$ in a forward-uniform family, and, respectively, for all $\bar{x}(0)$ in a backward-uniform family we have the *invariant projection* property

$$E^{X_-} x(0) = x_-(0) \tag{4.11}$$

$$E^{X_+} \bar{x}(0) = \bar{x}_+(0) \tag{4.12}$$

where $x_-(0)$ is the basis relative to the forward predictor space X_- in the first family and $\bar{x}_+(0)$ is the basis relative to the backward predictor space X_+ in the second family.

Proposition 4.2. *[50] A forward-uniform choice of bases in $\mathfrak{X}$ establishes a* lattice isomorphism *between $\mathfrak{X}$ and the corresponding family of state covariance matrices $\mathcal{P} := \{P = \mathrm{E}x(0)x(0)' \,|\, x(0)$ basis in $X\}$, the latter set being endowed with the natural partial order of positive semidefinite matrices. This is equivalent to saying that $X_1 \prec X_2 \Leftrightarrow P_1 \leq P_2$.*

In a backward-uniform choice the ordering of the corresponding state covariance matrices $\bar{\mathcal{P}} := \{\bar{P} = \mathrm{E}\bar{x}(0)\bar{x}(0)' \,|\, \bar{x}(0)$ basis in $X\}$ is reversed, namely, $X_1 \prec X_2 \Leftrightarrow \bar{P}_2 \leq \bar{P}_1$.

Proof. We shall prove only the implication $\Rightarrow$. For the reverse consult [50], p. 282-283. Everything descends from the orthogonality

$$x_2(0) - \mathrm{E}^{X_1} x_2(0) = x_2(0) - x_1(0) \perp x_1(0)$$

which is equivalent to the defining (4.9). From this it follows that the covariance matrix of $x_2(0) - x_1(0)$ is equal to $P_2 - P_1$ and hence $P_2 \geq P_1$. The backward case follows by symmetry. □

In particular, the bases $x_-(0)$ in X_- and $x_+(0)$ in X_+ in a forward-uniform choice, will have the smallest and, respectively, the largest state covariance matrices P_- and P_+ in $\mathcal{P}$. For a backward-uniform family it will instead happen that $\bar{P}_-$ is maximal and $\bar{P}_+$ is minimal.

That P_- and $\bar{P}_+$ are the covariance matrices of bases in the forward and backward predictor spaces can be seen also directly from the invariant projection property (4.11, 4.12). In fact $x_-(0)$ and $\bar{x}_+(0)$ are essentially the "forward and backward steady–state Kalman filter" estimates of any $x(0)$ (respectively, $\bar{x}(0)$) in a uniform family the bases. For, the first vector can be expressed as $x_-(0) := \mathrm{E}^{X_-} x(0)$ which is in turn equal to $\mathrm{E}^{H^-} x(0)$ by the splitting property. Hence $x(0) - x_-(0) \perp x_-(0) \in H^-$ so that $P - P_- \geq 0$ for all $P \in \mathcal{P}$. A completely analogous argument shows that $\bar{x}_+(0) = \mathrm{E}^{H^+} \bar{x}(0)$ and that $\bar{P} - \bar{P}_+ \geq 0$ for all $\bar{P} \in \bar{\mathcal{P}}$.

The most useful properties of uniform bases are summarized below.

Proposition 4.3. *[50] If $\{x(0)\}$ is a forward-uniform family of bases in $\mathfrak{X}$ then, for each $x(0)$, the corresponding* dual basis $\bar{x}(0)$, *uniquely defined in X by the condition $Ex(0)\bar{x}(0)' = I$, defines a backward-uniform family in $\mathfrak{X}$.*

Proposition 4.4. *[50] All* causal *minimal realizations of y corresponding to a forward-uniform family of bases are described by the same (A, C) pair. Conversely, a choice of bases in the subspaces $X \in \mathfrak{X}$ yielding state equations with the same (A, C) pairs is forward-uniform.*

Likewise, all anticausal *realizations corresponding to a backward-uniform family of bases are described by the same $(\bar{A}, \bar{C})$ parameters and conversely.*

Therefore choosing the realizations of spectral factors in $\mathcal{W}$ in such a way that the A and C matrices are the same (Theorem 4.1) is exactly the same thing as choosing a forward uniform basis in $\mathfrak{X}$. Dually, keeping the $\bar{A}, \bar{C}$ matrices invariant over the family of all minimal realizations of spectral factors $\bar{W} \in \bar{\mathcal{W}}$ is the same thing as having the corresponding state vectors $\bar{x}(0)$ related as in a backward uniform choice of basis.

The only backward-uniform family of bases we shall encounter in the following will be the *dual bases of a forward family.* Since these are related by the transformation $\bar{x}(0) = P^{-1}x(0)$ so that

$$\bar{P} = \mathrm{E}\bar{x}(0)\bar{x}(0)' = P^{-1} \tag{4.13}$$

it follows immediately that

Proposition 4.5. *The solution sets $\mathcal{P}$, $\bar{\mathcal{P}}$ of the Linear Matrix Inequality and of its dual (4.4) are related by the transformation $\bar{P} = P^{-1}$. In particular, the maximal element in $\mathcal{P}$ is $P_+ = \bar{P}_+^{-1}$.*

The following result says that choosing a uniform basis is after all as easy as choosing one basis in a particular X.

Proposition 4.6. *An arbitrary basis $x(0)$ in a minimal splitting subspace X can be uniquely extended to the whole family $\mathcal{X}$ in a uniform way (either in the forward or backward sense).*

Proof. First, compute $\bar{x}_+(0)$ by projecting $\bar{x}(0) := P^{-1}x(0)$ onto X_+, i.e., $\bar{x}_+(0) := \mathrm{E}^{X_+}P^{-1}x(0)$ and then go back to the "primal", i.e. let $x_+(0) = P_+^{-1}\bar{x}_+(0)$. Note that the projection of $x(0)$ onto X_+ is *not invariant* in a forward uniform basis.

Once $x_+(0)$ and (by a dual argument) $\bar{x}_-(0)$ are found, they can be used to generate two (dual families of uniform bases in $\mathcal{X}$, say $z(0)$ and $\bar{z}(0)$, by setting $z(0) := \mathrm{E}^X x_+(0)$, and $\bar{z}(0) := \mathrm{E}^X \bar{x}_-(0)$. It is immediate to check that $z(0)$ and $\bar{z}(0)$ are indeed dual bases and are related by the transformation $\bar{z}(0) = P^{-1}z(0)$ where $P = \mathrm{E}z(0)z(0)'$. □

5. Finite-Interval Realizations of a Stationary Process

Since in identification data are always finite we need to examine the problem of modeling the process y (which will still be assumed stationary and with a rational spectral density as in the previous two sections) on a *finite interval.* There seems to be no reason to worry about this since a stationary realization of the type (2.1) does of course describe the process also on any finite interval $[0, T]$. In this case however we must specify the *initial conditions vector* $x(0)$ at time zero, which is an essential extra parameter for a complete description of the process. The initial condition must in principle be estimated if the model class is chosen to consist of stationary realizations (2.1).

As we shall see there are *non-stationary* finite-interval realizations where the initial state is automatically fixed to zero and we do not need to worry about its estimation.

5.1 Forward and Backward Kalman Filtering and the Family of Minimal Stationary Realizations of y

Let us recall the structure of the Kalman filter for a minimal (forward) stochastic realization (2.1). Assume we are observing the output of the system starting at some finite time t_0 and define

$$H_{[t_0,t)} = \mathrm{span}\{a'y(s); a \in \mathbb{R}^m, t_0 \le s < t\}.$$

Then, the Kalman estimate

$$\hat{x}(t) = \mathrm{E}^{H_{[t_0,t)}}x(t) \tag{5.1}$$

is given by[12]

[12] Because of joint stationarity of x, y all time-varying quantities below depend on the time difference $t - t_0$. To keep notations simple we write this difference simply as t.

$$\hat{x}(t+1) = A\hat{x}(t) + K(t)[y(t) - C\hat{x}(t)]; \quad \hat{x}(t_0) = 0, \tag{5.2}$$

where the *Kalman gain matrix* $K(t)$ is given by

$$K(t) = [AQ(t)C' + BD'][CQ(t)C' + DD']^{-1}$$

$Q(t)$ being the error covariance matrix

$$Q(t) = \mathrm{E}\{[x(t) - \hat{x}(t)][x(t) - \hat{x}(t)]'\}$$

which is the solution of the *matrix Riccati equation*

$$\begin{aligned} Q(t+1) &= AQ(t)A' - \\ &- [AQ(t)C' + BD'][CQ(t)C' + DD']^{-1}[AQ(t)C' + BD']' + BB' \\ Q(0) &= P = \mathrm{E}\{x(0)x(0)'\} \end{aligned} \tag{5.3}$$

This Riccati equation depends on P, B and D, and consequently it varies with different realizations Σ, i.e. with different $P \in \mathcal{P}$. We shall now replace this Riccati equation with an invariant version, which depends only on the invariant parameters $A, C, \bar{C}$ and $\Lambda(0)$ of the spectral density of y. In this way we shall prove that the estimate (5.1) *is the same for all forward stationary realizations of* y.

To this end, note that from the orthogonality $x(t) - \hat{x}(t) \perp \hat{x}(t)$ it follows that $Q(t) = P - \Pi(t)$, where

$$\Pi(t) := \mathrm{E}\{\hat{x}(t)\hat{x}(t)'\}.$$

Then the Riccati equation (5.3) can be written

$$\begin{aligned} P - \Pi(t+1) &= APA' - A\Pi(t)A' \\ &- [\bar{C}' - A\Pi(t)C'][\Lambda(0) - C\Pi(t)C']^{-1}[\bar{C}' - A\Pi(t)C']' + BB' \end{aligned}$$

because $\bar{C}' = APC' + BD'$ and $\Lambda(0) = CPC' + DD'$. But P satisfies the Lyapunov equation

$$P = APA' + BB',$$

and therefore

$$\Pi(t+1) = \Pi(t) + \Lambda(\Pi(t)) \quad \Pi(0) = 0, \tag{5.4}$$

where $\Lambda : \mathbb{R}^{n\times n} \to \mathbb{R}^{n\times n}$ is given by

$$\Lambda(P) = A'PA - P + (\bar{C}' - APC')(\Lambda(0) - CPC')^{-1}(\bar{C}' - APC')'.$$

This is precisely the same quadratic function Λ introduced in the previous section. Moreover we can write,

$$K(t) = [\bar{C}' - A\Pi(t)C'][\Lambda(0) - C\Pi(t)C']^{-1},$$

which also depends only on $(A, C, \bar{C}, \Lambda(0))$ and on the solution of the Riccati equation (5.4). The transient (forward) innovation process

$$e(t) := y(t) - C\hat{x}(t) \quad t \geq t_0,$$

is also invariant. It is in fact a white noise process of covariance matrix

$$R(t) = \Lambda(0) - C\Pi(t)C',$$

i.e. $\mathrm{E}\{e(t)e(s)'\} = R(t)\delta_{ts}$. Hence we have shown the following fact.

Proposition 5.1. *Let $(A, C, \bar{C}, \Lambda(0))$ be a minimial realization of the spectral density matrix of the stationary process y. Then y has a non-stationary realization*

$$\hat{x}(t+1) = A\hat{x}(t) + B(t)\epsilon(t) \quad \hat{x}(t_0) = 0, \tag{5.5}$$
$$y(t) = C\hat{x}(t) + D(t)\epsilon(t) \tag{5.6}$$

on $\{t \geq t_0\}$, where the state $\hat{x}(t)$ is the orthogonal projection onto $H_{[t_0,t)}$ (the minimum variance one-step ahead estimate) of the state $x(t)$ of any minimal stationary realization of y in the uniform basis induced by (A, C), $B(t) := K(t)R(t)^{1/2}$, $D(t) := R(t)^{1/2}$ and $\epsilon(t) := -R(t)^{-1/2}e(t)$ is a normalized m-dimensional white noise process: the normalized transient innovation process *of y on $\{t \geq t_0\}$.*

The state of the *Kalman Filter realization* (5.5, 5.6) is a *non-stationary* process. In effect, since $\mathrm{E}^{H_{[t_0,t)}}y(t+k) = \mathrm{E}^{H_{[t_0,t)}}CA^k x(t) = CA^k\hat{x}(t)$ for all $k \geq 0$, the components of $\hat{x}(t)$ span the "finite-memory" predictor space

$$\check{X}_{t-} := \mathrm{E}^{H_{[t_0,t)}} H_t^+. \tag{5.7}$$

Relation with the forward stationary innovation model. This is well-known. Since each realization Σ is a minimal realization, by standard Kalman Fitering theory $\Pi(t) \to \Pi_\infty > 0$ as $t \to \infty$. Then Π_∞ satisfies the algebraic Riccati equation

$$\Lambda(P) = 0$$

so since it is also symmetric and positive definite, $\Pi_\infty \in \mathcal{P}_0 \subset \mathcal{P}$. Moreover $Q(t) \to Q_\infty \geq 0$, where $Q_\infty = P - \Pi_\infty$, and hence

$$P \geq \Pi_\infty. \tag{5.8}$$

for all P and therefore we see that $P_- = \Pi_\infty$. Now let $t_0 \to -\infty$ in the Kalman filter. Then $\hat{x}(t)$ converges by the (wide-sense) martingale convergence theorem [18] to the *steady state Kalman filter* estimate $x_-(t) = \mathrm{E}^{H_t^-}x(t)$. Moreover both $B(t)$ and $D(t)$ converge to constant matrices which indeed satisfy the Positive-Real lemma equations for $P = P_-$.

Dually, we construct a *backward Kalman filter* based on a minimal backward model

$$\begin{aligned} \bar{x}(t) &= A'\bar{x}(t+1) + \bar{B}\bar{w}(t) \\ y(t) &= \bar{C}\bar{x}(t+1) + \bar{D}\bar{w}(t) \end{aligned} \tag{5.9}$$

assuming we can observe $y(t)$ only in the interval $\{t \leq T\}$. Define the finite future space

$$H_{[t,T]} = \text{span}\{a'y(s); a \in \mathbb{R}^m, t \leq s \leq T\}.$$

and the backward Kalman estimate,

$$\hat{\bar{x}}(t) = \mathrm{E}^{H_{[t,T]}}\bar{x}(t) \tag{5.10}$$

Then an analysis that is completely symmetric to the one presented above, projecting over the future, yields a backward Kalman filter which can be written as a backward non-stationary realization of the stationary process y,

$$\begin{aligned} \hat{\bar{x}}(t) &= A'\hat{\bar{x}}(t+1) + \bar{B}(t)\bar{\epsilon}(t) \quad \hat{\bar{x}}(T) = 0, \\ y(t) &= \bar{C}\hat{\bar{x}}(t+1) + \bar{D}(t)\bar{\epsilon}(t) \end{aligned} \tag{5.11}$$

on $\{t \leq T\}$, where $\bar{B}(t) := \bar{K}(t)\bar{R}(t)^{1/2}$, $\bar{D}(t) := \bar{R}(t)^{1/2}$ and $\bar{\epsilon}(t) := -\bar{R}(t)^{-1/2}\bar{e}(t)$ is a normalized m-dimensional white noise process: the *normalized backward transient innovation process* of y on $\{t \leq T\}$.

The *backward Kalman gain*,

$$\bar{K}(t) = [C' - A'\bar{\Pi}(t)\bar{C}'][\Lambda(0) - \bar{C}\bar{\Pi}(t)\bar{C}']^{-1}, \tag{5.12}$$

is now computed from the solution of the *dual Riccati equation*

$$\bar{\Pi}(t-1) = \bar{\Pi}(t) + \bar{\Lambda}(\bar{\Pi}(t)) \quad \bar{\Pi}(T) = 0, \tag{5.13}$$

where $\bar{\Lambda} : \mathbb{R}^{n\times n} \to \mathbb{R}^{n\times n}$ is defined as

$$\bar{\Lambda}(\bar{P}) = A'\bar{P}A - \bar{P} + (C' - A'\bar{P}\bar{C}')(\Lambda(0) - \bar{C}\bar{P}\bar{C}')^{-1}(C' - A'\bar{P}\bar{C}')'.$$

which is again the same function $\bar{\Lambda}$ introduced in the previous section.

The state $\hat{\bar{x}}(t)$ is the orthogonal projection onto $H_{[t,T]}$ (the minimum error-variance "filter" estimate) of the state $\bar{x}(t)$ of any minimal stationary backward realization of y in the uniform basis induced by $(A', \bar{C})$. As $T \to +\infty$ $\hat{\bar{x}}(t)$ converges to $\bar{x}_+(t)$ and the backward transient model tends to the steady state backward model,

$$\begin{cases} \bar{x}_+(t-1) &= A'\bar{x}_+(t) + \bar{B}_+\bar{w}_+(t-1) \\ y(t-1) &= \bar{C}\bar{x}_+(t) + \bar{D}_+\bar{w}_+(t-1) \end{cases}$$

with state covariance $\bar{P}_+$.

In analogy to (5.7) the state $\hat{\bar{x}}(t)$ now spans the backward predictor space

$$\check{X}_{t+} := \mathrm{E}^{H_{[t,T]}}H_t^- . \tag{5.14}$$

The backward form of the "forward Kalman Filter" and the "forward form" of the backward Kalman Filter. Exactly as it happens in the stationary setup, the joint processes $[\hat{x}(t)'\ y(t-1)']'$ and $[\hat{\bar{x}}(t)'\ y(t)]'$, being Markov processes, admit both a *causal* or *forward* and an *anticausal* or *backward* representation. Unfortunately the name of "backward Kalman Filter" given to the representation (5.11) refers to both circumstances that $\bar{x}(t)$ is a basis in the backward predictor space (i.e. is an estimate based on the future of y) and to the anticausal structure of the realization. These two facts are strictly speaking unrelated and, although it may look a bit unnatural, there is also a *causal* form of the backward Kalman Filter representation (5.11),

$$\begin{aligned} \hat{\bar{x}}(t+1) &= \bar{\Pi}(t+1)A\bar{\Pi}(t)^{-1}\hat{\bar{x}}(t) + \bar{B}_+(t)\epsilon_+(t) \\ y(t) &= [\bar{C}\bar{\Pi}(t+1)A + \bar{D}(t)\bar{B}(t)']\bar{\Pi}(t)^{-1}\hat{\bar{x}}(t) + \bar{D}_+(t)\epsilon_+(t) \end{aligned} \tag{5.15}$$

where $\bar{B}_+(t), \bar{D}_+(t), \epsilon_+(t)$ have expressions which may be derived by the standard procedure explained e.g. in [48]. Of course the validity of the causal form above is subjected to the existence of the inverse $\bar{\Pi}(t)^{-1}$ which requires t "far enough" from the endpoint $t = T$. As shown in [48] and is visible in the last statement of Theorem 2.2, whenever we change causality structure it is convenient to go to the dual basis, namely define

$$\hat{x}_+(t) := \bar{\Pi}(t)^{-1}\hat{\bar{x}}(t) \tag{5.16}$$

and substitute into (5.15), which assume the simpler form

$$\begin{aligned} \hat{x}_+(t+1) &= A\hat{x}_+(t) + B_+(t)\epsilon_+(t) \\ y(t) &= C\hat{x}_+(t) + D_+(t)\epsilon_+(t). \end{aligned} \tag{5.17}$$

The appearence of the C matrix in the output equation is due to the fact that

$$\begin{aligned} [\bar{C}\bar{\Pi}(t+1)A + \bar{D}(t)\bar{B}(t)']\,\hat{x}_+(t) &= \mathrm{E}[\,y(t)\,|\,\hat{x}_+(t)\,] = \\ \mathrm{E}(y(t)\hat{x}_+(t)')[\mathrm{E}\hat{x}_+(t)\hat{x}_+(t)']^{-1}\hat{x}_+(t) &= \mathrm{E}(y(t)\hat{\bar{x}}(t)')\,\hat{x}_+(t), \end{aligned}$$

where, after inserting (5.10), the last member can be computed easily as

$$\begin{aligned} \mathrm{E}y(t)(\mathrm{E}^{H_{[t,T]}}\bar{x}(t))' &= \mathrm{E}y(t)\bar{x}(t)' = \mathrm{E}y(t)(A\bar{x}(t+1) + \bar{B}\bar{w}(t))' \\ &= \bar{C}\bar{P}A' + \bar{D}\bar{B}' = C. \end{aligned}$$

Note the last identity is identical to (2.17).

5.2 Finite-Interval Realizations

The situation of interest is really when both the past and the future spaces are finitely generated. To discuss this context it is better to recall some general facts from the geometric theory.

Consider the class of *Markovian Splitting Subspaces* at time t, X_t, for the process y on the interval $[0,T]$. These subspaces make the joint finite past and future spaces conditionally orthogonal (i.e. uncorrelated) given X_t, i.e.,

$$H_{[0,t)} \vee X_{[0,t]} \perp H_{[t,T]} \vee X_{[t,T]} | X_t. \tag{5.18}$$

where the symbols have an obvious meaning.

It is standard [49, 50] to show that the forward and backward *finite-memory* predictor spaces,

$$\hat{X}_{t-} := \mathrm{E}^{H_{[0,t)}} H_{[t,T]} \quad \text{and} \quad \hat{X}_{t+} := \mathrm{E}^{H_{[t,T]}} H_{[0,t)},$$

are such Markovian splitting subspaces. In fact these predictor spaces are also *minimal splitting*, exactly as in the stationary theory. The following representation theorem (proven in [55]) relates the finite predictor spaces to Kalman filter realizations.

Theorem 5.1. *Let (2.1) be any minimal stationary realization of the process y and let X be the relative state space. Then*

$$X_t := U^t X, \quad 0 \le t \le T \tag{5.19}$$

is a Markovian splitting subspace for the process subordinated by y on the interval $[0,T]$ which is minimal if t is far enough from the endpoints of the interval. Here "far enough" means $t \ge \nu_c$ and $T - t \ge \nu_o$ where ν_c and ν_o are the reachability and observability indices of the pairs $(A, \bar{C}')$ and (A, C) respectively.

Under the same conditions on t and $T-t$, the finite memory predictor spaces $\hat{X}_{t-}$ and $\hat{X}_{t+}$ coincide with $\check{X}_{t-}$ and $\check{X}_{t+}$ and in fact we have the representation formulas

$$\hat{X}_{t-} = \mathrm{E}^{H_{[0,t)}} X_t \quad \textit{and} \quad \hat{X}_{t+} = \mathrm{E}^{H_{[t,T]}} X_t. \tag{5.20}$$

which hold for any stationary splitting subspace X.

One consequence of the representation formulas (5.19) is that for any choice of basis $x(0)$ in a minimal stationary state space X, the process

$$\hat{x}(t) := \mathrm{E}^{H_{[0,t)}} x(t), \quad t \in [0,T] \tag{5.21}$$

is the state process of a finite-interval realization of y which is minimal for all times t "far enough" from the endpoints of the interval. Of course the relative state equations, either in the forward or in the backward representation, are nothing else but the Kalman filter equations for $\{y(t)\}$, described previously. This in a sense is not big surprise.

What is more interesting for identification is the following converse statement. It says that every basis in a finite memory predictor space is the Kalman filter of a uniformly ordered family of minimal stationary realizations. Hence picking a basis in the finite memory predictor space is the same thing as picking a legitimate (A, C) pair of a minimal stationary realization of y.

Theorem 5.2. *Fix any minimal stationary splitting subspace X and assume that t is far enough from the endpoints of the interval $[0,T]$ in the sense explained before. Then, to a basis vector $\hat{x}(t)$ in $\hat{X}_{t-}$ there corresponds a* unique *basis $x(t)$ in X_t for which (5.21) holds. As X varies in the family of all minimal stationary splitting subspaces $\mathcal{X}$ the bases $x(t)$ corresponding to a fixed $\hat{x}(t)$ describe a (forward) uniformly ordered family.*

Consequently a choice of basis in $\hat{X}_{t-}$ defines uniquely a matrix pair (A, C) of a minimal stationary realization of the process y.

A dual statement holds for the backward predictor space: any choice of basis $\hat{\bar{x}}(t)$ in $\hat{X}_{t+}$ defines uniquely a matrix pair $(\bar{A}, \bar{C})$ corresponding to a uniform family of minimal backward stationary realization of y of which $\hat{\bar{x}}(t)$ is the Kalman Filter estimate given the finite future of y.

Proof. The statement follows from the representation (5.20) and the minimality of X_t as a splitting subspace for $H_{[0,t)}$ and $H_{[t,T]}$. Minimality in this sense in particular implies that the constructibility operator,

$$\mathcal{C}_t := \mathrm{E}^{H_{[0,t)}}_{|X_t} : X_t \to \hat{X}_{t-}$$

is *injective* [49]. In other words there are unique random variables $x_k(t) \in X_t$ which are projected onto the components $\hat{x}_k(t)$; $k = 1, \ldots, n$.

The uniform order follows from the identity

$$\hat{x}(t) := \mathrm{E}^{H_{[0,t)}} x(t) = \mathrm{E}^{H_{[0,t)}} \mathrm{E}^{H_t^-} x(t).$$

It follows from this identity and from injectivity of $\mathcal{C}_t$ that all $x(t)$'s must have the same projection onto the infinite past $\mathrm{E}^{H_t^-} x(t)$. But this projection is the same as $\mathrm{E}^{X_{t-}} x(t)$ and so the bases $x(t)$ have the invariant projection property (4.11). The rest is immediate. □

The following concept will play a role in identification.

Definition 5.1. *Two basis vectors $\hat{x}(t)$ in $\hat{X}_{t-}$ and $\hat{\bar{x}}(t)$ in $\hat{X}_{t+}$ are* coherent *if the corresponding (A, C) and $(\bar{A}, \bar{C})$ pairs are such that $\bar{A} = A'$ and $(A, C, \bar{C}, \Lambda(0))$ is a minimal realization of the spectral density matrix of y.*

Proposition 5.2. *Two basis vectors $\hat{x}(t)$ in $\hat{X}_{t-}$ and $\hat{\bar{x}}(t)$ in $\hat{X}_{t+}$ are coherent if and only if*

$$\hat{x}(t) := \mathrm{E}^{\hat{X}_{t-}} \hat{x}_+(t) \tag{5.22}$$

where $\hat{x}_+(t)$ is the dual basis of $\hat{\bar{x}}(t)$ defined in (5.16).

Proof. Consider the causal version, $\hat{x}_+(t)$, of the (unique) backward Kalman Filter corresponding to the uniform family of stationary backward models attached to the pair $(\bar{A}, \bar{C})$ (or to $\hat{\bar{x}}(t)$). Clearly $\hat{\bar{x}}(t)$ is coherent with $\hat{x}(t)$ if and only if this causal nonstationary model has a time-invariant (A, C) pair coincident with the (A, C) pair of $\hat{x}(t)$. But then by the uniqueness theorem 5.2 $\hat{x}_+(t)$ must coincide with $\hat{x}(t)$. □

Remark. As we shall see later, the main idea of "subspace methods" identification is to recapture the *stationary* A, C and $\bar{A}, \bar{C}$ parameters of the process from the dynamic equations satisfied by the bases $\hat{x}(t)$ and $\hat{\bar{x}}(t)$ chosen in the finite-memory predictor spaces. As we have seen before these equations can be written as the Kalman Filter realizations (5.5, 5.6) or 5.11) where A, C and $\bar{A}, \bar{C}$ appear explicitly. However it should be stressed that the stationary parameters A, C and $\bar{A}, \bar{C}$ appear in the Kalman-Filter equations exactly because the bases $\hat{x}(t)$ and $\hat{\bar{x}}(t)$ have been obtained by projection of some stationary state $x(t)$ of the process. In identification, where we are actually attempting to recover the stationary dynamics of y, we do not have a stationary state-space model for y at our disposal. Instead we can pick the bases in the predictor spaces $\hat{X}_{t-}, \hat{X}_{(t+1)-}, \ldots$ at different time instants and compute the difference equations relating the bases at different time instants.

Of course one could in principle pick a basis arbitrarily at different time instants but this would naturally yield *time-varying* A and C parameters in the state equations. In fact, picking bases arbitrarily at each instant t yields time-varying A and C matrices which are not even similar to the stationary parameters we are looking for. So the question arises of choosing bases $\hat{x}(t)$ and $\hat{\bar{x}}(t)$ at different instants $t \in [0, T]$ in such a way that their time evolution is of the Kalman Filter type encountered so far, in particular described by difference equations with *constant* matrices A and C. In this case, by Theorem 5.2, A and C will be equal to the corresponding parameters of a stationary model (in fact, of the whole uniformly ordered family of stationary models corresponding to the basis, see Theorem 5.2). □

Now the Kalman Filter equations describe the propagation in time of the projection $\hat{x}(t)$ onto the past $H_{[0,t)}$, of a stationarily time-shifted state variable $x(t) = U^t x(0)$. Recall that the maps,

$$\mathcal{C}_t := \mathrm{E}^{H_{[0,t)}}_{|X_t} : X_t \to \hat{X}_{t-}, \quad \mathcal{O}_t := \mathrm{E}^{H_{[t,T]}}_{|X_t} : X_t \to \hat{X}_{t+}$$

are the *constructibility and observability operators* of X_t [49, 50] and that minimality of X_t implies that $\mathcal{C}_t$, $\mathcal{O}_t$ are invertible in their respective codomains (we have used this argument already in the proof of Theorem 5.2 above). We may define then a forward and backward *conditional shift operator* $\hat{U}(t)$ and $\hat{\bar{U}}(t)$, by setting

$$\hat{U}(t) := \mathcal{C}_{t+1} U \mathcal{C}_t^{-1} : \hat{X}_{t-} \to \hat{X}_{(t+1)-} \tag{5.23}$$

and

$$\hat{\bar{U}}(t) := \mathcal{O}_{t-1} U^{-1} \mathcal{O}_t^{-1} : \hat{X}_{t+} \to \hat{X}_{(t-1)+} \tag{5.24}$$

so that

$$\begin{aligned} \hat{x}(t+1) = \mathrm{E}^{H_{[0,t+1)}} x(t+1) &= \mathrm{E}^{H_{[0,t+1)}} U\, x(t) = \hat{U}(t)\hat{x}(t) \\ \hat{\bar{x}}(t-1) = \mathrm{E}^{H_{[t-1,T]}} \bar{x}(t-1) &= \mathrm{E}^{H_{[t-1,T]}} U^{-1}\, \bar{x}(t) = \hat{\bar{U}}(t)\hat{\bar{x}}(t). \end{aligned}$$

It is not difficult to show (but we shall not do it here) that the definition of $\hat{U}$ and $\hat{\bar{U}}$ is independent of the minimal stationary splitting subspace X entering in the equations (5.23, 5.24). This specific form of propagation in time, say for the basis $\hat{x}(t)$, is equivalent to the relative state equations being of the Kalman filter type, in particular involving A and C matrices which stay constant in time.

Proposition 5.3. *Let $\hat{x}(t)$ be a basis in $\hat{X}_{t-}$ and $\hat{z}(t+1)$ be an arbitrary basis in $\hat{X}_{(t+1)-}$. Then $\hat{z}(t+1)$ is the conditional shift of $\hat{x}(t)$, i.e. $\hat{z}(t+1) = \hat{x}(t+1) = \hat{U}(t)\hat{x}(t)$ if and only if*

$$\mathrm{E}[\hat{z}(t+1) \,|\, \hat{x}(t)\,] = A\hat{x}(t), \quad \mathrm{E}y(t)\hat{z}(t+1)' = \bar{C}$$

where A is the state-transition matrix of the uniform choice of bases determined by $\hat{x}(t)$ and $\bar{C}$ is the corresponding backward state-output matrix.

Dually, let $\hat{\bar{x}}(t)$ be a basis in $\hat{X}_{t+}$ and $\hat{\bar{z}}(t-1)$ be an arbitrary basis in $\hat{X}_{(t-1)+}$. Then $\hat{\bar{z}}(t-1)$ is the conditional shift of $\hat{\bar{x}}(t)$, i.e. $\hat{\bar{z}}(t-1) = \hat{\bar{x}}(t-1) = \hat{\bar{U}}(t)\hat{\bar{x}}(t)$ if and only if

$$\mathrm{E}[\hat{\bar{z}}(t-1) \,|\, \hat{\bar{x}}(t)\,] = A'\hat{\bar{x}}(t), \quad \mathrm{E}y(t)\hat{\bar{z}}(t-1)' = C$$

where A' is the state-transition matrix of the (backward) uniform choice of bases determined by $\hat{\bar{x}}(t)$ and C is the corresponding forward state-output matrix.

Proof. (if). By the Markovian splitting property every family of bases $\{\hat{z}(t); t \in [0,T]\,\}$ in the predictor spaces $\{\hat{X}_{t-}; t \in [0,T]\,\}$ forms a Markov process. It is also easy to see that the past space span$\{\hat{z}(s); s < t\,\}$ coincides with $H_{[0,t)}$ for all $t \geq 0$. Hence

$$\mathrm{E}[\hat{z}(t+1) \,|\, H_{[0,t)}\,] = \mathrm{E}[\hat{z}(t+1) \,|\, \hat{x}(t)\,] = A\hat{x}(t)$$

which is by assumption equal to

$$\mathrm{E}[\hat{x}(t+1) \,|\, \hat{x}(t)\,] = \mathrm{E}[\hat{x}(t+1) \,|\, H_{[0,t)}\,].$$

Therefore $\hat{z}(t+1)$ and $\hat{x}(t+1)$ have the same orthogonal projection onto $H_{[0,t)}$ i.e.

$$\mathrm{E}^{H_{[0,t)}}(\hat{z}(t+1) - \hat{x}(t+1)) = 0$$

so that, letting $Y_t^- := [y(t)'y(t-1)' \ldots y(0)']'$ and taking into account also the second condition in the statement of the proposition involving $\bar{C} = \mathrm{E}y(t)\hat{x}(t+1)'$ (this identity follows since $\bar{C}\hat{\bar{x}}_-(t+1) = \mathrm{E}[y(t)\,|\hat{\bar{x}}_-(t+1)\,]$ where $\hat{\bar{x}}_-(t+1)$ is the dual basis of $\hat{x}(t+1)$), we have

$$\mathrm{E}Y_t^- \hat{z}(t+1)' = \mathrm{E}Y_t^- \hat{x}(t+1)' = \begin{bmatrix} \bar{C} \\ \bar{C}A' \\ \vdots \\ \bar{C}(A')^t \end{bmatrix}.$$

We denote the constructibilty matrix on the right by $\bar{\Omega}$. Now clearly $\hat{z}(t+1) = M\hat{x}(t+1)$ for some nonsingular matrix M and by substituting in the first member of the equality yields $\bar{\Omega}M' = \bar{\Omega}$ which is equivalent to $M = I$ since (for t large enough) $\bar{\Omega}$ has independent columns. So $\hat{z}(t+1)$ coincide with the conditionally shifted basis $\hat{x}(t+1)$.

(only if). Pick any $x(t)$ in the stationary uniform family of bases corresponding to $\hat{x}(t)$ ($x(t) = C_t^{-1}\hat{x}(t)$, see Theorem 5.2). Then

$$\hat{x}(t+1) = \hat{U}(t)\hat{x}(t) = \mathrm{E}[x(t+1)|\, H_{[0,t+1)}\,]$$

is precisely the Kalman Filter estimate of the stationary state process x at time $t+1$. So the formulas follow from the derivations of the Kalman Filter equations at the beginning of the section.

The proof of the second half of the proposition involves completely similar arguments and is skipped. □

6. Estimation, Partial Realization and Balancing

We shall now concentrate on the statistical problem of describing an observed m-dimensional time series

$$\{y_0, y_1, y_2, \dots y_T\}, \tag{6.1}$$

by a finite-dimensional state-space model of the type (2.1) studied in the previous sections.

To put the methods discussed in this paper into perspective we should say that there are also different choices of the model class which are widely used in identification. One may choose,

1. A parametric class of spectral density functions; say all the rational spectra $\Phi(z)$ of fixed McMillan degree n.
2. A parametric class of (rational) minimal *shaping filter* representations, in other words models consisting of a pair: minimal spectral factor W, plus input white noise w. Expressing W as a polynomial matrix fraction,

$$W(z) = A(z^{-1})^{-1}B(z^{-1})$$

gives the model the familiar form of a linear difference equation

$$y(t) + \sum_{k=1}^{\nu} A_k y(t-k) = \sum_{k=0}^{\nu} B_k w(t-k) \tag{6.2}$$

i.e. an "ARMA" model, parametrized by the coefficients $\{A_k, B_k\}$ of the matrix polynomials $(A(z^{-1}), B(z^{-1}))$. As we have seen at the end of section 4, for square W's the input noise is uniquely determined by the output signal y.

3. Minimal state-space realizations of the type (2.1). These objects are the most "structured" kind of representation of the signal and can be reduced to the previous kind of models by eliminating the auxiliary variables (x and w). They will be our primary object of interest.

For each model class there is a problem of *unique parametrization*, i.e. of making the correspondence: parameter → model, generically bijective. The solution of this problem via the theory of canonical forms constitutes an important chapter of identification theory which has attracted much interest in the early seventies but is now a bit obsolete since *balanced canonical forms* [58], [59], which will be introduced later, are a much simpler and robust alternative.

Moreover, while a spectral density is a unique (wide-sense) probabilistic description of a signal, a family of different minimal spectral factors or state-space models (neglecting the indeterminacy inherent in the choice of basis) give rise to the same spectrum. For this reason when the model classes (2) and (3) are used it is necessary to specify a *representative* factor or minimal realization to get a 1:1 correspondence with the spectrum. Normally one chooses to describe a spectrum by its (unique) minimum phase spectral factor or *forward innovation models* i.e. or the corresponding causal "steady state Kalman Filter" realization. These models are 1:1 with the spectrum if we disregard the intrinsic indeterminacy in the input white noise (which is only defined modulo constant real orthogonal transformations) and the arbitrariness in the choice of basis in the relative state space X_-.

The model classes described above are wide-sense. In case the signal y is believed to be *Gaussian* they can equivalently be interpreted as defining the spectrum or the covariance function of a family of Gaussian probability laws for the underlying stochastic process. These probabiliy laws are uniquely determined by a corresponding model and are then also parametrized by the parameters $\{A, C, \bar{C}, \Lambda(0)\}$, $\{A_k, B_k\}$ and (A, B, C, D) respectively.

We shall consider two conceptually different approaches that are used to fit models to the data,

- The "direct" approach, based on the principle of minimizing a suitable function which measures the distance between the data [13] and the probability law induced by the model class. Well-known and widely accepted examples of distance functions are the *likelihood function* of the data according to the particular model, or the average squared *prediction-error* of the observed data corresponding to a particular choice of a model in the model class. Minimization of these criteria can (except in trivial cases) only

[13] This terminology is a bit misleading. In reality one minimizes a suitable "finite sample" approximation of a distance function between the *true law* of the data and the law induced by the model class. An example of distance function between probability measures which can be used to this purposse is the Kullback-Leibler distance.

be done numerically and hence the direct methods lead to iterative optimization algorithms in the space of the parameters, say the space of minimal (A, B, C, D) matrix quadruples, which parametrize the chosen model class.

– A two steps procedure which in principle can be described as identification of a rational model for the spectrum (or covariance) of the observed data, followed by stochastic realization. Here the first step is estimation of the parameters $(A, C, \bar{C})$ of a minimal realization of the spectral density matrix of the process.
From the spectral density matrix a state-space model (typically the forward innovation model) is then computed by solving the Linear Matrix Inequality, or the Riccati equation as seen in section 4.
The difference with the first approach is that the estimation of $(A, C, \bar{C})$ is *not* done by optimizing a likelihood or other distance functions but simply by *matching second order moments.* In other words, let

$$\{\Lambda_0, \Lambda_1, \dots, \Lambda_\nu\} \tag{6.3}$$

be a finite set of sample $m \times m$ covariance matrices estimated in some (as yet unspecified) way from the m-dimensional sequence of observations (6.1). The problem is of finding a minimal value of n and a minimal[14] triplet of matrices $(A, C, \bar{C})$, of dimensions $n \times n$, $m \times n$ and $m \times n$ respectively, such that

$$CA^{k-1}\bar{C}' = \Lambda_k \quad k = 1, 2, \dots, \nu \tag{6.4}$$

This is an instance of *estimation by the method of moments* described in the statistical textbooks e.g. [13, p. 497], which is a very old idea used extensively by K. Pearson in the beginning of the century. The underlying priciple is close in spirit to the wide-sense setting that we are working in. It does not necessarily guarantee minimal distance between the "true" and the model distributions but rather imposes that the parameters to be estimated match exactly the sample second order moments. These can easily be chosen at least "consistent" (i.e. tending to the true second order moments as the sample size goes to infinity) so the method gives consistent estimates in the sense that ν true moments $\Lambda_0(\tau)$ $\tau = 1, 2, \dots, \nu$ will be described exactly as $T \to \infty$. In other words the first ν lag values of the true covariance function will be matched exactly.

Some may argue that estimation by the method of moments is in general "non-efficient" and it is generally claimed in the literature that one should expect better results (in the sense of smaller asymptotic variance of the estimates) by direct methods. In practice this is true only to a point since the likelihood function or the average prediction error are computable only if we

[14] Recall that $(A, C, \bar{C})$ is minimal if (A, C) is completely observable and $(A, \bar{C}')$ is completely reachable.

assume Gaussian models (or linear predictors which amounts to the same) and this in the long run is equivalent to matching covariances anyway. In addition there is the structural handicap of iterative optimization methods which may get stuck in local minima and hence provide sub-optimal parameter estimates, a rather hard phenomenon to detect. The two-steps approach offers in this respect the major advantage of converting the nonlinear parameter estimation phase which is necessary in maximum-likelihood or prediction-error model identification into a partial realization problem, involving essentially the factorization of a Hankel matrix of estimated covariances, and the solution of a Riccati equation, both much better understood problems for which efficient numerical solution techniques are available.

6.1 Positivity

A warning is in order concerning the implementation of the method of moments described above in that it introduces some nontrivial mathematical questions related to positivity of the estimated spectrum.

In determining a minimal triplet $(A, C, \bar{C})$ interpolating the partial sequence (6.3) so that $CA^{k-1}\bar{C}' = \Lambda_k \quad k = 1, 2, \ldots, \nu$, we also completely determine the infinite sequence

$$\{\Lambda_0, \Lambda_1, \Lambda_2, \Lambda_3, \ldots\} \tag{6.5}$$

by setting $\Lambda_k = CA^{k-1}\bar{C}'$ for $k = \nu + 1, \nu + 2, \ldots$. This sequence is called a *minimal rational extension* of the finite sequence (6.3). The attribute "rational" is due to the fact that

$$Z(z) := \frac{1}{2}\Lambda_0 + \Lambda_1 z^{-1} + \Lambda_2 z^{-2} + \ldots = \frac{1}{2}\Lambda_0 + C(zI - A)^{-1}\bar{C}' \tag{6.6}$$

is a rational function. In order for (6.5) to be a bona fide covariance sequence, however, it is necessary, but *not* sufficient, that the Toeplitz matrix

$$T = \begin{bmatrix} \Lambda_0 & \Lambda_1 & \Lambda_2 & \cdots & \Lambda_\nu \\ \Lambda_1' & \Lambda_0 & \Lambda_1 & \cdots & \Lambda_{\nu_1} \\ \vdots & \vdots & \vdots & \ddots & \vdots \\ \Lambda_\nu' & \Lambda_{\nu-1}' & \Lambda_{\nu-2} & \cdots & \Lambda_0 \end{bmatrix} \tag{6.7}$$

be nonnegative definite. In fact, it is required that the function (the spectral density corresponding to (6.5))

$$\Phi(z) = \Lambda_0 + \sum_{k=1}^{\infty} \Lambda_k (z^k + z^{-k}) = Z(z) + Z(z^{-1})' \tag{6.8}$$

be nonnegative on the unit circle. This is equivalent to the function $Z(z)$ being *positive real*. Consequently, the interpolation needs to be done subject to the extra constraint of positivity.

The constraint of positivity is a rather tricky one and in all identification methods which are directly or indirectly, as the subspace methods described below, based on the interpolation condition (6.4) it is normally disregarded. For this reason these methods may fail to provide a positive extension and hence may lead to data $(A, C, \bar{C})$ for which there are no solutions of the LMI and hence to totally inconsistent results.

It is important to appreciate the fact that the problem of positivity of the extension has little to do with the "noise" or "sample variability" superimposed to the covariance data and is present equally well for (finite) data extracted from a true rational covariance sequence. For there is no guarantee that, even in this idealized situation, the order of a minimal rational extension 6.5 of the first ν covariance matrices of the sequaence, would be sufficiently high to equal the order of the infinite sequence and hence to generate a positive extension. A minimal partial realization may well fail to be positive because its order is too low to guarantee positivity.

Neglecting the positivity constraint amounts to tacitly assuming that

Assumption 6.1. The covariance data (6.3) can be generated exactly by some (unknown) stochastic system whose dimension is equal to the rank of the block Hankel matrix

$$H_\mu = \begin{bmatrix} \Lambda_1 & \Lambda_2 & \Lambda_3 & \cdots & \Lambda_\mu \\ \Lambda_2 & \Lambda_3 & \Lambda_4 & \cdots & \Lambda_{\mu+1} \\ \vdots & \vdots & \vdots & \ddots & \vdots \\ \Lambda_\mu & \Lambda_{\mu+1} & \Lambda_{\mu+2} & \cdots & \Lambda_{2\mu-1} \end{bmatrix}, \qquad (6.9)$$

where $\mu = [\frac{\nu}{2}]$.

This assumption is not "generically satisfied" and it can be shown that there are relatively "large" sets of data (6.3) for which it does not hold. It is not even enough to assume that the data is generated from a "true" finite-dimensional stochastic system: the rank condition is also necessary. Otherwise, for a minimal triplet $(A, C, \bar{C})$ which satisfies the interpolation condition (6.4), the positivity condition will not be automatically fulfilled, and the matrix A may even fail to be stable [10].

Following [55], we define the *algebraic degree* of the sequence (6.3) to be the minimal degree of any realization (6.6) satisfying (6.4) and the *positive degree* to be the minimal degree of a rational extension (6.6) for which, in addition, $Z(z)$ is positive real. Then Assumption 6.1 can also be written in the following equivalent form.

Assumption 6.1′. The positive degree of (6.3) is equal to the algebraic degree.

The fact that this equality cannot a priori be assumed to hold for generic covariance data can now be illustrated by the following fact.

Theorem 6.1. *[11] In the case $m = 1$, the generic value of the algebraic degree of (6.3) is $[\frac{\nu+1}{2}]$, whereas there is no generic value for the positive degree. In fact, for each $p = [\frac{\nu+1}{2}], [\frac{\nu+1}{2}]+1, \dots, \nu$, there is a nonempty open set of covariance data for which the positive degree is precisely p.*

The correct approach would in principle require to compute a rational *positive extension* of the finite covariance sequence (6.3), of minimal McMillan degree. Although there are methods to compute positive extensions, the most famous of which is the so-called "maximum-entropy" extension, based on the Levinson algorthm, these methods produce functions of very high complexity, in fact generically of the highest possible degree (ν in the case $m = 1$). Unfortunately there are no algorithms so far which compute positive extensions of minimal degree. A stochastic model reduction step would then be necessary but this is again, a rather underdeveloped area of system theory. For a discussion of these matters see [55].

In these notes we shall be content with discussing the deterministic partial realization aspect of the method, therefore tacitly assuming that the conditions described in Assumption 6.1 hold. There is standard software available for checking positivity (i.e. solvability of the LMI) of the partial realization. Whenever positivity fails one may try to add more covariance data so as to allow for an increase of the order (algebraic degree) of the partial realization. Once a positive triple $(A, C, \bar{C})$ is estimated, the computation of a state-space model is in principle just a matter of solving the LMI or the appropriate Riccati equation, as seen in section 4..

Historical remarks. The two-steps procedure was apparently first advocated in a systematic way by Faurre [21]; see also [22, 23]. More recent work is based on Singular Value Decomposition and canonical correlation analysis [2] and is due to Aoki [9], and van Overschee and De Moor [60]. There are versions of the algorithms based on canonical correlation analysis which apply directly to the observed data without even computing the covariance estimates [60].

The work of van Overschee and De Moor introduces an interesting "geometric" approach based on state-space construction and on the choice of particular bases in the state space. The system matrices are computed after the choice of basis by formulas analog to (2.12). This procedure on one hand makes very close contact with the geometric state-space construction ideas discussed in sections5.2 and 2. On the other hand it seems completely unrelated to the partial realization and covariance extension approach mentioned above.

In the rest of this paper we shall study the geometric "Subspace-methods" approach of [60] and show that it is very much related to the basic partial realization plus stochastic realization idea. In fact we shall show that the two approaches are equivalent and lead to exactly the same formulas.

6.2 The Hilbert Space of a Stationary Signal

In section 2. we have described an abstract model-building procedure based on geometric operations on certain subspaces of random variables constructed from linear statistics of the present and past histories of a stochastic process y. In practice instead one has just a collection of observed data,

$$\{y_0, y_1, \ldots, y_t, \ldots, y_T\} \tag{6.10}$$

with $y_t \in \mathbb{R}^m$, measured during an experiment. We shall assume that the sample size T is very large and that the data have been preprocessed so as to be compatible with the basic assumption of (wide-sense) stationarity and zero mean of the previous sections. This in particular means that we can pick N large enough so that the time averages

$$\frac{1}{N+1} \sum_{t=t_0}^{N+t_0} y_{t+\tau} y_t' \qquad \tau \geq 0 \tag{6.11}$$

are practically independent of the initial time t_0 and arbitrarily close to a *bona-fide* stationary covariance matrix sequence (the "true" covariance of the signal).

Under these assumptions on the data, the stochastic state-space theory of the sections 2.-5. can be translated into an isomorphic geometrical setup based on linear operations on the observed time series and can then applied to the problem of state-space modeling of the data.

In this section we shall briefly review the basic ideas behind this correspondence. For clarity of exposition we shall initially assume that $T = \infty$ and that the data collection has started in the infinitely remote past (so that the time series is actually doubly infinite).

For each $t \in \mathbb{Z}$ define the $m \times \infty$ matrices

$$\mathbf{y}(t) := [y_t, y_{t+1}, y_{t+2}, \ldots] \tag{6.12}$$

and consider the sequences $\mathbf{y} := \{\mathbf{y}(t) \mid t \in \mathbb{Z}\}$. This sequence will play a very similar role to the stationary processes y of the previous sections.

Define the vector space $\mathcal{Y}$ of all finite linear combinations

$$\mathcal{Y} := \{\sum a_k' \mathbf{y}(t_k) \qquad a_k \in \mathbb{R}^m,\ t_k \in \mathbb{Z}\} \tag{6.13}$$

Note that the vector space $\mathcal{Y}$ is just the row spaces of the family of semi-infinite matrices (6.12) or, equivalently the rowspace of the infinite Hankel matrix

$$Y_\infty := \begin{bmatrix} \vdots \\ \mathbf{y}(t) \\ \mathbf{y}(t+1) \\ \mathbf{y}(t+2) \\ \vdots \end{bmatrix}$$

This vector space of scalar semi-infinite sequences (rows) can be equipped with an inner product, which is first defined on the generators by the bilinear form

$$\langle a'\mathbf{y}(k), b'\mathbf{y}(j)\rangle := \lim_{T\to\infty} \frac{1}{T+1}\sum_{t=0}^{T} a' y_{t+k} y'_{t+j} b = a' \Lambda_0(k-j) b, \qquad (6.14)$$

and then extended by linearity to all finte linear combinations of elements of $\mathcal{Y}$. This inner product is nondegenerate if the Toeplitz matrix T_k, constructed with the true covariances $\{\Lambda_0(0), \Lambda_0(1), \dots, \Lambda_0(k)\}$, is a positive definite symmetric matrix for all k [55]. Note also that the limit does not change if in the limits of the sum (6.14) $t = 0$ is replaced by an arbitrary initial instant t_0, so that

$$\langle a'\mathbf{y}(k), b'\mathbf{y}(j)\rangle = \langle a'\mathbf{y}(t_0 + k), b'\mathbf{y}(t_0 + j)\rangle$$

for all t_0 (wide-sense stationarity). We also define a *shift operator* $\mathbf{U}$ on the family of semi-infinite matrices (6.12), by setting

$$\mathbf{U}a'\mathbf{y}(t) := a'\mathbf{y}(t+1) \quad t \in \mathbb{Z}, \quad a \in \mathbb{R}^m,$$

defining a linear map which is isometric with respect to the inner product (6.14) and extendable by linearity to all of $\mathcal{Y}$.

By closing the vector space $\mathcal{Y}$ with respect to convergence in the norm induced by the inner product (6.14), we obtain a Hilbert space [15] $\bar{\mathcal{Y}} =$ closure$\{\mathcal{Y}\}$ to which the shift operator $\mathbf{U}$ is extended by continuity as a unitary operator.

As explained in more detail in [55], this Hilbert space framework is isometrically isomorphic to the abstract "stochastic" geometric setup used in the previous sections. Now as stated formally in Proposition1.1 and in the subsequent generalization, we can formally think of the observed (infinitely long) time series as a regular sample path of a wide-sense stationary stochastic process $\mathbf{y}$, having covariance matrix equal to the true covariance function $\Lambda_0(.)$, equal to the limit of the sum (6.11) as $N \to \infty$. Then, at least as far as first and second order moments are concerned, the sequence of "tails" $\mathbf{y}$ defined in (6.12) behaves exactly like the abstract stochastic counterpart y. In particular all second order moments of the random process can equivalently be calculated in terms of the tail sequence $\mathbf{y}$ provided we substitute expectations with ergodic limits of the type (6.14). Since we only worry about second order properties in this paper, we may even formally *identify* the tail sequence $\mathbf{y}$ of (6.12) with the underlying stochastic process y. This requires just thinking of "random variables" as being semi-infinte strings of numbers and the expectation of products $\mathrm{E}\{\xi\eta\}$ as being the inner product of the

[15] Note that the symbol $\mathcal{Y}$ denotes a real inner-product space which need not be closed with respect to the inner product structure defined by (6.14). Since we will not have much use for the completed space in the following, we shall not introduce special symbols for it.

corresponding rows ξ and η. For reasons of uniformity of notation the inner product 6.14 will then be denoted

$$\langle \xi, \eta \rangle = \mathrm{E}\{\xi\eta\}, \tag{6.15}$$

Here as usual we allow $\mathrm{E}\{\cdot\}$ to operate on matrices, taking inner products row by row.

Hence all defintions and results in the geometric theory of stochastic realization can be carried over to the present framework. The orthogonal projection of ξ onto a subspace $\mathcal{H}$ of the space $\mathcal{Y}$ will still be denoted $\mathrm{E}[\xi\,|\mathcal{H}]$. Whenever $\mathcal{H}$ is given as the rowspace of some matrix of generators H, we shall write $\mathrm{E}[\xi \mid H]$ to denote the projection expressed (perhaps nonuniquely) in terms of the generators. It is clear that for finitely generated subspaces we have the representation formula

$$\mathrm{E}[\xi \mid H] = \mathrm{E}(\xi H')[\mathrm{E}(HH')]^{\sharp} H \tag{6.16}$$

and in case of linearly independent rows we can substitute the pseudoinverse $\sharp$ with a true inverse.

A (stationary) stochastic realization of $\mathbf{y}$ is a representation of the type

$$\left\{ \begin{array}{lcl} \mathbf{x}(t+1) & = & A\mathbf{x}(t) + B\mathbf{w}(t) \\ \mathbf{y}(t) & = & C\mathbf{x}(t) + D\mathbf{w}(t) \end{array} \right. \tag{6.17}$$

where $\{\mathbf{w}(t)\}$ is p-dimensional normalized white noise , i.e. $\mathrm{E}\{\mathbf{w}(t)\mathbf{w}(s)'\} = I\delta_{ts}$ $\mathrm{E}\{\mathbf{w}(t)\} = 0$, etc.

Remark 6.1. It should be kept in mind that the various linear operations in (6.17) hold in the sense of the metric of the space $\bar{\mathcal{Y}}$ defined above and are to be understood as "asymptotic equalities" between *sequences.* In particular, nothing can be said about the particular sample values, say y_t, x_t, w_t taken on by the time series involved in the model at a specific instant of time. This is similar to the interpretation that is given to the model (2.1) in case of *bona fide* stochastic processes, where the linear model can be expected to hold for each particular sample value only with probability one.

6.3 Identification Based on Finite Data

For data of finite lenght T the inner product (6.15) must be approximated by a finite sum

$$E\{\xi\eta\} \cong \frac{1}{T+1} \sum_{t=0}^{T} \xi_t \eta_t \tag{6.18}$$

which makes the "expectation" operator E essentially the same thing as ordinary Euclidean inner product in $\mathbb{R}^T$.

Assume $N \leq T$ is large enough for the time average in the ergodic limit (6.11) to be sufficiently close to the true covariance and for all subscripts below to make sense. Fix a "present" time $t = k$ and define the two mk-dimensional "random vectors" (i.e. block Hankel matrices of dimension $mk \times (N+1)$) formed by stacking the output data as

$$\mathbf{Y}_k^- = \begin{bmatrix} \mathbf{y}(0) \\ \mathbf{y}(1) \\ \vdots \\ \mathbf{y}(k-1) \end{bmatrix} = \begin{bmatrix} y_0 & y_1 & \cdots & y_N \\ y_1 & y_2 & \cdots & y_{N+1} \\ \vdots & \vdots & & \vdots \\ y_{k-1} & y_k & \cdots & y_{k+N-1} \end{bmatrix} \tag{6.19}$$

$$\mathbf{Y}_k^+ = \begin{bmatrix} \mathbf{y}(k) \\ \mathbf{y}(k+1) \\ \vdots \\ \mathbf{y}(2k-1) \end{bmatrix} = \begin{bmatrix} y_k & y_{k+1} & \cdots & y_{k+N} \\ y_{k+1} & y_{k+2} & \cdots & y_{k+N+1} \\ \vdots & \vdots & & \vdots \\ y_{2k-1} & y_{2k} & \cdots & y_{2k+N-1} \end{bmatrix} \tag{6.20}$$

The relative rowspaces $\mathbf{Y}_k^-, \mathbf{Y}_k^+$ generated by the rows of the $m \times (N+1)$ matrices $\mathbf{y}(t)$ for $0 \leq t < k$, and $k \leq t < 2k$ respectively, are the "past" and "future" spaces of the data at time k. Since the tail matrix sequences we can form with the observed signal are necessarily finite, these vector spaces can describe in reality only *finite* past and future histories of the signal $\mathbf{y}$ at time k. For simplicity of notations we use symbols that are not informative of this fact[16].

For later use let us define also the "augmented" future at time k (a $m(k+1) \times (N+1)$ block Hankel matrix)

$$\mathbf{Y}_{[k,2k]}^+ := \begin{bmatrix} \mathbf{Y}_k^+ \\ \mathbf{y}(2k) \end{bmatrix},$$

the relative rowspace will be denoted $\mathbf{Y}_{[k,2k]}$.

6.4 The Partial Realization Problem

In order to avoid trivial difficulties having to do with the fact that the rank of a finite Hankel matrix with too few rows or columns need not be equal to the algebraic degree of a finite sequence (6.3), we shall assume that the index k is chosen far enough from the endpoints $k = 0$ or $k = \nu$. There is in fact no loss of generality in assuming that we have $\nu = 2k+1$ sample covariance estimates,

$$\{\Lambda_0, \Lambda_1, \ldots, \Lambda_{2k}\}. \tag{6.21}$$

[16] More accurate notations would be,

$$\mathbf{Y}_k^- := \mathbf{Y}_{[0,k)} \quad \mathbf{Y}_k^+ := \mathbf{Y}_{[k,2k)}$$

and that the present time k has been chosen to be the "middle point" of the lag sequence of the covariance estimates (6.21).

A block Hankel matrix of stationary covariances can always be given the meaning of cross covariance matrix of the finite future and past of the underlying signal at time $t = k$,

$$H_k := \begin{bmatrix} \Lambda_1 & \Lambda_2 & \cdots & \Lambda_k \\ \Lambda_2 & \Lambda_3 & \cdots & \Lambda_{k+1} \\ \vdots & \vdots & \ddots & \vdots \\ \Lambda_k & \Lambda_{k+1} & \cdots & \Lambda_{2k-1} \end{bmatrix} = \frac{1}{N} \begin{bmatrix} \mathbf{y}(k) \\ \mathbf{y}(k+1) \\ \vdots \\ \mathbf{y}(2k-1) \end{bmatrix} \begin{bmatrix} \mathbf{y}(k-1) \\ \mathbf{y}(k-2) \\ \vdots \\ \mathbf{y}(0) \end{bmatrix}'$$

$$= \mathrm{E}\mathbf{Y}_k^+ (\bar{\mathbf{Y}}_k^-)' \tag{6.22}$$

where $\bar{\mathbf{Y}}_k^-$ is the "time reversal" of the vector $\mathbf{Y}_k^-$. The subscript k is attached to denote the "present" time. Similarly using the available covariance data we can form

$$H_{k+1} := \begin{bmatrix} \Lambda_1 & \Lambda_2 & \cdots & \Lambda_k & \Lambda_{k+1} \\ \Lambda_2 & \Lambda_3 & \cdots & \Lambda_{k+1} & \Lambda_{k+2} \\ \vdots & \vdots & \ddots & \vdots & \vdots \\ \Lambda_k & \Lambda_{k+1} & \cdots & \Lambda_{2k-1} & \Lambda_{2k} \end{bmatrix} = \begin{bmatrix} \Lambda_1 & \\ \Lambda_2 & \\ \vdots & \sigma H_k \\ \Lambda_k & \end{bmatrix}$$

$$= \mathrm{E}\{U\mathbf{Y}_k^+ (\bar{\mathbf{Y}}_{k+1}^-)'\}, \tag{6.23}$$

and

$$\bar{H}_{k+1} := \begin{bmatrix} \Lambda_1 & \Lambda_2 & \cdots & \Lambda_k \\ \Lambda_2 & \Lambda_3 & \cdots & \Lambda_{k+1} \\ \vdots & \vdots & \ddots & \vdots \\ \Lambda_k & \Lambda_{k+1} & \cdots & \Lambda_{2k-1} \\ \Lambda_{k+1} & \Lambda_{k+2} & \cdots & \Lambda_{2k} \end{bmatrix} = \begin{bmatrix} \Lambda_1 \quad \Lambda_2 \quad \cdots \quad \Lambda_k \\ \sigma H_k \end{bmatrix}$$

$$= \mathrm{E}\{\mathbf{Y}_{[k,2k]}^+ (\bar{\mathbf{Y}}_k^-)'\} \tag{6.24}$$

where σH_k is the *shifted Hankel matrix*, of the same dimension as H_k but with all entries shifted by one time unit i.e. with Λ_{i+1} replacing Λ_i everywhere.

We quote the following uniqueness result of partial realizations from [69].

Lemma 6.1. *The sequence (6.21) has a unique rational extension of minimal degree if and only if*

$$\mathrm{rank} H_k = \mathrm{rank} H_{k+1} = \mathrm{rank} \bar{H}_{k+1} := n \tag{6.25}$$

Uniqueness is understood in the sense that if $(A_1, C_1, \bar{C}_1)$ *and* $(A_2, C_2, \bar{C}_2)$ *both define minimal rational extensions of (6.21), then there is a nonsingular* $n \times n$ *matrix* T *such that*

$$A_2 = T^{-1} A_1 T, \quad C_2 = C_1 T, \quad \bar{C}_2' = T^{-1} \bar{C}_1'. \tag{6.26}$$

Computing a minimal partial realization can be done essentially via a rank factorization of the Hankel matrix H_k. The prototype algorithm, called the *Ho-Kalman* algorithm is reviwed below.

The Ho-Kalman Algorithm. Start by a *rank factorization* of H_k,

$$H_k = \Omega_k \bar{\Omega}_k' \tag{6.27}$$

where both factors Ω_k, $\bar{\Omega}_k$ have n linearly independent columns. Since by (6.25) column $-$ spanH_k = column $-$ spanH_{k+1} and, dually, row $-$ spanH_k = row $-$ span$\bar{H}_{k+1}$ there exist matrices $\bar{C}, \bar{\Delta}, C, \Delta$ such that

$$\begin{bmatrix} \Lambda_1 \\ \Lambda_2 \\ \vdots \\ \Lambda_k \end{bmatrix} = \Omega_k \bar{C}', \quad \sigma H_k = \Omega_k \Delta \tag{6.28}$$

and

$$\begin{bmatrix} \Lambda_1 & \Lambda_2 & \cdots & \Lambda_k \end{bmatrix} = C \bar{\Omega}_k', \quad \sigma H_k = \Delta \bar{\Omega}_k' \tag{6.29}$$

It is obvious from the last two equalities on the right that there must exist a *unique* matrix A of dimension $n \times n$ such that

$$\sigma H_k = \Omega_k A \bar{\Omega}_k'.$$

In conclusion, the matrices

$$\begin{aligned} A &= \Omega_k^{-L} \sigma H_k (\bar{\Omega}_k')^{-R} & (6.30) \\ C &= \begin{bmatrix} \Lambda_1 & \Lambda_2 & \cdots & \Lambda_k \end{bmatrix} (\bar{\Omega}_k')^{-R} & (6.31) \\ \bar{C} &= \begin{bmatrix} \Lambda_1' & \Lambda_2' & \cdots & \Lambda_k' \end{bmatrix} (\Omega_k')^{-R} & (6.32) \end{aligned}$$

are independent of the choice of the left- or right-inverses (denoted $^{-L}$ or $^{-R}$ respectively) and propagate the factorization (6.27) uniquely to H_{k+1} and $\bar{H}_{k+1}$ according to the formulas,

$$H_{k+1} = \begin{bmatrix} \Omega_k \bar{C}' & \Omega_k A \bar{\Omega}_k' \end{bmatrix} = \Omega_k \begin{bmatrix} \bar{C}' & A \bar{\Omega}_k' \end{bmatrix} := \Omega_k \bar{\Omega}_{k+1} \tag{6.33}$$

and

$$\bar{H}_{k+1} = \begin{bmatrix} C \bar{\Omega}_k \\ \Omega_k A \bar{\Omega}_k' \end{bmatrix} = \begin{bmatrix} C \\ \Omega_k A \end{bmatrix} \bar{\Omega}_k' := \Omega_{k+1} \bar{\Omega}_k'. \tag{6.34}$$

From these we obtain the following updating equations for the factors Ω_{k+1}, $\bar{\Omega}_{k+1}$,

$$\Omega_{k+1} = \begin{bmatrix} C \\ \Omega_k A \end{bmatrix}, \quad \bar{\Omega}_{k+1} = \begin{bmatrix} \bar{C} \\ \bar{\Omega}_k A' \end{bmatrix}. \tag{6.35}$$

Now once (6.28, 6.29) hold for some $(A, C, \bar{C})$ and k big enough, they must hold with the same $(A, C, \bar{C})$ *for all* $k = 1, \dots$ and then (6.35) can be interpreted as bona-fide recursions in k. From this we obtain precisely the classical structure of the observabililty and reconstructability matrices

$$\Omega_k = \begin{bmatrix} C \\ CA \\ \vdots \\ CA^{k-1} \end{bmatrix} \quad \bar{\Omega}_k = \begin{bmatrix} \bar{C} \\ \bar{C}A' \\ \vdots \\ \bar{C}(A')^{k-1} \end{bmatrix}', \tag{6.36}$$

seen in the literature.

It is important to note that under the equal ranks assumption (6.25), to each rank factorization (6.27) there corresponds a *unique triplet* $(A, C, \bar{C})$. In a sense fixing a rank factorization fixes the basis in the (deterministic) state space of the partial realization. Actually we may amplify this statement in the following way.

Theorem 6.2. *Each rank factorization (6.27) of the finite Hankel matrix H_k satisfying the equal ranks assumption (6.25), determines a unique partial realization $(A, C, \bar{C}, \Lambda(0))$ of the corresponding covariance sequence. Under Assumption 6.1 this realization defines a positive-real function $Z(z)$ and hence each factorization (6.27) determines also a unique uniformly ordered family of stationary realizations of the process* $\mathbf{y}$, *having the (same) $(A, C, \bar{C})$ parameters given in (6.30, 6.31, 6.32).*

The uniformly ordered family of minimal realizations of y corresponding to a given positive-real quadruple $(A, C, \bar{C}, \Lambda(0))$ was discussed in detail in sections 4. and 4.1.

The "subspace" identification procedure also produces uniformly ordered families of stationary realizations by choosing bases in the finite-memory predictor spaces.

6.5 Partial Realization via SVD

One particularly convenient choice of the rank factorization of the Hankel matrix, suggested in [76], later popularized in [44] and refined in [15, 16] is a normalized *Singular-Valuc factorization.*

Let L_k^- and L_k^+ be the lower triangular Cholesky factors of the block Toeplitz matrices

$$T_k - := \mathrm{E}\{\mathbf{Y}_k^-(\mathbf{Y}_k^-)'\} = L_k^-(L_k^-)' \qquad T_k^+ := \mathrm{E}\{\mathbf{Y}_k^+(\mathbf{Y}_k^+)'\} = L_k^+(L_k^+)'$$

and let

$$\mathbf{e}_k^- := (L_k^-)^{-1}\mathbf{Y}_k^- \qquad \bar{\mathbf{e}}_k^+ := (L_k^+)^{-1}\mathbf{Y}_k^+ \tag{6.37}$$

be the corresponding orthonormal bases in $\mathbf{Y}_k^-$, $\mathbf{Y}_k^+$ respectively (i.e. the finite interval forward and backward innovation vectors).

Introduce the *normalized Hankel matrix*:

$$\hat{H}_k := (L_k^+)^{-1} H_k (L_k^-)^{-T} = \mathrm{E}\,\bar{\mathbf{e}}_k^+(\mathbf{e}_k^-)'$$

and consider the Singular-Value decomposition (SVD) of $\hat{H}_k$,

$$\hat{H}_k = \hat{U}_k \hat{\Sigma}_k \hat{V}_k' \tag{6.38}$$

where $\hat{U}_k$, $\hat{V}_k$ are $mk \times mk$ orthogonal matrices and $\hat{\Sigma}_k$ is diagonal with nonnegative elements[17]

$$1 \geq \sigma_1 \geq \sigma_2 \geq \ldots \geq \sigma_{mk} \geq 0$$

We now do a numerical "rank determination" step which consists in setting equal to zero the canonical correlation coefficients which are smaller than a predetermined "noise treshold level". In this way we substitute for $\hat{\Sigma}_k$ a diagonal matrix of rank n,

$$\hat{\Sigma}_k \simeq \begin{bmatrix} \Sigma_k & 0 \\ 0 & 0 \end{bmatrix}$$

where

$$\Sigma_k = \operatorname{diag}\{\sigma_1 \geq \sigma_2 \geq \ldots \geq \sigma_n\}$$

where the σ_k's are significantly non-zero, and write (with some misuse of noation)

$$\hat{H}_k = U_k \Sigma_k V_k' \tag{6.39}$$

where now U_k and V_k are $mk \times n$ with orthonormal columns.

It is well-known that the "truncated" matrix on the right-hand side of (6.39) provides a best approximation of rank n of H_k in a variety of matrix norms [32]. It is however to be stressed also that this approximation is *no longer Hankel*, or if we prefer the euphemism, only "approximately Hankel".

Since the application of a rigorous Hankel approximation theory [1] would lead to complications, this difficulty is ignored in the following. An analysis of the additional errors involved in this approximation seems still to be an open problem.

From (6.39) a rank factorization of H_k is naturally,

$$\Omega_k := L_k^+ U_k \Sigma_k^{1/2}, \qquad \bar{\Omega}_k := L_k^- V_k \Sigma_k^{1/2}$$

which produces

$$A = \Sigma_k^{-1/2} U_k' (L_k^+)^{-1} \sigma H_k (L_k^-)^{-T} V_k \Sigma_k^{-1/2} \tag{6.40}$$

$$C = \begin{bmatrix} \Lambda_1 & \Lambda_2 & \cdots & \Lambda_k \end{bmatrix} (L_k^-)^{-T} V_k \Sigma_k^{-1/2} \tag{6.41}$$

$$\bar{C} = \begin{bmatrix} \Lambda_1' & \Lambda_2' & \cdots & \Lambda_k' \end{bmatrix} (L_k^+)^{-T} U_k \Sigma_k^{-1/2} \tag{6.42}$$

These formulas provide a partial realization of the sequence (6.21) enjoying special properties. Before turning to the analysis of these properties we remark that it may be desirable to rewrite them in a way which is more

[17] These are the well-known sample *canonical correlation coefficients* of the two random vectors $\mathbf{Y}_k^+$ and $\mathbf{Y}_k^-$. In geometric terms they are the cosines of the the *principal angles* between the subspaces $\mathbf{Y}_k^-$ and $\mathbf{Y}_k^+$.

convenient from the numerical point of view, where the explicit computation of the sample covariance matrices is not needed. The SVD computations can be done directly on a suitable QR-type factorization of the Hankel matrices representing the data. A number of other improvements can be introduced for problems of high dimension making (6.40)-(6.42) a quite reliable and fast computational scheme.

6.6 Stochastic Balanced Realizations: the Stationary Setting

We shall momentarily return to the abstract probabilistic setting of sections 2 and 3. Consider a stationary random process defined on the whole time axis $\mathbb{Z}$, with a rational spectral density $\Phi(z)$ represented as in (3.7). The following definition has been introduced by Desai and Pal in [14].

Definition 6.1. *A minimal realization* $(A, C, \bar{C}, \Lambda(0))$ *of a* $m \times m$ *positive real matrix is called* Stochastically Balanced[18] *if the minimal solutions* P_-, $\bar{P}_+$ *of the dual Linear Matrix Inequalities (4.1), (4.4) are both equal to the same diagonal matrix, i.e.*

$$P_- = \Sigma = \bar{P}_+$$

where $\Sigma = \text{diag}\{\sigma_1, \sigma_2, \ldots, \sigma_n\}$. *Whitout loss of generality we shall assume that the* σ_k*'s are ordered in decreasing magnitude, i.e.* $\sigma_{k+1} \geq \sigma_k$.

The motivation of this definition looks rather obscure at this stage and it is not really clear what balanced realizations should be good for. Below we shall provide an explanation based on [67].

Consider a minimal realization of y of the form (2.1) with (minimal) state space X. We shall start our discussion by attaching to each random variable ξ in X a pair of indices which quantify "how well" ξ can be estimated on the basis of the past or future history of the output process. We shall then define a choice of basis in X which has some "canonical" desirable properties in this respect. Initially our discussion will be completely coordinate-free.

For a random variable $\xi \in X$ we define the numbers,

$$\eta_+(\xi) := \frac{\|\mathrm{E}^{H^+}\xi\|^2}{\|\xi\|^2} \qquad \eta_-(\xi) := \frac{\|\mathrm{E}^{H^-}\xi\|^2}{\|\xi\|^2} \tag{6.43}$$

called the *future-* and, respectively, the *past- relative efficiency* of ξ. The numbers $\eta_\pm(\xi)$ are nonnegative and ≤ 1 and in the statistical literature are commonly referred to as the "percentage of explained variance" (of the random variable being estimated). Clearly, the larger $\eta_\pm(\xi)$, the better (in the sense of smaller estimation error variance) will be the corresponding estimate $\mathrm{E}^{H^\pm}\xi$.

[18] Or *Positive-Real Balanced.*

The relative efficiency indices have also a direct system theoretic interpretation in terms of the *observability* and *constructibility* operators associated to X [49] [50], defined respectively as

$$\mathcal{O} : X \to H^+, \qquad \mathcal{O}\xi := \mathrm{E}^{H^+}\xi \tag{6.44}$$

$$\mathcal{C} : X \to H^-, \qquad \mathcal{C}\xi := \mathrm{E}^{H^-}\xi \tag{6.45}$$

In terms of $\mathcal{O}$ and $\mathcal{C}$ the indices $\eta_+(\xi)$ and $\eta_-(\xi)$ may be interpreted as relative "degree of observability" or as relative "degree of constructibility" of $\xi \in X$.

Recall that the observability ad constructibility operators, introduced in geometric realization theory [49] play a somewhat similar role to the observability and reachability operators in deterministic systems theory to characterize minimality of a state space. In fact the splitting property of a subspace X can be shown to be equivalent to a factorization of the *Hankel operator* of the process y,

$$\mathbb{H} := \mathrm{E}^{H^-}|_{H^+} : H^+ \to H^-$$

through the space X, as [49]

$$\mathbb{H} = \mathcal{C}\mathcal{O}^* \tag{6.46}$$

a fundamental characterization of minimality being that X is a minimal splitting subspace if and only if the factorization (6.46) is canonical, i.e. $\mathcal{C}$ and $\mathcal{O}$ are both *injective* operators. Equivalently (in the finite dimensional case) $\mathcal{O}^* = \mathrm{E}^X|_{H^+}$ is *surjective.* Hence, for a minimal splitting subspace, both $\mathcal{C}^*\mathcal{C}$ and $\mathcal{O}^*\mathcal{O}$ are invertible maps $X \to X$.

It follows that in a minimal splitting subspace X there are two distinct orthonormal bases of eigenvectors, say $(\xi_1^+, \dots \xi_n^+)$ and $(\xi_1^-, \dots \xi_n^-)$ in which the operators $\mathcal{O}^*\mathcal{O}$ and $\mathcal{C}^*\mathcal{C}$ diagonalize, i.e.

$$\mathcal{O}^*\mathcal{O} = \mathrm{diag}\{\lambda_1^+ \dots \lambda_n^+\}, \qquad 1 \geq \lambda_1^+ \geq \dots \geq \lambda_n^+ > 0 \tag{6.47}$$

$$\mathcal{C}^*\mathcal{C} = \mathrm{diag}\{\lambda_1^- \dots \lambda_n^-\}, \qquad 1 \geq \lambda_1^- \geq \dots \geq \lambda_n^- > 0 \tag{6.48}$$

the statistical interpretation being that the states in X can be *ordered* in two different ways according to the magnitude of their future- and, respectively, past- relative efficiency indices. It is in fact immediate from the definition (6.43) that, in the ordering according to the index η_+ the "most observable" states are just those which lie parallel to the vector ξ_1^+, having maximal index $\eta_+(\xi) = \lambda_1^+$ while the "least observable" states ξ being those parallel to the direction ξ_n^+, having the smallest possible relative efficiency $\eta_+(\xi) = \lambda_n^+$. Of course a completely similar picture corresponds to the ordering induced by past–relative efficiency.

Assume for a moment that $H^+ \cap H^- = 0$ (which will be the case if, say, the spectrum of the process is coercive [50]). Then a direction of "very observable" states in X, being at a small relative angle with the future subspace H^+, will generally form a "large" angle with the past subspace H^- and hence give rise to projections onto H^- of small relative norm i.e. to small $\eta_-(\xi)$. The opposite phenomenon is of course to be expected in case a direction "very close" to H^- is selected. The idea of balancing in the stochastic framework has to do with a choice of basis which roughly speaking, is meant to "balance" i.e. to make equal (if possible), the two ordered sets of efficiency indices. This is meant to reduce as far as possible bad conditioning of the model in the same sense as in determinstic balancing. There is here a substantial difference with the deterministic case however, in that we have now *a whole family of minimal* X which need to be considered simultaneously for the choice of a balanced basis. For this reason the stochastic procedure will necessarily be somehow less obvious than in the deterministic case.

In order to analyze the effects of choosing a basis $x(0)$ in a minimal splitting subspace X, we shall introduce the linear map $T_{x(0)} : \mathbb{R}^n \to X$, defined by $T_{x(0)}a := a'x(0)$. Note that if $\mathbb{R}^n$ is equipped with the inner product $< a, b >_P := a'Pb$, where P is the covariance matrix of $x(0)$, then $T_{x(0)}$ becomes an isometry. From this observation it is not hard to check that $T_{x(0)}$ has the following properties,

Lemma 6.2. *Let P be the covariance matrix of the basis $x(0)$ in X. Then,*

$$T_{x(0)}^{-1} = P^{-1}T_{x(0)}^{*} \qquad\qquad T_{\bar{x}(0)} = T_{x(0)}P^{-1} \tag{6.49}$$

where $\bar{x}(0)$ is the dual basis of $x(0)$.

Obviously the efficiency indices (6.43) can be expressed in terms of the coordinates a, b, once a specific basis has been chosen. In particular the expressions of the numerators will be quadratic forms described by certain symmetric positive–definite matrices which we shall call, respectively, *Observability* and *Constructibility gramians* (relative to that particular basis). Provided they are expressed in dual bases, the two gramians, have a particularly simple expression that will be given in the Proposition below. Recall (Proposition 4.6) that a basis in an arbitrary X can be extended together with its dual, to the whole family of minimal splitting subspaces $\mathfrak{X}$ in such a way as to form a *uniform basis.*

Proposition 6.1. *Let $x(0)$ be a basis in the minimal splitting subspace X and $\bar{x}(0)$ be its dual basis. Then the constructibility and observability gramians relative to the bases $x(0)$ and $\bar{x}(0)$ respectively, are given by*

$$\hat{\mathcal{C}^*\mathcal{C}} := T_{x(0)}^{*}\mathcal{C}^*\mathcal{C}T_{x(0)} = P_- \tag{6.50}$$

$$\hat{\mathcal{O}^*\mathcal{O}} := T_{\bar{x}(0)}^{*}\mathcal{O}^*\mathcal{O}T_{\bar{x}(0)} = \bar{P}_+ = P_+^{-1} \tag{6.51}$$

where P_- and $\bar{P}_+$ are the covariance matrices of $x_-(0)$ and $\bar{x}_+(0)$ in the uniform basis induced by $x(0)$.

In particular the two gramians are invariant over $\mathfrak{X}$, i.e. do not depend on the particular minimal splitting subspace X.

Proof. The formulas follow from the orthogonality of any minimal splitting subspace to the so called "junk" spaces, N^-, N^+ (the subspace of H^- orthogonal to the future and, respectively, the subspace of H^+ orthogonal to the past), see e.g. [50], Corollary 4.9. This leads to the identities

$$\mathcal{C}\xi := \mathrm{E}^{H^-}\xi = \mathrm{E}^{X_-}\xi, \qquad \mathcal{O}\xi = \mathrm{E}^{H^+}\xi = \mathrm{E}^{X_+}\xi \tag{6.52}$$

the first of which, in force of (4.11), can be rewritten as $\mathcal{C}\xi = a'x_-(0)$ and immediately leads to (6.50). The second follows by a similar computation, using the dual invariant projection property (4.12). □

Note that in the forward basis induced by $x(0)$, the expression of the observability gramian would instead be

$$\hat{\mathcal{O}^*\mathcal{O}} := T^*_{x(0)}\mathcal{O}^*\mathcal{O}T_{x(0)} = P\,P_+^{-1}\,P \tag{6.53}$$

which is no longer invariant.

The invariance of the two Gramians with respect to the particular state space of the realization, pointed out in the proposition above, clarifies that the notion of balanced realization given by Desai and Pal in terms of covariance matrices turns (luckily) out to be the correct generalization of the deterministic idea to stochastic systems.

Theorem 6.3. *There is a choice of basis $\hat{x}(0) := [\hat{x}_1, \dots, \hat{x}_n]'$ in X, such that both the constructibility and observability gramians are represented by a diagonal matrix. In fact, there is a diagonal matrix Σ, with positive entries*

$$\Sigma = \mathrm{diag}\{\sigma_1 \dots \sigma_n\}, \qquad 1 \geq \sigma_1 \geq \dots \geq \sigma_n > 0 \tag{6.54}$$

such that, in the uniform basis induced by $\hat{x}(0)$ in $\mathfrak{X}$, one has

$$\hat{\mathcal{C}^*\mathcal{C}} = \Sigma = \hat{\mathcal{O}^*\mathcal{O}}, \tag{6.55}$$

where $\hat{\mathcal{C}^*\mathcal{C}}$ is the constructibility gramian relative to the basis $\hat{x}(0)$ and $\hat{\mathcal{O}^*\mathcal{O}}$ is the observability gramian relative to the dual basis *of $\hat{x}(0)$.*

If the numbers σ_k are all distinct, this choice of basis is unique modulo sign, i.e. for any other basis $\tilde{x}(0) := [\tilde{x}_1, \dots \tilde{x}_n]'$ leading to a diagonal structure of the form (6.55), one has $\tilde{x}_k = \pm\hat{x}_k, k = 1, ..., n$.

Recall, as observed in subsection 4.1, that choosing bases uniformly in the family of minimal state spaces $\mathfrak{X}$ is equivalent to fixing a (minimal) realization $(A, C, \bar{C}, \Lambda(0))$ of the spectral density matrix. It follows that balanced realizations are generically *canonical forms* with respect to system similarity for (deterministic) realizations $(A, C, \bar{C}, \Lambda(0))$ of the spectrum.

Corollary 6.1. *There always exists a similarity transformation which brings a minimal (positive-real) quadruple $(A, C, \bar{C}, \Lambda(0))$ into balanced form. If the numbers $\{\sigma_1, \sigma_2, \dots, \sigma_n\}$ are all distinct then the balanced realization is unique up to a signature matrix (i.e. any two balanced realizations differ by a change of basis given by a signature matrix).*

For a much deeper discussion of balanced canonical forms see [58, 59].

Algorithm for computing the change of basis matrix bringing a minimal positive realization $(A, C, \bar{C}, \Lambda(0))$ into balanced canonical form.

1. Compute a square factorization of P_-, i.e. let $P_- = RR^*$ where R is square nonsingular, e.g. a Cholesky factor.
2. Do Singular Value Decomposition of $R^*\bar{P}_+R$, i.e. compute the factorization $R^*\bar{P}_+R = U\Sigma^2U^*$ where U is an orthogonal matrix and Σ^2 is diagonal with positive entries ordered by magnitude in the decreasing sense.
3. Define $T := \Sigma^{1/2}U^*R^{-1}$. The matrix T is the desired basis transformation matrix.
4. Check: Compute

$$TP_-T^* = \Sigma^{1/2}U^*R^{-1}P_-R^{-*}U\Sigma^{1/2} = \Sigma$$

$$T^{-*}\bar{P}_+T^{-1} = \Sigma^{-1/2}U^*R^*\bar{P}_+RU\Sigma^{-1/2} = \Sigma$$

The meaning of the diagonal matrix Σ. Note that in force of (6.50),(6.51) and (6.55), the numbers $\{\sigma_1^2, \dots, \sigma_n^2\}$ can be computed directly as the the (ordered) eigenvalues of the ratio $P_- P_+^{-1}$.

The following statement, which brings up the meaning of the entries of Σ as the (nonzero) singular values of the Hankel operator of the process y, will be reported here for completeness. It has been known for a long time [15], [63]. The proof in the present setup is particularly simple.

Proposition 6.2. *The entries of $\Sigma = \text{diag}\{\sigma_1, \dots, \sigma_n\}$ are invariants of the process y, equal to the nonzero singular values of the Hankel operator $\mathbb{H}$. They coincide with the canonical correlation coefficients of the past and future spaces H^-, H^+ of the process y.*

Proof. One just needs to notice that $\{\sigma_1^2, \dots, \sigma_n^2\}$ are the eigenvalues of the operator $\mathcal{C}^*\mathcal{C}\mathcal{O}^*\mathcal{O}$, since

$$P_- P_+^{-1} = T^*_{x(0)}\mathcal{C}^*\mathcal{C}T_{x(0)}T^*_{\bar{x}(0)}\mathcal{O}^*\mathcal{O}T_{\bar{x}(0)} = T^*_{x(0)}\mathcal{C}^*\mathcal{C}\mathcal{O}^*\mathcal{O}T_{\bar{x}(0)} \qquad (6.56)$$

and by (6.49) $T_{\bar{x}(0)}T^*_{x(0)} = I$. On the other hand, the square of the nonzero singular values of $\mathbb{H}$ are the nonzero eigenvalues of $\mathbb{H}^*\mathbb{H}$, and it follows from the factorization (6.46) that the non–zero eigenvalues of $\mathbb{H}^*\mathbb{H}$ are indeed equal to those of $\mathcal{C}^*\mathcal{C}\mathcal{O}^*\mathcal{O}$.

That the singular values of the Hankel operator $\mathbb{H}$ coincide with the canonical correlation coefficients of the process is also quite standard. A formal verification can be found in [55]. □

In conclusion, in this section we have shown that the concept of stochastic balancing can be seen as a natural generalization of the deterministic idea of balancing of *stable systems*. In the geometric setting however the "stability" (which is necessary for deterministic balancing) of the model does not enter at all, as the choice of a particular state vector $x(0)$ has nothing to do with the choice of a particular stability (causality) structure of the corresponding realization. The particular causality structure of the model influences instead the computation of the "balancing" basis transformation $\hat{x}(0) = Tx(0)$ of Theorem 6.3. This aspect is discussed in [67].

6.7 Stochastic Balanced Realizations: the Case of Finite Data

The theory presented above only refers to the stationary case. The concept of balancing which applies to identification, where data are always finite, is that of *finite time balancing.*

Definition 6.2. *A minimal realization $(A, C, \bar{C}, \Lambda(0))$ of a $m \times m$ positive real matrix is called* Stochastically Balanced at time k, *if the solutions $P_-(k)$, $\bar{P}_+(k)$ of the dual Riccati equations (5.4), (5.13) started respectively at $t = 0$ with initial condition $\Pi(0) = 0$, and at time T with $\bar{\Pi}(T) = 0$, are both equal to the same diagonal matrix, i.e.*

$$P_-(k) = \Sigma(k) = \bar{P}_+(k)$$

where $\Sigma(k) = \operatorname{diag}\{\sigma_1(k), \sigma_2(k), \ldots, \sigma_n(k)\}$. Whitout loss of generality we assume that the $\sigma(k)$'s are ordered in decreasing magnitude, i.e. $\sigma_{i+1}(k) \geq \sigma_i(k)$.

The normalized SVD factorization (6.39) of a *finite* Hankel matrix leads precisely to a finite-time balanced realization.

Proposition 6.3. *The triple (6.40) is stochastically balanced* at time k.

The proof will result from the discussion presented at the end of the next section.

7. The "Subspace Methods" Identification Algorithm of Van Overschee and DeMoor

The general idea of the so-called "subspace methods" for identification of stochastic systems [60], is to operate directly on vector spaces generated by the data.

The system-theoretical background which explains the procedure in (isomorphic) probabilistic terms is exposed in section 5., see in particular Theorem 5.2. The procedure proposed in [60] consists of a number of steps which conceptually can be described as follows.

Given the past and future data spaces $\mathcal{Y}_k^-$, $\mathcal{Y}_k^+$,

1. Form the sample finite-memory predictor spaces $\hat{X}_{k-} = \mathrm{E}^{\mathcal{Y}_k^-} \mathcal{Y}_k^+$ and $\hat{X}_{k+} = \mathrm{E}^{\mathcal{Y}_k^+} \mathcal{Y}_k^-$.

2. Pick *coherent* bases $\hat{\mathbf{x}}(k), \hat{\bar{\mathbf{x}}}(k)$ in $\hat{X}_{k-}$ and $\hat{X}_{k+}$. These bases will define the state at time k of two finite-interval Kalman filter realizations of $\mathbf{y}$.
3. Repeat step n.2 to get coherent bases $\check{\mathbf{x}}(k+1), \check{\bar{\mathbf{x}}}(k-1)$for $\hat{X}_{(k+1)-}$ and $\hat{X}_{(k-1)+}$.
4. Multiply the bases computed in step 3. by a suitable transformation matrix so as they will correspond to $\hat{\mathbf{x}}(k), \hat{\bar{\mathbf{x}}}(k)$ conditionally shifted by one time step (see section 5. for the definition of the conditional shift).
5. Estimate the matrices $(A, C, \bar{C})$ by formulas of the type (2.12, 2.15). These formulas hold for the finite-interval Kalman filter realizations also, see Theorem 7.2 below.

To compute a stationary state-space model, say a forward stationary innovation model (A, C, B_-, D_-), starting from a realization of the spectrum $(A, C, \bar{C}, \Lambda_0)$, the following additional steps are needed.

6. Check $(A, C, \bar{C}, \Lambda_0)$ for positivity. If positivity is not satisfied one may try to re-run the algorithm by varying k and/or n.
7. If $(A, C, \bar{C}, \Lambda_0)$ is positive solve the Algebraic Riccati equation $\Lambda(P) = 0$ and find the unique stabilizing positive-definite solution P_-.
8. Compute (B_-, D_-) by the formulas

$$D_- = (\Lambda_0 - CP_-C')^{1/2}, \quad B_- = (\bar{C}' - AP_-C')(\Lambda_0 - CP_-C')^{-1/2}. \tag{7.1}$$

For pedagogical reasons we have chosen to follow closely the line of thought of [60] albeit, as argued in [55] this procedure involves some redundant computations which can be avoided. In the following sections we shall discuss in detail the basic steps listed above and explain the reasons of the redundancy.

The present time k will be assumed large enough throughout.

7.1 Choosing Bases in the Predictor Spaces

We shall show that there is a one-to-one correspondence between full rank factorizations of the Hankel matrix H_k and coherent choice of bases in the finite-memory predictor spaces $\hat{X}_{k-}$ and $\hat{X}_{k+}$. This correspondence relates the geometric approach of "subspace methods" to the partial realization approach discussed in section 6.4.

Theorem 7.1. *Let $\hat{\mathbf{x}}(k)$, $\hat{\bar{\mathbf{x}}}(k)$ be n-dimensional bases for the finite memory predictor spaces $\hat{X}_{k-}$ and $\hat{X}_{k+}$ and let*

$$\Omega_k \hat{\mathbf{x}}(k) := \mathrm{E}[\mathbf{Y}_k^+ | \hat{\mathbf{x}}(k)], \quad \bar{\Omega}_k \hat{\bar{\mathbf{x}}}(k) := \mathrm{E}[\bar{\mathbf{Y}}_k^- | \hat{\bar{\mathbf{x}}}(k)]. \tag{7.2}$$

Then H_k has the corresponding rank factorizations

$$H_k = \Omega_k \bar{\Delta}_k' = \Delta_k \bar{\Omega}_k'$$

for some $mk \times n$ matrices $\bar{\Delta}_k$, Δ_k with linearly independent columns. If $\hat{\mathbf{x}}(k)$, $\hat{\bar{\mathbf{x}}}(k)$ are a coherent pair, *then $\bar{\Delta}_k = \bar{\Omega}_k$ and $\Delta_k = \Omega_k$.*

Conversely, for each rank factorization (6.27) of the finite Hankel matrix H_k, the n-vectors

$$\hat{\mathbf{x}}(k) = \bar{\Omega}_k{}'(T_k^-)^{-1}\bar{\mathbf{Y}}_k^- \tag{7.3}$$

$$\hat{\bar{\mathbf{x}}}(k) = \Omega_k'(T_k^+)^{-1}\mathbf{Y}_k^+ \tag{7.4}$$

are coherent bases for the finite-memory forward and backward predictor spaces $\hat{X}_{k-}$ and $\hat{X}_{k+}$ respectively.

Proof. That the factorizations of H_k follow from (7.2) is a consequence of the splitting property (5.18) at time k of $\hat{X}_{k-}$ and $\hat{X}_{k+}$. In particular,

$$\mathcal{Y}_k^+ \perp \mathcal{Y}_k^- \,|\, \hat{X}_{k-}$$

which can be rewritten as,

$$\mathrm{E}\mathbf{y}(t)\mathbf{y}(s)' = \mathrm{E}\{\mathrm{E}[\mathbf{y}(t)|\hat{\mathbf{x}}(k)]\,\mathrm{E}[\mathbf{y}(s)|\hat{\mathbf{x}}(k)]'\}$$

for $t = k, \dots, 2k-1$ and $s = k-1, \dots, 0$. This relation arranged in matrix form is the same as $H_k = \Omega_k P(k)\tilde{\Delta}_k'$ where $P(k) := \mathrm{E}\hat{\mathbf{x}}(k)\hat{\mathbf{x}}(k)' > 0$ and $\tilde{\Delta}_k\hat{\mathbf{x}}(k) = \mathrm{E}[\bar{\mathbf{Y}}_k^-|\hat{\mathbf{x}}(k)]$. Letting $\bar{\Delta}_k := \tilde{\Delta}_k P(k)$ yields the first factorization of H_k. The fact that Ω_k and $\bar{\Delta}_k$ are full rank is implied by observability and constructibility (i.e. minimality) of $\hat{X}_{k-}$, since $\hat{\mathbf{x}}(k)$ is a basis. Naturally an analogous reasonig yields the other factorization.

Let $\hat{\mathbf{x}}_+(k) := \bar{P}(k)^{-1}\hat{\bar{\mathbf{x}}}(k)$ be the dual basis of $\hat{\bar{\mathbf{x}}}(k)$ and assume that $\mathrm{E}[\hat{\mathbf{x}}_+(k)|\hat{\mathbf{x}}(k)] = \hat{\mathbf{x}}(k)$ (Proposition 5.2). Since the components of $\hat{\mathbf{x}}_+(k)$ belong to the future we have $\bar{\mathbf{Y}}_k^- \perp \hat{\mathbf{x}}_+(k) \,|\, \hat{\mathbf{x}}(k)$ so that

$$\begin{aligned}\tilde{\Delta}_k P(k) &= \mathrm{E}\{\mathrm{E}[\bar{\mathbf{Y}}_k^-|\hat{\mathbf{x}}(k)]\hat{\mathbf{x}}(k)'\} = \mathrm{E}\{\mathrm{E}[\bar{\mathbf{Y}}_k^-|\hat{\mathbf{x}}(k)]\mathrm{E}[\hat{\mathbf{x}}_+(k)|\hat{\mathbf{x}}(k)]'\} \\ &= \mathrm{E}\{\bar{\mathbf{Y}}_k^-\hat{\mathbf{x}}_+(k)'\} = \mathrm{E}\{\bar{\mathbf{Y}}_k^-\hat{\bar{\mathbf{x}}}(k)'\}\bar{P}(k)^{-1} = \bar{\Omega}_k.\end{aligned}$$

To show the converse, take any random variable in $\mathcal{Y}_k^+$, i.e. any linear combination of the form $a'\mathbf{Y}_k^+, a \in \mathbb{R}^{mk}$ and project it onto $\mathcal{Y}_k^-$. Expressing the projection in terms of the generators $\bar{\mathbf{Y}}_k^-$ of $\mathcal{Y}_k^-$, we obtain

$$\mathrm{E}^{\mathcal{Y}_k^-} a'\mathbf{Y}_k^+ = a'H_k(T_k^-)^{-1}\bar{\mathbf{Y}}_k^- = a'\Omega_k\hat{\mathbf{x}}(k)$$

and since the columns of Ω_k are linearly independent it follows that the minimal splitting subspace $\hat{X}_{k-}$ is spanned by the scalar components of $\hat{\mathbf{x}}(k)$. These are also linearly independent since $\hat{\mathbf{x}}(k)$ has a positive definite variance matrix. A dual reasoning for $\hat{\bar{\mathbf{x}}}(k)$ leads to the same conclusion.

That the two bases are coherent is shown in the proposition below. □

The variance matrices $P(k) := \mathrm{E}\hat{\mathbf{x}}(k)\hat{\mathbf{x}}(k)'$ and $\bar{P}(k) := \mathrm{E}\hat{\bar{\mathbf{x}}}(k)\hat{\bar{x}}(k)'$ are given by

$$P(k) = \bar{\Omega}_k{}'(T_k^-)^{-1}\bar{\Omega}_k, \quad \bar{P}(k) = \Omega_k'(T_k^+)^{-1}\Omega_k$$

For future use we record also the formula,

$$\mathrm{E}\hat{\bar{\mathbf{x}}}(k)\hat{\mathbf{x}}(k)' = \bar{P}(k)P(k). \tag{7.5}$$

Proposition 7.1. *The two bases (7.3) and (7.4) are coherent in the sense explained in section 5. i.e. belong to the same uniform choice of bases, or, which is the same, to the same triple* $(A, C, \bar{C})$.

Proof. We shall interpret (with some foresight) $\hat{\bar{\mathbf{x}}}(k)$ as a *dual basis* in $\hat{X}_{k+}$ and $\hat{\mathbf{x}}(k)$ as a "primal basis" in $\hat{X}_{k-}$, the "primal" and dual corresponding bases being,

$$\hat{\mathbf{x}}_+(k) := \bar{P}(k)^{-1}\hat{\bar{\mathbf{x}}}(k), \quad \hat{\bar{\mathbf{x}}}_-(k) := P(k)^{-1}\hat{\mathbf{x}}(k)$$

respectively. Using (7.5) we compute

$$\mathrm{E}^{\hat{X}_{k-}}\hat{\mathbf{x}}_+(k) = \mathrm{E}\{\hat{\mathbf{x}}_+(k)\hat{\mathbf{x}}(k)'\}P(k)^{-1}\hat{\mathbf{x}}(k) = \bar{P}(k)^{-1}\bar{P}(k)P(k)P(k)^{-1}\hat{\mathbf{x}}(k) = \hat{\mathbf{x}}(k)$$

which is the projection condition in Proposition 5.2. So $\hat{\mathbf{x}}(k)$ and $\hat{\bar{\mathbf{x}}}(k)$ are coherent. □

Recall that in order to define the same triplet $(A, C, \bar{C})$, the two state vectors (7.3) and (7.4) must "match" i.e. be coherent in the sense explained in section 5..

As we have just seen, a choice of basis (state) vectors in the two spaces $\hat{X}_{k-}$ and $\hat{X}_{k+}$ is related in a one-to-one way to rank-factorizations of the Hankel matrix H_k. Note that by stationarity $\mathbf{y}$ admits also stationary realizations of dimension n of the standard structure (2.1) (see Theorem 5.1) and hence its spectrum is represented by some (minimal) triplet $(A, C, \bar{C})$ of degree n. Information about this triplet is encoded in the bases $\hat{\mathbf{x}}(k)$ and $\hat{\bar{\mathbf{x}}}(k)$, see Theorem 5.2.

We shall now describe a (conceptual) procedure to determine the triplet $(A, C, \bar{C})$ corresponding to an arbitrary choice of bases in the finite memory predictor spaces $\hat{X}_{k-}$, $\hat{X}_{k+}$ as operated above.

The basic idea to compute the dynamics, and in particular the A matrix, is to select bases in the "updated" predictor spaces $\hat{X}_{(k+1)-}$ and $\hat{X}_{(k-1)+}$ constructed with one more observation in the past and one more observation in the future, respectively.

Note however that this further basis selection must be done in such a way as to keep (A, C) and $(\bar{A}, \bar{C})$ constant in time. This is the same as *conditional shifting* defined in section 5.2. Once we know how to do this, the computation of $(A, C, \bar{C})$ is easy.

Theorem 7.2. *Let $\hat{\mathbf{x}}(k), \hat{\bar{\mathbf{x}}}(k)$ be coherent bases in $\hat{X}_{k-}$ and $\hat{X}_{k+}$ and let $\hat{\mathbf{x}}(k+1), \hat{\bar{\mathbf{x}}}(k-1)$ be the corresponding conditionally shifted bases in $\hat{X}_{(k+1)-}$ and $\hat{X}_{(k-1)+}$. The corresponding minimal triple $(A, C, \bar{C})$ can be computed by the following formulas,*

$$\begin{aligned} A &= \mathrm{E}\hat{\mathbf{x}}(k+1)\hat{\mathbf{x}}(k)'\hat{P}(k)^{-1} && (7.6)\\ C &= \mathrm{E}\mathbf{y}(k)\hat{\mathbf{x}}(k)'\hat{P}(k)^{-1} && (7.7)\\ A' &= \mathrm{E}\hat{\bar{\mathbf{x}}}(t-1)\hat{\bar{\mathbf{x}}}(k)'\hat{\bar{P}}(k)^{-1} && (7.8)\\ \bar{C} &= \mathrm{E}\mathbf{y}(k-1)\hat{\bar{\mathbf{x}}}(k)'\hat{\bar{P}}(k)^{-1} && (7.9) \end{aligned}$$

where $\hat{P}(k) = \mathrm{E}\hat{\mathbf{x}}(k)\hat{\mathbf{x}}(k)'$ and $\hat{\bar{P}}(k) = \mathrm{E}\hat{\bar{\mathbf{x}}}(k)\hat{\bar{\mathbf{x}}}(k)'$.

Proof. The formulas follow readily from the finite interval Kalman filter realizations corresponding to $\hat{\mathbf{x}}(k), \hat{\bar{\mathbf{x}}}(k)$. The fact that $\hat{\bar{\mathbf{x}}}(k)$ and $\hat{\mathbf{x}}(k)$ are coherent bases serves precisely the purpose of extracting the parameters $(A', \bar{C})$ from the backward Kalman filter corresponding to $\hat{\bar{\mathbf{x}}}(k)$. □

How do we select conditionally shifted bases? It is obvious that the statement of Theorem 7.1 applies as well to *any* block Hankel matrix constructed with the available covariance data and in particular to the "shifted" Hankel matrices H_{k+1} and $\bar{H}_{k+1}$ defined in (6.23) and (6.24). Assume the rank condition (6.25) holds and consider the Hankel factorizations (6.33, 6.34), namely $H_{k+1} = \Omega_k \bar{\Omega}'_{k+1}$, $\bar{H}_{k+1} = \Omega_{k+1}\bar{\Omega}'_k$ induced by the factorization (6.27) at time k. Corresponding to these factorizations introduce the n-dimensional vectors,

$$\begin{aligned} \hat{\mathbf{x}}(k+1) &= \bar{\Omega}'_{k+1}(T^-_{k+1})^{-1}\bar{\mathbf{Y}}^-_{k+1}, \quad \hat{\bar{\mathbf{x}}}(k+1) = \Omega_k(T^+_k)^{-1}\mathbf{U}\mathbf{Y}^+_k && (7.10)\\ \hat{\mathbf{x}}(k-1) &= \bar{\Omega}_k{}'(T^-_k)^{-1}\mathbf{U}^{-1}\bar{\mathbf{Y}}^-_k, \quad \hat{\bar{\mathbf{x}}}(k-1) = \Omega'_{k+1}(T^+_{k+1})^{-1}\mathbf{U}^{-1}\mathbf{Y}_{[k,2k]}. && (7.11) \end{aligned}$$

Now it follows from Theorem 7.1 above that (7.10) are basis vectors for the forward predictor space at time $k+1$ with memory $k+1$: $\hat{X}_{(k+1)-} := \mathrm{E}[\mathbf{U}\mathcal{Y}_{[k,2k-1]} | \mathcal{Y}^-_{k+1}]$, and respectively for the backward predictor space at time $k+1$ with memory of lenght k (in the future), defined as the orthogonal projection $\mathrm{E}[\mathcal{Y}^-_{k+1} | \mathbf{U}\mathbf{Y}_{[k,2k-1]}]$. By Theorem5.1 this projection is actually identical to $\mathrm{E}[\mathbf{U}\mathcal{Y}^-_k | \mathbf{U}\mathcal{Y}_{[k,2k-1]}] = \mathbf{U}\mathrm{E}[\mathcal{Y}^-_k | \mathbf{Y}_{[k,2k-1]}] = \mathbf{U}\hat{X}_{k+}$.

Dually (7.11) are basis vectors, respectively, for the forward predictor space $\mathbf{U}^{-1}\hat{X}_{k-}$ and for the backward predictor space with memory $k+1$ in the future, $\hat{X}_{(k-1)+} := \mathrm{E}[\mathcal{Y}^-_k | \mathcal{Y}_{[k,2k]}]$.

Proposition 7.2. *The random vectors $\hat{\mathbf{x}}(k+1)$, $\hat{\bar{\mathbf{x}}}(k-1)$ defined in (7.10) and (7.11) are the conditionally shifted versions of (7.3) one step forward in time and of (7.4) one step backwards in time.*

Proof. Let A and $\bar{C}$ be the $n \times n$ and $n \times m$ matrices in (6.30), (6.32). We proceed to show directly that

$$\mathrm{E}[\hat{\mathbf{x}}(k+1) | \hat{\mathbf{x}}(k)] = A\hat{\mathbf{x}}(k) \tag{7.12}$$

$$\mathrm{E}[\mathbf{y}(k)\hat{\mathbf{x}}(k+1)'] = \bar{C} \tag{7.13}$$

so that by Proposition 5.3 and by Theorem 6.2 our claim will follow. In fact,

$$\begin{aligned}
&\mathrm{E}\hat{\mathbf{x}}(k+1)\hat{\mathbf{x}}(k)'\hat{P}(k)^{-1} = \\
&\bar{\Omega}'_{k+1}(T^-_{k+1})^{-1}\mathrm{E}\bar{\mathbf{Y}}^-_{k+1}(\bar{\mathbf{Y}}^-_k)'(T^-_k)^{-1}\bar{\Omega}_k P(k)^{-1} = \\
&\bar{\Omega}'_{k+1}(T^-_{k+1})^{-1}\mathrm{E}\bar{\mathbf{Y}}^-_{k+1}(\bar{\mathbf{Y}}^-_k)'(\bar{\Omega}'_k)^{-R}
\end{aligned}$$

where the last equality follows since from the expression of $P(k)$ given above,

$$\bar{\Omega}_k{}'(T^-_k)^{-1}\bar{\Omega}_k P(k)^{-1} = I.$$

Now since

$$(T^-_{k+1})^{-1}\begin{bmatrix} \Lambda_0 & \Lambda_1 & \dots & \Lambda_k \\ \Lambda'_1 & \Lambda_0 & \dots & \Lambda_{k-1} \\ \vdots & \vdots & \ddots & \vdots \\ \Lambda'_k & \Lambda'_{k-1} & \dots & \Lambda_o \end{bmatrix} = \begin{bmatrix} I_m & 0 \\ 0 & I_{(k-1)m} \end{bmatrix}$$

we get,

$$(T^-_{k+1})^{-1}\mathrm{E}\bar{\mathbf{Y}}^-_{k+1}(\bar{\mathbf{Y}}^-_k)' = \begin{bmatrix} 0 \\ I_{(k-1)m} \end{bmatrix}$$

and finally,

$$\begin{aligned}
&\bar{\Omega}'_{k+1}\begin{bmatrix} 0 \\ I_{(k-1)m} \end{bmatrix}(\bar{\Omega}'_k)^{-R} = \Omega_k^{-L}H_{k+1}\begin{bmatrix} 0 \\ I_{(k-1)m} \end{bmatrix}(\bar{\Omega}'_k)^{-R} = \\
&\Omega_k^{-L}\sigma H_k(\bar{\Omega}'_k)^{-R} = A.
\end{aligned}$$

The verification of (7.13) can be done along similar lines. □

From the proof we obtain the following interesting statement,

Corollary 7.1. *The Kalman-Filter realizations having as state vectors the bases (7.3) and (7.4) in the finite-memory predictor spaces $\hat{X}_{k-}$ and $\hat{X}_{k+}$, have the same $A, C, \bar{C}$ parameters as those computed by the partial realization formulas (6.30, 6.31, 6.32) corresponding to the rank-factorization $H_k = \Omega_k\bar{\Omega}_k{}'$ induced by (7.3) and (7.4) in the sense described in Theorem 7.1.*

In other words, the "subspace methods" algorithm described at the beginning of this section is *equivalent to partial realization*, for the formulas (7.6, 7.7, 7.9) produce exactly the same $(A, C, \bar{C})$ matrices as the partial realization formulas (6.30, 6.31, 6.32) applied to the corresponding Hankel factorization.

Note that the conditionally shifted bases $\hat{\mathbf{x}}(k+1)$ and $\hat{\bar{\mathbf{x}}}(k-1)$ can be computed from the sole factorization (6.27) since $\bar{\Omega}_{k+1}$ and Ω_k are uniquely determined from (6.27) as

$$\bar{\Omega}'_{k+1} = \Omega_k^{-L} H_{k+1}, \quad \Omega_{k+1} = \bar{H}_{k+1}(\bar{\Omega}'_k)^{-R}$$

so that

$$\hat{\mathbf{x}}(k+1) = \Omega_k^{-L} H_{k+1}(T^-_{k+1})^{-1}\bar{\mathbf{Y}}^-_{k+1} \tag{7.14}$$

$$\hat{\bar{\mathbf{x}}}(k-1) = (\bar{\Omega}_k)^{-L}\bar{H}'_{k+1}(T^+_{k+1})^{-1}\mathbf{U}^{-1}\mathbf{Y}_{[k,2k]}. \tag{7.15}$$

Change of basis. If we pick arbitrarily an n-dimensional basis $\mathbf{s}(k+1)$ in $\hat{X}_{(k+1)-}$ the basis transformation matrix M taking $\mathbf{s}(k+1)$ into the conditionally shifted basis at time $k+1$ can be obtained by the following reasoning.

First notice that the first members of both expressions

$$\mathrm{E}[\mathbf{U}\mathbf{Y}^+_k | \hat{\mathbf{x}}(k+1)] = \Omega_k \hat{\mathbf{x}}(k+1)$$

$$\mathrm{E}[\mathbf{U}\mathbf{Y}^+_k | \mathbf{s}(k+1)] := \tilde{\Omega}_k \mathbf{s}(k+1),$$

are equal to $\mathrm{E}[\mathbf{U}\mathbf{Y}^+_k | \bar{\mathbf{Y}}^-_{k+1}]$ by the splitting property. Obviously they must be equal so that $\Omega_k \hat{\mathbf{x}}(k+1) = \tilde{\Omega}_k \mathbf{s}(k+1)$ and

$$\hat{\mathbf{x}}(k+1) = (\Omega_k)^{-L}\tilde{\Omega}_k \mathbf{s}(k+1). \tag{7.16}$$

which provides the change of basis formula in $\hat{X}_{(k+1)-}$. A similar formula can be derived easily for the change of basis in the backward predictor space. □

7.2 Skipping some Redundant Steps

As we have already warned the reader, the procedure for computing $(A, C, \bar{C})$ given so far is vastly redundant from a computational point of view. In principle we can eliminate the computation of the backward bases completely and reduce everything just to finding a basis $\hat{\mathbf{x}}(k)$ in $\hat{X}_{k-}$.

Also there is no need to pick a basis at time $k+1$ in $\hat{X}_{k+1}$ and to convert it to the conditionally shifted basis of $\hat{\mathbf{x}}(k)$, since the conditionally shifted basis $\hat{\mathbf{x}}(k+1)$ can be computed explicitely via formula (7.14). For, choosing a basis $\hat{\mathbf{x}}(k)$ induces a rank factorization (6.27) where the matrix Ω_k is determined by $\hat{\mathbf{x}}(k)$ as shown in (7.2) of Theorem 7.1 above.

Reduced "subspace" Algorithm.

1. Choose a basis $\hat{\mathbf{x}}(k)$ in $\hat{X}_{k-}$.
2. Compute the corresponding observability matrix Ω_k by (7.2).
3. Solve $H_k = \Omega_k \bar{\Omega}_k{}'$ to get (a unique) $\bar{\Omega}_k$.
4. Compute the conditionally shifted basis $\hat{\mathbf{x}}(k+1)$ by (7.14).
5. Compute $(A, C, \bar{C})$ by the following formulas,

$$A = \mathrm{E}\hat{\mathbf{x}}(k+1)\hat{\mathbf{x}}(k)'\hat{P}(k)^{-1} \tag{7.17}$$

$$C = \mathrm{E}\mathbf{y}(k)\hat{\mathbf{x}}(k)'\hat{P}(k)^{-1} \tag{7.18}$$

$$\bar{C} = \mathrm{E}\mathbf{y}(k-1)\hat{\mathbf{x}}(k)' \tag{7.19}$$

where $\hat{P}(k) = \mathrm{E}\hat{\mathbf{x}}(k)\hat{\mathbf{x}}(k)' = \bar{\Omega}_k{}'(T_k^-)^{-1}\bar{\Omega}_k$

□

Note that (7.19), which formally is derived from the backward (or anti-causal) form of the Kalman Filter realization with state $\hat{\mathbf{x}}(k)$, can be rewritten directly in terms of the dual basis $\hat{\bar{\mathbf{x}}}_-(k) = \hat{P}(k)^{-1}\hat{\mathbf{x}}(k)$ whereby,

$$\mathbf{y}(k-1) = \bar{C}\hat{\bar{\mathbf{x}}}_-(k) + \bar{D}_-(k)\bar{\epsilon}_-(k-1).$$

This reduced procedure should lead to a more effective numerical algorithm than the variants of the original subspace algorithm of [60] which have recently appeared in the literature.

7.3 The Least Squares Implementation

If the "expectation" operator E is written explicitely as in (6.18), then the formulas for $(A, C, \bar{C})$ of Theorem 7.2 express exactly the solution of the two dual *least squares problems,*

$$\min_{A,C} \left\| \begin{bmatrix} \hat{\mathbf{x}}(k+1) \\ \mathbf{y}(k) \end{bmatrix} - \begin{bmatrix} A \\ C \end{bmatrix} \hat{\mathbf{x}}(k) \right\|^2 \tag{7.20}$$

$$\min_{A',\bar{C}} \left\| \begin{bmatrix} \hat{\bar{\mathbf{x}}}(k-1) \\ \mathbf{y}(k-1) \end{bmatrix} - \begin{bmatrix} A' \\ \bar{C} \end{bmatrix} \hat{\bar{\mathbf{x}}}(k) \right\|^2 \tag{7.21}$$

where the norm is now ordinary Euclidean norm in $\mathbb{R}^N$. This equivalence can be used in the actual computation of $(A, C, \bar{C})$ requiring just a least-squares equation solver. Good numerical implementations for lest-squares problems are easily available. However we should notice that in this formulation we need to compute explicitely *all* the basis vectors $\hat{\mathbf{x}}(k), \hat{\bar{\mathbf{x}}}(k), \hat{\mathbf{x}}(k+1), \hat{\bar{\mathbf{x}}}(k-1)$.

This rephrasing of the formulas of Theorem 7.2 is used in commercially available codes. The appearence of least squares looks appealing to many and there have been attempts to use the reformulation above also for theoretical purposes. In this respect, there seems to be some confusion in the literature regarding the role played by the estimation residues of the least-squares solution, say

$$\begin{bmatrix} \hat{\mathbf{x}}(k+1) \\ \mathbf{y}(k) \end{bmatrix} - \begin{bmatrix} A \\ C \end{bmatrix} \hat{\mathbf{x}}(k) := \begin{bmatrix} \hat{\mathbf{e}}_{\mathbf{x}}(k) \\ \hat{\mathbf{e}}(k) \end{bmatrix}$$

in "proving" positive-realness of the estimated triple $(A, C, \bar{C})$.

Although it is easy to check that

$$\mathrm{E} \begin{bmatrix} \hat{\mathbf{e}}_{\mathbf{x}}(k) \\ \hat{\mathbf{e}}(k) \end{bmatrix} [\hat{\mathbf{e}}_{\mathbf{x}}(k)' \hat{\mathbf{e}}(k)'] = \begin{bmatrix} P(k+1) - AP(k)A' & \bar{C}' - AP(k)C' \\ \bar{C} - CP(k)A' & \Lambda(0) - CP(k)C' \end{bmatrix} \geq 0$$

there is obviously no guarantee that some $P \geq 0$ will satisfy the stationary matrix inequality $M(P) \geq 0$. To draw this conclusion from the previous expression requires existence of a positive limit of $P(k)$ as $k \to \infty$ which, as is well known, is equivalent to assuming positivity of $(A, C, \bar{C})$ from the beginning.

7.4 Use of the SVD

Of course determinig rank and "picking bases" in practice is a numerically nontrivial affair. The basic numerical tool which helps in this respect is the SVD. In particular the truncated SVD derived from (6.39) of the previous section leads to the choice

$$\Omega_k = L_k^+ U_k \Sigma_k^{1/2}, \quad \bar{\Omega}_k = L_k^- V_k \Sigma_k^{1/2} \tag{7.22}$$

These expressions are meant to be substituted for $\Omega_k, \bar{\Omega}_k$ everywhere in the formulas above in this section wehnever the purpose is to do actual computations.

For the sake of clarity of exposition, in this section we shall assume that the factorization (6.39) is *exact* i.e. that Σ_k is made of the n nonzero singular values of $\hat{H}_k$. From this particular choice of the factorization, the n-dimensional bases for the finite-memory forward and backward predictor spaces $\hat{X}_{k-}$ and $\hat{X}_{k+}$ are seen to be

$$\mathbf{z}(k) = \Sigma_k^{1/2} V_k' (L_k^-)^{-1} \mathbf{Y}_k^- \tag{7.23}$$

$$\bar{\mathbf{z}}(k) = \Sigma_k^{1/2} U_k' (L_k^+)^{-1} \mathbf{Y}_k^+ \tag{7.24}$$

We note immediately that in this basis the variance matrices $P(k)$ and $\bar{P}(k)$ are equal and diagonal,

$$\mathrm{E}\mathbf{z}(k)\mathbf{z}(k)' = \Sigma_k = \mathrm{E}\bar{\mathbf{z}}(k)\bar{\mathbf{z}}(k)'. \tag{7.25}$$

In fact we shall see shortly that $\mathbf{z}(k)$ and $\bar{\mathbf{z}}(k)$ are a (finite-time) *balanced basis.*

Moreover, since $U_k' U_k = I_n = V_k' V_k$ by orthonormality of the columns of U_k and V_k we see that the bases are diagonally correlated i.e.

$$\mathrm{E}\bar{\mathbf{z}}(k)\mathbf{z}(k)' = \Sigma_k^{1/2} U_k' \hat{H}_k V_k \Sigma_k^{1/2} = \Sigma_k^2. \tag{7.26}$$

Hence the vectors $\mathbf{z}(k)$ and $\bar{\mathbf{z}}(k)$ are essentially the so-called *canonical variates* of canonical correlation analysis [37]). The elements of Σ_k are the sample *canonical correlation coefficients* of the finite past and future spaces $\mathcal{Y}_k^-$, $\mathcal{Y}_k^+$. Their dimension n, i.e. the dimension of the predictor spaces, is in reality determined numerically (or statistically) in the truncation step leading to (6.39), by discarding the canonical correlation coefficients which are smaller than a certain "significance level". So the statement about n-dimensional realizability of $\mathbf{y}$ is really "approximate".

Proof of Proposition 6.3. According to the standard notations used in the stationary setting $\hat{\mathbf{x}}(k) = \hat{\mathbf{x}}_-(k)$ and $\hat{\bar{\mathbf{x}}}(k) = \hat{\bar{\mathbf{x}}}_+(k)$ .For any choice of bases $\hat{\mathbf{x}}(k), \hat{\bar{\mathbf{x}}}(k)$ it follows from the Kalman-Filter representations that the relative variance matrices

$$P(k) = \bar{\Omega}_k{}'(T_k^-)^{-1}\bar{\Omega}_k = P_-(k), \quad \bar{P}(k) = \Omega_k'(T_k^+)^{-1}\Omega_k = \bar{P}_+(k)$$

are the solutions of the (Finite-interval) Riccati equations (5.4), (5.13). From (7.25) we see that $P(k) = \Sigma_k = \bar{P}(k)$, so, as announced in Proposition 6.3, $\mathbf{z}(k), \bar{\mathbf{z}}(k)$ define a *balanced realization* at time k.

It should be stressed however that the formulas of Theorem 7.2 for $\mathbf{z}(k), \bar{\mathbf{z}}(k)$, analog to (6.40), *will not* yield the stationary balanced triple $(A, C, \bar{C})$ described in section 6.6. For getting the system matrices in this form we first need to solve the steady state Algebraic Riccati Equation obtained with the estimated coefficients $(A, C, \bar{C}, \Lambda(0))$, compute the maximal and minimal solutions P_-, P_+ and then apply to $(A, C, \bar{C})$ the balancing algorithm seen in section 6.6.

Acknowledgments

Discussions of G. Picci with colleagues (M. Deistler, J. van Schuppen, J.C. Willems) which took place at the Como NATO-ASI school are gratefully acknowledged.

References

1. Adamjan V.M., D. Z. Arov, and M. G. Krein, "Analytic properties of Schmidt pairs for a Hankel operator and the generalized Schur-Takagi problem", *Math. USSR Sbornik* **15**, 31–73,1971.
2. Akaike H., "Markovian representation of stochastic processes by canonical variables", *SIAM J. Control* **13**, 162–173,1975.
3. Akaike H., "Canonical Correlation Analysis of Time Series and the use of an Information Criterion", in *System Identification, Advances and case studies*, R.K. Mehra and D.L. Lainiotis eds. Academic Press, 1976.
4. Akhiezer N.I.,I. M. Glazman, *Theory of Linear Operators in Hilbert Space*, Ungar, 1966.
5. Anderson B.D.O., "The inverse problem of stationary covariance generation", *J. Statistical Physics* 1:133–147, 1969.
6. Anderson B.D.O., "A System Theory Criterion for Positive–Real Matrices", *SIAM Journal on Control*, 5, 2:171–182, 1967.
7. Anderson T.W., *Introduction to Multivariate Statistical Analysis*, John Wiley, 1958.
8. Arun K.S., and S.Y. Kung, "Balanced approximation of stochastic systems", *SIAM Journal on Matrix Analysis and Applications*, **11**: 42–68, 1990.
9. Aoki M., *State Space Modeling of Time Series*, 2nd edition, Springer-Verlag, 1991.
10. Byrnes C.I., and A. Lindquist, "The stability and instability of partial realizations", *Systems and Control Letters*, **2** , 2301–2312,1982.
11. Byrnes C.I., and A. Lindquist, "On the partial stochasic realization problem", to appear.
12. Caines P.E., *Linear Stochastic Systems*, Wiley, 1988.
13. Cramer H., *Mathematical Methods of Statistics*, Princeton, 1949.
14. Desai U.B., and D. Pal, "A realization approach to stochastic model reduction and balanced stochastic realization", *Proc 16th Annulal Conference on on Information Sciences and Sytems*, Princeton Univ, pp. 613–620, 1982, also in *Proc 21st Conference on Decision and Control*, Orlando, FL, pp.1105–1112, 1982.
15. Desai U.B., and D. Pal, "A realization approach to stochastic model reduction", *IEEE Transactions Automatic Control*,**AC-29**: 1097–1100, 1984.
16. Desai U.B., D. Pal, and R.D. Kikpatrick, "A realization approach to stochastic model reduction", *International Journal of Control*, 42: 821–838, 1985.
17. Desai U.B., *Modeling and Application of Stochastic Processes*, Kluwer Academic Publishers, 1986.
18. Doob J.L., "The Elementary Gaussian Processes", *Annals of Math. Statistics*, 15: 229–282, 1944.
19. Doob J.L. *Stochastic Processes*, Wiley, 1953.
20. Ekstrom M.P., "A spectral characterization of the ill-conditioning in numerical deconvolution", *IEEE Trans, Audio Electroacustics*, **AU-21**, pp. 344-348, 1973.
21. Faurre P., "Identification par minimisation d'une representation Markovienne de processus aleatoires", *Symposium on Optimization*, Nice, 1969.
22. Faurre P., and P. Chataigner, "Identification en temp reel et en temp differee par factorisation de matrices de Hankel", *Proc. French-Swedish colloquium on process control*, IRIA Roquencourt, 1971.
23. Faurre P., and J. P. Marmorat, "Un algorithme de réalisation stochastique", *C. R. Academie Sciences Paris* , **268**,1969.

24. Faurre P., "Representation Markovienne des processus stochastiques stationnaires", *INRIA Report de recherche*, 1973.
25. Faurre P., M. Clerget, and F. Germain, *Opérateurs Rationnels Positifs*, Dunod, 1979.
26. Gantmacher F.R., *Matrix Theory*, Vol. I, Chelsea, New York, 1959.
27. Gauss K.F., "Theoria Motus Corporum Coelestium", Liber II, in *Werke*, Julius Springer, Berlin, 1901.
28. Georgiou T.T., "Realization of power spectra from partial covariance sequences", *IEEE Transactions Acoustics, Speech and Signal Processing* **ASSP-35**, 438–449,1987.
29. Gevers M.R., and B.D.O. Anderson, "Representation of jointly stationary feedback free processes", *Intern. Journal of Control* **33**, pp.777-809,1981.
30. Gevers M.R., and B.D.O. Anderson, "On jointly stationary feedback free stochastic processes", *IEEE Trans. Automatic Control* **AC-27**, pp.431-436,1982.
31. Glover K., "All optimal Hankel norm approximations of linear multivariable systems and their L^{∞} error bounds", *International Journal of Control*, **39**, 6:1115–1193,1984.
32. Golub G.H., and C. R. Van Loan, *Matrix Computations* (2nd ed.). The Johns Hopkins Univ. Press,1989.
33. Granger C.W.J., "Economic processes involving feedback", *Information and Control* **6**, pp. 28-48,1963.
34. Green M., "Balanced stochastic realizations", *Linear Algebra and its Applications*, 98:211–247, 1988.
35. Hunt B.R., "A theorem on the difficulty of numerical deconvolution", *IEEE Trans, Audio Electroacustics*, **AU-20**, March 1972.
36. Heij Ch., T. Kloek, and A. Lucas, "Positivity conditions for stochastic state space modelling of time series", Reprint Series 695, Erasmus University Rotterdam.
37. Hotelling H., "Relations between two sets of variables", *Biometrica*, **28**, pp. 321–377,1936.
38. Harshavaradhana P., E. A. Jonckheere, and L. M. Silverman, "Stochastic balancing and approximation-stability and minimality", *IEEE Trans. Automatic Control*, **AC-29**, 744–746,1984.
39. Harshavadana P., and E.A. Jonckheere, "Spectral factor reduction by phase–matching, the continuous–time case", *International Journal of Control*, 42: 43–63, 1985.
40. Opdenacker P., and E.A. Jonckheere, "A state space approach to to approximation by phase–matching", in *Modelling, Identification and Robust Control* (C. I. Byrnes and A. Lindquist eds), Elsevier, 1986.
41. Kalman R.E., "Realization of covariance sequences", *Proc. Toeplitz Memorial Conference*, Tel Aviv, Israel, 1981.
42. Kalman R.E., P.L.Falb, and M.A.Arbib, *Topics in Mathematical Systems Theory*, McGraw-Hill, 1969.
43. Kimura H., "Positive partial realization of covariance sequences", *Modelling, Identification and Robust Control* (C. I. Byrnes and A. Lindquist, eds.), North-Holland, pp. 499–513,1987.
44. Kung S.Y., "A new identification and model reduction algoritm via singular value decomposition", *Proc. 12th Asilomar Conf. Circuit, Systems and Computers*, pp. 705–714,1978.
45. Larimore W.E., "System identification, reduced-order filtering and modeling via canonical variate analysis", *Proc. American Control Conference*, pp. 445–451,1990.

46. Larimore W.E., "Canonical Variate Analysis in Identification, Filtering, and Adaptive Control", *Proc. 29th IEEE Conference on Decision and Control*, pp. 596–604,1990.
47. Lindquist A., G. Picci, and G. Ruckebusch, "On minimal splitting subspaces and Markovian representation", *Math. System Theory*, **12**: 271-279, 1979.
48. Lindquist A., and G. Picci, "On the stochastic realization problem", *SIAM J. Control and Optimization*, **17**: 365–389, 1979.
49. Lindquist A., and G. Picci, "Realization theory for multivariate stationary Gaussian processes", *SIAM J. Control and Optimization*, **23**:809–857, 1985.
50. Lindquist A., and G. Picci, "A geometric approach to modelling and estimation of linear stochastic systems", *Journal of Mathematical Systems, Estimation and Control*, **1**:241–333, 1991.
51. Lindquist A., G. Michaletzky, and G. Picci, "Zeros of Spectral Factors, the Geometry of Splitting Subspaces, and the Algebraic Riccati Inequality", *SIAM J. Control & Optimization* (March 1995).
52. Lindquist A., and G. Michaletzky, "Output-induced subspaces,invariant directions and interpolation in linear discrete-time stochastic systems", *Tech Report TRITA/MAT-94-20*, Royal Institute of Technology, Stockholm,1994.
53. Lindquist A., and G. Picci, "On "subspace methods" identification", in *Systems and Networks: Mathematical Theory and Applications II*, U. Hemke, R. Mennicken and J Saurer, eds., Akademie Verlag, pp. 315–320,1994.
54. Lindquist A., and G. Picci, "On "subspace methods" identification and stochastic model reduction", *Proceedings 10th IFAC Symposium on System Identification*, Copenhagen, June 1994, Volume 2, pp. 397–403.
55. Lindquist A., and G. Picci, "Canonical Correlation Analysis Approximate Covariance Extension and Identification of Stationary Time Series", *Tech Report TRITA/MAT-94-32*, Royal Institute of Technology, Stockholm. (submitted to *Automatica*).
56. Molinari B.P., "The time-invariant linear-quadratic optimal-control problem", *Automatica*, 13:347–357, 1977.
57. Molinari B.P., "The stabilizing solution of the discrete algebraic Riccati equation", *IEEE Trans. Automatic Control*, **20**, 396–399,1975.
58. Ober R., "Balanced realizations: canonical forms, parametrization, model reduction", *International Journal of Control* **46** , pp. 643-670,1987.
59. Ober R., "Balanced parametrization of a class of linear systems", *SIAM Journal on Control & Optimization*, 29, 6:1251–1287, 1991.
60. van Overschee P., and B. De Moor, "Subspace algorithms for stochastic identification problem", *Automatica* **3**, 649-660,1993.
61. van Overschee P., and B. De Moor, "Two subspace algorithms for the identification of combined deterministic-stochastic systems", preprint.
62. van Overschee P., and B. De Moor, "A unifying theorem for subspace identification algorithms and its interpretation", *Proceedings 10th IFAC Symposium on System Identification*, Copenhagen, June Volume 2, pp. 145–156,1994.
63. Pavon M., "Canonical Correlations of past inputs and future outputs for linear stochastic systems", *Systems and Control Letters*, **4**: 209–215, 1984.
64. Pernebo L., and L. M. Silverman, "Model reduction via balanced state space representations", *IEEE Trans. Automatic Control*, **AC-27**, 382–387,1982.
65. Phillips D.L., "A technique for the numerical solution of certain integral equations of the first kind", *Journal of the Assoc. Comput. Mach.*, **9** pp. 97-101, 1962.
66. Picci G., "Stochastic realization of Gaussian Processes", *Proceedings of the IEEE*, **64**, pp. 112-122, 1976.

67. Picci G., and S. Pinzoni, "Acausal models and balanced realizations of stationary processes", *Linear Algebra and its Applications*, **205-206**, 957-1003,1994.
68. Rozanov N.I., *Stationary Random Processes*, Holden Day, 1963.
69. Tether A., "Construction of minimal state-variable models from input-output data", *IEEE Trans. Automatic Control* **AC-15**, pp. 427-436,1971.
70. Twomey S., "The application of nunerical filtering to the solution of integral equations of the first kind encountered in indirect sensing measurements", *Journal of the Franklin Institute*, **279**, pp. 95-109, 1965.
71. Vaccaro R.J., and T. Vukina, "A solution to the positivity problem in the state-space approach to modeling vector-valued time series", *J. Economic Dynamics and Control* **17**, pp. 401–421,1993.
72. Weiland S., "Theory of Approximation and disturbance attenuation for linear systems", *Doctoral Thesis*, University of Groningen, Jan 1991.
73. Wiener N., "Generalized Harmonic Analysis", in *The Fourier Integral and Certain of its Applications*, Cambridge Univ. Press 1933.
74. Wiener N., and P. Masani, "The prediction theory of multivariate stationary stochastic processes", I, *Acta Mathematica***98**, 11-150, (1957); II, *ibidem*, **99** 93-137, 1958.
75. Willems J.C., "Least squares stationary optimal control and the algebraic Riccati equation", *IEEE Trans. Automatic Control* **AC-16**, pp. 621–634,1971.
76. Zeiger H.P., and A. J. McEwen, "Approximate linear realization of given dimension via Ho's algorithm", *IEEE Trans. Automatic Control* **AC-19** , p. 153, 1974.
77. Youla D.C., "On The Factorization of Rational Matrices", *IRE Transactions PGIT* , **7**: 1961.

Parameter Estimation of Multivariable Systems Using Balanced Realizations

J. M. Maciejowski
Cambridge University
Engineering Dept.
Cambridge CB2 1PZ
England.

1. Introduction

The identification of multivariable systems is of extreme importance in practice. The majority of systems being manipulated or controlled, whether they are industrial processes, vehicles, aircraft, consumer products or, in a different arena, markets and economies, have more than one input and more than one output, and these usually exhibit significant interactions amongst each other. The very widespread tendency to control each variable separately, pretending that the others do not exist, is a compromise which to some extent has been driven by the difficulty of obtaining multivariable models.

Currently commercial pressures in the process industries put a considerable premium on the achievement of maximal efficiency and quality, and a key to achieving this is to implement multivariable control, which exploits the interactions between variables. It is now not uncommon, for example in the petrochemical sector, to find multivariable controllers handling 30 or 40 variables simultaneously. An essential ingredient of such a control scheme is a multivariable model of the process being controlled.

An important requirement in some applications is that the identification algorithm should, as far as possible, work routinely, automatically, and reliably, with no human intervention. Current methods can not yet deliver this requirement, but the methods presented in this chapter seem to be more promising in this respect than others, at least for multivariable problems. One can expect that it will always be necessary to monitor the results of an identification algorithm, either by a human or automatically. But it should be possible to avoid the need for human decision making as part of the identification algorithm.

Conceptually, the identification problem for multivariable systems is very similar to that for single-variable systems. But system theory tell us that, whereas in the single-variable case there is only one 'structural' decision to be made, namely the order of the model, in the multivariable case a relatively large number of other 'structural' decisions is required. Making these decisions has always been a stumbling block for multivariable identification, and has certainly made it difficult to move towards methods which could be applied 'routinely and automatically'. The major advance of the past 10 years has been the development of so-called 'subspace methods' for system identification. These appear to be very successful at finding multivariable models without making such decisions explicitly. As will be seen later in the chapter, we use such methods to obtain initial models, and these methods undoubtedly make a great contribution to the success of the complete identification scheme which we present here.

In the first few sections of this chapter, we motivate estimation of parameters in 'balanced' parametrizations. This is essentially the same as conventional parametric identification, except that we advocate an unusual choice of parameters. Then a substantial section is devoted to a description of the subspace methods mentioned above, which we use to obtain initial models. Estimation of parameters involves the optimization of the parameters of these initial models. Finally we present three examples. These are chosen to illustrate the use of balanced parametrizations for various classes of models. It is assumed that the reader is familiar with the basics of conventional parametric identification, as described in books such as [13, 30, 42, 49].

2. Problem Setting

Our objective is to find linear time-invariant dynamic models for multivariate data. This data may be available in various forms, depending on the problem being tackled. Most conventionally, the data is available in the form of an input vector time series $\{u(t) : u(t) \in \mathbf{R}^m, t = 1, \ldots, N\}$ and an output vector time series $\{y(t) : y(t) \in \mathbf{R}^p, t = 1, \ldots, N\}$. This corresponds to measurements taken from some process which is to be identified. Sometimes only the output measurements are available, and the objective is to find a model which produces an output with the same spectrum, or second-order statistics, as the observed output, when the model is driven by white noise. A variation of this problem occurs when the output data itself is not available, but its spectrum or autocovariance (either exact or approximate) is given; this is called the 'stochastic realization' problem. A relatively novel use of identification is as a way of obtaining approximate linear models from complex, detailed, nonlinear models. Such nonlinear models are being built increasingly frequently for industrial processes, and have long been used in macroeconomics. The idea is to apply identification methods to data generated by the nonlinear model, as if it were data generated by some industrial plant. A very important difference, however, is that the experimental conditions can be chosen freely; in particular, the input signal can be chosen to be an impulse, which is not usually possible with a real system.

Linear time-invariant multivariable systems can be represented by models in various forms. We review briefly the most common ones, which are:

Transfer function models:

$$y(t) = G(z)u(t) + \nu(t) \tag{2.1}$$

where $G(z)$ is a matrix of transfer functions, z is the forward shift (so that $zu(t) = u(t+1)$, $z^{-1}u(t) = u(t-1)$, etc), and $G(z)$ is related to the impulse response sequence of the model $\{g(t) : g(t) \in \mathbf{R}^{p\times m}, t = 0, 1, \ldots\}$ by:

$$G(z) = \sum_{t=0}^{\infty} g(t)z^{-t}. \tag{2.2}$$

$\nu(t) \in \mathbf{R}^p$ is an 'error', 'disturbance', or 'noise' term, which appears either as an admission that we cannot hope to model real data exactly without such terms, or as a representation of some additional influences on the output $y(t)$. If a stochastic setting is adopted and $\nu(t)$ is regarded as a stochastic process, then frequently a further model $\nu(t) = H(z)e(t)$ is introduced, in which $H(z)$ is another transfer function matrix, and $e(t)$ is a multivariate white noise process. An important reason for doing this is that the 'efficiency' of estimation of $G(z)$, namely the statistical accuracy of the estimates, can be increased by estimating $H(z)$ [30].

ARMAX models:

$$A(z)y(t) = B(z)u(t) + C(z)e(t) \tag{2.3}$$

Here $A(z)$, $B(z)$ and $C(z)$ are matrices of polynomials in z, and this is just a shorthand way of writing the model as a set of difference equations relating $y(t)$, $u(t)$ and $e(t)$. This is equivalent to the transfer function model if $G(z) = A(z)^{-1}B(z)$ and $H(z) = A(z)^{-1}C(z)$. Note that this model imposes 'common dynamics' — the matrix $A(z)$ — on the effects of the input $u(t)$ and the noise $e(t)$. This is not a restriction, since one can accommodate completely different dynamics by making $A(z)$ a 'common denominator', but it is frequently economical, in terms of model complexity, since real noise and disturbance effects frequently act on the output through the same mechanisms as the inputs do.

State-space models:

$$x(t+1) = A_d x(t) + B_d u(t) + K_d e(t) \tag{2.4}$$

$$y(t) = C_d x(t) + D_d u(t) + e(t). \tag{2.5}$$

An additional variable, the 'state vector' $x(t) \in \mathbf{R}^n$, appears in this model. When models are built using first principles mechanistic reasoning, the elements of the state vector usually represent 'physical' variables such as current, velocity, temperature, or, in a macroeconomic model, the foreign exchange rate. But when, as in this chapter, one considers 'black-box' identification, namely one assumes that nothing is known about the internal structure of the model, the state vector which results is an (approximation to) some unknown linear combination of such physical variables, and can not usually be given a physical interpretation. One frequently sees state-space models in which different noise variables are introduced in the two equations (2.4) and (2.5). But this is not necessary if the internal structure of the model is not known. Indeed the use of the same term $e(t)$, as here, is always enough to represent stochastic effects upon the output $y(t)$, at least up to second-order statistics (*ie* this is enough to reproduce any desired mean, covariance, and spectral properties). We have used the subscript 'd' to emphasise that the model description assumes discrete time. Later in the chapter we shall be jumping between discrete and continuous time representations, and this notation will help to avoid confusion there.

Note that frequently one assumes that the input $u(t)$ does not influence $y(t)$, but only future outputs $y(t+k)$ with $k > 0$, so that $g(0) = 0$. This is represented by $D_d = 0$ in (2.5). In the other model forms this shows up as the 'strictly proper' condition: the degree of each numerator in $G(z)$ is strictly smaller than the degree of the corresponding denominator. In ARMAX models the corresponding condition is more complicated. If $m = 1$ and $p = 1$ then the degree of $A(z)$ must be strictly greater than that of $B(z)$, but the generalisation of this to the multivariable case is not straightforward: if $A(z)$ is 'row reduced', then the degree of each row of $B(z)$ must be strictly smaller than the degree of the corresponding row of $A(z)$ [22, 26]. We shall not explain here what is meant by 'row reduced', or the degree of a row of a polynomial matrix; definitions can be found in the cited references.

We will assume throughout this chapter that we are only interested in *asymptotically stable* models, namely those for which the impulse responses satisfy

$$\max_{1\le i\le p}\sum_{j=1}^{m}\sum_{k=0}^{\infty}|g_{ij}(k)| \quad < \quad \infty \tag{2.6}$$

$$\max_{1\le i\le p}\sum_{j=1}^{p}\sum_{k=0}^{\infty}|h_{ij}(k)| \quad < \quad \infty \tag{2.7}$$

where $g(k)$ is the impulse response matrix for the input $u(t)$, as defined earlier, and $h(k)$ is the impulse response matrix for the noise $e(t)$. This requires that all the eigenvalues of A_d should lie strictly inside the unit disk, or all the poles of $G(z)$ and $H(z)$ should lie strictly inside the unit disk, or all the roots of $\det A(z)$ should lie strictly inside the unit disk. This assumption of stability is usually required either because we know *a priori* that the system being modelled is stable, or because we make some further assumptions (usually: that $\{u(t)\}$ and $\{y(t)\}$ are stationary stochastic processes, and that the effects of any initial conditions are negligible) which imply stability. In applications we can sometimes allow marginally stable models (distinct eigenvalues, poles, roots may be allowed on the unit circle), or know *a priori* that the model should be of this type, if it is known that the system includes integrators, or undamped resonant modes.

One can always obtain transfer function and ARMAX models corresponding to a given state-space model. For transfer functions the relationships are given by:

$$G(z) = C_d(zI - A_d)^{-1}B_d + D_d \tag{2.8}$$

$$H(z) = C_d(zI - A_d)^{-1}K_d + I\,. \tag{2.9}$$

It is not useful to write down a closed-form representation for the equivalent ARMAX model, but standard algorithms exist for this [9, 22, 26]. Going in the opposite direction, there are many possible state-space models which correspond to a given transfer function or ARMAX model. It is possible to obtain a state-space model from an ARMAX model in a straightforward way, (for example, take $x(t) = [y(t-1)^T, y(t-2)^T, \ldots, u(t-1)^T, u(t-2)^T, \ldots, e(t-1)^T, e(t-2)^T, \ldots]^T$), but it becomes considerably more complicated if care is taken to obtain a minimal model, namely one with the smallest possible state dimension n. But again standard algorithms are available.

We see therefore that all three models are equivalent to each other; given any one of them, models in the other two forms can be obtained. It therefore does not matter which form of model we obtain as the result of applying identification to the data, at least as far as *using* the models is concerned. On the other hand, there may be good reasons for preferring one of these forms of the model, when considering the identification problem.

In this chapter we shall use the state-space form (2.4) and (2.5). There are several reasons for this. Transfer function models are very natural and convenient in the 'SISO' (Single-Input, Single-Output) case, namely when $m = p = 1$. But in the multivariable case they have the drawback that each element of the transfer function matrix is independent of all the others. This means that it is difficult to exploit any common behaviour that may be present in the real system — which is usually the case — when identifying a model. Each element has to be identified separately from the others, in essence. (It should be noted, however, that if this is not an objection in a particular application, then there is a lot to be said for identifying a transfer function model, element-by-element, particularly if a noise model is not required or the input signals are uncorrelated with each other.) ARMAX models do not suffer from this problem, since the matrix $A(z)$ can represent dynamic behaviour which is common to the various input-output paths, as well as any dynamics common to the 'input-output' and the 'noise-output' behaviour. But again in the multivariable case, the use of ARMAX models becomes quite complicated. We have already seen that the enforcement of the 'strictly proper' condition becomes rather involved, for example. Even ensuring causality ($g(t) = 0$, $h(t) = 0$ for $t < 0$) is much more complicated than in the SISO case.

Another problem which appears with ARMAX models is that of needing an 'identifiable parametrization'. The observed input-output behaviour of a system can only determine the transfer function of a model, not a particular representation of that transfer function. But there are infinitely many different ARMAX models which correspond to the same transfer function. It is usually considered essential to fix a representation in such a way that the identification algorithm is prevented from wandering over a family of ARMAX models, all of whom correspond to the same transfer function. (The same is true for state-space models, as we shall see.) This is certainly necessary for practical reasons: otherwise the algorithm may wander into representations involving enormous numerical values for the coefficients, for instance, and run into numerical difficulties caused by the limitations of finite precision. It is more controversial whether an identifiable parametrization is needed for statistical reasons. It is undoubtedly necessary if the statistical properties of parameter estimates (namely, the coefficients in the model) are to be analysed. But it is questionable whether such properties are of interest. Statistical properties of the *transfer function estimates* are certainly important, but there is a growing body of opinion which claims that these properties do not depend on the particular parametrization, or even on whether a parametrization is identifiable or not. This issue is not yet resolved, or thoroughly understood.

State-space models share with ARMAX models the benefit of representing common dynamics, and the problem of requiring an identifiable parametrization. They are no better than ARMAX models in this respect, but also no worse in the multivariable case. (Whereas in the SISO case the problems with ARMAX models can be dealt with more easily.) State-space models have some mild advantages:

1. Causality is built into the model structure.
2. The 'strictly proper' property can be determined by inspection.
3. In general the best and most reliable numerical algorithms for system transformation, simplification, and analysis, as well as the exploitation of models in the synthesis of filters and control systems, assume that the model is given in state-space form.

A much stronger advantage of state-space models, as we shall show later in this chapter, is that there exist very nice identifiable parametrizations of them. In particular, it is known how to parametrize some specific classes of models which are of interest in applications. But the strongest argument in favour of state-space models is that we can assemble identification procedures for them which work very well and reliably.

It is usually desirable to find as simple a model as possible for the data. This means first, that our final state-space model should be minimal, namely reachable and observable (no redundant states which make no contribution to the input-output behaviour). It also means that the minimal state dimension (McMillan degree), n, should be as small as possible, given the data. Actually, there is an implicit (sometimes explicit) trade-off between the acceptable model complexity and how well it matches the data; clearly we get a very low complexity, though useless, model by setting $y(t) = 0u(t) + \nu(t)$, for example.

It is worth repeating that we shall assume in this chapter that we have no information on the internal structure of the real system which generated the data. In other words, we do not know, from physics for example, that certain elements of A_d should be zero, or what the appropriate value of n is, or that some element of $H(z)$ should be of a particular degree, and so on. If such information is available then the methods presented here are not appropriate, because they do not preserve structural information of this kind.

3. Identifiable Parametrizations

We say that a parametrization of a model is *identifiable* if only one parameter (vector) corresponds to each transfer function. In other words, if the map from parameters to input-output behaviour is injective. This property is necessary for an identification algorithm to have the possibility of converging (in some sense) on a parameter value. A necessary condition for parameter identifiability is that the degrees of freedom between the parameters (*ie* elements of the parameter vector) should be sufficiently restricted.

Consider the 'purely deterministic' state-space model (2.4), (2.5) when there is no noise input $e(t)$, and when $D_d = 0$:

$$x(t+1) = A_d x(t) + B_d u(t) \tag{3.1}$$

$$y(t) = C_d x(t)\,. \tag{3.2}$$

It is known that the set of all possible input-output behaviours of this system is a manifold of dimension $(m+p)n$ [12]. However, the number of elements in the A_d, B_d and C_d matrices is $(n+m+p)n$, so it is clear that n^2 degrees of freedom must be removed if parameter identifiability is to be achieved. (There are lower-dimensional submanifolds on which more than n^2 degrees of freedom must be removed, but that is not of importance here.) It is most often suggested that this restriction be achieved by setting most of the parameters to 0 or 1, and using one of the standard reachable or observable forms [22, 26]. For example, if $n = 4, m = 2, p = 3$, then an observable 'canonical form' (which can describe almost all, but not all, 2-input, 3-output systems of McMillan degree 4) is:

$$A = \begin{bmatrix} a_{11} & a_{12} & a_{13} & a_{14} \\ 0 & 0 & 1 & 0 \\ a_{31} & a_{32} & a_{33} & a_{34} \\ a_{41} & a_{42} & a_{43} & a_{44} \end{bmatrix} \tag{3.3}$$

$$B = \begin{bmatrix} b_{11} & b_{12} \\ b_{21} & b_{22} \\ b_{31} & b_{32} \\ b_{41} & b_{42} \end{bmatrix} \tag{3.4}$$

$$C = \begin{bmatrix} 1 & 0 & 0 & 0 \\ 0 & 1 & 0 & 0 \\ 0 & 0 & 0 & 1 \end{bmatrix}. \tag{3.5}$$

Observe that $n^2 = 16$ parameters have been fixed to be either 0 or 1, so that the degrees of freedom have been reduced as expected.

As the parameters a_{ij} and b_{ij} of this canonical form range over the real numbers, both unstable and unreachable systems will be encountered, and identification algorithms which use this form must take care to avoid such systems. This can be done by constraining the parameters, either explicitly or implicitly ('implicitly' here includes procedures such as ignoring the constraints unless they are seen to be violated, and then taking some *ad hoc* action). Avoiding unstable systems is not usually a problem with algorithms which rely on minimizing some cost function, because the value of the cost function would usually become very large for an unstable system. But avoiding unreachable systems (in general, with other parametrizations: avoiding non-minimal systems) involves the avoidance of an algebraic set in the $(m+p)n$-dimensional parameter space.

The use of this particular canonical form for identification is discussed further in [30, Appendix 4A], and the references given there.

The issue of which identifiable form to use, and even of whether to use one at all, is far from settled. True canonical forms (in the algebraic sense) partition the set of systems, so that each system can be represented by exactly one set of parameter values, but several (many!) structural forms are needed to cover the set of all systems. In the example above, other forms are needed to represent those systems not covered by the given form. Admittedly there are 'not many' such systems, in the sense that they all lie on submanifolds of smaller dimension than the dimension of the input-output behaviour manifold, but still there is a potential problem there. Glover and Willems [16] have argued that this is an avoidable problem: one can use 'overlapping' or 'pseudo-canonical' forms which have the advantage that each form can represent the majority of systems — typically, each form can represent 'almost all' systems, in the measure-theoretic sense — but now each system can be represented in several such forms. Each system still has a unique representation in each form, however, which is all that matters for identification. Hanzon and Ober [23] have proposed an overlapping form which bears some resemblance to the balanced form which will be introduced in the next section, and which is designed to represent systems in the neighbourhood of those at which the structure of the balanced form changes. McKelvey [37] has argued that it is unnecessary to use an explicit parametrization at all: it is enough to 'regularise' the parameter estimation problem. In fact the particular scheme he proposes estimates a system which is close to being balanced. A nice discussion and resume of much of the relevant literature is given by Veres [55].

It is worth mentioning that much of the discussion which appears in the literature about parametrizations of systems is concerned with the difficulties of estimating the correct 'structure', namely which particular form of a parametrization is most appropriate, given a set of data. We believe that currently available mehods for obtaining initial models, such as those described later in section 8., help to reduce the severity of these problems. Evidence which we have for this to date is that if we deliberately generate data using a model which does not have the generic structure, and then apply our identification methods to the data under the assumption that the structure *is* generic, then the algorithms seem to operate successfully, and exhibit no difficulties [11]. The probable explanation for this is that in the multivariable case the space of all stable systems of given input and output dimensions and a given McMillan degree is connected (in contrast to the SISO case) [8, 17, 44], so that the algorithms have no problem with working in the immediate neighbourhood of systems with non-generic structure. (This should be equally true for all algorithms.) When there are several generic structures to choose from, as is the case with balanced parametrizations, then the correct choice of such a structure is important; it appears that this can be done successfully by examining the structure of initial models which are obtained using the methods discussed in section 8..

4. Balanced Parametrization

Let a continuous-time linear system have state-space form:

$$\dot{x}(t) = A_c x(t) + B_c u(t) \tag{4.1}$$

$$y(t) = C_c x(t) + D_c u(t) \tag{4.2}$$

If this system is asymptotically stable then its reachability Gramian, G_r, and observability Gramian, G_o, are defined to be the solutions of the two Lyapunov equations:

$$A_c G_r + G_r A_c^T + B_c B_c^T = 0 \tag{4.3}$$

$$G_o A_c + A_c^T G_o + C_c^T C_c = 0\,. \tag{4.4}$$

The system is said to be *Lyapunov balanced* [40] if it is such that

$$G_r = \Sigma = G_o \tag{4.5}$$

and Σ is a diagonal matrix. It is always possible to find a state coordinate transformation which makes a stable system balanced in this sense.

The constraints implicit in equations (4.3) and (4.4) reduce the degrees of freedom between the elements of A_c, B_c, C_c by n^2, which is exactly what is needed to obtain an identifiable parametrization. It has been discovered that stable, minimal, continuous-time balanced systems can be parametrized explicitly [25, 43, 58], and that the parametrization is in fact a canonical form (which is therefore identifiable, *a fortiori*). An example of this parametrization, again for the case $n = 4$, $m = 2$, $p = 3$, is [45, 11] (A_c and C_c each have 4 columns):

$$A_c = \begin{bmatrix} -\frac{\|B_1\|^2}{2\sigma_1} & \frac{\sigma_2 B_1 B_2^T - \sigma_1 C_1^T C_2}{\sigma_1^2 - \sigma_2^2} & \cdots \\ \frac{\sigma_1 B_2 B_1^T - \sigma_2 C_2^T C_1}{\sigma_2^2 - \sigma_1^2} & -\frac{\|B_2\|^2}{2\sigma_2} & \cdots \\ \frac{\sigma_1 B_3 B_1^T - \sigma_3 C_3^T C_1}{\sigma_3^2 - \sigma_1^2} & \frac{\sigma_2 B_3 B_2^T - \sigma_3 C_3^T C_2}{\sigma_3^2 - \sigma_2^2} & \cdots \\ \frac{\sigma_1 B_4 B_1^T - \sigma_4 C_4^T C_1}{\sigma_4^2 - \sigma_1^2} & \frac{\sigma_2 B_4 B_2^T - \sigma_4 C_4^T C_2}{\sigma_4^2 - \sigma_2^2} & \cdots \end{bmatrix}$$

$$\begin{bmatrix} \frac{\sigma_3 B_1 B_3^T - \sigma_1 C_1^T C_3}{\sigma_1^2 - \sigma_3^2} & \frac{\sigma_4 B_1 B_4^T - \sigma_1 C_1^T C_4}{\sigma_1^2 - \sigma_4^2} \\ \frac{\sigma_3 B_2 B_3^T - \sigma_2 C_2^T C_3}{\sigma_2^2 - \sigma_3^2} & \frac{\sigma_4 B_2 B_4^T - \sigma_2 C_2^T C_4}{\sigma_2^2 - \sigma_4^2} \\ -\frac{\|B_3\|^2}{2\sigma_3} & \frac{\sigma_4 B_3 B_4^T - \sigma_3 C_3^T C_4}{\sigma_3^2 - \sigma_4^2} \\ \frac{\sigma_3 B_4 B_3^T - \sigma_4 C_4^T C_3}{\sigma_4^2 - \sigma_3^2} & -\frac{\|B_4\|^2}{2\sigma_4} \end{bmatrix} \tag{4.6}$$

$$B_c = \begin{bmatrix} b_{11} & b_{12} \\ b_{21} & b_{22} \\ b_{31} & b_{32} \\ b_{41} & b_{42} \end{bmatrix} \tag{4.7}$$

$$C_c = \begin{bmatrix} \cos(\phi_{21})\cos(\phi_{11})\|B_1\| & \cos(\phi_{22})\cos(\phi_{12})\|B_2\| & \cdots \\ \cos(\phi_{21})\sin(\phi_{11})\|B_1\| & \cos(\phi_{22})\sin(\phi_{12})\|B_2\| & \cdots \\ \sin(\phi_{21})\|B_1\| & \sin(\phi_{22})\|B_2\| & \cdots \end{bmatrix}$$

$$\begin{bmatrix} \cos(\phi_{23})\cos(\phi_{13})\|B_3\| & \cos(\phi_{24})\cos(\phi_{14})\|B_4\| \\ \cos(\phi_{23})\sin(\phi_{13})\|B_3\| & \cos(\phi_{24})\sin(\phi_{14})\|B_4\| \\ \sin(\phi_{23})\|B_3\| & \sin(\phi_{24})\|B_4\| \end{bmatrix} \tag{4.8}$$

Here $\sigma_1 > \ldots > \sigma_4 > 0$, $b_{i1} > 0$, $b_{ij} \in \mathbf{R}$ for $j > 1$, $-\pi/2 < \phi_{ij} < \pi/2$ (with additionally $\pi/2 < \phi_{1j} < 3\pi/2$ being allowed), $B_i = [b_{i1}, b_{i2}]$, C_i is the i'th column of C_c, and $\|B_i\| = \sqrt{B_i B_i^T}$. The σ_i parameters are called the *Hankel singular values* of the system, and they are the non-zero elements of the common reachability and observability Gramian defined in (4.5):

$$\Sigma = \operatorname{diag}(\sigma_1, \sigma_2, \sigma_3, \sigma_4). \tag{4.9}$$

This form looks a lot more complicated than the observable canonical one, but in fact it contains exactly the same number of parameters. Furthermore, as the parameters range over their allowed values, only asymptotically stable, minimal systems are generated. Note that the constraints on the parameters, required to achieve this, are much simpler than for the observable canonical form. In fact the parameters are almost unconstrained; since each one can vary over an open subset of the real line, it is possible to replace it by another unconstrained parameter. For example, the inequality constraints between the Hankel singular values (the σ_i parameters) can be avoided by using the alternative parameters $\tau_i = \sigma_i - \sigma_{i+1}, (i = 1, 2, 3), \tau_4 = \sigma_4$, and even the positivity constraint could be removed by using $\log \tau_i$. (Admittedly this advantage was not apparent when the use of balanced forms was originally advocated for parameter estimation, since explicit parametrizations were not known at the time, and it was in fact envisaged that complicated algebraic constraints would be required [31].)

The example above does not show the most general form of the parametrization, but is a 'generic' one, which is indicated by the fact that the σ_i parameters are distinct. Almost all systems of the given dimensions can be parametrized by a generic form of the parametrization. (For full details, and a more precise presentation of balanced canonical forms, see [45, 46, 11].)

As mentioned above, this balanced form parametrizes stable continuous-time systems, whereas we are really interested in stable discrete-time systems. But there is a standard bijection between stable continuous-time and stable discrete-time systems, and we can therefore parametrize stable discrete-time systems by means of this bijection. What is more, this bijection preserves the 'balanced' property [18, 45]. The bijection maps the left half complex plane (associated with continuous-time systems) to the unit disk (associated with discrete-time systems) according to:

$$z = \frac{1+s}{1-s}, \tag{4.10}$$

$$s = \frac{z-1}{z+1}. \tag{4.11}$$

Corresponding transformations of state-space models are well known [18, 45], and are given later in this chapter.

The observable canonical form is known to suffer from numerical problems for some parameter values. The general experience is, however, that fewer numerical problems arise with the balanced form. One well known property of the balanced canonical form is that, when this form is used, the transfer function is less sensitive to perturbations in the *elements* of the matrices A, B, and C than when any other state coordinate system is used [41]. (Note that this is not a statement about the sensitivity to changes in the parameters σ_i, b_{ij}, ϕ_{ij}.) This, however, is not a particular recommendation from the point of view of identification. Indeed, if the transfer function were completely insensitive to perturbations in these elements then one would expect very poor conditioning of the parameter estimation, because very slight changes in the input-output data could only be explained by enormous changes in the parameters. (The statistical consequence would be a large sample variance of the parameter estimates.) This is precisely the situation which is approached if one is estimating a system which is close to lack of minimality in some sense. But if one measures the 'distance' of a state-space model to lack of reachability by the smallest singular value of the reachability Gramian, $\underline{\sigma}(G_r)$, the distance to lack of observability by the smallest singular value of the observability Gramian, $\underline{\sigma}(G_o)$, and the distance to loss of minimality by $\min[\underline{\sigma}(G_r), \underline{\sigma}(G_o)]$, then for any transfer function matrix the balanced form is the one which maximises this distance [40]. The use of the balanced form should therefore avoid poor conditioning due to this source, to the maximum extent possible. This is another advantage of the balanced canonical form.

5. Some Useful Classes of Models

If we assume that $K_d = 0$ in (2.4), so that we are attempting to model only the relationship between $u(t)$ and $y(t)$, without trying to model the 'error' $\nu(t)$ (which now becomes the same as $e(t)$), then it is appropriate to look for models in the class of stable minimal systems, namely the class which we considered in the previous section. But more restricted classes of systems are also of interest in identification.

5.1 Minimum-Phase Models

If we want to model the error or noise, however, then it is usually appropriate to consider a smaller class, that of 'minimum-phase' models, namely those which are stable and whose inverses are stable. Since we don't know what $e(t)$ is, it is usual to assume that it is a 'white' stochastic process. 'Prediction error' methods of system identification require the one-step-ahead prediction of the output $y(t)$, namely the expected value of $y(t)$, conditional on the known previous values $y(t-i), u(t-i)$ for $i > 0$, and the known next value $u(t)$. It can be shown [22, 30] that this prediction $\hat{y}(t)$ is generated by the equations

$$\hat{x}(t+1) = A_d\hat{x}(t) + B_d u(t) + K_d[y(t) - \hat{y}(t)] \quad (5.1)$$

$$\hat{y}(t) = C_d\hat{x}(t), \quad (5.2)$$

if we continue to assume that $D_d = 0$, and with a suitable initial condition $\hat{x}(0)$. Substituting (5.2) in (5.1) gives

$$\hat{x}(t+1) = (A_d - K_d C_d)\hat{x}(t) + B_d u(t) + K_d y(t). \quad (5.3)$$

We require this equation to be stable, and hence we require that the eigenvalues of $A_d - K_d C_d$ should lie inside the unit circle. But (5.3) is just the state equation of the *inverse* system which has $y(t)$ as an input and $e(t)$ as its output, the output equation of this inverse system being

$$e(t) = -C_d\hat{x}(t) + y(t). \quad (5.4)$$

So we require this inverse system to be stable. Thus we arrive at the requirement that the system (A_d, K_d, C_d, I) should both be stable and have a stable inverse (equivalently, all the poles and transmission zeros of its transfer function should lie strictly inside the unit disk), namely that it be 'minimum-phase'. It can be shown that the second-order statistics of any wide-sense stationary stochastic process can be reproduced by a minimum-phase model.

A minimum-phase system is said to be *minimum-phase balanced* if its reachability Gramian is diagonal and equal to the observability Gramian of the inverse system [36]. (Other, slightly different definitions are also possible.) Furthermore any minimum-phase system can be made minimum-phase balanced by a suitable state coordinate transformation. It turns out that an explicit parametrization of minimum-phase balanced systems of given input and output dimensions and a given McMillan degree can be obtained [45, 46]. As before, the parametrization is in fact of continuous-time systems, but since the bijection (4.10), (4.11) preserves the minimum-phase property, we can again use this bijection to give us a parametrization of discrete-time minimum-phase systems. (This was first advocated in [35].)

For the case $m = 2, p = 3, n = 4$ which we considered earlier, a generic form of this parametrization is similar to the one presented above in (4.6), (4.7) and (4.8), but a little more complicated. Suppose that the 'continuous-time image', under the bilinear transformation (4.11), of the discrete-time model (A_d, K_d, C_d, I) — note that we omit the part associated with the input $u(t)$ — is

$$\dot{x}(t) = A_c x(t) + K_c e(t) \tag{5.5}$$

$$y(t) = C_c x(t) + E_c e(t). \tag{5.6}$$

The A_c, K_c, C_c, E_c matrices of a continuous-time minimum-phase system are parametrized as follows, where we have used a version of the parametrization which appears in [11, 36].

$$K_c = \begin{bmatrix} k_{11} & k_{12} \\ k_{21} & k_{22} \\ k_{31} & k_{32} \\ k_{41} & k_{42} \end{bmatrix} \tag{5.7}$$

where $k_{i1} > 0$, $k_{ij} \in \mathbf{R}$ for $j > 1$. The j'th column of C_c is given by

$$C_j = E_c \tilde{C}_j \tag{5.8}$$

where

$$\tilde{C}_j = \left(\sigma_j K_j^T + \sqrt{1 + \sigma_j^2} U_j \sqrt{K_j K_j^T} \right), \tag{5.9}$$

K_j is the j'th row of K_c and $U_j \in \mathbf{R}^p$ is a unit vector, which can be parametrized by two angles ϕ_{j1}, ϕ_{j2}, as was done in (4.8). The matrix A_c has elements

$$a_{ii} = -\frac{1}{2\sigma_i} K_i K_i^T \tag{5.10}$$

$$a_{ij} = \frac{1}{\sigma_i^2 - \sigma_j^2} [\sigma_j (1+\sigma_i^2) K_i K_j^T - -\sigma_i \sqrt{1+\sigma_i^2} \sqrt{1+\sigma_j^2} \sqrt{K_i K_i^T} \sqrt{K_j K_j^T} U_i U_j^T] \quad \text{if } i \neq j. \tag{5.11}$$

As before, $\sigma_1 > \ldots > \sigma_4 > 0$. The continuous-time minimum-phase system also has the matrix E_c which contains no new parameters, but takes the value

$$E_c = I - \tilde{C}_c (I - A_c + K_c \tilde{C}_c)^{-1} K_c \tag{5.12}$$

in order that the corresponding discrete-time system should have I as the direct feedthrough matrix (coefficient of $e(t)$ in (2.5)).

The matrix B_d which appears in (2.4) can now be allowed to have arbitrary real entries. Note that this allows nonminimum-phase transfer function matrices relating $u(t)$ to $y(t)$ to be identified, the minimum-phase 'restriction' applying only to the transfer function between $e(t)$ and $y(t)$. It also allows the possibility that the pair (A_d, B_d) may be unreachable, which would be quite reasonable if the real transfer function from $u(t)$ to $y(t)$ were of McMillan degree smaller than n.

The possibility of parametrising minimum-phase systems with almost no constraints on the parameters is a major advantage for identification.

A special case occurs when we have no input $u(t)$, but just a measured process $y(t)$, for which we would like to identify a model of the form

$$x(t+1) = A_d x(t) + K_d e(t) \tag{5.13}$$

$$y(t) = C_d x(t) + e(t). \tag{5.14}$$

If $e(t)$ is a white noise process then this gives a model of the stochastic process $y(t)$ as the output of a filter driven by white noise. The transfer function of this filter is a square root of the spectral density of $y(t)$, and is therefore a 'spectral factor' of the spectral density of $y(t)$. The model (5.13), (5.14) is called the 'innovations representation' of $y(t)$, providing that it is minimal, stable, and minimum-phase. It is apparent that this case can be treated using the parametrization just described, the only difference being that there is no 'B_d' matrix to estimate.

5.2 Positive-Real Models

A different situation occurs when we wish to obtain a model of the kind given by (5.13) and (5.14), but instead of the measurements $y(t)$ we have only estimates of the *autocovariance* function

$$R(\tau) = E\{y(t+\tau)y(t)\} \qquad (\tau = 0, 1, 2, \ldots, N) \tag{5.15}$$

(where $E\{.\}$ denotes the expectation operator), or of the spectral density of $y(t)$, which is the discrete Fourier transform of the autocovariance. We still want an innovations representation of $y(t)$, but now the requirement is that its autocovariance should match the given one (approximately). This is called the 'stochastic realization' problem.

Assume that the model (5.13),(5.14) holds and is stable, and let $P = E\{x(t)x(t)^T\}$ and $Q = E\{e(t)e(t)^T\}$. Then it is easy to show that

$$P = A_d P A_d^T + K_d Q K_d^T \tag{5.16}$$

and

$$R(0) = C_d P C_d^T + Q. \tag{5.17}$$

Also, we have

$$\begin{aligned} E\{x(t+\tau)x(t)^T\} &= A_d E\{x(t+\tau-1)x(t)^T\} + \dots \\ &\quad K_d E\{e(t+\tau-1)x(t)^T\} && (5.18) \\ &= \dots \\ &= A_d^\tau P && (5.19) \end{aligned}$$

where we have used the fact that $E\{e(t+k)x(t)^T\} = 0$ for $k > 0$. Hence for $\tau > 0$ we have

$$\begin{aligned} R(\tau) &= E\{[C_d x(t+\tau) + e(t+\tau)][C_d x(t) + e(t)]^T\} && (5.20) \\ &= C_d E\{x(t+\tau)x(t)^T\}C_d^T + E\{e(t+\tau)x(t)^T\}C_d^T + \dots \\ &\quad + C_d E\{x(t+\tau)e(t)^T\} + E\{e(t+\tau)e(t)^T\} && (5.21) \\ &= C_d A_d^\tau P C_d^T + C_d A_d^{\tau-1} K_d Q && (5.22) \\ &= C_d A_d^{\tau-1}(A_d P C_d^T + K_d Q) && (5.23) \end{aligned}$$

where we have used the fact that $E\{x(t+\tau)e(t)^T\} = A_d^{\tau-1} K_d Q$.

Now the system

$$x(t+1) = A_d x(t) + B_d u(t) \tag{5.24}$$

$$y(t) = C_d x(t) + D_d u(t) \tag{5.25}$$

has the impulse response sequence

$$g(t) = \begin{cases} C_d A_d^{t-1} B_d & (t > 0) \\ D_d & (t = 0) \end{cases} \tag{5.26}$$

so if we set

$$B_d = A_d P C_d^T + K_d Q \tag{5.27}$$

$$D_d = R(0) \tag{5.28}$$

then, from (5.23), the impulse response sequence will be the same as the given autocovariance sequence:

$$g(\tau) = R(\tau) \qquad (\tau \geq 0). \tag{5.29}$$

Hence the standard approach [4, 15] to the stochastic realization problem is:

1. Find a system (A_d, B_d, C_d, D_d) which is stable and has the sequence $\{R(\tau) : \tau \geq 0\}$ as its impulse response.
2. Assume Q is invertible (a reasonable assumption), and substitute from (5.27) and (5.17) into (5.16). This gives the Riccati equation

$$\begin{aligned} P &= A_d P A_d^T + \ldots \\ & (B_d - A_d P C_d^T)[R(0) - C_d P C_d^T]^{-1}(B_d - A_d P C_d^T)^T. \end{aligned} \tag{5.30}$$

3. Solve equation (5.30) for P, then find K_d from (5.27).

The solution of (5.30) is not unique, but it can be shown that if the smallest nonnegative solution is taken, then the resulting triple (A_d, K_d, C_d) gives an innovation representation: it is minimum-phase as well as stable [4, 15].

Of course, in the context of system identification we expect to find, in step 1, a system whose impulse response matches the given covariance sequence only approximately. But now we have a complication: if the impulse response of (A_d, B_d, C_d, D_d) is not a possible autocovariance sequence then the rest of the theory will not apply — we shall not be able to find a solution of (5.30), or the resulting K_d will not give a minimum-phase system. Now a sequence $\{R(\tau) : \tau \geq 0\}$ is an autocovariance sequence if and only if it is positive-definite, namely if the Toeplitz matrix built from it is positive definite:

$$\begin{bmatrix} R(0) & R(1)^T & R(2)^T & \ldots \\ R(1) & R(0) & R(1)^T & \ldots \\ R(2) & R(1) & R(0) & \ldots \\ \vdots & \vdots & \vdots & \ddots \end{bmatrix} > 0. \tag{5.31}$$

(An equivalent characterisation is that the corresponding spectral density is positive for all frequencies.) A stable system which has a positive definite impulse response is called (strictly) *positive-real.* To solve the approximate stochastic realization problem we need to identify a positive-real system in step 1.

A positive-real system is said to be *positive-real balanced* if the minimal nonnegative solution of (5.30) is diagonal, and equal to the minimal nonnegative solution to a dual Riccati equation; any positive-real system can be made positive-real balanced by a suitable state coordinate transformation. Once again, an explicit parametrization, similar to those for stable and for minimum-phase systems, is available [45, 46]. So again we can estimate parameters, with only mild constraints upon them, in the knowledge that we are not straying outside the class of systems which is of interest. As before, we only have an explicit parametrization of continuous-time positive-real systems, but the standard bijection (4.10), (4.11) between continuous-time and discrete-time systems preserves the positive-real property, so that we have an implicit parametrization of discrete-time positive-real systems. A generic form of the continuous-time parametrization for the case $m = 2, p = 3, n = 4$ is:

$$B_c = \begin{bmatrix} b_{11} & b_{12} \\ b_{21} & b_{22} \\ b_{31} & b_{32} \\ b_{41} & b_{42} \end{bmatrix} \tag{5.32}$$

$$C_j^T = \sqrt{B_j(D_c + D_c^T)^{-1}B_j^T}U_j(D_c + D_c^T)^{1/2}. \tag{5.33}$$

The matrix A_c has elements

$$a_{ii} = B_i(D_c + D_c^T)^{-1}C_i - \frac{1+p_i^2}{2p_i}B_i(D_c + D_c^T)^{-1}B_i^T \tag{5.34}$$

$$\begin{aligned} a_{ij} = {} & B_i(D_c + D_c^T)^{-1}C_j - \frac{1}{p_j^2 - p_i^2}[p_j(1-p_i^2)B_i(D_c + D_c^T)^{-1}B_j^T - \\ & p_i(1-p_j^2)C_i^T(D_c + D_c^T)^{-1}C_j \quad \text{if } i \neq j \end{aligned} \tag{5.35}$$

where $1 > p_1 > \ldots > p_4 > 0$, and all other terms have the meanings defined earlier. (This version of the parametrization is taken from [36].)

The parameters p_i are the *canonical correlation coefficients.* They are a little more constrained than the parameters σ_i in the earlier parametrizations, since they are bounded above as well as below, but each of them still ranges over a connected, bounded, open subset of $\mathbf{R}$, so this is still an easy constraint to handle. Remarkably, when a positive-real system is parametrized in this way, the solution to the continuous-time positive-real Riccati equation which corresponds to (5.30) is

$$P = \mathrm{diag}(p_1, p_2, \ldots, p_n) \tag{5.36}$$

and *this solution is also the solution to the discrete-time Riccati equation (5.30)*, since solutions to these two equations remain invariant under the bijection (4.10),(4.11). So if we identify a positive-real system in this parametrization then we can obtain K_d immediately from (5.27), since we already know P.

6. Outline of Parameter Estimation

System identification can be carried out in the following manner, which is standard except for the use of balanced parametrizations to restrict the optimization to the class of systems which is of interest. Choose a discrete-time model class. If this is the class of asymptotically stable systems, or of minimum-phase systems, or of positive-real systems, then parametrize it using a balanced parametrization, together with (4.10) and (4.11). Let θ denote the vector of parameters which appear in this parametrization.

Next, choose a criterion function $V(\theta)$ which measures the difference between the available data and the performance of the model in some way. (The model gets better as $V(\theta)$ decreases.)

Now estimate the parameters as follows:

Step 1. Set $i := 1$. Obtain an initial model somehow. From this initial model obtain an initial 'structure' of the parametrization (see [45, 46] for details), and an initial value of the parameter vector, θ_1. We shall present some effective ways of finding such an initial model later, although there are other possibilities [30, 60].

Step 2. Evaluate $V(\theta_i)$. This requires using (4.10) to find the discrete-time system corresponding to θ_i, followed by some kind of simulation (usually).

Step 3. Set $i := i + 1$. Use an optimization algorithm to find a parameter vector θ_i such that $V(\theta_i) < V(\theta_{i-1})$. If possible, increase the efficiency of this step by computing the gradient $\nabla_\theta V(\theta)$. Section 7. shows how this can be done.

Step 4. If successfully converged, stop. Otherwise return to Step 2.

Step 5. If the optimization diverges, or if it converges to an unsatisfactory model, return to Step 1 and try to find a better initial model. (A nonlinear optimization is usually involved, and the results can be very sensitive to the choice of initial model.)

This is a very general outline, which omits many details, but all the details which are not specific to the use of balanced parametrizations can be found in the literature [30, 42, 49].

We have referred above to the 'structure' of the parametrization which is to be used. By this we mean a number of integer-valued parameters which determine the details of the parametrization. An obvious one is the McMillan degree, n. Each singular value parameter σ_i may appear in the common Gramian ($G_r = G_o$ in the Lyapunov-balanced case) with some multiplicity. The 'generic' structures are those in which there are n distinct σ_i parameters. Our experience to date indicates that one can always assume such a generic structure. If there is only one output ($p = 1$) the $n(p-1)$ parameters ϕ_{ij} degenerate to n parameters s_j, each one taking the value ± 1; thus in this case there are n further integer parameters to estimate. Experience indicates that it *is* necessary to estimate these correctly, and the initial models obtained by the methods of section 8. usually yield the correct estimates. Strictly speaking, in the multi-output case ($p > 1$) these n 'structural' parameters do not disappear; each of the n parameters ϕ_{1j} may lie in either of the unconnected regions $-\pi/2 < \phi_{1j} < \pi/2$ and $\pi/2 < \phi_{1j} < 3\pi/2$. But in practice an optimization algorithm has no difficulty with 'stepping over' 0 to move this parameter into the correct region, so separate estimation of this aspect of the structure is not necessary. (A good initial estimate will put it into the correct region anyway, unless it is close to 0.)

7. Gradient Calculations

Parameter estimation algorithms can be made more efficient, both in terms of computation speed and of accuracy, if the gradient of the objective function $V(\theta)$ with respect to the parameter vector θ is supplied to the algorithm. To compute the gradient, 3 calculations are required:

1. The derivatives of the objective function must be obtained in terms of the derivatives of the discrete-time model matrices.
2. The derivatives of the discrete-time model matrices must be obtained in terms of the derivatives of the continuous-time model matrices which appear in the parametrizations.
3. The derivatives of the continuous-time matrices must be computed.

The first of these depends on the particular identification problem being solved, since it depends on the objective function being used. We will therefore not address this point in this chapter, but confine ourselves to indicating how the other calculations can be performed.

The formulae for state-space transformations which correspond to the bijection (4.10) are:

$$A_d(\theta) = (I + A_c(\theta))(I - A_c(\theta))^{-1} \tag{7.1}$$

$$B_d(\theta) = \sqrt{2}(I - A_c(\theta))^{-1}B_c(\theta) \tag{7.2}$$

$$C_d(\theta) = \sqrt{2}C_c(\theta)(I - A_c(\theta))^{-1} \tag{7.3}$$

$$D_d(\theta) = D_c(\theta) + C_c(\theta)(I - A_c(\theta))^{-1}B_c(\theta) \tag{7.4}$$

By using these, the derivatives of the discrete time state space matrices can be related to the derivatives of the continuous time state space matrices as follows:

$$\frac{\partial(A_d(\theta))}{\partial\theta_i} = (I + A_d(\theta))\frac{\partial(A_c(\theta))}{\partial\theta_i}(I - A_c(\theta))^{-1} \tag{7.5}$$

$$\begin{aligned}\frac{\partial(B_d(\theta))}{\partial\theta_i} &= (I - A_c(\theta))^{-1}\frac{\partial(A_c(\theta))}{\partial\theta_i}B_d(\theta) + \dots \\ &\quad \sqrt{2}(I - A_c(\theta))^{-1}\frac{\partial(B_c(\theta))}{\partial\theta_i}\end{aligned} \tag{7.6}$$

$$\begin{aligned}\frac{\partial(C_d(\theta))}{\partial\theta_i} &= \sqrt{2}\frac{\partial(C_c(\theta))}{\partial\theta_i}(I - A_c(\theta))^{-1} + \dots \\ &\quad C_d(\theta)\frac{\partial(A_c(\theta))}{\partial\theta_i}(I - A_c(\theta))^{-1}\end{aligned} \tag{7.7}$$

$$\begin{aligned}\frac{\partial(D_d(\theta))}{\partial\theta_i} &= \frac{\partial(D_c(\theta))}{\partial\theta_i} + \frac{1}{\sqrt{2}}\frac{\partial(C_c(\theta))}{\partial\theta_i}B_d(\theta) + \frac{1}{2}C_d(\theta)\frac{\partial(A_c(\theta))}{\partial\theta_i}B_d(\theta) + \\ & \frac{1}{\sqrt{2}}C_d(\theta)\frac{\partial(B_c(\theta))}{\partial\theta_i}\end{aligned} \tag{7.8}$$

The formulae for the derivatives of SISO continuous time balanced state space matrices were given in [32]. A difference between the SISO case and MIMO case is that in the MIMO case closed form formulae for the derivatives are not always available; it may be necessary to solve a few sets of linear equations and a few matrix Lyapunov equations in order to obtain the derivatives [33]. However, even in the MIMO case, closed form formulae are available in the generic case. As before, we shall only consider this case.

Suppose that we are identifying a model over the set of asymptotically stable models, so that we use the Lyapunov balanced parametrization given in section 4. (We shall continue to assume $n = 4, m = 2, p = 3$ for the sake of example.) The model is parametrized by the parameter vector $\theta = [\sigma_1, \ldots, \sigma_4, b_{11}, \ldots, b_{42}, \phi_{11}, \ldots, \phi_{24}]$.

The formulae for the derivatives of the state space matrices are as follows.

1. Derivatives of B_c
   We have
   a) $\frac{\partial B_c}{\partial x} = 0$ if $x = \sigma_i$ or $x = \phi_{ij}$.
   b) Let $\frac{\partial B_c}{\partial b_{ij}} = [\tilde{b}_{kl}]$, then

$$\tilde{b}_{kl} = \begin{cases} 1 & \text{if } k = i \text{ and } l = j \\ 0 & \text{otherwise} \end{cases} \tag{7.9}$$

2. Derivatives of C_c
   The matrix C_c depends on b_{ij} and ϕ_{ij} only.
   a) $\frac{\partial C_c}{\partial x} = 0$ if $x = \sigma_i$.
   b) Let $\frac{\partial C_c}{\partial b_{ij}} = [\tilde{c}_{kl}]$, then

$$\tilde{c}_{kl} = \begin{cases} c_{ij}\frac{b_{ij}}{||B_i||^2} & \text{if } k = i \\ 0 & \text{otherwise} \end{cases} \tag{7.10}$$

   In other words,

$$\frac{\partial C_c}{\partial b_{ij}} = \begin{bmatrix} 0_{p\times(i-1)} & C_i\frac{b_{ij}}{||B_i||^2} & 0_{p\times(n-i)} \end{bmatrix} \tag{7.11}$$

   c) Before we give an expression for the derivative $\frac{\partial C}{\partial \phi_{ij}}$, we define the following function:

$$\begin{aligned} g_q : \quad \mathbf{R}\times\mathbf{R}^q &\rightarrow \mathbf{R}^{q+1} \\ (K,\beta) &\mapsto w \end{aligned}$$

   such that

$$\begin{cases} w_1 &= Kcos(\beta_1)cos(\beta_2)cos(\beta_3)\dots cos(\beta_{q-1})cos(\beta_q) \\ w_2 &= Kcos(\beta_1)cos(\beta_2)cos(\beta_3)\dots cos(\beta_{q-1})sin(\beta_q) \\ w_3 &= Kcos(\beta_1)cos(\beta_2)cos(\beta_3)\dots sin(\beta_{q-1}) \\ \vdots & \quad \vdots \\ w_q &= Kcos(\beta_1)sin(\beta_2) \\ w_{q+1} &= Ksin(\beta_1) \end{cases} \tag{7.12}$$

Also let

$$K_j = \begin{cases} 1 & \text{if } j = 1 \\ cos(\phi_{i,p-1})cos(\phi_{i,p-2})\dots cos(\phi_{ij}) & \text{if } j > 1 \end{cases}$$

$$\gamma_j = \begin{bmatrix} \phi_{i,j-1} + \frac{\pi}{2} & \phi_{i,j-2} & \dots & \phi_{i1} \end{bmatrix}$$

The derivative $\frac{\partial C_c}{\partial \phi_{ij}}$ is

$$\frac{\partial C_c}{\partial \phi_{ij}} = ||B_i|| \begin{bmatrix} 0_{p\times(i-1)} & \begin{matrix} g_{j-1}(K_j, \gamma_j) \\ 0_{(p-j)\times 1} \end{matrix} & 0_{p\times(n-i)} \end{bmatrix} \tag{7.13}$$

3. Derivatives of A_c

a) Let $\frac{\partial A_c}{\partial \sigma_i} = [\tilde{a}_{kl}]$, then

$$\tilde{a}_{kl} = \begin{cases} -\frac{a_{kk}}{\sigma_k} & (k = l = i) \\ \frac{-2\sigma_k\sigma_l}{(\sigma_k^2-\sigma_l^2)^2} B_k B_l^T + \frac{(\sigma_k^2+\sigma_l^2)}{(\sigma_k^2-\sigma_l^2)^2} C_k^T C_l & (k = i, l \neq i) \\ \frac{(\sigma_k^2+\sigma_l^2)}{(\sigma_k^2-\sigma_l^2)^2} B_k B_l^T - \frac{2\sigma_k\sigma_l}{(\sigma_k^2-\sigma_l^2)^2} C_k^T C_l & (l = i, k \neq i) \\ 0 & \text{otherwise} \end{cases} \tag{7.14}$$

b) Let $\frac{\partial A_c}{\partial b_{ij}} = [\bar{a}_{kl}]$, then

$$\bar{a}_{kl} = \begin{cases} -\frac{b_{kl}}{\sigma_k} & (k = l = i) \\ \frac{1}{\sigma_k^2-\sigma_l^2}(\sigma_l b_{lj} - \frac{1}{||B_k||^2}\sigma_k b_{kj} C_k^T C_l) & (k = i, l \neq i) \\ \frac{1}{\sigma_k^2-\sigma_l^2}(\sigma_l b_{kj} - \frac{1}{||B_l||^2}\sigma_k b_{lj} C_k^T C_l) & (l = i, k \neq i) \\ 0 & \text{otherwise} \end{cases} \tag{7.15}$$

c) Let $\frac{\partial A_c}{\partial \phi_{ij}} = [\hat{a}_{kl}]$, then

$$\hat{a}_{kl} = \begin{cases} -\frac{\sigma_k}{\sigma_k^2-\sigma_l^2}(\frac{\partial C_k}{\partial \phi_{kj}})^T C_l & (k = i, l \neq i) \\ -\frac{\sigma_k}{\sigma_k^2-\sigma_l^2} C_k^T (\frac{\partial C_l}{\partial \phi_{lj}}) & (l = i, k \neq i) \\ 0 & \text{otherwise} \end{cases} \tag{7.16}$$

Note that $\frac{\partial C_k}{\partial \phi_{ij}}$ is the kth column of $\frac{\partial C_c}{\partial \phi_{ij}}$.

8. Finding an Initial Model

8.1 Available Methods

The material presented so far assumes that an initial linear model is available for subsequent parameter estimation. In fact, this is a universal assumption in the parametric approach to system identification, and is not confined to the use of balanced parametrizations. A real issue, which is particularly acute in the multivariable case, is how to get started: how to find a useful initial model.

Several possible methods exist, such as the multi-stage instrumental variable method suggested in [30] or the estimation of a high-order ARX model (namely (2.3) with $C(z) = I$) by linear regression followed by model reduction, as in [60]. But two particular classes of methods, based on linear system theory, seem to be particularly effective for multivariable systems. These will therefore be described in some detail in this section.

System-theoretic solutions to the identification problem have been available, in principle, since the 1960's, [24, 20, 10, 26]. These assume, however, that data has been generated by a finite-dimensional linear system, and is represented perfectly accurately — with no error due to noise or even finite-precision representation. Of course, real data is never generated by a finite-dimensional, time-invariant linear model, so this approach did not appear to be practical for some time. In the last few years, however, two classes of algorithms have emerged, which in effect provide numerically regularised versions of these classical systems theoretic solutions. (Actually, some of them have gone beyond the classical solutions, for example in their ability to model systems with both measured inputs and unmeasured noise.) These have proved to be very successful and convenient to use, largely because they can be applied to multivariable data in an almost automatic way.

The basic ideas of these algorithms are the following. According to linear systems theory, the state summarises the effect of past inputs on future outputs. 'Realization methods' exploit the fact that the map from past inputs to future outputs has finite rank (equal to the McMillan degree n), and a structure which allows it to be factored into two rank-n maps from which the state-space model can be recovered. The 'numerically regularised' versions of this procedure find low-rank approximate factors of this map, and then extract an approximate state-space model. 'Subspace methods' take as their starting point the fact that the state space can be given a geometric interpretation as the projection of the space of past inputs onto the space of future outputs (if future inputs are zero). With real data the dimension of this projection grows as fast as the number of data points, and does not stabilise. But practical versions of subspace algorithms identify a low-dimensional subspace of this projection which in some sense contains most of the information in the projection, and take that subspace to be the state space. A basis for this subspace provides a sequence of state estimates, and the matrices of the state-space model are found by using this sequence. This idea of estimating the state sequence first, and the matrices later, was introduced in [28, 61], and goes significantly beyond the original system theoretic approaches. There are many variations of these algorithms, some of which are somewhere between 'realization' and 'subspace' methods as defined here. Typically these are 'subspace' methods in the sense that they rely on projections of data onto low-dimensional subspaces, but they use the same approach to extracting model matrices as the 'realization' methods do — namely, without estimating a state sequence first.

Both classes of algorithms use techniques which are widely used in numerical analysis, particularly the singular value decomposition and the QR factorization. A very important feature of these techniques is that they provide means of estimating a suitable state dimension n.

Recent surveys of both classes of algorithms are given in [51] and [59]. The account of subspace methods given in this section has been heavily influenced by the recent thesis of van Overschee [54].

8.2 Realization Methods

Let the impulse response sequence of a causal, stable, multivariable, linear time-invariant system be $\{g(0), g(1), g(2), \ldots\}$ ($g(t) \in \mathbf{R}^{p \times m}$). Suppose that some input sequence $\{\ldots, u(t-2), u(t-1)\}$ is applied from the remote past up to time $t-1$, and that the input is then set to zero:

$$u(t+k) \quad = \quad 0 \qquad \text{for } k \geq 0. \tag{8.1}$$

Then

$$\begin{bmatrix} y(t) \\ y(t+1) \\ \vdots \end{bmatrix} = \begin{bmatrix} g(1) & g(2) & \ldots \\ g(2) & g(3) & \ldots \\ \vdots & \vdots & \vdots \end{bmatrix} \begin{bmatrix} u(t-1) \\ u(t-2) \\ \vdots \end{bmatrix}. \tag{8.2}$$

Let us write this as

$$Y^+ \quad = \quad H\, U^- \tag{8.3}$$

so that Y^+ is a vector of future outputs, U^- is a vector of past inputs, and H is a matrix with a block-Hankel structure (the matrix in position (i,j) is $g(i+j-1)$). Now suppose that the system in fact has a state-space realization (5.24),(5.25)) with McMillan degree n, so that its impulse response is given by (5.26). Then (writing A for A_d, etc)

$$H \quad = \quad \begin{bmatrix} CB & CAB & \ldots \\ CAB & CA^2B & \ldots \\ \vdots & \vdots & \vdots \end{bmatrix} \tag{8.4}$$

$$= \begin{bmatrix} C \\ CA \\ CA^2 \\ \vdots \end{bmatrix} [B,\ AB,\ A^2B,\ \ldots] \tag{8.5}$$

$$= \Omega\Gamma \tag{8.6}$$

where Ω and Γ are the extended observability and reachability matrices, respectively. Since each of these has rank n, it follows that $\text{rank}(H) = n$.

If we define

$$\Omega^{\uparrow} = \begin{bmatrix} CA \\ CA^2 \\ \vdots \end{bmatrix} \tag{8.7}$$

$$\Gamma^{\leftarrow} = [AB,\ A^2B,\ \ldots] \tag{8.8}$$

then important (*shift invariance*) properties are:

$$\Omega^{\uparrow} = \Omega A \tag{8.9}$$

$$\Gamma^{\leftarrow} = A\Gamma. \tag{8.10}$$

From this development the first approach to finding a state-space model follows, assuming that the impulse response sequence is available:

Algorithm 1. *1. Use the impulse response sequence to form the Hankel matrix H.*

2. Determine the rank n of H, and factorise it into two rank n factors Ω and Γ (with n columns and rows, respectively).

3. Take B to be the first m columns of Γ and C to be the first p rows of Ω, and form $A = \Omega^{\dagger}\Omega^{\uparrow}$ (where $\Omega^{\dagger}\Omega = I$) or $A = \Gamma^{\leftarrow}\Gamma^{\dagger}$ (where $\Gamma\Gamma^{\dagger} = I$).

This requires modification to obtain a practical algorithm. Firstly, the need for infinite matrices must be removed. Secondly, the fact that with real data, the rank of H will increase without bound as rows and columns are added, must be accommodated somehow. Note that equations (8.9) and (8.10) will be overdetermined and inconsistent in practice. In Step 3 one of them is solved in a least-squares sense, if $(.)^{\dagger}$ denotes the Moore-Penrose pseudo-inverse. It is possible to consider alternative solution strategies, such as Total Least Squares [19].

A successful practical algorithm, based on Algorithm 1, has been given by Zeiger and McEwen [62] and Kung [27]. The key to its success is the use of the singular value decomposition (SVD) for the estimation of the rank of H, and for factorising H.

Kung's algorithm is the following. Form the finite block-Hankel matrix

$$H = \begin{bmatrix} g(1) & g(2) & \cdots & g(N) \\ g(2) & g(3) & \cdots & g(N+1) \\ \vdots & \vdots & \vdots & \vdots \\ g(N) & g(N+1) & \cdots & g(2N-1) \end{bmatrix} \quad (8.11)$$

and find its singular value decomposition [19]

$$H = U\Sigma V^T \quad (8.12)$$

$$= [U_n \bar{U}_n] \begin{bmatrix} \Sigma_n & 0 \\ 0 & \bar{\Sigma}_n \end{bmatrix} \begin{bmatrix} V_n^T \\ \bar{V}_n^T \end{bmatrix} \quad (8.13)$$

where U_n,V_n have dimensions $pN \times n$ and $mN \times n$, respectively, for some $n \leq \min(mN, pN)$. Now partition U_n as

$$U_n = \begin{bmatrix} U_n^1 \\ U_n^2 \\ \vdots \\ U_n^N \end{bmatrix} \quad (8.14)$$

where each U_n^i has dimensions $p \times n$, and partition V_n similarly. Finally, define $U_n^{\uparrow}$ and $V_n^{\uparrow}$ to be the same as U_n and V_n, but 'shifted up' by p rows and m rows, respectively.

Then approximate realizations (A, B, C, D) of the sequence $(g(0), g(1), \ldots, g(2N-1))$ are obtained by taking $\Omega_N = U_n \Sigma_n^{1/2}$ and $\Gamma_N = \Sigma_n^{1/2} V_n^T$, which gives

$$A = \Sigma_n^{-1/2} U_n^T U_n^{\uparrow} \Sigma_n^{1/2} \quad (8.15)$$

$$B = \Sigma_n^{1/2} (V_n^1)^T \quad (8.16)$$

$$C = U_n^1 \Sigma_n^{1/2} \quad (8.17)$$

$$D = g(0) \quad (8.18)$$

or

$$A = \Sigma_n^{1/2} (V_n^{\uparrow})^T V_n \Sigma_n^{-1/2} \quad (8.19)$$

with B, C, D as above.

Note that [27] describes the algorithm assuming that an infinite Hankel matrix is available. With finite data there are several alternative ways of defining U_n, $U_n^{\uparrow}$, etc. One popular choice is to take

$$U_n = \begin{bmatrix} U_n^1 \\ U_n^2 \\ \vdots \\ U_n^{N-1} \end{bmatrix} \quad (8.20)$$

and

$$U_n^{\uparrow} = \begin{bmatrix} U_n^2 \\ U_n^3 \\ \vdots \\ U_n^N \end{bmatrix}. \tag{8.21}$$

An alternative is

$$U_n = \begin{bmatrix} U_n^1 \\ U_n^2 \\ \vdots \\ U_n^N \end{bmatrix} \tag{8.22}$$

and

$$U_n^{\uparrow} = \begin{bmatrix} U_n^2 \\ U_n^3 \\ \vdots \\ U_n^N \\ 0 \end{bmatrix} \tag{8.23}$$

which guarantees stability (see Theorem 8.1 below).

If $g(t) = 0$ for $t > N$ and the second choice is made for U_n and $U_n^{\uparrow}$ then the two alternative realizations are the same. If in addition the maximal possible McMillan degree is chosen, $n = \min(mN, pN)$, then they are exact realizations (in fact they are then 'Finite Impulse Response' realizations) [11], namely

$$CA^{t-1}B = \begin{cases} g(t) & (t = 1, 2, \dots, N) \\ 0 & (t > N) \end{cases} \tag{8.24}$$

and they are exactly (discrete-time) Lyapunov-balanced [40]:

$$A\Sigma A^T + BB^T = \Sigma \tag{8.25}$$
$$A^T \Sigma A + C^T C = \Sigma \tag{8.26}$$

(The exactness obtained in practice is close to the limits of machine precision; with IEEE standard arithmetic relative errors of the order of 10^{-14} are typical.) The usefulness of these realizations is due both to their guaranteed stability, and to the model approximation properties of balanced realizations: for any $\nu < n$, an approximate realization is obtained by extracting the $\nu \times \nu$ top left-hand corner of A, the top ν rows of B, and the leftmost ν columns of C (assuming the usual ordering of the entries of Σ, namely $\Sigma = \text{diag}\{\sigma_1, \sigma_2, \dots, \sigma_{Nm}\}$, with $\sigma_1 \geq \sigma_2 \geq \cdots \geq \sigma_{Nm} \geq 0$). In fact, the realization obtained in this way (by 'truncation') is the same as that obtained by applying Kung's algorithm with ν in place of n. If $g(t) \neq 0$ for $t > N$ or $n < \min(mN, pN)$ then the approximation error, and the distance of the realization from a balanced one, both depend on $||\bar{\Sigma}_n||$. They also depend, of course, on how close the 'tail' of the impulse response sequence is to zero; one can make it arbitrarily close by using a long enough sequence (large enough value of N), but at the cost of much greater computational expense — the bottleneck of the algorithm is computing the SVD of H. Vaccaro [50] has proposed scaling the impulse response before finding a realization (replacing $g(t)$ by $k^t g(t)$ with $0 < k < 1$), and 'undoing' the scaling afterwards.

An alternative way of finding a lower-order approximation from the one obtained with $n = \min(mN, pN)$ has been suggested in [3]. There the corresponding continuous-time system is found, using the bijection (4.10),(4.11), reduced by truncation, and then the discrete-time system corresponding to the reduced system is found, again using the bijection. If $n = \min(mN, pN)$ then the approximate model obtained in this way remains balanced, and its approximation error can be estimated. It is possible to follow this procedure with a slightly different bijection, which ensures that the frequency response of the approximate system remains invariant at frequency 0 (namely, step responses are preserved).

Realization-based methods can also be applied to the approximate stochastic realization problem which was defined in section 5.2. Recall that in this problem one starts with an estimated initial segment of an autocovariance sequence, and the objective is to find a model whose output has approximately the same autocovariance when it is driven by white noise. An obvious approach to solving this problem is to assemble the estimated autocovariances into a Hankel matrix, and apply Kung's algorithm to this matrix. This will yield a system whose impulse response matches the given autocovariance estimate, and one can then solve the positive-real Riccati equation (5.30), and hence find the required model. This sometimes works. The main problem with it is that if the estimated autocovariance sequence is not positive definite then the Riccati equation may not converge, and even if it does the resulting model may not be minimum-phase. In this case the model obtained from Kung's algorithm will not be positive-real, and will therefore not provide initial parameters for optimization over the space of positive-real systems.

Many variations of this idea have been proposed for the approximate stochastic realization problem, with good surveys available in [5, 6, 7]. A typical one is due to Desai *et al* [14]. From the estimated autocovariance sequence $\{\hat{R}(t)\}$ form the Toeplitz matrices

$$\mathcal{T}_N^+ = \begin{bmatrix} \hat{R}(0) & \hat{R}(1)^T & \cdots & \hat{R}(N-1)^T \\ \hat{R}(1) & \hat{R}(0) & \cdots & \hat{R}(N-2)^T \\ \hat{R}(2) & \hat{R}(1) & \cdots & \hat{R}(N-3)^T \\ \vdots & \vdots & \vdots & \vdots \\ \hat{R}(N-1) & \hat{R}(N-2) & \cdots & \hat{R}(0) \end{bmatrix} \tag{8.27}$$

$$\mathcal{T}_N^- = \begin{bmatrix} \hat{R}(0) & \hat{R}(1) & \cdots & \hat{R}(N-1) \\ \hat{R}(1)^T & \hat{R}(0) & \cdots & \hat{R}(N-2) \\ \hat{R}(2)^T & \hat{R}(1)^T & \cdots & \hat{R}(N-3) \\ \vdots & \vdots & \vdots & \vdots \\ \hat{R}(N-1)^T & \hat{R}(N-2)^T & \cdots & \hat{R}(0) \end{bmatrix} \tag{8.28}$$

and obtain (full column-rank) factorizations

$$\mathcal{T}_N^+ = L^+(L^+)^T \tag{8.29}$$

$$\mathcal{T}_N^- = L^-(L^-)^T. \tag{8.30}$$

Partition L^- as

$$L^- = \begin{bmatrix} L_1^- \\ L_2^- \\ \vdots \\ L_N^- \end{bmatrix} \tag{8.31}$$

where each L_i^- has p rows, and partition L^+ similarly. Define $L^{-\uparrow}$ as before (again, the original algorithm given in [14] assumes infinite data, and is not specific on how to do this). Now form the matrix

$$\bar{\mathcal{H}} = (L^+)^{-1}H(L^-)^{-T} \tag{8.32}$$

where H is the Hankel matrix with $\hat{R}(i+j-1)$ as its (i,j) entry. The algorithm is based on the fact that, with exact autocovariance information and $N = \infty$, the singular values of $\bar{\mathcal{H}}$ approach the canonical correlation coefficients between the past and the future of the output process which is being modelled, as $N \to \infty$. They are therefore parameters which appear in the balanced parametrization of positive-real systems (see section 5.2). Find the SVD

$$\bar{\mathcal{H}} = U\Sigma V^T \tag{8.33}$$

$$= [U_n \bar{U}_n] \begin{bmatrix} \Sigma_n & 0 \\ 0 & \bar{\Sigma}_n \end{bmatrix} \begin{bmatrix} V_n^T \\ \bar{V}_n^T \end{bmatrix} \tag{8.34}$$

and define

$$A = \Sigma_n^{1/2} V_n^T (L^{-\uparrow})^T (L^-)^{-T} V_n \Sigma_n^{-1/2} \tag{8.35}$$

$$B = \Sigma_n^{1/2} V_n^T (L_1^-)^T \tag{8.36}$$

$$C = L_1^+ U_n \Sigma_n^{1/2} \tag{8.37}$$

$$R = \hat{R}(0) - C\Sigma_n C^T \tag{8.38}$$

$$S = B - A\Sigma_n C^T \tag{8.39}$$

$$K = SR^{-1} \tag{8.40}$$

Then (A, K, C, I) is an approximate innovations representation (namely, stable and minimum-phase model) of the original covariance data, if the input white noise process has covariance matrix R. Again, this algorithm works sometimes, not always, but probably more often than the more primitive one based on Kung's algorithm. As before, it is possible to ensure that the resulting model is asymptotically stable, by introducing a block of zeros when defining $L^{-\uparrow}$ [34]. It appears always to give minimum-phase models when this is done, though to the author's knowledge this has not been proved. But it can fail to give initial parameter values for optimization, because occasionally the singular values of $\bar{\mathcal{H}}$ are larger than 1, which is inconsistent with the parametrization.

Of course, if the estimated covariance data sequence is not nonnegative then the factorizations (8.29)(8.30) do not exist, and the algorithm cannot even get started. (Note that estimating the Toeplitz matrices as $\mathcal{T}_N^- = YY^T/N$ for a suitable data matrix Y does not help — although this guarantees nonnegativity, the Toeplitz structure of $\mathcal{T}_N^-$ is lost, and so the autocovariance estimates, which are needed for constructing H, cannot be recovered.)

8.3 Subspace Methods

In the context of system identification we do not usually have impulse response information available, but an arbitrary sequence of inputs, and a corresponding sequence of outputs. In this case the approach of the previous subsection would require preliminary estimation of an impulse response matrix or of an autocovariance sequence from the data. Proceeding directly from the data, without such preliminary estimation, avoids the introduction of unnecessary approximation and computation. Subspace methods provide an approach to doing this.

In this subsection the input and output data will frequently be arranged into block Hankel matrices, for which we will use the following notation. Let

$$U_j^i = \begin{bmatrix} u(j) & u(j+1) & \dots & u(j+N-1) \\ u(j+1) & u(j+2) & \dots & u(j+N) \\ \vdots & \vdots & \vdots & \vdots \\ u(j+i-1) & u(j+i) & \dots & u(j+N+i-2) \end{bmatrix} \tag{8.41}$$

and let Y_j^i, be defined similarly. This notation is similar to that used in [52], except that here the superscript i denotes the number of block rows in the matrix, rather than the index of the last block row. The subscript j denotes the index of the first block row. For convenience, we shall define $U_p = U_0^M$, $U_f = U_M^M$, $Y_p = Y_0^M$, $Y_f = Y_M^M$, where M and N are positive integers such that $N > \max(pM, mM) > n$, and usually $N \gg M$. The suffices p and f are supposed to be mnemonic, representing 'past' and 'future', respectively. Also let

$$X_p = [x(0),\ x(1),\ \dots,\ x(N-1)] \quad \in \mathbf{R}^{n \times N} \tag{8.42}$$

and

$$X_f = [x(M),\ x(M+1),\ \dots,\ x(M+N-1)]. \tag{8.43}$$

Consider the deterministic case with no noise: suppose that the system is described by (5.24) and (5.25). Then

$$Y_p = \Omega_M X_p + T U_p \tag{8.44}$$
$$Y_f = \Omega_M X_f + T U_f \tag{8.45}$$
$$X_f = A^M X_p + \tilde{\Gamma}_M U_p \tag{8.46}$$

where Ω_M consists of the first M block rows of the extended observability matrix Ω, $T \in \mathbf{R}^{pM \times mM}$ is the matrix

$$T = \begin{bmatrix} D & 0 & 0 & \dots & 0 \\ CB & D & 0 & \dots & 0 \\ CAB & CB & D & \dots & 0 \\ \vdots & \vdots & \vdots & \ddots & \vdots \\ CA^{M-1}B & CA^{M-2}B & CA^{M-3}B & \dots & D \end{bmatrix} \tag{8.47}$$

and $\tilde{\Gamma}_M$ consists of the last M block columns of the reversed extended reachability matrix:

$$\tilde{\Gamma}_M = [A^{M-1}B, A^{M-2}B, \dots, B]. \tag{8.48}$$

Substituting (8.46) into (8.45), we see that

$$\text{row span}(Y_f) \ \subset \ \text{row span} \begin{bmatrix} X_p \\ U_p \\ U_f \end{bmatrix}. \tag{8.49}$$

But from (8.44) we have

$$\text{row span}(X_p) \ \subset \ \text{row span} \begin{bmatrix} Y_p \\ U_p \end{bmatrix}. \tag{8.50}$$

Hence

$$\text{row span}(Y_f) \ \subset \ \text{row span} \begin{bmatrix} Y_p \\ U_p \\ U_f \end{bmatrix} \tag{8.51}$$

which means that

$$Y_f = L_1 Y_p + L_2 U_p + L_3 U_f \tag{8.52}$$

for some matrices L_1, L_2 and L_3.

A *parallel projection* of a vector v along a subspace S_1, onto a subspace S_2, denoted by $v/_{S_1}S_2$, is defined by

$$v/_{S_1}S_2 = v \quad \text{if } v \in S_2 \tag{8.53}$$
$$v/_{S_1}S_2 = 0 \quad \text{if } v \in S_1 \tag{8.54}$$
$$v/_{S_1}S_2 = 0 \quad \text{if } v \in [S_1 \oplus S_2]^{\perp} \tag{8.55}$$

where $[S_1 \oplus S_2]^{\perp}$ denotes the orthogonal complement of $S_1 \oplus S_2$. In concrete terms, a parallel projector is a matrix which has S_1 and $[S_1 \oplus S_2]^{\perp}$ as eigenspaces with eigenvalues 0 and S_2 as an eigenspace with eigenvalue 1. The subspaces involved in the projections which we shall encounter will always be row spaces of certain matrices, and we shall use the abbreviated notation $v/_P Q$ to denote projection along the row space of P onto the row space of Q. We shall also use v/Q to denote the orthogonal projection onto the row space of Q.

Now from (8.52) it is clear that taking the parallel projection of (the row vectors in) Y_f along row span(U_f), onto row span$([Y_p^T, U_p^T]^T)$, annihilates the effect of U_f:

$$Y_f/_{U_f}[Y_p^T, U_p^T]^T = L_1 Y_p + L_2 U_p \,. \tag{8.56}$$

But from (8.45) we see that

$$Y_f/_{U_f}[Y_p^T, U_p^T]^T = \Omega_M X_f \,. \tag{8.57}$$

Now note that the left hand-side of this is computable; if we can factorise it into the product of two rank-n factors (of appropriate dimensions) then we can take these factors to be Ω_M and X_f, respectively. Let us suppose for the time being that we know the value of n, and that we can perform such a factorisation somehow. Note that precisely which rank-n factorisation is taken does not matter: the factorisation is unique up to a nonsingular transformation ($\Omega_M \to \Omega_M Z$, $X_f \to Z^{-1} X_f$) which corresponds to a change of state coordinates. Then we can recover the state vector sequence from

$$X_f = \Omega_M^{\dagger} Y_f/_{U_f}[Y_p^T, U_p^T]^T \,. \tag{8.58}$$

Now note that if we advance time by one step, by making the definitions:

$$X_f^+ = [x(M+1),\ x(M+2),\ \ldots,\ x(M+N)] \tag{8.59}$$
$$U_p^+ = U_0^{M+1} \tag{8.60}$$
$$Y_p^+ = Y_0^{M+1} \tag{8.61}$$
$$U_f^- = U_{M+1}^{M-1} \tag{8.62}$$
$$Y_f^- = Y_{M+1}^{M-1} \tag{8.63}$$

then by similar reasoning we arrive at

$$X_f^+ = \Omega_{M-1}^\dagger Y_f^- / _{U_f^-} [Y_p^{+T}, U_p^{+T}]^T . \tag{8.64}$$

Now the matrices A,B,C,D of the state-space model can be recovered by solving the equation

$$\begin{bmatrix} X_f^+ \\ Y_M^1 \end{bmatrix} = \begin{bmatrix} A & B \\ C & D \end{bmatrix} \begin{bmatrix} X_f \\ U_M^1 \end{bmatrix} \tag{8.65}$$

which follows directly from (5.24) and (5.25).

So we have

Algorithm 2. *1. Assemble input and output data, and compute the parallel projections $Y_f/_{U_f}[Y_p^T, U_p^T]^T$ and $Y_f^-/_{U_f^-}[Y_p^{+T}, U_p^{+T}]^T$.*
2. Determine the rank, n, of these projections and factorise them (as in (8.57)) to get X_f and X_f^+.
3. Obtain the state-space matrices by solving (8.65).

To obtain a practical implementation of this, a means of determining the 'approximate rank' of the projections is required, and then of obtaining the factorisations. As before, the Singular Value Decomposition is the key to doing this. Consider the SVD:

$$Y_f/_{U_f}\begin{bmatrix} Y_p \\ U_p \end{bmatrix} = U\Sigma V^T \tag{8.66}$$

$$= [U_n \, \bar{U}_n] \begin{bmatrix} \Sigma_n & 0 \\ 0 & \bar{\Sigma}_n \end{bmatrix} \begin{bmatrix} V_n^T \\ \bar{V}_n^T \end{bmatrix} \tag{8.67}$$

where n is determined, for example, by inspection of the magnitudes of the singular values. Then one can take $\Omega_M = U_n \Sigma_n^{1/2}$, for example — note the similarity to Kung's algorithm in this case. Other choices are possible, provided that

$$\text{col span}(\Omega_M) = \text{col span}(Y_f/_{U_f}[Y_p^T, U_p^T]^T) . \tag{8.68}$$

Now Ω_{M-1} can be taken to be the first $p(M-1)$ rows of Ω_M (the procedure adopted in [54]), or it could be computed from an SVD of $Y_f^-/_{U_f^-}[Y_p^{+T}, U_p^{+T}]^T$. Equation (8.65), encountered in Step 3, will be inconsistent with real data, and so must be solved approximately. A least-squares solution is the usual, but not the only possible, choice.

We note, without going into details, that the projections required in this and other subspace algorithms can be implemented efficiently using the QR decomposition [19]. This is extremely important for the speed and memory requirements of these algorithms, and much space in the principal papers on the topic, such as [52, 56], is devoted to this question.

Another algorithm is obtained by finding the system matrices without finding the state sequence first:

Algorithm 3. *1. Assemble input and output data, and compute the parallel projection* $Y_f/_{U_f}[Y_p^T, U_p^T]^T$.

2. Determine the rank, n, of this projection and factorise it (as in (8.57)) to get Ω_M.

3. Determine the system matrices A and C as in Step 3 of Algorithm 1.

4. Determine the matrices B and D by solving an overdetermined set of linear equations, such as:

$$Y_f/\Omega_M^{\perp} = TU_f/\Omega_M^{\perp} \tag{8.69}$$

where $v/\Omega_M^{\perp}$ denotes the orthogonal projection of v onto the orthogonal complement of the subspace spanned by the rows of Ω_M, (8.69) follows from (8.45), and T is defined in (8.47). n and Ω_M can be determined as described above. Note that with this algorithm stability of the model can be ensured by applying Theorem 8.1 in Step 3.

In order to identify a stochastic model of the form

$$x(t+1) = Ax(t) + Bu(t) + Ke(t) \tag{8.70}$$

$$y(t) = Cx(t) + Du(t) + e(t) \tag{8.71}$$

the same general approach can be followed, though there are significant differences of detail. The conceptual difference is that geometrical notions of subspaces and projections are defined in terms of sample correlations. In particular, an inner product between two vectors $w \in \mathbf{R}^{1\times N}$ and $v \in \mathbf{R}^{1\times N}$ is defined as

$$< w, v > = \frac{1}{N} wv^T. \tag{8.72}$$

So if the vectors contain samples of stochastic processes:

$$w = [w(t), w(t+1), \ldots, w(t+N-1)] \tag{8.73}$$

$$v = [v(\tau), w(\tau+1), \ldots, v(\tau+N-1)] \tag{8.74}$$

then the inner product is a sample cross-correlation:

$$< w, v > = \frac{1}{N} \sum_{i=0}^{N-1} w(t+i)v(\tau+i) \tag{8.75}$$

and, if wide-sense stationarity of all stochastic processes is assumed, then with probability 1 this inner product converges to the ensemble cross-correlation of the two processes $E\{w(t)v(\tau)\}$. With this definition, two vectors are orthogonal, namely $< w, v >= 0$, if the corresponding two processes are uncorrelated, providing that N is adequately large. The geometry obtained in this way corresponds to the usual Hilbert space geometry of wide-sense stationary processes [9].

Now consider the computable orthogonal projection

$$\mathcal{Z}_M = Y_f / \begin{bmatrix} Y_p \\ U_p \\ U_f \end{bmatrix} \tag{8.76}$$

of Y_f onto the combined row span of Y_p, U_p and U_f. According to linear estimation theory (see [13, Theorem 3.1.1] for example),

$$\mathcal{Z}_M \rightarrow \hat{Y}_f \qquad \text{as } N \rightarrow \infty \tag{8.77}$$

where $\hat{Y}_f$ denotes the optimal (minimum mean-square error) linear estimate of Y_f, conditional on knowledge of Y_p,U_p, and U_f. But it is known from Kalman filtering theory that

$$\hat{Y}_f = \Omega_M \hat{X}_f + TU_f \tag{8.78}$$

where

$$\hat{X}_f = [\hat{x}(0), \hat{x}(1), \ldots, \hat{x}(N-1)] \tag{8.79}$$

and each $\hat{x}(i)$ is the prediction of $x(i)$, conditional on data $\{u(i-M), \ldots, u(i-1), u(i)\}$, $\{\ldots, y(i-1)\}$, as obtained from a Kalman filter with an appropriately chosen initial state $\hat{x}(0)$. (Ω_M and T are defined as above.) From (8.77) and (8.78) it is clear that

$$\mathcal{Z}_M /_{U_f} \begin{bmatrix} Y_p \\ U_p \end{bmatrix} \rightarrow \Omega_M \hat{X}_f /_{U_f} \begin{bmatrix} Y_p \\ U_p \end{bmatrix} \tag{8.80}$$

and note, since $(v/[L^T, M^T]^T)/_M L = v/_M L$, that from this we get

$$Y_f /_{U_f} \begin{bmatrix} Y_p \\ U_p \end{bmatrix} \rightarrow \Omega_M \hat{X}_f /_{U_f} \begin{bmatrix} Y_p \\ U_p \end{bmatrix} \qquad \text{as } N \rightarrow \infty. \tag{8.81}$$

This is, of course, analogous to (8.57), and we can recover $\hat{X}_f /_{U_f} \begin{bmatrix} Y_p \\ U_p \end{bmatrix}$ as in (8.58).

Now to proceed as in Algorithm 2 we would like to obtain an equation analogous to (8.64). We can, indeed, define

$$\hat{X}_f^+ /_{U_f} [Y_p^{+T}, U_p^{+T}]^T = \Omega_{M-1}^{\dagger} Y_f^- /_{U_f^-} [Y_p^{+T}, U_p^{+T}]^T \tag{8.82}$$

but unfortunately $\hat{X}_f /_{U_f} [Y_p^T, U_p^T]^T$ and $\hat{X}_f^+ /_{U_f} [Y_p^{+T}, U_p^{+T}]^T$ are not simply related. It turns out that $\hat{X}_f$ and $\hat{X}_f^+$ can be obtained as state estimate sequences, but each with a different initial condition. However, it is shown in [52] that the initial condition is the same if $\{u(t)\}$ is a white noise process (which is unlikely), and is approximately the same if M is sufficiently large. One can therefore proceed by assuming that

$$\begin{bmatrix} \hat{X}_f^+ \\ Y_M^1 \end{bmatrix} = \begin{bmatrix} A & B \\ C & D \end{bmatrix} \begin{bmatrix} \hat{X}_f \\ U_M^1 \end{bmatrix} + \begin{bmatrix} W_M^1 \\ E_M^1 \end{bmatrix} \tag{8.83}$$

where $W_M^1 = KE_M^1$ and E_M^1 is defined analogously to U_M^1 and Y_M^1, and accepting that subsequent results will be biased, with the bias decreasing as M increases. With this reservation, $\hat{X}_f$ and $\hat{X}_f^+$ can be estimated as was the case earlier with Algorithm 2, (8.83) can be solved for A, B, C, D in a least-squares sense, and W_f, E_f estimated as the residuals from this solution. Finally, and this step is new to the stochastic case, one can estimate the covariance matrix

$$\begin{bmatrix} KQK^T & KQ \\ QK^T & Q \end{bmatrix} \approx \begin{bmatrix} < W_f, W_f > & < W_f, E_f > \\ < E_f, W_f > & < E_f, E_f > \end{bmatrix} \tag{8.84}$$

from which it is possible to extract K. However, the resulting stochastic model (A, K, C, I) may not be minimum-phase; it is possible to ensure the minimum-phase condition by following a more complicated procedure: define

$$\begin{bmatrix} \Phi & S \\ S^T & Q \end{bmatrix} = \begin{bmatrix} < W_f, W_f > & < W_f, E_f > \\ < E_f, W_f > & < E_f, E_f > \end{bmatrix}. \tag{8.85}$$

Now following the development of (5.16),(5.17),(5.27), solve $\hat{P} = A\hat{P}A^T + \Phi$ for $\hat{P}$, $\Lambda_0 = C\hat{P}C^T + Q$ for Λ_0, and $G = A\hat{P}C^T + S$ for G, and find the minimal positive definite solution P of the Riccati equation (analogous to (5.30))

$$\begin{aligned} P &= A_d P A^T + \\ &\quad +(G - APC^T)[\Lambda_0 - CPC^T]^{-1}(G - APC^T)^T. \end{aligned} \tag{8.86}$$

Finally, set

$$K = (G - APC^T)(\Lambda_0 - CPC^T)^{-1}. \tag{8.87}$$

Note that with perfect covariance estimates P would be the same as $\hat{P}$.

It is possible to develop algorithms which avoid the introduction of bias, even for small M. These are more complicated, and follow the approach of Algorithm 3. See [52, 54] for details.

If a *weighted* parallel projection $W_1(Y_f/_{U_f}[Y_p^T, U_p^T]^T)W_2$ is used in (8.66), then it is shown in [53, 54] that some of the better-known versions of subspace algorithms, including in particular those of [29, 52, 56], can be obtained as special cases, each corresponding to a different choice of W_1 and W_2. The weight W_2 used in [29, 57] corresponds to a further orthogonal projection onto $U_f^\perp$, the orthogonal complement of U_f. This avoids the introduction of bias which was mentioned above.

8.4 Guaranteeing Stability

The following result gives a way of guaranteeing that the models obtained by the methods outlined above are stable. It can be applied to all the realization-based methods, and to those subspace methods in which the 'A' matrix is formed as the product of a shifted matrix with a pseudo-inverse. So it can be used for all the methods based on Algorithms 1 and 3. It should be used with caution, however, because a price to be paid for the guaranteed stability is some distortion of the results.

Suppose that $Z \in \mathbf{R}^{Np \times n}$ is a matrix which is partitioned into blocks, each of size $p \times n$:

$$Z = \begin{bmatrix} Z_1 \\ Z_2 \\ \vdots \\ Z_N \end{bmatrix} \tag{8.88}$$

and let $Z_o^{\uparrow}$ denote the matrix which is formed when each block is shifted upwards by p rows and a block of zeros is introduced at the bottom:

$$Z_o^{\uparrow} = \begin{bmatrix} Z_2 \\ Z_3 \\ \vdots \\ Z_N \\ 0 \end{bmatrix}. \tag{8.89}$$

Also let $Z^{\dagger}$ denote a pseudo-inverse of Z, so that $Z^{\dagger} Z = I_n$. (We assume that $Np \geq n$.)

Theorem 8.1. *If*

$$A = Z^{\dagger} Z_o^{\uparrow} \tag{8.90}$$

then

$$|\lambda(A)| \leq 1 \tag{8.91}$$

holds for any eigenvalue $\lambda(A)$ of A.

Proof: See [34].

Remark: A similar theorem holds if

$$Z = [Z_1, Z_2, \ldots, Z_N] \tag{8.92}$$

$$Z_o^{\leftarrow} = [Z_2, \ldots, Z_N, 0] \tag{8.93}$$

$$A = Z_o^{\leftarrow} Z^{\dagger}. \tag{8.94}$$

8.5 Estimating n

If the data were really generated by a linear system of McMillan degree n, and N were extremely large, and the data could be represented with infinite precision, then we would have $\bar{\Sigma}_n = 0$ in (8.12) or (8.66), and it would be very clear what the true value of n was. Of course none of these conditions hold in practice. Data usually comes from a nonlinear system, which is in addition very likely changing with time. In these circumstances it is not appropriate to think of finding the 'true' value of n. The best that can be hoped for is that one may find a value of n which is not too inconsistent with the data. Commonly-given advice is that the McMillan degree which is finally selected can be determined by examining the singular values in Σ, and choosing n such that the entries of $\bar{\Sigma}_n$ are much smaller than those of Σ_n. This advice is not very useful in practice, because the singular values tend to decrease gradually, with no clear division between 'large' and 'small' values. In practice a range of values of n must be tried, and the quality of approximation offered by each compared either qualitatively or in some quantitative manner. Some results are available which bound the error in various situations [3, 18, 21, 48]. Larimore [28, 29] advocates the use of the AIC criterion to select n (see [13, 55] for an account of this and other criteria). There is some evidence that Larimore's 'CVA' algorithm is particularly effective at determining a suitable McMillan degree.

If the models produced by subspace algorithms are to serve as initial models for parameter estimation using balanced parametrizations, then they can be put into balanced form, providing that they are stable, and the initial parameters can then be obtained easily. The balanced form can also be useful for reducing the McMillan degree, if desired, most easily by 'truncation' of the balanced state vector. This provides an alternative approach to determining the McMillan degree: the value chosen for n in the algorithms can be rather high (limited by computational resources, perhaps), and can then be reduced in a separate approximation stage, in a manner reminiscent of the approach taken by Wahlberg [60], who finds high-order ARX models and subsequently reduces the order. In [39] a subspace algorithm is proposed in which the resulting model is already balanced.

'Balanced truncation' is a relatively crude procedure for approximating a model, and more sophisticated possibilities exist. In particular, 'AAK' or optimal Hankel-norm methods are available [1, 18]. However, the added sophistication of these methods may not be needed in practice, particularly if the subspace methods are to be followed by further parameter estimation. On the other hand, the optimisation involved in parameter estimation is known to have many local minima, so that starting from a slightly better initial model may lead to a significantly better final model. It should be said, however, that in many applications the additional accuracy to be obtained from subsequent parameter estimation is simply not needed.

9. Examples

In this section we present three examples to show how balanced parametrisations can be used for estimating parameters of multivariable linear systems. The examples concern a distillation column, an industrial dryer, and the spectrum of sea waves, and they illustrate the estimation of stable systems, minimum phase systems, and positive real systems, respectively.

9.1 Distillation Column

This example concerns a fractional distillation column with 3 inputs (input cooling temperature, reboiling temperature, and pressure) and 2 outputs (top product flow rate, C4 concentration). A complex 'high-fidelity' nonlinear simulation of the column was available, constructed from physical and thermodynamic 'first-principles'. In order to exploit this model for control, an automatic way of approximating it by a low-order linear model was required. The quality of the approximation was to be judged by its ability to reproduce low-amplitude step responses, although this is not the best criterion for control applications. As is to be expected, and as we shall demonstrate later, a model which reproduced the step response very well did not reproduce the frequency response particularly well.

Since a simulation model was available, it was possible to produce impulse response data from it — in fact by differencing step response data — and therefore to use an approximate realization algorithm to obtain the initial model. The only property of the model known *a priori* was that it was stable. It was therefore appropriate to use the 'Lyapunov balanced' parametrisation to improve the initial model. Extremely good results were obtained using McMillan degrees as low as 6.

A very important part of multivariable system identification is appropriate scaling of the data. It is essential to 'tell' the identification algorithm what are comparable variations in each signal. Otherwise the algorithm will inevitably provide good models for those transmission paths which happen to have relatively large numerical signal values associated with them.

An initial indication of a suitable McMillan degree (state dimension) for the linearised model is given by examining the Hankel singular values of a high-order model obtained by Kung's [27] or Al-Saggaf's [3] method, as described in section 8.2. This high-order model is in fact an FIR model, whose impulse response matches the initial segment of the nonlinear model's impulse response exactly, and is zero subsequently. For this example we supplied the algorithm with an impulse response over 250 time steps, the interval between steps representing 10 seconds. Since there are 3 inputs and 2 outputs, the FIR model has $\min(2,3) \times 250 = 500$ states. The first 15 Hankel singular values (of 500) are shown in Fig. 9.1. Assessing these 'by eye', they indicate that about 8 states should be sufficient, and even 6 may be enough.

The parameter estimates were first improved by optimising the criterion

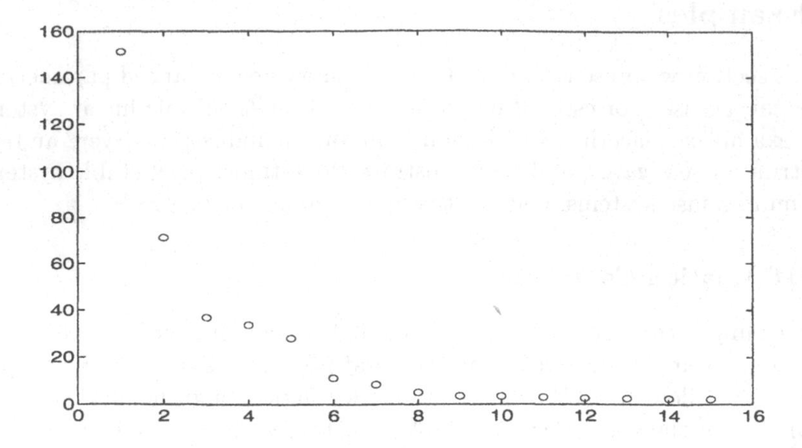

Fig. 9.1. The largest 15 singular values

$$V(\theta) = \sum_{t=0}^{N} ||D_o Y(t) D_i - \hat{Y}(t,\theta)||_F^2 \qquad (9.1)$$

where $Y(t)$ and $\hat{Y}(t,\theta)$ are the step response matrices, at time t, of the nonlinear model and of the linearised model, respectively, D_o and D_i are diagonal scaling matrices, and $||.||_F$ denotes the Frobenius norm.

Table 9.1 shows the values of $V(\theta)$ before and after optimization for 6'th and 8'th order models obtained using both Kung's and Al-Saggaf's algorithms. The fact that each optimization depends on the initial estimate, even though the behaviour of the initial Kung and Al-Saggaf models appears very similar, emphasises the susceptibility of parameter estimation to local optima. Fig. 9.2 shows the step responses of the original nonlinear model (the 'real data'), the initial 8-state model obtained from Kung's algorithm, and the final 8-state model obtained after parameter estimation. It can be seen that parameter optimisation gives a significant improvement on the initial model.

Table 9.1. The model accuracy before and after optimization

	Kung's		Al-Saggaf's	
Objective function $V(\theta)$	6th order	8th order	6th order	8th order
Before optimization	116.57	32.68	42.73	16.55
After optimization	27.54	9.94	23.84	13.68

Replicating the step response may not be particularly meaningful for control, of course, since much attention is paid to modelling low frequency behaviour, and it is precisely this behaviour that will be changed by any control

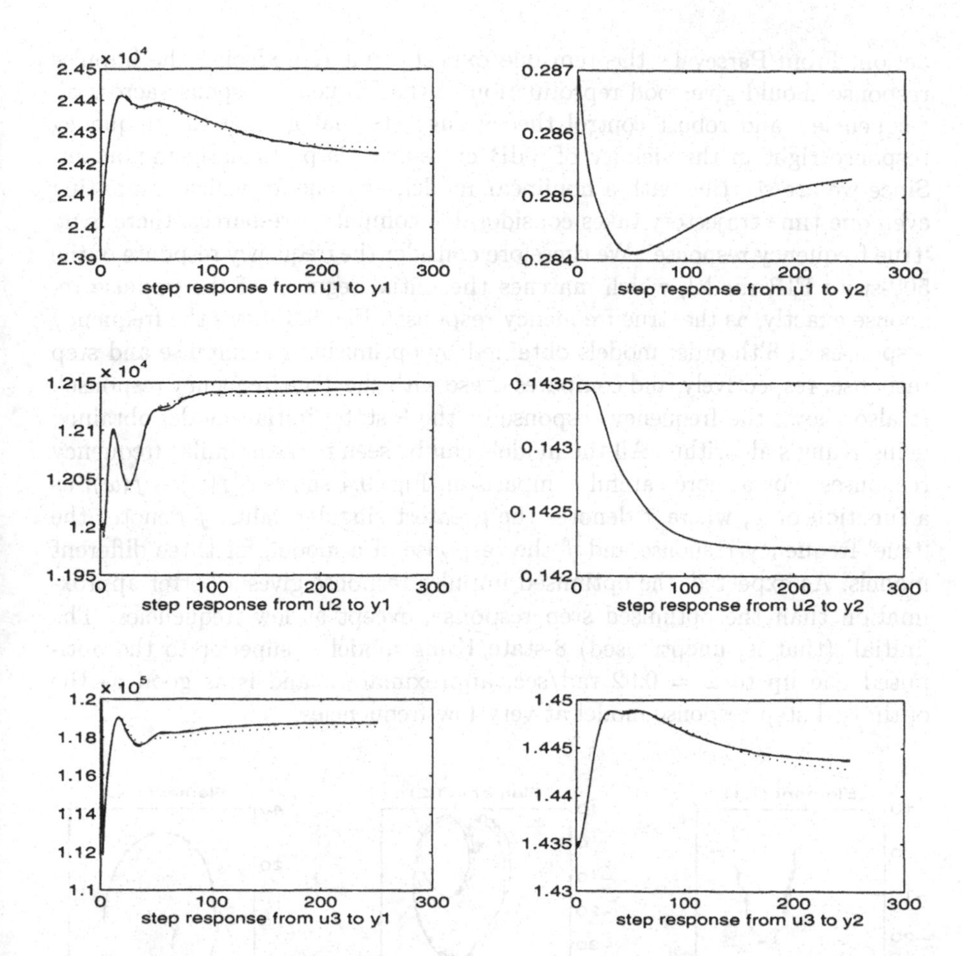

Fig. 9.2. Simulation model (solid). 8th order Kung's estimate obtained from scaled data (dotted). Optimised model (dashed). Unit time interval = 10s.

action. From Parseval's theorem one expects that reproducing the impulse response should give good reproduction of the frequency response across all frequencies, and robust control theory suggests that getting the frequency response right in the vicinity of '0dB cross-over' is particularly important. Since we are starting with a nonlinear model, and one for which computing even one time trajectory takes considerable computing resources, there is no 'true frequency response'. We therefore consider the frequency response of the 500-state FIR model, which matches the initial segment of the impulse response exactly, as the 'true frequency response'. Fig. 9.3 shows the frequency responses of 8'th order models obtained by optimising the impulse and step response, respectively, and compares these with the 'true frequency response'. It also shows the frequency response of the 8-state 'initial' model obtained using Kung's algorithm. All the models can be seen to have similar frequency responses. For a more careful comparison, Fig. 9.4 shows $\bar{\sigma}[f(\omega) - \hat{f}(\omega)]$ as a function of ω, where $\bar{\sigma}$ denotes the greatest singular value, f denotes the 'true' frequency response and $\hat{f}$ the response of a model, for three different models. As expected, the optimised impulse response gives a better approximation than the optimised step response, except at low frequencies. The 'initial' (that is, unoptimised) 8-state Kung model is superior to the optimised one up to $\omega = 0.02$ rad/sec, approximately, and is as good as the optimised step-response model at very low frequencies.

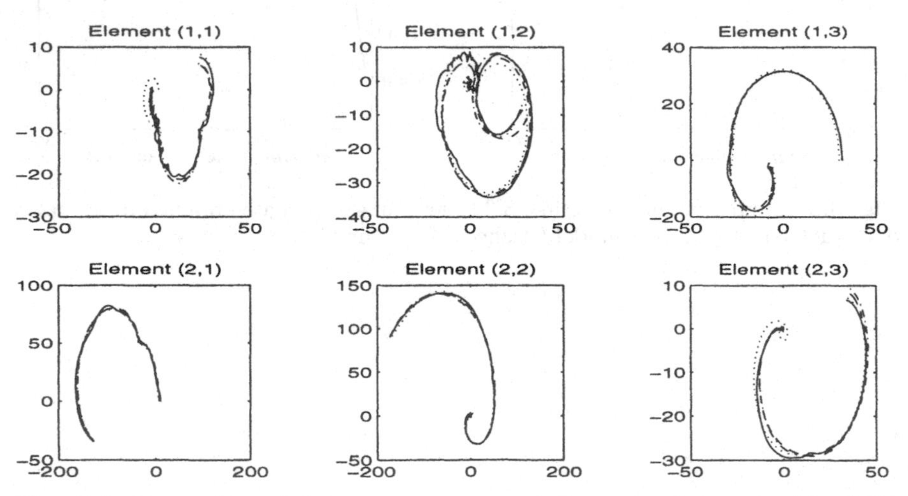

Fig. 9.3. Comparison of frequency responses of alternative 8-state models. Solid line: FIR model, Dashed line: 8-state, optimised impulse response, Chained line: 8-state initial model, Dotted line: 8-state, optimised step response.

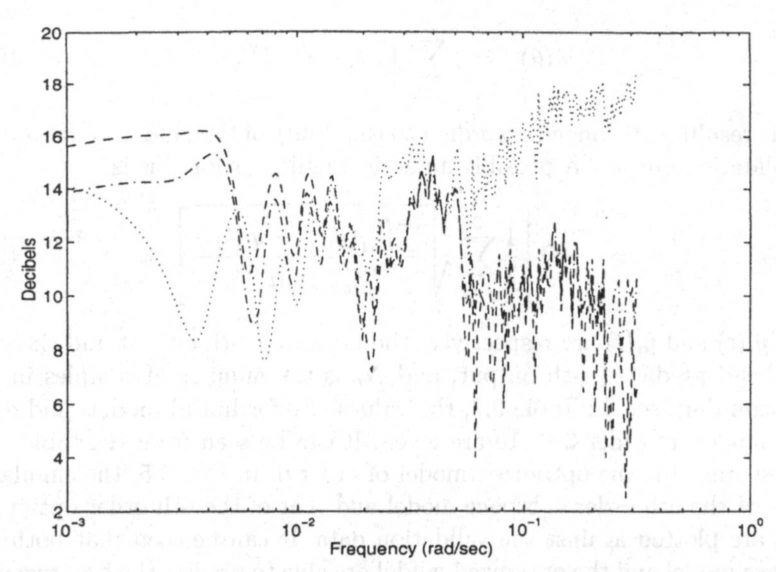

Fig. 9.4. Maximum singular value of difference from FIR model. Key as for Fig. 9.1.

9.2 Industrial Dryer

The second example concerns an industrial drying process, with 3 inputs and 3 outputs. The raw material enters the dryer continuously and the moisture in it is removed by passing hot gas over it. The efficiency of the process depends on both the temperature and the rate of flow of the hot gas. The process inputs are: the fuel flow rate, the hot gas exhaust fan speed, and the rate of flow of raw material. The three outputs are: dry bulb temperature, wet bulb temperature, and the moisture content of the raw material when it leaves the dryer.

The data of this example consists of measurements taken on a real process. The first two inputs could be manipulated, and PRBS signals were applied to these. The third input is an external disturbance which could not be manipulated, but could be measured, and its actual value was recorded. After a certain amount of pre-processing (detrending, pre-filtering and decimation), the data consisted of 867 samples for each input and output, with sampling period 10 seconds. The first 600 samples were used to obtain a model and the rest of the data was used for model validation.

In this case it was appropriate to obtain a model in the 'innovations' form (2.4),(2.5), with the noise model (A_d, K_d, C_d, I) being minimum-phase. We therefore used the N4SID subspace algorithm [52], which produces models of this form, to obtain an initial model. We then used the balanced parametrisation of minimum-phase systems (the parameters in B_d and D_d being unrestricted) to improve the model by prediction-error optimisation. The optimisation criterion was the unweighted prediction error:

$$V(\theta) = \sum_{i=0}^{N} \|y(t) - \hat{y}(t)\|^2 . \qquad (9.2)$$

The results were judged according to the ability of the models to reproduce the validation output. A possible numerical indicator for this is

$$\epsilon = 100 \left[\frac{1}{3} \sum_{i=1}^{3} \sqrt{\frac{\sum_{t=1}^{N_v} (y_i(t) - \hat{y}_i(t))^2}{\sum_{t=1}^{N_v} (y_i(t))^2}} \right] \% \qquad (9.3)$$

where $y_i(t)$ and $\hat{y}_i(t)$ are respectively the measured i-th output and the one-step ahead predicted i-th output, and N_v is the number of samples in the validation data set. In Table 9.2, the values of ϵ for initial models and optimized models of order 2 to 10 are given. It can be seen from the table that the best model is the optimized model of order 6. In Fig. 9.5, the simulated output of the 6th order subspace model and that of the 6th order optimized model are plotted against the validation data. It can be seen that both the subspace model and the optimized model are able to predict the first two outputs very well, but the optimized model predicts the 3rd output significantly better than the subspace model.

Table 9.2. Model error of subspace model and optimized model

model order	subspace models	optimized models
2	26.3406	15.6249
3	14.6125	9.2929
4	13.9288	9.6979
5	12.5058	9.1138
6	12.3669	* 8.7947
7	11.0762	9.0579
8	11.1219	9.6294
9	13.3647	9.1706
10	9.3996	9.0817

9.3 Sea Wave Spectrum

Hydrodynamicists derive spectra of ocean waves on the basis of 'first-principles' reasoning. A widely-used example is the Pierson-Moskwitz spectrum, which has the form

$$\Phi(\omega) = \frac{4\pi^3 h^2}{\omega^5 z^4} \exp\left(\frac{-16\pi^3}{\omega^4 z^4}\right) \qquad (9.4)$$

where h is the wave height, z is the zero-crossing period, and ω is the frequency [47]. In applications such as the control of ships, particularly for

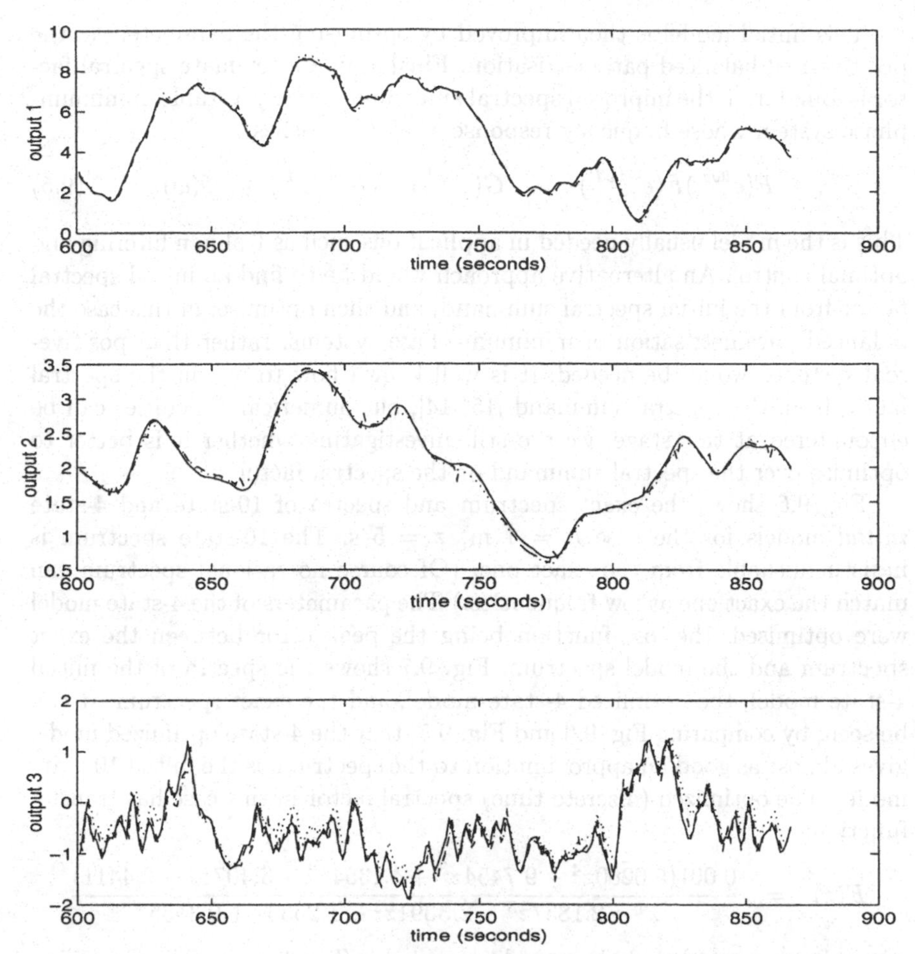

Fig. 9.5. Predicted output on the validation data. Measured output (solid), 6-th order subspace model (dotted), 6-th order optimized model (dashed).

station-keeping (eg when recovering sunken objects or surveying), it is usually necessary to have finite-dimensional models of disturbances such as waves. It is therefore of great importance to find finite-dimensional approximations of irrational spectra such as this one. It is also important to do so with techniques which can be generalised to multivariate spectra, since directional wave spectrum information is frequently available.

The approach we use is to obtain the autocovariance function $\phi(\tau)$ corresponding to the given spectrum $\Phi(\omega)$, and then to use the approximate stochastic realization of algorithm [14], as described in section 8.2, to find an initial positive-real spectral summand, namely a finite-dimensional system whose frequency response $G(e^{i\omega T})$ satisfies $\Phi(\omega) \approx G(e^{i\omega T}) + G(e^{-i\omega T}) \geq 0$.

This initial model is then improved by optimising the parameters in the positive-real balanced parametrisation. Finally an approximate spectral factor is found from the improved spectral summand, namely a stable, minimum-phase system whose frequency response $F(e^{i\omega T})$ satisfies

$$F(e^{i\omega T})F(e^{-i\omega T}) = G(e^{i\omega T}) + G(e^{-i\omega T}) \approx \Phi(\omega) \tag{9.5}$$

This is the model usually needed in applications such as Kalman filtering and optimal control. An alternative approach would be to find an initial spectral factor from the initial spectral summand, and then optimise; in this case the balanced parametrisation of minimum-phase systems, rather than positive-real systems, would be needed. It is well known how to obtain the spectral factor from the spectral summand [15, 14], but numerical difficulties can be encountered at this stage; we are still investigating whether it is better to optimise over the spectral summand or the spectral factor.

Fig. 9.6 shows the exact spectrum and spectra of 10-state and 4-state *initial* models for the case $h = 1$ m, $z = 5$ s. The 10-state spectrum is indistinguishable from the exact one. (Of course no rational spectrum can match the exact one at low frequencies.) The parameters of the 4-state model were optimised, the loss function being the peak error between the exact spectrum and the model spectrum. Fig. 9.7 shows the spectra of the initial 4-state model, the optimized 4-state model, and the exact spectrum. It can be seen, by comparing Fig. 9.6 and Fig. 9.7, that the 4-state optimized model gives almost as good an approximation to the spectrum as the initial 10-state model. The optimized (discrete-time) spectral factor in this case has transfer function:

$$F(z) = \frac{0.001(5.0260z^4 - 9.7454z^3 + 4.1354z^2 + 3.4071z - 2.4411)}{z^4 - 2.1837z^3 + 2.3391z^2 - 1.2533z + 0.3435}$$

where it is assumed that the sampling period is $T = 1$ s.

Although our example concerns a scalar spectrum, the algorithm extends to multivariate spectra with no modifications of the ideas, and only minor modifications of the computational steps.

10. Conclusions

Balanced parametrisations seem to offer an effective and convenient solution to the identifiability problem which occurs when identifying multivariable linear systems. The relatively recent development of subspace methods has already increased the attractiveness of using state-space models for multivariable identification, and we believe that the use of balanced parametrisations will reinforce this trend.

The ability to focus on specific sub-classes of linear systems is a very nice feature of balanced parametrisations. We have shown how this can be

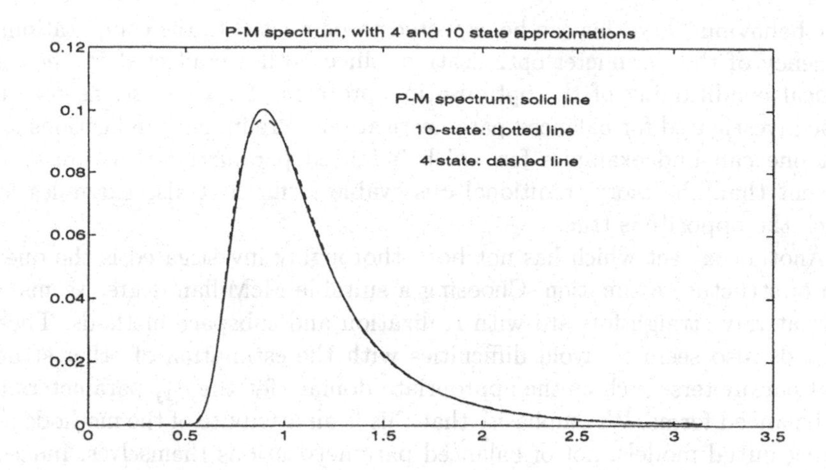

Fig. 9.6. Spectra of initial 4-state and 10-state models

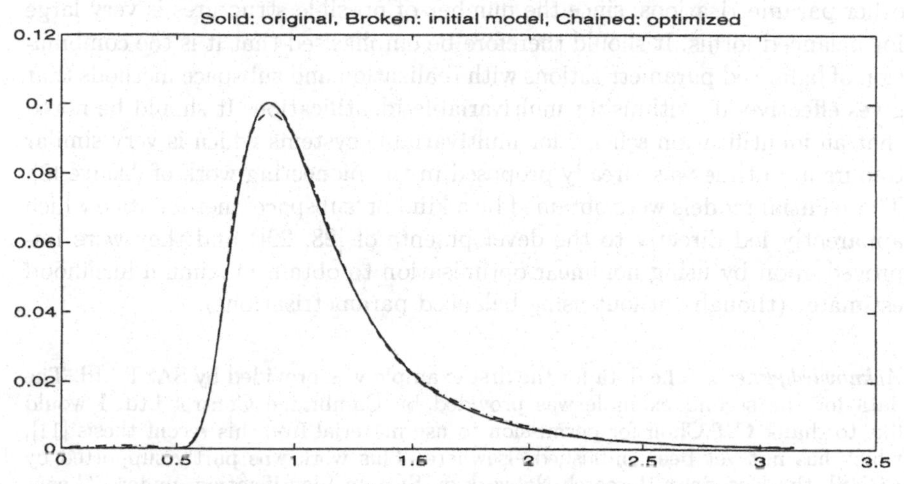

Fig. 9.7. Spectra of 4-state initial and optimised models

applied to particular examples. There is scope to exploit this ability further in applications such as identification of slowly time-varying systems, and adaptive signal processing and control schemes.

We have not discussed the statistical consequences of using balanced parametrisations. Asymptotically (as the amount of data grows), one would expect that the particular parametrisation used should not affect the statistics of the estimated transfer function. Furthermore, existing results on the asymptotic statistical behaviour of parameter estimates depend on the identification algorithm (such as the prediction error method), rather than on the particular parametrisation used (provided that it is identifiable) [30]. But in many applications small-sample behaviour is more important than asymptotic behaviour, and this can be expected to be related to the computational efficiency of the parameter optimisation, since both are affected by the numerical conditioning of the optimisation problem. These questions remain to be investigated for balanced parametrisations. Preliminary indications are that one can find examples for which balanced parametrisations are more efficient than the more traditional observable forms, but also examples for which the opposite is true.

Another aspect which has not been thoroughly investigated is the question of structure estimation. Choosing a suitable McMillan degree seems to be relatively straightforward with realization and subspace methods. These methods also seem to avoid difficulties with the estimation of other structural parameters, such as the appropriate domain for the ϕ_{1j} parameters in the balanced forms. We emphasise that this is an attribute of the methods for finding initial models, not of balanced parametrisations themselves. Indeed, without such effective methods for finding initial models, the structure estimation problem with balanced parametrisations may well be worse than with other parametrisations, since the number of possible structures is very large for balanced forms. It should therefore be emphasised that it is the combination of balanced parametrisations with realization and subspace methods that gives effective algorithms for multivariable identification. It should be noted that an identification scheme for multivariable systems which is very similar to ours in outline was already proposed in the pioneering work of Akaike [2]. There initial models were obtained by a kind of 'subspace' method (one which apparently led directly to the developments of [28, 29]) and they were improved upon by using nonlinear optimisation to obtain maximum likelihood estimates (though without using balanced parametrisations).

Acknowledgement. The data for the first example was provided by SAST Ltd. The data for the second example was provided by Cambridge Control Ltd. I would like to thank C.T.Chou for permission to use material from his recent thesis [11], which has not yet been published elsewhere. This work was partly supported by ERNSI, the European Research Network on System Identification, under EC contract SC1*CT920779.

References

1. Adamjan V.M, D.Z. Arov, and M.G. Krein, "Analytic properties of Schmidt pairs for a Hankel operator and the generalised Schur-Takagi problem", *Math. USSR Sbornik*, 15, 31-73, 1971.
2. Akaike H., "Canonical correlation analysis of time series and the use of an information criterion", in: Mehra,R.K, and Lainiotis,D.G, (eds), *System Identification Advances and Case Studies*, New York: Academic Press 1976.
3. Al-Saggaf U.M., and G.F. Franklin, "Model reduction via balanced realizations: an extension and frequency weighted techniques", *IEEE Trans. Auto. Contr.*, AC-33, 687-692, 1988.
4. Anderson B.D.O., "The inverse problem of stationary covariance generation", *J.Stat.Phys.*, 1, 133-147, 1969.
5. Aoki M., *State Space Modeling of Time Series*, Berlin: Springer 1987.
6. Arun K.S., and S.Y. Kung, "Generalized principal component analysis and its application in approximate stochastic realization", in: Desai,U.B, (ed), *Modelling and Application of Stochastic Processes*, Boston: Kluwer 1986.
7. Arun K.S., and S.Y. Kung, "Balanced approximation of stochastic systems", *SIAM J.Matrix Anal. Appl.*, 11, 42-68, 1990.
8. Brockett R.W., "Some geometric questions in the theory of linear systems", *IEEE Trans.Auto.Contr.*, 21, 449-454, 1976.
9. Caines P.E., *Linear Stochastic Systems*, New York: Wiley 1988.
10. Chen C.T., *Linear System Theory and Design*, New York: Holt, Rinehart and Winston 1984.
11. Chou C.T., "Geometry of Linear Systems and Identification", *PhD Thesis*, Cambridge University, 1994.
12. Clark J.M.C., "The consistent selection of parametrizations in systems identification", *Proc. Joint Amer.Contr.Conf.*, 1976.
13. Davis M.H.A., and R.B. Vinter, *Stochastic Modelling and Control*, London: Chapman and Hall 1985.
14. Desai U.B., D. Pal, and R.D. Kirkpatrick, "A realization approach to stochastic model reduction", *Int. J.Contr.*, 42, 821-838, 1985.
15. Faurre P., "Stochastic realization algorithms", in: Mehra,R.K, and Lainiotis,D.G, (eds), *System Identification: Recent Advances and Case Studies*, New York: Academic Press 1976.
16. Glover K., and J.C. Willems, "Parametrization of linear dynamical systems: canonical forms and identifiability", *IEEE Trans. Auto. Contr.*, AC-19, 640-645, 1974.
17. Glover K., "Some geometrical properties of linear systems with implications in identification", *Proc. IFAC World Congress*, Boston, 1975.
18. Glover K., "All optimal Hankel-norm approximations of linear multivariable systems and their L^{∞} error bounds", *Int. J.Contr.*, 39, 1115-1193, 1984.
19. Golub G., and C. Van Loan, *Matrix Computations*, Baltimore: Johns Hopkins University Press 1983.
20. Gopinath B., "On the identification of linear time-invariant systems from input-output data", *Bell System Tech. Jnl.*, 48, 1101-1113, 1969.
21. Green M., and B.D.O. Anderson, "Generalized balanced stochastic truncation", *Proc. 29th IEEE Conf. on Decision and Control*, Hawaii, 1990.
22. Hannan E.J., and Deistler M., *The Statistical Theory of Linear Systems*, New York: Wiley 1988.

23. Hanzon B., and R.J. Ober, "Overlapping block-balanced canonical forms and parametrizations: the stable SISO case", *Research memorandum 1992-29*, Vrije Universiteit, Amsterdam, 1992. Also submitted to SIAM J.Contr. and Optim.
24. Ho B.L., and R.E. Kalman, "Effective construction of linear, state variable models from input-output functions", *Regelungstechnik*, 14, 545-548, 1966.
25. Kabamba P.T., "Balanced forms: canonicity and parametrization", *IEEE Trans. Auto. Contr.*, AC-30, 1106-1109, 1985.
26. Kailath T., *Linear Systems*, Englewood Cliffs: Prentice-Hall 1980.
27. Kung S.Y., "A new low-order approximation algorithm via singular value decomposition", *Proc. 12th Asilomar Conf. on Circuits*, Systems and Computers, 1978.
28. Larimore W.E., "System identification, reduced-order filtering and modeling via canonical variate analysis", *Proc. American Contr.Conf.*, San Francisco, 1983.
29. Larimore W.E., "Canonical variate analysis in identification, filtering, and adaptive control", *Proc. 29th IEEE Conf. on Decision and Control*, Hawaii, 1990.
30. Ljung L., *System Identification: Theory for the User*, Englewood Cliffs: Prentice-Hall 1987.
31. Maciejowski J.M., "Balanced realizations in system identification", *Proc. 7'th IFAC Symp. Identification and Parameter Estimation*, York, UK, 1985.
32. Maciejowski J.M., and R.J. Ober, "Balanced parametrizations and canonical forms for system identification", *Proc. 8'th IFAC Symp. Identification and Parameter Estimation*, Beijing, China, 1988.
33. Maciejowski J.M., and C.T. Chou, "Identification for control using balanced parametrizations", *Proc. 2nd European Control Conf.*, Groningen, 1993.
34. Maciejowski J.M., "Guaranteed stability with subspace algorithms", Submitted to *Syst. and Contr.Lett.*, 1995.
35. McGinnie B.P., R.J. Ober, and J.M. Maciejowski, "Balanced parametrizations in time-series identification", *Proc. 29th IEEE Conf. on Decision and Control*, Hawaii, 1990.
36. McGinnie B.P., "A Balanced View of System Identification", *PhD Thesis*, Cambridge University, 1994.
37. McKelvey T., "Identification of State-Space Models from Time and Frequency Data", *PhD Thesis*, Linkoping University, 1994.
38. Moonen M., B. De Moor, L. Vandenberghe, and J. Vandewalle, "On and off line identification of linear state-space models", *Int. J. Contr.*, 49, 219-232, 1989.
39. Moonen M., and J. Ramos, "A subspace algorithm for balanced state space system identification", *IEEE Trans. Auto. Contr.*, 38, 1727-1729, 1993.
40. Moore B.C., "Principal component analysis in linear systems: controllability, observability and model reduction", *IEEE Trans. Auto. Contr.*, 26, 17-32, 1989.
41. Mullis C.T., and R.A. Roberts, "Synthesis of minimum roundoff noise fixed point digital filters", *IEEE Trans. Circuits and Systems*, 23, 551-562, 1976.
42. Norton J.P., *An Introduction to Identification*, New York: Academic Press 1986.
43. Ober R.J., "Topology of the set of asymptotically stable minimal systems", *Int.J.Contr.*, 46, 263-280, 1987.
44. Ober R.J., "Connectivity properties of classes of linear systems", *Int.J.Contr.*, 50, 2049-2073, 1989.
45. Ober R.J., "Balanced parametrization of classes of linear systems", *SIAM J.Contr. and Optim.*, 29, 1251-1287, 1991.
46. Ober R.J., "Balanced realizations and canonical forms", this volume.
47. Patel M.H., *Dynamics of Offshore Structures*, London: Butterworths 1989.

48. Safonov M., and R.Y. Chiang, "Model reduction for robust control: a Schur relative error method", *Int.J.Adaptive Contr. and Signal Proc.*, 2, 259-272, 1988.
49. Söderström T., and P. Stoica, *System Identification*, New York: Prentice Hall 1989.
50. Vaccaro R.J., "Finite-data algorithms for approximate stochastic realization", in: Desai,U.B, (ed), *Modelling and Application of Stochastic Processes*, Boston: Kluwer 1986.
51. Van Der Veen A-J, E.F. Deprettere, and A.L. Swindlehurst, "Subspace-based signal analysis using singular value decomposition", *Proc. IEEE*, 81, 1277-1308, 1993.
52. van Overschee P., and B. DeMoor, "N4SID: Subspace algorithms for the identification of combined deterministic stochastic systems", *Automatica*, 30, 75-93, 1994.
53. van Overschee P., and B. DeMoor, "A unifying theorem for three subspace system identification algorithms", *Proc. 10'th IFAC Symp. on System Identification*, Copenhagen, (1994).
54. van Overschee P., "Subspace Identification", *PhD Thesis*, Katholieke Universiteit Leuven, Belgium, 1995.
55. Veres S., *Structure Selection of Stochastic Dynamic Systems*, New York: Gordon and Breach 1991.
56. Verhaegen M., and P.M. Dewilde, "Subspace model identification", Parts 1 and 2, *Int. J. Contr.*, 56, 1187-1241, 1992.
57. Verhaegen M., "Identification of the deterministic part of MIMO state space models given in innovations form from input-output data", *Automatica*, 30, 61-74, 1994.
58. Verriest E., and T. Kailath, "On generalised balanced realizations", *IEEE Trans. Auto. Contr.*, AC-28, 833-844, 1983.
59. Viberg M., "On subspace-based methods for the identification of linear time-invariant systems", *Technical Report CTH-TE-22*, Chalmers University of Technology, Sweden, 1994.
60. Wahlberg B., "On model reduction in system identification", *Proc. American Control Conf*, Seattle, 1986.
61. Willems J.C., "From time series to linear systems, part II", *Automatica*, 22, 675-694, 1986.
62. Zeiger H.P., and A.J. McEwen, "Approximate linear realization of given dimension via Ho's algorithm", *IEEE Trans. Auto. Contr.*, AC 19, 153, 1974.

Balanced Canonical Forms*

Raimund J. Ober

Center for Engineering Mathematics
University of Texas at Dallas
Richardson, TX75083, USA
E-mail: ober@utdallas.edu

1. Introduction

Canonical forms for linear systems are of importance since they provide a unique state space representation of linear systems. They therefore play a major role in system identification where a unique parametrization of the systems in the model set is essential to avoid identifiability problems. Various types of canonical forms for linear systems have been introduced and studied (see e.g. [31], [8], [13], [9]). Most of these canonical forms for multi-variable systems are generalizations of the observer or controller canonical form for single-input single-output systems. The purpose of this paper is to review canonical forms that are based on balanced realizations.

The usefulness of a canonical form depends on its properties. One of the standard canonical forms, the controller canonical form, is of particular significance since the parameters of the canonical form have an immediate interpretation as the coefficients of the transfer function. Moreover, this canonical form permits a straightforward proof of the pole-shifting theorem. There are, however, drawbacks of the controller canonical form especially concerning the resulting parametrization of linear systems. The set of parameters in the controller canonical form that lead to a minimal system is very complicated. This makes it difficult to use this canonical form in cases where it is important to have a geometrically well-behaved parameter space, e.g. in some optimization tasks. One of the main advantages of the balanced canonical forms that are discussed here, is that the parameter space has some desirable geometric properties. This is at the expense that even for single input single output systems discrete structural parameters have to be introduced.

The results presented in this paper extend the previous results ([21], [28], [26]) also to the case of systems with coefficients in the complex field $\mathcal{C}$. In the multivariable case we use a modification due to Hanzon ([10]) of the previously published canonical forms for systems with identical singular values. The approach to the proofs of the canonical form for minimal systems, bounded real systems and positive real systems is also new and based on the bijections that were introduced in ([27]). The objective of this paper is to present these results in a comprehensive form and to derive them in a unified

---

* Dedicated to the memory of Ted Hannan.

way. Because of this tutorial nature of the presentation proofs are given of all essential results even if the proofs have been published already elsewhere.

For each of the classes of transfer functions which we consider we will now define a specific type of balanced realization. To simplify the presentation we introduce the following notation. The set of all p-dimensional output and m-dimensional input minimal continuous-time systems of McMillan degree n is denoted by $L_n^{p,m}$. We call a continuous-time system (A, B, C, D) *stable* if all the eigenvalues of A are in the open left half plane. The subset of $L_n^{p,m}$ of all stable systems is denoted by $S_n^{p,m}$. The subsets of $S_n^{p,m}$ of inner systems and bounded real systems are denoted by $I_n^{p,m}$ and $B_n^{p,m}$. A system (A, B, C, D) in $S_n^{p,m}$ with transfer function G is called *inner* if $(G(s))^*G(s) = I$ for all $s \in \mathcal{C}$. It is called *bounded-real* if $I - (G(i\omega))^* + G(i\omega) > 0$ for all $\omega \in \Re \cup \{\pm\infty\}$. If $p = m$, then P_n^m stands for the subset of $S_n^{m,m}$ of positive real systems. A system (A, B, C, D) in $S_n^{m,m}$ with transfer function G is called *positive-real* if $(G(i\omega))^* + G(i\omega) > 0$ for all $\omega \in \Re \cup \{\pm\infty\}$. In this paper we will study systems with coefficients in the real field $\Re$ and in the complex field $\mathcal{C}$. If a statement is valid for both situations we will use the symbol $\mathcal{K}$ to denote either $\Re$ or $\mathcal{C}$. One of our aims is to study *canonical forms* for these classes of systems in terms of balanced realizations. Two systems $(A_i, B_i, C_i, D_i) \in L_n^{p,m}$, $i = 1, 2$, are called equivalent if there exists a nonsingular matrix $T \in \mathcal{K}^{n \times n}$ such that $(A_1, B_1, C_1, D_1) = (TA_2T^{-1}, TB_2, C_2T, D_2)$.

Definition 1.1. *A canonical form for system equivalence on a subset $\mathcal{A} \subseteq L_n^{p,m}$ is a map*

$$\Gamma : \mathcal{A} \to \mathcal{A},$$

such that

1. *$\Gamma(a) \sim a$ for all $a \in \mathcal{A}$.*
2. *if $a, b \in \mathcal{A}$, and $a \sim b$ then $\Gamma(a) = \Gamma(b)$.*

We also refer to $\Gamma(a)$ as the canonical form of $a \in \mathcal{A}$.

We now introduce the different types of balancing. The principle behind the definition of the different types of balancing is that associated with each class of systems there is a natural pair of Riccati or Lyapunov equations. A system is then called balanced if specified solutions of each of the two equations are identical and diagonal.

Definition 1.2. *1.* (**LQG-balancing**) *The system $(A, B, C, D) \in L_n^{p,m}$ is called* LQG-balanced *if the stabilizing solutions Y and Z to the control and filter algebraic Riccati equations,*

$$0 = A_L^*Y + YA_L - YBR^{-1}B^*Y + C^*S^{-1}C,$$

$$0 = A_LZ + ZA_L^* - ZC^*S^{-1}CZ + BR^{-1}B^*,$$

where $A_L := A - BR^{-1}D^*C$, $R = I + D^*D$, $S = I + DD^*$, are such that

$$Y = Z = diag(\sigma_1, \sigma_2, \dots, \sigma_n) := \Sigma_L,$$

with $\sigma_1 \geq \sigma_2 \geq \dots \geq \sigma_n > 0$. *The matrix* Σ_L *is called the* LQG-grammian *of the system. The positive numbers* $\sigma_1, \sigma_2, \dots, \sigma_n$ *are called the* LQG-singular values *of the system.*

2. (**Lyapunov-balancing**) *The system* $(A, B, C, D) \in S_n^{p,m}$ *is called* Lyapunov-balanced *if the positive definite solutions* Y *and* Z *to the Lyapunov equations,*

$$0 = AZ + ZA^* + BB^*,$$

$$0 = A^*Y + YA + C^*C,$$

are such that

$$Y = Z = diag(\sigma_1, \sigma_2, \dots, \sigma_n) := \Sigma_S,$$

with $\sigma_1 \geq \sigma_2 \geq \dots \geq \sigma_n > 0$. *The matrix* Σ_S *is called the* Lyapunov-grammian *of the system. The positive numbers* $\sigma_1, \sigma_2, \dots, \sigma_n$ *are called the* Lyapunov-singular values *of the system.*

3. (**Bounded-real-balancing**) *The system* $(A, B, C, D) \in B_n^{p,m}$ *is called* bounded-real-balanced *if the stabilizing solutions* Y *and* Z *to the control and filter bounded-real Riccati equations,*

$$0 = A_B^*Y + YA_B + YBR^{-1}B^*Y + C^*S^{-1}C,$$

$$0 = A_BZ + ZA_B^* + ZC^*S^{-1}CZ + BR^{-1}B^*,$$

where $A_B := A + BR^{-1}D^*C$, $R = I - D^*D$, $S = I - DD^*$, *are such that*

$$Y = Z = diag(\sigma_1, \sigma_2, \dots, \sigma_n) := \Sigma_B,$$

with $\sigma_1 \geq \sigma_2 \geq \dots \geq \sigma_n > 0$. *The matrix* Σ_B *is called the* bounded-real-grammian *of the system. The positive numbers* $\sigma_1, \sigma_2, \dots, \sigma_n$ *are called the* bounded-real-singular values *of the system.*

4. (**Positive-real-balancing**) *The system* $(A, B, C, D) \in P_n^m$ *is called* positive-real-balanced *if the stabilizing solutions* Y *and* Z *to the control and filter positive-real Riccati equations,*

$$0 = A_P^*Y + YA_P + YBR^{-1}B^*Y + C^*R^{-1}C,$$

$$0 = A_PZ + ZA_P^* + ZC^*R^{-1}CZ + BR^{-1}B^*,$$

where $A_P := A - BR^{-1}C$, $R = D + D^*$, *are such that*

$$Y = Z = diag(\sigma_1, \sigma_2, \dots, \sigma_n) =: \Sigma_P,$$

with $\sigma_1 \geq \sigma_2 \geq \dots \geq \sigma_n > 0$. *The matrix* Σ_P *is called the* positive-real-grammian *of the system. The positive numbers* $\sigma_1, \sigma_2, \dots, \sigma_n$ *are called the* positive-real-singular values *of the system.*

The definition of LQG balancing is due to Jonckheere and Silverman ([14]) and Verriest ([33]). The notion of Lyapunov balancing has been introduced by the author to distinguish this type of balancing from others that were defined later. Balancing for stable systems has been considered by many authors for different purposes. Roberts and Mullis ([20]) studied these types of realizations because of their good sensitivity properties for the implementation of filters. The good behavior of Lyapunov balanced realizations from the point of view of model reduction has apparently first been recognized by Moore ([19]). For the definition of positive-real balancing see the work by Desai and Pal (e.g [2]). Bounded-real balancing has been considered by Opdenacker and Jonckheere ([29]).

The existence of these various types of balanced realizations is established in the following theorem.

Theorem 1.1. *Let G be a proper rational (stable, antistable, bounded-real, positive-real) function. Then G has a LQG- (Lyapunov-, bounded-real-, positive-real-) balanced realization.*

If (A, B, C, D) is a LQG- (Lyapunov-, bounded-real-, positive-real-) balanced realization of G with grammian $\Sigma = diag(\sigma_1 I_{n_1}, \sigma_2 I_{n_2}, \dots, \sigma_k I_{n_k})$, $\sigma_1 > \sigma_2 > \dots > \sigma_k > 0$, then all other LQG- (Lyapunov-, bounded-real-, positive-real-) balanced realizations over $\mathcal{K}$ are given by

$$(QAQ^*, QB, CQ^*, D),$$

where Q is a constant unitary matrix over $\mathcal{K}$ with $Q = diag(Q_1, Q_2, \dots, Q_k)$, with $Q_j \in \mathcal{K}^{n_j \times n_j}$, $j = 1, \dots, k$. Moreover, all other LQG- (Lyapunov-, bounded-real-, positive-real-) balanced realizations of G have the grammian Σ.

Proof. Let $(A, B, C, D) \in L_n^{p,m}$ be a realization of the proper rational function G. Let Y, Z be the stabilizing solutions to the control and filter algebraic Riccati equations. Assume that T is a state space transformation of the system (A, B, C, D). It is easily verified that the stabilizing solutions to the control and filter algebraic Riccati equations of the system $(TAT^{-1}, TB, CT^{-1}, D)$ are given by TZT^* and $T^{-*}YT^{-1}$. To show that there exists an invertible T which simultaneously diagonalizes Z and Y, first note that since Z is positive and hermitian there exists T_1 invertible, such that

$$T_1 Z T_1^* = I.$$

Since Y is positive and hermitian there exists T_2 unitary such that

$$T_2^{-*} T_1^{-*} Y T_1^{-1} T_2^{-1} =: \Sigma^2 =: diag(\sigma_1^2, \sigma_2^2, \dots, \sigma_n^2),$$

for some $\sigma_1 \geq \sigma_2 \geq \dots \geq \sigma_n > 0$. Now with

$$T := \Sigma^{1/2} T_2 T_1$$

we have

$$TZT^* = \Sigma^{1/2} T_2 T_1 Z T_1^* T_2^* \Sigma^{1/2} = \Sigma^{1/2} T_2 I T_2^* \Sigma^{1/2} = \Sigma^{1/2} I \Sigma^{1/2} = \Sigma$$

and

$$T^{-*} Y T^{-1} = \Sigma^{-1/2} T_2^{-*} T_1^{-*} Y T_1^{-1} T_2^{-1} \Sigma^{-1/2} = \Sigma^{-1/2} \Sigma^2 \Sigma^{-1/2} = \Sigma,$$

which shows the first statement.

Let $(A_1, B_1, C_1, D) \in L_n^{p,m}$ be another LQG-balanced realization of G with LQG-grammian Σ_1. Then there exists a state space transformation T such that

$$(A_1, B_1, C_1, D) = (TAT^{-1}, TB, CT^{-1}, D)$$

and

$$\Sigma_1 = T\Sigma T^* = T^{-*} \Sigma T^{-1}.$$

Then $\Sigma_1^2 = T\Sigma^2 T^{-1}$. Hence Σ_1^2 and Σ^2 are equivalent and since both are diagonal with decreasing diagonal entries we have that $\Sigma_1^2 = \Sigma^2$. Since $\Sigma_1 = T\Sigma T^*$ is diagonal we have that T is unitary. As $\Sigma_1^2 T = T\Sigma^2$ it is easily verified that T has the required structure. If a state space transformation Q is given as in the statement of the Theorem it can be checked in a straightforward way that the transformed system is again LQG-balanced.

The statements for the other types of balancing follow analogously. □

Since balanced realizations are not unique they do not define a canonical form. Much of this paper will be devoted to the introduction of further constraints on balanced realizations to obtain canonical forms.

One of the interesting facts of the canonical forms that are presented here is that they all have a very similar structure, called the balanced form. This is irrespective of the class of systems for which they are derived. It will be shown that balanced forms also provide parametrizations of the respective classes of systems. For example, a canonical form will be derived for the class of stable minimal systems $S_n^{p,m}$ which is given in terms of the so-called Lyapunov balanced form. Conversely, it will be shown that each system which is in Lyapunov balanced form is necessarily minimal and stable. This provides for a parametrization of the class $S_n^{p,m}$ using Lyapunov balanced realizations. Analogous results will be derived for the classes $L_n^{p,m}$, $B_n^{p,m}$ and P_n^m using the corresponding balanced systems.

In many areas of applications models of dynamic processes are given in terms of high dimensional linear systems. Often however the dimension of the model is too high for an efficient analysis of the system. If a high dimensional model is obtained e.g. of an electrical circuit cost considerations may prohibit the implementation. In these and similar situations the question arises whether the high dimensional system can be approximated well by a low dimensional system. The 'goodness' of an approximation will of course depend on the application and a norm-based criterion will be discussed later.

A principal requirement of a model reduction scheme should be that important qualitative properties of the high order system be retained in the approximant. If for example stability is an important feature of the system, then the approximant should also be stable. Balanced realizations provide model reduction schemes that have such properties. For example the balanced model reduction scheme for positive real systems guarantees that the approximating system is again positive real. The *balanced model reduction scheme* is defined as follows. Let (A, B, C, D) be a LQG (Lyapunov, bounded-real, positive-real) system over $\mathcal{K}$ that is partitioned as follows,

$$A = \begin{pmatrix} A_{11} & A_{12} \\ A_{21} & A_{22} \end{pmatrix}, \quad B = \begin{pmatrix} B_1 \\ B_2 \end{pmatrix},$$

$$C = \begin{pmatrix} C_1 & C_2 \end{pmatrix},$$

with $A_{11} \in \mathcal{K}^{r\times r}$, $B_1 \in \mathcal{K}^{r\times m}$, $C_1 \in {}^{p\times r}$ and $r < n$. The r-dimensional system (A_{11}, B_1, C_1, D) is then called the *r-dimensional LQG (Lyapunov, bounded-real, positive-real) balanced approximant* of (A, B, C, D).

One of the reasons for the interest that balanced realizations received is due their model reduction properties which will be discussed in later sections and their connection to Hankel norm approximation (see e.g. [7], [5]).

Lyapunov balanced realizations play a particularly important role in our development. Many properties of the other types of balanced realizations can be deduced from properties of Lyapunov balanced realizations. The necessary machinery for this process will be introduced in Sect. 3.. Model reduction properties of Lyapunov balanced realizations will be investigated in Sect. 2.. The introduction of canonical forms for various classes of systems is the topic of Sections 4. and 5..

Having analyzed a canonical form, parametrization and model reduction for stable minimal systems in Sect. 3., a bijection between $S_n^{p,m}$ and $L_n^{p,m}$ will be introduced in Sect. 4.. This bijection will be used to carry the results for $S_n^{p,m}$ over to $L_n^{p,m}$. In a similar way canonical forms, parametrizations and model reduction results will be derived for bounded-real and positive real systems in Sect. 5..

If the system (A, B, C, D) is a realization of the proper rational function G, i.e. $G(s) = C(sI - A)^{-1}B + D$, for $s \in \mathcal{C}$, then we write $G \stackrel{r}{=} (A, B, C, D)$. Occasionally we will also write $\left(\begin{array}{c|c} A & B \\ \hline C & D \end{array}\right)$ to denote the system (A, B, C, D).

This work was supported by NSF Grant DMS-9304696.

2. Lyapunov Balanced Realizations and Model Reduction

In this section we are going to introduce basic facts concerning model reduction properties of Lyapunov balanced realizations. We are going to prove a result by Pernebo and Silverman [30] that shows that if (A, B, C, D) is a Lyapunov balanced system and (A_{11}, B_1, C_1, D) is its r dimensional balanced approximant then the approximant is stable, minimal and again Lyapunov balanced. This is the case if a mild condition on the singular values is satisfied. In Theorem 3.3, it will be shown that this condition can be dropped if the system is in Lyapunov balanced canonical form.

The following Lemma will be needed frequently.

Lemma 2.1. *Let (A, B, C, D) be a continuous-time system and let P (Q) be a positive definite solution to the Lyapunov equation,*

$$AP + PA^* = -BB^* \quad (A^*Q + QA = -C^*C),$$

Then the system is stable if and only if it is reachable (observable).

Proof. That reachability (observability) implies stability is a standard result. Let now the system be stable and let x be an eigenvector of A^* with eigenvalue λ, i.e. $A^*x = \lambda x$. Then

$$0 \geq -< BB^*x, x >=< (AP + PA^*)x, x >=< Px, A^*x > + < PA^*x, x >$$

$$= \bar{\lambda} < Px, x > +\lambda < Px, x >= 2Re(\lambda) < Px, x > .$$

Since P is positive definite $< Px, x >> 0$. The stability of the system implies that

$$Re(\lambda) < 0.$$

Hence $< B^*x, B^*x >> 0$ which implies that $B^*x \neq 0$. This implies the reachability of the system. The corresponding implication on observability of the system follows analogously. □

Theorem 2.1. **(Pernebo-Silverman)** *Let $(A, B, C, D) \in S_n^{p,m}$ be a n-dimensional Lyapunov balanced system with Lyapunov grammian $\Sigma = diag(\sigma_1, \sigma_2, \ldots, \sigma_n)$, $\sigma_1 \geq \sigma_2 \geq \cdots \geq \sigma_n \geq 0$. Let $r < n$ be such that $\sigma_{r+1} \neq \sigma_n$. Then the r-dimensional balanced approximant (A_{11}, B_1, C_1, D) of (A, B, C, D) is stable, minimal and Lyapunov balanced with Lyapunov grammian $\Sigma_1 = diag(\sigma_1, \sigma_2, \ldots, \sigma_r)$.*

Proof. Let $\Sigma_1 = diag(\sigma_1, \ldots, \sigma_k)$. Then it is easily observed that Σ_1 solves the Lyapunov equations for the approximant (A_{11}, B_1, C_1, D), i.e.

$$A_{11}\Sigma_1 + \Sigma_1 A_{11}^* = -B_1 B_1^*,$$

$$A_{11}^*\Sigma_1 + \Sigma_1 A_{11} = -C_1^* C_1.$$

We need to show that A_{11} is stable. Let λ be an eigenvalue of A_{11}^* with eigenvector x, i.e.

$$A_{11}^* x = \lambda x.$$

Then

$$\begin{aligned} 0 \geq - < B_1^* x, B_1^* x > &= - < B_1 B_1^* x, x > = < (A_{11}\Sigma_1 + \Sigma_1 A_{11}^*)x, x > \\ &= \lambda < \Sigma_1 x, x > + \bar{\lambda} < \Sigma_1 x, x > \\ &= 2Re(\lambda) < \Sigma_1 x, x >. \end{aligned}$$

Since $< \Sigma_1 x, x > \geq 0$ this shows that $Re(\lambda) \leq 0$. Assume now that $\lambda = iy$, $y \in \Re$. Then the above calculation shows that

$$- < B_1^* x, B_1^* x > = 2Re(\lambda) < \Sigma_1 x, x > = 0,$$

and therefore that $B_1^* x = 0$. Multiplying the Lyapunov equation on the right by x we obtain,

$$0 = -B_1 B_1^* x = A_{11}\Sigma_1 x + \Sigma_1 A_{11}^* x = A_{11}\Sigma_1 x + \lambda \Sigma_1 x.$$

Hence,

$$A_{11}\Sigma_1 x = -\lambda \Sigma_1 x.$$

Using the equation

$$A_{11}^*\Sigma_1 + \Sigma_1 A_{11} = -C_1^* C_1,$$

we have,

$$\begin{aligned} &- < C_1\Sigma_1 x, C_1\Sigma_1 x > = - < C_1^* C_1 \Sigma_1 x, \Sigma_1 x > \\ &= < (A_{11}^*\Sigma_1 + \Sigma_1 A_{11})\Sigma_1 x, \Sigma_1 x > = < \Sigma_1^2 x, A_{11}\Sigma_1 x > + < \Sigma_1 A_{11}\Sigma_1 x, \Sigma_1 x > \\ &= -\bar{\lambda} < \Sigma_1^2 x, \Sigma_1 x > - \lambda < \Sigma_1^2 x, \Sigma_1 x > = 0, \end{aligned}$$

since $\lambda + \bar{\lambda} = 0$. Therefore $C_1\Sigma_1 x = 0$. Multiplying the Lyapunov equation $A_{11}^*\Sigma_1 + \Sigma_1 A_{11} = -C_1^* C_1$ on the right by $\Sigma_1 x$ we have

$$0 = -C_1^* C_1 \Sigma_1 x = (A_{11}^*\Sigma_1 + \Sigma_1 A_{11})\Sigma_1 x = A_{11}^*\Sigma_1^2 x - \lambda \Sigma_1^2 x,$$

i.e.

$$A_{11}^*\Sigma_1^2 x = \lambda \Sigma_1^2 x.$$

This shows that the eigenspace of A_{11}^* with eigenvalue λ is invariant under Σ_1^2. If $\Sigma_1 = diag(\tilde{\sigma}_1 I_{n_1}, \tilde{\sigma}_2 I_{n_2}, \ldots, \tilde{\sigma}_l I_{n_l})$, where $\tilde{\sigma}_1 > \tilde{\sigma}_2 > \cdots > \tilde{\sigma}_l > 0$, then the invariant subspaces of Σ_1^2 are subspaces of

$$E_i := diag(0I_{n_1+\cdots+n_{i-1}}, I_{n_i}, 0I_{n_{i+1}+\cdots+n_l})\mathcal{K}^r, \quad 1 \le i \le l.$$

Hence $x \in E_{i_0}$ for some $1 \le i_0 \le l$, i.e. $\Sigma_1^2 x = \sigma_{i_0}^2 x$. Now consider the $(2,1)$ blocks of the Lyapunov equations of the original system (A, B, C, D), i.e.

$$A_{21}\Sigma_1 + \Sigma_2 A_{12}^* = -B_2 B_1^*,$$

$$A_{12}^*\Sigma_1 + \Sigma_2 A_{21} = -C_2^* C_1.$$

Multiplying the first equation on the right by x, on the left by Σ_2, and the second equation on the right by $\Sigma_1 x$ we obtain,

$$\Sigma_2 A_{21}\Sigma_1 x + \Sigma_2^2 A_{12}^* x = -\Sigma_2 B_2 B_1^* x = 0$$

$$A_{12}^*\Sigma_1^2 x + \Sigma_2 A_{21}\Sigma_1 x = -C_2^* C_1 \Sigma_1 x = 0.$$

Therefore,

$$\Sigma_2^2 A_{12}^* x = -\Sigma_2 A_{21}\Sigma_1 x = A_{12}^*\Sigma_1^2 x = \sigma_{i_0}^2 A_{12}^* x.$$

Since by assumption the diagonal entries of Σ_2^2 are distinct from $\sigma_{i_0}^2$ we have that $A_{12}^* x = 0$. Then with $\tilde{x} = (x^*, 0I_{n-r})^*$ we have

$$A^*\tilde{x} = \begin{pmatrix} A_{11}^* & A_{21}^* \\ A_{21}^* & A_{22}^* \end{pmatrix} \begin{pmatrix} x \\ 0 \end{pmatrix} = iy \begin{pmatrix} x \\ 0 \end{pmatrix},$$

which is a contradiction to the stability of A^* and hence of A.

Reachability and observability of the system now follow from Lemma 2.1.

□

The following result will give a quantitative bound on the size of the error that is incurred by the approximation process. The error is in term of the so-called H^∞ norm of the error function. The H^∞-norm of a stable proper rational transfer function F is given by

$$\|F\|_\infty = \sup_{\omega \in \Re} \|F(i\omega)\|.$$

This norm is of particular relevance in robust control (see e.g. [4]). This result was derived independently by Enns ([3]) and Glover ([7]).

Theorem 2.2. *Let G be the transfer function of a continuous-time n-dimensional Lyapunov balanced system with Lyapunov grammian $\Sigma = diag(\tilde{\sigma}_1 I_{n_1}, \tilde{\sigma}_2 I_{n_2}, \ldots, \tilde{\sigma}_k I_{n_k})$, with $\tilde{\sigma}_1 > \tilde{\sigma}_2 > \cdots > \tilde{\sigma}_k$. Let $r := n_1 + n_2 + \cdots + n_l$, $1 \le l \le k$. Let $\hat{G}_r$ be the transfer function of the r-dimensional balanced approximant. Then*

$$\|G - \hat{G}_r\|_\infty \le 2(\tilde{\sigma}_{l+1} + \cdots + \tilde{\sigma}_{n_k}).$$

3. A Lyapunov Balanced Canonical Form for Stable Continuous-Time Systems

In this section we derive a canonical form for the set $S_n^{p,m}$ of stable minimal continuous-time systems of fixed McMillan degree. As in the previous sections we only consider continuous-time systems and refer to the Concluding Remarks for a discussion of the discrete-time case. The canonical form of a stable minimal system will be Lyapunov balanced. Before addressing the derivation of a canonical form for the whole set of systems we will consider the subset of systems with identical singular values. A canonical form for this subset will serve as a building block for the canonical form for the full set $S_n^{p,m}$.

The following type of matrix will be used repeatedly. A matrix $M \in \mathcal{K}^{n\times l}$ with $rank(M) = n \leq l$ is called *positive upper triangular* if it is of the form

$$M_0 := Q_0 M =$$

$$\begin{pmatrix} 0 & \dots & 0 & m_{1i_1} & * & \dots & * & * & * & \cdots & * & * & * & \cdots & * \\ 0 & \dots & 0 & 0 & 0 & \dots & 0 & m_{2i_2} & * & \dots & * & * & * & \cdots & * \\ \vdots & & \vdots & \vdots & \vdots & & \vdots & & & & & & \vdots & & \vdots \\ 0 & \dots & 0 & 0 & 0 & \dots & 0 & 0 & 0 & \dots & 0 & m_{ni_n} & * & \dots & * \end{pmatrix},$$

with $m_{ji_j} \in \Re, m_{ji_j} > 0$ for all $j = 1, 2, \dots, n$. The indices $i_1, \dots, i_n$ are called the *independence indices*.

Note that positive upper triangular matrices are of full rank. Also, if $M_1 \in \mathcal{K}^{m\times n}$ and $M_2 \in \mathcal{K}^{n\times r}$ are both positive upper triangular then M_1M_2 is positive upper triangular.

The following Lemma will be of importance.

Lemma 3.1. *Let $M \in \mathcal{K}^{n\times l}$ with $n = rank(M) \leq l$. Then there exists a unitary matrix $Q_0 \in \mathcal{K}^{n\times n}$ such that $M_0 := Q_0M$ is positive upper triangular. The matrices M_0 and Q_0 are unique, i.e. if $\tilde{M}_0 = \tilde{Q}_0M$ is also positive upper triangular and $\tilde{Q}_0$ unitary then $\tilde{M}_0 = M_0$ and $\tilde{Q}_0 = Q_0$.*

Proof. Write $M = (m_1, m_2, \dots, m_l)$, $m_j \in \mathcal{K}^n$, $1 \leq j \leq l$. Let i_1 be such that $m_{i_1} \neq 0$ and $m_j = 0$ for all $1 \leq j < i_1$. Then there exists a unitary matrix $Q_1 \in \mathcal{K}^{n\times n}$, such that

$$Q_1 m_{i_1} = \begin{pmatrix} m_{1i_1} \\ 0 \\ \vdots \\ 0 \end{pmatrix},$$

with $m_{1i_1} > 0$. Hence

$$M_1 := Q_1 M = \left(\begin{array}{cccc|c} 0 & \cdots & 0 & m_{1i_1} & \\ 0 & \cdots & 0 & 0 & * \ \cdots \ * \\ \vdots & & \vdots & \vdots & \\ 0 & \cdots & 0 & 0 & \tilde{M}_2 \end{array}\right),$$

with $\tilde{M}_2 \in \mathcal{K}^{(n-1)\times(l-i_1)}$ and $rank(\tilde{M}_2) = n-1$. It follows in a straightforward way that all unitary Q such that QM_1 has the same structure as M_1 are given by $Q = diag(1, \tilde{Q}_2)$, with $Q_2 \in \mathcal{K}^{(n-1)\times(n-1)}$ unitary and otherwise arbitrary. Find now in the same way as above $\tilde{Q}_{2,0}$ such that with $Q_2 = diag(1, \tilde{Q}_{2,0})$,

$$M_2 := Q_2Q_1M = Q_2M_1 = \left(\begin{array}{cccc|ccc} 0 & \cdots & 0 & m_{1i_1} & * & \cdots & * \\ 0 & \cdots & 0 & 0 & & & \\ \vdots & & \vdots & \vdots & & \tilde{Q}_{2,0}\tilde{M}_2 & \\ 0 & \cdots & 0 & 0 & & & \end{array}\right),$$

$$= \begin{pmatrix} 0 & \cdots & 0 & m_{1i_1} & * & \cdots & * & * & * & \cdots & * \\ 0 & \cdots & 0 & 0 & 0 & \cdots & 0 & m_{2i_2} & * & \cdots & * \\ \vdots & & \vdots & \vdots & \vdots & & \vdots & \vdots & \vdots & & \vdots \\ 0 & \cdots & 0 & 0 & 0 & \cdots & 0 & 0 & * & \cdots & * \end{pmatrix}.$$

Proceeding inductively we can obtain the desired unique structure.

□

A m-input p-output system (A, B, C, D) of dimension n is said to be in *σ-block form* if there exist integers, the so-called *step sizes* of the system, $m = \tau_0 \geq \tau_1 \geq \tau_2 \geq \cdots \geq \tau_l > 0$ with $\sum_{i=1}^{l} \tau_i = n$ and

1. D is an arbitrary matrix in $\mathcal{K}^{p\times m}$,
2. $B = \begin{pmatrix} \overline{B} \\ 0 \end{pmatrix}$, where $\overline{B}$ is a $\tau_1 \times \tau_0$ positive upper triangular matrix,
3.

$$A = \begin{pmatrix} \tilde{A} + \mathcal{S}_1 & -\mathcal{A}_1^* & 0 & 0 & 0 & \cdots & 0 \\ \mathcal{A}_1 & \mathcal{S}_2 & -\mathcal{A}_2^* & 0 & 0 & \cdots & 0 \\ 0 & \mathcal{A}_2 & \mathcal{S}_3 & \ddots & \ddots & & \vdots \\ 0 & 0 & \ddots & \ddots & \ddots & \ddots & \vdots \\ \vdots & \vdots & \ddots & \ddots & \ddots & -\mathcal{A}_{l-2}^* & 0 \\ \vdots & \vdots & & 0 & \mathcal{A}_{l-2} & \mathcal{S}_{l-1} & -\mathcal{A}_{l-1}^* \\ 0 & 0 & \cdots & 0 & 0 & \mathcal{A}_{l-1} & \mathcal{S}_l \end{pmatrix},$$

   where
   a) $\mathcal{S}_i$ is a $\tau_i \times \tau_i$ skew-hermitian matrix, $i = 1, 2, \ldots, l$,
   b) $\mathcal{A}_i$ is a positive upper triangular $\tau_{i+1} \times \tau_i$ matrix, $i = 1, \ldots, l-1$.
   c) $\tilde{A} \in \mathcal{K}^{\tau_1 \times \tau_1}$, is a function of σ, $\overline{B}$ and D, with $\sigma > 0$.
4. $C = (\overline{C}, 0)$ where $\overline{C} \in \mathcal{K}^{p\times\tau_1}$ is a function of $\overline{B}, D$ and the matrix $U \in \mathcal{K}^{p\times\tau_1}$ is such that $U^*U = I_{\tau_1}$.

Depending on the class of systems that we consider and the type of balancing that we study a different function will be chosen in 3.c.) and 4.). If the system (A, B, C, D) is in σ-block form with $\tilde{A} = -\frac{1}{2\sigma}\overline{BB}^*$ and $\overline{C} = U(\overline{BB}^*)^{\frac{1}{2}}$ then the system is said to be in *Lyapunov σ-block form.* Clearly a system in Lyapunov σ-block form is uniquely specified by

1. the *discrete parameters* $n, p, m, l, \tau_1, \tau_2, \cdots, \tau_l$.
2. the *continuous parameters* σ, $\overline{B}$, U, D, $\mathcal{A}_1, \ldots, \mathcal{A}_{l-1}, \mathcal{S}_1, \ldots, \mathcal{S}_l$. As specified before the fine structure of the matrices $\mathcal{A}_1, \ldots, \mathcal{A}_{l-1}$ is determined by the independence indices.

The relevance of the above systems becomes clear in the following results. The next proposition shows that a system in Lyapunov σ-block form is necessarily stable and minimal. A square transfer function G is called σ- *inner*, $\sigma > 0$, if $\frac{1}{\sigma}G$ is inner, i.e. G is σ-inner if it is stable and $G^*G = \sigma^2 I$. Similarly, a minimal system is called σ-inner if its transfer function is σ-inner.

We will need the following characterization of inner functions (see e.g. [6]).

Lemma 3.2. *Let G be a not necessarily square proper rational function. Then the following two statements are equivalent.*

1. G is inner,
2. If $G \stackrel{r}{=} (A, B, C, D)$ is a minimal state space realization, then there exists $Y = Y^* > 0$ such that
   a) $A^*Y + YA = -C^*C$,
   b) $C^*D + YB = 0$,
   c) $D^*D = I$.

Proof. Consider first the case of inner functions. Assume that 1.) is true. Let $G \stackrel{r}{=} (A, B, C, D)$ be a minimal realization of the inner function G. Then $G^* \stackrel{r}{=} (-A^*, -C^*, B^*, D^*)$ is also a minimal realization of G^*. Since G is inner

$$I = G^*G \stackrel{r}{=} \left(\begin{array}{cc|c} -A^* & -C^*C & -C^*D \\ 0 & A & B \\ \hline B^* & D^*C & D^*D \end{array}\right).$$

This implies that $D^*D = I$. Since (A, B, C, D) is stable there exists $Y = Y^* > 0$ such that

$$A^*Y + YA = -C^*C.$$

Performing a state space transformation of this realization of G^*G with $\begin{pmatrix} I & -Y \\ 0 & I \end{pmatrix}$ we obtain

$$I = G^*G \stackrel{r}{=} \left(\begin{array}{cc|c} -A^* & -A^*Y - YA - C^*C & -C^*D - YB \\ 0 & A & B \\ \hline B^* & B^*Y + D^*C & I \end{array}\right)$$

$$= \left(\begin{array}{cc|c} -A^* & 0 & -C^*D - YB \\ 0 & A & B \\ \hline B^* & B^*Y + D^*C & I \end{array} \right)$$

$$= \left(\begin{array}{c|c} -A^* & -C^*D - YB \\ \hline B^* & I \end{array} \right) + \left(\begin{array}{c|c} A & B \\ \hline B^*Y + D^*C & 0 \end{array} \right).$$

Since the first system in this decomposition is antistable and the second system is stable, the addition of these systems can only be I if the first system is I and the second system is the 0 system. Since (A, B, C, D) is minimal it follows that $(A, B, B^*Y + D^*C, 0)$ is reachable. Hence this system is the 0 system only if $B^*Y + D^*C = 0$. But if $B^*Y + D^*C = 0$ then also $(-A^*, -C^*D - YB, B^*, I) \stackrel{r}{=} I$. This shows 2.).

That 2.) implies 1.) is easily verified. □

Proposition 3.1. *Let (A, B, C, D) be a m-input, p-output continuous-time system in Lyapunov σ-block form, then $(A, B, C, D) \in S_n^{p,m}$, i.e. the system is minimal and stable. Moreover, it is Lyapunov balanced with Lyapunov grammian $\Sigma_S = \sigma I$. If $p \geq m$ and if D is such that $D^*D = \sigma^2 I$ and $C^*D + \Sigma_S B = 0$, then the system is σ-inner.*

Proof. Let (A, B, C, D) be a n-dimensional system in Lyapunov σ-block form. The specific structures of A and B imply that $\begin{bmatrix} B & AB & \cdots & A^{n-1}B \end{bmatrix}$ is positive upper triangular and hence of rank n. Therefore the system is reachable. It is also easily checked that the Lyapunov equation

$$AP + PA^* = -BB^*,$$

has the positive definite solution $P = \sigma I$. Hence the reachability of the system implies by Lemma 2.1 that the system is stable. As by construction $BB^* = C^*C$, the Lyapunov equation

$$A^*Q + QA = -C^*C$$

has the solution $Q = \sigma I$. Applying Lemma 2.1 again we have that the stability of the system implies that it is observable. Hence $(A, B, C, D) \in S_n^{p,m}$. Since $P = Q = \sigma I$ the system is Lyapunov balanced with Lyapunov grammian $\Sigma_S = \sigma I$.

If G is the transfer function of the system, then $(A, \frac{1}{\sigma}B, C, \frac{1}{\sigma}D)$ is a minimal realization of $\frac{1}{\sigma}G$. It is checked easily that this system satisfies the conditions of Lemma 3.2. Hence $\frac{1}{\sigma}G$ is inner. □

In the previous Proposition it was shown that a system in Lyapunov σ-block form is Lyapunov balanced and its grammian is a multiple of the identity matrix. We now need to show a converse result which states that if a stable minimal system has a Lyapunov grammian that is a multiple of the identity matrix then there exists an equivalent system which is in Lyapunov σ-block form. The following Lemma will be needed to show that the 'step sizes' decrease.

Lemma 3.3. *Let (A, B, C, D) be a n-dimensional system. Let $t_i := rank([\ B \quad AB \quad \cdots \quad A^{i-1}B\ ])$, $i = 2, 3, \ldots, n$; $t_1 := rank(B)$. Let $\tau_i = t_i - t_{i-1}$, $i = 1, \ldots, n$, with $t_0 = 0$. Then*

$$\tau_0 \geq \tau_1 \geq \cdots \geq \tau_n \geq 0 \text{ and } \sum_{i=1}^{n} \tau_i = n.$$

Proof. For some r such $1 \leq r \leq n$, let $c_1, \ldots, c_k$ be the columns of $A^{r-1}B$ that are linearly dependent on the columns of $[\ B \quad AB \quad \cdots A^{r-2}B\ ]$. Then $Ac_1, Ac_2, \ldots, Ac_k$ are linearly dependent on the columns of $A[\ B \quad AB \quad \cdots A^{r-2}B\ ]$ and therefore are linearly dependent on the columns of $[\ B \quad AB \quad \cdots A^{r-1}B\ ]$. Hence

$$\begin{aligned} 0 \leq \tau_r &= rank[\ B \quad AB \quad \cdots A^{r}B\ ] - rank[\ B \quad AB \quad \cdots A^{r-1}B\ ] \\ &\leq rank[\ B \quad AB \quad \cdots A^{r-1}B\ ] - rank[\ B \quad AB \quad \cdots A^{r-2}B\ ] \\ &= \tau_{r-1}. \end{aligned}$$

Moreover

$$\sum_{i=1}^{n} \tau_i = \sum_{i=1}^{n} t_i - t_{i-1} = t_n - t_0 = n.$$

□

Theorem 3.1. *Let $(A, B, C, D) \in S_n^{p,m}$ and let Y, Z be the positive definite solutions to the Lyapunov equations*

$$0 = AZ + ZA^* + BB^*, \quad 0 = A^*Y + YA + C^*C.$$

Assume that $ZY = \sigma^2 I$, $\sigma > 0$. Then there exists a state space transformation T such that

$$(A_b, B_b, C_b, D_b) := (TAT^{-1}, TB, CT^{-1}, D)$$

is in Lyapunov σ-block form. Therefore (A_b, B_b, C_b, D_b) is Lyapunov balanced with Lyapunov grammian $\Sigma_S = \sigma I$. Moreover, the map Γ_S that assigns to such a system (A, B, C, D) the system (A_b, B_b, C_b, D_b) is a canonical form.

Proof. Since $ZY = \sigma^2 I$ we can assume that (A, B, C, D) is Lyapunov balanced with Lyapunov grammian $\Sigma_S = \sigma I$. By Theorem 1.1 all other Lyapunov balanced realizations are given by a state space transformation with an orthogonal state space transformation Q. By the minimality of the system the $n \times nm$ matrix $R := \begin{bmatrix} B & AB & \dots & A^{n-1}B \end{bmatrix}$ has rank n. Let now Q_0 be the unique unitary matrix such that $Q_0 R$ is positive upper triangular (Lemma 3.1). Consider the Lyapunov balanced system

$$(A_b, B_b, C_b, D_b) := (Q_0 A Q_0^*, Q_0 B, C Q_0^*, D).$$

Since the matrix

$$R_b := Q_0 R = \begin{bmatrix} B_b & A_b B_b & \cdots & A_b^{n-1} B_b \end{bmatrix}$$

is positive upper triangular we can write it as

$$R_b = \begin{pmatrix} R_{11} & * & * & \cdots & * & \cdots & * \\ 0 & R_{22} & * & \cdots & * & \cdots & * \\ \vdots & \ddots & \ddots & \ddots & & & \vdots \\ 0 & \cdots & 0 & R_{ll} & * & \cdots & * \end{pmatrix},$$

where R_{ii} is a $\tau_i \times m$, $1 \leq s_i \leq m$, $i = 1, \dots, l$, positive upper triangular full rank matrix. The indices τ_i, $i = 1, \dots, l$, are all strictly larger than 0 because of the special structure of the matrix R_b. That $\tau_{i+1} \leq \tau_i$, $i = 1, \dots, l-1$, and $\sum_{i=1}^{l} \tau_i = n$, follows from Lemma 3.3.

Set $\overline{B} := R_{11}$. Then it can be seen that

$$B_b = \begin{pmatrix} \overline{B} \\ 0 \end{pmatrix} = \begin{pmatrix} R_{11} \\ 0 \end{pmatrix}$$

which has the required structure. The second block column of R_b is given by

$$\begin{pmatrix} * \\ R_{22} \\ 0 \end{pmatrix} = A_b B_b = A_b \begin{pmatrix} \overline{B} \\ 0 \end{pmatrix}.$$

This implies that A_b is of the form

$$A_b = \left(\begin{array}{c|c} A_{11} & * \\ A_{21} & * \\ 0 & * \end{array} \right),$$

where $A_{11} \in \mathcal{K}^{\tau_0 \times \tau_0}$, $A_{21} \in \mathcal{K}^{\tau_1 \times \tau_0}$. As $R_{22} = A_{21} B_1$ we have that necessarily A_{21} is positive upper triangular, since both R_{22} and B_1 are positive upper triangular. Considering stepwise all other block columns of R_b shows that A_b has the structure,

$$A_b = \begin{pmatrix} A_{11} & * & * & \cdots & * & \cdots & * \\ A_{21} & * & * & \cdots & * & \cdots & * \\ 0 & A_{32} & \ddots & & \vdots & & \vdots \\ \vdots & \ddots & \ddots & \ddots & \vdots & & \vdots \\ 0 & \cdots & 0 & A_{l,l-1} & * & \cdots & * \end{pmatrix},$$

where $A_{i+1,i} \in \mathcal{K}^{\tau_{i+1} \times \tau_i}$ and positive upper triangular, $i = 1, \ldots, l-1$.

Since the system (A_b, B_b, C_b, D_b) is Lyapunov balanced with Lyapunov grammian $\Sigma_S = \sigma I$ we have that

$$A_b + A_b^* = \frac{-1}{\sigma} B_b B_b^* = \frac{-1}{\sigma} diag(\overline{BB}^*, 0, \ldots, 0).$$

This implies immediately the required structure for A_b. Also, $A_b + A_b^* = \frac{-1}{\sigma} B_b B_b^* = \frac{-1}{\sigma} C_b^* C_b$. Hence $B_b B_b^* = C_b^* C_b$ and therefore $C_b = (\overline{C}, 0)$, for some $\overline{C} \in \mathcal{K}^{p \times \tau_1}$. Thus $\overline{BB}^* = \overline{C}^* \overline{C}$. Writing $U = \overline{C}(\overline{BB}^*)^{-\frac{1}{2}}$, we have

$$U^* U = (\overline{BB}^*)^{-\frac{1}{2}} \overline{C}^* \overline{C} (\overline{BB}^*)^{-\frac{1}{2}} = (\overline{BB}^*)^{-\frac{1}{2}} \overline{BB}^* (\overline{BB}^*)^{-\frac{1}{2}} = I_{\tau_1}.$$

Therefore $\overline{C} = U(\overline{BB}^*)^{\frac{1}{2}}$ has the required structure.

It remains to show that the map Γ is a canonical form. By construction, Definition 1.1, part 1 is satisfied. It remains to show Definition 1.1, part 2. This is the case if we can show that two equivalent systems (A_i, B_i, C_i, D_i), $i = 1, 2$, in Lyapunov σ-block are identical. Both systems (A_i, B_i, C_i, D_i), $i = 1, 2$, are Lyapunov balanced and since both systems are equivalent they have the same Lyapunov grammian $\Sigma_S = \sigma I$ and there exists an orthogonal Q such that $(A_1, B_1, C_1, D_1) = (Q A_2 Q^*, Q B_2, C_2 Q^*, D_2)$. Hence for $R_i = \begin{bmatrix} B_i & A_i B_i & \cdots & A_i^{n-1} B_i \end{bmatrix}$, $i = 1, 2$, we have $R_1 = Q R_2$. Note that by construction both R_i, $i = 1, 2$, are positive upper triangular. The uniqueness statement of Lemma 3.1, therefore implies that $R_1 = R_2$ and $Q = I$, and hence both systems are identical. □

In ([21]) a similar result was given for systems with real coefficients. The work by Hanzon ([11]) contains this result in its version for systems with real coefficients. As a Corollary to Proposition 3.1 and to Theorem 3.1 we obtain the following state space characterization of square σ-inner systems.

To prove the Corollary we need the following Lemma.

Lemma 3.4. *Let (A, B, C, D) be a minimal realization of the square inner rational function G. Let Y, Z be such that*

$$A^* Y + Y A = -C^* C, \quad A Z + Z A^* = -B B^*.$$

Then $YZ = I$.

Proof. Since G is inner, by Lemma 3.2, $C^*D+YB=0$. Since $DD^*=D^*D=I$, $CY^{-1}=-DB^*$. As $A^*Y+YA=-C^*C$ we have

$$AY^{-1}+Y^{-1}A^* = -Y^{-1}C^*CY^{-1} = -(CY^{-1})^*(CY^{-1})$$
$$= -BD^*DB^* = -BB^*.$$

By the uniqueness of the solution to this Lyapunov equation it follows that $Y^{-1}=Z$, which implies the claim. □

Corollary 3.1. *Let $(A_\sigma, B_\sigma, C_\sigma, D_\sigma)$ be a m-input, m-output continuous-time system of dimension n. The following statements are equivalent.*

1. *$(A_\sigma, B_\sigma, C_\sigma, D_\sigma)$ is a minimal σ-inner system.*
2. *$(A_\sigma, B_\sigma, C_\sigma, D_\sigma)= (TAT^{-1}, TB, CT^{-1}, D)$ for some invertible matrix T, where (A,B,C,D) is in the following Lyapunov σ-block form: there exist indices $m=\tau_0\geq\tau_1\geq\cdots\geq\tau_l>0$, with $\sum_{i=1}^{l}\tau_i=n$, $\sigma>0$, such that*

   a) $B=\begin{pmatrix}\overline{B}\\ 0\end{pmatrix}$, *where $\overline{B}$ is a $\tau_1\times\tau_0$ positive upper triangular matrix,*

   b)

$$A=\begin{pmatrix}
\tilde{A}+\mathcal{S}_1 & -\mathcal{A}_1^* & 0 & 0 & 0 & \cdots & 0\\
\mathcal{A}_1 & \mathcal{S}_2 & -\mathcal{A}_2^* & 0 & 0 & \cdots & 0\\
0 & \mathcal{A}_2 & \mathcal{S}_3 & \ddots & \ddots & & \vdots\\
0 & 0 & \ddots & \ddots & \ddots & \ddots & \vdots\\
\vdots & \vdots & \ddots & \ddots & \ddots & -\mathcal{A}_{l-2}^* & 0\\
\vdots & \vdots & & 0 & \mathcal{A}_{l-2} & \mathcal{S}_{l-1} & -\mathcal{A}_{l-1}^*\\
0 & 0 & \cdots & \cdots & 0 & \mathcal{A}_{l-1} & \mathcal{S}_l
\end{pmatrix},$$

   where

$$\mathcal{S}_i,\ \textit{is a } \tau_i\times\tau_i \textit{ skew-hermitian matrix}, i=1,2,\ldots,l,$$

$$\mathcal{A}_i \textit{ is a positive upper triangular } \tau_{i+1}\times\tau_i \textit{ matrix}, i=1,\ldots,l-1;$$

$$\tilde{A}=-\frac{1}{2\sigma}\overline{B}\overline{B}^*,$$

   c)

$$D \quad \textit{is a } m\times m\textit{-matrix such that } D^*D=\sigma^2 I$$

   d)

$$C=-\frac{1}{\sigma}DB^*.$$

Moreover, the system (A,B,C,D) as defined in 2.) is Lyapunov balanced with Lyapunov grammian $\Sigma_S = \sigma I$. The map Γ that assigns to each σ-inner system $(A_\sigma, B_\sigma, C_\sigma, D_\sigma)$ in 1.) the realization in 2.) is a canonical form.

Proof. Let $(A_\sigma, B_\sigma, C_\sigma, D_\sigma)$ is a minimal square σ-inner system. Let Y, Z be the solutions to the Lyapunov equations

$$A_\sigma^* Y + Y A_\sigma = -C_\sigma^* C_\sigma, \; A_\sigma Z + Z A_\sigma^* = -B_\sigma B_\sigma^*.$$

Lemma 3.4 implies that $YZ = \sigma^2 I$. Hence by the Theorem there exists a unique equivalent system (A,B,C,D) that is in Lyapunov σ-block form. This implies that A and B have the stated structure. Since the system is square and σ-inner, it follows by Lemma 3.2 that $D^*D = DD^* = \sigma^2 I$ and $C\Sigma + DB^* = 0$, where $\Sigma = \sigma I$ is the Lyapunov grammian of the system. Hence $C = -\frac{1}{\sigma} DB^*$ and therefore 2.)

Let now (A,B,C,D) be as in 2.). It is necessary to show that the system is in Lyapunov σ-block form. Since $C = -\frac{1}{\sigma} DB^*$, it follows that $C = (\overline{C}, 0)$, where $\overline{C} \in \mathcal{K}^{p \times \tau_1}$ is such that $\overline{C} = -\frac{1}{\sigma} D\overline{B}^* = U(\overline{B}\overline{B}^*)^{\frac{1}{2}}$, with $U := -\frac{1}{\sigma} DB^*(\overline{B}\overline{B}^*)^{-\frac{1}{2}}$ such that $U^*U = I_{\tau_1}$. Hence (A,B,C,D) is in Lyapunov σ-block form and by Proposition 3.1 the system is minimal and σ-inner. This implies 1.) The remaining statements follow from the Theorem. □

In the single-input single-output case this representation simplifies substantially.

Corollary 3.2. *Let $(A_\sigma, b_\sigma, c_\sigma, d_\sigma)$ be a single-input single-output continuous-time system of dimension n. The following statements are equivalent:*

1. *$(A_\sigma, b_\sigma, c_\sigma, d_\sigma)$ is a minimal σ-inner system.*
2. *$(A_\sigma, b_\sigma, b_\sigma, d_\sigma) = (TAT^{-1}, Tb, cT^{-1}, d)$ for some invertible matrix T, where (A,b,c,d) is in the following σ-inner form:*

$$A = \begin{pmatrix} a_{11} + i\beta_1 & -\alpha_1 & 0 & \cdots & \cdots & 0 \\ \alpha_1 & i\beta_2 & -\alpha_2 & 0 & & \vdots \\ 0 & \alpha_2 & i\beta_3 & \ddots & \ddots & \vdots \\ \vdots & \ddots & \ddots & \ddots & -\alpha_{n-2} & 0 \\ \vdots & & 0 & \alpha_{n-2} & i\beta_{n-1} & -\alpha_{n-1} \\ 0 & \cdots & \cdots & 0 & \alpha_{n-1} & i\beta_n \end{pmatrix},$$

$$b = \begin{pmatrix} b_1 \\ 0 \\ \vdots \\ 0 \end{pmatrix}, \quad c = \begin{pmatrix} s_1 b_1, 0, \cdots, 0 \end{pmatrix}, \quad d = -s_1 \sigma,$$

with $\sigma > 0$, $b_1 > 0$, $s_1 \in \mathcal{K}, |s_1| = 1$, $\alpha_i > 0$, $i = 1, \ldots, n-1$, *and* $a_{11} := -\frac{b_1^2}{2\sigma}$. *For* $\mathcal{K} = \Re$, $\beta_i = 0$, $i = 1, 2, \ldots, n$, *and* $s_1 = \pm 1$.

Moreover, the system (A, b, c, d) *as defined in 2.) is Lyapunov balanced with Lyapunov grammian* $\Sigma_S = \sigma I$. *The map that assigns to each* σ*-inner system* $(A_\sigma, b_\sigma, c_\sigma, d_\sigma)$ *in 1.) the realization in 2.) is a canonical form.*

Having analyzed in some depth the canonical form for systems with identical Lyapunov singular values we can now proceed to develop a canonical form for the class $S_n^{p,m}$ of stable minimal systems. The approach will be to reduce the canonical form problem for general systems to the canonical form problem for subsystems with identical singular values. To this end we need to introduce the following definitions.

An m-input p-output system (A, B, C, D) of dimension n is said to be in *balanced form* if there exist so-called *block indices* $n_1, n_2, \ldots, n_k$, $\sum_{i=1}^k n_i = n$ such that if (A, B, C, D) is partitioned as

$$A = \begin{pmatrix} A(1,1) & \cdots & A(1,j) & \cdots & A(1,k) \\ \vdots & & \vdots & & \vdots \\ A(i,1) & \cdots & A(i,j) & \cdots & A(i,k) \\ \vdots & & \vdots & & \vdots \\ A(k,1) & \cdots & A(k,j) & \cdots & A(k,k) \end{pmatrix}, \quad B = \begin{pmatrix} B(1) \\ \vdots \\ B(i) \\ \vdots \\ B(k) \end{pmatrix},$$

$$C = \begin{pmatrix} C(1) & \cdots & C(i) & \cdots & C(k) \end{pmatrix},$$

where $A(i,j) \in \mathcal{K}^{n_i \times n_j}$, $B(j) \in \mathcal{K}^{n_j \times m}$ and $C(i) \in \mathcal{K}^{p \times n_i}$, $1 \le i, j \le k$, we have that

1. the block diagonal systems $(A(i,i), B(i), C(i), D)$ are in σ_i-block form with $\sigma_1 > \cdots > \sigma_i > \cdots > \sigma_k > 0$, and step sizes $m = \tau_0^j \ge \tau_1^j \ge \tau_2^j \ge \cdots \ge \tau_{l_j}^j > 0$, $\sum_{i=1}^{l_j} \tau_i^j = n_j$, $1 \le j \le k$.
2. the block entries $A(i,j)$, $1 \le i, j \le k$, $i \ne j$, are given by

$$A(i,j) = \begin{pmatrix} \tilde{A}_{ij} & 0 \\ 0 & 0 \end{pmatrix},$$

where $\tilde{A}_{ij} \in \mathcal{K}^{\tau_1^i \times \tau_1^j}$ is a function of σ_i, σ_j, $U(i)$, $U(j)$, $\overline{B}(i)$, $\overline{B}(j)$ and D, $1 \le i, j \le k$, $i \ne j$.

The specific function that is chosen in 2.) will depend on the type of balancing that we study. If a system (A, B, C, D) is in balanced form such that the block diagonal systems $(A(i,i), B(i), C(i), D)$ are in Lyapunov σ_i-block form and

$$\tilde{A}_{ij} = \frac{1}{\sigma_i^2 - \sigma_j^2}\left(\sigma_j \overline{B}_i \overline{B}_j^* - \sigma_i (\overline{B}_i \overline{B}_i^*)^{\frac{1}{2}} U_i^* U_j (\overline{B}_j \overline{B}_j^*)^{\frac{1}{2}}\right),$$

for $1 \leq i,j \leq k$, $i \neq j$, then the system is said to be in *Lyapunov balanced form* with Lyapunov singular values $\sigma_1, \ldots, \sigma_k$.

The following proposition states an interesting *augmentation property* of linear systems whose corresponding Lyapunov equations have positive definite solutions. If the system is partitioned into two sets of states and the corresponding block diagonal subsystems are stable and minimal, then the system itself is stable and minimal provided a weak technical condition is satisfied.

Note that if the system (A, B, C, D) is in Lyapunov balanced form then the block diagonal subsystems $(A(i,i), B(i), C(i), D)$ are stable, minimal and Lyapunov balanced with Lyapunov grammian $\sigma_i I_{n_i}$, $i = 1, \ldots, k$. As will be shown in the proof of Theorem 3.2, the definition of Lyapunov balanced form is such that the Proposition can be immediately applied to a system being in Lyapunov balanced form.

Proposition 3.2. *(***Kabamba [15]***) Let*

$$(A, B, C, D) = \left(\begin{pmatrix} A_{11} & A_{12} \\ A_{21} & A_{22} \end{pmatrix}, \begin{pmatrix} B_1 \\ B_2 \end{pmatrix}, \begin{pmatrix} C_1 & C_2 \end{pmatrix}, D \right),$$

be a n-dimensional linear system that is conformally partitioned, i.e. (A_{11}, B_1, C_1, D) is a k-dimensional continuous-time system, $0 < k < n$. Assume that there exist positive definite $n \times n$ matrices $P = diag(P_1, P_2)$, $Q = diag(Q_1, Q_2)$, where P_1, Q_1 are $k \times k$, such that the sets of eigenvalues of $P_1 Q_1$ and of $P_2 Q_2$ have zero intersection. If further

$$AP + PA^* = -BB^*, \quad A^*Q + QA = -C^*C,$$

and (A_{ii}, B_i, C_i, D), $i = 1, 2$, are minimal then (A, B, C, D) is stable and minimal.

Proof. Note that for $i = 1, 2$,

$$A_{ii}P_i + P_i A_{ii}^* = -B_i B_i^*,$$

and therefore by Lemma 2.1 the reachability of (A_{ii}, B_i, C_i, D), $i = 1, 2$, also implies the stability of the systems.

Assume now that A is not stable. Therefore there exists $x \in \mathcal{K}^n$, $x \neq 0$, $\lambda \in \mathcal{C}$, with $Re(\lambda) \geq 0$ such that

$$A^*x = \lambda x.$$

Now consider

$$0 \geq - < BB^*x, x > = - < B^*x, B^*x >$$

$$= < (AP + PA^*)x, x > = < PA^*x, x > + < x, PA^*x >$$

$$= 2Re(\lambda) < x, Px > .$$

Since $P > 0$ this inequality can only hold if $Re(\lambda) = 0$ and $B^*x = 0$. Hence $\lambda = i\omega$ for some $\omega \in \Re$. Multiplying the Lyapunov equation on the right by x we obtain

$$0 = BB^*x = APx + PA^*x = APx + \lambda x.$$

Hence

$$APx = -\lambda Px,$$

i.e. Px is an eigenvector of A with eigenvalue $-\lambda$. Using the second Lyapunov equation the above reasoning applied to (A, C) shows that $C^*Px = 0$ and

$$A^*QPx = -QAPx = \lambda QPx,$$

i.e. QPx is also an eigenvector of A^* with the same eigenvalue λ. Hence the eigenspace of A^* corresponding to the eigenvalue λ is invariant under the transformation $QP = diag(Q_1P_1, Q_2P_2)$.

Let $\mathcal{K}^n = X_1 \oplus X_2$ be the orthogonal decomposition of the state-space that gives rise to the block partitioning of the state-space matrices and the matrices P and Q. Due to the block structure of QP and the fact that the eigenvalues of Q_1P_1 and of Q_2P_2 are different, we have that if E is an invariant subspace of QP, then either $E \subseteq X_1$ or $E \subseteq X_2$. Therefore also the eigenspace of A^* with eigenvalue λ has this form. Hence $x = \begin{pmatrix} x_1 \\ 0 \end{pmatrix}$, for some $x_1 \in X_1$, $x_1 \neq 0$, or $x = \begin{pmatrix} 0 \\ x_2 \end{pmatrix}$, for some $x_2 \in X_2$, $x_2 \neq 0$. If $x = \begin{pmatrix} x_1 \\ 0 \end{pmatrix}$ then

$$A^*x = \begin{pmatrix} A_{11}^* & A_{21}^* \\ A_{12}^* & A_{22}^* \end{pmatrix} \begin{pmatrix} x_1 \\ 0 \end{pmatrix} = \begin{pmatrix} A_{11}^* x_1 \\ A_{12}^* x_1 \end{pmatrix} = \begin{pmatrix} \lambda x_1 \\ 0 \end{pmatrix}$$

and hence $A_{11}^* x_1 = \lambda x_1$, which is a contradiction to the stability of A_{11}. Similarly if $x = \begin{pmatrix} 0 \\ x_2 \end{pmatrix}$ we obtain a a contradiction to the stability of A_{22}. Hence we have the stability of A. Since the Lyapunov equations corresponding to the system have the positive definite solutions P and Q we therefore also have by Lemma 2.1 the reachability and the observability of the system (A, B, C, D). □

We can now give a canonical form and a parametrization result for minimal and stable systems of dimension n, i.e. for systems in the class $S_n^{p,m}$. This canonical form is called *Lyapunov balanced canonical form.* The other implication of the Theorem shows the parametrization result that each system in Lyapunov balanced form is automatically stable, minimal and Lyapunov balanced.

Theorem 3.2. *Let (A_s, B_s, C_s, D_s) be a m-input p-output continuous-time system of dimension n. Then the following are equivalent:*

1. (A_s, B_s, C_s, D_s) *is a stable and minimal system, i.e. is in* $S_n^{p,m}$.
2. $(A_s, B_s, C_s, D_s) = (TAT^{-1}, TB, CT^{-1}, D)$ *for some invertible* T, *where* (A, B, C, D) *is in Lyapunov balanced form, i.e. there exist block indices* $n_1, \ldots, n_k$, $\sum_{j=1}^{k} n_j = n$, *parameters* $\sigma_1 > \sigma_2 > \cdots > \sigma_k > 0$ *and* k *families of step sizes* $m = \tau_0^j \geq \tau_1^j \geq \tau_2^j \geq \cdots \geq \tau_{l_j}^j > 0$, $\sum_{i=1}^{l_j} \tau_i^j = n_j$, $1 \leq j \leq k$, *such that*
   a) $B = (\underbrace{\overline{B}_1^T, 0}_{n_1}, \underbrace{\overline{B}_2^T, 0}_{n_2}, \ldots, \underbrace{\overline{B}_k^T, 0}_{n_k})^T$, *where* $\overline{B}_j \in \mathcal{K}^{\tau_1^j \times \tau_0^j}$ *is positive upper triangular,* $1 \leq j \leq k$.
   b) $C = (\underbrace{U_1(\overline{B}_1\overline{B}_1^*)^{\frac{1}{2}}, 0}_{n_1}, \underbrace{U_2(\overline{B}_2\overline{B}_2^*)^{\frac{1}{2}}, 0}_{n_2}, \ldots, \underbrace{U_k(\overline{B}_k\overline{B}_k^*)^{\frac{1}{2}}, 0}_{n_k})$, *where* $U_j \in \mathcal{K}^{p \times \tau_1^j}$, $U_j^* U_j = I_{\tau_1^j}$, $1 \leq j \leq k$.
   c)

$$A = \left(\begin{array}{c|c|c|c|c} A(1,1) & \cdots & A(1,i) & \cdots & A(1,k) \\ \hline \vdots & \ddots & \vdots & \vdots & \vdots \\ \hline A(i,1) & \cdots & A(i,i) & \cdots & A(i,k) \\ \hline \vdots & & \vdots & \ddots & \vdots \\ \hline A(k,1) & \cdots & A(k,i) & \cdots & A(k,k) \end{array}\right),$$

where for $1 \leq i \leq k$

$$A(i,i) = \begin{pmatrix} \tilde{A}_{ii} + \mathcal{S}_1^i & (-\mathcal{A}_1^i)^* & & 0 \\ \mathcal{A}_1^i & \mathcal{S}_2^i & \ddots & \\ & \ddots & \ddots & (-\mathcal{A}_{l_i-1}^i)^* \\ 0 & & \mathcal{A}_{l_i-1}^i & \mathcal{S}_{l_i}^i \end{pmatrix},$$

for $1 \leq i, j < k$, $i \neq j$

$$A(i,j) = \begin{pmatrix} \tilde{A}_{ij} & 0 & \cdots & 0 \\ 0 & 0 & \cdots & 0 \\ \vdots & \vdots & & \vdots \\ 0 & 0 & \cdots & 0 \end{pmatrix},$$

and

i. $\mathcal{S}_i^j$ *is a* $\tau_i^j \times \tau_i^j$ *skew-hermitian matrix,* $i = 1, 2, \ldots, l_j$, $1 \leq j \leq k$.
ii. $\mathcal{A}_i^j$ *is a positive upper triangular* $\tau_{i+1}^j \times \tau_i^j$ *matrix,* $i = 1, 2, \ldots, l_j - 1$, $1 \leq j \leq k$.
iii. $\tilde{A}_{ij} \in \mathcal{K}^{\tau_1^i \times \tau_1^j}$ *is given by*

$$\tilde{A}_{ij} = \frac{1}{\sigma_i^2 - \sigma_j^2}(\sigma_j \overline{B}_i \overline{B}_j^* - \sigma_i (\overline{B}_i \overline{B}_i^*)^{\frac{1}{2}} U_i^* U_j (\overline{B}_j \overline{B}_j^*)^{\frac{1}{2}}),$$

$$\textit{for } 1 \leq i,j \leq k, \quad i \neq j, \quad \textit{and}$$
$$\tilde{A}_{ii} = -\frac{1}{2\sigma_i}\overline{B_i}\overline{B}_i^* \quad \textit{for } 1 \leq i \leq k.$$

Moreover, the system (A,B,C,D) as defined in (2) is Lyapunov balanced with Lyapunov grammian $\Sigma_S = diag(\sigma_1 I_{n_1},\ldots,\sigma_k I_{n_k})$, where $n_1,\ldots,n_k$ are the block indices and $\sigma_1 > \cdots > \sigma_k > 0$ are the Lyapunov singular values of the system in (2). The map Γ_S that assigns to each system in $S_n^{p,m}$ the realization in (2) is a canonical form.

Proof. Assume that (A_s,B_s,C_s,D_s) is a stable and minimal system. We can assume without loss of generality that the system is Lyapunov balanced with Lyapunov grammian

$$\Sigma_S = diag(\sigma_1 I_{n_1}, \sigma_2 I_{n_2}, \ldots, \sigma_k I_{n_k}), \quad \sigma_1 > \sigma_2 > \cdots > \sigma_k > 0.$$

By Theorem 1.1 all other Lyapunov balanced realizations can be obtained by a state space transformation of the form $Q = diag(Q_1, Q_2, \ldots, Q_k)$, $Q_j \in \mathcal{K}^{n_j \times n_j}, j = 1,2,\ldots,k$.

Let (A_s,B_s,C_s,D_s) be partitioned according to the block indices $n_1, n_2, \ldots, n_k$, such that

$$\begin{aligned} A_s &= (A_s(i,j))_{1\leq i,j\leq k}, & A_s(i,j) &\in \mathcal{K}^{n_i \times n_j}, \ 1 \leq i,j \leq k, \\ B_s &= (B_s(1)^T, B_s(2)^T, \cdots, B_s(k)^T)^T, & B_s(j) &\in \mathcal{K}^{n_j \times m}, \ 1 \leq j \leq k, \\ C_s &= (C_s(1), C_s(2), \cdots, C_s(k)), & C_s(i) &\in \mathcal{K}^{p \times n_i}, \ 1 \leq i \leq k. \end{aligned}$$

By Theorem 2.1 the block diagonal subsystems $(A_s(j,j), B_s(j), C_s(j), D_s)$ are stable, minimal and Lyapunov balanced with Lyapunov grammian $\sigma_j I_{n_j}$, $1 \leq j \leq k$. By Theorem 3.1 there exists a unique unitary $Q_j \in \mathcal{K}^{n_j \times n_j}$, such that

$$(A(j,j), B(j), C(j), D) := (Q_j A_s(j,j) Q_j^*, Q_j B_s(j), C_s(j) Q_j^*, D_s)$$

is in Lyapunov σ_j-block form, $1 \leq j \leq k$. Then $(A,B,C,D) := (QA_sQ^*, QB_s, C_sQ^*, D_s)$ where $Q := diag(Q_1, Q_2, \ldots, Q_k)$ is uniquely determined, and the block diagonal subsystems have the desired structure. The system is Lyapunov balanced with Lyapunov grammian $\Sigma = \Sigma_S$. To conclude this part of the proof it remains to be shown that the off-diagonal block matrices of A have the stated representation. Let (A,B,C,D) be partitioned according to the block indices $n_1, n_2, \ldots, n_k$, i.e.

$$\begin{aligned} A &= (A(i,j))_{1\leq i,j\leq k}, & A(i,j) &\in \mathcal{K}^{n_i \times n_j}, \ 1 \leq i,j \leq k, \\ B &= (B(1)^T, B(2)^T, \cdots, B(k)^T)^T, & B(j) &\in \mathcal{K}^{n_j \times m}, \ 1 \leq j \leq k, \\ C &= (C(1), C(2), \cdots, C(k)), & C(i) &\in \mathcal{K}^{p \times n_i}, \ 1 \leq i \leq k. \end{aligned}$$

Let $1 \leq i,j \leq k, i \neq j$, and consider the (i,j) block entry of the Lyapunov equations

$$A^*\Sigma + \Sigma A = -C^*C, \qquad A\Sigma + \Sigma A^* = -BB^*,$$

i.e.

$$A(j,i)^*\sigma_j + \sigma_i A(i,j) = -C(i)^*C(j), \quad A(i,j)\sigma_j + \sigma_i A(j,i)^* = -B(i)B(j)^*,$$

or

$$\begin{bmatrix} \sigma_i & \sigma_j \\ \sigma_j & \sigma_i \end{bmatrix} \begin{bmatrix} A(i,j) \\ A(i,j)^* \end{bmatrix} = - \begin{bmatrix} C(i)^*C(j) \\ B(i)B(j)^* \end{bmatrix}.$$

Since by assumption $\sigma_i \neq \sigma_j$, this equation can be solved to give

$$A(i,j) = \frac{1}{\sigma_i^2 - \sigma_j^2}[\sigma_j B(i)B(j)^* - \sigma_i C(i)^*C(j)].$$

The structure of $B(i), B(j), C(i)$ and $C(j)$ shows that

$$A(i,j) = \begin{bmatrix} \tilde{A}_{ij} & 0 \\ 0 & 0 \end{bmatrix},$$

where

$$\tilde{A}_{ij} := \frac{1}{\sigma_i^2 - \sigma_j^2}[\sigma_j \overline{B}_i \overline{B}_j^* - \sigma_i \overline{C}_i^* \overline{C}_j] \in \mathcal{K}^{\tau_1^i \times \tau_1^j}.$$

Hence 2.)

Now assume 2.). By construction of (A, B, C, D), $\Sigma_S = diag(\sigma_1 I_{n_1}, \sigma_2 I_{n_2}, \cdots, \sigma_k I_{n_k})$ solves the Lyapunov equations

$$A^*\Sigma + \Sigma A = -C^*C, \qquad A\Sigma + \Sigma A^* = -BB^*.$$

Let (A, B, C, D) be partitioned according to the block indices. By construction the block diagonal subsystems are in Lyapunov σ_i-block form, $1 \leq i \leq k$. Hence Proposition 3.1 implies that they are stable and minimal. Hence the system (A, B, C, D) is stable and minimal by Proposition 3.2. Since Σ_S solves the two Lyapunov equations the system is balanced with Lyapunov grammian Σ_S.

□

Specialization of the theorem to the single input single output case gives the following Corollary.

Corollary 3.3. *Let (A_s, b_s, c_s, d_s) be a single-input single-output continuous-time system of dimension n. Then the following statements are equivalent:*

1. *(A_s, b_s, c_s, d_s) is a stable minimal system.*
2. *$(A_s, b_s, c_s, d_s) = (TAT^{-1}, Tb, cT^{-1}, d)$ for some invertible matrix T, where (A, b, c, d) is in the following Lyapunov balanced form with block indices $n_1, n_2, \ldots, n_k$, $\sum_{i=1}^k n_i = n$.*

a) $b = (\underbrace{b_1, 0, \ldots, 0}_{n_1}, \underbrace{b_2, 0, \ldots, 0}_{n_2}, \ldots, \underbrace{b_k, 0, \ldots, 0}_{n_k})^T$, *where* $b_i > 0$, $1 \leq i \leq k$.

b) $c = (\underbrace{s_1 b_1, 0, \ldots, 0}_{n_1}, \underbrace{s_2 b_2, 0, \ldots, 0}_{n_2}, \ldots, \underbrace{s_k b_k, 0, \ldots, 0}_{n_k})$, *where* $s_i \in \mathcal{K}$, $|s_i| = 1$, $1 \leq i \leq k$.

c)

$$A = \left(\begin{array}{c|c|c|c} A(1,1) & A(1,2) & \cdots & A(1,k) \\ \hline A(2,1) & A(2,2) & \cdots & A(2,k) \\ \hline \vdots & & \ddots & \vdots \\ \hline A(k,1) & A(k,2) & \cdots & A(k,k) \end{array} \right),$$

where for $1 \leq j \leq k$,

$$A(j,j) = \begin{pmatrix} a_{jj} + i\beta_1^j & -\alpha_1^j & & 0 \\ \alpha_1^j & i\beta_2^j & \ddots & \\ & \ddots & \ddots & -\alpha_{n_j-1}^j \\ 0 & & \alpha_{n_j-1}^j & i\beta_{n_j}^j \end{pmatrix},$$

for $1 \leq i, j \leq k$, $i \neq j$,

$$A(i,j) = \begin{pmatrix} a_{ij} & 0 & \cdots & 0 \\ 0 & 0 & \cdots & 0 \\ \vdots & \vdots & & \vdots \\ 0 & 0 & \cdots & 0 \end{pmatrix},$$

and

$$\alpha_j^i > 0 \text{ for } 1 \leq j \leq n_i - 1, \quad 1 \leq i \leq k,$$

$$a_{ij} = \frac{\sigma_j - \bar{s}_i s_j \sigma_i}{\sigma_i^2 - \sigma_j^2} b_i b_j \text{ for } 1 \leq i, j \leq k, \quad i \neq j,$$

$$a_{ii} = -\frac{1}{2\sigma_i} b_i^2 \text{ for } 1 \leq i \leq k.$$

$$\sigma_1 > \sigma_2 > \cdots > \sigma_k > 0.$$

If $\mathcal{K} = \Re$, *then*

$$s_i = \pm 1, \text{ for } i = 1, 2, \ldots, k,$$

$$\beta_j^i = 0, \text{ for } 1 \leq j \leq n_i, \quad 1 \leq i \leq k,$$

$$a_{ij} = \frac{-b_i b_j}{s_i s_j \sigma_i + \sigma_j}, \text{ for } 1 \leq i, j \leq k.$$

Moreover, the system (A, b, c, d) *as defined in* (2) *is Lyapunov balanced with Lyapunov grammian* $\Sigma_S = diag(\sigma_1 I_{n_1}, \sigma_2 I_{n_2}, \ldots, \sigma_k I_{n_k})$. *The map* Γ_S *that assigns to each system in* $S_n^{1,1}$ *the realization in* (2) *is a canonical form.*

Proof. The proof follows by specializing the previous theorem to the case $m = p = 1$ and renaming the parameters $\overline{B}_i$ by b_i, $1 \le i \le k$; S_i^j by $\sqrt{-1}\beta_i^j$, $i = 1, \ldots, l_j$, $1 \le j \le k$, and U_i by S_i, $i = 1, \ldots, k$. □

We can now reconsider the model reduction problem for Lyapunov balanced systems. In Theorem 2.1 we showed that a k-dimensional balanced approximant of a Lyapunov balanced system with Lyapunov grammian $\Sigma = diag(\sigma_1, \ldots, \sigma_r, \sigma_{r+1}, \ldots, \sigma_n)$ is stable minimal and Lyapunov balanced if $\sigma_r > \sigma_{r+1}$. If the system is in Lyapunov balanced canonical form the same result holds without the condition on the point of truncation.

Theorem 3.3. *Let $(A, B, C, D) \in S_n^{p,m}$ be in Lyapunov balanced canonical form with Lyapunov grammian Σ. Let $1 \le r < n$. Then the r-dimensional balanced approximant (A_1, B_1, C_1, D) is in $S_r^{p,m}$ and is in Lyapunov balanced canonical form with Lyapunov grammian Σ_1 where $\Sigma = diag(\Sigma_1, \Sigma_2)$, $\Sigma_1 \in \mathcal{K}^{r \times r}$.*

Proof. Let $(\hat{A}, \hat{B}, \hat{C}, \hat{D})$ be the r-dimensional balanced approximant of (A, B, C, D). Let $\hat{\Sigma}$ be the $r \times r$ principal submatrix of Σ. It is easily checked that

$$\hat{A}\hat{\Sigma} + \hat{\Sigma}\hat{A}^* = -\hat{B}\hat{B}^*,$$
$$\hat{A}^*\hat{\Sigma} + \hat{\Sigma}\hat{A} = -\hat{C}^*\hat{C}.$$

Let $\Sigma = diag(\sigma_1 I_{n_1}, \sigma_2 I_{n_2}, \ldots, \sigma_k I_{n_k})$, $\sigma_1 > \sigma_2 > \cdots > \sigma_k$. If $r = n_1 + \cdots + n_l$ for some $1 \le l < k$, then the result follows by Theorem 2.1 or the immediately verified fact that $(\hat{A}, \hat{B}, \hat{C}, \hat{D})$ admits a Lyapunov balanced parametrization. If $n_1 + \cdots + n_l < r < n_1 + \cdots + n_{l+1}$, for some $1 \le l < k - 1$, we will also show that $(\hat{A}, \hat{B}, \hat{C}, \hat{D})$ admits a Lyapunov balanced parametrization.

Partition $(\hat{A}, \hat{B}, \hat{C}, \hat{D})$ such that

$$\hat{A} = \begin{pmatrix} \hat{A}_{11} & \hat{A}_{12} \\ \hat{A}_{21} & \hat{A}_{22} \end{pmatrix}, \quad \hat{B} = \begin{pmatrix} \hat{B}_1 \\ \hat{B}_2 \end{pmatrix}, \quad \hat{C} = \begin{pmatrix} \hat{C}_1 & \hat{C}_2 \end{pmatrix},$$

$$\hat{\Sigma} = diag(\hat{\Sigma}_1, \hat{\Sigma}_2),$$

such that $(\hat{A}_{11}, \hat{B}_1, \hat{C}_1, \hat{D}_1)$ is a $n_1 + \cdots + n_l$-dimensional system and $\hat{\Sigma}_2 = \sigma_{l+1} I_{r-(n_1+\cdots+n_l)}$. Then

$$\hat{A}_{ii}\hat{\Sigma}_i + \hat{\Sigma}_i\hat{A}_{ii}^* = -\hat{B}_i\hat{B}_i^*,$$
$$\hat{A}_{ii}^*\hat{\Sigma}_i + \hat{\Sigma}_i\hat{A}_{ii} = -\hat{C}_i^*\hat{C}_i,$$

for $i = 1, 2$, and $(\hat{A}_{11}, \hat{B}_1, \hat{C}_1, \hat{D})$ admits a Lyapunov balanced parametrization. If we show that $(\hat{A}_{22}, \hat{B}_2, \hat{C}_2, \hat{D})$ is in Lyapunov σ_{l+1}-block form, then $(\hat{A}, \hat{B}, \hat{C}, \hat{D})$ is in Lyapunov balanced form since $\hat{\Sigma}$ solves the observability and reachability Lyapunov equation for $(\hat{A}, \hat{B}, \hat{C}, \hat{D})$. Then the result will follow from Theorem 3.2.

It follows by inspection that $\hat{B}_2$ has the structure of a B-matrix of a system in Lyapunov σ_{l+1}-block form. Also by inspection and since

$$\hat{A}_{22}\hat{\Sigma}_2 + \hat{\Sigma}_2\hat{A}_{22}^* = -\hat{B}_2\hat{B}_2^*$$

it follows that $\hat{A}_{22}$ has the structure of a A-matrix of a system in Lyapunov σ_{l+1}-block form. Hence $(\hat{A}_{22}, \hat{B}_2, \hat{C}_2, \hat{D})$ is stable and reachable, by Proposition 3.1. Since

$$\hat{A}_{22}^*\hat{\Sigma}_2 + \hat{\Sigma}_2\hat{A}_{22} = -\hat{C}_2\hat{C}_2$$

it follows by Corollary 2.1 that $(\hat{A}_{22}, \hat{B}_2, \hat{C}_2, \hat{D})$ is also observable. Hence $(\hat{A}_{22}, \hat{B}_2, \hat{C}_2, \hat{D})$ is stable, minimal and in Lyapunov balanced with Lyapunov grammian $\hat{\Sigma}_2$. Moreover, because of the structure of $\hat{A}_{22}$ and $\hat{B}_2$, the system is in Lyapunov σ_{l+1}-block form.

□

4. L-Characteristic, LQG-Balanced Canonical Form and Model Reduction for Minimal Systems

In this section a canonical form will be given for the class $L_n^{p,m}$ of minimal systems of fixed McMillian degree. The canonical form is defined in terms of LQG-balanced realizations. This canonical form is derived from the Lyapunov-balanced canonical form for stable systems using the L-characteristic, a bijection between the class $L_n^{p,m}$ of minimal systems of McMillian degree n and the class $S_n^{p,m}$ of stable minimal systems of McMillian degree n. A substantial part of the section will be devoted to the introduction and analysis of this bijection. This bijection is here introduced using a state space formulation. The analysis of the L-characteristic will be followed by the derivation of the LQG-balanced canonical form, a parametrization result for $L_n^{p,m}$ and an investigation of the model reduction properties of LQG balanced systems.

In the following definition the L-characteristic is introduced for a system in $L_n^{p,m}$. For a system in $S_n^{p,m}$ the inverse L-characteristic will be defined. Part of this section is devoted to show that the L-characteristic defines a bijection from $L_n^{p,m}$ to $S_n^{p,m}$ whose inverse is the inverse L-characteristic.

Throughout this section we will use the following abbreviations. Let (A, B, C, D) be a system, then set $R_L := I + D^*D$, $S_L := I + DD^*$ and $A_L := A - BR_L^{-1}D^*C$.

Definition 4.1. *1. Let (A, B, C, D) be a minimal system. Let Y be the stabilizing solution of the control algebraic Riccati equation and let Z be the stabilizing solution of the filter algebraic Riccati equation, i.e.*

$$0 = A_L^* Y + Y A_L - Y B R_L^{-1} B^* Y + C^* S_L^{-1} C$$
$$0 = A_L Z + Z A_L^* - Z C^* S_L^{-1} C Z + B R_L^{-1} B^*$$

with $A_L - BR_L^{-1}B^*Y$ and $A_L - ZC^*S_L^{-1}C$ stable. Then the system

$$\begin{aligned}(\mathcal{A}, \mathcal{B}, \mathcal{C}, \mathcal{D}) &:= \chi_L((A, B, C, D)) \\ &:= (A_L - BR_L^{-1}B^*Y, BR_L^{-\frac{1}{2}}, S_L^{-\frac{1}{2}}C(I + ZY), D) \\ &= ((I + ZY)^{-1}(A_L - ZC^*S_L^{-1}C)(I + ZY), BR_L^{-\frac{1}{2}}, S_L^{-\frac{1}{2}}C(I + ZY), D)\end{aligned}$$

is called the L-characteristic *of the system.*

2. Let $(\mathcal{A}, \mathcal{B}, \mathcal{C}, \mathcal{D})$ be a stable minimal system and let P and Q be the solutions to the Lyapunov equations

$$\mathcal{A}P + P\mathcal{A}^* = -\mathcal{B}\mathcal{B}^*, \quad \mathcal{A}^*Q + Q\mathcal{A} = -\mathcal{C}^*\mathcal{C}.$$

Then the system

$$\begin{aligned}(A, B, C, D) &:= I\chi_L((\mathcal{A}, \mathcal{B}, \mathcal{C}, \mathcal{D})) \\ &= (\mathcal{A} + \mathcal{B}(\mathcal{B}^*Q + \mathcal{D}^*\mathcal{C})(I + PQ)^{-1}, \mathcal{B}R_L^{\frac{1}{2}}, S_L^{\frac{1}{2}}\mathcal{C}(I + PQ)^{-1}, \mathcal{D}) \\ &= \Big((I + PQ)(\mathcal{A} + (I + PQ)^{-1}(P\mathcal{C}^* + \mathcal{B}\mathcal{D}^*)\mathcal{C})(I + PQ)^{-1}, \\ &\qquad \mathcal{B}R_L^{\frac{1}{2}}, S_L^{\frac{1}{2}}\mathcal{C}(I + PQ)^{-1}, \mathcal{D}\Big)\end{aligned}$$

is called the inverse L-characteristic *of the system $(\mathcal{A}, \mathcal{B}, \mathcal{C}, \mathcal{D})$. Here $R_L := I + \mathcal{D}^*\mathcal{D}$ and $S_L := I + \mathcal{D}\mathcal{D}^*$.*

Note that both expressions for the L-characteristic are identical because of the Bucy relations given in the following Lemma.

Lemma 4.1. *Let (A, B, C, D) be a minimal system. Then*

$$(I + ZY)(A_L - BR^{-1}B^*Y) = (A_L - ZC^*S^{-1}C)(I + ZY),$$

where Y is the stabilizing solution to the control algebraic Riccati equation and Z is the stabilizing solution to the filter algebraic Riccati equation.

Proof. Consider the two Riccati equation,

$$0 = A_L^* Y + Y A_L - Y B R^{-1} B^* Y + C^* S^{-1} C,$$
$$0 = A_L Z + Z A_L^* - Z C^* R^{-1} C Z + B S^{-1} B^*.$$

Multiplying the first equation on the left by Z and the second equation on the right by Y, equating both equations and adding A_L to both sides we obtain

$$A_L + Z A_L^* Y + Z Y A_L - Z Y B R^{-1} B^* Y + Z C^* S^{-1} C$$
$$= A_L + A_L Z Y + Z A_L^* Y - Z C^* S^{-1} C Z Y + B R^{-1} B^* Y.$$

Canceling the term $Z A_L^* Y$ from either side and collecting terms, we obtain

$$(I + ZY)(A_L - BR^{-1}B^*Y) = (A_L - ZC^*S^{-1}C)(I + ZY).$$

□

That the two expressions for the inverse L-characteristic are identical follows from the following Lemma.

Lemma 4.2. *Let* $(\mathcal{A}, \mathcal{B}, \mathcal{C}, \mathcal{D}) \in S_n^{p,m}$ *and let* P *and* Q *be such that*

$$\mathcal{A}P + P\mathcal{A}^* = -\mathcal{B}\mathcal{B}^*, \quad \mathcal{A}^*Q + Q\mathcal{A} = -\mathcal{C}^*\mathcal{C}.$$

Then

$$[\mathcal{A} + \mathcal{B}(\mathcal{B}^*Q + \mathcal{D}^*\mathcal{C})(I + PQ)^{-1}][I + PQ] =$$
$$[I + PQ][\mathcal{A} + (I + PQ)^{-1}(P\mathcal{C}^* + \mathcal{B}\mathcal{D}^*)\mathcal{C}]$$

and

$$[\mathcal{A} + \mathcal{B}\mathcal{B}^*Q(I + PQ)^{-1}][I + PQ] = [I + PQ][\mathcal{A} + (I + PQ)^{-1}P\mathcal{C}^*\mathcal{C}]$$

Proof. We have

$$[\mathcal{A} + \mathcal{B}(\mathcal{B}^*Q + \mathcal{D}^*\mathcal{C})(I + PQ)^{-1}][I + PQ] = \mathcal{A}(I + PQ) + \mathcal{B}(\mathcal{B}^*Q + \mathcal{D}^*\mathcal{C})$$
$$= \mathcal{A} + \mathcal{A}PQ + \mathcal{B}\mathcal{B}^*Q + \mathcal{B}\mathcal{D}^*C = \mathcal{A} + (\mathcal{A}P + \mathcal{B}\mathcal{B}^*)Q + \mathcal{B}\mathcal{D}^*C$$
$$= \mathcal{A} + (-P\mathcal{A}^*)Q + \mathcal{B}\mathcal{D}^*\mathcal{C} = \mathcal{A} - P(-Q\mathcal{A} - \mathcal{C}^*\mathcal{C}) + \mathcal{B}\mathcal{D}^*\mathcal{C}$$
$$= \mathcal{A} + PQ\mathcal{A} + (P\mathcal{C}^* + \mathcal{B}\mathcal{D}^*)\mathcal{C}$$
$$= [I + PQ][\mathcal{A} + (I + PQ)^{-1}(P\mathcal{C}^* + \mathcal{B}\mathcal{D}^*)\mathcal{C}].$$

which shows the first identity. Subtracting $\mathcal{B}\mathcal{D}^*\mathcal{C}$ from either side implies the second identity. □

The following Lemma shows that both the L-characteristic and inverse L-characteristic preserve the minimality and equivalence of systems.

Lemma 4.3. *1. The L-characteristic of a minimal system is stable and minimal. The L-characteristics of two equivalent systems are equivalent.*
2. The inverse L-characteristic of a stable minimal system is minimal. The inverse L-characteristics of two equivalent systems are equivalent.

Proof. 1.) Since Y is the stabilizing solution of the control algebraic Riccati equation the matrix $A_L - BR_L^{-1}B^*Y$ is stable by definition. It is easily seen that the characteristic system is reachable. The observability of the system follows by the second representation of the characteristic.

Let $(A, B, C, D) \in L_n^{p,m}$. If Z is the stabilizing solution to the Riccati equation,

$$A_L Z + ZA_L^* - ZC^*S_L^{-1}CZ + BR_L^{-1}B^* = 0,$$

then TZT^* is the stabilizing solution to this Riccati equation for the system $(TAT^{-1}, TB, CT^{-1}, D)$, where T is non-singular. Similarly, if Y is the stabilizing solution to

$$A_L^*Y + YA_L - YBR_L^{-1}B^*Y + C^*S_L^{-1}C = 0,$$

then $T^{-*}YT^{-1}$ is the stabilizing solution to this Riccati equation for the system $(TAT^{-1}, TB, CT^{-1}, D)$. Using this fact it is easily seen that the L-characteristics of two equivalent systems are equivalent.

2.) The proof is similar to the proof of 1.). □

The L-characteristic and the inverse L-characteristic have interesting properties concerning the way solutions of Riccati equations respectively Lyapunov equations are mapped under the characteristic maps.

Proposition 4.1. *1. Let (A, B, C, D) be a minimal system and let Y be the stabilizing solution to the control algebraic Riccati equation and let Z be the stabilizing solution to the filter algebraic Riccati equation. Let $(\mathcal{A}, \mathcal{B}, \mathcal{C}, \mathcal{D})$ be the L-characteristic of (A, B, C, D). Then the Lyapunov equations*

$$\mathcal{A}P + P\mathcal{A}^* = -\mathcal{B}\mathcal{B}^*, \quad \mathcal{A}^*Q + Q\mathcal{A} = -\mathcal{C}^*\mathcal{C}.$$

have solutions

$$P := (I + ZY)^{-1}Z = Z(I + YZ)^{-1}, \quad Q := Y + YZY.$$

2. Let $(\mathcal{A}, \mathcal{B}, \mathcal{C}, \mathcal{D})$ be a stable minimal system. Let P, Q be the positive definite solutions to the Lyapunov equations

$$\mathcal{A}P + P\mathcal{A}^* = -\mathcal{B}\mathcal{B}^*, \quad \mathcal{A}^*Q + Q\mathcal{A} = -\mathcal{C}^*\mathcal{C}.$$

Let (A, B, C, D) be the inverse L-characteristic of $(\mathcal{A}, \mathcal{B}, \mathcal{C}, \mathcal{D})$. Then

$$Y := Q(I + PQ)^{-1} = (I + QP)^{-1}Q$$

is the stabilizing solution to the control algebraic Riccati equation and

$$Z := P + PQP$$

is the stabilizing solution to the filter algebraic Riccati equation of the system (A, B, C, D).

Proof. 1.) We want to show that with $P = (I + ZY)^{-1}Z = Z(I+YZ)^{-1}$ we have,

$$\mathcal{A}P + P\mathcal{A}^* = -\mathcal{B}\mathcal{B}^*.$$

To do this consider

$$\begin{aligned}
&(I + ZY)[\mathcal{A}P + P\mathcal{A}^*](I + YZ)\\
&= (I + ZY)[(A_L - BR_L^{-1}B^*Y)P + P(A_L - BR_L^{-1}B^*Y)^*](I + YZ)\\
&= (I + ZY)(A_L - BR_L^{-1}B^*Y)Z + Z(A_L - BR_L^{-1}B^*Y)^*(I + YZ)\\
&= A_LZ + ZA_L^* + Z(YA_L + A_L^*Y)Z\\
&-2ZYBR_L^{-1}B^*YZ - BR_L^{-1}B^*YZ - ZYBR_L^{-1}B^*.
\end{aligned}$$

Using the two Riccati equations this gives,

$$\begin{aligned}
&(I + ZY)[\mathcal{A}P + P\mathcal{A}^*](I + YZ)\\
&= ZC^*S_L^{-1}CZ - BR_L^{-1}B^* + Z[YBR_L^{-1}B^*Y - C^*S_L^{-1}C)]Z\\
&-2ZYBR_L^{-1}B^*YZ - BR_L^{-1}B^*YZ - ZYBR_L^{-1}B^*\\
&= -(I + ZY)BR_L^{-1}B^*(I + YZ)\\
&= -(I + ZY)\mathcal{B}\mathcal{B}^*(I + YZ),
\end{aligned}$$

which implies the claim. Now with $Q = Y + YZY$, we have

$$\begin{aligned}
&\mathcal{A}^*Q + Q\mathcal{A} = \mathcal{A}^*Y(I + ZY) + (I + YZ)Y\mathcal{A}\\
&= (A_L^*Y - YBR_L^{-1}B^*Y)(I + ZY) + (I + YZ)(YA_L - YBR_L^{-1}B^*Y).
\end{aligned}$$

Using the Riccati equation, we have

$$\begin{aligned}
&\mathcal{A}^*Q + Q\mathcal{A} = (-YA_L - C^*S_L^{-1}C)(I + ZY) + (I + YZ)(-A_L^*Y - C^*S_L^{-1}C)\\
&= -C^*S_L^{-1}C(I + ZY) - (I + YZ)C^*S_L^{-1}C - A_L^*Y - YA_L - Y(A_LZ + ZA_L^*)Y\\
&= -C^*S_L^{-1}C(I + ZY) - (I + YZ)C^*S_L^{-1}C\\
&-YBR_L^{-1}B^*Y + C^*S_L^{-1}C - Y(ZC^*S_L^{-1}CZ)Y + YBR_L^{-1}B^*Y\\
&= -(I + YZ)C^*S_L^{-1}C(I + ZY)\\
&= -\mathcal{C}^*\mathcal{C}.
\end{aligned}$$

2.) First note that

$$A_L = A - BR_L^{-1}D^*C$$

$$= \mathcal{A} + \mathcal{B}(\mathcal{B}^*Q + \mathcal{D}^*\mathcal{C})(I + PQ)^{-1} - \mathcal{B}R_L^{1/2}R_L^{-1}\mathcal{D}^*S_L^{1/2}\mathcal{C}(I + PQ)^{-1}$$
$$= \mathcal{A} + \mathcal{B}\mathcal{B}^*Q(I + PQ)^{-1},$$

where we have used that $R_L^{1/2}\mathcal{D}^* = \mathcal{D}^*S_L^{1/2}$. Since,

$$\begin{aligned}
& (I + QP)[A_L^*Y + YA_L - YBR_L^{-1}B^*Y + C^*S_L^{-1}C](I + PQ) \\
= \; & (I + QP)[(\mathcal{A} + \mathcal{B}\mathcal{B}^*Q(I + PQ)^{-1})^*Q(I + PQ)^{-1} \\
& +(I + QP)^{-1}Q(\mathcal{A} + \mathcal{B}\mathcal{B}^*Q(I + PQ)^{-1}) \\
& -(I + QP)^{-1}Q\mathcal{B}R_L^{1/2}R_L^{-1}R_L^{1/2}\mathcal{B}^*Q(I + PQ)^{-1} \\
& +(I + PQ)^{-*}\mathcal{C}^*S_L^{1/2}S_L^{-1}S_L^{1/2}\mathcal{C}(I + PQ)^{-1}](I + PQ) \\
= \; & (I + QP)\mathcal{A}^*Q + Q\mathcal{B}\mathcal{B}^*Q + Q\mathcal{A}(I + PQ) + Q\mathcal{B}\mathcal{B}^*Q - Q\mathcal{B}\mathcal{B}^*Q + \mathcal{C}^*\mathcal{C} \\
= \; & \mathcal{A}^*Q + Q\mathcal{A} + \mathcal{C}^*\mathcal{C} + Q(P\mathcal{A}^* + \mathcal{A}P + \mathcal{B}\mathcal{B}^*)Q \\
= \; & 0,
\end{aligned}$$

we have verified the first identity. Now with $Z := P(I + QP)$ we have

$$\begin{aligned}
& A_LZ + ZA_L^* - ZC^*S_L^{-1}CZ + BR_L^{-1}B^* \\
= \; & (\mathcal{A} + \mathcal{B}\mathcal{B}^*Q(I + PQ)^{-1})(I + PQ)P + P(I + QP)(\mathcal{A} + \mathcal{B}\mathcal{B}^*Q(I + PQ)^{-1})^* \\
& -P(I + QP)(I + PQ)^{-*}\mathcal{C}^*S_L^{1/2}S_L^{-1}S_L^{1/2}\mathcal{C}(I + PQ)^{-1}(I + PQ)P \\
& +\mathcal{B}R_L^{1/2}R_L^{-1}R_L^{1/2}\mathcal{B}^* \\
= \; & \mathcal{A}P + \mathcal{A}PQP + \mathcal{B}\mathcal{B}^*QP + P\mathcal{A}^* + PQP\mathcal{A}^* + PQ\mathcal{B}\mathcal{B}^* - P\mathcal{C}^*\mathcal{C}P + \mathcal{B}\mathcal{B}^* \\
= \; & \mathcal{A}^*P + P\mathcal{A}^* + \mathcal{B}\mathcal{B}^* + (\mathcal{A}P + \mathcal{B}\mathcal{B}^*)QP - P\mathcal{C}^*\mathcal{C}P + PQ(P\mathcal{A}^* + \mathcal{B}\mathcal{B}^*) \\
= \; & 0 - P\mathcal{A}^*QP - P\mathcal{C}^*\mathcal{C}P + PQ(P\mathcal{A}^* + \mathcal{B}\mathcal{B}^*) \\
= \; & -P(\mathcal{A}^*Q + \mathcal{C}^*\mathcal{C})P + PQ(P\mathcal{A}^* + \mathcal{B}\mathcal{B}^*) \\
= \; & P(Q\mathcal{A})P + PQ(P\mathcal{A}^* + \mathcal{B}\mathcal{B}^*) \\
= \; & PQ(\mathcal{A}P + P\mathcal{A}^* + \mathcal{B}\mathcal{B}^*) \\
= \; & 0,
\end{aligned}$$

which shows the second identity. Since

$$A_L - BR_L^{-1}B^*Y$$
$$= \mathcal{A} + \mathcal{B}\mathcal{B}^*Q(I + PQ)^{-1} - \mathcal{B}R_L^{1/2}R_L^{-1}R_L^{1/2}\mathcal{B}^*Q(I + PQ)^{-1}$$
$$= \mathcal{A},$$

which is stable and

$$A_L - ZC^*S_L^{-1}C$$
$$= \mathcal{A} + \mathcal{B}\mathcal{B}^*Q(I + PQ)^{-1} - P(I + QP)(I + PQ)^{-*}\mathcal{C}^*S_L^{1/2}S_L^{-1}S_L^{1/2}\mathcal{C}(I + PQ)^{-1}$$
$$= \mathcal{A} + \mathcal{B}\mathcal{B}^*Q(I + PQ)^{-1} - P\mathcal{C}^*\mathcal{C}(I + PQ)^{-1}$$
$$= (I + PQ)[\mathcal{A} + (I + PQ)^{-1}P\mathcal{C}^*\mathcal{C}](I + PQ)^{-1} - P\mathcal{C}^*\mathcal{C}(I + PQ)^{-1}$$

$$= (I + PQ)\mathcal{A}(I + PQ)^{-1},$$

is stable, where we have used Lemma 4.2, we have shown that Z, Y are the stabilizing solutions to the Riccati equations. □

That the L-characteristic indeed induces a bijection between $L_n^{p,m}$ and $S_n^{p,m}$ is established in the following theorem.

Theorem 4.1. *The map*

$$\chi_L : L_n^{p,m} \to S_n^{p,m}$$

is a bijection that preserves system equivalence, with inverse $I\chi_L$.

Proof. It is first shown that χ_L is injective with left inverse $I\chi_L$, i.e. that $I\chi_L \cdot \chi_L$ is the identity map on $L_n^{p,m}$.

Let $(A, B, C, D) \in L_n^{p,m}$ and let $(\mathcal{A}, \mathcal{B}, \mathcal{C}, \mathcal{D}) \in S_n^{p,m}$ be its L-characteristic, i.e.

$$(\mathcal{A}, \mathcal{B}, \mathcal{C}, \mathcal{D}) = (A_L, -BR_L^{-1}B^*Y, BR_L^{-1/2}, S_L^{-1/2}C(I + ZY), D),$$

where Y and Z are the stabilizing solutions to the respective Riccati equations. We know by Proposition 4.1 that the solutions to the Lyapunov equations

$$\mathcal{A}P + P\mathcal{A}^* = -\mathcal{B}\mathcal{B}^*, \quad \mathcal{A}^*Q + Q\mathcal{A} = -\mathcal{C}^*\mathcal{C}$$

are given by $P = (I+ZY)^{-1}Z = Z(I+YZ)^{-1}$, $Q = Y+YZY$. Hence we can see that $PQ = ZY$. Now apply $I\chi_L$ to $(\mathcal{A}, \mathcal{B}, \mathcal{C}, \mathcal{D})$ and set $(A_1, B_1, C_1, D_1) := I\chi_L((\mathcal{A}, \mathcal{B}, \mathcal{C}, \mathcal{D}))$, then $D_1 = D$ and

$$B_1 = \mathcal{B}R_L^{1/2} = BR_L^{-1/2}R_L^{1/2} = B,$$

$$C_1 = S_L^{1/2}\mathcal{C}(I + PQ)^{-1} = S_L^{1/2}S_L^{-1/2}C(I + ZY)(I + ZY)^{-1} = C,$$

$$\begin{aligned} A_1 &= \mathcal{A} + \mathcal{B}\mathcal{B}^*Q(I + PQ)^{-1} + \mathcal{B}\mathcal{D}^*\mathcal{C}(I + PQ)^{-1} \\ &= A - BR_L^{-1}(D^*C + B^*Y) + BR_L^{-1}B^*Y(I + ZY)(I + ZY)^{-1} \\ &\quad + BR_L^{-1}D^*C(I + ZY)(I + ZY)^{-1} \\ &= A, \end{aligned}$$

i.e. $I\chi_L \cdot \chi_L((A, B, C, D)) = (A, B, C, D)$ for $(A, B, C, D) \in L_n^{p,m}$.

It is now shown that χ_L is surjective with right inverse $I\chi_L$, i.e. that $\chi_L \cdot I\chi_L$ is the identity map on $S_n^{p,m}$.

Let $(\mathcal{A}, \mathcal{B}, \mathcal{C}, \mathcal{D}) \in S_n^{p,m}$ and let $(A, B, C, D) = (\mathcal{A} + \mathcal{B}(\mathcal{B}^*Q + \mathcal{D}^*\mathcal{C})(I + PQ)^{-1}, \mathcal{B}R_L^{1/2}, S_L^{1/2}\mathcal{C}(I+PQ)^{-1}, \mathcal{D})$ be its inverse L-characteristic. Now consider

$$(\mathcal{A}_1, \mathcal{B}_1, \mathcal{C}_1, \mathcal{D}) := \chi_L \cdot I\chi_L((\mathcal{A}, \mathcal{B}, \mathcal{C}, \mathcal{D}))$$

$$= \chi_L((A,B,C,D)) = (A_L - BR_L^{-1}B^*Y, BR_L^{-1/2}, S_L^{-1/2}C(I+ZY), D),$$

where Y, Z are the stabilizing solutions to the control respectively filter algebraic Riccati equations of the system (A,B,C,D). By Proposition 4.1

$$Y = Q(I+PQ)^{-1} = (I+QP)^{-1}Q, \quad Z = P + PQP.$$

where Q, P are the positive definite solutions to the Lyapunov equations

$$\mathcal{A}P + P\mathcal{A}^* = -\mathcal{B}\mathcal{B}^*, \; \mathcal{A}^*Q + Q\mathcal{A} = \mathcal{C}^*\mathcal{C}.$$

Now, using that $A_L = \mathcal{A} + \mathcal{B}\mathcal{B}^*Q(I+PQ)^{-1}$, we have $\mathcal{D}_1 = D$, and

$$\mathcal{A}_1 = A_L - BR_L^{-1}B^*Y$$

$$= \mathcal{A} + \mathcal{B}\mathcal{B}^*Q(I+QP)^{-1} - \mathcal{B}R_L^{1/2}R_L^{-1}R_L^{1/2}\mathcal{B}Q(I+QP)^{-1}$$

$$= \mathcal{A},$$

$$\mathcal{B}_1 = BR_L^{-1/2} = \mathcal{B}R_L^{1/2}R_L^{-1/2} = \mathcal{B},$$

$$\mathcal{C}_1 = S_L^{-1/2}C(I+ZY) = S_L^{-1/2}S_L^{1/2}\mathcal{C}(I+PQ)^{-1}(I+PQ) = \mathcal{C},$$

which shows the claim that $\chi_L \cdot I\chi_L$ is the identity map. Therefore χ_L is a bijection with inverse $\chi_L^{-1} = I\chi_L$. That χ_L preserves system equivalence was shown in Lemma 4.3. □

This theorem was first shown in ([27]) where it was used to show that the manifolds $L_n^{p,m}/\sim$ and $S_n^{p,m}/\sim$ are diffeomorphic.

In the following corollary it is shown that the L-characteristic maps LQG balanced systems to stable minimal systems whose reachability and observability grammians are diagonal.

Corollary 4.1. *Let $(A,B,C,D) \in L_n^{p,m}$ and let $(\mathcal{A},\mathcal{B},\mathcal{C},\mathcal{D})$ be its L-characteristic. Let $Y(Z)$ be the stabilizing solution to the control (filter) algebraic Riccati equation and let P, Q be the solutions to the Lyapunov equations*

$$\mathcal{A}P + P\mathcal{A}^* = -\mathcal{B}\mathcal{B}^*, \; \mathcal{A}^*Q + Q\mathcal{A} = -\mathcal{C}^*\mathcal{C}.$$

Then

1. *if (A,B,C,D) is LQG balanced with LQG grammian Σ_L then*

$$P = \Sigma_L(I+\Sigma_L^2)^{-1}, \; Q = \Sigma_L(I+\Sigma_L^2).$$

2. *if $(\mathcal{A},\mathcal{B},\mathcal{C},\mathcal{D})$ is Lyapunov balanced with Lyapunov grammian Σ_S, then*

$$Y = \Sigma_S(I+\Sigma_S^2)^{-1}, \; Z = \Sigma_S(I+\Sigma_S^2).$$

Proof. The statements follow immediately from Proposition 4.1. □

If $\Gamma_S: \quad S_n^{p,m} \to S_n^{p,m}$ is a canonical form, then a canonical form can be defined on $L_n^{p,m}$ using the bijection χ_L by setting

$$\Gamma := \chi_L^{-1} \circ \Gamma_S \circ \chi_L.$$

If Γ_S is the Lyapunov balanced canonical form, it follows from Corollary 4.1 that a system in the Γ canonical form is close to being LQG-balanced. In fact a simple diagonal state space transformation will LQG-balance the system.

Lemma 4.4. *1. If $(\mathcal{A}, \mathcal{B}, \mathcal{C}, \mathcal{D}) \in S_n^{p,m}$ is such that the Lyapunov equations*

$$\mathcal{A}P + P\mathcal{A}^* = -\mathcal{B}\mathcal{B}^*, \ \mathcal{A}^*Q + Q\mathcal{A} = -\mathcal{C}^*\mathcal{C}.$$

have solutions

$$P = \Sigma(I + \Sigma^2)^{-1}, \ Q = \Sigma(I + \Sigma^2)$$

for some positive diagonal matrix Σ, then

$$\Delta_S((\mathcal{A}, \mathcal{B}, \mathcal{C}, \mathcal{D})) := (T\mathcal{A}T^{-1}, T\mathcal{B}, \mathcal{C}T^{-1}, \mathcal{D})$$

with $T = (I + \Sigma^2)^{\frac{1}{2}}$ is Lyapunov balanced with Lyapunov grammian Σ.
2. If $(A, B, C, D) \in L_n^{p,m}$ is such that the control (filter) algebraic Riccati equation has the stabilizing solution $Y(Z)$ with

$$Y = \Sigma(I + \Sigma^2)^{-1}, \ Z = \Sigma(I + \Sigma^2)$$

for some positive diagonal matrix Σ, then

$$\Delta_L((A, B, C, D)) := (TAT^{-1}, TB, CT^{-1}, D)$$

with $T = (I + \Sigma^2)^{\frac{1}{2}}$ is LQG balanced with LQG grammian Σ.

Proof. 1.) If the system $(\mathcal{A}, \mathcal{B}, \mathcal{C}, \mathcal{D})$ has reachability grammian P and observability grammian Q then the system $(T\mathcal{A}T^{-1}, T\mathcal{B}, \mathcal{C}T^{-1}, \mathcal{D})$ has reachability grammian TPT^* and observability grammian $T^{-*}QT^{-1}$. With P, Q and T as in the statement of the Lemma we therefore have

$$TPT^* = (I + \Sigma^2)^{\frac{1}{2}} \Sigma (I + \Sigma^2)^{-1} (I + \Sigma^2)^{\frac{1}{2}} = \Sigma,$$

$$T^{-*}QT^{-1} = (I + \Sigma^2)^{-\frac{1}{2}} \Sigma (I + \Sigma^2)(I + \Sigma^2)^{-\frac{1}{2}} = \Sigma,$$

which implies 1.).
2.) This is shown analogously to 1.). □

With the diagonal scaling map Δ_L as defined in the previous Lemma set

$$\Gamma_L := \Delta_L \circ \chi_L^{-1} \circ \Gamma_S \circ \chi_L.$$

The diagonal scaling now assures that Γ_L is a canonical form for $L_n^{p,m}$ in terms of LQG balanced systems. This fact will be used in the following Theorem to derive a canonical form and parametrization for $L_n^{p,m}$. This canonical form is called the *LQG balanced canonical form.*

Theorem 4.2. *Let (A_l, B_l, C_l, D_l) be a m-input p-output continuous-time system of dimension n. Then the following are equivalent:*

1. *(A_l, B_l, C_l, D_l) is a minimal system, i.e in $L_n^{p,m}$.*
2. *$(A_l, B_l, C_l, D_l) = (TAT^{-1}, TB, CT^{-1}, D)$ for some invertible T, where (A, B, C, D) is in LQG balanced form, i.e. there exist block indices $n_1, \dots, n_k$, $\sum_{j=1}^k n_j = n$, parameters $\sigma_1 > \sigma_2 > \dots > \sigma_k > 0$ and k families of step sizes $m = \tau_0^j \geq \tau_1^j \geq \tau_2^j \geq \dots \geq \tau_{l_j}^j > 0$, $\sum_{i=1}^{l_j} \tau_i^j = n_j$, $1 \leq j \leq k$, such that*
   a) *$B = (\underbrace{\overline{B}_1^T, 0}_{n_1}, \underbrace{\overline{B}_2^T, 0}_{n_2}, \dots, \underbrace{\overline{B}_k^T, 0}_{n_k})^T R_L^{\frac{1}{2}}$, where $\overline{B}_j \in \mathcal{K}^{\tau_1^j \times \tau_0^j}$ is positive upper triangular, $1 \leq j \leq k$.*
   b) *$C = S_L^{\frac{1}{2}}(\underbrace{U_1(\overline{B}_1\overline{B}_1^*)^{\frac{1}{2}}, 0}_{n_1}, \underbrace{U_2(\overline{B}_2\overline{B}_2^*)^{\frac{1}{2}}, 0}_{n_2}, \dots, \underbrace{U_k(\overline{B}_k\overline{B}_k^*)^{\frac{1}{2}}, 0}_{n_k})$, where $U_j \in \mathcal{K}^{p \times \tau_1^j}$, $U_j^* U_j = I_{\tau_1^j}$, $1 \leq j \leq k$.*
   c)

$$A = \left(\begin{array}{c|c|c|c|c} A(1,1) & \cdots & A(1,i) & \cdots & A(1,k) \\ \hline \vdots & \ddots & \vdots & \vdots & \vdots \\ \hline A(i,1) & \cdots & A(i,i) & \cdots & A(i,k) \\ \hline \vdots & & \vdots & \ddots & \vdots \\ \hline A(k,1) & \cdots & A(k,i) & \cdots & A(k,k) \end{array}\right),$$

where for $1 \leq i \leq k$

$$A(i,i) = \begin{pmatrix} \tilde{A}_{ii} + \mathcal{S}_1^i & (-\mathcal{A}_1^i)^* & & 0 \\ \mathcal{A}_1^i & \mathcal{S}_2^i & \ddots & \\ & \ddots & \ddots & (-\mathcal{A}_{l_i-1}^i)^* \\ 0 & & \mathcal{A}_{l_i-1}^i & \mathcal{S}_{l_i}^i \end{pmatrix},$$

for $1 \leq i, j \leq k$, $i \neq j$

$$A(i,j) = \begin{pmatrix} \tilde{A}_{ij} & 0 & \cdots & 0 \\ 0 & 0 & \cdots & 0 \\ \vdots & \vdots & & \vdots \\ 0 & 0 & \cdots & 0 \end{pmatrix},$$

and

i. S_i^j *is a* $\tau_i^j \times \tau_i^j$ *skew-hermitian matrix,* $i = 1, 2, \ldots, l_j$, $1 \le j \le k$.

ii. A_i^j *is a positive upper triangular* $\tau_{i+1}^j \times \tau_i^j$ *matrix,* $i = 1, 2, \ldots, l_j - 1$, $1 \le j \le k$.

iii. $\tilde{A}_{ij} \in \mathcal{K}^{\tau_1^i \times \tau_1^j}$ *is given by*

$$\tilde{A}_{ij} = \frac{1}{\sigma_i^2 - \sigma_j^2}(\sigma_j(1+\sigma_i^2)\overline{B}_i\overline{B}_j^* - \sigma_i(1+\sigma_j^2)(\overline{B}_i\overline{B}_i^*)^{\frac{1}{2}}U_i^*U_j(\overline{B}_j\overline{B}_j^*)^{\frac{1}{2}})$$

$$+\overline{B}_iD^*U_j(\overline{B}_j\overline{B}_j^*)^{\frac{1}{2}},$$

$$\text{for} \quad 1 \le i, j \le k, \quad i \ne j, \quad \text{and}$$

$$\tilde{A}_{ii} = -\frac{1}{2\sigma_i}(1 - \sigma_i^2)\overline{B}_i\overline{B}_i^* + \overline{B}_iD^*U_i(\overline{B}_i\overline{B}_i^*)^{\frac{1}{2}} \quad \text{for} \quad 1 \le i \le k.$$

Moreover, the system (A, B, C, D) *as defined in* (2) *is LQG balanced with LQG grammian* $\Sigma_L = diag(\sigma_1 I_{n_1}, \ldots, \sigma_k I_{n_k})$*, where* $n_1, \ldots, n_k$ *are the block indices and* $\sigma_1 > \cdots > \sigma_k > 0$ *are the LQG singular values of the system in* (2)*. The map* Γ_L *that assigns to each system in* $L_n^{p,m}$ *the realization in* (2) *is a canonical form.*

Proof. Let $(A_l, B_l, C_l, D_l) \in L_n^{p,m}$. Then $\chi_L((A_l, B_l, C_l, D_l))$ is in $S_n^{p,m}$. Let $(\mathcal{A}, \mathcal{B}, \mathcal{C}, \mathcal{D}) := \Gamma_S(\chi_L((A_l, B_l, C_l, D_l)))$ be the Lyapunov balanced canonical form of $\chi_L((A_l, B_l, C_l, D_l))$. Since χ_L and χ_L^{-1} respects system equivalence the system

$$(A, B, C, D) = \Delta_L(\chi_L^{-1}(\Gamma_S(\chi_L((A_l, B_l, C_l, D_l)))))$$

is equivalent to (A_l, B_l, C_l, D_l). Moreover the system is LQG balanced. It is straightforward to check that $\Gamma_L := \Delta_L \circ \chi_L^{-1} \circ \Gamma_L \circ \chi_L$ defines a canonical form for $L_n^{p,m}$.

It is necessary to show that (A, B, C, D) admits the stated parametrization. Consider now the Lyapunov balanced parametrization of $(\mathcal{A}, \mathcal{B}, \mathcal{C}, \mathcal{D})$, i.e. let $n_1, \ldots, n_k$, $\sum_{j=1}^k n_j = n$ be the block sizes, let $\sigma_1 > \sigma_2 > \cdots > \sigma_k > 0$ the Lyapunov singular values, let $m = \tau_0^j \ge \tau_1^j \ge \tau_2^j \ge \cdots \ge \tau_{l_j}^j > 0$, $\sum_{i=1}^{l_j} \tau_i^j = n_j$, $1 \le j \le k$ the families of step sizes such that

1. $\mathcal{B} = (\underbrace{\overline{\mathcal{B}}_1^T, 0}_{n_1}, \underbrace{\overline{\mathcal{B}}_2^T, 0}_{n_2}, \ldots, \underbrace{\overline{\mathcal{B}}_k^T, 0}_{n_k})^T$,

   where $\overline{\mathcal{B}}_j \in \mathcal{K}^{\tau_1^j \times \tau_0^j}$ is positive upper triangular, $1 \le j \le k$.
2. $\mathcal{C} = (\underbrace{\mathcal{U}_1(\overline{\mathcal{B}}_1\overline{\mathcal{B}}_1^*)^{\frac{1}{2}}, 0}_{n_1}, \underbrace{\mathcal{U}_2(\overline{\mathcal{B}}_2\overline{\mathcal{B}}_2^*)^{\frac{1}{2}}, 0}_{n_2}, \ldots, \underbrace{\mathcal{U}_k(\overline{\mathcal{B}}_k\overline{\mathcal{B}}_k^*)^{\frac{1}{2}}, 0}_{n_k})$,

   where $\mathcal{U}_j \in \mathcal{K}^{p \times \tau_1^j}, \mathcal{U}_j^*\mathcal{U}_j = I_{\tau_1^j}$, $1 \le j \le k$.

3.

$$A = \left(\begin{array}{c|c|c|c|c} A(1,1) & \cdots & A(1,i) & \cdots & A(1,k) \\ \hline \vdots & \ddots & \vdots & \vdots & \vdots \\ \hline A(i,1) & \cdots & A(i,i) & \cdots & A(i,k) \\ \hline \vdots & & \vdots & \ddots & \vdots \\ \hline A(k,1) & \cdots & A(k,i) & \cdots & A(k,k) \end{array}\right),$$

where for $1 \leq i \leq k$

$$A(i,i) = \begin{pmatrix} \tilde{\mathcal{A}}_{ii} + \mathcal{S}_1^i & (-\mathcal{A}_1^i)^* & & 0 \\ \mathcal{A}_1^i & \mathcal{S}_2^i & \ddots & \\ & \ddots & \ddots & (-\mathcal{A}_{l_i-1}^i)^* \\ 0 & & \mathcal{A}_{l_i-1}^i & \mathcal{S}_{l_i}^i \end{pmatrix},$$

for $1 \leq i,j \leq k$, $i \neq j$

$$A(i,j) = \begin{pmatrix} \tilde{\mathcal{A}}_{ij} & 0 & \cdots & 0 \\ 0 & 0 & \cdots & 0 \\ \vdots & \vdots & & \vdots \\ 0 & 0 & \cdots & 0 \end{pmatrix},$$

and

a) $\mathcal{S}_i^j$ is a $\tau_i^j \times \tau_i^j$ skew-hermitian matrix, $i = 1, 2, \ldots, l_j$, $1 \leq j \leq k$.

b) $\mathcal{A}_i^j$ is a positive upper triangular $\tau_{i+1}^j \times \tau_i^j$ matrix, $i = 1, 2, \ldots, l_j - 1$, $1 \leq j \leq k$.

c) $\tilde{\mathcal{A}}_{ij} \in \mathcal{K}^{\tau_1^i \times \tau_1^j}$ is given by

$$\tilde{\mathcal{A}}_{ij} = \frac{1}{\sigma_i^2 - \sigma_j^2}\left(\sigma_j \overline{\mathcal{B}}_i \overline{\mathcal{B}}_j^* - \sigma_i (\overline{\mathcal{B}}_i \overline{\mathcal{B}}_i^*)^{\frac{1}{2}} \mathcal{U}_i^* \mathcal{U}_j (\overline{\mathcal{B}}_j \overline{\mathcal{B}}_j^*)^{\frac{1}{2}}\right),$$

$$\text{for } 1 \leq i,j \leq k, \quad i \neq j, \quad \text{and } \tilde{\mathcal{A}}_{ii} = -\frac{1}{2\sigma_i} \overline{\mathcal{B}}_i \overline{\mathcal{B}}_i^* \quad \text{for } 1 \leq i \leq k.$$

The system $(\mathcal{A}, \mathcal{B}, \mathcal{C}, \mathcal{D})$ is Lyapunov balanced with Lyapunov grammian $\Sigma = diag(\sigma_1 I_{n_1}, \ldots, \sigma_k I_{n_k})$. Then by Corollary 4.1 and Lemma 4.4

$$\begin{aligned} (A, B, C, D) \;\; = \;\; & ((I + \Sigma^2)^{-\frac{1}{2}} (\mathcal{A} + \mathcal{B}(\mathcal{B}^* \Sigma + \mathcal{D}^* \mathcal{C})(I + \Sigma^2)^{-1})(I + \Sigma^2)^{\frac{1}{2}}, \\ & (I + \Sigma^2)^{-\frac{1}{2}} \mathcal{B} R_L^{\frac{1}{2}}, S_L^{\frac{1}{2}} \mathcal{C}(I + \Sigma^2)^{-\frac{1}{2}}, \mathcal{D}). \end{aligned}$$

Setting

$$\overline{B_j} := \frac{1}{\sqrt{1 + \sigma_j^2}} \overline{\mathcal{B}}_j, \quad 1 \leq i \leq k,$$

it follows that $B = (I + \Sigma^2)^{-\frac{1}{2}} \mathcal{B} R_L^{\frac{1}{2}}$ has the required structure. Since with $U_j := \mathcal{U}_j$

$$S_L^{\frac{1}{2}} \mathcal{U}_j (\overline{\mathcal{B}}_j \overline{\mathcal{B}}_j^*)^{\frac{1}{2}} \frac{1}{\sqrt{1+\sigma_j^2}} = S_L^{\frac{1}{2}} U_j (\overline{B}_j \overline{B}_j^*)^{\frac{1}{2}}$$

for $1 \le i \le k$, it follows that $C = S_L^{\frac{1}{2}} \mathcal{C} (I + \Sigma^2)^{-\frac{1}{2}}$ also has the required structure. Note that

$$\begin{aligned} A &= (I+\Sigma^2)^{-\frac{1}{2}} (\mathcal{A} + \mathcal{B}(\mathcal{B}^* \Sigma + \mathcal{D}^* \mathcal{C})(I+\Sigma^2)^{-1})(I+\Sigma^2)^{\frac{1}{2}} \\ &= (I+\Sigma^2)^{-\frac{1}{2}} \mathcal{A} (I+\Sigma^2)^{\frac{1}{2}} + BR_L^{-1} B^* \Sigma + BR_L^{-1} D^* BS_L^{-1} C. \end{aligned}$$

If (A, B, C, D) is partitioned according to the block indices $n_1, \ldots, n_k$, then for $1 \le i, j \le k$, $i \ne j$,

$$A_{ij} = \frac{\sqrt{1+\sigma_j^2}}{\sqrt{1+\sigma_i^2}} \mathcal{A}_{ij} + B_i R_L^{-1} B_j^* \sigma_j + B_j R_L^{-1} D^* S_L^{-1} C_j.$$

This shows that all entries of A_{ij} are zero with the exception of the principal $\tau_1^i \times \tau_1^j$-subblock $\tilde{A}_{ij}$ which is given by

$$\begin{aligned} \tilde{A}_{ij} &= \frac{\sqrt{1+\sigma_j^2}}{\sqrt{1+\sigma_i^2}} \tilde{\mathcal{A}}_{ij} + \overline{B}_i \overline{B}_j^* \sigma_j + \overline{B}_i D^* U_j (\overline{B}_j \overline{B}_j^*)^{\frac{1}{2}} \\ &= \frac{\sqrt{1+\sigma_j^2}}{\sqrt{1+\sigma_i^2}} \left(\frac{1}{\sigma_i^2 - \sigma_j^2} (\sigma_j \overline{\mathcal{B}}_i \overline{\mathcal{B}}_j^* - \sigma_i (\overline{\mathcal{B}}_i \overline{\mathcal{B}}_i^*)^{\frac{1}{2}} \mathcal{U}_i^* \mathcal{U}_j (\mathcal{B}_j \overline{\mathcal{B}}_j^*)^{\frac{1}{2}}) \right) \\ &\quad + \overline{B}_i \overline{B}_j^* \sigma_j + \overline{B}_i D^* U_j (\overline{B}_i \overline{B}_j^*)^{\frac{1}{2}} \\ &= \frac{1+\sigma_j^2}{\sigma_i^2 - \sigma_j^2} \left(\sigma_j \overline{B}_i \overline{B}_j^* - \sigma_i (\overline{B}_i \overline{B}_i^*)^{\frac{1}{2}} U_i^* U_j (\overline{B}_j \overline{B}_j^*)^{\frac{1}{2}} \right) \\ &\quad + \overline{B}_i \overline{B}_j^* \sigma_j + \overline{B}_i D^* U_j (\overline{B}_j \overline{B}_j^*)^{\frac{1}{2}} \\ &= \frac{1}{\sigma_i^2 - \sigma_j^2} \left(\sigma_j (1+\sigma_i^2) \overline{B}_i \overline{B}_j^* - \sigma_i (1+\sigma_j^2) (\overline{B}_i \overline{B}_i^*)^{\frac{1}{2}} U_i^* U_j (\overline{B}_j \overline{B}_j^*)^{\frac{1}{2}} \right) \\ &\quad + \overline{B}_i D^* U_j (\overline{B}_j \overline{B}_j^*)^{\frac{1}{2}} \end{aligned}$$

for $1 \le i, j \le k$, $i \ne j$. For $1 \le i \le k$, we have

$$A_{ii} = \mathcal{A}_{ii} + B_i R_L^{-1} (B_i^* \sigma_i + D^* S_L^{-1} C_i).$$

The principal $\tau_1^i \times \tau_1^i$ submatrix of A_{ii}, $1 \le i \le k$, is given by

$$\tilde{\mathcal{A}}_{ii} + \mathcal{S}_1^i + \overline{B}_i R_L^{\frac{1}{2}} R_L^{-1} (R_L^{\frac{1}{2}} \overline{B}_i^* \sigma_i + D^* S_L^{\frac{1}{2}} U_i (\overline{B}_i \overline{B}_i^*)^{\frac{1}{2}})$$

$$= -\frac{1}{2\sigma_i}\overline{\mathcal{B}}_i\overline{\mathcal{B}}_i^* + \mathcal{S}_1^i + \overline{B}_i\overline{B}_i^*\sigma_i + \overline{B}_iD^*U_i(\overline{B}_i\overline{B}_i^*)^{\frac{1}{2}}$$

$$= -\frac{1}{2\sigma_i}(1+\sigma_i^2)\overline{B}_i\overline{B}_i^* + S_1^i + \overline{B}_i\overline{B}_i^*\sigma_i + \overline{B}_iD^*U_i(\overline{B}_i\overline{B}_i^*)^{\frac{1}{2}}$$

$$= -\frac{1}{2\sigma_i}(1-\sigma_i^2)\overline{B}_i\overline{B}_i^* + \overline{B}_iD^*U_i(\overline{B}_i\overline{B}_i^*)^{\frac{1}{2}} + S_1^i,$$

where we set $S_i^j = \mathcal{S}_i^j$ for $i = 1, 2, \ldots, l_j$, $1 \le j \le k$. This shows that A has the stated form. Therefore (A, B, C, D) is in the stated canonical form. Hence Γ_L has the claimed properties.

To complete the proof it remains to be shown that if a system (A, B, C, D) has the parametrization stated in 2.) then the system is minimal and LQG balanced with LQG grammian $\Sigma_L = diag(\sigma_1 I_{n_1}, \ldots, \sigma_k I_{n_k})$. Let

$$\begin{aligned}(\mathcal{A}, \mathcal{B}, \mathcal{C}, \mathcal{D}) \; := \; & ((I+\Sigma^2)^{\frac{1}{2}}(A_L - BR_L^{-1}B^*\Sigma)(I+\Sigma^2)^{-\frac{1}{2}}, \\ & (I+\Sigma^2)^{\frac{1}{2}}BR_L^{-\frac{1}{2}}, S_L^{-\frac{1}{2}}C(I+\Sigma^2)^{-\frac{1}{2}}, D).\end{aligned}$$

Straightforward algebraic manipulations, similar to those in the first part of the proof, show that $(\mathcal{A}, \mathcal{B}, \mathcal{C}, \mathcal{D})$ admits the parametrization for Lyapunov balanced systems as given in Theorem 3.2. Hence $(\mathcal{A}, \mathcal{B}, \mathcal{C}, \mathcal{D}) \in S_n^{p,m}$ is Lyapunov balanced with Lyapunov grammian Σ_L. Then it follows from the construction of $(\mathcal{A}, \mathcal{B}, \mathcal{C}, \mathcal{D})$ as the 'formal' pre-image of (A, B, C, D) under $\Delta_L \circ \chi_L^{-1}$, that $(\mathcal{A}, \mathcal{B}, \mathcal{C}, \mathcal{D}) = \chi_L(\Delta_L^{-1}((A, B, C, D)))$. Hence (A, B, C, D) is minimal and LQG balanced with LQG grammian Σ_L. □

It is worth commenting that as was established in the proof of the theorem, if $(\mathcal{A}, \mathcal{B}, \mathcal{C}, \mathcal{D})$ is a Lyapunov balanced canonical form and $(A, B, C, D) = \Delta_L(\chi_L^{-1}((\mathcal{A}, \mathcal{B}, \mathcal{C}, \mathcal{D})))$ then up to some scaling of the $\overline{B}_i$ parameters, the parameters of (A, B, C, D) are the same as those for $(\mathcal{A}, \mathcal{B}, \mathcal{C}, \mathcal{D})$. The only essential difference between the parametrization of $S_n^{p,m}$ given in Theorem 3.2 and the parametrization of $L_n^{p,m}$ given in the previous Theorem is the way in which the system parameters go into the entries $\tilde{A}_{ij}$, $1 \le i, j \le k$.

Specialization of the theorem to the single input single output case gives the following corollary.

Corollary 4.2. *Let (A_l, b_l, c_l, d_l) be a single-input single-output continuous-time system of dimension n. Then the following statements are equivalent:*

1. *(A_l, b_l, c_l, d_l) is a minimal system.*
2. *$(A_l, b_l, c_l, d_l) = (TAT^{-1}, Tb, cT^{-1}, d)$ for some invertible matrix T, where (A, b, c, d) is in the following LQG balanced form with block indices $n_1, n_2, \ldots, n_k$, $\sum_{i=1}^k n_i = n$:*
   a) $b = (\underbrace{b_1, 0, \ldots, 0}_{n_1}, \underbrace{b_2, 0, \ldots, 0}_{n_2}, \ldots, \underbrace{b_k, 0, \ldots, 0}_{n_k})^T$,
   where $b_i > 0$, $1 \le i \le k$.

b) $c = (\underbrace{s_1 b_1, 0, \ldots, 0}_{n_1}, \underbrace{s_2 b_2, 0, \ldots, 0}_{n_2}, \ldots, \underbrace{s_k b_k, 0, \ldots, 0}_{n_k}),$
where $s_i \in \mathcal{K}$, $|s_i| = 1$, $1 \leq i \leq k$.

c)

$$A = \left(\begin{array}{c|c|c|c} A(1,1) & A(1,2) & \cdots & A(1,k) \\ \hline A(2,1) & A(2,2) & \cdots & A(2,k) \\ \hline \vdots & & \ddots & \vdots \\ \hline A(k,1) & A(k,2) & \cdots & A(k,k) \end{array} \right),$$

where for $1 \leq j \leq k$

$$A(j,j) = \begin{pmatrix} a_{jj} + i\beta_1^j & -\alpha_1^j & & 0 \\ \alpha_1^j & i\beta_2^j & \ddots & \\ & \ddots & \ddots & -\alpha_{n_j-1}^j \\ 0 & & \alpha_{n_j-1}^j & i\beta_{n_j}^j \end{pmatrix},$$

for $1 \leq i, j \leq k$, $i \neq j$,

$$A(i,j) = \begin{pmatrix} a_{ij} & 0 & \cdots & 0 \\ 0 & 0 & \cdots & 0 \\ \vdots & \vdots & & \vdots \\ 0 & 0 & \cdots & 0 \end{pmatrix},$$

and

$$\alpha_j^i > 0 \text{ for } 1 \leq j \leq n_i - 1, \quad 1 \leq i \leq k,$$

$$a_{ij} = \tfrac{-b_i b_j}{1+|d|^2} \left[\tfrac{\bar{s}_i s_j \sigma_i (1+\sigma_j^2) - \sigma_j (1+\sigma_i^2)}{\sigma_i^2 - \sigma_j^2} - s_j d^* \right]$$
$$\text{for } 1 \leq i, j \leq k, \quad i \neq j,$$

$$a_{ii} = -\frac{b_i^2}{1+|d|^2} \left[\frac{1}{2\sigma_i}(1 - \sigma_i^2) - s_i d^* \right], \text{ for } 1 \leq i \leq k,$$

$$\sigma_1 > \sigma_2 > \cdots > \sigma_k > 0.$$

If $\mathcal{K} = \Re$, *then*

$$s_i = \pm 1, \quad 1 \leq i \leq k,$$

$$\beta_j^i = 0, \text{ for } 1 \leq j \leq n_i, \quad 1 \leq i \leq k,$$

$$a_{ij} = \frac{-b_i b_j}{1 + d^2} \left[\frac{1 - s_i s_j \sigma_i \sigma_j}{s_i s_j \sigma_i + \sigma_j} - s_j d \right], \text{ for } 1 \leq i, j \leq k.$$

Moreover, the system (A, b, c, d) *as defined in* (2) *is LQG balanced with LQG grammian* $\Sigma_L = diag(\sigma_1 I_{n_1}, \sigma_2 I_{n_2}, \ldots, \sigma_k I_{n_k})$. *The map* Γ_L *that assigns to each system in* $L_n^{1,1}$ *the realization in* (2) *is a canonical form.*

Proof. The result follows immediately from the previous theorem by absorbing $R_L^{\frac{1}{2}} = S_L^{\frac{1}{2}}$ into $\overline{B}_i$ parameters and setting $b_i := \overline{B}_i R_L^{\frac{1}{2}}$, $1 \le i \le k$. The remaining parameter replacements are as in the proof of Corollary 3.3. □

Having established a parametrization of $L_n^{p,m}$ in terms of LQG balanced realizations, LQG balanced model reduction can now be analyzed.

Theorem 4.3. *Let $(A, B, C, D) \in L_n^{p,m}$ be in LQG balanced canonical form with LQG grammian Σ. Let $1 \le r < n$. Then the r-dimensional balanced approximant (A_{11}, B_1, C_1, D) is in $L_r^{p,m}$ and is in LQG balanced canonical form with LQG grammian Σ_1, where $\Sigma = diag(\Sigma_1, \Sigma_2)$, $\Sigma_1 \in \mathcal{K}^{r\times r}$.*

Proof. Let $(\mathcal{A}, \mathcal{B}, \mathcal{C}, \mathcal{D}) = \Delta_L(\chi_L((A, B, C, D)))$. It follows by Corollary 4.1 and Lemma 4.4 that if $(A, B, C, D) \in L_n^{p,m}$ is in LQG balanced canonical form with LQG grammian Σ, then $(\mathcal{A}, \mathcal{B}, \mathcal{C}, \mathcal{D})$ is in Lyapunov balanced canonical form with Lyapunov grammian Σ.

Let $P_r \in \mathcal{K}^{r\times n}$ be the projection matrix given by

$$P_r = (\delta_{ij})_{1\le i\le r, 1\le j\le n},$$

then

$$(A_{11}, B_1, C_1, D) = (P_r A P_r^*, P_r B, C P_r^*, D).$$

If $(\mathcal{A}_{11}, \mathcal{B}_1, \mathcal{C}_1, \mathcal{D})$ is the r-dimensional balanced approximant of $(\mathcal{A}, \mathcal{B}, \mathcal{C}, \mathcal{D})$ then

$$\begin{aligned}
B_1 &= P_r B = P_r(I + \Sigma^2)^{-\frac{1}{2}} \mathcal{B} R_L^{\frac{1}{2}} = (I + \Sigma_1^2)^{-\frac{1}{2}} P_r \mathcal{B} R_L^{\frac{1}{2}} = (I + \Sigma_1^2)^{-\frac{1}{2}} \mathcal{B}_1 R_L^{\frac{1}{2}},\\
C_1 &= C P_r^* = S_L^{\frac{1}{2}} \mathcal{C}(I + \Sigma^2)^{-\frac{1}{2}} P_r^* = S_L^{\frac{1}{2}} \mathcal{C} P_r^* (I + \Sigma_1^2)^{-\frac{1}{2}} = S_L^{\frac{1}{2}} \mathcal{C}_1 (I + \Sigma_1^2)^{-\frac{1}{2}},\\
A_{11} &= P_r A P_r^* = P_r (I + \Sigma^2)^{-\frac{1}{2}} (\mathcal{A} + \mathcal{B}(\mathcal{B}^* \Sigma + \mathcal{D}^* \mathcal{C})(I + \Sigma^2)^{-1})(I + \Sigma^2)^{\frac{1}{2}} P_r^*\\
&= (I + \Sigma_1^2)^{-\frac{1}{2}} (P_r \mathcal{A} P_r^* + P_r \mathcal{B}(\mathcal{B}^* P_r^* \Sigma_1 + \mathcal{D}^* \mathcal{C} P_r)(I + \Sigma_1^2)^{-1})(I + \Sigma_1^2)^{\frac{1}{2}}\\
&= (I + \Sigma_1^2)^{-\frac{1}{2}} (\mathcal{A}_{11} + \mathcal{B}_1(\mathcal{B}_1^* \Sigma_1 + \mathcal{D}^* \mathcal{C}_1)(I + \Sigma_1^2)^{-1})(I + \Sigma_1^2)^{\frac{1}{2}}.
\end{aligned}$$

Since Σ_1 is the Lyapunov grammian of the system $(\mathcal{A}_{11}, \mathcal{B}_1, \mathcal{C}_1, \mathcal{D})$ which is in Lyapunov balanced canonical form, this shows that

$$(A_{11}, B_1, C_1, D) = \Delta_L(\chi_L^{-1}((\mathcal{A}_{11}, \mathcal{B}_1, \mathcal{C}_1, \mathcal{D})))$$

and hence by the properties of $\Delta_L \circ \chi_L^{-1}$, the system (A_{11}, B_1, C_1, D) is minimal and in LQG balanced canonical form with LQG grammian Σ_1. □

As a corollary we also obtain a model reduction result for general LQG balanced systems ([14]). Analogously to Theorem 2.1 a condition has to be placed on the point at which truncation occurs.

Corollary 4.3. *Let (A,B,C,D) be a n-dimensional LQG balanced system in $L_n^{p,m}$ with LQG grammian $\Sigma = diag(\sigma_1 I_{n_1}, \sigma_2 I_{n_2}, \dots, \sigma_k I_{n_k})$, $\sigma_1 > \sigma_2 > \dots > \sigma_k > 0$. Let $r = n_1 + n_2 + \dots + n_l$ for some $1 \leq l \leq k$. Then the r-dimensional balanced approximant (A_{11}, B_1, C_1, D) of (A,B,C,D) is minimal and LQG balanced with LQG grammian $\Sigma_1 = diag(\sigma_1 I_{n_1}, \sigma_2 I_{n_2}, \dots, \sigma_l I_{n_l})$.*

Proof. Let (A,B,C,D) be as in the statement. By Theorem 1.1 all equivalent LQG balanced systems are given by (QAQ^*, QB, CQ^*, D) with $Q = diag(Q_1, Q_2, \dots, Q_k)$, $Q_i \in \mathcal{K}^{n_i \times n_i}$ unitary. Let $Q_0 = diag(Q_1^0, Q_2^0, \dots, Q_k^0)$, $Q_i^0 \in \mathcal{K}^{n_i \times n_i}$ unitary, be such that $(Q_0 A Q_0^*, Q_0 B, C Q_0^*, D)$ is in LQG balanced canonical form. By Theorem 4.3 the r-dimensional balanced approximant of this system is in $L_r^{p,m}$ and LQG balanced with LQG grammian $\Sigma_1 = diag(\sigma_1 I_{n_1}, \sigma_2 I_{n_2}, \dots, \sigma_l I_{n_l})$. Because of the block diagonal structure of Q_0 and the assumption on r, the approximant can be written as $(\hat{Q}_0 A_{11} \hat{Q}_0^*, \hat{Q}_0 B_1, C_1 \hat{Q}_0^*, D)$ where $\hat{Q}_0 = (\hat{Q}_1^0, \hat{Q}_2^0, \dots, \hat{Q}_l^0)$. Since $\hat{Q}_0$ is unitary this shows that (A_{11}, B_1, C_1, D) has the claimed properties. □

5. Characteristics, Canonical Forms and Model Reduction for Bounded-Real and Positive-Real Systems

The purpose of this section is to derive canonical forms for bounded-real and positive-real systems in terms of bounded-real respectively positive-real balanced systems. The approach that is taken is analogous to the one used in the previous section to derive the LQG balanced canonical form for minimal systems in $L_n^{p,m}$.

Analogously to the discussion in the previous section we will introduce characteristic maps χ_B and χ_P for bounded-real and positive-real systems. These characteristic maps are used to carry the Lyapunov balanced canonical form for systems in $S_n^{p,m}$ over to $B_n^{p,m}$ respectively P_n^m to introduce a canonical form for bounded real and positive real systems. In contrast to the characteristic map χ_L, the range space of χ_B and χ_P will no longer be the set $S_n^{p,m}$ of stable minimal systems of given McMillan degree, but rather the subsets $US_{n,B}^{p,m}$ and $US_{n,P}^{m,m}$. A system $(\mathcal{A}, \mathcal{B}, \mathcal{C}, \mathcal{D}) \in S_n^{p,m} (S_n^{m,m})$ is in $US_{n,B}^{p,m}$ $(US_{n,P}^{m,m})$ if

1.) $\lambda_{max}(PQ) < 1$, where P, Q are the solutions to the Lyapunov equation

$$\mathcal{A}P + P\mathcal{A}^* = -\mathcal{B}\mathcal{B}^*, \quad \mathcal{A}^*Q + Q\mathcal{A} = -\mathcal{C}^*\mathcal{C}.$$

2.) $I - \mathcal{D}^*\mathcal{D} > 0 \quad (\mathcal{D} + \mathcal{D}^* > 0)$.

The results on the parameterization of bounded-real/positive-real systems will be used to analyze the balanced model reduction method for bounded-real and positive-real systems.

Throughout this section we will use the following abbreviations. Let (A, B, C, D) be a system. If $I - D^*D > 0$ set $R_B := I - D^*D$, $S_B := I - DD^*$ and $A_B := A + BR_B^{-1}D^*C$. If the system has equal input and output dimensions and $D + D^* > 0$ set $R_P := D + D^*$ and $A_P := A - BR_P^{-1}C$.

The characteristic map and inverse characteristic map for bounded-real systems is defined as follows.

Definition 5.1. *1.) Let (A, B, C, D) be a bounded-real system. Let Y be the stabilizing solution of the control bounded-real Riccati equation and let Z be the stabilizing solution of the filter bounded real Riccati equation, i.e.*

$$0 = A_B^*Y + YA_B + YBR_B^{-1}B^*Y + C^*S_B^{-1}C$$

$$0 = A_BZ + ZA_B^* + ZC^*S_B^{-1}CZ + BR_B^{-1}B^*$$

with $A_B + BR_B^{-1}B^*Y$ and $A_B + ZC^*S_B^{-1}C$ stable. Then the system

$$\chi_B((A,B,C,D)) := (A_B + BR_B^{-1}B^*Y, BR_B^{-\frac{1}{2}}, S_B^{-\frac{1}{2}}C(I - ZY), D)$$

$$= ((I - ZY)^{-1}(A_B + ZC^*S_B^{-1}C)(I - ZY), BR_B^{-\frac{1}{2}}, S_B^{-\frac{1}{2}}C(I - ZY), D)$$

is called the B-characteristic *of the system (A, B, C, D).*

2.) Let $(\mathcal{A}, \mathcal{B}, \mathcal{C}, \mathcal{D})$ be a stable minimal system in $US_{n,B}^{p,m}$ and let P and Q be the solutions of the Lyapunov equations

$$\mathcal{A}P + P\mathcal{A}^* = -\mathcal{B}\mathcal{B}^*, \quad \mathcal{A}^*Q + Q\mathcal{A} = -\mathcal{C}^*\mathcal{C}.$$

Then the system

$$I\chi_B((\mathcal{A},\mathcal{B},\mathcal{C},\mathcal{D})) := (\mathcal{A}-\mathcal{B}(\mathcal{B}^*Q+\mathcal{D}^*\mathcal{C})(I-PQ)^{-1}, \mathcal{B}R_B^{\frac{1}{2}}, S_B^{\frac{1}{2}}\mathcal{C}(I-PQ)^{-1}, \mathcal{D})$$

$$= ((I-PQ)(\mathcal{A}-(I-PQ)^{-1}(P\mathcal{C}^*+\mathcal{B}\mathcal{D}^*)\mathcal{C})(I-PQ)^{-1}, \mathcal{B}R_B^{\frac{1}{2}}, S_B^{\frac{1}{2}}\mathcal{C}(I-PQ)^{-1}, \mathcal{D})$$

is called the inverse B-characteristic *of the system $(\mathcal{A}, \mathcal{B}, \mathcal{C}, \mathcal{D})$. Here $R_B := I - \mathcal{D}^*\mathcal{D}$ and $S_B := I - \mathcal{D}\mathcal{D}^*$.*

Since the analysis of the characteristic map for bounded-real systems is quite similar to that of the characteristic map for positive-real systems, the definition of the positive real characteristic will be given now.

Definition 5.2. *1.) Let (A, B, C, D) be a positive-real system. Let Y be the stabilizing solution of the control positive real Riccati equation and let Z be the stabilizing solution of the filter positive real Riccati equation, i.e.*

$$0 = A_P^*Y + YA_P + YBR_P^{-1}B^*Y + C^*R_P^{-1}C$$

$$0 = A_PZ + ZA_P^* + ZC^*R_P^{-1}CZ + BR_P^{-1}B^*$$

with $A_P + BR_P^{-1}B^*Y$ and $A_P + ZC^*R_P^{-1}C$ stable. Then the system

$$\chi_P((A,B,C,D)) := (A_P + BR_P^{-1}B^*Y, BR_P^{-\frac{1}{2}}, R_P^{-\frac{1}{2}}C(I - ZY), D)$$

$$= ((I - ZY)^{-1}(A_P + ZC^*R_P^{-1}C)(I - ZY), BR_P^{-\frac{1}{2}}, R_P^{-\frac{1}{2}}C(I - ZY), D)$$

is called the P-characteristic *of the system* (A, B, C, D).

2.) Let $(\mathcal{A}, \mathcal{B}, \mathcal{C}, \mathcal{D})$ *be a stable minimal system in* $US_{n,P}^{m,m}$ *and let* P *and* Q *be the solutions of the Lyapunov equations*

$$\mathcal{A}P + P\mathcal{A}^* = -\mathcal{B}\mathcal{B}^*, \quad \mathcal{A}^*Q + Q\mathcal{A} = -\mathcal{C}^*\mathcal{C}.$$

Then the system

$$I\chi_P((\mathcal{A}, \mathcal{B}, \mathcal{C}, \mathcal{D})) := (\mathcal{A} - \mathcal{B}(\mathcal{B}^*Q - \mathcal{C})(I - PQ)^{-1}, \mathcal{B}R_P^{\frac{1}{2}}, R_P^{\frac{1}{2}}\mathcal{C}(I - PQ)^{-1}, \mathcal{D})$$

$$= ((I - PQ)(\mathcal{A} - (I - PQ)^{-1}(P\mathcal{C}^* + \mathcal{B})\mathcal{C})(I - PQ)^{-1}, \mathcal{B}R_P^{\frac{1}{2}}, R_P^{\frac{1}{2}}\mathcal{C}(I - PQ)^{-1}, \mathcal{D})$$

is called the inverse P-characteristic *of the system* $(\mathcal{A}, \mathcal{B}, \mathcal{C}, \mathcal{D})$. *Here* $R_P = \mathcal{D} + \mathcal{D}^*$.

Note that both expressions for the B-characteristic and P-characteristic are identical because of the following Lemma.

Lemma 5.1. *1. With the assumptions as in the Definition 5.1,*

$$(I - ZY)(A_B + BR_B^{-1}B^*Y) = (A_B + ZC^*S_B^{-1}C)(I - ZY),$$

where Y *is the stabilizing solution to the control bounded-real Riccati equation and* Z *is the stabilizing solution to the filter bounded-real Riccati equation.*

2. With the assumptions as in the Definition 5.2,

$$(I - ZY)(A_P + BR_P^{-1}B^*Y) = (A_P + ZC^*S_P^{-1}C)(I - ZY),$$

where Y *is the stabilizing solution to the control positive-real Riccati equation and* Z *is the stabilizing solution to the filter positive-real Riccati equation.*

Proof. 1.) Consider the two Riccati equations,

$$0 = A_B^*Y + YA_B - YBR_B^{-1}B^*Y + C^*S_B^{-1}C,$$

$$0 = A_BZ + ZA_B^* - ZC^*R_B^{-1}CZ + BS_B^{-1}B^*.$$

Multiplying the first equation on the left by $-Z$ and the second equation on the right by $-Y$, equating both equations and adding A_B to both sides we obtain

$$A_B - ZA_B^*Y - ZYA_B - ZYBR_B^{-1}B^*Y - ZC^*S_B^{-1}C$$
$$= A_B - A_BZY - ZA_B^*Y - ZC^*S_B^{-1}CZY - BR_B^{-1}B^*Y.$$

Canceling the term ZA_B^*Y from either side and collecting terms, we obtain

$$(I - ZY)(A_B + BR_B^{-1}B^*Y) = (A_B + ZC^*S_B^{-1}C)(I - ZY).$$

2.) The statement follows analogously to 1.).

□

That $I - ZY$ is invertible follows by general results on the *gap* between the maximal and minimal solution to the bounded-real Riccati equation (see e.g. [34]). The analogous argument shows that the two expressions for the positive real characteristic are identical and well-defined.

It is now necessary to analyze the bounded-real and positive-real characteristic. The main result is Theorem 5.1 in which it will be shown that these two characteristic maps are bijections whose inverses are the corresponding inverse characteristic maps. To this end the following Lemmas and Proposition need to be established.

That the two expressions for the inverse B-characteristic and the inverse P-characteristic are identical follows from the following Lemma.

Lemma 5.2. *Let $(\mathcal{A}, \mathcal{B}, \mathcal{C}, \mathcal{D})$ be a stable minimal system and let P and Q be such that*

$$\mathcal{A}P + P\mathcal{A}^* = -\mathcal{B}\mathcal{B}^*, \quad \mathcal{A}^*Q + Q\mathcal{A} = -\mathcal{C}^*\mathcal{C}.$$

1. *If $(\mathcal{A}, \mathcal{B}, \mathcal{C}, \mathcal{D}) \in US^{p,m}_{n,B}$ then*

$$[\mathcal{A} - \mathcal{B}(\mathcal{B}^*Q + \mathcal{D}^*\mathcal{C})(I - PQ)^{-1}][I - PQ]$$
$$= [I - PQ][\mathcal{A} - (I - PQ)^{-1}(P\mathcal{C}^* + \mathcal{B}\mathcal{D}^*)\mathcal{C}],$$

and

$$[\mathcal{A} - \mathcal{B}\mathcal{B}^*Q(I - PQ)^{-1}][I - PQ] = [I - PQ][\mathcal{A} - (I - PQ)^{-1}P\mathcal{C}^*\mathcal{C}].$$

2. *If $(\mathcal{A}, \mathcal{B}, \mathcal{C}, \mathcal{D}) \in US^{m,m}_{n,P}$ then*

$$[\mathcal{A} - \mathcal{B}(\mathcal{B}^*Q + \mathcal{C})(I - PQ)^{-1}][I - PQ] = [I - PQ][\mathcal{A} - (I - PQ)^{-1}(P\mathcal{C}^* + \mathcal{B})\mathcal{C}],$$

and

$$[\mathcal{A} - \mathcal{B}\mathcal{B}^*Q(I - PQ)^{-1}][I - PQ] = [I - PQ][\mathcal{A} - (I - PQ)^{-1}P\mathcal{C}^*\mathcal{C}].$$

Proof. 1.) We have

$$[\mathcal{A} - \mathcal{B}(\mathcal{B}^*Q + \mathcal{D}^*\mathcal{C})(I - PQ)^{-1}][I - PQ]$$
$$= \mathcal{A}(I - PQ) - \mathcal{B}(\mathcal{B}^*Q + \mathcal{D}^*\mathcal{C}) = \mathcal{A} - \mathcal{A}PQ - \mathcal{B}\mathcal{B}^*Q - \mathcal{B}\mathcal{D}^*\mathcal{C}$$
$$= \mathcal{A} - [\mathcal{A}P + \mathcal{B}\mathcal{B}^*]Q - \mathcal{B}\mathcal{D}^*\mathcal{C}$$
$$= \mathcal{A} - [-P\mathcal{A}^*]Q - \mathcal{B}\mathcal{D}^*\mathcal{C} = \mathcal{A} - PQ\mathcal{A} - P\mathcal{C}^*\mathcal{C} - \mathcal{B}\mathcal{D}^*\mathcal{C}$$
$$= [I - PQ][\mathcal{A} - (I - PQ)^{-1}(P\mathcal{C}^* + \mathcal{B}\mathcal{D}^*)\mathcal{C}].$$

Adding $\mathcal{B}\mathcal{D}^*\mathcal{C}$ to both sides gives the second identity.

2.) The proof is analogous to the proof of 1.). □

In the following lemma it is shown that the characteristic and inverse characteristic maps preserve minimality and respect system equivalence.

Lemma 5.3. *1. The B-characteristic (P-characteristic) of a bounded real (positive real) system is stable and minimal. The B-characteristics (P-characteristics) of two equivalent systems are equivalent.*

2. The inverse B-characteristic (inverse P-characteristic) of a system in $US^{p,m}_{n,B}$ ($US^{m,m}_{n,P}$) is minimal. The inverse B-characteristics (inverse P-characteristics) of two equivalent systems are equivalent.

Proof. The proof is analogous to the proof of Lemma 4.3. □

The following Proposition shows important connections between the solutions of Riccati equations corresponding to a bounded-real (positive-real) system and the solutions to the Lyapunov equations of its bounded-real (positive-real) characteristic.

Proposition 5.1. *1. Let (A, B, C, D) be a bounded-real (positive-real) system and let Y be the stabilizing solution to the control bounded-real (positive-real) Riccati equation and let Z be the stabilizing solution to the filter bounded-real (positive-real) Riccati equation. Let $(\mathcal{A}, \mathcal{B}, \mathcal{C}, \mathcal{D})$ be its B-characteristic (P-characteristic). Then the Lyapunov equations*

$$\mathcal{A}P + P\mathcal{A}^* = -\mathcal{B}\mathcal{B}^*, \quad \mathcal{A}^*Q + Q\mathcal{A} = -\mathcal{C}^*\mathcal{C}.$$

have solutions

$$P = (I - ZY)^{-1}Z = Z(I - YZ)^{-1}, \quad Q = Y - YZY.$$

2. Let $(\mathcal{A}, \mathcal{B}, \mathcal{C}, \mathcal{D})$ be a stable minimal system in $US^{p,m}_{n,B}$ ($US^{m,m}_{n,P}$). Let P, Q be the solutions to the Lyapunov equations

$$\mathcal{A}P + P\mathcal{A}^* = -\mathcal{B}\mathcal{B}^*, \quad \mathcal{A}^*Q + Q\mathcal{A} = -\mathcal{C}^*\mathcal{C}.$$

Let (A, B, C, D) be its inverse B-characteristic (P-characteristic). Then

$$Y = Q(I - PQ)^{-1} = (I - QP)^{-1}Q$$

is the stabilizing solution to the control bounded-real (positive-real) Riccati equation and $Z = P - PQP$ is the stabilizing solution to the filter bounded-real (positive-real) Riccati equation.

Proof. We only consider the case for bounded-real systems. The proof of the results for positive-real systems is analogous.

1.) We first show that $\mathcal{A}P + P\mathcal{A}^* = -\mathcal{B}\mathcal{B}^*$ with $P = (I - ZY)^{-1}Z = Z(I - YZ)^{-1}$. Since $I - ZY$ is invertible this follows from,

$$\begin{aligned}
(I - ZY)(\mathcal{A}P + P\mathcal{A}^*)(I - YZ) &= (I - ZY)\mathcal{A}Z + Z\mathcal{A}^*(I - YZ) \\
&= (I - ZY)[A_B + BR_B^{-1}B^*Y]Z + Z[A_B + BR_B^{-1}B^*Y]^*(I - YZ) \\
&= A_BZ + ZA_B^* - Z(YA_B + A_B^*Y)Z \\
&\quad +(I - ZY)BR_B^{-1}B^*YZ + ZYBR_B^{-1}B^*(I - YZ) \\
&= -ZC^*S_B^{-1}CZ - BR_B^{-1}B^* - Z[-YBR_B^{-1}B^*Y - C^*S_B^{-1}C]Z \\
&\quad +(I - ZY)BR_B^{-1}B^*YZ + ZYBR_B^{-1}B^*(I - YZ) \\
&= -BR_B^{-1}B^* + ZYBR_B^{-1}B^*YZ \\
&\quad +(I - ZY)BR_B^{-1}B^*YZ + ZYBR_B^{-1}B^*(I - YZ) \\
&= -(I - ZY)BR_B^{-1}B^*(I - YZ) \\
&= -(I - ZY)\mathcal{B}\mathcal{B}^*(I - YZ).
\end{aligned}$$

Let now $Q = Y - YZY$. We are going to show that $\mathcal{A}^*Q + Q\mathcal{A} = -\mathcal{C}^*\mathcal{C}$. We consider

$$\begin{aligned}
&\mathcal{A}^*Q + Q\mathcal{A} \\
&= [A_B + BR_B^{-1}B^*Y]^*Y(I - ZY) + (I - YZ)Y[A_B + BR_B^{-1}B^*Y] \\
&= [A_B^*Y + YBR_B^{-1}B^*Y](I - ZY) + (I - YZ)[YA_B + YBR_B^{-1}B^*Y].
\end{aligned}$$

Using the bounded-real Riccati equation this gives,

$$\begin{aligned}
\mathcal{A}^*Q + Q\mathcal{A} &= [-YA_B - C^*S_B^{-1}C](I - ZY) + (I - YZ)[-A_B^*Y - C^*S_B^{-1}C] \\
&= -C^*S_B^{-1}C(I - ZY) - (I - YZ)C^*S_B^{-1}C - YA_B - A_B^*Y + Y(A_BZ + ZA_B^*)Y \\
&= C^*S_B^{-1}(I - ZY) - (I - YZ)C^*S_B^{-1}C + YBR_B^{-1}B^*Y + C^*S_B^{-1}C \\
&\quad +Y[-ZC^*S_B^{-1}CZ - BR_B^{-1}B^*]Y \\
&= -(I - YZ)C^*S_B^{-1}C(I - ZY) \\
&= -\mathcal{C}^*\mathcal{C}.
\end{aligned}$$

2.) First note that

$$\begin{aligned}
A_B &= A + BR_B^{-1}D^*C \\
&= \mathcal{A} - \mathcal{B}(\mathcal{B}^*Q + \mathcal{D}^*\mathcal{C})(I - PQ)^{-1} + \mathcal{B}R_B^{1/2}R_B^{-1}\mathcal{D}^*S_B^{1/2}\mathcal{C}(I - PQ)^{-1} \\
&= \mathcal{A} - \mathcal{B}\mathcal{B}^*Q(I - PQ)^{-1}.
\end{aligned}$$

Since $I - QP$ is invertible, we have for $Y = Q(I - PQ)^{-1} = (I - QP)^{-1}Q$,

$$\begin{aligned}
&(I - QP)[A_B^*Y + YA_B + YBR_B^{-1}B^*Y + C^*S_B^{-1}C](I - PQ) \\
&= (I - QP)[(\mathcal{A} - \mathcal{B}\mathcal{B}^*Q(I - PQ)^{-1})^*Q(I - PQ)^{-1}
\end{aligned}$$

$$+(I-QP)^{-1}Q(\mathcal{A}-\mathcal{B}\mathcal{B}^*Q(I-PQ)^{-1})$$
$$+(I-QP)^{-1}Q\mathcal{B}R_B^{1/2}R_B^{-1}R_B^{1/2}\mathcal{B}^*Q(I-PQ)^{-1}$$
$$+(I-PQ)^{-*}\mathcal{C}^*S_B^{1/2}S_B^{-1}S_B^{1/2}\mathcal{C}(I-PQ)^{-1}](I-PQ)$$
$$=(I-QP)\mathcal{A}^*Q-Q\mathcal{B}\mathcal{B}^*Q+Q\mathcal{A}(I-PQ)-Q\mathcal{B}\mathcal{B}^*Q+Q\mathcal{B}\mathcal{B}^*Q+\mathcal{C}^*\mathcal{C}$$
$$=\mathcal{A}^*Q+Q\mathcal{A}+\mathcal{C}^*\mathcal{C}-Q(P\mathcal{A}^*+\mathcal{A}P+\mathcal{B}\mathcal{B}^*)Q$$
$$=0,$$

we have verified the first identity. Now with $Z=P(I-QP)$ we have

$$A_BZ+ZA_B^*+ZC^*S_B^{-1}CZ+BR_B^{-1}B^*$$
$$=(\mathcal{A}-\mathcal{B}\mathcal{B}^*Q(I-PQ)^{-1})(I-PQ)P+P(I-QP)(\mathcal{A}-\mathcal{B}\mathcal{B}^*Q(I-PQ)^{-1})^*$$
$$+P(I-QP)(I-PQ)^{-*}\mathcal{C}^*S_B^{1/2}S_B^{-1}S_B^{1/2}\mathcal{C}(I-PQ)^{-1}(I-PQ)P$$
$$+\mathcal{B}R_B^{1/2}R_B^{-1}R_B^{1/2}\mathcal{B}^*$$
$$=\mathcal{A}P-\mathcal{A}PQP-\mathcal{B}\mathcal{B}^*QP+P\mathcal{A}^*-PQP\mathcal{A}^*-PQ\mathcal{B}\mathcal{B}^*+P\mathcal{C}^*\mathcal{C}P+\mathcal{B}\mathcal{B}^*$$
$$=\mathcal{A}^*P+P\mathcal{A}^*+\mathcal{B}\mathcal{B}^*-(\mathcal{A}P+\mathcal{B}\mathcal{B}^*)QP+P\mathcal{C}^*\mathcal{C}P-PQ(P\mathcal{A}^*+\mathcal{B}\mathcal{B}^*)$$
$$=0+P\mathcal{A}^*QP+P\mathcal{C}^*\mathcal{C}P-PQ(P\mathcal{A}^*+\mathcal{B}\mathcal{B}^*)$$
$$=P(\mathcal{A}^*Q+\mathcal{C}^*\mathcal{C})P-PQ(P\mathcal{A}^*+\mathcal{B}\mathcal{B}^*)$$
$$=-P(Q\mathcal{A})P-PQ(P\mathcal{A}^*+\mathcal{B}\mathcal{B}^*)$$
$$=-PQ(\mathcal{A}P+P\mathcal{A}^*+\mathcal{B}\mathcal{B}^*)$$
$$=0,$$

which shows the second identity. Since

$$A_B+BR_B^{-1}B^*Y=\mathcal{A}-\mathcal{B}\mathcal{B}^*Q(I-PQ)^{-1}+\mathcal{B}R_B^{1/2}R_B^{-1}R_B^{1/2}\mathcal{B}^*Q(I-PQ)^{-1}$$
$$=\mathcal{A},$$

is stable and

$$A_B+ZC^*S_B^{-1}C$$
$$=\mathcal{A}-\mathcal{B}\mathcal{B}^*Q(I-PQ)^{-1}+P(I-QP)(I-PQ)^{-*}\mathcal{C}^*S_B^{1/2}S_B^{-1}S_B^{1/2}\mathcal{C}(I-PQ)^{-1}$$
$$=\mathcal{A}-\mathcal{B}\mathcal{B}^*Q(I-PQ)^{-1}+P\mathcal{C}^*\mathcal{C}(I-PQ)^{-1}$$
$$=(I-PQ)[\mathcal{A}-(I-PQ)^{-1}P\mathcal{C}^*\mathcal{C}](I-PQ)^{-1}+P\mathcal{C}^*\mathcal{C}(I-PQ)^{-1}$$
$$=(I-PQ)\mathcal{A}(I-PQ)^{-1},$$

is stable, where we have used Lemma 5.2, we have shown that Z, Y are the stabilizing solutions to the bounded-real Riccati equations.

□

We are now in a position to show that the B-characteristic and P-characteristic are bijections.

Theorem 5.1. *The maps*

$$\chi_B : \quad B_n^{p,m} \to US_{n,B}^{p,m}$$

$$\chi_P : \quad P_n^m \to US_{n,P}^{m,m}$$

are bijections that preserve system equivalence with inverses $\chi_B^{-1} = I\chi_B$ *and* $\chi_P^{-1} = I\chi_P$.

Proof. We are only going to consider the B-characteristic. The proof for the P-characteristic is analogous. We first show that $\chi_B(B_n^{p,m}) \subseteq US_{n,B}^{p,m}$. Let $(A,B,C,D) \in B_n^{p,m}$. It follows that $\lambda_{max}(ZY) < 1$, where Z, Y are the stabilizing solutions to the bounded-real Riccati equations (see e.g. [34]). If $(\mathcal{A},\mathcal{B},\mathcal{C},\mathcal{D}) = \chi_B((A,B,C,D))$ and P, Q are the positive definite solutions to the Lyapunov equations

$$\mathcal{A}P + P\mathcal{A}^* = -\mathcal{B}\mathcal{B}^*, \quad \mathcal{A}^*Q + Q\mathcal{A} = -\mathcal{C}^*\mathcal{C},$$

then as a consequence of Proposition 5.1 we have $PQ = ZY$ and therefore that $\lambda_{max}(PQ) < 1$. Clearly, $I - \mathcal{D}^*\mathcal{D} = I - D^*D > 0$ and hence $(\mathcal{A},\mathcal{B},\mathcal{C},\mathcal{D}) \in US_{n,B}^{p,m}$.

We show next that $I\chi_B(US_{n,B}^{p,m}) \subseteq B_n^{p,m}$. Let $(\mathcal{A},\mathcal{B},\mathcal{C},\mathcal{D}) \in US_{n,B}^{p,m}$ and let $(A,B,C,D) = I\chi_B((\mathcal{A},\mathcal{B},\mathcal{C},\mathcal{D}))$. By Lemma 5.3 the system (A,B,C,D) is minimal. In Proposition 5.1 it was shown that the two bounded real Riccati equations for the system (A,B,C,D) have positive definite stabilizing solutions Y, Z. This together with the fact that $\lambda_{max}(ZY) = \lambda_{max}(PQ) < 1$ implies (see e.g. [34]) that (A,B,C,D) is bounded-real.

That χ_B preserves system equivalence was established in Lemma 5.3. We are next going to show that χ_B is injective, or more precisely that $I\chi_B \cdot \chi_B$ is the identity map. Let $(A,B,C,D) \in B_n^{p,m}$, let $(\mathcal{A},\mathcal{B},\mathcal{C},\mathcal{D}) = \chi_B((A,B,C,D))$ and set $(A_1,B_1,C_1,D_1) := I\chi_B((\mathcal{A},\mathcal{B},\mathcal{C},\mathcal{D}))$. We have

$$D_1 = \mathcal{D},$$

$$B_1 = \mathcal{B}R_B^{1/2} = BR_B^{-1/2}R_B^{1/2} = B,$$

$$C_1 = S_B^{1/2}\mathcal{C}(I - PQ)^{-1} = S_B^{-1/2}S_B^{1/2}C(I - ZY)(I - ZY)^{-1} = C,$$

$$\begin{aligned} A_1 &= \mathcal{A} - \mathcal{B}(\mathcal{B}^*Q + \mathcal{D}^*\mathcal{C})(I - PQ)^{-1} \\ &= A_B + BR_B^{-1}B^*Y - BR_B^{-1/2}R_B^{-1/2}B^*Y \\ &\quad -BR_B^{-1/2}D^*S_B^{-1/2}C(I - ZY)(I - ZY)^{-1} \\ &= A + BR_B^{-1}D^*C - BR_B^{-1}D^*C \\ &= A, \end{aligned}$$

which shows that $I\chi_B \cdot \chi_B((A,B,C,D)) = (A,B,C,D)$ and hence that χ_B is injective. We now show that χ_B is surjective by showing that $\chi_B \cdot I\chi_B$ is the identity map. Let $(\mathcal{A},\mathcal{B},\mathcal{C},\mathcal{D}) \in US^{p,m}_{n,B}$, let $(A,B,C,D) := I\chi_B((\mathcal{A},\mathcal{B},\mathcal{C},\mathcal{D}))$ and set $(\mathcal{A}_1,\mathcal{B}_1,\mathcal{C}_1,\mathcal{D}_1) := \chi_B((A,B,C,D))$. Then

$$\mathcal{D}_1 = \mathcal{D},$$
$$\mathcal{C}_1 = R_B^{-1/2}C(I-ZY) = S_B^{-1/2}S_B^{1/2}\mathcal{C}(I-PQ)^{-1}(I-PQ) = \mathcal{C},$$
$$\mathcal{B}_1 = BR_B^{-1/2} = \mathcal{B}R_B^{1/2}R_B^{-1/2} = \mathcal{B},$$
$$\mathcal{A}_1 = A_B + BR_B^{-1}B^*Y$$
$$= \mathcal{A} - \mathcal{B}\mathcal{B}^*Q(I-PQ)^{-1} + \mathcal{B}R_B^{1/2}R_B^{-1}R_B^{1/2}\mathcal{B}^*Q(I-PQ)^{-1}$$
$$= \mathcal{A}.$$

This shows that χ_B is surjective. Hence we have that χ_B is bijective with inverse $\chi_B^{-1} = I\chi_B$. □

Analogous to the situation for the L-characteristic we can also show that the B-characteristic (P-characteristic) maps bounded-real balanced (positive-real balanced) systems to stable minimal systems whose reachability and observability grammians are diagonal.

Corollary 5.1. *Let $(A,B,C,D) \in B^{p,m}_n$ (P^m_n) and let $(\mathcal{A},\mathcal{B},\mathcal{C},\mathcal{D})$ be its B-characteristic (P-characteristic). Let Y be the stabilizing solution to the control bounded-real (positive-real) Riccati equation and let Z be the stabilizing solutions to the filter bounded-real (positive-real) Riccati equation. If P, Q are the solutions to the Lyapunov equations*

$$\mathcal{A}P + P\mathcal{A}^* = -\mathcal{B}\mathcal{B}^*, \qquad \mathcal{A}^*Q + Q\mathcal{A} = -\mathcal{C}^*\mathcal{C},$$

and

1. *if (A,B,C,D) is bounded-real (positive-real) balanced with bounded-real (positive-real) grammian Σ, then*

$$P = \Sigma(I-\Sigma^2)^{-1}, \qquad Q = \Sigma(I-\Sigma^2).$$

2. *if $(\mathcal{A},\mathcal{B},\mathcal{C},\mathcal{D})$ is Lyapunov balanced with Lyapunov grammian Σ, then*

$$Y = \Sigma(I-\Sigma^2)^{-1}, \qquad Z = \Sigma(I-\Sigma^2).$$

If $\Gamma_S : S^{p,m}_n \to S^{p,m}_n$ is the Lyapunov balanced canonical form, then

$$\Gamma_B := \chi_B^{-1} \circ \Gamma_S \circ \chi_B, \qquad \Gamma_P := \chi_P^{-1} \circ \Gamma_S \circ \chi_P$$

define canonical forms for $B^{p,m}_n$ and P^m_n. But the canonical forms are not in terms of the respective balanced realizations as is clear from the previous corollary. Analogously to the construction of the LQG-balanced canonical form diagonal scaling will produce the desired result.

Lemma 5.4. *1. If $(\mathcal{A}, \mathcal{B}, \mathcal{C}, \mathcal{D}) \in S_n^{p,m}$ is such that the Lyapunov equations*

$$\mathcal{A}P + P\mathcal{A}^* = -\mathcal{B}\mathcal{B}^*, \qquad \mathcal{A}^*Q + Q\mathcal{A} = -\mathcal{C}^*\mathcal{C}$$

have solutions

$$P = \Sigma(I - \Sigma^2)^{-1}, \; Q = \Sigma(I - \Sigma^2)$$

for some positive diagonal matrix $\Sigma < 1$, then

$$\Delta_S((\mathcal{A}, \mathcal{B}, \mathcal{C}, \mathcal{D})) := (T\mathcal{A}T^{-1}, T\mathcal{B}, \mathcal{C}T^{-1}, \mathcal{D})$$

with $T = (I - \Sigma^2)^{\frac{1}{2}}$ is Lyapunov balanced with Lyapunov grammian Σ.

2. If $(A, B, C, D) \in B_n^{p,m}$ is such that the control (filter) bounded-real Riccati equation has the stabilizing solution $Y(Z)$ with

$$Y = \Sigma(I - \Sigma^2)^{-1}, \; Z = \Sigma(I - \Sigma^2)$$

for some positive diagonal matrix $\Sigma < 1$, then

$$\Delta_B((A, B, C, D)) := (TAT^{-1}, TB, CT^{-1}, D)$$

with $T = (I - \Sigma^2)^{\frac{1}{2}}$ is bounded-real balanced with bounded-real grammian Σ.

3. If $(A, B, C, D) \in P_n^m$ is such that the control (filter) positive-real Riccati equation has the stabilizing solution $Y(Z)$ with

$$Y = \Sigma(I - \Sigma^2)^{-1}, \; Z = \Sigma(I - \Sigma^2)$$

for some positive diagonal matrix $\Sigma < 1$, then

$$\Delta_P((A, B, C, D)) := (TAT^{-1}, TB, CT^{-1}, D)$$

with $T = (I - \Sigma^2)^{\frac{1}{2}}$ is positive-real balanced with positive-real grammian Σ.

Proof. The proof is analogous to the proof of Lemma 4.4. □

With the diagonal scaling maps Δ_B, Δ_P set

$$\Gamma_B := \Delta_B \circ \chi_B^{-1} \circ \Gamma_S \circ \chi_B,$$

$$\Gamma_P := \Delta_P \circ \chi_P^{-1} \circ \Gamma_S \circ \chi_P.$$

Then Γ_B (Γ_P) defines a canonical form for bounded-real (positive-real) systems in terms of bounded-real (positive-real) realizations.

The *bounded-real balanced canonical form* and *bounded-real balanced parametrization* result is given in the following Theorem.

Theorem 5.2. *Let (A_b, B_b, C_b, D_b) be a m-input p-output continuous-time system of dimension n. Then the following are equivalent:*

1. (A_b, B_b, C_b, D_b) *is a bounded-real system, i.e in* $B_n^{p,m}$.
2. $(A_b, B_b, C_b, D_b) = (TAT^{-1}, TB, CT^{-1}, D)$ *for some invertible* T, *where* (A, B, C, D) *is in bounded-real balanced form, i.e. there exist block indices* $n_1, \ldots, n_k$, $\sum_{j=1}^k n_j = n$, *parameters* $1 > \sigma_1 > \sigma_2 > \cdots > \sigma_k > 0$ *and* k *families of step sizes* $m = \tau_0^j \geq \tau_1^j \geq \tau_2^j \geq \cdots \geq \tau_{l_j}^j > 0$, $\sum_{i=1}^{l_j} \tau_i^j = n_j$, $1 \leq j \leq k$, *such that*
   a) $B = (\underbrace{\overline{B}_1^T, 0}_{n_1}, \underbrace{\overline{B}_2^T, 0}_{n_2}, \ldots, \underbrace{\overline{B}_k^T, 0}_{n_k})^T R_B^{\frac{1}{2}}$, *where* $\overline{B}_j \in \mathcal{K}^{\tau_1^j \times \tau_0^j}$ *is positive upper triangular,* $1 \leq j \leq k$.
   b) $C = S_B^{\frac{1}{2}}(\underbrace{U_1(\overline{B}_1\overline{B}_1^*)^{\frac{1}{2}}, 0}_{n_1} \underbrace{U_2(\overline{B}_2\overline{B}_2^*)^{\frac{1}{2}}, 0}_{n_2}, \ldots, \underbrace{U_k(\overline{B}_k\overline{B}_k^*)^{\frac{1}{2}}, 0}_{n_k})$, *where* $U_j \in \mathcal{K}^{p \times \tau_1^j}$, $U_j^* U_j = I_{\tau_1^j}$, $1 \leq j \leq k$.
   c)

$$A = \left(\begin{array}{c|c|c|c|c} A(1,1) & \cdots & A(1,i) & \cdots & A(1,k) \\ \hline \vdots & \ddots & \vdots & \vdots & \vdots \\ \hline A(i,1) & \cdots & A(i,i) & \cdots & A(i,k) \\ \hline \vdots & & \vdots & \ddots & \vdots \\ \hline A(k,1) & \cdots & A(k,i) & \cdots & A(k,k) \end{array}\right),$$

where for $1 \leq i \leq k$

$$A(i,i) = \begin{pmatrix} \tilde{A}_{ii} + \mathcal{S}_1^i & (-\mathcal{A}_1^i)^* & & 0 \\ \mathcal{A}_1^i & \mathcal{S}_2^i & \ddots & \\ & \ddots & \ddots & (-\mathcal{A}_{l_i-1}^i)^* \\ 0 & & \mathcal{A}_{l_i-1}^i & \mathcal{S}_{l_i}^i \end{pmatrix},$$

for $1 \leq i, j \leq k$, $i \neq j$

$$A(i,j) = \begin{pmatrix} \tilde{A}_{ij} & 0 & \cdots & 0 \\ 0 & 0 & \cdots & 0 \\ \vdots & \vdots & & \vdots \\ 0 & 0 & \cdots & 0 \end{pmatrix},$$

and

   i. $\mathcal{S}_i^j$ *is a* $\tau_i^j \times \tau_i^j$ *skew-hermitian matrix,* $i = 1, 2, \ldots, l_j$, $1 \leq j \leq k$.
   ii. $\mathcal{A}_i^j$ *is a positive upper triangular* $\tau_{i+1}^j \times \tau_i^j$ *matrix,* $i = 1, 2, \ldots, l_j - 1$, $1 \leq j \leq k$.

iii. $\tilde{A}_{ij} \in \mathcal{K}^{\tau_1^i \times \tau_1^j}$ *is given by*

$$\tilde{A}_{ij} = \frac{1}{\sigma_i^2 - \sigma_j^2}(\sigma_j(1-\sigma_i^2)\overline{B}_i\overline{B}_j^* - \sigma_i(1-\sigma_j^2)(\overline{B}_i\overline{B}_i^*)^{\frac{1}{2}}U_i^*U_j(\overline{B}_j\overline{B}_j^*)^{\frac{1}{2}})$$

$$- \overline{B}_iD^*U_j(\overline{B}_j\overline{B}_j^*)^{\frac{1}{2}},$$

for $1 \leq i,j \leq k, \quad i \neq j$ *and*

$$\tilde{A}_{ii} = -\frac{1}{2\sigma_i}(1+\sigma_i^2)\overline{B}_i\overline{B}_i^* - \overline{B}_iD^*U_i(\overline{B}_i\overline{B}_i^*)^{\frac{1}{2}}, \quad \text{for } 1 \leq i \leq k.$$

Moreover, the system (A,B,C,D) *as defined in* (2) *is bounded-real balanced with bounded-real grammian* $\Sigma_B = diag(\sigma_1 I_{n_1},\ldots,\sigma_k I_{n_k})$, *where* $n_1,\ldots,n_k$ *are the block indices and* $1 > \sigma_1 > \cdots > \sigma_k > 0$ *are the bounded-real singular values of the system in* (2). *The map* Γ_B *that assigns to each system in* $B_n^{p,m}$ *the realization in* (2) *is a canonical form.*

Proof. Let $(A_b,B_b,C_b,D_b) \in B_n^{p,m}$. Then $\chi_B((A_b,B_b,C_b,D_b))$ is in $US_n^{p,m}$. Let $(\mathcal{A},\mathcal{B},\mathcal{C},\mathcal{D}) := \Gamma_S(\chi_B((A_b,B_b,C_b,D_b)))$ be the Lyapunov balanced canonical form of $\chi_B((A_b,B_b,C_b,D_b))$. Since $\Gamma_S := \chi_B^{-1} \circ \Gamma_S \circ \chi_B$ defines a canonical form for $B_n^{p,m}$, it follows that

$$(A,B,C,D) := \Delta_B(\chi_B^{-1}((\mathcal{A},\mathcal{B},\mathcal{C},\mathcal{D})))$$

is a bounded-real balanced system which is equivalent to (A_b,B_b,C_b,D_b). To check that (A,B,C,D) is in the stated form, consider the Lyapunov balanced parametrization of $(\mathcal{A},\mathcal{B},\mathcal{C},\mathcal{D})$. Assume that the parametrization of $(\mathcal{A},\mathcal{B},\mathcal{C},\mathcal{D})$ is given using the notation that was introduced in the proof of Theorem 4.2. The system $(\mathcal{A},\mathcal{B},\mathcal{C},\mathcal{D})$ is Lyapunov balanced with Lyapunov grammian $\Sigma = diag(\sigma_1 I_{n_1}, \sigma_2 I_{n_2},\ldots,\sigma_k I_{n_k})$ with $1 > \sigma_1 > \sigma_2 > \cdots > \sigma_k > 0$. By Corollary 5.1 and Lemma 5.3

$$\begin{aligned}(A,B,C,D) \;=\; & ((I-\Sigma^2)^{-\frac{1}{2}}(\mathcal{A} - \mathcal{B}(\mathcal{B}^*\Sigma - \mathcal{C})(I-\Sigma^2)^{-1})(I-\Sigma^2)^{\frac{1}{2}}, \\ & (I-\Sigma^2)^{-\frac{1}{2}}\mathcal{B}R_B^{\frac{1}{2}}, S_B^{\frac{1}{2}}\mathcal{C}(I-\Sigma^2)^{-\frac{1}{2}}, \mathcal{D}))\end{aligned}$$

Setting $\overline{B}_j := \frac{1}{\sqrt{1-\sigma_j^2}}\mathcal{B}_j$, $1 \leq j \leq k$, it follows that $B = (I-\Sigma^2)^{-\frac{1}{2}}\mathcal{B}R_B^{\frac{1}{2}}$ has the required structure. Since with $U_j := \mathcal{U}_j$,

$$S_B^{\frac{1}{2}}\mathcal{U}_j(\overline{\mathcal{B}}_j\overline{\mathcal{B}}_j^*)^{\frac{1}{2}}\frac{1}{\sqrt{1-\sigma_j^2}} = S_B^{\frac{1}{2}}U_j(\overline{B}_j\overline{B}_j^*)^{\frac{1}{2}}$$

for $1 \leq j \leq k$, it follows that $C = S_B^{\frac{1}{2}}\mathcal{C}(I-\Sigma^2)^{-\frac{1}{2}}$ has the required structure. Note that

$$A = (I-\Sigma^2)^{-\frac{1}{2}}(\mathcal{A} - \mathcal{B}(\mathcal{B}^*\Sigma + \mathcal{D}^*\mathcal{C})(I-\Sigma^2)^{-1})(I-\Sigma^2)^{\frac{1}{2}}$$

$$= (I-\Sigma^2)^{-\frac{1}{2}}\mathcal{A}(I-\Sigma^2)^{\frac{1}{2}} - (I-\Sigma^2)^{-\frac{1}{2}}\mathcal{B}(\mathcal{B}^*\Sigma+\mathcal{D}^*\mathcal{C})(I-\Sigma^2)^{-1})(I-\Sigma^2)^{\frac{1}{2}}$$

$$= (I-\Sigma^2)^{-\frac{1}{2}}\mathcal{A}(I-\Sigma^2)^{\frac{1}{2}} - BR_B^{-\frac{1}{2}}R_B^{-\frac{1}{2}}B^*\Sigma - BR_B^{-\frac{1}{2}}D^*S_B^{-\frac{1}{2}}C$$

$$= (I-\Sigma^2)^{-\frac{1}{2}}\mathcal{A}(I-\Sigma^2)^{\frac{1}{2}} - BR_B^{-1}B^*\Sigma - BR_B^{-\frac{1}{2}}D^*S_B^{-\frac{1}{2}}C.$$

If (A,B,C,D) is partitioned according to the block indices $n_1, n_2, \ldots, n_k$, then for $1 \le i,j \le k$, $i \ne j$,

$$A_{ij} = \frac{\sqrt{1-\sigma_j^2}}{\sqrt{1-\sigma_i^2}}\mathcal{A}_{ij} - B_iR_B^{-1}B_j^*\sigma_j - B_iR_B^{-\frac{1}{2}}D^*S_B^{-\frac{1}{2}}C_j.$$

This show that all entries A_{ij} are zero with the exception of the principal $\tau_1^i \times \tau_1^j$-subblock $\tilde{A}_{ij}$ which is given by

$$\tilde{A}_{ij} = \frac{\sqrt{1-\sigma_j^2}}{\sqrt{1-\sigma_i^2}}\tilde{\mathcal{A}}_{ij} - \overline{B}_i\overline{B}_j^*\sigma_j - \overline{B}_iD^*U_j(\overline{B}_j\overline{B}_j^*)^{\frac{1}{2}}$$

$$= \frac{\sqrt{1-\sigma_j^2}}{\sqrt{1-\sigma_i^2}}\left(\frac{1}{\sigma_i^2-\sigma_j^2}\left(\sigma_j\overline{\mathcal{B}}_i\overline{\mathcal{B}}_j^* - \sigma_i(\overline{\mathcal{B}}_i\overline{\mathcal{B}}_i^*)^{\frac{1}{2}}\mathcal{U}_i^*\mathcal{U}_j(\mathcal{B}_j\overline{\mathcal{B}}_j^*)^{\frac{1}{2}}\right)\right)$$

$$-\overline{B}_i\overline{B}_j^*\sigma_j - \overline{B}_iD^*U_j(\overline{B}_j\overline{B}_j^*)^{\frac{1}{2}}$$

$$= \frac{1-\sigma_j^2}{\sigma_i^2-\sigma_j^2}\left(\sigma_j\overline{B}_i\overline{B}_j^* - \sigma_i(\overline{B}_i\overline{B}_i^*)^{\frac{1}{2}}U_i^*U_j(\overline{B}_j\overline{B}_j^*)^{\frac{1}{2}}\right)$$

$$-\overline{B}_i\overline{B}_j^*\sigma_j - \overline{B}_iD^*U_j(\overline{B}_j\overline{B}_j^*)^{\frac{1}{2}}$$

$$= \frac{1}{\sigma_i^2-\sigma_j^2}\left(\sigma_j(1-\sigma_i^2)\overline{B}_i\overline{B}_j^* - \sigma_i(1-\sigma_j^2)(\overline{B}_i\overline{B}_i^*)^{\frac{1}{2}}U_i^*U_j(\overline{B}_j\overline{B}_j^*)^{\frac{1}{2}}\right)$$

$$-\overline{B}_iD^*U_j(\overline{B}_j\overline{B}_j^*)^{\frac{1}{2}}$$

for $1 \le i,j \le k$, $i \ne j$. For $1 \le i \le k$, we have

$$A_{ii} = \mathcal{A}_{ii} - B_iR_B^{-1}B_i^*\sigma_i - B_iR_B^{-\frac{1}{2}}D^*S_B^{-\frac{1}{2}}C_i$$

The principal $\tau_1^i \times \tau_1^i$ submatrix, $1 \le i \le k$, is given by

$$\tilde{\mathcal{A}}_{ii} + \mathcal{S}_1^i - \overline{B}_i\overline{B}_i^*\sigma_i - \overline{B}_iD^*U_i(\overline{B}_i\overline{B}_i^*)^{\frac{1}{2}}$$

$$= -\frac{1}{2\sigma_i}\overline{\mathcal{B}}_i\overline{\mathcal{B}}_i^* - \mathcal{S}_1^i - \overline{B}_i\overline{B}_i^*\sigma_i - \overline{B}_iD^*U_i(\overline{B}_i\overline{B}_i^*)^{\frac{1}{2}}$$

$$= -\frac{1-\sigma_i^2}{2\sigma_i}\overline{B}_i\overline{B}_i^* + S_1^i - \overline{B}_i\overline{B}_i^*\sigma_i - \overline{B}_iD^*U_i(\overline{B}_i\overline{B}_i^*)^{\frac{1}{2}}$$

$$= -\frac{1+\sigma_i^2}{2\sigma_i}\overline{B}_i\overline{B}_i^* - \overline{B}_iD^*U_i(\overline{B}_i\overline{B}_i^*)^{\frac{1}{2}} + S_1^i,$$

where we set $S_i^j := \mathcal{S}_i^j$, for $i = 1, 2, \ldots, l_j$, $1 \le j \le k$. This show that A has the stated form. Hence (A, B, C, D) admits the stated canonical form. The parameterization part of the theorem is shown similar to the analogous part of the proof of Theorem 4.2. □

Specializing the previous result to the single-input single-output case results in the following corollary.

Corollary 5.2. *Let (A_b, b_b, c_b, d_b) be a single-input single-output continuous-time system of dimension n. Then the following statements are equivalent:*

1. *(A_b, b_b, c_b, d_b) is a stable minimal system.*
2. *$(A_b, b_b, c_b, d_b) = (TAT^{-1}, Tb, cT^{-1}, d)$ for some invertible matrix T, where (A, b, c, d) is in the following bounded-real balanced form with block indices $n_1, n_2, \ldots, n_k$, $\sum_{i=1}^k n_i = n$, $|d| < 1$:*

   a) $b = (\underbrace{b_1, 0, \ldots, 0}_{n_1}, \underbrace{b_2, 0, \ldots, 0}_{n_2}, \ldots, \underbrace{b_k, 0, \ldots, 0}_{n_k})^T$, *where* $b_i > 0$, $1 \le i \le k$.

   b) $c = (\underbrace{s_1 b_1, 0, \ldots, 0}_{n_1}, \underbrace{s_2 b_2, 0, \ldots, 0}_{n_2}, \ldots, \underbrace{s_k b_k, 0, \ldots, 0}_{n_k})$, *where* $s_i \in \mathcal{K}$, $|s_i| = 1$, $1 \le i \le k$.

   c)

$$A = \left(\begin{array}{c|c|c|c} A(1,1) & A(1,2) & \cdots & A(1,k) \\ \hline A(2,1) & A(2,2) & \cdots & A(2,k) \\ \hline \vdots & & \ddots & \vdots \\ \hline A(k,1) & A(k,2) & \cdots & A(k,k) \end{array}\right),$$

where for $1 \le j \le k$,

$$A(j,j) = \begin{pmatrix} a_{jj} + i\beta_1^j & -\alpha_1^j & & 0 \\ \alpha_1^j & i\beta_2^j & \ddots & \\ & \ddots & \ddots & -\alpha_{n_j-1}^j \\ 0 & & \alpha_{n_j-1}^j & i\beta_{n_j}^j \end{pmatrix},$$

for $1 \le i, j \le k$, $i \ne j$,

$$A(i,j) = \begin{pmatrix} a_{ij} & 0 & \cdots & 0 \\ 0 & 0 & \cdots & 0 \\ \vdots & \vdots & & \vdots \\ 0 & 0 & \cdots & 0 \end{pmatrix},$$

and

$$\alpha_j^i > 0 \; for \; 1 \le j \le n_i - 1, \;\; 1 \le i \le k,$$

$$a_{ij} = \tfrac{-b_i b_j}{1-|d|^2} \left[\tfrac{\bar{s}_i s_j \sigma_i (1-\sigma_j^2) - \sigma_j (1-\sigma_i^2)}{\sigma_i^2 - \sigma_j^2} + s_j d^* \right] \quad for \; 1 \le i, j \le k, \;\; i \ne j,$$

$$a_{ii} = -\frac{b_i^2}{1-|d|^2} \left[\frac{1}{2\sigma_i}(1+\sigma_i^2) + s_i d^* \right] \; for \; 1 \le i \le k,$$

$$1 > \sigma_1 > \sigma_2 > \cdots > \sigma_k > 0.$$

If $\mathcal{K} = \Re$, *then*

$$s_i = \pm 1, \; for \; 1 \le i \le k,$$

$$\beta_j^i = 0, \; for \; 1 \le j \le n_i, \;\; 1 \le i \le k,$$

$$a_{ij} = \frac{-b_i b_j}{1-d^2} \left[\frac{1 + s_i s_j \sigma_i \sigma_j}{s_i s_j \sigma_i + \sigma_j} + s_j d \right], \; for \; 1 \le i, j \le k.$$

Moreover, the system (A, b, c, d) *as defined in* (2) *is bounded-real balanced with bounded-real grammian* $\Sigma_B = diag(\sigma_1 I_{n_1}, \sigma_2 I_{n_2}, \ldots, \sigma_k I_{n_k})$. *The map* Γ_B *that assigns to each system in* $B_n^{1,1}$ *the realization in* (2) *is a canonical form.*

The *positive-real balanced canonical form* and *positive-real balanced parametrization* result is given in the following Theorem.

Theorem 5.3. *Let* (A_p, B_p, C_p, D_p) *be a* m*-input* m*-output continuous-time system of dimension* n. *Then the following are equivalent:*

1. (A_p, B_p, C_p, D_p) *is a positive-real system, i.e in* P_n^m.
2. $(A_p, B_p, C_p, D_p) = (TAT^{-1}, TB, CT^{-1}, D)$ *for some invertible* T, *where* (A, B, C, D) *is in positive-real balanced form, i.e. there exist block indices* $n_1, \ldots, n_k$, $\sum_{j=1}^k n_j = n$, *parameters* $1 > \sigma_1 > \sigma_2 > \cdots > \sigma_k > 0$ *and* k *families of step sizes* $m = \tau_0^j \ge \tau_1^j \ge \tau_2^j \ge \cdots \ge \tau_{l_j}^j > 0$, $\sum_{i=1}^{l_j} \tau_i^j = n_j$, $1 \le j \le k$, *such that*
   a) $B = (\underbrace{\overline{B}_1^T, 0}_{n_1}, \underbrace{\overline{B}_2^T, 0}_{n_2}, \ldots, \underbrace{\overline{B}_k^T, 0}_{n_k})^T R_P^{\frac{1}{2}}$, *where* $\overline{B}_j \in \mathcal{K}^{\tau_1^j \times \tau_0^j}$ *is positive upper triangular,* $1 \le j \le k$.
   b) $C = R_P^{\frac{1}{2}}(\underbrace{U_1(\overline{B}_1 \overline{B}_1^*)^{\frac{1}{2}}, 0}_{n_1} \underbrace{U_2(\overline{B}_2 \overline{B}_2^*)^{\frac{1}{2}}, 0}_{n_2}, \ldots, \underbrace{U_k(\overline{B}_k \overline{B}_k^*)^{\frac{1}{2}}, 0}_{n_k})$, *where* $U_j \in \mathcal{K}^{p \times \tau_1^j}$, $U_j^* U_j = I_{\tau_1^j}$, $1 \le j \le k$.
   c)

$$A = \left(\begin{array}{c|c|c|c|c} A(1,1) & \cdots & A(1,i) & \cdots & A(1,k) \\ \hline \vdots & \ddots & \vdots & \vdots & \vdots \\ \hline A(i,1) & \cdots & A(i,i) & \cdots & A(i,k) \\ \hline \vdots & & \vdots & \ddots & \vdots \\ \hline A(k,1) & \cdots & A(k,i) & \cdots & A(k,k) \end{array} \right),$$

where for $1 \le i \le k$

$$A(i,i) = \begin{pmatrix} \tilde{A}_{ii} + \mathcal{S}_1^i & (-\mathcal{A}_1^i)^* & & 0 \\ \mathcal{A}_1^i & \mathcal{S}_2^i & \ddots & \\ & \ddots & \ddots & (-\mathcal{A}_{l_i-1}^i)^* \\ 0 & & \mathcal{A}_{l_i-1}^i & \mathcal{S}_{l_i}^i \end{pmatrix},$$

for $1 \le i,j \le k,\ i \ne j$

$$A(i,j) = \begin{pmatrix} \tilde{A}_{ij} & 0 & \cdots & 0 \\ 0 & 0 & \cdots & 0 \\ \vdots & \vdots & & \vdots \\ 0 & 0 & \cdots & 0 \end{pmatrix},$$

and

i. $\mathcal{S}_i^j$ *is a* $\tau_i^j \times \tau_i^j$ *skew-hermitian matrix,* $i = 1, 2, \ldots, l_j$, $1 \le j \le k$.

ii. $\mathcal{A}_i^j$ *is a positive upper triangular* $\tau_{i+1}^j \times \tau_i^j$ *matrix,* $i = 1, 2, \ldots, l_j - 1$, $1 \le j \le k$.

iii. $\tilde{A}_{ij} \in \mathcal{K}^{\tau_1^i \times \tau_1^j}$ *is given by*

$$\tilde{A}_{ij} = \frac{1}{\sigma_i^2 - \sigma_j^2}(\sigma_j(1-\sigma_i^2)\overline{B}_i\overline{B}_j^* - \sigma_i(1-\sigma_j^2)(\overline{B}_i\overline{B}_i^*)^{\frac{1}{2}}U_i^*U_j(\overline{B}_j\overline{B}_j^*)^{\frac{1}{2}})$$

$$+ \overline{B}_iU_j(\overline{B}_j\overline{B}_j^*)^{\frac{1}{2}},$$

for $1 \le i,j \le k$, $i \ne j$ *and*

$$\tilde{A}_{ii} = -\frac{1}{2\sigma_i}(1+\sigma_i^2)\overline{B}_i\overline{B}_i^* + \overline{B}_iU_i(\overline{B}_i\overline{B}_i^*)^{\frac{1}{2}}, \quad \textit{for } 1 \le i \le k.$$

Moreover, the system (A, B, C, D) *as defined in* (2) *is positive-real balanced with positive-real grammian* $\Sigma_P = diag(\sigma_1 I_{n_1}, \ldots, \sigma_k I_{n_k})$*, where* $n_1, \ldots, n_k$ *are the block indices and* $1 > \sigma_1 > \cdots > \sigma_k > 0$ *are the positive-real singular values of the system in* (2)*. The map* Γ_P *that assigns to each system in* P_n^p *the realization in* (2) *is a canonical form.*

Proof. The proof is analogous to the proof of Theorem 5.2. □

Specializing the previous result to the single-input single-output case results in the following corollary.

Corollary 5.3. *Let* (A_p, b_p, c_p, d_p) *be a single-input single-output continuous-time system of dimension* n*. Then the following statements are equivalent:*

1. (A_p, b_p, c_p, d_p) *is a stable minimal system.*

2. $(A_p, b_p, c_p, d_p) = (TAT^{-1}, Tb, cT^{-1}, d)$ *for some invertible matrix* T, *where* (A, b, c, d) *is in the following positive-real balanced form with block indices* $n_1, n_2, \ldots, n_k$, $\sum_{i=1}^k n_i = n$, $Re(d) > 0$:

   a) $b = (\underbrace{b_1, 0, \ldots, 0}_{n_1}, \underbrace{b_2, 0, \ldots, 0}_{n_2}, \ldots, \underbrace{b_k, 0, \ldots, 0}_{n_k})^T$, *where* $b_i > 0$, $1 \le i \le k$.

   b) $c = (\underbrace{s_1 b_1, 0, \ldots, 0}_{n_1}, \underbrace{s_2 b_2, 0, \ldots, 0}_{n_2}, \ldots, \underbrace{s_k b_k, 0, \ldots, 0}_{n_k})$, *where* $s_i \in \mathcal{K}$, $|s_i| = 1$, $1 \le i \le k$.

   c)
$$A = \left(\begin{array}{c|c|c|c} A(1,1) & A(1,2) & \cdots & A(1,k) \\ \hline A(2,1) & A(2,2) & \cdots & A(2,k) \\ \hline \vdots & & \ddots & \vdots \\ \hline A(k,1) & A(k,2) & \cdots & A(k,k) \end{array}\right),$$

   where for $1 \le j \le j$
$$A(j,j) = \begin{pmatrix} a_{jj} + i\beta_1^j & -\alpha_1^j & & 0 \\ \alpha_1^j & i\beta_2^j & \ddots & \\ & \ddots & \ddots & -\alpha_{n_j-1}^j \\ 0 & & \alpha_{n_j-1}^j & i\beta_{n_j}^j \end{pmatrix},$$

   for $1 \le i, j \le k$, $i \ne j$,
$$A(i,j) = \begin{pmatrix} a_{ij} & 0 & \cdots & 0 \\ 0 & 0 & \cdots & 0 \\ \vdots & \vdots & & \vdots \\ 0 & 0 & \cdots & 0 \end{pmatrix},$$

   and
$$\alpha_j^i > 0 \text{ for } 1 \le j \le n_i - 1, \quad 1 \le i \le k,$$
$$a_{ij} = \frac{-b_i b_j}{2Re(d)} \left[\frac{\bar{s}_i s_j \sigma_i (1-\sigma_j^2) - \sigma_j (1-\sigma_i^2)}{\sigma_i^2 - \sigma_j^2} - s_j \right] \quad \text{for } 1 \le i, j \le k, \quad i \ne j,$$
$$a_{ii} = -\frac{b_i^2}{2Re(d)} \left[\frac{1}{2\sigma_i}(1 + \sigma_i^2) - s_i \right] \text{ for } 1 \le i \le k,$$
$$1 > \sigma_1 > \sigma_2 > \cdots > \sigma_k > 0.$$

   If $\mathcal{K} = \Re$, *then*
$$s_i = \pm 1, \quad \text{for } 1 \le i \le k,$$
$$\beta_j^i = 0, \text{ for } 1 \le j \le n_i, \quad 1 \le i \le k,$$
$$a_{ij} = \frac{-b_i b_j}{2d(s_i s_j \sigma_i + \sigma_j)}(1 - s_i\sigma_i)(1 - s_j\sigma_j), \text{ for } 1 \le i, j \le k.$$

Moreover, the system (A, b, c, d) *as defined in* (2) *is positive-real balanced with positive-real grammian* $\Sigma_P = diag(\sigma_1 I_{n_1}, \sigma_2 I_{n_2}, \ldots, \sigma_k I_{n_k})$. *The map* Γ_P *that assigns to each system in* P_n^1 *the realization in* (2) *is a canonical form.*

We can now use the previous parametrization results to prove a model reduction result for bounded-real (positive-real) balanced systems.

Theorem 5.4. *Let* $(A, B, C, D) \in B_n^{p,m}$ (P_n^m) *be in bounded-real (positive-real) balanced canonical form with bounded-real (positive-real) grammian* Σ. *Let* $1 \leq r < n$. *Then* r*-dimensional balanced approximant* (A_{11}, B_1, C_1, D) *is in* $B_r^{p,m}$ (P_r^m) *and is in bounded-real (positive-real) balanced canonical form with bounded-real (positive-real) grammian* Σ_1, *where* $\Sigma = diag(\Sigma_1, \Sigma_2)$, $\Sigma_1 \in \mathcal{K}^{r \times r}$.

Proof. The proof of this theorem is analogous to the proof of Theorem 4.3. □

In the following Corollary a model reduction result is obtained for bounded-real (positive-real) balanced systems which are not necessarily in the respective canonical form.

Corollary 5.4. *Let* (A, B, C, D) *be a* n*-dimensional bounded-real (positive-real) balanced system in* $B_n^{p,m}$ (P_n^m) *with bounded-real (positive-real) grammian* $\Sigma = diag(\sigma_1 I_{n_1}, \sigma_2 I_{n_2}, \ldots, \sigma_k I_{n_k})$, $\sigma_1 > \sigma_2 > \cdots > \sigma_k > 0$. *Let* $r = n_1 + n_2 + \cdots + n_l$ *for some* $1 \leq l \leq k$. *Then the* r*-dimensional balanced approximant* (A_{11}, B_1, C_1, D) *of* (A, B, C, D) *is in* $B_r^{p,m}$ (P_r^m) *and is bounded-real (positive-real) with bounded-real (positive-real) grammian* $\Sigma_1 = diag(\sigma_1 I_{n_1}, \sigma_2 I_{n_2}, \ldots, \sigma_l I_{n_l})$.

Proof. The proof is analogous to the proof of Corollary 4.3. □

6. Concluding Remarks

In this paper we discussed canonical forms and parametrization results for minimal, stable, bounded-real and positive-real continuous-time systems. In many applications, such as system identification, it is however desirable to have the analogous results for discrete-time systems. Balanced realizations for corresponding classes of discrete-time systems can be defined in a completely analogous way to the continuous-time setting, by balancing solutions to the respective discrete-time Riccati equations (see e.g. [19], [2]). If $DS_n^{p,m}$ is the set of discrete-time stable minimal n-dimensional systems with p-dimensional output space and m-dimensional input space, the *bilinear transform*

$$M: \quad S_n^{p,m} \to DS_n^{p,m}$$
$$(A_c, B_c, C_c, D_c) \mapsto (A_d, B_d, C_d, D_d)$$

where

$$A_d = (I - A_c)^{-1}(I + A_c)$$
$$B_d = \sqrt{2}(I - A_c)^{-1}B_c$$
$$C_d = \sqrt{2}C_c(I - A_c)^{-1}$$
$$D_d = C_c(I - A_c)^{-1}B_c + D_c$$

is a bijection which preserves system equivalence. It also maps bijectively continuous-time bounded-real (positive-real) systems to discrete-time bounded-real (positive-real) systems. Since the bilinear transform also preserves the various notations of balancing, it can be used to carry the continuous-time results over to the discrete-time case to define balanced canonical forms for discrete-time systems (see [21], [26]). In particular

$$D\Gamma_S := M \circ \Gamma_S \circ M^{-1}$$

defines a balanced canonical form for the class of stable discrete-time systems $DS_n^{p,m}$ and

$$D\Gamma_B := M \circ \Gamma_B \circ M^{-1}$$
$$D\Gamma_P := M \circ \Gamma_P \circ M^{-1}$$

define canonical forms for the class $DB_n^{p,m}$ respectively DP_n^m of discrete-time bounded-real systems in $DS_n^{p,m}$ and for the class DP_n^m of discrete-time positive-real systems in $DS_n^{m,m}$.

In this paper only we did not present any results on canonical forms for minimum phase systems. Such results are, however, easily derived from those for positive-real systems using the state space formulae that relate a positive real system to the associated spectral factor ([26]). Such a parametrization for minimum phase systems is of importance in time series analysis, where based on an observed time series an innovative model is to be identified (see e.g. [1], [9], [18], [17]). That balanced realizations may provide a good canonical forms for system identification has for the first time been suggested by Maciejowski ([16]).

The parametrization results for $S_n^{p,m}$, $L_n^{p,m}$, $B_n^{p,m}$ and P_n^m allow an analysis of the number of connected components of the associated manifolds of linear systems ([24], [25]). A disadvantage of the balanced parametrization is, however, that it does not induce an atlas for the manifold of systems. This property of the canonical form is not ideal for the implementation in system identification algorithms where overlapping charts are of importance ([10]). That the canonical forms can be changed to lead to an overlapping parametrization was shown in ([12]).

The importance of Lyapunov balanced realizations in the theory of Hankel operators and H^∞ control is due to the interpretation of the Lyapunov singular values as the singular values of the Hankel operator whose kernel is given by the impulse response of the system. Let $(A, B, C, D) \in S_n^{p,m}$ and set $H(t) := Ce^{tA}B$, $t \geq 0$. Then the *Hankel operator* with kernel H is defined by

$$\mathcal{H}: \; L^2([0,\infty]) \rightarrow L^2([0,\infty])$$

$$u \mapsto (\mathcal{H}(u))(\cdot) = \int_0^\infty H(t+\cdot)u(t)dt$$

This operator has rank n and we have that the singular values of $\mathcal{H}$ are given by

$$\sigma((\mathcal{H}^*\mathcal{H})^{\frac{1}{2}}) = \{\sigma_1, \sigma_2, \ldots, \sigma_n\},$$

where $\sigma_1, \sigma_2, \ldots, \sigma_n$ are the Lyapunov singular values of (A,B,C,D). Using Theorem 3.2 it is therefore quite straightforward to construct finite rank Hankel operators with prescribed singular values. This system theoretic approach was used in ([22], [23], [32]) to solve the inverse spectral problem for Hankel operators.

References

1. Caines P.E., *Linear Stochastic Systems.* John Wiley, New York, 1988.
2. Desai U.B., and D. Pal, "A transformation approach to stochastic model reduction", *IEEE Transactions on Automatic Control*, 29:1097–1100, 1984.
3. Enns D., "Model reduction for control system design", *PhD thesis*, Department of Aeronautics and Astronautics, Stanford University, 1984.
4. Francis B., *A course in H^∞ control theory.* Lecture Notes in Control and Information Sciences, volume 88. Springer, 1987.
5. Fuhrmann P.A., and R. J. Ober, "A functional approach to LQG balancing", *International Journal of Control*, 57:627–741, 1993.
6. Genin Y.P., P. V. Dooren, and T. Kailath, "On Σ-lossless transfer functions and related questions", *Linear Algebra and its Applications*, 50:251–275, 1983.
7. Glover K., "All optimal Hankel-norm approximations of linear multivariable systems and their L^∞-error bounds", *International Journal of Control*, 39:1115–1193, 1984.
8. Guidorzi R.P., "Invariants and canonical forms for systems: structural and parametric identification", *Automatica*, 17:117–133, 1981.
9. Hannan E.J., and M. Deistler, *The statistical theory of linear systems.* John Wiley, New York, 1988.
10. Hanzon B., "Recursive identification and spaces of linear dynamical systems", CWI Tracts 63, 64, CWI Amsterdam, 1989.
11. Hanzon B., "A new balanced canonical form for stable multivariable systems", *Research Memorandum*, Vakgroep Econometrie 1993-13, Free University Amsterdam, 1993.
12. Hanzon B., and R. J. Ober, "Overlapping block-balanced canonical forms and parametrizations: the SISO case", *SIAM Journal on Control and Optimization.* To appear.
13. Hinrichsen D., "Canonical forms and parametrization problems in linear system theory", in P. Cook, editor, *Fourth IMA International Conference on Control Theory.* Academic Press, New York, 1986.
14. Jonckheere E.A., and L. M. Silverman, "A new set of invariants for linear systems–application to reduced order compensator design", *IEEE Transactions on Automatic Control*, 28:953–964, 1983.
15. Kabamba P.T., "Balanced forms: canonicity and parametrization", *IEEE Transactions on Automatic Control*, 30:1106–1109, 1985.
16. Maciejowski J.M., "Balanced realizations in system identification", in *IFAC Identification and System Parameter Estimation, York, U.K.*, pages 1823–1827. Pergamon Press, Oxford, 1985.
17. McGinnie B.P., J. M. Maciejowski, and R. J. Ober, "Balanced realizations in system identification", Monograph, in preparation.
18. McGinnie B.P., R. J. Ober, and J. M. Maciejowski, "Balanced parametrizations in time series identification", in *Proceedings Conference on Decision on Control*, 1990.
19. Moore B.C., "Principal component analysis in linear systems: controllability, observability and model reduction", *IEEE Transactions on Automatic Control*, 26:17–32, 1981.
20. Mullis C.T., and R. A. Roberts, "Synthesis of minimum roundoff noise fixed point digital filters", *IEEE Transactions on Circuits and Systems*, CAS 23:551–562, 1976.
21. Ober R.J., "Balanced realizations: canonical forms, parametrizations, model reduction", *International Journal of Control*, 46:643–670, 1987.

22. Ober R.J., "A note on a system theoretic approach to a conjecture by Peller-Khrushchev", *Systems and Control Letters*, 8:303–306, 1987.
23. Ober R.J., "A note on a system theoretic approach to a conjecture by Peller-Khrushchev: the general case", *IMA Journal of Mathematical Control and Information*, 4:263–279, 1987.
24. Ober R.J., "Topology of the set of asymptotically stable systems", *International Journal of Control*, 46:263–280, 1987.
25. Ober R.J., "Connectivity properties of classes of linear systems", *International Journal of Control*, 50:2049–2073, 1989.
26. Ober R.J., "Balanced parametrization of classes of linear systems", *SIAM Journal on Control and Optimization*, 29:1251–1287, 1991.
27. Ober R.J., and P. A. Fuhrmann, "Diffeomorphisms between sets of linear systems", in P. V. Dooren and B. Wyman, editors, *Linear Algebra for Control Theory*, IMA Volumes in Mathematics and Its Applications, volume 62, pages 117–157. Springer, 1994.
28. Ober R.J., and D. McFarlane, "Balanced canonical forms for minimal systems: a normalized coprime factorization approach", *Linear Algebra and its Applications*, 122-124:23–64, 1989.
29. Opdenacker P.C., and E. A. Jonckheere. "A contraction mapping preserving balanced reduction scheme and its infinity norm error bounds", *IEEE Transactions on Circuits and Systems*, 35:184–189, 1988.
30. Pernebo L., and L. M. Silverman, "Model reduction via balanced state space representations", *IEEE Transactions on Automatic Control*, 37:382–387, 1982.
31. Rissanen J., "Basics of invariants and canonical forms for linear dynamic systems", *Automatica*, 10:175–182, 1974.
32. Treil S.R., "The inverse spectral problem for the modulus of a Hankel operator and balanced realizations", *Leningrad Mathematical Journal*, 2:353–375, 1991.
33. Verriest E.I., "Low sensitivity design and optimal order reduction for the LQG-problem", in *Proceedings 24th Midwest Symposium on Circuits and Systems Albuquerque, New Mexico*, June 1981.
34. Willems J.C., "Least squares stationary optimal control and algebraic Riccati equation", *IEEE Transactions on Automatic Control*, 16:621–634, 1971.

From Data to State Model

Paolo Rapisarda[1] and Jan C. Willems[2]

[1] Dipartimento di Ingegneria Elettronica, Elettrotecnica ed Informatica
Universita' di Trieste
34127 Trieste
Italy
Email: rapisard@univ.trieste.it

[2] Institute of Mathematics and Computing Science
University of Groningen
9700 AV Groningen
The Netherlands
Email: willems@math.rug.nl

1. Introduction

System identification deals with the problem of constructing models of systems, based on observed or experimental data. The models are chosen in a model class, a set of candidate models, which reflects the *a priori* available knowledge about the system and the assumptions the modeler is willing to make on the nature of the system under study. Given some measurements of the relevant characteristic of the system under study, the choice of a model in the model class is based on a criterion, which determines the quality of the model with respect to the available data. On the basis of this criterion, the identification procedure selects the model among those available.

In the field of engineering and physics, a mathematical model consists of a number of relations among the variables describing the system under study. In the case of dynamical systems, these relations are usually given as difference or differential equations. In this case, the procedure for identifying a dynamical system consists in choosing a suitable model class (for example, that of linear systems), and to select, on the basis of the data and the identification criterion, the "best" model in the class - for example, the model which reproduces the data most accurately. This identification procedure typically involves specifying a model structure as a starting point. This consists of imposing an input-output structure on the system, reflecting the cause-effect relationships which the modeler believes are present in the system, and of setting other structural characteristics, such as the orders of the difference or differential equations describing the system. Moreover, a number of unknown parameters are specified, which are to be determined or adjusted on the basis of the available data. In some cases, an *identification experiment* is set up, in which the system is excited by means of known "input signals", and its corresponding responses measured.

In general, no matter how accurately the identification experiment or the measurements have been set up, and how carefully the structural characteristics of the model have been chosen to start with, the model obtained by the identification procedure will not fit the data exactly. It is common practice, for solving this problem, to adopt the philosophy of statistics, and to include randomness in the model, in the form of additional, uncontrollable inputs, and as measurement noise. In this way a stochastic model is obtained from the data, which is thus considered to be a realization of a stochastic process.

The approach to system identification described in this paper, first put forward in [13, 14, 15], departs from the classical one.

To begin with, no stochastic assumptions will be made about the nature of the data generating mechanism: the approach is *completely deterministic.* There are many good reasons to avoid stochastic assumptions when modeling. First, very often the problem is that of modeling a single time series, which cannot be considered as a sample from a population of time series. Therefore, unless additional assumptions are made, this renders a stochastical analysis of the given time series impossible. A second reason is that, more often than not, the lack of fit between data and model should be attributed to the fact that the system is too complex for the model class chosen. Hence, approximation will be better suited than stochasticity, in order to penetrate the cause why the model is unable to explain the data exactly.

The deterministic point of view of [14] considers as the very first problem of identification that of finding models which explain the data *exactly.* This leads to the development of a language to formulate and to analyse deterministic identification problems. The philosophy underlying the approach is to let the data speak as much as possible, that is, to restrain from making unnecessary assumptions on the nature of the system. This implies, for example, that no *a priori* assumptions on the causality structure of the system are made, and that the input-output structure is entirely deduced from the data. Moreover, other structural issues, as the number of equations of the model, are settled on the basis of the data, rather than being postulated arbitrarily.

The problem we are concerned with in this paper is the identification of real vector time series, and the model class is that of discrete time, finite dimensional, time-invariant linear systems. The elements of this class can be represented in many different ways, which have been discussed at length in [13]; among these different forms, those in which we will be most interested in are *kernel representations*, which correspond to systems of homogeneous difference equations, and *state space representations.*

The exact identification procedures which we will describe are those set up in [14], which pass from a time series $\tilde{w}$, consisting of measurements of the relevant attributes of a phenomenon, to a model which explains $\tilde{w}$ and as little else as possible. The algorithms compute a kernel representation or a state space representation of the model. Our aim will be that of deriving a state space model from the data. In order to do this, two paths are available. The first one consists in computing a kernel representation of the model, and deriving from this representation a state space model. This can be done by determining a set of state variables for the system, and computing on their basis the state space equations. The paper [7] has dealt with the problem of obtaining state variables from different kinds of representations of a system. In this paper we will summarize those results, and we will describe how to pass from a set of state variables and a given representation to a state space model. This is a rather indirect way of solving the identification of a state space model, since it requires first setting up a representation in terms of higher order difference equations. The second procedure consists of computing a state model directly from the data. Such algorithms have been derived in [14]. They will be described at length in this paper, and applied to numerical data in a series of simulations.

Recently considerable attention has been directed towards *subspace identification methods* ([6], [10], [11], [12]). These algorithms are based on the computation of a subspace, defined by the span of the rows of matrices derived from the input-output data. In this paper we will illustrate the relationships between these methods and the exact identification algorithms of [14]. The concept of *common features* between two sets of observations will be used to illustrate these relationships.

Exact identification is a topic which, although constituting the first step towards a rigorous foundation of identification, remains mainly of theoretical interest. In fact, in real applications the phenomenon to be modeled is too complex to be described within the model class, and there will always be a lack of fit between the model and the data. Rather than invoking randomness and attributing the lack of fit to "noise", it may be more sound to reformulate the identification problem as an approximation one: given some data, produced by a system which cannot be represented exactly by an element of the model class, find a model which gives a "good", but not necessarily exact, fit on the data. The paper [15] is devoted to the foundation of approximate modeling. Two methodologies are set up for solving this problem, and we will briefly discuss them in a section of this paper.

The paper is organized as follows.

In section 2. an overview of the behavioral approach to dynamical systems is given, and some basic notions are introduced. Among these, the different kind of representations available for linear, time-invariant, finite dimensional systems are discussed.

Section 3. discusses the problem of computing a state space representation of a system. For this purpose, the concept of *state map* is introduced, and procedures are described for obtaining state equations.

In section 4. the general framework of exact identification is described, and the crucial notion of the most powerful unfalsified model is introduced. This is the model which explains the data and as little else as possible. The main ideas underlying the procedures of [14] for state space modeling are discussed, and classical realization theory is cast into the exact identification framework of [14].

Section 5. illustrates the algorithms to derive a state model from time series. The interpretation of the state as the variables which capture the "common features" of past and present is discussed. The connections among the procedures of [14] for state space modeling, and what has been lately known as *subspace identification*, are described.

In section 6. a brief description of the main features of the approximate modeling procedures of [15] is given. Two methodologies are described for obtaining an optimal approximate model.

Section 7. contains some numerical experiments, illustrating the application of the procedures of [14] to the problem of modeling a noncausal system.

A section devoted to notation concludes the paper.

2. Background

In this section we will give an introduction to the behavioral framework to system theory, emphasizing the notions which are more closely related to the problem of obtaining models from time series; we refer the reader to [13, 16, 17] for a thorough exposition. Following [13, 16, 17], a dynamical system is defined as a family of trajectories, and a characterization of linear, time invariant and complete systems is given. This definition is generalized to the case in which the description of the system involves two sets of variables, the first corresponding to the evolution of the manifest variables, the variables of direct interest, and the second corresponding to the evolution of the latent variables, introduced mainly for ease of modeling. Among latent variables, a special role will be given to state variables.

2.1 Discrete Time Systems

Let us first formalize the very basic notion of *system*.

Definition 2.1. *A* discrete time dynamical system *is a triple* $\Sigma = (\mathbb{Z}, \mathbb{R}^q, \mathcal{B})$ *with* $\mathbb{Z}$ *the* time set, $\mathbb{R}^q$ *the* signal space, *and* $\mathcal{B}$ *the* behavior *of the system.*

The *behavior* of the system is the central object of interest in the approach to system theory put forward in [13, 16, 17]. It consists of all trajectories $w : \mathbb{Z} \to \mathbb{R}^q$ which are compatible with the laws governing the system.

Example 2.1. (Samuelson's multiplier-accelerator model) The typical multiplier-accelerator model of income determination and of the business cycle is the following. Let Y be the national income, and C the national consumption. Denote with β the marginal average propensity to consume, $0 < \beta < 1$; the relation between Y and C is

$$C(t+1) = \beta Y(t). \tag{2.1}$$

Denote with I the total amount of money invested at time t in the whole country. It is the sum of the private investments I' and of the public investments I'':

$$I(t) = I'(t) + I''(t). \tag{2.2}$$

I' varies according to the demand for consumption goods:

$$I'(t+1) = \gamma(C(t+1) - C(t)) \tag{2.3}$$

with γ a constant. The final equation of the model is

$$Y(t) = C(t) + I(t). \tag{2.4}$$

These equations define a discrete-time system as follows. The variables involved are Y, C, I', I'', I, that is, $q = 5$. The behavior $\mathcal{B}$ is defined as

$$\mathcal{B} = \{(Y, C, I', I'', I) \mid eqs.\ (2.1) - (2.4)\ hold\}. \tag{2.5}$$

□

Definition 2.1 considers all variables involved in the description of the system at the same level: no structure (for example, input-output, i.e., cause-effect relationship among the variables) is postulated *a priori*. This is particularly important in identification and modeling, in which a clear *a priori* distinction between cause and effect is in many cases impossible.

In this paper we deal with linear, time-invariant, and complete systems over $\mathbb{Z}$.

Definition 2.2. *$\Sigma = (\mathbb{Z}, \mathbb{R}^q, \mathcal{B})$ is* linear *if $\mathcal{B}$ is a linear subspace of $(\mathbb{R}^q)^{\mathbb{Z}}$, and* time invariant *if, $\forall\ \tau \in \mathbb{Z}$, $\sigma^\tau \mathcal{B} = \mathcal{B}$.*

Here σ^τ denotes the *τ-shift*: for $w \in (\mathbb{R}^q)^{\mathbb{Z}}$, $\sigma^\tau w$ is defined as $(\sigma^\tau w)(t) = w(t+\tau)$.

Therefore in a linear system any linear combination of trajectories is still an admissible trajectory of the system. Time invariance, on the other hand, states that the laws governing the system do not change with time.

Completeness is defined as follows:

Definition 2.3. *Let $\Sigma = (\mathbb{Z}, \mathbb{R}^q, \mathcal{B})$ be a dynamical system. Then it is* complete *if*

$$\{w \in \mathcal{B}\} \iff \{w_{|[t_0,t_1]\bigcap \mathbb{Z}} \in \mathcal{B}_{|[t_0,t_1]\bigcap \mathbb{Z}} \quad \forall\ -\infty < t_0 \leq t_1 < +\infty\} \tag{2.6}$$

Completeness of a dynamical system is equivalent to the fact that the properties of a trajectory at plus or minus infinity are not important to decide whether the trajectory belongs to the behavior or not [1].

The following mathematical characterization of linear, time-invariant, complete systems holds:

---

[1]

Willst du ins Unendliche schreiten
Geh nur im Endlichen nach allen Seiten
(J.W. Goethe, *Sprüche in Reimen: Gott, Gemüth und Welt*).

Theorem 2.1. *([13]) Let $\Sigma = (\mathbb{Z}, \mathbb{R}^q, \mathcal{B})$. Then it is linear and complete if and only if $\mathcal{B}$ is a linear subspace of $(\mathbb{R}^q)^{\mathbb{Z}}$ closed in the topology of pointwise convergence.*

A concrete characterization of linear, time invariant and complete systems can be given in terms of polynomial operators in the shift.

Theorem 2.2. *([13]) The system $\Sigma = (\mathbb{Z}, \mathbb{R}^q, \mathcal{B})$ is linear, time invariant and complete if and only if $\mathcal{B}$ is the kernel of a polynomial operator in the shift.*

Polynomial operators in the shift have the following representation. Let $R_i \in \mathbb{R}^{g\times q}$, $i = -l, -l+1, \ldots, 0, \ldots, L$, and let a matrix dipolynomial be associated with the matrices R_i's as

$$R(\xi) := R_{-l}\xi^{-l} + R_{-l+1}\xi^{-l+1} + \ldots + R_0 + R_1\xi + \ldots + R_L\xi^L \in \mathbb{R}^{g\times q}[\xi, \xi^{-1}]; \tag{2.7}$$

$R(\xi, \xi^{-1})$ defines a polynomial operator in the shift

$$\begin{aligned} R(\sigma, \sigma^{-1}) &: (\mathbb{R}^q)^{\mathbb{Z}} \to (\mathbb{R}^g)^{\mathbb{Z}} \\ R(\sigma, \sigma^{-1})(w) &:= R_{-l}\sigma^{-l}w + \ldots + R_0 w + R_1\sigma w + \ldots + R_L\sigma^L w. \end{aligned}$$

Theorem 2.2 states that Σ is linear, time invariant, and complete if and only if there exists a dipolynomial matrix R such that $w \in \mathcal{B}$ if and only if

$$R(\sigma, \sigma^{-1})w = 0. \tag{2.8}$$

By time invariance, (2.8) is equivalent to

$$\sigma^l R(\sigma, \sigma^{-1})w = 0 \tag{2.9}$$

that is, to

$$R'(\sigma)w = 0 \tag{2.10}$$

with $R'(\xi) = R_{-l} + \ldots + R_0\xi^l + \ldots + R_L\xi^{l+L}$. Therefore the behavior of a dscrete time system with time axis $\mathbb{Z}$ can be represented as the kernel of a polynomial operator in the shift defined by a *polynomial* matrix, that is, by an element of $\mathbb{R}^{\bullet\times\bullet}[\xi]$. In the sequel of the paper we will assume that this is the case, and through the paper we will use only operators in the shift defined by matrices in $\mathbb{R}^{\bullet\times\bullet}[\xi]$.

For obvious reasons, a representation (2.10) of $\mathcal{B}$ will be called a *kernel representation.* The result of theorem 2.2 is of great importance, given the fact that many real-world phenomena are effectively described by linear relations among the variables of interest and their lags.

Example 2.2. (Example 1 - continued) Samuelson's multiplier-accelerator model defines a linear, time invariant, and complete system. Its kernel representation is

$$\begin{aligned} \sigma C &= \beta Y \\ I &= I' + I'' \\ \sigma I &= \gamma(\sigma - 1)C \\ Y &= C + I \end{aligned} \tag{2.11}$$

so that, with the ordering (Y, C, I', I'', I) of the variables, the polynomial matrix R associated with (2.11) is

$$R(\xi) = \begin{pmatrix} \beta & -\xi & 0 & 0 & 0 \\ 0 & 0 & 1 & 1 & -1 \\ 0 & -\gamma(\xi - 1) & 0 & 0 & \xi \\ 1 & -1 & 0 & 0 & -1 \end{pmatrix} \tag{2.12}$$

□

Note that a system has an infinite number of (equivalent) kernel representations. Given any matrix R such that $\mathcal{B} = Ker\ R(\sigma)$, one can obtain a different representation of $\mathcal{B}$ stacking R on top of some polynomial combinations of its rows. Even when restricting attention to full row rank representations, any matrix $R' = UR$, with U unimodular, represents $\mathcal{B}$ in kernel form (see [16]).

In definition 2.1, all variables w are treated on an equal footing. A special case of kernel representation occurs when the variables w are split in two sets of components. Let $P \in \mathbb{R}^{p \times p}[\xi]$ and $Q \in \mathbb{R}^{p \times m}[\xi]$ be given, with $det(P) \neq 0$ and $P^{-1}Q$ proper. Now consider the system $\Sigma = (\mathbb{Z}, \mathbb{R}^q, \mathcal{B})$, with $\mathcal{B}$

$$\mathcal{B} = \{w = col(u, y) \mid P(\sigma)y = Q(\sigma)u\} \tag{2.13}$$

with $u \in (\mathbb{R}^m)^{\mathbb{Z}}$, $y \in (\mathbb{R}^p)^{\mathbb{Z}}$. This will be called an *input-output representation* of $\mathcal{B}$, with u the *inputs*, and y the *outputs*, of the system. Note that (2.13) is a kernel representation of $\mathcal{B}$, with the matrix R defined as $R = \begin{pmatrix} Q & -P \end{pmatrix}$.

On the other hand, every kernel representation may be transformed, by appropriate selection of components in w, in an input-output representation. That is, given any kernel representation associated with $R \in \mathbb{R}^{g \times q}[\xi]$, there exists a permutation matrix $T \in \mathbb{R}^{q \times q}$ and polynomial matrices P and Q, with P nonsingular and $P^{-1}Q$ proper, such that

$$col(u, y) = Tw \in \{col(u, y) \mid P(\sigma)y = Q(\sigma)u\} \iff w \in KerR(\sigma). \tag{2.14}$$

That is, given a model in the class of linear, time invariant, and complete systems, one can always obtain a partition of the variables in inputs and outputs, without having to assume any causality structure *a priori*; this is one of the most distinctive and innovative aspects of the behavioral framework.

The notion of input in the behavioral framework involves three properties of a variable: it must determine, along with the initial conditions and the laws of the system, the evolution of the output variables (processing); it can be chosen freely, i.e., it can be any trajectory in $\mathbb{R}^{\mathbb{Z}}$; it does not anticipate the output, i.e., its past does not contain information about the future of the output other than the information already contained in the dynamical laws. From a modeling point of view, inputs are to be interpreted as signals which are unexplained, but imposed by the environment. The outputs, on the other hand, are to be interpreted as determined by the inputs and the initial conditions. We refer the reader to [16] for the details.

It is important to note that the selection of P and Q of (2.13) is not unique, in general. This implies that for a system whose behavior is described by (2.10), different selections of inputs and outputs can be given. However, it is possible to prove (see [13]) that the number of outputs in any representation (2.10) of the behavior of the system is unique, and coincides with $rank(R)$.

The componentwise partition between inputs and outputs happens at the level of the variables w. Another way of distinguishing between different kinds of variables is to introduce latent variables, besides the manifest variables. This reflects a common situation in modeling, where the need arises to introduce *auxiliary variables*, in addition to those constituting the attributes of the system which we are trying to model. For example, when considering electrical circuits and their behavior, one is forced to introduce currents and voltages in the internal branches in order to describe the voltages and the currents at the external ports. In the next section we introduce the notion of system with latent variable, which formalizes situations like this one.

2.2 Latent Variables

The following definition formalizes the notion of system with auxiliary variables.

Definition 2.4. *A* discrete-time system with latent variables *is a quadruple* $\Sigma_f = (\mathbb{Z}, \mathbb{R}^q, \mathbb{R}^d, \mathcal{B}_f)$ *with* $\mathbb{Z}$ *the* time set, $\mathbb{R}^q$ *the* manifest signal space, $\mathbb{R}^d$ *the* latent signal space, *and* $\mathcal{B}_f \subseteq (\mathbb{R}^q \times \mathbb{R}^d)^{\mathbb{Z}}$ *the* full behavior.

A system with latent variables defines a latent variable representation of the *external* or *manifest system* $\Sigma = (\mathbb{Z}, \mathbb{R}^q, \mathcal{B})$ with $\mathcal{B}$, the *manifest behavior*, defined as

$$\mathcal{B} = \{\, w : \mathbb{Z} \to \mathbb{R}^q \mid \exists\, \ell : \mathbb{Z} \to \mathbb{R}^d \;\; s.t. \;\; (w, \ell) \in \mathcal{B}_f \}. \tag{2.15}$$

Equivalently, considering the projection on the external variable π_w, defined as $\pi_w((w, \ell)) = w$, one can interpret $\mathcal{B}$ as the result of the projection of $\mathcal{B}_f$: $\mathcal{B} = \pi_w(\mathcal{B}_f)$.

In the context of linear, time-invariant, and complete systems, the full behavior of a system Σ_f with latent variable, can be represented by equations of the form

$$R(\sigma)w = M(\sigma)\ell, \tag{2.16}$$

where $R \in \mathbb{R}^{\bullet \times q}[\xi]$, $M \in \mathbb{R}^{\bullet \times d}[\xi]$. We call (2.16) a *hybrid representation* of the full behavior of the system.

The external system associated with a linear, time-invariant, and complete system with latent variable Σ_f is itself linear, time invariant and complete, as the following result shows.

Theorem 2.3. *([13]) Let $\Sigma_f = (\mathbb{Z}, \mathbb{R}^q, \mathbb{R}^d, \mathcal{B}_f)$ be described by (2.16). Then there exists a polynomial matrix $\bar{R} \in \mathbb{R}^{\bullet \times q}[\xi]$ such that the external system Σ_e of Σ_f has the kernel representation*

$$\bar{R}(\sigma)w = 0. \tag{2.17}$$

This theorem will be referred to in the following as the *latent variable elimination theorem.* It states that no matter how many auxiliary variables are introduced in the modeling process, the model corresponding to the variables of interest will consist, after elimination of the latent variable, of a kernel representation. The proof of theorem 2.3 actually suggests an algorithm to perform the elimination of the latent variables. We summarize this procedure as follows: given a full row rank hybrid representation

$$R(\sigma)w = M(\sigma)\ell, \tag{2.18}$$

find a unimodular matrix U such that

$$U\,(\,R \quad -M\,) = \begin{pmatrix} R_1' & 0 \\ R_2' & -M_2' \end{pmatrix}, \tag{2.19}$$

with M_2' of full row rank. The fact that U is unimodular implies that the full behavior of

$$\begin{aligned} R_1'(\sigma)w &= 0 && (2.20)\\ R_2'(\sigma)w &= M_2'(\sigma)\ell && (2.21) \end{aligned}$$

is the same as that of (2.18). It turns out that (2.20) is the desired kernel description of the external behavior of the original system (2.18).

Example 2.3. (Example 1 - continued) An economist may be interested mainly in the evolution of national income, and interpret Y as the external manifest variable, while consumption, and induced, total and autonomous investment, are considered as auxiliary latent variables. The model can then be written as

$$\begin{aligned} \beta Y &= \sigma C \\ 0 &= I - I' - I'' \\ Y &= C + I \\ 0 &= \sigma I - \gamma(\sigma - 1)C. \qquad (2.22) \end{aligned}$$

After elimination of the auxiliary variables, the manifest behavior is represented as

$$\beta(-\sigma + \gamma)Y + \sigma^2 Y = 0. \tag{2.23}$$

□

The notions of controllability and observability emerge in the behavioral framework as follows. The time invariant system $(\mathbb{Z}, \mathbb{R}^q, \mathcal{B})$ is said to be *controllable* if for all w_1, w_2 in $\mathcal{B}$, there exists a $T \geq 0$ and a $w \in \mathcal{B}$ such that $w(t) = w_1(t)$ for $t < 0$ and $w(t+T) = w_2(t)$ for $t \geq 0$. The notion of observability deals with latent variable systems, and refers to the possibility of deducing the latent variables from the manifest ones. Thus (2.18) defines an *observable* system if there exists a map $F : (\mathbb{R}^q)^{\mathbb{Z}} \mapsto (\mathbb{R}^d)^{\mathbb{Z}}$ such that $((w, \ell) \in \mathcal{B}_f) \Longrightarrow (\ell = F(w))$. For linear latent variable systems this is equivalent to $((0, \ell) \in \mathcal{B}_f) \Longrightarrow (\ell = 0)$.

The question when a system in kernel form is controllable can be answered effectively in terms of R. Indeed, Σ described by $R(\sigma)w = 0$ is controllable if and only if $rank(R(\lambda)) = rank(R)$ for all $\lambda \in \mathbb{C}$ (here one should view $R(\lambda)$ as a matrix over the field of complex numbers and R as a matrix over the field of real rational functions). Analogously, (2.18) will be observable if and only if $M(\lambda)$ is right prime (equivalently, if and only if $M(\lambda)$ is of full column rank for all $\lambda \in \mathbb{C}$). A proof of these claims can be found in [16].

Actually, controllability can also be characterized in terms of (2.18). Take $R = I$ in (2.18), yielding

$$w = M(\sigma)\ell. \tag{2.24}$$

Let $\mathcal{B}$ be the manifest behavior of (2.24). This yields the dynamical system $(\mathbb{Z}, \mathbb{R}^q, \mathcal{B})$, and for obvious reasons we will call (2.24) an *image representation* of $\mathcal{B}$. By the latent variable elimination theorem already mentioned, $\mathcal{B}$ admits a kernel representation. However, not every system in kernel form has an image representation. As proved in [16], this is the case if and only if the system is controllable!

One may interpret the elimination of the latent variable as a step towards simplification of a model. However, this point of view does not take into account the fact that a model not involving latent variables is not necessarily simpler than one that it does. In fact, the elimination process will usually increase the lags of the equations relating the manifest variables among each other, making the computation of system trajectories a complex task; or, a "good" choice of latent variable could display the structure of the memory of the system. These considerations bring us to the concept of state variable.

2.3 State Models

To formalize the concept of state of a discrete-time system, we introduce the following

Definition 2.5. *A (discrete-time)* state space dynamical system *is a system with latent variables* $\Sigma_s = (\mathbb{Z}, \mathbb{R}^q, \mathbb{R}^d, \mathcal{B}_s)$ *in which the latent variable, customarily denoted with* x*, satisfies the* axiom of state*: if* (w_1, x_1), (w_2, x_2) *belong to* $\mathcal{B}_s$*, and* $x_1(t') = x_2(t')$ *for some* $t' \in \mathbb{Z}$*, then*

$$(w, x) := (w_1, x_1) \wedge_{t'} (w_2, x_2) \tag{2.25}$$

the concatenation of (w_1, x_1) and (w_2, x_2) at time $\bar{t}$, *defined as*

$$((w_1, x_1) \wedge_{t'} (w_2, x_2))(t) = \begin{cases} (w_1(t), x_1(t)) & t < t' \\ (w_2(t), x_2(t)) & w \geq t', \end{cases}$$

belongs to $\mathcal{B}_s$.

This definition captures the state as a variable which summarizes all information from the past relevant to the determination of the future behavior of the system. In other words, the state variable *splits* (in a technical sense, see [16]) past and future of the behavior; equivalently, the state variable captures the *common features* of past and future: any trajectory arriving at a given state has enough in common with any other trajectory emanating from that state, that an admissible trajectory can be composed out of the two by concatenation.

The importance of state space systems lies in the fact that they admit representations which are of first order in x, and zeroth order in w. We state the result for the class of linear time-invariant complete systems, the one we are interested in this paper, although the result is more general, see [16].

Theorem 2.4. $\Sigma_s = (\mathbb{Z}, \mathbb{R}^q, \mathbb{R}^d, \mathcal{B}_s)$ *is a linear time-invariant and complete state space system if and only if it admits a kernel representation of the form*

$$E\sigma x + Fx + Gw = 0. \tag{2.26}$$

Note that (2.26) may involve algebraic relations among the variables. In particular, implicit systems are part of the class of state space systems considered in this paper.

The simplicity of representations of the form (2.26) makes state space models a natural choice when considering the simulation of a dynamical system, since it allows easy updating of the values of x and easy computation of the values of w, and it provides a framework in which analysis and control of physical systems are addressed in a very effective way, as shown by the history of our discipline.

By theorem 2.3, the external system Σ of a state space system Σ_s represented by (2.26) is linear time-invariant and complete. We will say say that Σ_s *represents* Σ *in state space form.*

The question arises whether *every* system represented in kernel form allows a representation in state form by a judicious choice of state variable. This is the *state space representation problem.* Connected to this problem, specific questions arise regarding the uniqueness of state space representations, and regarding the minimality of these representations. We address these questions in the next section.

2.4 Existence and Uniqueness of State Space Models

Denote with $\mathcal{L}^q$ the family of all linear, time-invariant, closed subspaces of $(\mathbb{R}^q)^{\mathbb{Z}}$, equipped with the topology of pointwise convergence. By theorem 2.1 and theorem 2.2, $\mathcal{L}^q$ consists of all linear time-invariant and complete systems, or equivalently of all systems whose behavior is the kernel of a polynomial operator in the shift.

The question whether an element in $\mathcal{L}^q$ admits a state space representation is addressed in

Theorem 2.5. *([16]) Every element of $\mathcal{L}^q$ is the external system of a linear time-invariant complete state space system.*

Together with theorem 2.3, the result of theorem 2.5 shows that the class of linear time-invariant and complete systems is exactly the one which has an underlying finite dimensional state space, and provides an alternative way of representing elements of $\mathcal{L}^q$, besides kernel and hybrid representations.

Let us now examine the issues regarding uniqueness and minimality of a state space representation of an element of $\mathcal{L}^q$.

In the context of state space models, *minimality* is introduced at two levels.

Definition 2.6. *The system $\Sigma_s = (\mathbb{Z}, \mathbb{R}^q, \mathbb{R}^d, \mathcal{B}_s)$ represented by*

$$E\sigma x + Fx + Gw = 0 \tag{2.27}$$

with E, $F \in \mathbb{R}^{f\times n}$, $G \in \mathbb{R}^{f\times q}$, with external behavior $\mathcal{B}$, is said to be minimal *if, for every other state space system $\Sigma_s' = (\mathbb{Z}, \mathbb{R}^q, \mathbb{R}^d, \mathcal{B}_s')$ represented by*

$$E'\sigma x + F'x + G'w = 0 \tag{2.28}$$

with E', $F' \in \mathbb{R}^{f'\times n'}$, $G \in \mathbb{R}^{f'\times q}$, and with the same external behavior $\mathcal{B}$, there holds $f' \geq f$ and $n' \geq n$.

Σ_s is said to be state minimal *if, for any Σ_s' as above, there holds $n' \geq n$.*

Thus, minimal state representations are characterized by as few as possible equations and as few as possible state variables among the state space systems representing a given behavior, while state minimal state representations have a minimal number of state variables among the state space systems with the same external behavior. In the context of the class of all minimal state systems representing the same behavior $\mathcal{B}$, the question of uniqueness of representation is answered as follows.

Theorem 2.6. *([16]) Let Σ_s be a minimal state space representation of the external behavior $\mathcal{B}$, associated with*

$$E\sigma x + Fx + Gw = 0, \tag{2.29}$$

E, $F \in \mathbb{R}^{f\times n}$, $G \in \mathbb{R}^{f\times q}$. Then Σ_s' associated with

$$E'\sigma x' + F'x' + G'w = 0, \tag{2.30}$$

is another minimal representation of $\mathcal{B}$ if and only if there exist nonsingular matrices $T \in \mathbb{R}^{n\times n}$ and $V \in \mathbb{R}^{f\times f}$ such that

$$\begin{aligned} E' &= VET \\ F' &= VFT \\ G' &= VG. \end{aligned} \tag{2.31}$$

In other words, minimal state space representations are unique, modulo a choice of basis in the state space and in the equation space.

An analogous result holds for state minimality:

Theorem 2.7. *([16]) Let Σ_s be a state minimal state space representation of the external behavior $\mathcal{B}$, with full behavior $\mathcal{B}_f$ described by*

$$E\sigma x + Fx + Gw = 0, \tag{2.32}$$

E, $F \in \mathbb{R}^{f\times n}$, $G \in \mathbb{R}^{f\times q}$. Then Σ_s' is another state minimal state representation of $\mathcal{B}$, with full behavior $\mathcal{B}_f'$ described by

$$E'\sigma x' + F'x' + G'w = 0, \tag{2.33}$$

E', $F' \in \mathbb{R}^{f'\times n}$, $G' \in \mathbb{R}^{f'\times q}$, if and only if there exists a nonsingular matrix $T \in \mathbb{R}^{n\times n}$ such that

$$\{(w,x) \in \mathcal{B}_f\} \Longleftrightarrow \{(w,Tx) \in \mathcal{B}_f'\} \tag{2.34}$$

Therefore, all state minimal state representations of the same external behavior are unique, modulo a choice of basis in the state space.

The state space representation problem has been addressed with reference to representations like (2.26). Introducing more structure on w, and combining the input-output framework with the notion of state, yields the class of input/state/output systems, while considering a free latent variable v besides the latent variable x yields the class of state space with driving variable systems.

2.5 Input/State/Output, Output Nulling, and Driving Variable Representations

The definition of input/state/output system is given below only in the linear, time-invariant, and complete case; the reader is referred to [17] for a more general setting.

Definition 2.7. *A linear time-invariant complete* input/state/output system *is a quintuple* $\Sigma_{i/s/o} = (\mathbb{Z}, \mathbb{R}^m, \mathbb{R}^p, \mathbb{R}^n, \mathcal{B}_f)$ *with* $\mathbb{R}^m$ *the* input space, *the space in which the input variable* u *takes its values,* $\mathbb{R}^p$ *the* output space, *the space in which the output variable* y *takes its values,* $\mathbb{R}^n$ *the* state space, *and* $\mathcal{B}_f$ *the* full behavior, *described by*

$$\begin{aligned} \sigma x &= Ax + Bu \\ y &= Cx + Du. \end{aligned} \tag{2.35}$$

An input/state/output system is a special case of state space system (2.26) in which the matrix E has a special form, and the external variables have been partitioned in inputs and outputs: (2.35) is equivalent to

$$\begin{pmatrix} I_n \\ 0 \end{pmatrix} \sigma x + \begin{pmatrix} -A \\ -C \end{pmatrix} x + \begin{pmatrix} -B & 0 \\ -D & I_n \end{pmatrix} \begin{pmatrix} u \\ y \end{pmatrix} = 0. \tag{2.36}$$

In the previous section we have discussed the state representation problem for systems in state space form. On the other hand, section 2.1 discussed the derivation of an input-output system from a kernel representation, by componentwise partition of the external variables in inputs and outputs. The following result should therefore come as no surprise.

Theorem 2.8. *([17]) For any element* Σ *of* $\mathcal{L}^q$ *there exists an input/state/output system with the behavior of* Σ *as external behavior.*

Theorem 2.8 settles the question of existence of input/state/output representations and provides the class $\mathcal{L}^q$ with an additional kind of representation besides those already discussed.

Minimality of input/state/output systems is defined analogously to definition 2.6. Due to the special structure of the equations (2.35), for this class of systems minimality is equivalent to state minimality (see [17]).

We address the problem of uniqueness of input/state/output representations in the context of minimal representations.

Theorem 2.9. *([17]) Let* Σ_s *be a minimal input/state/output representation of the external behavior* $\mathcal{B}$*, associated with the equations*

$$\begin{aligned} \sigma x &= Ax + Bu \\ y &= Cx + Du. \end{aligned} \tag{2.37}$$

Then Σ_s'*, associated with the equations*

$$\begin{aligned} \sigma x' &= A'x' + B'u \\ y &= C'x' + D'u, \end{aligned} \tag{2.38}$$

is another minimal input/state/output representation of $\mathcal{B}$*, if and only if there exists a nonsingular matrix* T *such that*

$$\begin{aligned} A' &= TAT^{-1} \\ B' &= TB \\ C' &= CT^{-1} \\ D' &= D. \end{aligned} \tag{2.39}$$

Let us conclude this section with two additional ways of representing elements of $\mathcal{L}^q$, akin to input/state/output systems: *state systems with driving variable*, and *output nulling representations.*

Definition 2.8. *A linear, time -invariant, and complete* state space with driving variable system *is a quintuple* $\Sigma = (\mathbb{Z}, \mathbb{R}^q, \mathbb{R}^n, \mathbb{R}^d, \mathcal{B}_f)$ *with* $\mathbb{R}^q$ *the external signal space,* $\mathbb{R}^d$ *the* driving variable space, *in which the latent* driving variable v *takes its values, and* $\mathcal{B}_f$ *the full behavior, described by the equations*

$$\begin{aligned} \sigma x &= Ax + Bv \\ w &= Cx + Dv \end{aligned} \tag{2.40}$$

Analogously to theorem 2.5 and 2.8, the following result allows to consider state space systems with driving variable as a possible choice of representation of elements of the class $\mathcal{L}^q$.

Theorem 2.10. *([17]) For every element* Σ *of* $\mathcal{L}^q$ *there exists a state space system with driving variable having the behavior of* Σ *as external behavior.*

Minimality of driving variable representations is characterized in terms of the dimensions of the state and of the driving variable being the smallest possible among those of driving variable systems describing the same external behavior. It can be shown (see [17]) that then the number of equations is also minimal, and that all minimal driving variable representations of a given external behavior can be generated from one such representation (2.40) by considering the matrices

$$\begin{aligned} A' &= S(A + BF)S^{-1} \\ B' &= SBR \\ C' &= (C + DF)S^{-1} \\ D' &= DR \end{aligned} \tag{2.41}$$

for S nonsingular in $\mathbb{R}^{n \times n}$, $R \in \mathbb{R}^{m \times m}$ nonsingular, and $F \in \mathbb{R}^{m \times n}$.

Output nulling representations are defined as follows.

Definition 2.9. *A linear, time-invariant, complete state space system* $\Sigma = (\mathbb{Z}, \mathbb{R}^q, \mathbb{R}^n, \mathcal{B}_f)$, *is said to have an* output nulling representation *if and only if there exist* $A \in \mathbb{R}^{n \times n}$, $B \in \mathbb{R}^{n \times q}$, $C \in \mathbb{R}^{\bullet \times n}$, $D \in \mathbb{R}^{\bullet \times q}$, *such that* $\mathcal{B}_f$ *is represented as*

$$\begin{aligned} \sigma x &= Ax + Bw \\ 0 &= Cx + Dw. \end{aligned} \tag{2.42}$$

The following result allows to consider output nulling models as possible representations of elements of $\mathcal{L}^q$.

Theorem 2.11. *([4]) For every element $\Sigma \in \mathcal{L}^q$ with full behavior $\mathcal{B}_f$, there exist matrices $A \in \mathbb{R}^{n\times n}$, $B \in \mathbb{R}^{n\times q}$, $C \in \mathbb{R}^{\bullet\times n}$, $D \in \mathbb{R}^{\bullet\times q}$, such that $\mathcal{B}_f$ is represented as*

$$\begin{aligned} \sigma x &= Ax + Bw \\ 0 &= Cx + Dw. \end{aligned} \tag{2.43}$$

An output nulling representation is termed *minimal* if all other output nulling representations of the same external behavior have greater or equal number of states and number of equations. All minimal output nulling representations of a system can be obtained from one such representation, by considering the matrices

$$\begin{aligned} A' &= S(A + LC)S^{-1} \\ B' &= S(B + LD) \\ C' &= PCS^{-1} \\ D' &= PD \end{aligned} \tag{2.44}$$

where L is any matrix, and P and S are invertible matrices ([4]).

2.6 Recapitulation

In this section we have given definitions and results to which we will frequently refer in the sequel. The notions of system, of manifest and latent variable, of polynomial operator in the shift have been given. The class $\mathcal{L}^q$ of linear time-invariant complete systems has been defined. A number of different representations for elements of $\mathcal{L}^q$ have been introduced. Their properties and some issues regarding their structure have been discussed. The notions of controllability and observability have been introduced.

From this overview of the behavioral approach to system theory, two things should be apparent: the strong link between this approach and polynomial matrix algebra, and the importance of the notion of state, and of state representations of elements of $\mathcal{L}^q$. These two aspects are further connected by means of the concept of *state map* which is the subject of the next section.

3. From Difference Equation to State Models

In this section we introduce the notion of state map, and show how state maps can be computed starting from representations of elements of $\mathcal{L}^q$ in kernel, image, or hybrid form. We will see that a state space representation and a state space with driving variable representation can be readily computed from a kernel, image, or hybrid representation. This section is based on the paper [7], which dealt with the continuous time case; anyway, the results apply, with the obvious modifications, to the discrete time case as well.

3.1 Basic Notions

This section is devoted to the introduction of some notation related to polynomials and rational functions, which will be extensively used in the following.

Any rational function can be written in a unique way as the sum of a polynomial and of a strictly proper rational function: given $q \in \mathbb{R}(\xi)$, there exist unique $p \in \mathbb{R}[\xi]$ and $s \in \mathbb{R}_+(\xi)$, the set of strictly proper rational functions, such that $q = p + s$. This defines the map

$$(\ )_+ : \mathbb{R}(\xi) \mapsto \mathbb{R}[\xi] \tag{3.1}$$

by

$$(q(\xi))_+ := p(\xi). \tag{3.2}$$

On the set of rational functions in the indeterminate ξ, multiplication by ξ^{-1} defines a map $\xi^{-1} : \mathbb{R}(\xi) \mapsto \mathbb{R}(\xi)$ in the obvious way.

Definition 3.1. *The* shift-and-cut *map σ_+ is defined as*

$$\begin{aligned} \sigma_+ \ &: \ \mathbb{R}(\xi) \mapsto \mathbb{R}[\xi] \\ \sigma_+ \ &:= \ (\ )_+ \circ \xi^{-1}. \end{aligned} \tag{3.3}$$

This definition is extended to vectors and matrices of rational functions in a componentwise manner.

Iterated application of σ_+ will be denoted as $\sigma_+^k := \overbrace{\sigma_+ \circ \sigma_+ \circ \cdots \circ \sigma_+}^{k-times}$.

In the following, special importance will be given to the action of σ_+ on vector polynomials. Therefore, let us examine in detail what is the result of the application of σ_+ to a vector polynomial $p \in \mathbb{R}^{1\times q}[\xi]$. Write

$$p(\xi) := p_\delta \xi^\delta + p_{\delta-1}\xi^{\delta-1} + \cdots + p_1\xi + p_0. \tag{3.4}$$

Then

$$\sigma_+(p(\xi)) = p_\delta \xi^{\delta-1} + p_{\delta-1}\xi^{\delta-2} + \cdots + p_1, \tag{3.5}$$

that is,

$$\sigma_+(p(\xi)) = \xi^{-1}(p(\xi) - p_0). \tag{3.6}$$

As we will see, the shift-and-cut map plays an important role in the computation of state maps, in connection with the notions of *Ξ-matrix* and of *Ξ-space.*

Definition 3.2. *Let* $R := R_0 + R_1\xi + \cdots + R_L\xi^L \in \mathbb{R}^{g\times q}[\xi]$. *Define* R^k, $k = 0,\ldots,L$, *as* $R^0 := R$, *and* $R^k := \sigma_+^k R = \sigma_+ R^{k-1}$, $k = 1,\ldots,L$. *The* Ξ-matrix *of* R, *denoted with* R_Ξ, *is*

$$R_\Xi := col(R^k)_{k=1,\ldots,L} = \begin{pmatrix} R^1 \\ R^2 \\ \vdots \\ R^L \end{pmatrix}. \tag{3.7}$$

The Ξ-space *of* R, *denoted with* Ξ_R, *is the vector space over* $\mathbb{R}$ *generated by the rows of* R_Ξ.

Introduce now on $\mathbb{R}^{1\times q}[\xi]$ the equivalence relation $\overset{R}{\sim}$ defined as follows: $p, q \in \mathbb{R}^{1\times q}[\xi]$ are *equivalent modulo* R, written $p \overset{R}{\sim} q$, if and only if there exists $r(\xi) \in \mathbb{R}^{1\times q}[\xi]$ such that $p(\xi) - q(\xi) = r(\xi)R(\xi)$.

The equivalence relation $\overset{R}{\sim}$ is well defined also on Ξ_R; moreover, the vector space structure on Ξ_R induces in the natural way a vector space structure on the set of equivalence classes induced by $\overset{R}{\sim}$ on Ξ_R. We will denote this set of equivalence classes with Ξ_R $(mod\ R)$. That is,

$$\Xi_R\ (mod\ R) = \{[p] \in 2^{\Xi_R} \mid q \in [p]\ iff\ \exists\, r \in \mathbb{R}^{1\times q}[\xi]\ s.t.\ p = q + rR\}. \tag{3.8}$$

The following example illustrates the above notions.

Example 3.1. Let

$$R = \begin{pmatrix} \xi^2 + 2\xi - 1 & \xi + 1 \\ \xi - 1 & \xi^2 - 3 \end{pmatrix} \tag{3.9}$$

and consider its first row, $(\xi^2 + 2\xi - 1 \quad \xi + 1)$. The shift and cut operator acts on this row as

$$\begin{aligned} \sigma_+(\xi^2 + 2\xi - 1 \quad \xi + 1) &= (\xi + 2 \quad 1) \\ \sigma_+^2(\xi^2 + 2\xi - 1 \quad \xi + 1) &= (1 \quad 0) \\ \sigma_+^3(\xi^2 + 2\xi - 1 \quad \xi + 1) &= (0 \quad 0). \end{aligned} \tag{3.10}$$

Hence the Ξ_R space is the vector space spanned by

$$(\xi + 2 \quad 1), (1 \quad 0), (1 \quad \xi), (0 \quad 1), \tag{3.11}$$

and these vectors already form a basis for this space. It is easily verified that the vectors (3.11), interpreted as representing elements of Ξ_R $(mod\ R)$, are linearly independent as well, and therefore form a basis of Ξ_R $(mod\ R)$. □

Note that selecting from the rows of R_Ξ a maximal set of linearly independent rows, and considering these as representatives of elements of Ξ_R $(mod\ R)$, does not necessarily yield a basis for Ξ_R $(mod\ R)$, as made explicit by the following example.

Example 3.2. Let

$$R = \begin{pmatrix} 1 & 0 & -1 \\ 0 & 1 & \xi^3 \\ 0 & 0 & \xi \end{pmatrix} . \tag{3.12}$$

A maximal set of linearly independent rows of R_Ξ is

$$(\,0 \quad 0 \quad \xi^2\,),\ (\,0 \quad 0 \quad \xi\,),\ (\,0 \quad 0 \quad 1\,), \tag{3.13}$$

but the first and the second element of this set are equivalent to zero modulo R, since $(\,0 \quad 0 \quad \xi^2\,) = (\,0 \quad 0 \quad \xi\,)R$ and $(\,0 \quad 0 \quad \xi\,) = (\,0 \quad 0 \quad 1\,)R$. Therefore $\Xi_R(mod\ R)$ has a basis consisting of $(\,0 \quad 0 \quad 1\,)$ only. □

The notions of shift-and-cut map, of Ξ-matrix and of Ξ-space associated with a polynomial matrix R, and the equivalence relation $\overset{R}{\sim}$, will play an important role in the solution of the following

Problem 3.1. Given a system Σ represented in kernel form, determine an integer n and a polynomial matrix $X \in \mathbb{R}^{n \times q}[\xi]$ such that

$$\begin{aligned} R(\sigma)w &= 0 \\ x &= X(\sigma)w \end{aligned} \tag{3.14}$$

defines a (minimal) state space system with external behavior $Ker\ R(\sigma)$.

We will call such a shift operator $X(\sigma)$ a *state map* for the system described by $R(\sigma)w = 0$.

A problem statement analogous to this one can be given in case the system is described in image or hybrid form; in these cases, one looks for a shift operator $X(\sigma)$ which maps full trajectories (w, ℓ) to state variable trajectories for a state space representation of the external behavior of the system.

In the following we will assume that the external behavior of the system under study is a proper subset of $(\mathbb{R}^q)^\mathbb{Z}$, and that it does not consist of the zero trajectory only. In fact, in cases such as these, the problem of computing a state map is trivial. In case the external behavior coincides with $(\mathbb{R}^q)^\mathbb{Z}$, the only possible state map is the zero one. In the second case, any polynomial matrix X defines an admissible state map.

The following two sections will be devoted to the characterization of state maps for systems in kernel, hybrid and image form, respectively.

3.2 From Kernel Representations to $X(\xi)$

In this section we deal with the problem of characterizing state maps for systems described in kernel form. In dealing with this characterization, we will make use of the following, equivalent definition of state space system, which uses a version of the axiom of state modified in the case of linear, time-invariant systems.

Definition 3.3. *Let Σ be a linear time-invariant complete system with latent variable x. Σ is a state space system if, whenever (w, x) is an admissible full trajectory, and $x(0) = 0$, also $(0 \wedge_0 w, 0 \wedge_0 x)$ is an admissible full trajectory.*

Note that, due to time invariance, the choice of $t = 0$ as the istant of concatenation is not restrictive. In the following we will assume that concatenation always occurs at time 0, and we will use, instead of $\wedge_0$, the notation $\wedge$.

As a preliminary result, we examine the conditions under which a trajectory is concatenable with the zero trajectory. These conditions correspond to a system of linear equations in w and its lags, and define a polynomial operator in the shift $X(\sigma)$ which, in fact, corresponds to a state map. It turns out that the rows of $X(\xi)$ defined in this way, have an immediate interpretation in terms of the matrix R_{Ξ} introduced in the previous section.

Proposition 3.1. *([7]) Let a kernel representation*

$$R(\sigma)w = 0 \tag{3.15}$$

of the external behavior $\mathcal{B}$ of a linear, time-invariant, and complete system be given. A trajectory $w \in \mathcal{B}$ is concatenable with the zero trajectory, that is, $0 \wedge w \in \mathcal{B}$, if and only if

$$(R_{\Xi}(\sigma)w)(0) = 0. \tag{3.16}$$

This result yields the desired characterization of state maps.

Theorem 3.1. *([7]) Let $R(\sigma)w = 0$ describe the external behavior of a system $\Sigma \in \mathcal{L}^q$. The polynomial matrix $X \in \mathbb{R}^{\bullet \times q}[\xi]$ defines a state map $X(\sigma)$, that is,*

$$\begin{aligned} R(\sigma)w &= 0 \\ X(\sigma)w &= x \end{aligned} \tag{3.17}$$

defines a state space system, if and only if there exist matrices A, $A' \in \mathbb{R}^{\bullet\times\bullet}$ and polynomial matrices B, $B' \in \mathbb{R}^{\bullet\times\bullet}$ such that

$$\begin{aligned} R_{\Xi}(\xi) &= AX(\xi) + B(\xi)R(\xi) \\ X(\xi) &= A'R_{\Xi}(\xi) + B'(\xi)R(\xi). \end{aligned} \tag{3.18}$$

If the rows of $X(\xi)$ are interpreted as representative of elements of $\Xi_R(modR)$, theorem 3.1 can be restated saying that X defines a state map if and only if its rows form a set of generators of $\Xi_R(modR)$. This yields the following corollary to theorem 3.1, which addresses the problem of state minimality.

Corollary 3.1. *([7]) The polynomial matrix $X(\xi)$ defines a minimal state variable x for*

$$\begin{aligned} R(\sigma)w &= 0 \\ X(\sigma)w &= x \end{aligned} \tag{3.19}$$

if and only if the rows of $X(\xi)$ form a basis for $\Xi_R(mod\ R)$.

Example 3.3. Consider the system with behavior described by

$$\sigma^n w_1 + \cdots + p_1 \sigma w_1 + p_0 w_1 = q_n \sigma^n w_2 + \cdots + q_1 \sigma w_2 + q_0 w_2 \tag{3.20}$$

where w_i, $i = 1, 2$, are scalar functions. Defining

$$p^i := \sigma_+^i(p) = \xi^{n-i} + p_{n-1}\xi^{n-i-1} + \cdots + p_i \tag{3.21}$$

and analogously for q^i, it is easy to see that

$$f_i := \begin{pmatrix} p^i & -q^i \end{pmatrix} \tag{3.22}$$

$i = 1, \ldots, n$, form a basis for $\Xi_{\begin{pmatrix} p & -q \end{pmatrix}}$ $(mod\ \begin{pmatrix} p & -q \end{pmatrix})$. Stacking the f^i vectors yields a minimal state inducing polynomial matrix. □

Theorem 3.1 characterizes the polynomial shift operators which induce a state variable for systems described in kernel form. As seen in section 2, a frequent situation in modeling is the need to introduce auxiliary variables besides the manifest ones, and hybrid representations result from this procedure. The following section examines the problem of computing state variables for systems in hybrid form.

3.3 From Hybrid Representation to $X(\xi)$

In this section we will discuss the characterization of state maps for the external behavior of a system described in hybrid form.

We will focus on state maps which map *full* trajectories (w, ℓ) to state trajectories for the external system, that is, (w, ℓ)-induced state maps, for two reasons. The first one is that manifest and latent variables enter the description of the system on an equal footing; this is exemplified by the fact that a hybrid representation can be considered a kernel description of the full behavior. A state function is therefore most naturally chosen as a function of both the manifest and the latent variables. A second reason for not considering state maps induced only by the external variable, is the following. Such state map would be computed by first eliminating the latent variable, therefore modifying the original equations. Such a modification, however, is not a desirable feature of a state construction procedure, since a state variable should reflect as much as possible the physical structure of the system, as put in evidence by the original equations.

For these reasons, we will construct (w, ℓ)-induced state maps. This choice makes available as a starting point for the computation of a state map for the external behavior, a state map for the full behavior. In fact, it is easily seen from the definitions that any state map for the full behavior (which can be computed according to theorem 3.1, since the full behavior is represented in kernel form) is a state map for the external behavior as well. It is interesting to characterize the situations in which the converse holds, that is, to determine for which systems in hybrid form a state map for the external behavior is also a state map for the full behavior.

Consider a hybrid representation (2.20), (2.21), which can be obtained by unimodular premultiplication of (2.18) and is equivalent to the original equations. A state map for the full behavior is obtained by applying the shift-and-cut operator to the rows of $\begin{pmatrix} R_1' & 0 \\ R_2' & -M_2' \end{pmatrix}$, thus computing $\begin{pmatrix} R_1' & 0 \\ R_2' & -M_2' \end{pmatrix}_{\Xi}$. From this Ξ-matrix the Ξ-matrix $(R_1' \quad 0)_{\Xi} = (R_{1\Xi}' \quad 0)$ can be extracted as a submatrix. This implies that a state map for the external behavior is a state map for the full behavior if and only if the vector space spanned by the rows of $(R_{1\Xi}' \quad 0)$, which in general is a proper subspace of the vector space generated by the rows of $\begin{pmatrix} R_1' & 0 \\ R_2' & -M_2' \end{pmatrix}_{\Xi}$, coincides with that space. This happens if and only if the dimension of the minimal state spaces for the external and the full behavior are the same. An efficient way of checking this is given in the following proposition.

Proposition 3.2. *([7]) Let a system be described in hybrid form as in (2.18), and let ℓ be observable from w. Then the dimensions of the minimal state spaces for the external and the full behavior, respectively, are the same, if and only if there exists an input-output selection in (w, ℓ) such that the variables ℓ are all outputs for the full behavior.*

Remark 3.1. The above proposition implies that, if there exists an input-output selection in (w, ℓ) such that the latent variables can all be chosen as outputs, a polynomial operator in the shift $X(\sigma)$ is a state map for the external behavior if and only if it is a state map for the full behavior.

Remark 3.2. Existence of an input-output selection on (w, ℓ) such that ℓ is entirely composed of outputs can be checked efficiently, cf. [7].

Assume now that the latent variable ℓ is observable, but that in any selection of inputs and outputs for the full system, some components of ℓ have to be chosen as inputs. From proposition 3.2 it follows that the dimension of the minimal state space for the full behavior is greater than that of the minimal state space for the external behavior. This is a consequence of the fact that a state inducing map for the full behavior corresponds to concatenability conditions that involve both w *and* ℓ, while concatenability in the external variable only is a weaker concept. That is, some of the concatenability conditions imposed by a state variable for the full behavior become too restrictive when considering concatenation of the external variables only: even if the state variable for the full behavior is not zero at $t = 0$, it could still be possible to concatenate the external trajectory with the zero trajectory. The idea we pursue in the following is to derive from the state variable for the full behavior, a state variable for the external behavior by projecting it down with a suitably defined linear map. We call the computation of a state for the external behavior from a state for the full behavior via this projection the *reduction* of the state variable for $\mathcal{B}_f$ to a state variable for $\mathcal{B}_{ext}$.

The reduction process involves introducing some new concepts.

Assume that the full row rank polynomial matrix

$$(R \mid -M)(\xi) = \sum_{j=0}^{L} (R_j \mid -M_j) \xi^j \tag{3.23}$$

has g rows. Consider

$$\begin{pmatrix} (R \mid -M)_{\Xi} \\ 0_{g\times(d+q)} \end{pmatrix} = col(\sigma_+^k((R \mid -M)))_{k=1,\dots,L+1} \tag{3.24}$$

and the matrix $T := col(M_i)_{i=0,\dots,L}$. Define $E := \{r \in \mathbb{R}^{1\times(L+1)g} \mid rT = 0\}$, the vector space of constant left annihilators of T.

As was done for the case of kernel representations, we first consider the conditions under which a trajectory $(w,\ell) \in \mathcal{B}_f$ is externally concatenable with zero, that is, $0 \wedge w \in \mathcal{B}_{ext}$. These conditions correspond to a system of linear equations involving w, ℓ, and their derivatives, and define a polynomial operator in the shift which is in fact a state map. The rows of the corresponding polynomial matrix turn out to have an interpretation in terms of a set of generators of E, and of the matrix $\begin{pmatrix} (R \mid -M)_{\Xi} \\ 0_{g\times(d+q)} \end{pmatrix}$.

Proposition 3.3. *([7]) Let a hybrid representation (2.18) be given, with ℓ observable from w. Assume that for every input-output partition of (w,ℓ) at least one component of ℓ is chosen as an input.*

A trajectory $(w,\ell) \in \mathcal{B}_f$ is externally concatenable with zero, that is, $0 \wedge w \in \mathcal{B}_{ext}$, if and only if, given any set $\{v_1,\dots,v_s\}$ of generators of E, there holds

$$(v_i \begin{pmatrix} (R \mid -M)_{\Xi} \\ 0_{g\times(d+q)} \end{pmatrix} (\sigma) \begin{pmatrix} w \\ \ell \end{pmatrix})(0) = 0 \tag{3.25}$$

$i = 1,\dots,s$.

We can now state the main result regarding systems in hybrid form with ℓ observable from w:

Theorem 3.2. *([7]) Let a system be described in hybrid form as in (2.18), and let ℓ be observable from w. The matrix $X \in \mathbb{R}^{\bullet\times(q+d)}[\xi]$ defines a (w,ℓ)-induced state map for the external system corresponding to (2.18), that is,*

$$\begin{aligned} R(\sigma)w &= M(\sigma)\ell \\ x &= X(\sigma)\begin{pmatrix} w \\ \ell \end{pmatrix}, \end{aligned} \tag{3.26}$$

defines a state model for the external behavior corresponding to (2.18), if and only if either it defines a state for the full behavior, or for each constant *matrix V whose rows generate E, there exist constant matrices A, A', and polynomial matrices B, B', such that*

$$V\begin{pmatrix} (R \mid -M)_{\Xi} \\ 0 \end{pmatrix} = AX + B(R \mid -M) \tag{3.27}$$

and

$$X = A'V\begin{pmatrix} (R \mid -M)_{\Xi} \\ 0 \end{pmatrix} + B'(R \mid -M). \tag{3.28}$$

Remark 3.3. Theorem 3.2 may be restated as follows: X defines a state inducing map if and only if either it defines a state for the full behavior, or its rows form a set of generators for the vector space

$$\{r\begin{pmatrix} col(\sigma_+^i(R \mid -M))_{i=1,\dots,L} \\ 0_{g\times(q+d)} \end{pmatrix} \mid r \in E\} \ (mod\ (R \mid -M)), \tag{3.29}$$

defined as the set of equivalence classes determined by the equivalence $\underset{\sim}{(R \mid -M)}$ on the vector space

$$\{r\begin{pmatrix} col(\sigma_+^i(R \mid -M))_{i=1,\dots,L} \\ 0_{g\times(q+d)} \end{pmatrix} \mid r \in E\}. \tag{3.30}$$

This equivalent formulation yields the following characterization of minimal state inducing maps:

Corollary 3.2. *([7]) X defines a (w,ℓ)-induced minimal state map for the external system corresponding to (2.18) if and only if either there exists an input-output selection in (w,ℓ) in which ℓ is entirely composed of outputs for the full behavior, and the rows of X form a basis of*

$$(R \mid -M)_{\Xi} \ (mod\ (R\ -M)), \tag{3.31}$$

or the rows of X form a basis of the vector space

$$\{r\begin{pmatrix} col(\sigma_+^i(R \mid -M))_{i=1,\dots,L} \\ 0_{g\times(q+d)} \end{pmatrix} \mid r \in E\} \ (mod\ (R \mid -M)). \tag{3.32}$$

The results exposed in this section can be adapted to the case in which the latent variable ℓ is not observable from w. This has been discussed in [7].

3.4 From Image Representation to $X(\xi)$

Image representations

$$w = M(\sigma)\ell, \tag{3.33}$$

where $w \in (\mathbb{R}^q)^{\mathbb{Z}}$, $\ell \in (\mathbb{R}^d)^{\mathbb{Z}}$, $M \in \mathbb{R}^{q\times d}[\xi]$, of the behavior of a linear time invariant complete system, have been introduced in section 2. in connection with the notion of controllability: the behavior of a system has an image representation if and only if the system is controllable.

In this section we consider the computation of state maps for the external behavior of systems whose full behavior is described by (3.33). These may be considered to be a special case of systems representable in hybrid form, with $R = I_q$. We will restrict attention to the case in which $M(\xi)$ of (3.33) is right prime, i.e. the latent variable ℓ is observable from w, referring the reader to [7] for a discussion of the general case.

It is a matter of straightforward verification that in a system whose behavior is described by (3.33), the latent variables can be chosen as playing the role of outputs in the full behavior. This is most easily seen by considering that, for full column rank M, a suitable subset R_1 of the columns of the $q \times q$ identity matrix exists such that

$$\begin{pmatrix} R_1 & M \end{pmatrix} \tag{3.34}$$

is nonsingular, and arranging the columns of a complementary canonical basis of R_1 in $\mathbb{R}^q$ in a matrix R_2, we have

$$\begin{pmatrix} R_1 & M \end{pmatrix}^{-1} R_2 \tag{3.35}$$

proper. Then the external variables corresponding to R_2 can be chosen as inputs, while those corresponding to R_1, and the latent variable ℓ, can be chosen as outputs for the full behavior.

This result, together with proposition 3.2, allows us to conclude that, for observable image representations, the dimensions of the minimal state space of the external and full behavior are equal.

Consider now the problem of determining a state inducing map for an observable image representation. The following theorem is a consequence of the considerations made above:

Theorem 3.3. *([7]) Let a system be represented in image form with ℓ observable from w. $X \in \mathbb{R}^{\bullet\times(q+d)}$ defines a state map for the system (3.33) (i.e. $w = M(\sigma)\ell$, $x = X(\sigma)\begin{pmatrix} w \\ \ell \end{pmatrix}$ defines a state system) if and only if there exists constant matrices A, $A' \in \mathbb{R}^{\bullet\times\bullet}$ and polynomial matrices B, $B' \in \mathbb{R}^{\bullet\times q}$ such that*

$$\begin{aligned} \begin{pmatrix} I_q & -M \end{pmatrix}_{\Xi} &= AX + B\begin{pmatrix} I_q & -M \end{pmatrix} \\ X &= A'\begin{pmatrix} I_q & -M \end{pmatrix}_{\Xi} + B'\begin{pmatrix} I_q & -M \end{pmatrix}. \end{aligned} \tag{3.36}$$

Note that $\sigma_+^k \,(\, I_q \quad | \; -M\,) = (\, 0_{q\times q} \quad -\sigma_+^k M\,)$, $k \in \mathbb{N}$, and therefore any state map for the system (3.33) is ℓ-induced.

The results shown in this section and in the preceding two have shown that state variables for systems of $\mathcal{L}^q$ represented in kernel, hybrid, or image form, are computed in a straightforward way from the equations describing the system. An appealing feature of the notion of state map is that state space equations, in their various forms, are computed directly from the equations of the system, and from a state map. The following section will illustrate this point.

3.5 From $X(\xi)$ to State Equation

In this section we illustrate, by means of some examples, how to compute state space representations, starting from a representation of the system in kernel or hybrid form, and a state map.

Let us first consider the problem of passing from a kernel representation and a corresponding state map, to a state space representation, by means of the following example. Let the kernel representation

$$\sigma^2 w_1 + \sigma w_2 - w_2 = 0, \tag{3.37}$$

be given, and assume that the state map $X(\xi)$

$$X(\xi) = \begin{pmatrix} \xi & 1 \\ 1 & 0 \\ 2 & 0 \end{pmatrix} + \begin{pmatrix} \xi^2 & \xi - 1 \\ \xi^3 & \xi^2 - \xi \\ 2\xi^2 & 2\xi - 2 \end{pmatrix}, \tag{3.38}$$

has been computed. Note that $x := X(\sigma)w$ is not a minimal state variable; in fact, it can be verified that $x' := \begin{pmatrix} \sigma w_1 + w_2 \\ w_1 \end{pmatrix}$ is a minimal state variable.

It is easy to check that

$$\begin{aligned} E &= \begin{pmatrix} 1 & 0 & 0 \\ 0 & 1 & 0 \\ 0 & -2 & 1 \\ 0 & 0 & 0 \end{pmatrix} \\ F &= \begin{pmatrix} 0 & 0 & 0 \\ 1 & 0 & 0 \\ 0 & 0 & 0 \\ 0 & 1 & 0 \end{pmatrix} \\ G &= \begin{pmatrix} 0 & -1 \\ 0 & -1 \\ 0 & 0 \\ -1 & 0 \end{pmatrix} \\ H(\xi) &= \begin{pmatrix} 1-\xi \\ -\xi^2 - 1 \\ 0 \\ -\xi \end{pmatrix} \end{aligned} \tag{3.39}$$

are such that $(\xi E + F)X(\xi) + G = H(\xi)R(\xi)$, and therefore

$$\begin{aligned} \sigma x_1 - w_2 &= 0 \\ \sigma x_2 + x_1 - w_2 &= 0 \\ -2\sigma x_1 + x_3 &= 0 \\ x_2 - w_1 &= 0 \end{aligned} \tag{3.40}$$

hold for every $\begin{pmatrix} w_1 \\ w_2 \end{pmatrix} \in Ker\,(\,\sigma^2 \quad \sigma - 1\,)$. By eliminating the state variables, equation (3.37) is obtained. Therefore (3.40) is a state space representation of the system (3.37).

Let us now consider the procedure for passing from a hybrid representation, and a corresponding state map, to a state space representation of the external behavior. Let a hybrid representation

$$\begin{aligned} \sigma w_1 + w_2 &= \ell \\ \sigma w_1 + \sigma^2 w_2 + w_2 &= \sigma\ell - \ell \end{aligned} \tag{3.41}$$

of a behavior $\mathcal{B}$ be given. A state map for the external system can be computed as

$$X(\sigma) = \begin{pmatrix} 2 & \sigma & -1 \\ 1 & -1 & 0 \end{pmatrix}. \tag{3.42}$$

It is easy to see that

$$\begin{aligned} E &= \begin{pmatrix} 1 & 0 \\ 0 & 1 \\ 0 & 0 \end{pmatrix} \\ F &= \begin{pmatrix} 0 & 0 \\ 1 & 0 \\ 0 & 1 \end{pmatrix} \\ G &= \begin{pmatrix} 0 & 2 \\ -2 & 1 \\ -1 & 1 \end{pmatrix} \\ H &= \begin{pmatrix} 1 & 1 \\ 1 & 0 \\ 0 & 0 \end{pmatrix} \end{aligned} \tag{3.43}$$

are such that $(\xi E + F)X(\xi) + (\,G \quad 0_{3\times 1}\,) = H(\xi)\,(\,R \quad -M\,)\,(\xi)$. Therefore

$$\begin{aligned} \sigma x_1 + 2w_2 &= 0 \\ \sigma x_2 + x_1 - 2w_1 + w_2 &= 0 \\ x_2 - w_1 + w_2 &= 0 \end{aligned} \tag{3.44}$$

for every $\begin{pmatrix} w \\ \ell \end{pmatrix}$ in the full behavior of (3.41). Moreover, by eliminating the state variables from the equations (3.44), the equation

$$\sigma^2 w_1 - 2\sigma w_1 - \sigma^2 w_2 + \sigma w_2 - 2w_2 = 0 \tag{3.45}$$

which describes the external behavior of the system (3.41), is obtained. Therefore (3.44) is a state space representation of the external behavior of (3.41).

Procedures analogous to the above ones can be applied to obtain state space representations from image representations of the external behavior of a system.

As seen from the above examples, the procedures for obtaining a state space representation, starting from the equations of the system and a state map, involve the solution of a system of polynomial equations, and are therefore of standard implementation.

3.6 $V(\xi)$

Computation of state space with driving variable models starting from a kernel or a hybrid representation of a behavior, involves the construction of a driving variable v, induced by the variables describing the system. That is, an integer m and a polynomial matrix $V(\xi) \in \mathbb{R}^{m\times\bullet}$ are to be determined, such that $V(\sigma)w$, or $V(\sigma)\begin{pmatrix} w \\ \ell \end{pmatrix}$ in case a hybrid representation is considered, is a latent driving variable for the system. In this section we will show how to compute such a polynomial matrix $V(\xi)$ by means of some examples.

Let us first consider the problem for the case of a kernel representation $R(\sigma)w = 0$, for which a state map $X(\sigma)$ is available. The procedure we will use to compute a driving variable representation of this system, is inspired by the procedure for computing a compoenetwise partition of the external variables in inputs and outputs, starting from a kernel representation, cfr. [13]. That is, given a kernel representation $R(\sigma)w = 0$ of $\mathcal{B}$, let $R(\xi) = R_L\xi^L + R_{L-1}\xi^{L-1} + \ldots + R_1\xi + R_0 \in \mathbb{R}^{g\times q}[\xi]$, and denote the i-th row of R_L by $R_{L,i}$. Define $m := q - dimspan_R\{R_{L,i}\}_{i=1,\ldots,g}$, and let $\{u_1, \ldots, u_m\}$ be a complementary basis for $span_{\mathbb{R}}\{R_{L,i}\}_{i=1,\ldots,g}$ in $\mathbb{R}^{1\times q}$. The matrix $col(u_i) =: V$ defines a set of m free variables. To write the state space equations, we compute matrices $A \in \mathbb{R}^{n\times n}$, $B \in \mathbb{R}^{n\times m}$, $C \in \mathbb{R}^{q\times n}$, $D \in \mathbb{R}^{q\times m}$, and $H \in \mathbb{R}^{n\times g}[\xi]$, $H' \in \mathbb{R}^{q\times g}[\xi]$, such that

$$\begin{aligned} \xi X(\xi) &= AX(\xi) + BV(\xi) + H(\xi)R(\xi) \\ I_q &= CX(\xi) + DV(\xi) + H'(\xi)R(\xi). \end{aligned} \tag{3.46}$$

As an application, consider the system

$$\sigma^2 w_1 + \sigma w_2 - w_2 = 0, \tag{3.47}$$

be given, and assume that the state map $X(\xi)$

$$X(\xi) = \begin{pmatrix} \xi & 1 \\ 1 & 0 \\ 2 & 0 \end{pmatrix} + \begin{pmatrix} \xi^2 & \xi - 1 \\ \xi^3 & \xi^2 - \xi \\ 2\xi^2 & 2\xi - 2 \end{pmatrix}, \tag{3.48}$$

has been computed. Since $R_L = (1 \quad 0)$, a complementary basis for $span_{\mathbb{R}} R_L$ is $(0 \quad 1)$. The reader can verify that the equations

$$\begin{aligned} \sigma x_1 &= v \\ \sigma x_2 &= -x_1 + v \\ \sigma x_3 &= -2x_1 + 2v \\ w_1 &= x_2 \\ w_2 &= v \end{aligned} \tag{3.49}$$

define a state space with driving variable representation of $Ker\,(\sigma^2 \quad \sigma - 1)$.

Let us now consider the computation of a driving variable for the external behavior of a system described in hybrid form. The underlying idea is the same as that used in the kernel representation case. That is, we compute a matrix $V(\xi)$ whose rows form a complementary basis for

$$span_{\mathbb{R}} rows X(\xi)(mod\,(R \quad -M))$$

in

$$span_{\mathbb{R}} rows \begin{pmatrix} \xi X(\xi) \\ (I_q \quad 0_{q\times d}) \end{pmatrix} (mod\,(R \quad -M)),$$

and then compute matrices A, B, C, D, $H(\xi)$, $H'(\xi)$ such that

$$\begin{aligned} \xi X(\xi) &= AX(\xi) + BV(\xi) + H(\xi)\,(R \quad -M) \\ (I_q \quad 0_{q\times d}) &= CX(\xi) + DV(\xi) + H'(\xi)\,(R \quad -M)\,. \end{aligned} \tag{3.50}$$

As an example, consider the system

$$\begin{aligned} \sigma w_1 + w_2 &= \ell \\ \sigma w_1 + \sigma^2 w_2 + w_2 &= \sigma\ell - \ell \end{aligned} \tag{3.51}$$

of a behavior $\mathcal{B}$ be given. A state map for the external system can be computed as

$$X(\sigma) = \begin{pmatrix} 2 & \sigma & -1 \\ 1 & -1 & 0 \end{pmatrix}. \tag{3.52}$$

It is easily seen that $span_{\mathbb{R}} rows X(\xi)(mod\;(R \;\mid\; -M))$ equals

$$< (2 \quad \xi \quad -1), (1 \quad -1 \quad 0) > . \tag{3.53}$$

Note that the first row of $\xi X(\xi)$ satisfies

$$(2\xi \quad \xi^2 \quad -\xi) = (0 \quad -2 \quad 0) + (1 \quad 1) \begin{pmatrix} \xi & 1 & -1 \\ \xi & \xi^2 & -\xi + 1 \end{pmatrix}, \tag{3.54}$$

and therefore $(2\xi \quad \xi^2 \quad -\xi)\,(mod\,(R \quad -M) = (0 \quad -2 \quad 0)$.

The second row of $\xi X(\xi)$ satisfies

$$\begin{aligned}(\xi \quad -\xi \quad 0) &= -(2 \quad \xi \quad -1) + (2 \quad -1 \quad 0) + (\xi \quad 1 \quad -1) = \\ &= -(2 \quad \xi \quad -1) + (2 \quad -1 \quad 0)(mod(R \quad -M)). \end{aligned} \tag{3.55}$$

Therefore

$$span_{\mathbb{R}} rows \begin{pmatrix} \xi X(\xi) \\ (I_2 \quad 0_{2\times 1}) \end{pmatrix} (mod(R \quad -M)) =$$

$$span_{\mathbb{R}}\{(0 \quad -2 \quad 0), -(2 \quad \xi \quad -1) + (2 \quad -1 \quad 0), (1 \quad 0 \quad 0)\}. \tag{3.56}$$

A complementary basis for $span_{\mathbb{R}} rows X(\xi)(mod(R \quad -M))$ in

$$span_{\mathbb{R}}\{(0 \quad -2 \quad 0), -(2 \quad \xi \quad -1) + (2 \quad -1 \quad 0)\}$$

can be chosen as $(0 \quad 1 \quad 0)$. Therefore $v = w_2$ is a driving variable for the external system. Computation of the system matrices A, B, C, D yields the following representation for the external behavior:

$$\begin{aligned} \sigma \begin{pmatrix} x_1 \\ x_2 \end{pmatrix} &= \begin{pmatrix} 0 & 0 \\ -1 & 2 \end{pmatrix} \begin{pmatrix} x_1 \\ x_2 \end{pmatrix} + \begin{pmatrix} -2 \\ 1 \end{pmatrix} v \\ \begin{pmatrix} w_1 \\ w_2 \end{pmatrix} &= \begin{pmatrix} 0 & 1 \\ 0 & 0 \end{pmatrix} \begin{pmatrix} x_1 \\ x_2 \end{pmatrix} + \begin{pmatrix} 1 \\ 1 \end{pmatrix} v. \end{aligned} \tag{3.57}$$

As seen from these examples, the procedures for computing state space with driving variable representations, starting from the equations of the system and a state map, involve standard polynomial computations. This has a special importance as for what regards applications.

3.7 Simulation

The notions introduced in this section are relevant especially for the problem of simulating a dynamical system. By *simulation* we mean a procedure for computing an arbitrary element of the behavior $\mathcal{B}$ of the system. Such a procedure can, in principle, start from any of the representations of elements of $\mathcal{L}^q$ introduced in section 2, but state space representations lend themselves better than any other to the problem of selecting and computing an element of $\mathcal{B}$, due to the easy updating of the state trajectory, from which the external trajectory is directly computable.

Given a representation of $\mathcal{B}$, in kernel, image or hybrid form, compute first a state space representation of $\mathcal{B}$. For purposes of simulation, a state space with driving variable representation is best suited. Assume for the moment that in

$$\begin{aligned} \sigma x &= Ax + Bv \\ w &= Cx + Dv \end{aligned} \tag{3.58}$$

the matrix A is nonsingular. A trajectory of $\mathcal{B}$ can then be computed as follows: choose $x(0) \in \mathbb{R}^n$ and $v \in (\mathbb{R}^d)^{\mathbb{Z}}$, and compute x and w via

$$x(k+1) = Ax(k) + Bv, \qquad k \geq 0, \tag{3.59}$$

and

$$x(k-1) = A^{-1}x(k) - A^{-1}Bv, \qquad k < 0. \tag{3.60}$$

w can be computed via v and x according to

$$w(k) = Cx(k) + Dv(k) \qquad k \in \mathbb{Z}. \tag{3.61}$$

In other words, once the "initial conditions" $x(0)$ and the driving input v have been chosen by the user (for example, with the purpose of generating a particular trajectory of the external behavior), the whole state trajectory x can be computed according to (3.59) and (3.60), and the external trajectory w can be read out from (3.61).

Some care has to be taken in case A is singular, since in this case the choice of the initial state $x(0)$ and of the driving variable v cannot be made independently. In fact, a part of x will be completely determined by v. We refer the reader to [17] for the details.

The simulation procedure illustrated above can be easily adapted to the case in which an input/state/output representation is available.

3.8 Recapitulation

In this section we have developed the notion of state map, introduced in section 3.1, and derived state models starting form various kind of representations of elements of $\mathcal{L}^q$, and we have shown how to use a state model in the simulation of the behavior of a system. The underlying assumption was, of course, that a model of the system was available to start with, as a set of equations derived from first principles, and possibly involving latent variables. In many cases, however, the physical *a priori* knowledge about the system is not sufficient to obtain model equations, and the need arises of *modeling from data*, that is, to obtain a model starting from observations of the phenomenon under study. In the next section, we will deal with such an approach.

4. The Most Powerful Unfalsified Model

In this section we deal with exact modeling of time series from the behavioral point of view, following [14]. We first introduce some notions which will be instrumental in the solution of the problem; of crucial importance among these is the concept of most powerful unfalsified model. Next, the question of existence of the most powerful model is addressed. The main ideas underlying the procedures for obtaining a state space model from time series are discussed, and classical realization theory is interpreted as a special case of the more general problem of modeling from data.

4.1 Basics

Identification is the process of obtaining dynamical models from data, and consists in finding a dynamical model which explains given measurements of a phenomenon. The measurements are assumed to be arranged in a q-dimensional vector and the problem is to fit these data to some model. In what sense the model should *explain* the data, what is the meaning of *fit*, and the very basic question of what a *model* is, are usually answered as follows. A model is considered to be a set of equations containing a number of unknown parameters. These parameters are then determined on the basis of the data. This procedure is exemplified in its simplest form by least squares regression. In general, the model obtained in this way does not explain the data exactly, in the sense that it is not capable of reproducing the data in an exact way. The standard way of dealing with this situation is to include randomness or noise in the model. The lack of fit between data and model is attributed to randomness or noise, and this assumption allows the use of models which, strictly speaking, are not supported by the observations.

This point of view, however, does not take into account the fact that, in most cases, the lack of fit between data and model is the result of having chosen a model which is unable to explain the observations. A typical example of this situation is when a linear time-invariant model is used to fit data produced by a nonlinear system.

These considerations suggest that modeling is better cast into an *approximation* framework, in which a "good" model is the one which "best" approximates the measurements.

The approximate modeling framework will be dealt with in another section of this paper. In this section, instead, we focus on the exact modeling of time series, as the logical starting point of identification. We first introduce some concepts which formalize the modeling process. A thorough discussion of this topic, on which the present treatment is based, can be found in [14].

A dynamical phenomenon is characterized by a certain set of measurable attributes. For the purposes of this paper, these attributes will consist of real numbers, collected in a q-dimensional vector. The phenomenon will produce *outcomes* which consist of q-dimensional vector time series; that is, the *phenomenon space* S is $(\mathbb{R}^q)^{\mathbb{Z}}$.

Definition 4.1. *A* model M *for a phenomenon with phenomenon space* S *is a subset* $M \subseteq S$*. A* model class $\mathcal{M}$ *is a subset* $\mathcal{M} \subseteq 2^S$.

The choice of the model class $\mathcal{M}$ embodies the *a priori* assumptions on the nature of the phenomenon, for example, linearity. On the other hand, a model is a set of possible outcomes of a phenomenon: according to the model M, only the outcomes in M can be produced by the phenomenon. In this sense, the choice of a model M *restricts* the possible outcomes to those in M.

Definition 4.2. *Let* M_1, M_2 *be models in the model class* $\mathcal{M}$. M_1 *is* more powerful *than* M_2 *if* $M_1 \subseteq M_2$.

This definition puts the prohibitive power of a model as the natural ordering criterion: the more a model forbids, restricting the possible outcomes of the phenomenon to a smaller set, the better it is.

A *measurement* of a phenomenon is a set of time series $Z \subset S$. The following definition gives a precise meaning to exact modeling of a time series.

Definition 4.3. *A model M is* unfalsified *by the measurements Z, if $Z \subseteq M$.*

Let us now introduce the important concept of most powerful unfalsified model.

Definition 4.4. *Let $\mathcal{M}$ be a model class, and Z a set of measurements. M^* is the* most powerful unfalsified model *in $\mathcal{M}$ if $Z \subseteq M^*$, and if for all $M \in \mathcal{M}$ such that $Z \subseteq M$, there holds $M^* \subseteq M$.*

Therefore, the most powerful unfalsified model is the more restrictive model in $\mathcal{M}$ among those not refuted by the data.

Example 4.1. Assume the phenomenon space S is $\mathbb{R}^q$. Therefore the observations are q-dimensional vectors, $Z = \{z_1, \ldots, z_r\}$, $z_i \in \mathbb{R}^q$. Assume that the model class $\mathcal{M}$ is chosen as that of all linear subspaces of $\mathbb{R}^q$. Then a most powerful model exists, and in fact for Z as above the most powerful model $M_Z^* = span_{\mathbb{R}}\{z_i\}_{i=1,\ldots,g}$. □

4.2 Existence of the Most Powerful Unfalsified Model

The most powerful unfalsified model introduced in definition 4.4 need not exist; as an example, take the case in which no unfalsified model exists in the model class $\mathcal{M}$. However, provided that the model class $\mathcal{M}$ enjoys the *intersection property*, the most powerful unfalsified model exists, as stated in the following

Proposition 4.1. *([14]) Assume that $\mathcal{M}$ enjoys the* intersection property, *that is, if $\mathcal{M}' \subseteq \mathcal{M}$, then $\bigcap_{M \in \mathcal{M}'} M \in \mathcal{M}$. Moreover, assume $S \in \mathcal{M}$. Then for every measurement Z, there exists a most powerful unfalsified model for Z in $\mathcal{M}$.*

Let us now specialize the definitions given in section 4.1 to the case of direct interest to us, that of modeling q-dimensional time series by subsets of $(\mathbb{R}^q)^{\mathbb{Z}}$. A model is a subset $\mathcal{B}$ of $(\mathbb{R}^q)^{\mathbb{Z}}$, and it identifies a dynamical system $\Sigma = (\mathbb{Z}, \mathbb{R}^q, \mathcal{B})$ in the obvious way. The model class we will consider in the following is $\mathcal{M} = \mathcal{L}^q$, that consisting of all linear, finite dimensional, time-invariant systems. Given a finite set Z of time series, corresponding to observations of a dynamical system, the model $\mathcal{B}$ is unfalsified by Z if $Z \subseteq \mathcal{B}$. One may wonder whether, given Z, the most powerful unfalsified model for Z exists in $\mathcal{M}$. The answer is positive, since the class $\mathcal{L}^q$ enjoys the intersection property.

Proposition 4.2. *([14]) The model class $\mathcal{L}^q$ enjoys the intersection property. Therefore, if the phenomenon space is $S = (\mathbb{R}^q)^{\mathbb{Z}}$, and $\mathcal{M} = \mathcal{L}^q$, for every finite set of data d_1, d_2, ..., d_r, the most powerful unfalsified model $\Sigma^* = (\mathbb{Z}, \mathbb{R}^q, \mathcal{B}^*_Z)$ exists.*

According to proposition 4.2, given a finite set of measurements Z, there exists a linear time-invariant complete system Σ^*, which can reproduce the time series in Z, and as little else as possible. In fact, given a set of measurements $Z = \{\tilde{w}_1, \tilde{w}_2, \ldots, \tilde{w}_r\}$, the computation of $\mathcal{B}^*_Z$ is immediate: $\mathcal{B}^*_Z$ is simply the closure in the topology of pointwise convergence in $(\mathbb{R}^q)^{\mathbb{Z}}$ of the span of the time series $\sigma^t \tilde{w}_i$, $t \in \mathbb{Z}$, $1 \leq i \leq r$; we will denote this set with

$$\mathcal{B}^*_Z = span\ \{\sigma^t \tilde{w}_i \mid t \in \mathbb{Z}, 1 \leq i \leq r\}. \tag{4.1}$$

It is clear that $\mathcal{B}^*_Z$ defined in this way contains Z (take $t = 0$ in (4.1)). On the other hand, any linear, time-invariant, and complete model unfalsified by Z, must contain the closure of the span of all shifts of $\tilde{w}_i$, $1 \leq i \leq r$, and therefore will have to contain $\mathcal{B}^*_Z$, it will be less powerful than $\mathcal{B}^*_Z$.

$\mathcal{B}^*_Z$ defined in (4.1) is a rather abstract way of answering the question of what is the most powerful unfalsified model of a set of measurements Z: more concrete representations of $\Sigma^* = (\mathbb{Z}, \mathbb{R}^q, \mathcal{B}^*_Z)$ are required. This need prompts the search for algorithms which, given as input data a set of time series Z, give as output a representation of $\mathcal{B}^*_Z$, in terms of a set of equations, in one of the forms introduced in section 2. In [14] this problem has been discussed for the case of modeling a time series with a kernel representation or a state space with driving variable representation. Therein, two algorithms for passing from a q-dimensional time series $\tilde{w}$ to a polynomial matrix $R^*(\xi)$ such that $R^*(\sigma)w = 0$ is a kernel representation of $\mathcal{B}^*_Z$, have been discussed, and we refer the interested reader to the original source. In the following section, and in section 5., instead, we will focus on the algorithms for computing a state space model.

4.3 From Time Series to State Space Model

In this section we state in a formal way the problem of passing from a time series to a state space model, and we outline the basic ideas underlying the algorithms illustrated in section 5. of this paper.

The problem we deal with in this section is the following: given a q-dimensional vector time series $\tilde{w}$, find a minimal state space with driving variable representation of $\mathcal{B}^*_{\tilde{w}} = span\ \{\sigma^t \tilde{w} \mid t \in \mathbb{Z}\}$. That is, given $\tilde{w}$, we look for matrices A, B, C, D, such that

$$\begin{aligned} \sigma x &= Ax + Bv \\ w &= Cx + Dv \end{aligned} \tag{4.2}$$

is a state space with driving variable representation of $\mathcal{B}^*_{\tilde{w}}$, with minimal dimension of the state space, and minimal dimension of the driving variable space. A more general version of the problem consists of modeling r time series $\{\tilde{w}_1, \ldots, \tilde{w}_r\}$.

Note that no information on the dimension of the state space and on the dimension of the driving variable space is available to start with. While the former is a common situation in identification, the latter is not: from realization theory, through the entire *corpus* of identification theory, the input space is assumed to be known to the modeler. The fact that the state space and the driving input space are simultaneously identified by the algorithms we will illustrate, is, in fact, one of the main novelties and points of interest in the approach put forward in [14].

The algorithms of section 5. examine the Hankel matrix of the data $\mathcal{H}(\tilde{w})$:

$$\begin{matrix} \vdots \\ row\text{ -t}\rightarrow \\ \vdots \\ row\text{ -1}\rightarrow \\ row\text{ 0}\rightarrow \\ row\text{ 1}\rightarrow \\ \vdots \\ row\text{ t}\rightarrow \\ \vdots \end{matrix} \begin{pmatrix} \cdots & \vdots & \vdots & \vdots & \cdots & \vdots & \cdots \\ \cdots & \tilde{w}(-t+1) & \tilde{w}(-t+2) & \tilde{w}(-t+3) & \cdots & \tilde{w}(-t+t') & \cdots \\ \cdots & \vdots & \vdots & \vdots & \cdots & \vdots & \cdots \\ \cdots & \tilde{w}(-1) & \tilde{w}(0) & \tilde{w}(1) & \cdots & \tilde{w}(t') & \cdots \\ \cdots & \tilde{w}(0) & \tilde{w}(1) & \tilde{w}(2) & \cdots & \tilde{w}(t'+1) & \cdots \\ \cdots & \tilde{w}(1) & \tilde{w}(2) & \tilde{w}(3) & \cdots & \tilde{w}(t'+2) & \cdots \\ \cdots & \vdots & \vdots & \vdots & \cdots & \vdots & \cdots \\ \cdots & \tilde{w}(t-1) & \tilde{w}(t) & \tilde{w}(t+1) & \cdots & \tilde{w}(t'+t) & \cdots \\ \cdots & \vdots & \vdots & \vdots & \cdots & \vdots & \cdots \end{pmatrix} \tag{4.3}$$

and its truncations $\mathcal{H}_t(\tilde{w})$, the submatrices consisting of the block rows of $\mathcal{H}(\tilde{w})$ indexed 0 to t. $\mathcal{H}(\tilde{w})$ provides a concrete representation of the most powerful unfalsified model $\mathcal{B}^*_{\tilde{w}}$: the columns of $\mathcal{H}(\tilde{w})$ represent the shifts $\sigma^t\tilde{w}$, and therefore $\mathcal{B}^*_{\tilde{w}}$ coincides with the column span of $\mathcal{H}(\tilde{w})$. As we will see, all information regarding the structure of the system is embedded in $\mathcal{H}(\tilde{w})$ and its truncations.

In the more general version of the problem, consisting of modeling r time series $\{\tilde{w}_1, \ldots, \tilde{w}_r\}$, the algorithms examine the block Hankel matrix

$$\mathcal{H}(\tilde{W}) := \begin{pmatrix} \cdots & \vdots & \vdots & \vdots & \cdots & \vdots & \cdots \\ \cdots & \tilde{W}(-t+1) & \tilde{W}(-t+2) & \tilde{W}(-t+3) & \cdots & \tilde{W}(-t+t') & \cdots \\ \cdots & \vdots & \vdots & \vdots & \cdots & \vdots & \cdots \\ \cdots & \tilde{W}(-1) & \tilde{W}(0) & \tilde{W}(1) & \cdots & \tilde{W}(t') & \cdots \\ \cdots & \tilde{W}(0) & \tilde{W}(1) & \tilde{W}(2) & \cdots & \tilde{W}(t'+1) & \cdots \\ \cdots & \tilde{W}(1) & \tilde{W}(2) & \tilde{W}(3) & \cdots & \tilde{W}(t'+2) & \cdots \\ \cdots & \vdots & \vdots & \vdots & \cdots & \vdots & \cdots \\ \cdots & \tilde{W}(t-1) & \tilde{W}(t) & \tilde{W}(t+1) & \cdots & \tilde{W}(t'+t) & \cdots \\ \cdots & \vdots & \vdots & \vdots & \cdots & \vdots & \cdots \end{pmatrix}, \tag{4.4}$$

with

$$\tilde{W}(t) := \begin{pmatrix} \tilde{w}_1(t) & \ldots & \tilde{w}_r(t) \end{pmatrix}. \tag{4.5}$$

The main ideas underlying the algorithms we will illustrate in section 5. are those of truncation, of simultaneous construction of the state space and of the input space, and that of splitting. We briefly discuss them in the following.

The first idea is that of *truncation*. In general, the subspace $\mathcal{B}^*_{\tilde{w}} = span\ \{\sigma^t \tilde{w} \mid t \in \mathbb{Z}\}$ is infinite dimensional; consider, for example, the case of a one-dimensional time series composed of a free input only: in this case $span\ \{\sigma^t \tilde{w} \mid t \in \mathbb{Z}\} = (\mathbb{R}^q)^{\mathbb{Z}}$. However, it will be shown in section 5. that examining the truncations of $\mathcal{B}^*_{\tilde{w}}$ to finite but "sufficiently large" time intervals, no information about $\mathcal{B}^*_{\tilde{w}}$ is lost in the truncation, and therefore the computations reduce to finite dimensional linear algebra. The truncation of $\mathcal{B}^*_{\tilde{w}}$ to $[0,t]$ consists of the restriction of the trajectories in $\mathcal{B}^*_{\tilde{w}}$ to the time interval $[0,t]$, and is denoted with $(\mathcal{B}^*_{\tilde{w}})_t$; note that $(\mathcal{B}^*_{\tilde{w}})_t$ coincides with the column span of $\mathcal{H}_t(\tilde{w})$.

The second idea underlying the identification procedures of section 5., is the simultaneous construction of the state space and of the input space. These spaces are computed from the truncations $(\mathcal{B}^*_{\tilde{w}})_t$ of $\mathcal{B}^*_{\tilde{w}}$ which contain all relevant information, via projections on appropriate subspaces of $(\mathcal{B}^*_{\tilde{w}})_t$. The computation of the input space, in particular, can be interpreted as the determination of the part of $\tilde{w}(t)$ left unexplained by the state at time t.

The third concept we will use in section 5., especially for the second identification algorithm we will illustrate, is that of the state as the quantity which *splits* past and future of the behavior of the system, or equivalently, which captures the *common features* of past and future. This point of view will be formalized with the notion of *relative row rank of a partitioned matrix*.

4.4 Realization Theory as a Special Case

Although it is common practice in identification to assume a stochastic point of view, that is, to introduce stochastics in order to explain the discrepancy between data and model, a special case of identifying a state space model, namely realization theory, has dealt extensively with the problem from a purely deterministic point of view.

The problem of impulse response realization can be considered as a special case of the problem stated in section 4.3; the time series $\tilde{w}_i$, $i = 1, \dots, m$, consist in this case of the impulse response of the system: $\tilde{w}_i(t) := \begin{pmatrix} \delta_i(t) \\ \tilde{g}_i(t) \end{pmatrix}$, where δ_i is the impulse in the i-th input channel:

$$\delta_i(t) = \begin{cases} e_i & t = 0 \\ 0 & t \neq 0 \end{cases}$$

and $\tilde{g}_i(t)$ is the vector of the impulse responses at time t due to an impulse in the i-th input channel.

In the classical input-output setting, many algorithms are known for passing from impulse response measurements to a corresponding minimal state space realization; among these, Ho-Kalman's [3] and Silverman's algorithm [9] are the best known. As we will see, the algorithms illustrated in section 5. have these classical realization theory algorithms as special cases. Of course, the fact that no *a priori* input-output structure is assumed, makes the algorithms of section 5. more than a simple extension or generalization of the impulse response realization algorithms.

4.5 Recapitulation

In this section we have introduced the notion of most powerful unfalsified model as the model which explains the observations but as little else as possible. We have investigated the question of existence of the most powerful unfalsified model for the problem of modeling q-dimensional time series, and we have discussed the basic concepts and techniques on which the state space identification algorithms illustrated in the following section are built. Before going further, let us remark that the notion of most powerful unfalsified model has been successfully applied to problems other than those discussed in this paper. For example, this notion can be applied to the case of modeling a finite time series; the interested reader is referred to [2], where a formalization of the process of modeling a finite amount of data from a time series has been undertaken, and the very important concepts of *simplicity* and *corroboration* have been introduced. In [1], instead, the concept of most powerful unfalsified model has been applied to the problem of modeling a finite set of polynomial exponential time series. Other directions of research stemming from the behavioral approach to modeling, have been pursued in [8].

5. Algorithms

In this section we illustrate the procedures of [14] for obtaining state space models from time series. We first address the abstract problem of how to pass from a given behavior $\mathcal{B}$ to a state space with driving variable model. This will be instrumental in introducing the concepts on which the modeling algorithms are based. The concept of relative row rank is discussed, and its connections with the problem of extracting common features from a set of observations are illustrated. The relationships of the identification procedures of [14], and the family of identification schemes currently known as *subspace identification methods* ([6, 10]), are discussed. Classical realization theory algorithms are shown to be special cases of the algorithms illustrated in this section.

5.1 From Behavior to State Space Model

This section illustrates how to compute a minimal state space with driving variable representation, having a given behavior $\mathcal{B}$ as external behavior. The discussion is based on sections 9 of [13], and 16 of [14], to which the reader is referred for a thorough exposition.

In determining the state space and the driving variable space of a state space with driving variable representation of $\mathcal{B}$, we will make use of the following three subspaces of $(\mathbb{R}^q)^{\mathbb{Z}_+}$:

$$\begin{aligned} \mathcal{B}^+ &:= \mathcal{B}_{|\mathbb{Z}_+} \\ \mathcal{B}^0 &:= \{w \in \mathcal{B}^+ \mid (\sigma^*)^t w \in \mathcal{B}^+ \; \forall\, t \in \mathbb{Z}_+\} \\ \mathcal{B}^1 &:= \{w \in \mathcal{B}^0 \mid w(0) = 0\}. \end{aligned} \tag{5.1}$$

Here σ^* denotes the forwards shift: for $w \in (\mathbb{R}^q)^{\mathbb{Z}_+}$, $(\sigma^* w)(t) := w(t-1)$, $t \geq 1$, and $(\sigma^* w)(0) := 0$.

$\mathcal{B}^+$ is the restriction of $\mathcal{B}$ to $\mathbb{Z}_+$; interpreting the time instant $t = 0$ as the current time, $\mathcal{B}^+$ is the set of all possible future trajectories of $\mathcal{B}$.

$\mathcal{B}^0 \subseteq \mathcal{B}^+$ is the subspace of future trajectories of $\mathcal{B}^+$ which can be preceded by an arbitrary number of zeros. This subspace consists of all trajectories which can be considered to be in the zero state at $t = 0$, since they are concatenable with the zero trajectory.

$\mathcal{B}^1$ is the subspace of $\mathcal{B}^0$ consisting of the trajectories which are zero at $t = 0$, that is, all trajectories of $\mathcal{B}$ which, besides being in the zero state at $t = 0$, can be considered to have had zero input at time $t = 0$.

Introduce now the following equivalence relation on $\mathcal{B}^+$:

$$w_1 \overset{\mathcal{B}^0}{\sim} w_2 \Longleftrightarrow w_1 - w_2 \in \mathcal{B}^0 \tag{5.2}$$

(read "w_1 and w_2 are equivalent modulo $\mathcal{B}^0$"). The set of equivalence classes $\mathcal{B}^+(mod\ \mathcal{B}^0)$ can be identified as the state space: any two trajectories of $\mathcal{B}^+$ equivalent modulo $\mathcal{B}^0$ have the same state at $t = 0$, since their difference can be preceded by any number of zeros.

Define an equivalence relation $\overset{\mathcal{B}^1}{\sim}$ on $\mathcal{B}^0$ as follows:

$$w_1 \overset{\mathcal{B}^1}{\sim} w_2 \Longleftrightarrow w_1 - w_2 \in \mathcal{B}^1. \tag{5.3}$$

$\mathcal{B}^0(mod\mathcal{B}^1)$ can be interpreted as the input space: all trajectories in $\mathcal{B}^0$ start in the zero state, and if they take the same value at $t = 0$, this is due to the action of the same input at $t = 0$.

Note that $\mathcal{B}^+(mod\ \mathcal{B}^1) \cong \mathcal{B}^+(mod\ \mathcal{B}^0) \oplus \mathcal{B}^0(mod\ \mathcal{B}^1)$, that is, we can identify $\mathcal{B}^+(mod\ \mathcal{B}^1)$ with $X \oplus U$, with $X = \mathbb{R}^n$, and $U = \mathbb{R}^m$, for suitable integers n and m. This yields the following commutative diagram:

$$\begin{array}{ccccc} \mathbb{R}^q & \overset{M_2}{\longleftarrow} & \mathcal{B}^+(mod\mathcal{B}^1) & \overset{M_1}{\longrightarrow} & \mathcal{B}^+(mod\mathcal{B}^0) \\ id \downarrow & & \uparrow\downarrow iso & & \uparrow\downarrow iso \\ \mathbb{R}^q & \overset{(C \quad D)}{\longleftarrow} & \mathbb{R}^n \oplus \mathbb{R}^m & \overset{(A \quad B)}{\longrightarrow} & \mathbb{R}^n \end{array}$$

in which the *forwards shift map* M_1 and the *read-out map* M_2 are shown. M_1 computes the state at time $t+1$, on the basis of the state and the input at time t. M_2 computes the value of w at time t, based on the value of the current state and of the current input. Existence of the maps M_1 and M_2 follows from

Theorem 5.1. *([14]) Let $\mathcal{B}$ be a linear, time-invariant, and complete subspace of $(\mathbb{R}^q)^{\mathbb{Z}}$. Then $\mathcal{B}^+(mod\mathcal{B}^0)$ and $\mathcal{B}^0(mod\mathcal{B}^1)$ are finite dimensional, and there exist linear maps M_1 and M_2 such that the diagram commutes. Moreover, having fixed a basis for $\mathcal{B}^+(mod\mathcal{B}^0) \cong \mathbb{R}^n$ and a basis for $\mathcal{B}^0(mod\mathcal{B}^1) \cong \mathbb{R}^m$, let M_1 be represented by the matrix $(A \quad B)$ with $A \in \mathbb{R}^{n\times n}$, $B \in \mathbb{R}^{n\times m}$, and let M_2 be represented by the matrix $(C \quad D)$, $C \in \mathbb{R}^{q\times n}$, $D \in \mathbb{R}^{q\times m}$. Then*

$$\begin{aligned} \sigma x &= Ax + Bv \\ w &= Cx + Dv \end{aligned} \tag{5.4}$$

is a minimal state space with driving variable representation of $\mathcal{B}$.

The spaces $\mathcal{B}^+$, $\mathcal{B}^0$, and $\mathcal{B}^1$ are, in general, infinite dimensional; however, by examining their truncations $\mathcal{B}_t^+ := \mathcal{B}^+_{|[0,t]\bigcap\mathbb{Z}_+}$, $\mathcal{B}_t^0 := \mathcal{B}^0_{|[0,t]\bigcap\mathbb{Z}_+}$, $\mathcal{B}_t^1 := \mathcal{B}^1_{|[0,t]\bigcap\mathbb{Z}_+}$, for a suitably large t, computation of the system matrices A, B, C, D can be performed using finite dimensional linear algebra. To determine how large the truncation interval must be, to ensure that no information is lost in the truncation process, let us introduce the sequence of *relative dimension indices* ρ_t (cfr. [13], section 7).

Definition 5.1. *Consider a linear, time-invariant, complete behavior $\mathcal{B}$, and the restrictions $\mathcal{B}_t := \mathcal{B}_{|[0,t]\bigcap\mathbb{Z}_+}$. The t-th* relative dimension index *ρ_t is*

$$\rho_t := \begin{cases} dim\mathcal{B}_t - dim\mathcal{B}_{t-1} & t \geq 1 \\ dim\mathcal{B}_0 & t = 0. \end{cases}$$

It is easily seen that

$$q \geq \rho_0 \geq \rho_1 \geq \ldots \geq 0, \tag{5.5}$$

so that

$$\rho_\infty := \lim_{t\to\infty} \rho_t \tag{5.6}$$

exists. It can be shown that $\rho_\infty = m$, the number of inputs in an input-output, or the minimal dimension of the driving variable in a driving variable representation of $\mathcal{B}$.

Let $t^* := min_{t \in \mathbb{Z}_+}\{t \text{ s.t. } \rho_t = \rho_\infty\}$. t^* is the minimal index such that the input-output and state space structure of $\mathcal{B}$, can be determined on the basis of $\mathcal{B}_t^+$, $\mathcal{B}_t^0$, and $\mathcal{B}_t^1$, as stated in the following

Proposition 5.1. *([14]) The following statements are equivalent:*

1. $t \geq t^*$
2. $dim\mathcal{B}_t^+(mod\mathcal{B}_t^0) = dim\mathcal{B}^+(mod\mathcal{B}^0)$
3. $dim\mathcal{B}_t^0(mod\mathcal{B}_t^1) = dim\mathcal{B}^0(mod\mathcal{B}^1)$

Therefore, the state space and the input space can be determined from $\mathcal{B}_t^+$, $\mathcal{B}_t^0$, $\mathcal{B}_t^1$, provided t is sufficiently large.

Note, however, that $\mathcal{B}_t^0$ and $\mathcal{B}_t^1$ cannot in general be determined on the basis of $\mathcal{B}_t^+$, respectively $\mathcal{B}_t^0$, alone. Let us therefore introduce the spaces:

$$\begin{aligned} \bar{\mathcal{B}}_t^0 &:= \bigcap_{0 \leq t' \leq t} (\sigma_t^*)^{-t'} \mathcal{B}_t^+ \\ \bar{\mathcal{B}}_t^1 &:= \bar{\mathcal{B}}_t^0 \bigcap Ker\pi^0. \end{aligned} \tag{5.7}$$

Here π^t is the projection on the interval $[0,t] \bigcap \mathbb{Z}_+$:

$$\pi^t(w) := w_{|[0,t] \bigcap \mathbb{Z}_+}; \tag{5.8}$$

and $\sigma_t^* := \pi^t \circ \sigma^*$. Note that $\bar{\mathcal{B}}_t^0$ consists of the restrictions of length $t+1$ of elements of $\mathcal{B}^+$, which can be preceded by any number of zeros. On the other hand, $\bar{\mathcal{B}}_t^1$ represents the sequences in $\bar{\mathcal{B}}_t^0$ which, in addition, are zero at $t = 0$. Of course, $\mathcal{B}_t^0 \subseteq \bar{\mathcal{B}}_t^0$ and $\mathcal{B}_t^1 \subseteq \bar{\mathcal{B}}_t^1$; however, it turns out that restrictions of $\mathcal{B}^+$ of length at least t^* guarantee equality.

Proposition 5.2. *([14]) The following statements are equivalent:*

1. $t \geq t^*$
2. $\mathcal{B}_{t+1}^0 = \bar{\mathcal{B}}_{t+1}^0$
3. $\mathcal{B}_{t+1}^1 = \mathcal{B}_{t+1}^1$

This result allows us to draw the following diagram for $t \geq t^*$:

$$\begin{array}{ccccc} \mathbb{R}^q & \xleftarrow{\bar{M}_2} & \mathcal{B}^+(mod\bar{\mathcal{B}}^1) & \xrightarrow{\bar{M}_1} & \mathcal{B}^+(mod\bar{\mathcal{B}}^0) \\ id \downarrow & & \uparrow\downarrow iso & & \uparrow\downarrow iso \\ \mathbb{R}^q & \xleftarrow{(C \quad D)} & \mathbb{R}^n \oplus \mathbb{R}^m & \xrightarrow{(A \quad B)} & \mathbb{R}^n \end{array}$$

which is the analogous of the previous one. The following theorem is a restatement of theorem 5.1 for the case at hand.

Theorem 5.2. *([14]) Let a linear, time-invariant, and complete behavior $\mathcal{B}$, and $t \geq t^*$, be given. There exist maps $\bar{M}_1$ and $\bar{M}_2$ such that the diagram commutes, and $\bar{M}_1$ and $\bar{M}_2$ yield a state space with driving variable representation of $\mathcal{B}$.*

5.2 From Time Series to State Space Model I

In this section we illustrate the algorithm of section 17 of [14]. Basically, the algorithm amounts to applying theorem 5.2 to $\mathcal{B}^*_{\tilde{w}} = span_{\mathbb{R}} columns \mathcal{H}(\tilde{w})$. To this purpose, the integer t^* is first computed. Then $\bar{\mathcal{B}}^0_t$ and $\bar{\mathcal{B}}^1_t$, are determined from the columns of $\mathcal{H}_t(\tilde{w})$, and they are used to compute a basis of the state space and a basis of the driving variable space. Finally, a matrix representation of the forwards shift map and of the read-out map is computed.

Algorithm 1

Data: ..., $\tilde{w}(-1)$, $\tilde{w}(0)$, $\tilde{w}(1)$, ..., $\tilde{w}(t)$, ...
Output: Matrices A, B, C, D of a state space with driving variable representation of $\mathcal{B}^*_{\tilde{w}}$

Step 1. Compute $\rho_0 := rank\mathcal{H}_0(\tilde{w})$, and $\rho_t := rank\mathcal{H}_t(\tilde{w}) - rank\mathcal{H}_{t-1}(\tilde{w})$, $t \geq 1$.
Choose t' be such that $\rho_t = \rho_{t'}$ $\forall t \geq t'$.
Step 2. Determine $\mathcal{B}_{t'} := span\{col(\tilde{w}, \ldots, \sigma^{t'}\tilde{w}\}$
Step 3. Define $\sigma^*_t : \mathbb{R}^{q(t'+1)} \to \mathbb{R}^{q(t'+1)}$, as

$$\sigma^*_t(col(a_0, a_1, \ldots, a_{t'})) = col(0, a_0, a_1, \ldots, a_{t'-1}),$$

and let $\pi^0 : \mathbb{R}^{q(t'+1)} \to \mathbb{R}^q$ be defined as

$$\pi^0(col(a_0, a_1, \ldots, a_{t'})) = a_0.$$

Compute

$$\begin{aligned} \bar{\mathcal{B}}^0_{t'} &:= \bigcap_{0 \leq t \leq t'} (\sigma^*_{t'})^{-t} \mathcal{B}^+_t \\ \bar{\mathcal{B}}^1_{t'} &:= \bar{\mathcal{B}}^0_{t'} \bigcap Ker\pi^0. \end{aligned} \tag{5.9}$$

Step 4. Determine a submatrix H of $\mathcal{H}_{t'}(\tilde{w})$ with n_2 columns, and matrices $P \in \mathbb{R}^{n \times (t'+1)q}$, $Q_x \in \mathbb{R}^{n_2 \times n}$, and $Q_u \in \mathbb{R}^{n_2 \times m}$, such that

$$\begin{aligned} \bar{\mathcal{B}}^0_{t'} \oplus ImHQ_x &= \mathcal{B}_{t'} \\ \bar{\mathcal{B}}^1_{t'} \oplus ImHQ_u &= \mathcal{B}^0_{t'} \\ PHQ_x &= I_n \\ P\bar{\mathcal{B}}^0_{t'} &= 0 \end{aligned} \tag{5.10}$$

Step 5. Let H consist of the k_1-th, k_2-th, ..., k_{n_2}-th column of $\mathcal{H}_{t'}(\tilde{w})$. Define σH to be the submatrix of $\mathcal{H}_{t'}(\tilde{w})$ consisting of the (k_1+1)-th, (k_2+1)-th,

..., $(k_{n_2}+1)$-th column of $\mathcal{H}_{t'}(\tilde{w})$. Let H^0 be the submatrix of H consisting of its first q rows. Compute

$$\begin{aligned} A &= P\sigma H Q x \\ B &= P\sigma H Q_u \\ C &= H^0 Q x \\ D &= H^0 Q_u. \end{aligned} \tag{5.11}$$

The matrices A, B, C, D, are the desired ones.

Example 5.1. (Example 5 of [14]) Consider the system described in kernel form as

$$\tilde{w}_2(t+1) = \tilde{w}_1(t) + \tilde{w}_2(t) \tag{5.12}$$

and the following time series, which belongs to the behavior of the system:

$$\begin{aligned} \tilde{w}_1 &= \{\ldots,1,0,1,0,0,1,0,0,0,1,0,0,0,1,0,\ldots\} \\ \tilde{w}_2 &= \{\ldots,0,1,1,2,2,2,3,3,3,3,4,4,4,4,4,\ldots\}. \end{aligned} \tag{5.13}$$

The matrix $\mathcal{H}(\tilde{w})$ is

$$\mathcal{H}(\tilde{w}) = \begin{pmatrix} \vdots & \vdots & \vdots & \vdots & \vdots & \vdots & \vdots & \vdots & \vdots & \vdots & \vdots & \vdots & \vdots & \vdots & \vdots & \vdots & \vdots \\ \ldots & 1 & 0 & 1 & 0 & 0 & 1 & 0 & 0 & 0 & 1 & 0 & 0 & 0 & 0 & 1 & \ldots \\ \ldots & 0 & 1 & 1 & 2 & 2 & 2 & 3 & 3 & 3 & 3 & 4 & 4 & 4 & 4 & 4 & \ldots \\ \ldots & 0 & 1 & 0 & 0 & 1 & 0 & 0 & 0 & 1 & 0 & 0 & 0 & 0 & 1 & 0 & \ldots \\ \ldots & 1 & 1 & 2 & 2 & 2 & 3 & 3 & 3 & 3 & 4 & 4 & 4 & 4 & 4 & 5 & \ldots \\ \ldots & 1 & 0 & 0 & 1 & 0 & 0 & 0 & 1 & 0 & 0 & 0 & 0 & 1 & 0 & 0 & \ldots \\ \ldots & 1 & 2 & 2 & 2 & 3 & 3 & 3 & 3 & 4 & 4 & 4 & 4 & 4 & 5 & 5 & \ldots \\ \vdots & \vdots & \vdots & \vdots & \vdots & \vdots & \vdots & \vdots & \vdots & \vdots & \vdots & \vdots & \vdots & \vdots & \vdots & \vdots & \vdots \end{pmatrix} \tag{5.14}$$

and $t^* = 2$. The reader can verify that

$$\mathcal{B}_2 = Im \begin{pmatrix} 1 & 0 & 1 & 0 \\ 0 & 1 & 1 & 2 \\ 0 & 1 & 0 & 0 \\ 1 & 1 & 2 & 2 \\ 1 & 0 & 0 & 1 \\ 1 & 2 & 2 & 2 \end{pmatrix}. \tag{5.15}$$

Its is easily verified that only the first column of the matrix in (5.15), together with its forwards shifts, belong to $\mathcal{B}_2$. Therefore

$$\mathcal{B}_2^0 = Im \begin{pmatrix} 1 & 0 & 0 \\ 0 & 0 & 0 \\ 0 & 1 & 0 \\ 1 & 0 & 0 \\ 1 & 0 & 1 \\ 1 & 1 & 0 \end{pmatrix} \tag{5.16}$$

and

$$\mathcal{B}_2^1 = Im \begin{pmatrix} 0 & 0 \\ 0 & 0 \\ 1 & 0 \\ 0 & 0 \\ 0 & 1 \\ 1 & 0 \end{pmatrix} . \tag{5.17}$$

One can take

$$H = \begin{pmatrix} 1 & 0 \\ 0 & 2 \\ 0 & 0 \\ 1 & 2 \\ 1 & 1 \\ 1 & 2 \end{pmatrix} , \tag{5.18}$$

that is, the first and fourth column of the part of $\mathcal{H}_2(\tilde{w})$ displayed in (5.14), and

$$Q_x = \begin{pmatrix} 0 \\ 1 \end{pmatrix}$$

$$Q_u = \begin{pmatrix} 1 \\ 0 \end{pmatrix} . \tag{5.19}$$

P can be chosen as

$$P = \left(-\tfrac{1}{4}, \tfrac{1}{4}, 0, \tfrac{1}{4}, 0, 0 \right) . \tag{5.20}$$

With

$$\sigma H = \begin{pmatrix} 0 & 0 \\ 1 & 2 \\ 1 & 1 \\ 1 & 2 \\ 0 & 0 \\ 2 & 3 \end{pmatrix} \tag{5.21}$$

and

$$H^0 = \begin{pmatrix} 1 & 0 \\ 0 & 2 \end{pmatrix} , \tag{5.22}$$

this yields $A = 1$, $B = \frac{1}{2}$, $C = \begin{pmatrix} 0 \\ 2 \end{pmatrix}$, $D = \begin{pmatrix} 1 \\ 0 \end{pmatrix}$ which is a minimal state space with driving variable representation of the system. □

5.3 Common Features and Relative Row Rank

In this section we will discuss the interpretation of the state as a quantity which extracts the common features of past and future of a behavior, a point of view put forward in [14].

This interpretation is better introduced by means of the finite measurements case. Assume that two (finite) sets of observations $c_k \in \mathbb{R}^{n_1}$, $k = 1, 2, \ldots, m$, and $d_k \in \mathbb{R}^{n_2}$, $k = 1, 2, \ldots, m$, are given. One can think of the c_k's and d_k's as measurements of different, but related, phenomena, for example, the marks on various tests of a set of m students in high school and in University, respectively. The problem is to extract the common features between the phenomena of which the c_k's and the d_k's are the measured quantities. This can formalized in a linear algebraic setting as follows. Let $\mathcal{C} = span_{\mathbb{R}}\{c_k\}_{k=1,\ldots,m}$, and $\mathcal{D} = span_{\mathbb{R}}\{d_k\}_{k=1,\ldots,m}$ be the subspaces generated by the observations. Determine a basis for $\mathcal{C}$, and a basis for $\mathcal{D}$, such that as many as possible components of the c_k's, represented by means of the basis chosen for $\mathcal{C}$, are equal to those of the d_k's, represented in the basis chosen for $\mathcal{D}$. In this way an optimal choice of representation is performed, which maximizes the display of common features among the two sets of measurements.

In this setting, computation of the number of common features among $\mathcal{C}$ and $\mathcal{D}$ can be performed as follows: find an integer n, the number of common features among the c_k's and the d_k's, enjoying the two properties:

1. There exist subspaces $\mathcal{C}^0$ and $\mathcal{D}^0$ of $\mathcal{C}$ and $\mathcal{D}$, respectively, such that

$$n = dim\mathcal{C}(mod\mathcal{C}^0) = \mathcal{D}(mod\mathcal{D}^0);$$

2. n is maximal among the integers satisfying property 1.

In practice, n is more easily computed resorting to the notion of *relative row rank* of a partitioned matrix.

Definition 5.2. *The* relative row rank *of a partitioned matrix* $M = \begin{pmatrix} M_1 \\ M_2 \end{pmatrix}$ *is*

$$rrr(M_1, M_2) := rankM_1 + rankM_2 - rankM.$$

Let now the observations c_k and d_k be arranged in two matrices

$$C = \begin{pmatrix} c_1 & c_2 & \ldots & c_m \end{pmatrix} \in \mathbb{R}^{n_1 \times m}$$

and

$$D = \begin{pmatrix} d_1 & d_2 & \ldots & d_m \end{pmatrix} \in \mathbb{R}^{n_1 \times m}.$$

From the above discussion, it follows that the number of common features among $\mathcal{C} = span_{\mathbb{R}}columnsC$ and $\mathcal{D} = span_{\mathbb{R}}columnsD$, equals $rrr(C, D)$.

The problem of computing the common features among two finite sets of observations is a natural starting point from which to deal with the problem of extracting the common features of the past and the future of a trajectory. To do this, we have first to generalize definition 5.2 for the case of infinite matrices. Let us first consider the case when $M = \begin{pmatrix} M_1 \\ M_2 \end{pmatrix}$ is a partitioned matrix with an infinite number of columns, and with a finite number of rows. Let $M_{1,t'}$ be the submatrix of M_1 consisting of its $(-t')$-th, $(-t'+1)$-th, ..., t'-th column, and let $M_{2,t'}$ be analogously defined. We define the relative row rank of M as

$$rrr(M_1, M_2) := \lim_{t' \to \infty} rrr(M_{1,t'}, M_{2,t'}). \tag{5.23}$$

Let now M have an infinite number of columns, and M_1 and M_2 have an infinite number of rows:

$$M = \begin{pmatrix} \vdots \\ m_{13} \\ m_{12} \\ m_{11} \\ m_{21} \\ m_{22} \\ m_{23} \\ \vdots \end{pmatrix}, \tag{5.24}$$

where m_{ij} is the j-th row of M_i, $i = 1, 2$, $j \in \mathbb{N}$. Denote with M_1^t the submatrix of M_1 consisting of its rows indexed 1, 2, ..., t. Analogously, let $M_2^{t'}$ be the submatrix of M_2 consisting of its t' lowest indexed rows. Define

$$rrr(M_1, M_2) := \lim_{t \to \infty, t' \to \infty} rrr(M_1^t, M_2^{t'}). \tag{5.25}$$

Finally, observe that augmenting M_1 or M_2 with additional rows never decreases the relative row rank. This yields the following

Definition 5.3. *Let $M = \begin{pmatrix} M_1 \\ M_2 \end{pmatrix}$ be a matrix with an infinite number of columns, and let M_1 and M_2 both have an infinite number of rows. With the notation introduced above,*

$$rrr(M_1, M_2) := \sup_{t', t''} rrr(M_1^{t'}, M_2^{t''}). \tag{5.26}$$

Let us now turn to the problem of extracting the common features of the past and of the future of a trajectory. Consider an observed time series $\tilde{w}$, and denote with $\mathcal{B}^*_{\tilde{w}|\mathbb{Z}_+}$ the set

$$\mathcal{B}^*_{\tilde{w}|\mathbb{Z}_+} := \{w_{|\mathbb{Z}_+} \mid w \in \mathcal{B}^*_{\tilde{w}}\} \tag{5.27}$$

and, similarly, let

$$\mathcal{B}^*_{\tilde{w}|\mathbb{Z}_-} := \{w_{|\mathbb{Z}_-} \mid w \in \mathcal{B}^*_{\tilde{w}}\}. \tag{5.28}$$

Note that $\mathcal{B}^*_{\tilde{w}|\mathbb{Z}_-}$ and $\mathcal{B}^*_{\tilde{w}|\mathbb{Z}_+}$ are the column span of the upper, respectively, lower, part of the infinite partitioned matrix

$$\mathcal{H}(\tilde{w}) := \begin{pmatrix} \cdots & \vdots & \vdots & \vdots & \cdots & \vdots & \cdots \\ \cdots & \tilde{w}(-2) & \tilde{w}(-1) & \tilde{w}(0) & \cdots & \tilde{w}(t'-1) & \cdots \\ \cdots & \tilde{w}(-1) & \tilde{w}(0) & \tilde{w}(1) & \cdots & \tilde{w}(t') & \cdots \\ - & - & - & - & - & - & - \\ \cdots & \tilde{w}(0) & \tilde{w}(1) & \tilde{w}(2) & \cdots & \tilde{w}(t'+1) & \cdots \\ \cdots & \tilde{w}(1) & \tilde{w}(2) & \tilde{w}(3) & \cdots & \tilde{w}(t'+2) & \cdots \\ \cdots & \vdots & \vdots & \vdots & \cdots & \vdots & \cdots \end{pmatrix} = \begin{pmatrix} \mathcal{H}_-(\tilde{w}) \\ \mathcal{H}_+(\tilde{w}) \end{pmatrix}. \tag{5.29}$$

Note that the matrix in (5.29) is the Hankel matrix of the data (4.3), in which past and future of the behavior have been separated explicitly. In fact, the number of common features between past and future has an interesting interpretation in terms of this partition.

Proposition 5.3. *(([14]) $rrr(\mathcal{H}_-(\tilde{w}), \mathcal{H}_+(\tilde{w}))$ is finite, and it equals the dimension of a minimal state space representation of $\mathcal{B}^*_{\tilde{w}}$.*

On the basis of this result, and of the theory of *splitting linear relations* put forward in [14], an algorithm for state space realization of time series can be formulated. We will illustrate this procedure in the next section.

5.4 From Time Series to State Model II

In this section an algorithm will be described for passing from a time series $\tilde{w}$ to a state space with driving variable model. Although the algorithm is based on the same ideas as the algorithm described in section 5.2, it differs from that one in that the state trajectory $\{x(\cdot)\}$, and the driving variable sequence $\{u(\cdot)\}$ are explicitly computed from the data. The system matrices are then determined on the basis of $\{x(\cdot)\}$ and of $\{u(\cdot)\}$, and not on the basis of the original data as done by the algorithm of section 5.2. As we will see in the next section, computation of the state sequence is a feature from which subspace identification methods, which have recently drawn much attention from the identification community, have been inspired.

The algorithm we will formally state in the sequel proceeds according to the ideas of section 5.1. A finite submatrix $\begin{pmatrix} H_1 \\ H_2 \end{pmatrix}$ of $\mathcal{H}(\tilde{w})$ is computed, with relative row rank equal to that of $\begin{pmatrix} \mathcal{H}_-(\tilde{w}) \\ \mathcal{H}_+(\tilde{w}) \end{pmatrix}$. Computation of the state sequence is performed by projecting the column space of H_2 onto a certain subspace of ImH_2, while computation of the input sequence is performed on the basis of the time series $\tilde{w}$ and of the state sequence. Finally, the system matrices are obtained by solving a linear system of equations.

Algorithm II

Data: $\ldots, \tilde{w}(-1), \tilde{w}(0), \tilde{w}(1), \ldots, \tilde{w}(t), \ldots$
Output: Matrices A, B, C, D of a state space with driving variable representation of $\mathcal{B}^*_{\tilde{w}}$

Step 1. Determine matrices H_- and H_+, consisting of rows of $\mathcal{H}_-(\tilde{w})$ and $\mathcal{H}_+(\tilde{w})$, respectively, such that

$$rrr(H_-, H_+) = (\mathcal{H}_-(\tilde{w}); \mathcal{H}_+(\tilde{w})) =: n.$$

Step 2. Determine a matrix $\begin{pmatrix} H_1 \\ H_2 \end{pmatrix}$ consisting of a finite set of columns of $\begin{pmatrix} H_- \\ H_+ \end{pmatrix}$ such that the columns of $\begin{pmatrix} H_1 \\ H_2 \end{pmatrix}$ span $Im \begin{pmatrix} H_- \\ H_+ \end{pmatrix}$.
Step 3. Compute $Ker H_1$, and

$$H_2 \; Ker H_1 := \{h_2 \in Im H_2 \mid h_2 = H_2 x, \; x \in Ker H_1\}.$$

Step 4. Let $h_+(t)$ denote the t-th column of H_+. Compute

$$x(t) := h_+(t)(mod \; H_2 Ker H_1)$$

for all columns of H_+. $\{x(t)\}$ is the state sequence, and it identifies the subspace $X := span\{x(t)\}$ of $Im H_+$, of dimension n. Therefore $X \cong \mathbb{R}^n$, and there exists a bijective map $\pi_x : Im H_+ \; (mod \; H_2 Ker H_1) \to \mathbb{R}^n$. For ease of notation, we will denote with $x(t) \in \mathbb{R}^n$ also the projection $\pi_x(h_+(t)(mod \; H_2 Ker H_1))$.
Step 5. Define $f(t) := \begin{pmatrix} \tilde{w}(t) \\ x(t) \end{pmatrix} \in \mathbb{R}^{q+n}$, and let $F = span_{\mathbb{R}}\{f(t)\}$. Let

$$u(t) := f(t) \; (mod \; span_{\mathbb{R}} x(t)).$$

This determines a subspace U of F such that $F = U \oplus span_{\mathbb{R}}\{x(t)\}$, and the integer $m := dim U$. The map $\pi_u : F \; (mod \; span_m R\{x(t)\} \to \mathbb{R}^m$ is surjective. For ease of notation, we will denote with $u(t) \in \mathbb{R}^m$ also the image of $f(t)$ $(mod \; span_{\mathbb{R}} x(t))$ under π_u. $\{u(\cdot)\}$ is the driving input sequence. *Step 6.* Determine $n+m$ time instants t_i such that $\{f(t_i)\}_{i=1,\ldots,n+m}$ is a basis for F, and let $\{u(t_i)\}_{i=1,\ldots,n+m}$ be the corresponding driving input sequence.
Step 7. Find $M \in \mathbb{R}^{(n+q)\times(n+m)}$ such that

$$\begin{pmatrix} x(t_i+1) \\ w(t_i) \end{pmatrix} = M \begin{pmatrix} x(t_i) \\ u(t_i) \end{pmatrix}.$$

Partition M as

$$\begin{array}{c} \\ n \\ q \end{array} \begin{array}{c} \begin{array}{cc} n & m \end{array} \\ \left[\begin{array}{cc} A & B \\ C & D \end{array}\right] \end{array}$$

The matrices A, B, C, D are the required ones.

Example 5.2. (Example 5 of [14]) Consider the sequence of data of example 5.1. Choose the matrix $H = \begin{pmatrix} H_- \\ H_+ \end{pmatrix}$, with

$$H_- = \begin{pmatrix} 1 & 0 & 1 & 0 & 0 & 1 & \dots \\ 0 & 1 & 1 & 2 & 2 & 2 & \dots \end{pmatrix}, \tag{5.30}$$

and

$$\begin{pmatrix} 0 & 1 & 0 & 0 & 1 & 0 & \dots \\ 1 & 1 & 2 & 2 & 2 & 3 & \dots \end{pmatrix}. \tag{5.31}$$

Choosing the matrices H_1 and H_2 as

$$\begin{aligned} H_1 &= \begin{pmatrix} 1 & 0 & 1 & 0 \\ 0 & 1 & 1 & 2 \end{pmatrix} \\ H_2 &= \begin{pmatrix} 0 & 1 & 0 & 0 \\ 1 & 1 & 2 & 2 \end{pmatrix}, \end{aligned} \tag{5.32}$$

it is easily verified that $\begin{pmatrix} H_1 \\ H_2 \end{pmatrix}$ has relative row rank equal to that of $\begin{pmatrix} H_- \\ H_+ \end{pmatrix}$. Note that

$$Ker H_1 = \begin{pmatrix} 1 & 0 \\ 1 & -2 \\ -1 & 0 \\ 0 & 1 \end{pmatrix} \tag{5.33}$$

and

$$H_2 Ker H_1 = Im \begin{pmatrix} 1 \\ 0 \end{pmatrix}. \tag{5.34}$$

This yields the state sequence x

$$x = \{1, 1, 2, 2, 2, 3, \dots\}. \tag{5.35}$$

The space F is spanned by

$$F =< \begin{pmatrix} 0 \\ 1 \\ 1 \end{pmatrix}, \begin{pmatrix} 1 \\ 1 \\ 1 \end{pmatrix} > \tag{5.36}$$

and this yields $\mathbb{R}$ as the input space. The projection π_u can be chosen as the projection of $f \in F$ on the first component, and this yields an input sequence

$$u = \{0, 1, 0, 0, 1, 0, \dots\}. \tag{5.37}$$

Observe now that

$$\begin{pmatrix} 1 & 2 \\ 0 & 1 \\ 1 & 1 \end{pmatrix} = M \begin{pmatrix} 1 & 1 \\ 0 & 1 \end{pmatrix}, \tag{5.38}$$

which yields

$$M = \begin{pmatrix} 1 & 1 \\ 0 & 1 \\ 1 & 0 \end{pmatrix}, \tag{5.39}$$

corresponding to the state space equations

$$\begin{aligned} \sigma x &= x + u \\ w &= \begin{pmatrix} 0 \\ 1 \end{pmatrix} x + \begin{pmatrix} 1 \\ 0 \end{pmatrix} u. \end{aligned} \tag{5.40}$$

□

The feature which distinguishes the algorithm illustrated in this section from the one illustrated in section 5.2, is the explicit computation of the driving input sequence $\{u(t_i)\}$ and of the state sequence $\{x(t_i)\}$. This latter computation lies at the core of the so called *subspace identification methods*. In the next section, we present the main features of the family of subspace identification schemes, and discuss their relationships with the approach of [14] which we have illustrated here.

5.5 Subspace Identification

Recently, *subspace identification* has drawn considerable attention from researchers in the identification field. According to [10], subspace identification algorithms "...formulate and solve a major part of the identification problem on a signal level ..." and "...the main characteristic of these schemes is the approximation of a subspace, defined by the span of the column or row space of matrices determined by the input-output data. The parametric time-invariant model, in this case, is calculated from these spans by exploiting their special structure, such as the shift-invariance property".

We can see from this description that subspace identification algorithms have much in common with the approach put forward in [14], of which we have illustrated the main features in the preceding sections. In fact, they formulate the problem of identification on a signal level, although they keep the distinction of the data in inputs and outputs. Moreover, they let the data speak, and formulate the identification problem in terms of the recorded signals and their relationships. In this sense, they depart sharply from the classical identification methods, in which an *a priori* choice of a model with a prespecified structure is made, and the parameters of the model are subsequently determined on the basis of the data.

Subspace identification methods have been applied to deterministic, stochastic, and combined deterministic-stochastic identification problems ([6, 5, 11, 12]). In the following we will focus on the deterministic identification scheme of [6], of which we will illustrate the similarities with the identification approach described in sections 5.1, 5.2, 5.4.

In [6] it is assumed that a linear, time-invariant, input-state- output system

$$\begin{aligned} \sigma x &= Ax + Bu \\ y &= Cx + Du \end{aligned} \tag{5.41}$$

with $A \in \mathbb{R}^{n\times n}$, $B \in \mathbb{R}^{n\times m}$, $C \in \mathbb{R}^{p\times n}$, $D \in \mathbb{R}^{p\times m}$, generates an input-output sequence $\begin{pmatrix} u(t) \\ y(t) \end{pmatrix}$, $t \in \mathbb{Z}_+$. The problem is that the matrices A, B, C, D, are unknown and should be deduced from the data. For the purposes of computation, finite subsequences of this input-output series are arranged in partitioned Hankel matrices of the form

$$H_{t|i-1} := \begin{pmatrix} Y_{t|i-1} \\ U_{t|i-1} \end{pmatrix} = \begin{pmatrix} y(t) & y(t+1) & \dots & y(t+j-1) \\ y(t+1) & y(t+2) & \dots & y(t+i) \\ \vdots & \vdots & \ddots & \vdots \\ y(t+i-1) & y(t+i) & \dots & y(t+i+j-2) \\ u(t) & u(t+1) & \dots & u(t+j-1) \\ u(t+1) & u(t+2) & \dots & u(t+j) \\ \vdots & \vdots & \ddots & \vdots \\ u(t+i-1) & u(t+i) & \dots & u(t+i+j-2) \end{pmatrix} \tag{5.42}$$

where i, and j, with $j >> i$, are assumed to be "sufficiently large", so that $H_{t|i-1}$ contains enough information on the system.

Under mild conditions on the input u (persistency of excitation), the following result holds:

Theorem 5.3. *([6]) Define $X \in \mathbb{R}^{n\times j}$ as*

$$X := \begin{pmatrix} x(i) & x(i+1) & \dots & x(i+j-1) \end{pmatrix}, \tag{5.43}$$

with $x(t)$ the state vector of the generating system (5.41), $k = i, \dots, i+j-1$. Let H denote

$$H = \begin{pmatrix} H_{0|i-1} \\ H_{i|2i-1} \end{pmatrix}. \tag{5.44}$$

Then:

1. $dim(span_{\mathbb{R}} rows H_{0|i-1} \bigcap span_{\mathbb{R}} rows H_{i|2i-1}) = n$
2. $span_{\mathbb{R}} rows H_{0|i-1} \bigcap span_{\mathbb{R}} rows H_{i|2i-1} = span_{\mathbb{R}} rows X$.

Theorem 5.3 constitutes the basis of the identification algorithm of [6]. It shows how to determine a state sequence corresponding to the input-output data, by computing the intersection of the row span of the "past" Hankel matrix $H_{0|i-1}$ and of the "future" Hankel matrix $H_{i|2i-1}$. Once a state sequence has been computed, the system matrices are computed solving the system of equations

$$\begin{pmatrix} x(i+1) & \dots & x(i+j-1) \\ y(i) & \dots & y(i+j-2) \end{pmatrix} = \begin{pmatrix} A & B \\ C & D \end{pmatrix} \begin{pmatrix} x(i) & \dots & x(i+j-2) \\ u(i) & \dots & u(i+j-2) \end{pmatrix}. \tag{5.45}$$

Let us now examine the common features of the algorithm of [6], and the algorithms of [14] which we have illustrated in sections 5.1, 5.2 and 5.4, starting from the statements of theorem 5.3, which we will interpret in the light of the notion of relative row rank of a partitioned matrix.

Let us first examine statement 1). By using Grassman's dimension theorem,

$$\begin{aligned} dim(span_{\mathbb{R}} rows \quad & H_{0|i-1} \bigcap span_{\mathbb{R}} rows H_{i|2i-1}) = dim(span_{\mathbb{R}} rows H_{0|i-1}) + \\ + \quad & dim(span_{\mathbb{R}} rows H_{i|2i-1}) - dim(span_{\mathbb{R}} rows H). \qquad (5.46) \end{aligned}$$

Note that the quantity on the right hand side of (5.46) is the relative row rank of the partitioned matrix H. Note also that H may be obtained from steps 1 and 2 of Algorithm II, section 5.4, with the following choice of H_-, H_+, H_1, H_2:

1. Let H_-, respectively H_+, be a set of t consecutive block rows of $\mathcal{H}_-(\tilde{w})$, respectively $\mathcal{H}_+(\tilde{w})$, with $t \geq t^* = \min_{t' \in \mathbb{Z}} \{t' \mid \rho_{t'} = \rho_\infty\}$;
2. Select H_1, respectively H_2, as a set of j consecutive columns of H_-, respectively H_+, determined as in 1), such that the column span of $\begin{pmatrix} H_1 \\ H_2 \end{pmatrix}$ equals that of $\begin{pmatrix} H_- \\ H_+ \end{pmatrix}$.

Statement 2) of the theorem suggests to perform the computation of a state sequence by determining a basis for the intersection of the row spaces of the past and of the future Hankel matrices. In Algorithm II, instead, the state trajectory was computed on the basis of the column span of the past Hankel matrix. This difference is due to the fact that the theoretical developments of section 5.1 have a natural preference for columnwise examination of $\mathcal{H}_t(\tilde{w})$, whose column span equals $\mathcal{B}_t(\tilde{w})$.

Of course, the algorithms of [14] depart sharply from the approach of deterministic subspace identification methods in that they make no *a priori* assumptions on the input-output structure of the system; it is reasonable, however, to view the exact identification procedures of [14] as subspace identification methods, even though the name came *en vogue* only later.

5.6 Realization Theory as a Special Case

As we have discussed in section 4.4, the impulse response realization problem can be cast into the identification framework put forward in [14]. In this section we will elaborate on the relationships existing between classical impulse response algorithms and the algorithms of sections 5.1, 5.2, and 5.4.

Classical impulse response algorithms operate on the sequence of matrices $\tilde{G}(0)$, $\tilde{G}(1)$, ..., where

$$\tilde{G}(t) = (\, \tilde{g}_1(t) \quad \tilde{g}_2(t) \quad \ldots \quad \tilde{g}_m(t)\,) \qquad (5.47)$$

is a $p \times m$ matrix whose i-th column, $i = 1, \ldots, m$, consists of the responses in the j-th output channel, $j = 1, \ldots, p$, to an impulse in the i-th input channel at time $t = 0$. The algorithms have in common the construction of the Hankel matrix

$$\mathcal{H}(\tilde{G}) := \begin{pmatrix} \tilde{G}(0) & \tilde{G}(1) & \tilde{G}(2) & \ldots \\ \tilde{G}(1) & \tilde{G}(2) & \tilde{G}(3) & \ldots \\ \tilde{G}(2) & \tilde{G}(3) & \tilde{G}(4) & \ldots \\ \vdots & \vdots & \vdots & \ddots \end{pmatrix}. \tag{5.48}$$

From this Hankel matrix, impulse response realization algorithms compute its (finite) rank n. This is instrumental in the selection of a suitable finite submatrix of rank n, which contains all information regarding the system, and from which the system matrices are determined. Each algorithm performs in a different way the selection of the submatrix: Ho-Kalman's algorithm selects a leading submatrix of rank n, while Silverman's algorithm selects a nonsingular $n \times n$ submatrix, of $\mathcal{H}(\tilde{G})$.

On the basis of the submatrix selected, the action of the shift on the output sequences is determined, and the system matrices are computed accordingly.

Let us cast this way of proceeding into the framework of section 5.4, considering for simplicity of exposition the case of a single-input, single-output system ($m = p = 1$). In this case, the data consists of a time series $\tilde{w} = \begin{pmatrix} \delta \\ y \end{pmatrix}$, with δ the unit impulse centered at $t = 0$, and y the corresponding response. In this case, the Hankel matrix $\mathcal{H}(\tilde{w})$ is

$$\begin{matrix} \vdots \\ \textit{input} \;\; \textit{row} \;\; \text{-1} \;\; \rightarrow \\ \textit{output} \;\; \textit{row} \;\; \text{-1} \;\; \rightarrow \\ \textit{input} \;\; \textit{row} \;\; 0 \;\; \rightarrow \\ \textit{output} \;\; \textit{row} \;\; 0 \;\; \rightarrow \\ \textit{input} \;\; \textit{row} \;\; 1 \;\; \rightarrow \\ \textit{output} \;\; \textit{row} \;\; 1 \;\; \rightarrow \\ \vdots \end{matrix} \begin{pmatrix} \ldots & \vdots & \vdots & \vdots & \vdots & \ldots & \vdots & \ldots \\ \ldots & 0 & 0 & 1 & 0 & \ldots & 0 & \ldots \\ \ldots & 0 & 0 & y(0) & y(1) & \ldots & y(t) & \ldots \\ \ldots & 0 & 1 & 0 & 0 & \ldots & 0 & \ldots \\ \ldots & 0 & y(0) & y(1) & y(2) & \ldots & y(t-1) & \ldots \\ \ldots & 1 & 0 & 0 & 0 & \ldots & 0 & \ldots \\ \ldots & y(0) & y(1) & y(2) & y(3) & \ldots & y(t-2) & \ldots \\ \ldots & \vdots & \vdots & \vdots & \vdots & \ldots & \vdots & \ldots \end{pmatrix}$$

Permuting the rows yields the following matrix, in which the input rows and the output rows have been permuted:

$$
\begin{array}{l}
\vdots \\
\textit{input row } {-2} \rightarrow \\
\textit{input row } {-1} \rightarrow \\
\vdots \\
\textit{output row } {-2} \rightarrow \\
\textit{output row } {-1} \rightarrow \\
\textit{output row } 0 \rightarrow \\
\textit{output row } 1 \rightarrow \\
\vdots \\
\textit{input row } 0 \rightarrow \\
\textit{input row } 1 \rightarrow \\
\vdots
\end{array}
\left(\begin{array}{cccccccc}
\cdots & \vdots & \cdots & \vdots & \vdots & \vdots & \vdots & \cdots \\
\cdots & 0 & \cdots & 0 & 1 & 0 & 0 & \cdots \\
\cdots & 0 & \cdots & 1 & 0 & 0 & 0 & \cdots \\
\cdots & \vdots & \cdots & \vdots & \vdots & \vdots & \vdots & \cdots \\
\cdots & 0 & 0 & 0 & y(0) & y(1) & y(2) & \cdots \\
\cdots & 0 & 0 & y(0) & y(1) & y(2) & y(3) & \cdots \\
\cdots & 0 & y(0) & y(1) & y(2) & y(3) & y(4) & \cdots \\
\cdots & y(0) & y(1) & y(2) & y(3) & y(4) & y(5) & \cdots \\
\cdots & \vdots & \vdots & \vdots & \vdots & \vdots & \cdots & \cdots \\
\cdots & 0 & 1 & 0 & 0 & \cdots & 0 & \cdots \\
\cdots & 1 & 0 & 0 & 0 & \cdots & 0 & \cdots \\
\cdots & \vdots & \vdots & \vdots & \vdots & \cdots & \vdots & \cdots
\end{array}\right)
$$

Let us apply Algorithm II to $\mathcal{H}(\tilde{w})$, and assume now that the relative row rank n of $\mathcal{H}(\tilde{w})$ has been computed. According to step 2, two submatrices H_- and H_+ of $\mathcal{H}(\tilde{w})$ have to be selected, so that $\begin{pmatrix} H_- \\ H_+ \end{pmatrix}$ has relative row rank n. This can be done as follows. Select a set of n independent rows corresponding to the output variable y in $\mathcal{H}_-(\tilde{w})$. Each of these rows corresponds to some shift $\sigma^{-t}y$, where $t \in I_-$, and I_- is a set of n indices $t_1, \ldots, t_n$. Let $m_- = maxI_-$, and define

$$
H_- := \begin{pmatrix} col(\sigma^{-t}\delta)_{1 \leq t \leq m_-} \\ col(\sigma^{-t}y)_{t \in I_-} \end{pmatrix}. \tag{5.49}
$$

Analogously, select a submatrix of n independent rows of $\mathcal{H}_+$ corresponding to the output component, and assume that they correspond to the shifts $\sigma^{-t}y$, $t \in I_+$. Let $m_+ = maxI_+$, and define

$$
H_+ := \begin{pmatrix} col(\sigma^{-t}\delta)_{1 \leq t \leq m_+} \\ col(\sigma^{-t}y)_{t \in I_+} \end{pmatrix}. \tag{5.50}
$$

With the matrices H_- and H_+ selected in this way, the relative row rank of $\begin{pmatrix} H_- \\ H_+ \end{pmatrix}$ equals n. Step 4 of Algorithm II requires the selection of the submatrices H_1 and H_2 of H_-, H_+, respectively, such that $\begin{pmatrix} H_1 \\ H_2 \end{pmatrix}$ has relative row rank n, and its column span equals that of $\begin{pmatrix} H_- \\ H_+ \end{pmatrix}$. This can be done as follows. Select the columns of $\begin{pmatrix} H_- \\ H_+ \end{pmatrix}$ corresponding to the nonzero value of the input, and complete this set with enough of columns, to span $span_{\mathbb{R}}\ columns \begin{pmatrix} H_- \\ H_+ \end{pmatrix}$. On the basis of the finite matrix $\begin{pmatrix} H_1 \\ H_2 \end{pmatrix}$ determined in this way, Algorithm II yields a minimal realization of the system having impulse response y.

Proposition 5.4. *([14]) Let H_1, H_2 be selected from $\mathcal{H}_-(\tilde{w})$, $\mathcal{H}_+(\tilde{w})$, following the above procedure. Then, Step 3 to Step 7 of Algorithm II compute a minimal realization of the impulse response y.*

In fact, in [14] it has ben proven that Ho-Kalman's and Silverman's algorithm are special cases of Algorithm II, each corresponding to a particular choice of the matrices H_- and H_+. The interested reader is referred to [14] for the details. The important difference of Algorithm II with classical impulse response realization, however, is that it does not assume a prespecified input-output structure, and therefore is more than a mere extension of this kind of algorithms.

6. Approximate Modeling

The concept of most powerful unfalsified model and the algorithms described in section 5. are concerned with exact modeling: a set of data is given, and a model is derived from the data, describing the data in an exact way. This approach to modeling, however, is not very realistic: in practice, very often the data are generated by a system which does not satisfy the assumptions of linearity or time invariance. In cases such as these, the algorithms of section 5. would in general produce useless models, which have no restrictive power. Take, for example, the case of a scalar time series corresponding to a free variable: in this case the most powerful model is simply $(\mathbb{R}^q)^{\mathbb{Z}}$, a model which describes every possible scalar time series and which has therefore no prohibitive power.

The problem of "real world" modeling is usually dealt taking randomness into the picture, for example assuming that the data are a realization of a stochastic process, and by using the concepts and methods of statistics to explain the data. This is questionable for several reasons. First, very often the problem is to analyze a *single* time series. No knowledge about the existence of other time series associated with the phenomenon under study can be assumed; actually, in many cases the existence of a population of time series can be *a priori* declared impossible, because the experiment which has given rise to the data is not replicable, or because it is impractical to replicate it. In these cases, invoking the philosophy of statistics is inappropriate, although under certain assumptions also a single time series can be stochastically analysed by looking at a population of its time windows. Moreover, the stochastic approach does not take into account the fact that usually the phenomenon to be modeled does not, in first place, belong to the model class being used, and therefore the lack of fit between the data and the model is not to be attributed to randomness, but rather to a fundamental inadequacy of the model class. This inadequacy is, of course, the price one has to pay to restrict oneself to a model class whose elements are easy to use and to understand, such as the model class of linear and time-invariant systems.

These considerations point out that modeling "real data" is foremost a matter of *approximating* a phenomenon with a model which is not able to capture its complexity: given a set of measurements, the modeling process consists of selecting an appropriate element of the model class which best approximates the data.

The modeling approach put forward in [15], of which we outline the main features in this section, is based on the idea of *approximate modeling*: on the basis of the data, a model is selected which does not necessarily explain the data exactly. Of course, such a model has to be simple, that is, easy to understand and to use; on the other hand, it has to be accurate, to explain the data as precisely as possible.

Simplicity of a model is measured by means of a complexity function $c : \mathcal{M} \to \mathcal{C}$, with $\mathcal{C}$ the complexity space, a partially ordered set. Accuracy is measured by means of a misfit function ϵ, which associates to each pair (data, model) an element $\epsilon(Z, M) \in \mathcal{E}$, the misfit level space, also a partially ordered space.

In general, simple models will give a poor fit, while a good fit usually requires complex models. Since simultaneous minimization of the complexity and of the misfit of a model is generally impossible, two approximate modeling methodologies are proposed in [15]. The first one assumes that a maximal admissible complexity c^{adm} is fixed, while the second one assumes a maximal tolerated misfit ϵ^{tol} is fixed. In both cases, the simplest admissible model which best fits the data is selected.

Let us formalize these two methodologies.

Methodology 1: Fix the *maximal admissible complexity* c^{adm}. $M^* \in \mathcal{M}$ is the *optimal approximate model* in the model class $\mathcal{M}$ for the measurements Z if

1. $c(M^*) \leq c^{adm}$
2. $\{M \in \mathcal{M},\ c(M) \leq c^{adm}\} \Longrightarrow \{\epsilon(Z, M^*) \leq \epsilon(Z, M)\}$
3. $\{M \in \mathcal{M},\ c(M) \leq c^{adm}, \epsilon(Z, M) = \epsilon(Z, M^*)\} \Longrightarrow \{c(M^*) \leq c(M)\}$.

Methodology 2: Fix the *maximal tolerated misfit* ϵ^{tol}. $M^* \in \mathcal{M}$ is the *optimal approximate model* in the model class $\mathcal{M}$ for the measurements Z if

1. $\epsilon(Z, M^*) \leq \epsilon^{tol}$
2. $\{M \in \mathcal{M},\ \epsilon(M) \leq \epsilon^{tol}\} \Longrightarrow \{c(M^*) \leq (M)\}$
3. $\{M \in \mathcal{M},\ \epsilon(Z, M) \leq \epsilon^{tol}, c(M) = c(M^*)\} \Longrightarrow \{\epsilon(Z, M^*) \leq \epsilon(Z, M)\}$.

Note that property 3 of the optimal approximate model often induces uniqueness, which is not guaranteed by properties 1 and 2. In [15] the problem of approximating in an optimal way a given time series $\tilde{w}$ with a model in $\mathcal{L}^q$ has been considered. We will not enter into the details of the solution, referring the interested reader to the original source for a thorough exposition. In the following, we will only introduce some of the concepts which lie at the core of the approximate modeling procedures of [15]. Let us first consider the definition of complexity of a dynamical system in the class $\mathcal{L}^q$.

Definition 6.1. *The* complexity *of a system* $\Sigma \in \mathcal{L}^q$, *with behavior* $\mathcal{B}$, *is the sequence*

$$c_t(\mathcal{B}) := \frac{dim\ \mathcal{B}_t}{q(t+1)} \tag{6.1}$$

This complexity measures the richness of the behavior $\mathcal{B}$ on finite time intervals. It has been shown in [15] that this complexity function is closely connected with the *equation structure* of the system, that is, with the set of indices representing the orders of the difference equations describing the system in kernel form. This connection allows to interpret complex systems as those described in kernel form by few and high order equations.

The misfit between a time series and a linear system has been defined in [15] as an *equation-error* oriented misfit measure. The lack of fit is measured between the data, and each equation of an appropriate kernel representation of the model. In this way, a sequence of misfits is defined. The misfits are computable in a straightforward way by performing a singular value decomposition of the empirical covariance matrix of the data. We refer the interested reader to [15], section 23, for the details of this procedure, and for a precise definition of the misfit measure.

With complexity and misfit defined as sequences of real numbers, a partial order has to be imposed on $(\mathbb{R})^{\mathbb{N}}$. In [17] the lexicographic ordering has been chosen, thus expressing a preference for models consisting of few and low order equations.

On the basis of the two methodologies described above, two algorithms are proposed in [15] to compute a kernel representation of the optimal approximate model. They consist in manipulations of the empirical covariance matrix of the data, based on singular value decompositions. These computations simultaneously determine the equations of the model and their misfit. We refer the interested reader to [15], sections 25.1 and 25.2, for a description of the algorithms, and to [2] for the case in which the data consist of a finite set of measurements.

7. Simulations

In this section we illustrate the application of the algorithm of section 5.4 to the problem of modeling a system with a noncausal impulse response. To this purpose, we will consider a system with $q = 2$, described in continuous time by

$$w_2(t) = -\frac{d^2}{dt^2}\int_{-\infty}^{+\infty} \phi(x) w_1(t-x)dx \tag{7.1}$$

where $\phi(x)$ is the filter $\frac{1}{\sqrt{2\pi}}e^{-x^2}$. This type of system arises in the problem of detecting abrupt changes in an observed signal w_1, which is first smoothed via the convolution.

In the simulations we will consider a discrete version of (7.1), with a time step of length 0.2:

$$w_2(t) = \sum_{-N}^{N} G_k w_1(t-k) \tag{7.2}$$

with $N = 40$ and $G_k = 0.2\frac{d^2}{dx^2}\phi(0.2k)$. Note that (7.2) describes a noncausal system. The impulse response of this system (the "Mexican hat") is depicted in Fig. 7.1. The identification experiment consists in convoluting a zero mean,

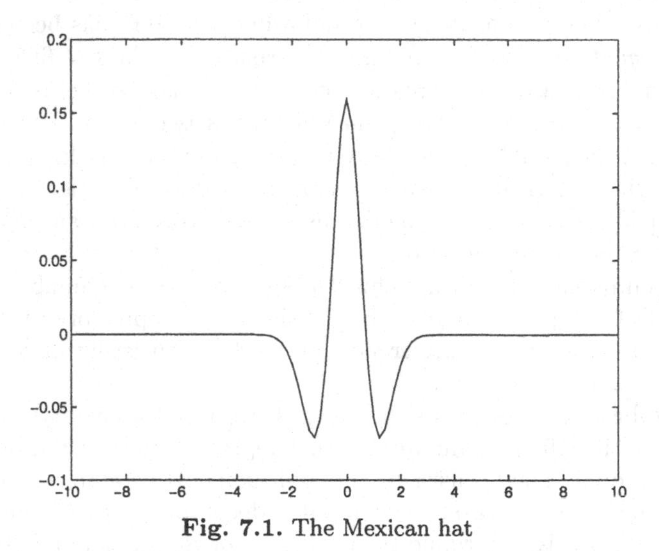

Fig. 7.1. The Mexican hat

Gaussian sequence $\tilde{w}_1$, with the causal part of the Mexican hat, to produce the signal $\tilde{w}_2$. $\tilde{w}_1$ and $\tilde{w}_2$ are collected in a vector $\tilde{w} = \begin{pmatrix} \tilde{w}_1 \\ \tilde{w}_2 \end{pmatrix}$, which forms the input data to a Matlab implementation of Algorithm II.

By carrying out the computation with different accuracy thresholds, models of different order are obtained. Figure 7.2 compares the simulated response of the model of order one obtained in this way, with the output data $\tilde{w}_2$. To evaluate the accuracy of the model of order one in a different way, let us consider the approximation of the Mexican hat obtained from the impulse response of the model; the anticausal part of the approximation is obtained by symmetry. The result is shown in Fig. 7.3. A model of order three yields a better fit on the data, as shown in Fig. 7.4, and a better approximation of the Mexican hat, as depicted in Fig. 7.5. Finally, Fig. 7.6 depicts the approximation of the Mexican hat obtained from the model of order five.

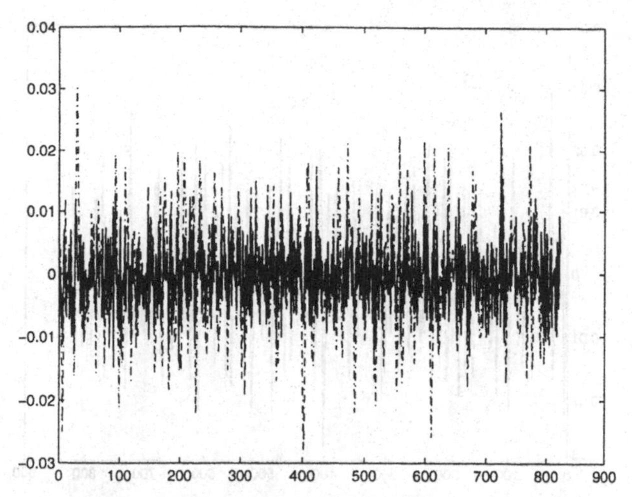

Fig. 7.2. The output data (dashed line) and the experimental data (solid line) for the model of order one

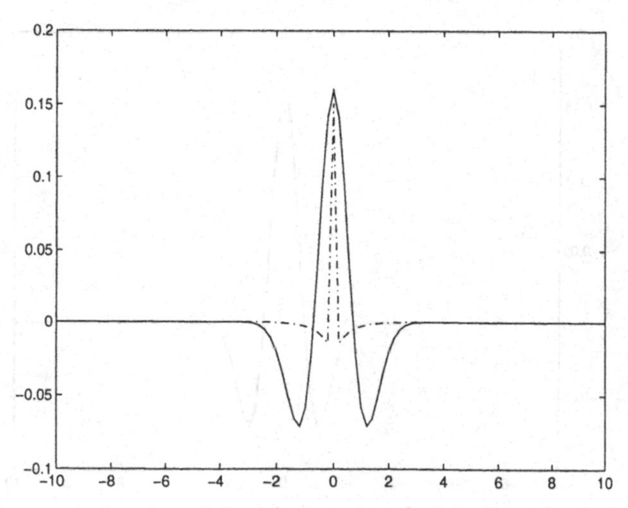

Fig. 7.3. The approximation of the Mexican hat (solid) obtained from the impulse response of the model of order one (dashed)

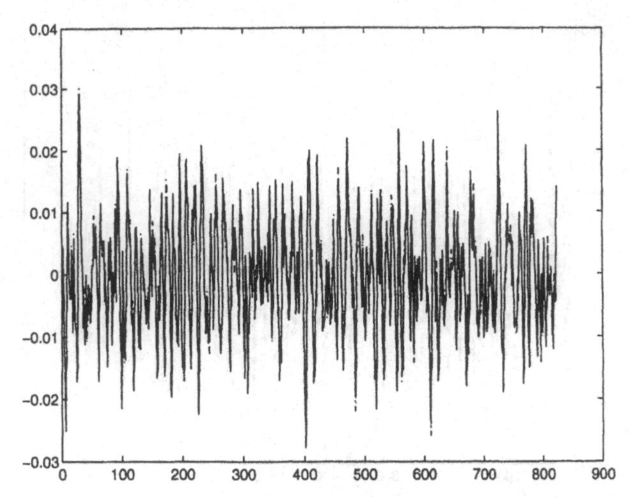

Fig. 7.4. The output data (dashed line) and the experimental data (solid line) for the model of order three

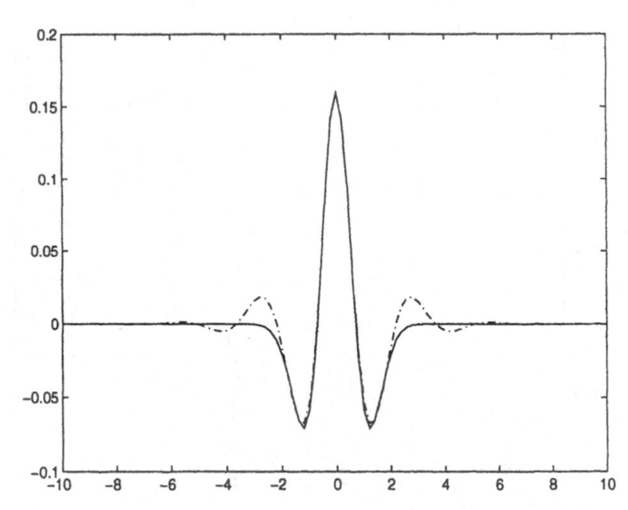

Fig. 7.5. The approximation of the Mexican hat (solid) obtained from the impulse response of the model of order three (dashed)

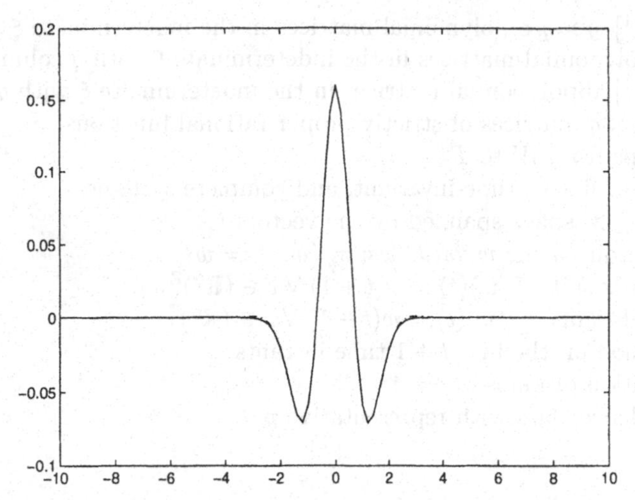

Fig. 7.6. The approximation of the Mexican hat (solid) obtained from the impulse response of the model of order five (dashed)

Acknowledgement. The research of the first author has been supported by NWO/CNR. The research of the second author has been supported by the SCIENCE project System Identification, with contract number SC1*-CT92-0779.

A. Notation

$\mathbb{N}$ natural numbers (0 is not included)
$\mathbb{Z}$ integers
$\mathbb{Z}_+$ nonnegative integers
$\mathbb{Z}_-$ negative integers
$\mathbb{R}$ real numbers
$\mathbb{C}$ complex numbers
$\mathbb{R}[\xi]$ polynomials with real coefficients
$\mathbb{R}[\xi\xi^{-1}]$ dipolynomials with real coefficients
$\mathbb{R}_+(\xi)$ proper rational functions
e_i the i-th vector of a canonical basis vector in $\mathbb{R}^{1\times\bullet}$
$\mathbb{R}^{g\times q}$ $g \times q$ real matrices
$\mathbb{R}^{\bullet\times q}$ real matrices with q columns

$col(r_1, \ldots, r_n)$ the matrix $\begin{pmatrix} r_1 \\ r_2 \\ \vdots \\ r_n \end{pmatrix}$

$diag(x_k)_{k=1,\ldots,r}$ $r \times r$ diagonal matrix with diagonal elements x_k
$\mathbb{R}^{g\times q}[\xi] \cdot g \times q$ polynomial matrices in the indeterminate ξ

$\mathbb{R}^{g \times q}[\xi, \xi^{-1}]$ $g \times q$ dipolynomial matrices in the indeterminate ξ
$\mathbb{R}^{\bullet \times q}[\xi]$ polynomial matrices in the indeterminate ξ with q columns
$\mathbb{R}^{\bullet \times q}[\xi, \xi^{-1}]$ dipolynomial matrices in the indeterminate ξ with q columns
$\mathbb{R}_{+}^{g \times q}(\xi)$ $g \times q$ matrices of strictly proper rational functions
$(W)^T$ maps from W to T
$\mathcal{L}^q$ family of linear, time-invariant, and complete systems
$< r_1, \ldots, r_n >$ space spanned by the vectors r_i
π_w projection on the w variables: $\pi_w(w, \ell) := w$
σ backwards shift: $(\sigma w)(t) = w(t+1)$ $\forall w \in (\mathbb{R}^q)^{\mathbb{Z}}$
σ^* forwards shift: $(\sigma^* w)(t) = w(t-1)$ $\forall w \in (\mathbb{R}^q)^{\mathbb{Z}}_{+}$
π^t projection on the first $t+1$ time instants
$\circ$ composition of maps
$[p]$ equivalence class with representative p

References

1. Antoulas A.C., and J.C. Willems, "A behavioral approach to linear exact modeling", *IEEE Transactions on Automatic Control*, vol. 38, pp. 1776–1802, 1993.
2. Heij C., "Deterministic Identification of Dynamical Systems", Lecture Notes in Control and Information Sciences, vol. 127. New York: Springer, 1989.
3. Ho B.L., and R.E. Kalman, "Effective Construction of Linear, State-Variable Models from Input/Output Functions", *Regelungstechnik*, vol. 14, pp. 545-548, 1966.
4. Kuijper M., "First order representations of linear systems". Basel: Birkhäuser, 1994.
5. Larimore W., "Canonical variate analysis in identification, filtering and adaptive control", *Proc. 29th IEEE Conf. Decision and Control*, pp. 596–604, 1990.
6. Moonen M., B. De Moor, L. Vanderberghe, and J. Vandewalle, "On- and off-line identification of linear state space models", *International Journal of Control*, vol. 49, pp. 219–232, 1989.
7. Rapisarda P., and J.C. Willems, "State Maps for Linear Systems", *SIAM Journal on Control and Optimization.* To appear (1994)
8. Roorda B., "Global Total Least Squares: a method for the construction of open approximate models from vector time series", Ph. D. Thesis, Erasmus University Rotterdam, The Netherlands, 1995.
9. Silverman L.M., "Realization of linear dynamical systems", *IEEE Transactions on Automatic Control*, vol. 16, pp. 554–567,1971.
10. Verhaegen M., and P. Dewilde, "Subspace model identification. Part I: The output-error state-space model identification class of algorithms", *International Journal of Control*, vol. 56, pp. 1187–1210,1992.
11. Van Overschee P., and B. De Moor, "Subspace Algorithms for the Stochastic Identification Problem", *Automatica*, vol. 29, pp. 649–660, 1993.
12. Van Overschee P., and B. De Moor, "N4SID: Subspace Algorithms for the Identification of Combined Deterministic-Stochastic Systems", *Automatica*, vol. 30, pp. 75–93, 1994.
13. Willems J.C., "From Time Series to Linear System. Part I - Finite Dimensional Linear Time Invariant Systems", *Automatica*, vol. 22, pp. 561–580, 1986.
14. Willems J.C., "From Time Series to Linear System. Part II - Exact Modeling", *Automatica*, vol. 22, pp. 675–694, 1986.
15. Willems J.C., "From Time Series to Linear System. Part III - Approximate Modeling", *Automatica*, vol. 23, pp. 87–115, 1987.
16. Willems J.C., "Models for Dynamics", *Dynamics Reported*, vol. 2, pp. 171–269, 1989.
17. Willems J.C., "Paradigms and Puzzles in the Theory of Dynamical Systems", *IEEE Transactions on Automatic Control*, vol. 36, pp. 259–294, 1991.

Identification of Linear Systems from Noisy Data

Manfred Deistler and Wolfgang Scherrer

Institute for Econometrics, Operations Research and Systems Theory
Technical University Vienna
Argentinierstraße 8
A-1040 Vienna
Austria.

1. Introduction

Systems identification is concerned with obtaining a "good" model from data. In many cases this is a fairly nontrivial problem and systematic approaches for solving this task have been developed. Usually, for identification, the following has to be specified:

(i) The *model class*, i.e. the class of all a priori feasible candidate models
(ii) The *class of observations*
(iii) The *identification procedure*, which is a rule (in the automatic case a function) attaching to every finite part $x_t, t = 1, \ldots, T$; $T \in \mathbb{N}$ of the observations a model (or a class of models) from the model class. Throughout this paper, x_t will denote the n-dimensional vector of observed variables at time t.

Typically, the data are noisy in the sense that a finite part of the observations does not completely reveal the underlying data generating mechanism. This is a main problem, both for the design and the evaluation of identification procedures.

The most common models for time series data are dynamic systems and stochastic processes. In this paper we restrict ourselves to (discrete, equidistant) time series, (wide sense) stationary processes and linear systems.

There are different philosophies in modeling with systems and in dealing with noise. One philosophy is to derive a deterministic model from the data by (at least conceptionally) separating an exact part from a noisy part of the data and by describing the exact linear relations among the exact part of the data. In this case one may or may not provide a (stochastic) model for the noisy part of the data.

Another philosophy – which is quite common – is to model the data as an ARMA process

$$a(\mathrm{z})x_t = b(\mathrm{z})\epsilon_t \tag{1.1}$$

where (ϵ_t) is n-dimensional white noise, z is the backward-shift of the integers $\mathbb{Z}$, i.e. $\mathrm{z}(x_t|t \in \mathbb{Z}) = (x_{t-1}|t \in \mathbb{Z})$ as well as a complex variable and

$$a(\mathrm{z}) = \sum_{j=0}^{p} A_j \mathrm{z}^j \quad ; \quad b(\mathrm{z}) = \sum_{j=0}^{p} B_j \mathrm{z}^j \quad ; \quad A_j, B_j \in \mathbb{R}^{n\times n}$$

are polynominal matrices. We assume

$$\det a(\mathrm{z}) \neq 0 \text{ for } |\,\mathrm{z}\,| \leq 1 \text{ and } \det b(\mathrm{z}) \neq 0 \text{ for } |\,\mathrm{z}\,| < 1$$

Then the Wold decomposition (see, e.g., [22]) is given as

$$x_t = k(\mathrm{z})\epsilon_t$$

where $k(\mathrm{z}) = a(\mathrm{z})^{-1}b(\mathrm{z})$ is the corresponding (rational) transfer function.

ARMA systems provide models for a stationary process with rational spectral density, based on a finite dimensional linear system with (unobserved) white noise inputs. Their advantage is that, for given p, only a finite number of parameters is needed for their description (or to be more precise, for the description of their second moments). On the other hand (for suitably chosen p) every linearly regular, stationary process can be approximated by an ARMA process with arbitrary accuracy. In other words estimation for ARMA systems (in a certain sense) is a parametric estimation problem and a wide class of stationary processes can be described this way. Clearly ARMA models are *symmetric* in the sense that we do not have to distinguish between inputs and outputs in x_t a priori. ARMA models have proved to be of great use for applications in particular for the univariate case; however, in the multivariate case, even for moderate n problems from the curse of dimensionality may arise. Consider, e.g., the case $p = 6$, $n = 5$ then "generically" (for more details see [14]), $5 \times 5 \times 6 \times 2 = 300$ parameters for $a(\mathrm{z}), b(\mathrm{z})$ have to be estimated.

In case the classification of the observations $x_t = (z_t', y_t')'$ into (observed) inputs z_t and outputs y_t is known a priori, we may use an ARMAX model

$$a(\mathrm{z})y_t = d(\mathrm{z})z_t + b(\mathrm{z})\epsilon_t \tag{1.2}$$

Here $d(\mathrm{z})$ is a, say, $m \times s$ polynomial matrix, $m + s = n$, and the other symbols have analogous meanings and properties as before. In addition it is assumed that observed inputs and the white noise are orthogonal, i.e.,

$$\mathrm{E}\, z_s \epsilon_t' = 0 \;\forall t, s \tag{1.3}$$

The solution of (1.2) is of the form

$$y_t = a(z)^{-1}d(z)z_t + a(z)^{-1}b(z)\epsilon_t \tag{1.4}$$

Equation (1.4) can be interpreted as a deterministic system relating "latent" (unobserved) outputs $\hat{y}_t = a(z)^{-1}d(z)z_t$ to the observed inputs z_t plus noise $a(z)^{-1}b(z)\epsilon_t$ added to the latent outputs $\hat{y}_t$ in order to obtain the observed outputs y_t. This is an *unsymmetric system model*, as by assumption inputs and outputs in x_t are distinguished a priori. In comparison to an ARMA model here no model for (z_t) is given and the dimension of the parameter space is smaller. Consider, e.g,. the "generic" case for $p = 6$, $m = 2$, $s = 3$, then $2 \times 2 \times 6 \times 2 + 2 \times 3 \times 6 = 84$ parameters have to be estimated for $a(z), b(z), d(z)$. Equation (1.4) also gives an *unsymmetric noise model*, since all noise is added to the outputs (or to the equations, which is equivalent from our point of view) and the inputs are assumed to be free of noise. In econometrics this is called *errors-in-equations* modeling.

In this paper we will adhere to the first philosophy described above and we will deal with a *symmetric noise model* where both, inputs and outputs may be subject to noise. Models of this kind are called *errors-in-variables* (EV) *models*, or – in a different but equivalent formulation – *factor models*. The usual ARMA and ARMAX modeling belongs to the *main stream* of system identification ([8]). The *alternative approach* of EV (or factor) modeling, at least for the static case, has a long history, in particular in areas like statistics, psychometrics and econometrics (see, e.g., [1], [27], [13], [12]). In the last two decades there has been a resurging interest in such models (see, e.g., [2], [6]). Recently, mainly triggered by Kalman's work ([17], [18]) EV models have also been analyzed in systems engineering. The dynamic case has been treated in, e.g., [4], [9], [24], [11], [25], [16]. The main stream approach of noise modeling is justified in a great number of applications dealing for instance with prediction where, e.g., (1.3) is a natural assumption. On the other hand, in a number of cases, e.g., the asymmetry in errors-in-equations modeling cannot be justified and may lead to *prejudiced* results (Kalman). For example, in sonar array processing, when an array of n sensors is assumed to receive noisy signals from $n - m$ sources ([15]) EV models arise in a natural way. More generally we can distinguish three main areas for EV modeling:

(i) If we are interested in the *true system* underlying the data (rather than, e.g., in prediction) and if we cannot be sure a priori that the inputs have been observed free of noise. This is the "classical" motivation for EV models e.g. in econometrics.
(ii) If we want to approximate a high dimensional data vector by a small number of factors. This is the "classical" motivation for factor analysis (e.g., in psychometrics, where an example would be to determine the intelligence factors underlying the test scores). A related issue is that EV modeling may considerably reduce the dimension of parameter spaces in comparison with multivariate AR or ARMA models.

(iii) In a number of cases no sufficient a priori information about the number of equations and/or about the classification of the variables into inputs and outputs is available. Then one has to use a *symmetric system model* which in turn demands a *symmetric noise model*. This point has been emphasized in particular by [17].

To a good part because of the technical complications, EV modeling is still far from being complete as far its theory is concerned and is still not a standard tool in system identification.

Now let us introduce some notation. For a matrix, A say, we use the corresponding lowercase letter a_{ij} to denote its i, j-th entry. The (left) kernel of a matrix A is denoted by $\ker(A)$; $\mathrm{rk}(A)$ and $\mathrm{crk}(A)$ respectively denote the rank and the corank respectively of A. If $A \in \mathbb{C}^{n\times n}$ is a Hermitean matrix then $\lambda_1(A) \leq \cdots \leq \lambda_n(A)$ denote its eigenvalues. For a vector v, $\mathrm{diag}(v)$ denotes the (square) diagonal matrix, whose diagonal elements are the corresponding entries of v. For a complex Matrix A the matrix A^* is the complex conjugate transposed matrix. For a rational matrix $A(\mathrm{z})$ (with real coefficients) we define $A^*(\mathrm{z}) = A'(1/\mathrm{z})$. For a subset of a topological space, $\mathcal{A}$ say, $\mathcal{A}^o$ denotes the interior of A and $\overline{\mathcal{A}}$ denotes the closure of $\mathcal{A}$.

2. The Model

The systems considered are of the form

$$w(\mathrm{z})\hat{x}_t = 0 \tag{2.1}$$

where $\hat{x}_t$ is an n-dimensional vector of latent (i.e. not necessarily observed) variables and where

$$w(\mathrm{z}) = \sum_{j=-\infty}^{\infty} W_j\, \mathrm{z}^j \quad ; \quad W_j \in \mathbb{R}^{m\times n}$$

We will call $w(\mathrm{z})$ the *relation function*; it represents an exact (i.e. deterministic) linear system of a very general form; no a priori classification of the components of x_t as inputs and outputs and no a priori information about causality is required. Systems of this form have been studied in detail in [29]. Here also the number m of equations in (2.1) is not assumed to be known a priori. Without restriction of generality we assume $1 \leq m \leq n$ and that $w(\mathrm{z})$ contains no linearly dependent rows.

We use a stochastic model for the observations of the form:

$$x_t = \hat{x}_t + u_t \tag{2.2}$$

where x_t is the vector of observed variables and u_t is the n-dimensional noise vector. Throughout we will assume

(a1) (x_t), $(\hat{x}_t)$ and (u_t) are stationary processes with absolutely summable autocovariance functions. Thus the spectral densities Σ, $\hat{\Sigma}$ and $\tilde{\Sigma}$ respectively are defined pointwise.
(a2) $\mathrm{E}\,\hat{x}_t = 0$ and $\mathrm{E}\,u_t = 0$.
(a3) $\mathrm{E}\,\hat{x}_s u_t' = 0\ \forall s, t$.
(a4) $\Sigma > 0$

Clearly stochastic models have certain limitations. In a number of cases assumptions such as stationarity and ergodicity can hardly be justified. However, most of the results of this paper can also be obtained without a stochastic setting from sample second moments rather than from their population counterparts. In other words the essential points of our analysis are "philosophy invariant" in the sense that they also hold in a nonstochastic setting The additional assumptions (a1)–(a4) are either natural or not very restrictive: Note that rational spectral densities satisfy (a1). Assumption (a3) is natural in a large number of applications.

In this paper, unless the contrary is stated explicitely, the spectral densities $\Sigma, \hat{\Sigma}, \tilde{\Sigma}$, as well as the relation function $w(e^{-i\lambda})$ are considered for fixed frequency λ; thus $\Sigma, \hat{\Sigma}, \tilde{\Sigma}$ and w are considered as constant matrices with complex entries. The results obtained thus may be interpreted pointwise in frequency, and in an identification context they relate naturally to nonparametric spectral estimation of Σ.

An alternative but equivalent formulation is a socalled factor model, where the linear restrictions on $(\hat{x}_t)$ are expressed as

$$\hat{x}_t = l(\mathrm{z}) f_t$$

where

$$l(\mathrm{z}) = \sum_{j=-\infty}^{\infty} L_j \, \mathrm{z}^j \quad ; \quad L_j \in \mathbb{R}^{n \times n-m}$$

and (f_t) is an $(n-m)$-dimensional full rank regular stationary process. (f_t) is called *factor process.* Assuming that (f_t) is white noise with covariance matrix unity, then

$$\hat{\Sigma}(\lambda) = l(e^{-i\lambda}) l(e^{-i\lambda})^* \tag{2.3}$$

If $\Sigma, \hat{\Sigma}, \tilde{\Sigma}$ and $w(e^{-i\lambda})$ are constant over frequency λ and have real entries (i.e. (x_t), $(\hat{x}_t)$ and (u_t) are white noise and $W_j = 0$, $j \neq 0$) we speak of a *static model.*

The main problem dealt with in this paper is to obtain the underlying system (e.g., represented by its relation function w or by the kernel of $\hat{\Sigma}$) –

or to be more precise, the set of all observationally equivalent systems – from the population second moments of the observations. In this paper no information contained in the observations except for second moments is used. For the use of higher order moments see, e.g., [7], [28]. Clearly starting from population rather than from sample second moments gives an idealized problem (a stochastic realization problem rather than an identification problem), and the effect of sampling variation is not investigated. However, by our analysis, main difficulties in EV model identification are addressed; *in particular we analyze how uncertainty about the noise structure leads to uncertainty about the underlying system*, an uncertainty which has nothing to do with sampling variation and thus remains in an infinite data record. A major part of this paper deals with the description of this uncertainty in terms of the resulting classes of observationally equivalent systems.

It should be stressed that this kind of uncertainty does not exist in the errors-in-equations approach, where under very general conditions (persistent excitation) the transfer function is uniquely determined from the second moments of the observations. Also for ARMA models, under fairly general conditions, the transfer function is unique. The main additional complication in EV model identification is caused by the uncertainty described above and from the point of view of mathematical analysis, by the more complicated relation between the second moments of the observations and the underlying systems and noise structures.

By (2.2) and by our assumptions, we obtain the equation

$$\Sigma = \hat{\Sigma} + \tilde{\Sigma} \tag{2.4}$$

for the spectral densities. By (2.1), the spectral density $\hat{\Sigma}$ is singular and

$$w\hat{\Sigma} = 0 \tag{2.5}$$

holds. For given Σ, a matrix $\hat{\Sigma}$ is called *compatible* (with Σ) if (2.4) is satisfied, where $\hat{\Sigma}$ and $\tilde{\Sigma}$ are spectral densities (and thus positive semidefinite) and where in addition $\hat{\Sigma}$ is singular and typically $\tilde{\Sigma}$ satisfies further assumptions such as (a5) or (a6) below. Thus the set of all matrices $\hat{\Sigma}$ compatible with Σ is the set of all observationally equivalent $\hat{\Sigma}$. A relation function w is called *compatible* if there exists a compatible $\hat{\Sigma}$ such that (2.5) holds, where the rows of w form a basis for the left kernel of $\hat{\Sigma}$. From (2.5) we see that a part of the nonuniqueness of w is trivial: Even if we commence from $\hat{\Sigma}$ rather than from Σ, then w is nonunique. Since we assume that the rows of w form a basis for the left kernel of $\hat{\Sigma}$, then for given $\hat{\Sigma}$, the matrix w is unique up to left multiplication by nonsingular matrices.

In order to remove this trivial part of uncertainty, we may consider equivalence classes $\{tw | t \in \mathbb{C}^{m\times m}, \det t \neq 0\}$, rather than relation functions w. Clearly such equivalence classes may be identified with the subspace $\ker \hat{\Sigma}$. Note that the set of all equivalence classes $\{tw | \det t \neq 0\}$, $w \in \mathbb{C}^{m\times n}$, $\operatorname{rk} w = m$, endowed with the quotient topology is a differentiable manifold of real dimension $2m(n-m)$, called the Grassmannian $\mathcal{G}(m,n)$. By eventually rearranging the components of $\hat{x}_t$ (and thus accordingly rearranging the columns of w) the relation function w can be written as $w = (w_1, w_2)$ where w_2 is square and nonsingular. Then $(-w_2^{-1}w_1, -I)$ is a normalized representation from the same equivalence class, where the first $n-m$ components of $\hat{x}_t$, $\hat{z}_t$ say, are a possible choice of inputs and $k = -w_2^{-1}w_1$ gives a transferfunction. Of course in general there are serveral possible choices for inputs.

For the sake of simplicity of presentation of theorems below, we also introduce the concept of an m-solution (for Σ), which is a matrix $w \in \mathbb{C}^{m\times n}$ with $\mathrm{rk}(w) = m$ such that there exists a compatible $\hat{\Sigma}$ satisfying

$$w\hat{\Sigma} = 0$$

Clearly every compatible relation function is an m-solution but the converse is not necessarily true since $\mathrm{rk}\, w \leq \mathrm{crk}\, \hat{\Sigma}$.

Without imposing additional a priori assumptions, every relation function w would be compatible with a given Σ; thus some additional structure has to be imposed.

A classical method of seperating noise from latent variables is to project the i-th component of x_t, x_t^i say, onto the space spanned by the x_s^j, $j \neq i$, $j = 1, \ldots, n$, $s \in \mathbb{Z}$. This is called the *i-th elementary regression*. Here $\hat{\Sigma}$ is unique and $\ker \hat{\Sigma}$ is one-dimensional. A corresponding $w \in \ker \hat{\Sigma}$, $w \neq 0$ is called an *i-th elementary solution*. This case is characterised by a $\tilde{\Sigma}$ of the form $\mathrm{diag}(0, \ldots, 0, \tilde{\sigma}_{ii}, 0, \ldots, 0)$. Clearly in many cases attaching all noise to the i-th component of x_t and assuming that all other components are free of noise is a prejudice ([19]). E.g. for the first elementary regression, $\hat{\Sigma}$ is given by the classical least squares formula

$$\hat{\Sigma} = \begin{pmatrix} \Sigma_{12}\Sigma_{22}^{-1}\Sigma_{21} & \Sigma_{12} \\ \Sigma_{21} & \Sigma_{22} \end{pmatrix}$$

where

$$\Sigma = \begin{pmatrix} \sigma_{11} & \Sigma_{12} \\ \Sigma_{21} & \Sigma_{22} \end{pmatrix}$$

In this paper, the two following (additional) alternative assumptions are considered

(a5) $\tilde{\Sigma}$ is diagonal

This is a fairly common assumption in factor analysis. The idea behind this assumption is to provide a decoupling of common and individual effects among the variables. The common effects are attributed to the system, the individual effects to the noise. In other words, the latent variables generate a splitting subspace which makes the components of x_t conditionally orthogonal ([21]). This case will be called the *Frisch case* ([17]), and (2.4) will be called a

Frisch decomposition then. As easiliy can be seen in this case a decomposition (2.4) (and thus an EV model) exists for arbitrarily chosen Σ.

The alternative assumption considered here is that the noise level is bounded, or to be more precise, that

(a6) $\lambda_n(\tilde{\Sigma}(\lambda)) \leq \epsilon > 0$

holds, where λ_n denotes the maximum eigenvalue of $\tilde{\Sigma}$ and ϵ is an a priori given bound. This case will be called the *bounded noise case.*

The following three problems are analyzed in the paper:

(i) Neither assumption (a5) nor (a6) will give in general a unique system for a given Σ. As has been indicated already, our basic philosophy is to identify a class of observationally equivalent systems rather than a single system; therefore one important problem is to describe for given Σ the class of all observationally equivalent systems, i.e. of all systems compatible with Σ. This class, in general, will contain systems with different numbers of outputs (i.e. with different m). In addition a description of the subclasses corresponding to a fixed number of outputs (and thus equations) is of interest. An important integer is the maximum corank of $\hat{\Sigma}$, denoted by $\mathrm{mc}(\Sigma)$, among all $\hat{\Sigma}$ compatible with given Σ. Clearly $\mathrm{mc}(\Sigma)$ is the maximum number of equations in the equivalence class and $n - \mathrm{mc}(\Sigma)$ is the minimum number of of factors in the equivalence class. The subclass corresponding to $\mathrm{mc}(\Sigma)$ is of special interest.

(ii) From the point of view of identification the continuity of the mapping attaching classes of observationally equivalent systems to Σ is important. This relates to consistency of estimation of these equivalence classes, since Σ can be consistently estimated under general assumptions.

(iii) Another important problem is estimation of $\mathrm{mc}(\Sigma)$. For this purpose some properties of the sets $\mathcal{S}_m$ of spectral densities Σ such that $\mathrm{mc}(\Sigma) = m$ holds (i.e. $\mathcal{S}_m = \{\Sigma |\, \mathrm{mc}(\Sigma) = m\}$) are analyzed.

Let us introduce some additional notation: The set of all m-solutions corresponding to a given Σ is called the *m-solution set* $\underline{\mathcal{L}}_m$ (of Σ). For $m = \mathrm{mc}(\Sigma)$, the set $\underline{\mathcal{L}}_m$ is the set of all compatible $m \times n$ relation functions. By $\mathcal{L}_m$ we denote the quotient space of $\underline{\mathcal{L}}_m$ with respect to left multiplication by nonsingular matrices. Clearly $\mathcal{L}_m$ is a subset of the Grassmanniann $\mathcal{G}(m, n)$. For some statements it is more convinient to discuss sets of compatible $\hat{\Sigma}$, rather than the corresponding kernels $\ker \hat{\Sigma}$. Since, for given Σ, the matrices $\hat{\Sigma}$ and $\tilde{\Sigma}$ are in a trivial one-to-one relation, and since for the Frisch case the $\tilde{\Sigma}$ can easily be identified with $(\tilde{\sigma}_{11}, \ldots, \tilde{\sigma}_{22})' \in \mathbb{R}^n$, we introduce the following notation: By $\mathcal{E}$ we denote the set of all $\Sigma - \hat{\Sigma} = \tilde{\Sigma}$ where $\hat{\Sigma}$ is compatible with Σ and by $\mathcal{E}_m$ we denote the subset of $\mathcal{E}$ where $\mathrm{crk}(\Sigma - \tilde{\Sigma}) = m$ holds. Of course $\mathcal{E} = \mathcal{E}_1 \cup \cdots \cup \mathcal{E}_n$. Sometimes we also use the symbol $\mathcal{E}_m(\Sigma)$ to make the dependence on Σ explicit. Analogously we use the notation $\underline{\mathcal{L}}_m(\Sigma)$ and $\mathcal{L}_m(\Sigma)$. By $\mathcal{S}$ we denote the set of positive definite matrices $\Sigma \in \mathbb{C}^{n \times n}$.

Let $d(x, y)$ denote a metric on a space $\mathcal{A}$; then for compact subsets such as $\mathcal{U}, \mathcal{V} \subseteq \mathcal{A}$ the socalled Hausdorff metric $d_H(\mathcal{U}, \mathcal{V})$ is defined as

$$d_H(\mathcal{U}, \mathcal{V}) = \min(\rho(\mathcal{U}, \mathcal{V}), \rho(\mathcal{V}, \mathcal{U}))$$

where

$$\rho(\mathcal{U}, \mathcal{V}) = \sup_{x \in \mathcal{U}} \inf_{y \in \mathcal{V}} d(x, y)$$

3. The Frisch Case, Bivariate Observations

It should be noted that our approach is genuine multivariate as it gives no sense for the univariate case $n = 1$. Thus the simplest case is the bivariate ($n = 2$) one, which is discussed in this section. A typical difficulty of the case $n > 2$, namely the determination of $\mathrm{mc}(\Sigma)$ from the observations, does not arise in the bivariate case, as in the bivariate case $\mathrm{mc}(\Sigma) = 1$ holds unless Σ is diagonal.

Let us consider the *static* case first, where Σ, $\hat{\Sigma}$ and w are constant with real entries. We assume for simplicity that $\sigma_{12} \neq 0$ holds. As is easily seen the set of all feasible $\hat{\Sigma}$ is characterized by

$$\begin{gathered} 0 = \det \hat{\Sigma} = \hat{\sigma}_{11}\hat{\sigma}_{22} - \sigma_{12}^2 \\ 0 \leq \hat{\sigma}_{11} \leq \sigma_{11} \\ 0 \leq \hat{\sigma}_{22} \leq \sigma_{22} \end{gathered}$$

In this case we are free to choose the inputs. Let $\hat{x}_t = (\hat{z}_t', \hat{y}_t')'$ and let us take $\hat{z}_t$ as the input. Then normalizing $w = (k, -1)$, the set of all observationally equivalent relation functions is described by the slopes k satisfying ([13]):

$$\left|\frac{\sigma_{21}}{\sigma_{11}}\right| \leq |k| \leq \left|\frac{\sigma_{22}}{\sigma_{12}}\right| \quad ; \quad \mathrm{sign}\, k = \mathrm{sign}\, \sigma_{21} \tag{3.1}$$

Note that the bounds in (3.1) correspond to the elementary regressions where either all noise has been added to $\hat{y}_t$ (i.e. $\hat{z}_t$ is observed free of noise) or to $\hat{z}_t$ respectively.

Completely analogous results can be obtained in a nonstochastic setting: If we fit a least squares line to a two-dimensional scatter plot, the resulting line of course depends on the way the distance of a point from the scatter plot to the line is measured. The *traditional* way is to take the distance parallel to the y- or to the z-axis respectively (which corresponds to the bounds in (3.1)). If we take all distances corresponding to unit balls which are ellipses with main axes parallel to the coordinate axes, then the corresponding least squares lines will have slope parameters as described by (3.1).

The above results can be generalized to the *dynamic Frisch case* (Anderson and Deistler (1984)): Under the assumption $\sigma_{12}(\lambda) \neq 0\ \forall \lambda \in [-\pi, \pi]$, again $\hat{z}_t$ may be taken as input and the set of all observationally equivalent relation functions $w = (k, -1)$ is characterized as the set of all transfer functions k satisfying

$$\left|\frac{\sigma_{21}(\lambda)}{\sigma_{11}(\lambda)}\right| \leq |k(e^{-i\lambda})| \leq \left|\frac{\sigma_{22}(\lambda)}{\sigma_{12}(\lambda)}\right| \quad ; \quad \arg k(e^{-i\lambda}) = \arg \sigma_{21}(\lambda) \tag{3.2}$$

In the dynamic case clearly k is in general complex valued and dependent on frequency λ. The phase of $k(e^{-i\lambda})$, $\arg k(e^{-i\lambda})$ is uniquely determined from Σ (as is the sign of k in the static case) but the gain of k is non-unique in general. Again the bounds $|\sigma_{21}/\sigma_{11}|$ and $|\sigma_{22}/\sigma_{12}|$ (which correspond to the elementary regressions or in other words to the Wiener-filters) correspond to the extreme cases where either $\hat{z}_t$ or $\hat{y}_t$ are observed without noise. (See Fig. 5.2 for a plot of the set of observationally equivalent transferfunctions k.) It should be noted that here k is not necessarily causal. For the case where k is a priori known to be causal or for questions relating to the existence of causal transfer functions in the equivalence class see [7], [3], [9].

4. The Frisch Case, General n

In this section we assume throughout that (a5) holds, i.e. that $\tilde{\Sigma}$ is diagonal. For general n, there is still a number of open problems. First we treat some special cases where rather complete results are available and then we present some general results, mainly relating to topological and geometrical properties, for instance of equivalence classes.

The first special case considered is the *static case where* $\mathrm{mc}(\Sigma) = 1$ *holds.* For this case, the following theorem (which has a long history see e.g. [12], [17], [20]) gives a complete picture, both for the characterization of the case $\mathrm{mc}(\Sigma) = 1$ in terms of Σ, and for the characterization of the solution set.

Theorem 4.1. *For the static case we have:*

(i) $\mathrm{mc}(\Sigma) = 1$ *iff there exists a diagonal matrix U with diagonal elements* 1 *or* -1 *such that $U\Sigma^{-1}U$ is a positive matrix (i.e. a matrix consisting of strictly positive entries only).*
(ii) For $\mathrm{mc}(\Sigma) = 1$, $\underline{\mathcal{L}}_1$ *is the union of the positive cone, C say, generated by the n rows of Σ^{-1} and* $-C = \{-w | w \in C\}$.

Let $S = \Sigma^{-1}$ and let s_i denote the i-th row of S. As easily can be seen from

$$s_i \Sigma = (0, \ldots, 0, 1, 0, \ldots, 0) = s_i \,\mathrm{diag}(0, \ldots, 0, s_{ii}^{-1}, 0, \ldots, 0)$$

s_i is an i-th elementary solution. For $\mathrm{mc}(\Sigma) = 1$ then no elementary solution s_i contains a zero component, thus without loss of generality we may introduce the normalized elementary solutions $(1, e_i) = s_{i1}^{-1} s_i$. Then, by theorem

4.1, $\mathrm{mc}(\Sigma) = 1$ is equivalent to saying that, eventually after changing signs in the elements of x_t, for all elementary regressions the corresponding e_i, $i = 1, \ldots, n$, are contained in the (strictly) positive orthant of $\mathbb{R}^{n-1}$ and that the set of all e such that $(1, e) \in \underline{\mathcal{L}}_1$ is the convex hull generated by the e_i, $i = 1, \ldots, n$. This is shown for the case $n = 3$ in Fig. 4.1. Figure 4.2, again for $n = 3$, gives an example of the static case $\mathrm{mc}(\Sigma) = 2$.

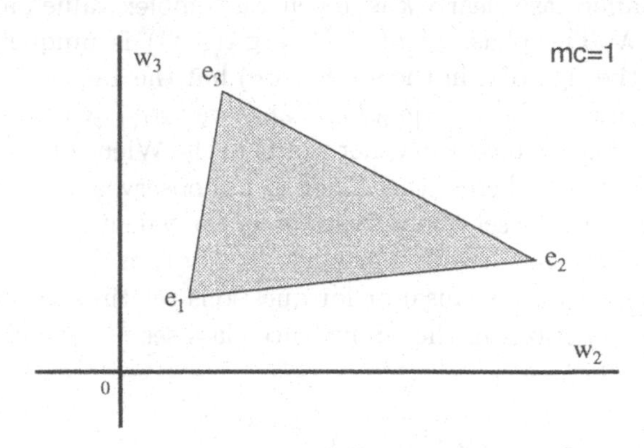

Fig. 4.1. The set of normalized 1-solutions $(1, w_2, w_3)$ for the static Frisch case, $n = 3$, $\mathrm{mc}(\Sigma) = 1$.

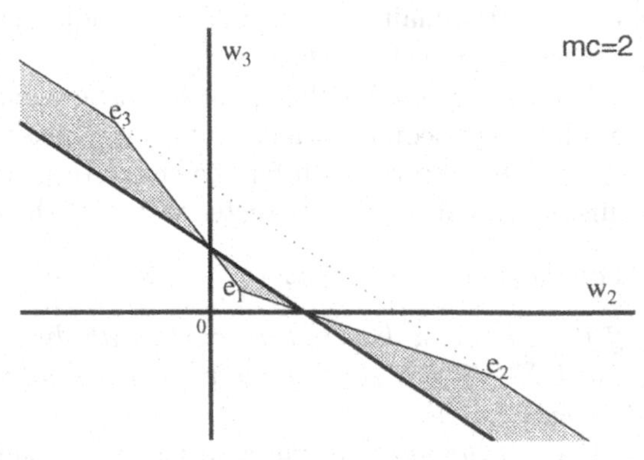

Fig. 4.2. The set of normalized 1-solutions $(1, w_2, w_3)$ for the static Frisch case, $n = 3$, $\mathrm{mc}(\Sigma) = 2$.

Figures 4.3 and 4.4 show sets $\mathcal{E}$ for the static case $n = 3$ for $\mathrm{mc}(\Sigma) = 1$ and $\mathrm{mc}(\Sigma) = 2$ respectively (compare theorem 4.2 below and the remarks following this theorem.)

For the dynamic $\mathrm{mc}(\Sigma) = 1$ case, things are much more complicated (see [10] and [26]). The condition $\mathrm{mc}(\Sigma) = 1$ is equivalent to saying that no $w \in \underline{\mathcal{L}}_1$ has a zero entry, but this does not seem to be a useful "test". As can

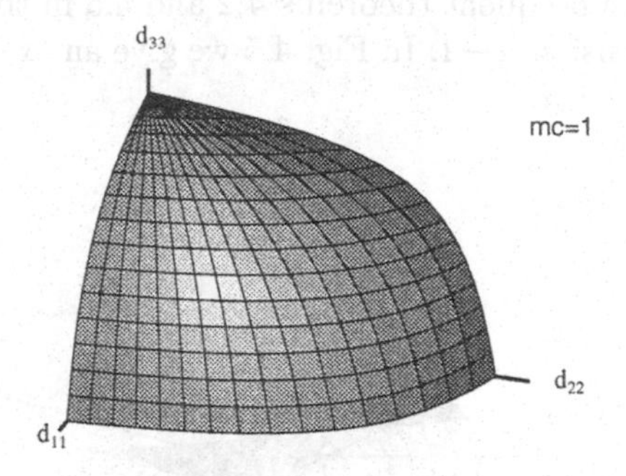

Fig. 4.3. The set $\mathcal{E}$ of compatible noise spectral densities for the Frisch case $n = 3$, $\mathrm{mc}(\Sigma) = 2$. Here $d_{ii} = \tilde{\sigma}_{ii}$ denote the i-th diagonal elements of $\tilde{\Sigma}$.

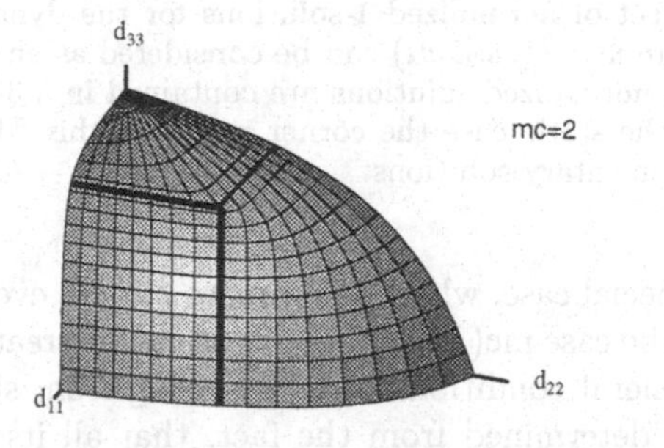

Fig. 4.4. The set $\mathcal{E}$ of compatible noise spectral densities for the Frisch case $n = 3$, $\mathrm{mc}(\Sigma) = 2$. Note that $\mathcal{E}$ contains segments of three lines parallel to the coordinate axes. The intersection point, $\tilde{\Sigma}_0$ say, of these segments is the unique $\tilde{\Sigma} \in \mathcal{E}$ with $\mathrm{crk}(\Sigma - \tilde{\Sigma}) = 2$. Thus $\mathcal{E}_2 = \{\tilde{\Sigma}_0\}$ is a singleton and $\mathcal{E}$ is not smooth at the point $\tilde{\Sigma}_0$.

be seen from the subsequent theorems 4.2 and 4.5 in this case $\mathcal{L}_1$ is compact and of (real) dimension $n-1$. In Fig. 4.5 we give an example for the dynamic case $n=3$, $\mathrm{mc}(\Sigma)=1$.

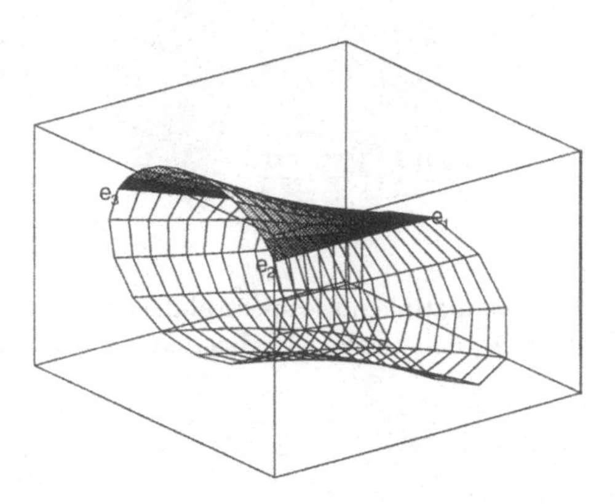

Fig. 4.5. The set of normalized 1-solutions for the dynamic Frisch case, $n=3$, $\mathrm{mc}(\Sigma)=1$. Here $w=(1,w_2,w_3)$ can be considered as an element of R^4. It can be shown that the normalized solutions are contained in a 3-dimensional subspace of this R^4. As in the static case the corner points of this "deformed" triangle correspond to the elementary solutions.

Another special case, where a complete picture even for the dynamic case is available is the case $\mathrm{mc}(\Sigma)=n-1$. This case is treated in detail in [5]. Here under very general conditions $\mathcal{E}_{n-1}$ is a singleton, since the corresponding $\hat{\Sigma}$ is uniquely determined from the fact, that all its 2×2 submatrices are singular and thus we can determine e.g. $\hat{\sigma}_{11}$ from $\hat{\sigma}_{11}\sigma_{13}-\sigma_{21}\sigma_{23}=0$. For this situation, both a characterization of the case $\mathrm{mc}(\Sigma)=n-1$ in terms of Σ and a complete description of the solution set are available.

Now we give a number of general results. We here describe basic features, for more details and proofs the reader is referred to [24], [11] and [25].

In a first step we describe some properties of the set $\mathcal{E}$ of all observationally equivalent $\tilde{\Sigma}$.

Theorem 4.2. *$\mathcal{E}$ is homeomorphic to the positive part of the unit sphere $\mathcal{K}^+=\{d\in\mathbb{R}^n\,|d_i\geq 0,\|d\|=1\}$. In particular $\mathcal{E}$ is compact and a topological manifold with boundary of (real) dimension n-1.*

Proof. Let $1/\lambda_0$ denote the largest eigenvalue of $\Sigma^{-1/2}\operatorname{diag}(d)\Sigma^{-*/2}$ for $d\in\mathcal{K}^+$. Then the matrix $\Sigma-\lambda\operatorname{diag}(d)$ is positive semidefinite and singular iff $\lambda=\lambda_0$ holds. Clearly $\hat{\Sigma}=\Sigma-\lambda_0\operatorname{diag}(d)$ is compatible with Σ. We now define a function on $\mathcal{K}^+$ by $d\mapsto\lambda_0\operatorname{diag}(d)$. This function is continuous, because the largest eigenvalue is a contiuous function of the matrix elements. As Σ is non-singular, the inverse mapping defined by $\tilde{\Sigma} \mapsto (\tilde{\sigma}_{11}, \ldots, \tilde{\sigma}_{11})/\|(\tilde{\sigma}_{11}, \ldots, \tilde{\sigma}_{11})\|$ is well defined and continuous too. The second statement of the proposition is an immediate consequence of the first.

As a trivial consequence of the theorem above we see that a Frisch decomposition always exists and that $\hat{\Sigma}$ is never unique without imposing further restrictions. By the above theorem $\mathcal{E}$ and $\mathcal{K}^+$ are topologically equivalent, however $\mathcal{E}$ may not be smooth. As has been shown in [23], $\mathcal{E}$ is smooth exactly at those points where $\operatorname{crk} \hat{\Sigma} = 1$ holds.

In a next step we analyze the subsets $\mathcal{E}_m$ of $\mathcal{E}$ corresponding to a given corank m of $\Sigma - \tilde{\Sigma}$, i.e. to the m-outputs systems. We have

Theorem 4.3. *For every $1 \leq m \leq n$, the set $\mathcal{E}_1 \cup \cdots \cup \mathcal{E}_m$ is open and dense in $\mathcal{E}$.*

Now by (2.3), we can write (2.4) in the form

$$\Sigma = ll^* + \tilde{\Sigma} \quad ; \quad l \in \mathbb{C}^{n \times n-m} \tag{4.1}$$

where we restrict ourselves to the case $\operatorname{crk} \hat{\Sigma} = \operatorname{crk} l = m$. Σ as a positive definite matrix with complex entries has n^2 free real parameters, $\tilde{\Sigma}$ has n free real parameters and $\hat{\Sigma} = ll^*$ has $n^2 - m^2$ free real parameters, in other words we have $n_L = n^2$ free parameters on the left hand side and $n_R = n^2 - m^2 + n$ on the right hand side of (4.1). Thus using a heuristic (and not exact) argument based on mere counting of free parameters, one could conclude that the socalled *Ledermann bound* $\mathrm{m}_L = \sqrt{n}$ plays an important role in our analysis. One might expect that for $m \geq \mathrm{m}_L$ (where $n_L \geq n_R$ holds), the set $\mathcal{E}_m$ is either empty or consists of a finite number of points, whereas for $m < \mathrm{m}_L$, the set $\mathcal{E}_m$ is either empty or has dimension $n - m^2$. By a similar argument one might expect that the sets $\mathcal{S}_m = \{\Sigma | \operatorname{mc}(\Sigma) = m\}$ are "thin" in $\mathcal{S}$ for $m > \mathrm{m}_L$ and "thick" otherwise. It should be noted, that the term Ledermann bound is mainly used for the static case where $\mathrm{m}_L = (-1 + \sqrt{1 + 8n})/2$. The heuristic reasoning above is made precise in the following theorems.

Theorem 4.4. *There exists a set $\mathcal{S}^r \subset \mathcal{S}$ with the following properties:*

(i) $\mathcal{S}^r$ is generic in $\mathcal{S}$ in the sense that its complement $\mathcal{S} \setminus \mathcal{S}^r$ is a set of Lebesque measure zero.
(ii) $\mathcal{S}_m \cap \mathcal{S}^r = \emptyset$ for all $m > \mathrm{m}_L$.
(iii) $\mathcal{S}_m \cap \mathcal{S}^r$ is open in $\mathcal{S}$ and nonvoid for all $m \leq \mathrm{m}_L$ and $\overline{(\mathcal{S}_m \cap \mathcal{S}^r)} \cap \mathcal{S}_m = \mathcal{S}_m$, i.e. $\mathcal{S}_m \cap \mathcal{S}^r$ is dense in $\mathcal{S}_m$, for all $m \leq \mathrm{m}_L$.
(iv) For $\Sigma \in \mathcal{S}^r$ and $m < \mathrm{m}_L$ the set $\mathcal{E}_m$ is either empty or a differentiable submanifold of $\mathbb{R}^n$ with boundaries of dimension $n - m^2$. For $\Sigma \in \mathcal{S}^r$ and $m = \mathrm{m}_L$ the set $\mathcal{E}_m$ contains only a finite number of points.

In many cases one is only interested in the class of all observationally equivalent systems with a maximum number of equations (i.e. in factor models with a minimum number of factors). The reason for this is that the corresponding systems explain as much as possible by the system, leaving as little as possible to the *outside world*. We have:

Theorem 4.5. *For $m = \mathrm{mc}(\Sigma)$, the sets $\mathcal{E}_m$ and $\mathcal{L}_m$ are homeomorphic. If in addition $\Sigma \in \mathcal{S}^r$ and $\mathrm{mc}(\Sigma) < \mathrm{m}_L$ hold then $\mathcal{L}_m$ is a differentiable submanifold of $\mathcal{G}(m,n)$ with boundaries of dimension $n - m^2$ and $\mathcal{E}_m$ and $\mathcal{L}_m$ are differeomorphic.*

From the point of view of identification an important integer is $\mathrm{mc}(\Sigma)$ (remember that $\mathrm{mc}(\Sigma)$ is the minimum number of factors). In order to estimate or test for $\mathrm{mc}(\Sigma)$, some properties of the sets $\mathcal{S}_m$ have to be investigated. In addition to the properties described in theorem 4.4 we have:

Theorem 4.6.

(i) $\mathcal{S}_m$ *is nonvoid for all* $1 \leq m \leq n$.
(ii) $\mathcal{S}_m \cup \cdots \cup \mathcal{S}_1$ *is open in* $\mathcal{S}$ *for all* $1 \leq m \leq n$
(iii) $\overline{\mathcal{S}_m} = \overline{\mathcal{S}_n \cup \cdots \cup \mathcal{S}_m}$ *for* $m \geq \mathrm{m}_L$.

In the last part of this section we deal with *continuity* of mappings relating sets of observationally equivalent systems to the spectral denstity of the observations. The motivation for this is the following: In a stationary context Σ can be estimated consistently by the usual spectral estimators. Continuity then implies also consistency for the estimators of the corresponding classes of observationally systems. In [10], [24], [11] and [25] a number of such continuity results has been proved; here we only present a part of these results.

Let the set $\mathcal{A}_m(\Sigma)$ be defined as $\mathcal{E}_m(\Sigma) \cup \cdots \cup \mathcal{E}_n(\Sigma)$; by theorems 4.2 and 4.3, $\mathcal{A}_m(\Sigma)$ is compact. We have

Theorem 4.7.

(i) *For $\Sigma^k, \Sigma^0 \in \mathcal{S}$, $\Sigma^k \to \Sigma^0$ we have*

$$\rho(\mathcal{A}_m(\Sigma^k), \mathcal{A}_m(\Sigma^0)) = \sup_{\tilde{\Sigma}^k \in \mathcal{A}_m(\Sigma^k)} \inf_{\tilde{\Sigma}^0 \in \mathcal{A}_m(\Sigma^0)} \|\tilde{\Sigma}^k - \tilde{\Sigma}^0\| \to 0$$

(ii) *If in addition $\Sigma^k, \Sigma^0 \in \mathcal{S}^r$, where $\mathcal{S}^r$ is the generic subset of $\mathcal{S}$ considered in theorem 4.4, then $\Sigma^k \to \Sigma^0$ implies*

$$\min\left(\rho(\mathcal{A}_m(\Sigma^k), \mathcal{A}_m(\Sigma^0)), \rho(\mathcal{A}_m(\Sigma^0), \mathcal{A}_m(\Sigma^k))\right) \to 0$$

i.e. the mapping $\Sigma \mapsto \mathcal{A}_m(\Sigma)$ on $\mathcal{S}^r$ is continuous with respect to the Hausdorff metric for the sets $\mathcal{A}_m$.

5. The Bounded Noise Case

Here again we only present results without proofs; for more results and proofs, the reader is refered to [25]. It should be noted that the word compatible now refers to assumption (a.5), and thus symbols such as e.g. $\mathcal{L}_1$ have a different meaning compared to section 4.

First, let us forget about (a.5) and let us consider the following problem: For a given Σ, we fix a $w \in \mathbb{C}^{m\times n}$ with $\operatorname{rk} w = m$. Then clearly always Hermitean positive semidefinite matrices $\hat{\Sigma}$ and $\tilde{\Sigma}$ satisfying (2.4) and (2.5) can be found. Now it can be shown (see [19] for the static case and [25]) that there is a unique minimal (w.r. to the semiordering given by semipositivity of matrices) element among all such $\tilde{\Sigma}$, given by

$$\tilde{\Sigma}_w = \Sigma w^*(w\Sigma w^*)^{-1} w\Sigma$$

Because of this minimality we call

$$\Sigma = (\Sigma - \tilde{\Sigma}_w) + \tilde{\Sigma}_w$$

a generalized least squares decomposition. From this result we can derive the following characterization of m-solutions and of $\operatorname{mc}(\Sigma)$.

Theorem 5.1.

(i) $w \in \mathbb{C}^{m\times n}$, $\operatorname{rk} w = m$ *is an* m*-solution iff* $w(\epsilon\Sigma - \Sigma\Sigma)w^* \geq 0$ *holds.*
(ii) $\operatorname{mc}(\Sigma) = m$ *iff* $\lambda_1(\Sigma) \leq \cdots \leq \lambda_m(\Sigma) \leq \epsilon < \lambda_{m+1}(\Sigma)$ *holds.*

We note that in the bounded noise case, unlike in the Frisch case, there may be no solutions if ϵ is too small and that the determination of $\operatorname{mc}(\Sigma)$, which is still an open problem for the general Frisch case, here is rather simple.

In the next theorem some properties of the m-solution set $\mathcal{L}_m$ are given:

Theorem 5.2.

(i) $\mathcal{L}_m$ *is a compact subset of* $\mathcal{G}(m,n)$.
(ii) For $\lambda_m(\Sigma) < \epsilon$, *the set* $\mathcal{L}_m$ *contains a nonvoid open subset of* $\mathcal{G}(m,n)$ *and* $\mathcal{L}_m = \overline{\mathcal{L}_m^o}$ *holds.*

In the next theorem the sets $\mathcal{S}_m$ of spectral densities are considered:

Theorem 5.3. *$\mathcal{S}_m$ contains a nonvoid open subset and $\mathcal{S}_n \cup \cdots \cup \mathcal{S}_m$ is closed in $\mathcal{S}$.*

By the theorem above all sets $\mathcal{S}_m$ are "thick" subsets of $\mathcal{S}$; for the Frisch case, on the contrary for $m > \mathrm{m}_L$ the sets $\mathcal{S}_m$ are "thin".

Finally we are concerned with continuity: Let $\mathcal{S}(\epsilon) \subset \mathcal{S}$ denote the set of all spectral densities which have no eigenvalue equal to ϵ. Note that a Hausdorff metric on $\mathcal{G}(m,n)$ can be introduced as follows: Let $\mathfrak{x}, \mathfrak{y} \in \mathcal{G}(m,n)$

(where $\mathcal{G}(m,n)$ is considered as the set of all m-dimensional subspaces of $\mathbb{C}^n$) then we define

$$\mathrm{d}_{\mathcal{G}}(\mathfrak{x},\mathfrak{y}) = \mathrm{d}_H(\{x \in \mathfrak{x}; \|x\| = 1\}, \{y \in \mathfrak{y}; \|y\| = 1\})$$

Then $\mathrm{d}_{\mathcal{G}}$ again can be used to define a Hausdorff metric on the set, $\mathcal{C}$ say, of all compact subsets of $\mathcal{G}(m,n)$. For this Hausdorff metric then:

Theorem 5.4. *The mapping*

$$\mathcal{S}(\epsilon) \to \mathcal{C} : \Sigma \mapsto \mathcal{L}_m(\Sigma)$$

is continuous for $1 \leq m \leq n$.

In Fig. 5.1 for $n = 2$, sets of observationally equivalent transferfunctions for the bounded noise case, for different noise levels ϵ, are shown, and a comparison with the Frisch case is given. Figures 5.2 and 5.3 show the set of all observationally equivalent transferfunctions $k(e^{-i\lambda})$ as a function of frequency λ for the Frisch and the bounded noise case respectively.

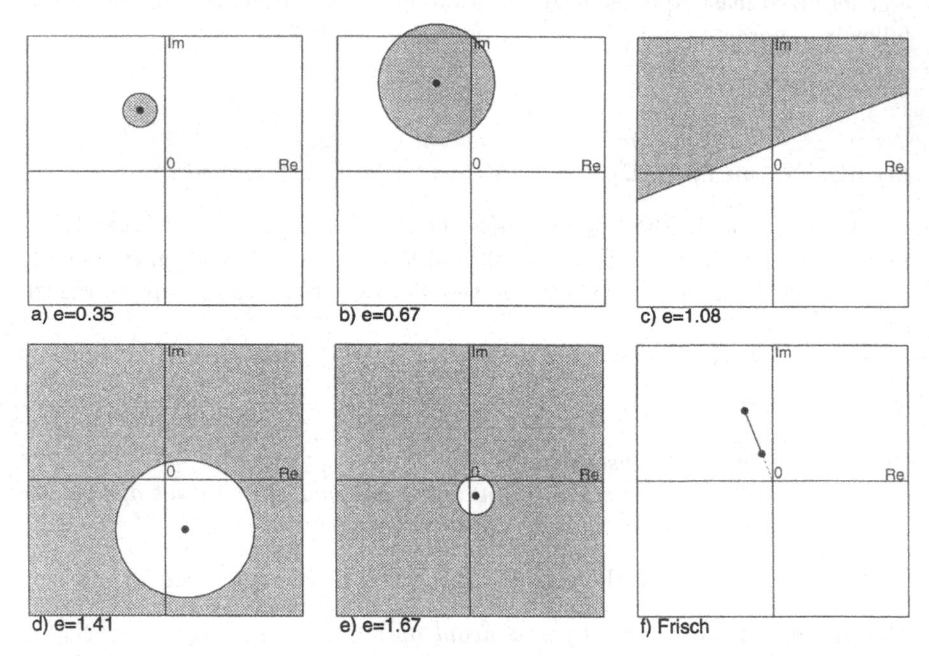

Fig. 5.1. Here for a given Σ, $n = 2$, sets of compatible transferfunctions k, evaluated at a fixed frequency, are represented in the complex plane (marked by dark areas). Figures a)-e) show the bounded noise case for a growing noise bound ϵ. Figure f) shows the Frisch case. Note that the sets are 2-dimensional for the bounded noise case and 1-dimensional for the Frisch case.

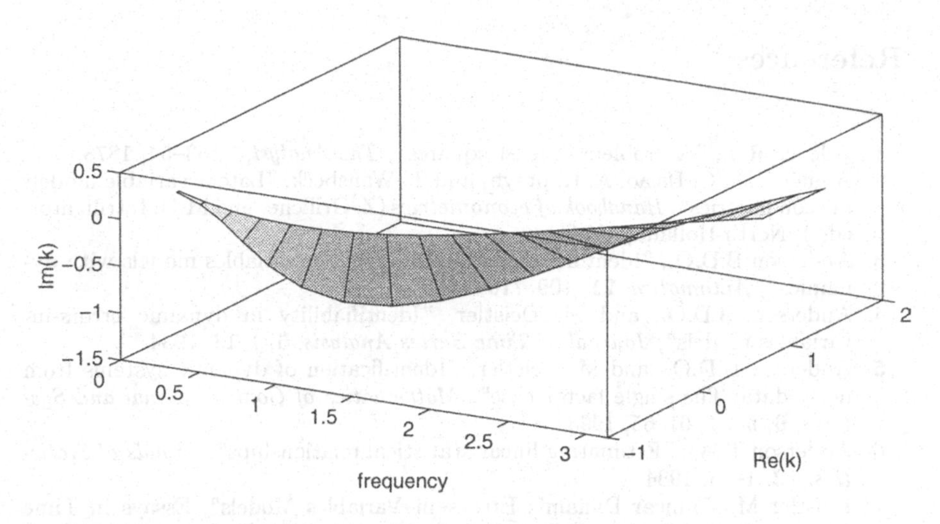

Fig. 5.2. Here for a given Σ, $n = 2$, the set of all observationally equivalent transferfunctions k as a function of frequency is shown for the Frisch case.

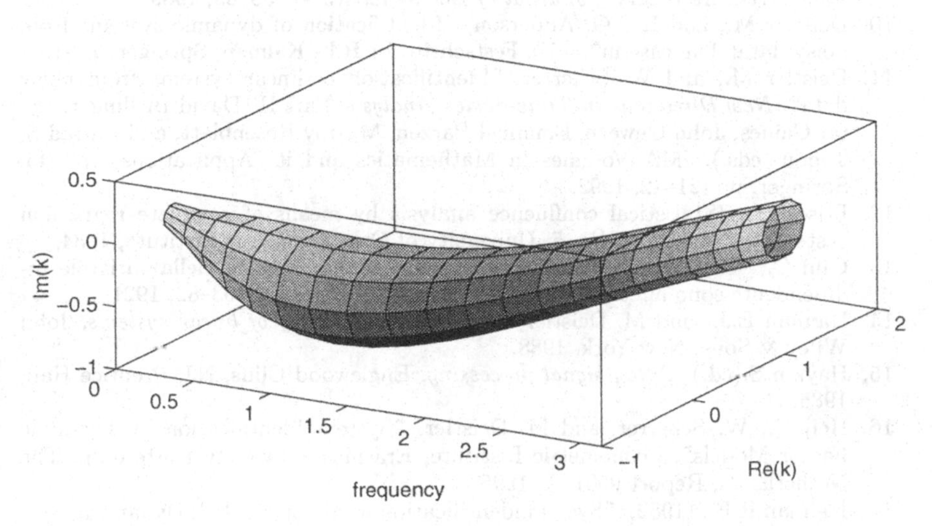

Fig. 5.3. Here for a given Σ, $n = 2$ and a given noise bound $\epsilon > 0$, the set of all observationally equivalent transferfunctions k as a function of frequency is shown for the bounded noise case.

References

1. Adcock R.J., "A problem in least squares", *The Analyst*, **5**, 53–54, 1878.
2. Aigner D.J., C. Hsiao, A. Kapteyn, and T. Wansbeck, "Latent variable models in econometrics", *Handbook of econometrics* (Z. Griliches and M.D. Intriligator, eds.), North-Holland, 1984.
3. Anderson B.D.O., "Identification of scalar errors-in-variables models with dynamics", *Automatica*, **21**, 709–716, 1985.
4. Anderson B.D.O., and M. Deistler, "Identifiability in dynamic errors-in-variables models", *Journal of Time Series Analysis*, **5**, 1–13, 1984.
5. Anderson B.D.O., and M. Deistler, "Identification of dynamic systems from noisy data: The single factor case", *Mathematics of Control, Signal and Systems*, **6**, no. 1, 61–65, 1993.
6. Anderson T.W., "Estimating linear statistical relationships", *Annals of Statistics*, **12**, 1–45, 1984.
7. Deistler M., "Linear Dynamic Errors–in–Variables Models", Essays in Time Series and Alied Processes, Festschrift in Honour of E.J. Hannan, (J. Gani and M. Priestly, ed.), Applied Probability Trust, pp. 23–39, 1986.
8. Deistler M., "Linear system identification - a survey", *From data to model* (J. Willems, ed.), Springer, Berlin, pp. 1–25, 1989.
9. Deistler M., and B.D.O. Anderson, "Linear dynamic errors-in-variables models, some structure theory", *Journal of Econometrics* **41**, 39–63, 1989.
10. Deistler M., and B.D.O. Anderson, "Identification of dynamic systems from noisy data: The case $m^* = 1$", Festschrift for R.E. Kalman, Springer, 1991.
11. Deistler M., and W. Scherrer, "Identification of linear systems from noisy data", *New Directions in Time-Series Analysis*, Part II (David Brillinger, Peter Caines, John Geweke, Emanuel Parzen, Murray Rosenblatt, and Murad S. Taqqu, eds.), IMA Volumes in Mathematics and its Applications, vol. 46, Springer, pp. 21–42, 1992.
12. Frisch R., "Statistical confluence analysis by means of complete regression systems", Publication No. 5, University of Oslo, Economic Institute, 1934.
13. Gini C., "Sull'interpolazione di una retta quando i valori della variabile indipendente sono affetti da errori accidentali", *Metron*, **1**, 63–82, 1921.
14. Hannan E.J., and M. Deistler, *The statistical theory of linear systems*, John Wiley & Sons, New York, 1988.
15. Haykin S.(ed.), *Array signal processing*, Englewood Cliffs, NJ, Prentice Hall, 1985.
16. Heij C., W. Scherrer, and M. Deistler, "System Identification by Dynamic Factor Models", Econometric Institute, Erasmus University Rotterdam, The Netherlands, Report 9501/A, 1995.
17. Kalman R.E., (1982): "System identification from noisy data", Dynamical Systems II, a University of Florida international Symposium (A. Bednarek and L. Cesari, eds.), Academic Press, New York , 1982.
18. Kalman R.E., "Identifiability and modeling in econometrics", Developments in statistics (Krishnaiah, ed.), vol. 4, Academic Press, New York, 1983.
19. Kalman R.E., *A theory for the identification of linear relations*, Colloques Lions (H. Brezis and P.G. Ciarlet ed.), 1989.
20. Klepper S., and E.E. Leamer (1984): "Consistent Sets of Estimates for Regressions with Errors in all Variables", *Econometrica*, **52**, No. 1, January, 1984.
21. Picci G., and S. Pinzoni, "Dynamic factor-analysis models for stationary processes", *IMA Journal of Mathematical Control and Information*, **3**, 185–210, 1986.

22. Rozanov Yu.A., *Stationary random processes*, Holden-Day, San Francisco, 1967.
23. Schachermayer W., and M. Deistler, *Identification of linear systems from noisy data with uncorrelated error components: some structure theory*, mimeo, 1990.
24. Scherrer W., "Strukturtheorie von linearen dynamischen Fehler–in–den–Variablen Modellen mit diagonalem Fehlerspektrum", Ph.D. thesis, Technical University Vienna, Austria, September 1991.
25. Scherrer W., and M. Deistler, *A structure theory for linear dynamic errors-in-variables models*, mimeo, 1993.
26. Scherrer W., M. Deistler, M. Kopel, and W. Reitgruber, "Solution sets for linear dynamic errors-in-variables models", *Statistica Neerlandica*, **45**, No. 4, pp. 391–404, 1991.
27. Spearman C., "General intelligence, objectively determined and measured", *Amer. Jour. Psych.*, **15**, 201–293, 1904.
28. Tugnait J.K., "Stochastic system identification with noisy input using cumulant statics", *IEEE Tr. AC*, **37**, 476–485, 1992.
29. Willems J.C., "From time series to linear systems", *Automatica*, **22**, 561–580, 1986.

Identification in $\mathcal{H}_\infty$: Theory and Applications

Pramod P. Khargonekar[1], Guoxiang Gu[2], and Jonathan Friedman[3]

[1] Dept. of Electrical Engineering and Computer Science
The University of Michigan
Ann Arbor, MI 48109-2122.

[2] Dept. of Electrical Engineering
Louisiana State University
Baton Rouge, LA 70803-5901.

[3] Dept. of Aerospace Engineering
The University of Michigan
Ann Arbor, MI 48109-2140.

1. Introduction

System modeling and identification play a central and critical role in the design of control systems. The classical approach to system identification is based on a time-domain stochastic system theory perspective. See the book by Ljung [27] for a comprehensive exposition. There have been many successful applications of this large body of knowledge.

In recent years a number of researchers have started to explore alternative (worst-case and/or deterministic and/or frequency domain) approaches [2, 3, 6, 7, 11, 13, 14, 17, 18, 19, 20, 21, 22, 23, 25, 24, 26, 28, 31, 29, 32, 33, 36, 38, 39, 40, 42, 43, 44, 45, 46, 47, 48, 50, 51]. This list is not meant to be complete but just to give a sampling of the literature in this active and growing area.

A significant motivation for this flurry of research has come from the developments in the field of robust control. Most of the modern robust control problems are formulated using worst-case deterministic criteria. Therefore, it is natural to demand that the modeling efforts produce a nominal model and guaranteed worst-case bounds on the modeling error which can be used for robust control analysis and design. While this motivation is quite reasonable, it is fair to say that the research efforts to-date have not resulted in a really successful merger of robust control and identification.

Among these new research directions, one particular avenue being explored is *identification in* $\mathcal{H}_\infty$. The basic idea is to perform input-output experiments using, for example, sinusoidal excitation, and extract frequency response information on the system under consideration. The resulting (noisy and partial) frequency response information is then mapped into a transfer function model for the system. Such an approach is by no means new – it is a very commonly used procedure. Some early references dealing with this frequency domain approach include [8, 41].

Frequency response based system identification is again being investigated very actively; see [2, 3, 13, 14, 19, 20, 21, 28, 33, 34, 35, 36, 32, 4] and the references cited there. A canonical problem, known as identification in $\mathcal{H}_\infty$, was formulated by Helmicki, Jacobson, and Nett [19]. The basic problem is to map noisy and partial frequency response data into a transfer function model for a stable linear system such that the worst case error measured in the $\mathcal{H}_\infty$ norm converges to zero as the number of data points goes to infinity and as the noise amplitude goes to zero. Additionally, it is required to produce guaranteed worst case errors on the modeling error.

Several researchers have been working on this problem for the last few years. We have also been working on this problem for the last few years and have developed quite promising algorithms for solving this problem. In this paper, we will give a brief summary of some of our work on this problem. The aim of the paper is to expose the potential reader to the basic concepts and results and give pointers to the recent literature.

Very recently, we have done some work on applying these algorithms to very challenging data sets on flexible systems [9, 15]. These results indicate that our algorithms have the potential of being quite useful in real engineering applications.

2. Problem Formulation

In this paper, we will focus attention on a specific frequency domain identification problem known as "identification in $\mathcal{H}_\infty$ ". The problem was posed in this particular form by Helmicki, Jacobson and Nett [19]. We briefly describe this problem formulation next.

2.1 Discrete-Time Systems

Suppose that the "true" unknown system to be identified is a causal linear exponentially stable discrete-time single-input single-output shift-invariant system with transfer function $\hat{h}$. Thus, we can express the transfer function $\hat{h}$ as a power series

$$\hat{h} = \sum_{k=0}^{\infty} h_k z^k .$$

As is well known, the sequence h_k is the unit pulse response of the system. Also, note that we have used positive powers of z to form the power series. Thus, for a rational transfer function $\hat{h}$, exponential stability of the system would be equivalent to requiring that $\hat{h}$ have no poles inside the closed unit disc.

It will be assumed that we have some prior knowledge concerning the transfer function $\hat{h}$. At a general abstract level, this will be expressed by the statement: $\hat{h} \in \mathcal{S} \subset \mathcal{A}$, where $\mathcal{A}$ is the disk algebra, i. e.,

$$\mathcal{A} := \left\{ \hat{f} \in \mathcal{H}_\infty : \hat{f} \text{ continuous on the unit circle} \right\}.$$

The nature of the subset $\mathcal{S}$, thus, reflects the prior knowledge of the properties of the unknown system to be identified.

A particularly important model set $\mathcal{S}$ is the collection of exponentially stable bounded systems

$$\mathcal{H}(D_\rho, M) = \left\{ \hat{f} : \hat{f} \text{ analytic in } D_\rho,\ |\hat{f}| \leq M\ \forall z \in D_\rho \right\}$$

where $\rho > 1$ and $D_\rho := \{z : |z| < \rho\}$ is the complex disc of radius ρ. It is a simple exercise to verify that if $\hat{h} \in \mathcal{H}(D_\rho, M)$, then

$$|h_k| \leq M\rho^{-k}.$$

Thus, ρ corresponds to the prior knowledge of the *relative stability* or the exponential decay rate of the unknown system. On the other hand, M is one measure of the maximum gain of the system.

Throughout this paper, we will only deal with single-input single-output systems, although most of the results carry over without much difficulty to multiple-input multiple-output systems as well.

2.2 Experimental Data

The experimental data on $\hat{h}$ will be taken to be noisy values of the frequency response of the system at a finite number of frequencies. Such frequency response information can be obtained by applying linear combinations of sinusoids of various frequencies as input excitation to the system to be identified. As this is quite well known in the standard linear systems literature, we will not dwell on this issue.

Let us now suppose that we are given a finite number N of possibly noisy experimental frequency response data

$$E_k^N(\hat{h}, \hat{\eta}) := \hat{h}(e^{j2k\pi/N}) + \hat{\eta}_k, \tag{2.1}$$

where $k = 0, 1, 2, \ldots, N-1$ and

$$\hat{\eta} \in B_N(\epsilon) := \left\{ (\hat{\eta}_0, \hat{\eta}_1, \ldots, \hat{\eta}_{N-1}) \in C^N : |\hat{\eta}_k| \leq \epsilon\ \ \forall k \right\}, \tag{2.2}$$

is the measurement noise in the frequency response.

We have taken a frequency independent bound ϵ on the noise $\hat{\eta}_k$. A more general case is to assume a noise bound of the form

$$\|\hat{w}\hat{\eta}\|_\infty \leq \epsilon$$

for some known weighting function $\hat{w}$. As pointed out in [19], the resulting identification problem can be transformed into an equivalent problem with

$\hat{w} \equiv 1$. Thus, for the sake of simplicity, we will assume that the bound on the noise is independent of frequency.

The bound ϵ quantifies the uncertainty in the estimation of the frequency response from the time-domain experiment. While extracting a good estimate of the transfer function at a given frequency using the time domain data from a sinusoidal excitation experiment is quite straightforward, the problem of extracting the noise bound ϵ from the experimental data is much harder. We will not deal with this issue here.

It should be noted that the experimental data E_k^N represents the frequency response of the unknown system at the frequency $z = e^{2jk\pi/N}$. Thus, as k ranges from 0 to N, these frequencies are *uniformly* spaced on the unit circle. This assumption is critical to this paper. The interested reader is referred to [3, 35] for the case of nonuniformly spaced frequencies. This issue is particularly important for continuous-time systems which are also treated in [2, 20, 5].

2.3 Identification in $\mathcal{H}_\infty$

The identification problem is to find an algorithm A_N which maps the given experimental information into an identified model $\hat{h}_{id}^N(z) \in \mathcal{A}$. The performance of the identification algorithm A_N is measured in a worst-case sense as follows:

$$e_N(\epsilon, \mathcal{S}) := \sup_{\hat{h} \in \mathcal{S},\, \hat{\eta} \in B_N(\epsilon)} \|\hat{h} - \hat{h}_{id}^N\|_\infty. \tag{2.3}$$

We will deal with families of algorithms A_N depending on N. With a slight abuse of terminology, we will refer to such a family of algorithms as an algorithm.

An algorithm is called **convergent** if

$$\lim_{\epsilon \to 0,\, N \to \infty} e_N(\epsilon, \mathcal{S}) = 0. \tag{2.4}$$

If in addition, the convergence of the algorithm is independent of the *a priori* information, $\epsilon and \mathcal{S}$, then it is called **robustly convergent and untuned.**

Under very mild assumptions on the set of systems $\mathcal{S}$, it can be shown that the worst case error

$$e_N(\epsilon, \mathcal{S}) \geq \epsilon.$$

Thus, ϵ is a lower bound on the performance of any algorithm. It will be seen that we have algorithms that produce models within a guaranteed asymptotic error of 2ϵ.

The efforts of many researchers in the last few years have focused on the analysis and construction of efficient algorithms for the problem of identification in $\mathcal{H}_\infty$. This has led to a very good understanding of the basic features of the problem, limits on performance of identification algorithms, and procedures for construction of very efficient algorithms for solving this problem. The main purpose of this paper is to give an exposition of some, but not all, of these developments.

3. Background Results

There is a close connection between the problem of identification in $\mathcal{H}_\infty$ and ideas from n-width approximation theory [37, 49]. This is quite natural since the identified model is, in some sense, an approximation of the true system. Here we will collect together some simple results from approximation theory. Needless to say, there is a vast amount of related literature on approximation of systems for which the reader is referred to [37, 52] and the references cited there.

Let $\mathcal{P}_m$ be the collection of all polynomials (or FIR models) of degree no greater than $m-1$:

$$\mathcal{P}_m = \left\{\hat{p} : \hat{p} = p_0 + p_1 z + \ldots + p_{m-1} z^{m-1}\right\}.$$

We shall specialize the definition of the n-width to $\mathcal{S}$ viewed as a subset of $\mathcal{H}_\infty$ as follows. The n-width of $\mathcal{S}$ with respect to $\mathcal{H}_\infty$ [37] is given by:

$$\delta_n(\mathcal{S}, \mathcal{H}_\infty) := \inf_{P_n} \sup_{x \in \mathcal{S}} \|x - P_n(x)\|_\infty \tag{3.1}$$

where the infimum is taken over all possible continuous linear operators P_n of $\mathcal{H}_\infty$ into $\mathcal{H}_\infty$ of rank n, i. e., those operators P_n whose range dimension is no greater than n. Any continuous linear operator P_n of rank at most n for which

$$\delta_n(\mathcal{S}, \mathcal{H}_\infty) = \sup_{x \in \mathcal{S}} \|x - P_n(x)\|_\infty$$

is called an *optimal linear operator* for $\delta_n(\mathcal{S}, \mathcal{H}_\infty)$.

The n-width of $\mathcal{S}$ is a measure of the complexity of the set $\mathcal{S}$. As a trivial example, if $\mathcal{S} = \mathcal{P}_N$, then $\delta_n = 0$ for all $n \geq N$.

For simplicity, we will denote $\delta_n(\mathcal{S}, \mathcal{H}_\infty)$ by merely δ_n for the special case $\mathcal{S} = \mathcal{H}(D_\rho, M)$. As a matter of fact, δ_n is a function of M and ρ. In the literature on n-widths, many results on δ_n can be found. The following lemma (Theorem 2.1 on page 250 and Proposition 2.5 on page 256 of [37]) will be used extensively for the quantification of the worst-case identification error.

Lemma 3.1. *Let $M > 0, \rho > 1$ and let $\mathcal{S} = \mathcal{H}(D_\rho, M)$. Then for $m = 0, 1, \ldots,$*

$$\delta_m(\mathcal{S}, \mathcal{H}_\infty) = \sup_{\hat{h} \in \mathcal{H}(D_\rho, M)} \inf_{\hat{p} \in \mathcal{P}_m} \|\hat{h} - \hat{p}\|_\infty = M\rho^{-m}.$$

An optimal linear operator for $\mathcal{H}(D_\rho, M)$ is given by

$$\hat{p}_m^*[\hat{h}] := \sum_{k=0}^{m-1} \left(1 - \rho^{2(k-m)}\right) h_k z^k. \tag{3.2}$$

Thus,

$$\sup_{\hat{h} \in \mathcal{H}(D_\rho, M)} \|\hat{h} - \hat{p}_m^*[\hat{h}]\|_\infty = M\rho^{-m}. \tag{3.3}$$

A key conclusion here is that the n-width of $\mathcal{H}(D_\rho, M)$ goes to zero exponentially as $n \to \infty$. More importantly, polynomials are "optimal approximants" for the case of $\mathcal{S} = \mathcal{H}(D_\rho, M)$.

For general sets $\mathcal{S}$, the n-widths, optimal linear operators, and optimal approximants can not be found as easily as above for the special case $\mathcal{S} = \mathcal{H}(D_\rho, M)$. We introduce the following definition which will be used for quantification of identification error:

$$\delta_n[\mathcal{S}] := \sup_{\hat{h} \in \mathcal{S}} \inf_{\hat{p}_n \in \mathcal{P}_n} \|\hat{h} - \hat{p}_n\|_\infty. \tag{3.4}$$

By the results described above, for the special case, $\mathcal{S} = \mathcal{H}(D_\rho, M)$, $\delta_m[\mathcal{S}] = \delta_m(\mathcal{S}, \mathcal{H}_\infty)$. This will not be true in general. However, it is easy to verify that $\delta_m[\mathcal{S}] \geq \delta_m(\mathcal{S}, \mathcal{H}_\infty)$.

Clearly, certain restrictions on the prior information set $\mathcal{S}$ are needed for the problem of identification in $\mathcal{H}_\infty$ to have a solution. We will assume throughout the paper that the model set $\mathcal{S}$ is admissible in the following sense.

Definition 3.1. *A subset $\mathcal{S} \subset \mathcal{A}$ is said to be admissible if*

$$1.\ M_s := \sup\left\{\|\hat{h}\|_\infty : \hat{h} \in \mathcal{S}\right\} < \infty; \qquad 2.\ \lim_{n\to\infty} \delta_n[\mathcal{S}] = 0.$$

The following proposition reveals an intrinsic property of admissible sets [3].

Proposition 3.1. *A model set $\mathcal{S} \subset \mathcal{A}$ is admissible if and only if $\mathcal{S}$ is totally bounded in $\mathcal{A}$.*

Thus, admissibility is a sort of compactness assumption on the set of unknown systems $\mathcal{S}$. In what follows we will consider classes of linear algorithms and nonlinear algorithms for identification in $\mathcal{H}_\infty$ and their engineering applications.

4. Linear Algorithms

For the experimental data in (2.1), define its (inverse) DFT coefficients by

$$f_i(E^N) = \frac{1}{N} \sum_{k=0}^{N-1} E_{k+1}^N e^{j2ik\pi/N}, \tag{4.1}$$

where i is any integer, and $j = \sqrt{-1}$. Intuitively, we think of f_i as an estimate of the pulse response coefficient h_i. If the noise is zero, i. e., $\epsilon = 0$ and as $N \to \infty$, then it follows that $f_i \to h_i$. These simple observations motivate some linear algorithms for the problem of identification in $\mathcal{H}_\infty$.

A class of linear algorithms consists of the identified model based on weighted partial summation

$$\hat{h}_{id}^{n} := \sum_{k=0}^{n-1} w_{n,k} f_k(E^N) z^k, \tag{4.2}$$

where $n \leq N$ and $w_n = \{w_{n,k}\}_{k=0}^{n-1}$ is a suitable "window function", independent of the *a priori* information. It is assumed that n is a function of N and satisfies $\lim_{N\to\infty} n(N) = \infty$. For simplicity this functional relation will not be shown explicitly. Basically, the window function scales our estimates f_i of the pulse response of the unknown system. The choice of the window function $w_{n,k}$ and $n(N)$ completely determines the algorithm. Moreover, it is obvious that the identified model is a **linear function** of the data E^N.

4.1 The Kernel Function

We associate for each window function a kernel

$$K_n(\omega) = \sum_{k=0}^{n-1} w_{n,k} e^{jk\omega}.$$

Define the *discretized convolution* in the frequency domain by

$$(K_n \circledast \hat{g})(\omega) := \frac{1}{N} \sum_{i=0}^{N-1} \hat{g}(e^{j2k\pi/N}) K_n(\omega - \frac{2i\pi}{N}).$$

Now some simple calculations show that

$$\hat{h}_{id}^{n}(e^{j\omega}) = K_n \circledast \left(\hat{h} + \hat{\eta}\right).$$

The linearity of the algorithm is manifested in the decomposition

$$\hat{h}_{id}^{n} = \hat{h}_{n,N} + \hat{\eta}_{n,N}, \tag{4.3}$$

where $\hat{h}_{n,N} = K_n \circledast \hat{h}$ and $\hat{\eta}_{n,N} = K_n \circledast \hat{\eta}$. As a simple consequence, we get the inequality

$$\|\hat{h} - \hat{h}_{id}^{n}\|_\infty \leq \|\hat{h} - \hat{h}_{n,N}\|_\infty + \|\hat{\eta}_{n,N}\|_\infty \tag{4.4}$$

which implies that the identification error for each $\hat{h} \in \mathcal{S}$ is bounded by a sum of two components: *the approximation error* $\|\hat{h} - \hat{h}_{n,N}\|_\infty$ and *the noise error* $\|\hat{\eta}_{n,N}\|_\infty$.

4.2 Error Analysis

Next we give a general procedure for analyzing the worst-case identification error defined in (2.3). This analysis is based on the concept of n-width explained previously.

Using the definition in (3.4), each $\hat{h} \in \mathcal{S}$ can be written as a sum of a polynomial approximation and a residual as follows:

$$\hat{h} = \hat{p}_n^*[\hat{h}] + \hat{\zeta}[\hat{h}], \;\; \hat{p}_n^*[\hat{h}] \in \mathcal{P}_n, \;\; \|\hat{\zeta}[\hat{h}]\|_\infty \leq \delta_n[\mathcal{S}]. \tag{4.5}$$

For each window function $w_{n,k}$, define

$$C_w^n = \frac{1}{N} \sup_{\omega \in [0,2\pi/N]} \sum_{i=0}^{N-1} |K_n(\omega = -2i\pi/N)|$$

and

$$C_w = \limsup \{C_w^n, n(N) > 0\}.$$

The following result demonstrates the impact of the window function on the worst-case identification error.

Theorem 4.1. *The worst-case identification error (2.3) for the linear algorithm represented by (4.2) satisfies*

$$e_N(\epsilon, \mathcal{S}) \leq (1 + C_w^n)\delta_n[\mathcal{S}] + C_w^n \epsilon + \sup_{\hat{h} \in \mathcal{S}} \|K_n \circledast \hat{p}_n^*[\hat{h}] - \hat{p}_n^*[\hat{h}]\|_\infty. \tag{4.6}$$

Proof. Using (4.4), for each $\hat{h} \in \mathcal{S}$, we have

$$\|\hat{h} - \hat{h}_{id}^n\|_\infty \leq \|\hat{h} - \hat{h}_{n,N}\|_\infty + \|\hat{\eta}_{n,N}\|_\infty.$$

Next, using (4.5), we can write

$$\|\hat{h} - \hat{h}_{n,N}\|_\infty \leq \|\hat{h} - \hat{p}_n^*[\hat{h}]\|_\infty + \|\hat{h}_{n,N} - \hat{p}_n^*[\hat{h}]\|_\infty.$$

Also,

$$\hat{h}_{n,N} = K_n \circledast \hat{p}_n^*[\hat{h}] + K_n \circledast \hat{\zeta}[\hat{h}].$$

Observe that

$$\|K_n \circledast \hat{g}\|_\infty \leq C_w^n \|\hat{g}\|_\infty.$$

Using the fact that $\|\hat{\zeta}[\hat{h}]\|_\infty \leq \delta_n[\mathcal{S}]$, and combining the above inequalities leads to

$$\|\hat{h} - \hat{h}_{n,N}\|_\infty \leq (1 + C_w^n)\delta_n[\mathcal{S}] + \|K_n \circledast \hat{p}_n^*[\hat{h}] - \hat{p}_n^*[\hat{h}]\|_\infty.$$

Taking supremum on both sides leads to the desired result.

Now let us consider the rectangular window function

$$w_{n,k} = 1, 0 \le k \le n, \quad \text{and} \quad 0 \quad \text{for} \quad k > n$$

as studied in [12, 32]. Then, it is not difficult to see that $K_n \circledast \hat{p}_n^*[\hat{h}] = \hat{p}_n^*[\hat{h}]$. The next result follows immediately from this observation.

Corollary 4.1. *Let the window function be given by* $w_{n,k} = 1$ *for* $0 \le k \le n-1$ *and zero elsewhere. Suppose* $\mathcal{S} = \mathcal{H}(D_\rho, M)$. *Then the linear algorithm in (4.2) gives worst case identification error*

$$e_N(\epsilon, \mathcal{S}) \le (1 + C_w^n) M \rho^{-n} + C_w^n \epsilon.$$

Next we give some bounds on C_w for the rectangular window. By [52, page 67 of vol. 1], we have that as $N \to \infty$, $C_w^n = L_n$, a Lebesgue's constant. Combining this result with Gronwall's inequality, we can conclude that [13]

$$C_w \le \alpha \log(n) + \beta, \quad \frac{4}{\pi^2} \le \alpha \le \frac{2}{\pi}$$

where β is an absolute constant. Thus a new error bound for the linear algorithm with rectangular window is given by

$$e_N(\epsilon, \mathcal{S}) \le (1 + \alpha \log(n) + \beta) M \rho^{-n} + (\alpha \log(n) + \beta) \epsilon. \tag{4.7}$$

Comparing this against the earlier error bound in [13]:

$$e_N(\epsilon, \mathcal{S}) \le \frac{M(\rho^{-n+1} + \rho^{-N+1})}{\rho - 1} + (\alpha \log(n) + \beta) \epsilon, \tag{4.8}$$

we conclude that the error bound in (4.7) is often more attractive than (4.8) especially if ρ is very close to one as in the case of lightly damped system.

A similar result can also be established for the triangular window as studied in [13].

Corollary 4.2. *Let* $w_{n,k} = 1 - \frac{k}{n}, k = 0, 1, \ldots, n; w_{n,k} = 0$ *for* $k \ge n$ *be the one-sided triangular window and* $\mathcal{S} = \mathcal{H}(D_\rho, M)$. *Then the linear algorithm in (4.2) gives worst case identification error*

$$\begin{aligned} e_N(\epsilon, \mathcal{S}) \quad \le \quad & (1 + C_w) M \rho^{-n} + C_w \epsilon \\ & + M \frac{\rho^2(1 - \rho^{-(n+1)}) - (n+1)(\rho - 1)\rho^{-n+1}}{n(\rho - 1)^2}, \end{aligned}$$

where $C_w = \frac{\log(n)}{\pi} + b$ *and* b *is an absolute constant.*

This result is obtained by combining Theorem 4.1 and the results in [13]. It is interesting to compare it with Corollary 4.1. While the noise error is smaller for the triangular window (the coefficient of $\log(n)$ is $1/\pi$ as compared with $2/\pi$), the last term of the worst-case identification error in (4.6) is nonzero thereby increasing the approximation error.

Using a rather clever argument, Partington [34] has shown that there does not exist a robustly convergent linear algorithm. Therefore, we turn our attention next to nonlinear algorithms.

5. Nonlinear Algorithms

A class of nonlinear algorithms for the problem of identification in $\mathcal{H}_\infty$ is based on modifications of the linear algorithm from last section. Let $\{f_i(E^N)\}$ be the inverse DFT coefficients as defined in (4.1). Obviously, this is a periodic sequence with period N. The following type of algorithms will be discussed in this section.

5.1 Two-Stage Nonlinear Algorithm

We begin by describing a family of nonlinear algorithms for solving the problem of identification in $\mathcal{H}_\infty$. All of these algorithms have a two-stage structure as explained below. Many of the algorithms in the existing literature fit this two-stage structure. This includes algorithms proposed by Helmicki, Jacobson, and Nett, Makila, Partington, and ourselves.

– Stage 1: Set pre-identified model $\hat{h}^n_{pi}$ by

$$\hat{h}^n_{pi}(\lambda) := \sum_{k=-n+1}^{n-1} w_{n,k} f_k(E^N) z^k \tag{5.1}$$

for some two-sided window function w_n.

– Stage 2: Take identified model $\hat{h}^n_{id}$ as

$$\hat{h}^n_{id} := \text{argmin}\left\{\|\hat{h}^n_{pi} - \hat{g}\|_\infty : \hat{g} \in \mathcal{H}_\infty(D)\right\}. \tag{5.2}$$

It is noted that Stage 2 involves solving the Nehari best approximation problem. This is necessitated by the fact that the pre-identified model $\hat{h}^n_{pi}$ is not necessarily analytic inside the unit disc as it may have nonzero negative Fourier coefficients. Using the standard Nehari theorem [30], the approximation error induced in the second stage is no larger than the Hankel norm of the anticausal part of $\hat{h}^n_{pi}$. Therefore the performance of the two-stage algorithm is critically dependent on the window function in the first stage which is the only *free design parameter* in the two-stage nonlinear algorithm.

We associate for each two-sided window function a kernel

$$K_n(\omega) = \sum_{k=-n+1}^{n-1} w_{n,k} e^{jk\omega}$$

and define C_w as in the previous section. Then, the pre-identified model in Stage 1 can also be written as the discretized convolution:

$$\hat{h}^n_{pi} = K_n \circledast \left(\hat{h} + \hat{\eta}\right).$$

Using the decomposition (4.5) for $\hat{h} \in \mathcal{S}$, we have that

$$\hat{h}_{pi}^n = K_n \circledast \hat{p}_n^*[\hat{h}] + K_n \circledast \left(\hat{\eta} + \hat{\zeta}\right).$$

If $2n \leq N+1$, the function $K_n \circledast \hat{p}_n^*[\hat{h}]$ is analytic, and thus the anticausal component of $\hat{h}_{pi}^n$ comes from the term $K_n \circledast \left(\hat{\eta} + \hat{\zeta}\right)$ and is insignificant if n is large and ϵ is small.

The next result indicates a general procedure for the quantification of the worst-case identification error.

Theorem 5.1. *Let the worst-case identification error at the first stage be defined by*

$$e_N^{pi}(\epsilon, \mathcal{S}) := \sup_{\hat{h} \in \mathcal{S},\, \hat{\eta} \in B_N(\epsilon)} \|\hat{h} - \hat{h}_{pi}^n\|_\infty.$$

Let $N+1 \geq 2n$. Then

$$e_N^{pi}(\epsilon, \mathcal{S}) \leq \delta_n[\mathcal{S}] + C_w\left(\epsilon + \delta_n[\mathcal{S}]\right) + \sup_{\hat{h} \in \mathcal{S}} \|K_n \circledast \hat{p}_n^*[\hat{h}] - \hat{p}_n^*[\hat{h}]\|_\infty.$$

Furthermore, the worst-case identification error for the two-stage algorithm satisfies

$$\begin{aligned} e_N(\epsilon, \mathcal{S}) &\leq e_N^{pi}(\epsilon, \mathcal{S}) + C_w\left(\epsilon + \delta_n[\mathcal{S}]\right) \\ &\leq \delta_n[\mathcal{S}] + 2C_w\left(\epsilon + \delta_n[\mathcal{S}]\right) + \sup_{\hat{h} \in \mathcal{S}} \|K_n \circledast \hat{p}_n^*[\hat{h}] - \hat{p}_n^*[\hat{h}]\|_\infty, \end{aligned}$$

where C_w is defined as in Theorem 4.1.

It is noted that the worst-case identification error for the class of nonlinear algorithms is similar to the class of linear algorithms except an additional term $C_w\left(\epsilon + \delta_n[\mathcal{S}]\right)$ due to the the anticausal component in $K_n \circledast \left(\hat{\eta} + \hat{\zeta}\right)$.

While C_w^n diverges as $n \to \infty$ for one-sided window functions, it can be made uniformly bounded if the window function is two-sided thereby guaranteeing the convergence of the nonlinear algorithm. In fact such convergence property is robust and independent of the *a priori* information. An important problem is thus the characterization of the robust convergence in terms of the window function. This question was resolved in [14, 16]. Our first result is taken from [14] where robust convergence is characterized in the time-domain.

Theorem 5.2. *Suppose that the window function $w_{n,k}$ is even symmetric with respect to k with $n < N/2$. Then, Stage 1 of the two-stage nonlinear algorithm is robustly convergent if the window function $w_{n,k}$ satisfies*

(i) $\lim_{n\to\infty} \Delta^2 w_{n,k} = 0$ *for* $k = 0, 1, \ldots$;

(ii) $N_s := \limsup \left\{ N_n := n|\Delta w_{n,n-1}| + \sum_{k=0}^{n-2} (k+1)|\Delta^2 w_{n,k}| : n \geq 0 \right\} < \infty$;

(iii) $\lim_{n\to\infty} w_{n,0} = 1$.

Consequently, if the above conditions hold, then the two-stage nonlinear identification algorithm is robustly convergent.

5.2 Convex and Concave Windows

It is not difficult to see that for symmetric windows, the inequality $C_w \leq N_s$ holds. For some commonly used window functions, the quantity N_s, an upper bound of C_w, can be explicitly calculated. The following notions of convex and concave windows are useful.

Definition 5.1. *Let the two-side window function be symmetric. The window function $w_{n,k}$ is called convex at k if $\Delta^2 w_{n,k} \geq 0$, and called a convex function if $\Delta^2 w_{n,k} \geq 0$ for all $k \geq 0$; The window function $w_{n,k}$ is called concave at k, if $\Delta^2 w_{n,k} \leq 0$ and called a concave function if $\Delta^2 w_{n,k} \leq 0$ for all $k \geq 0$; The window function $w_{n,k}$ is called non-increasing if $\Delta w_{n,k} \geq 0$ for all $k \geq 0$.*

With these notions of convex and concave window functions, we have the following result [14] which gives some simple and easy way to compute upper bounds.

Theorem 5.3. *Let the window function $w_{n,k}$ used in the two-stage algorithm be even symmetric which is non-increasing and $N > n > 0$. Then the following hold:*

(1) If $w_{n,k}$ is is a convex function, then

$$C_w \leq N_s = w_{n,0};$$

(2) If $w_{n,k}$ is is a concave function, then

$$C_w \leq N_s = 2nw_{n,n-1} - w_{n,0};$$

(3) If $w_{n,k}$ is convex for $k < m(< n)$ and concave for $k \geq m$, then

$$C_w \leq N_s \leq w_{n,0} - 2(m+1)w_{n,m} + 2mw_{n,m+1} + 2nw_{n,n-1};$$

(4) If $w_{n,k}$ is concave for $k < m(< n)$ and convex for $k \geq m$, then

$$C_w \leq N_s \leq 2(m+1)w_{n,m} - 2mw_{n,m+1} - w_{n,0}.$$

5.3 Frequency Domain Analysis

We comment that while $C_w \leq N_s$, in some cases even though C_w is finite, N_s is infinity. Hence the conditions in Theorem 5.2 are sufficient but not necessary for robust convergence of the two stage nonlinear algorithms. The next result [16] characterizes the robust convergence of the two-stage algorithm in frequency domain. In other words, conditions are given in terms of the Fourier transform of the window function.

Theorem 5.4. *Suppose that the window function $w_{n,k}$ is even symmetric with respect to k with $n < N/2$. Then, the two-stage nonlinear algorithm is robustly convergent if*

1. $\lim_{N\geq 2n\to\infty} \frac{1}{N}\sum_{i=0}^{N-1} K_n(2i\pi/N)\, e^{-jk\frac{2i\pi}{N}} = 1$ *for* $k = 0, \pm1, \pm2, \ldots$; *and*
2. $C_w < \infty$.

Conversely, if the first stage of the two-stage nonlinear algorithm is robustly convergent then conditions 1 and 2 are satisfied.

The proof of the converse statement involves construction of a noise sequence which is conjugate symmetric [16]. Although the class of window functions satisfying the conditions in Theorem 5.4 is larger than the corresponding class in Theorem 5.3, it is difficult to compute or estimate C_w. However, it will be seen in the next section, that for an important class of windows called trapezoidal windows, Theorem 5.4 gives a better estimate than Theorem 5.3.

5.4 Trapezoidal Window

In light of Theorem 5.1, the worst-case identification error bound can be written as

$$e_N(\epsilon, \mathcal{S}) \leq 2C_w\epsilon + 2\,(1 + C_w)\,\delta_n[\mathcal{S}] + \sup_{\hat{h}\in\mathcal{S}} \|K_n \circledast \hat{p}_n^*[\hat{h}] - \hat{h}\|_\infty. \tag{5.3}$$

The first term is the error induced by measurement noise and the rest of the terms grouped together will be called the approximation error.

In attempting to optimize the window function, we have found that there is a trade-off between the noise error and the approximation error. For example, consider the following facts:

- For the one-sided rectangular window, the last term in the error expression vanishes but the corresponding C_w is not bounded.
- On the other hand, the two-sided triangular window has the smallest possible value for C_w, namely one, but the corresponding approximation error is $\mathcal{O}(\frac{1}{n})$ [13].

Hence, an attractive choice is a *trapezoidal window:* a combination of the triangular and rectangular windows. A special trapezoidal window associated with the de la Vallee kernel [52] was used by Partington [34]. We will employ following parameterized trapezoidal window [14, 15]:

$$w_{n,k} = \begin{cases} 1 & 0 \leq k \leq m-1, \\ 1 + \frac{k}{n}, & -n \leq k \leq -1, \\ 1 - \frac{k-m+1}{n} & m \leq k \leq n+m-1, \\ 0 & \text{elsewhere}, \end{cases} \tag{5.4}$$

where $n+m < N$. Clearly $m = 1$ corresponds to two-sided triangular window while $n = 1$ corresponds to one-sided rectangular window. The above window has two additional advantages. First the free integer parameter m can be used to yield a right trade-off between the two error terms. Second it gives more weighting on the causal part of the pre-identified model. It is noted that if m is odd, then the shifted window

$$\tilde{w}_{n,k} = w_{n,k+(m-1)/2}$$

is symmetric about $k = 0$. Because shifting does not change the value of C_w, the worst-case identification error bound in (5.3) remains valid provided that $n + m < N$. Moreover, we have the following result.

Theorem 5.5. *For the trapezoidal window in (5.4), the pre-identified model is interpolatory if $N = n + m - 1$:*

$$\hat{h}_{pi}^n(W_N^k) = E_k^N, \quad k = 0, 1, ..., N-1.$$

Moreover $C_w \leq \sqrt{N/n}$ [16]. Thus the resulting worst-case identification error for the two-stage nonlinear algorithm is bounded by

$$e_N(\epsilon, \mathcal{S}) \leq 2\epsilon\sqrt{N/n} + \left(1 + 2\sqrt{N/n}\right)\delta_m[\mathcal{S}].$$

In particular if n, m are such that $N = 2n$ and $m = 1 + N - n = 1 + N/2$, and $\mathcal{S} = \mathcal{H}(D_\rho, M)$, then

$$e_N(\epsilon, \mathcal{S}) \leq 2\sqrt{2}\epsilon + \left(1 + 2\sqrt{2}\right) M\rho^{-(1+N/2)}.$$

Remark 5.1. It is known that the interpolation algorithm in [6, 17] is optimal within a factor of two. However it can only be shown in [17] that the resulting worst-case identification error for the interpolation algorithm is no larger than twice of $e_N^{pi}(\epsilon, \mathcal{S})$. In light of Theorem 5.1, the resulting identification error for the two-stage algorithm is strictly smaller than twice of $e_N^{m}(\epsilon, \mathcal{S})$.

6. Engineering Applications

While algorithms are often designed to have attractive theoretical properties, they often require some heuristics for successful application to real engineering problems. Identification in $\mathcal{H}_\infty$ is no exception to this. In two recent papers [9, 10], a set of heuristics which can guide the use of these algorithms by other engineers have been presented. These heuristics are built from the experience we gained while applying the algorithms for identification in $\mathcal{H}_\infty$ to experimental data sets, which are discussed below.

In [15], we attempted to apply the two stage nonlinear algorithms to a frequency response data set from a Jet Propulsion Laboratory flexible structure.

This data was supplied to us by Dr. David Bayard. Unfortunately, at first we could only obtain satisfactory results by using a combination of the classical Sanathanan-Koerner (SK) iteration followed by the two-stage nonlinear algorithm. SK iteration is an iterative procedure to solve a nonlinear least squares problem of fitting the frequency response data by a parameterized transfer function.

However in [9, 10], we developed some new heuristics and re-applied the two-stage nonlinear algorithm alone to the same JPL data set and to the frequency response data of a flexible pointing system test bed at the Automation and Robotics Laboratory at the Armament Research Development and Engineering Center (ARDEC), Picatinny Arsenal, NJ. (This test bed was the subject of two invited sessions at the 1993 American Control Conference.) This time the results were much more promising. The interested reader should consult [9, 10] for a detailed discussion of these applications.

The results of the identification of these two flexible systems are shown in Figures 6.1 - 6.4. Since the ARDEC system has only 3 or 4 moderately damped modes, compared to 13 lightly damped modes in the JPL structure, we considered it to be the easier of the two structures to identify and we will begin our discussion with it.

We begin by examining the inverse Fourier coefficients of the ARDEC structure computed in (4.1) to make an educated guess at the window parameters m and n. The identified model using this approach was roughly $670th$ order, which is too large to be of an practical use since the true system only has 3 or 4 modes. Thus we use the combination of FIR balanced model reduction procedure to arrive at a 13th order model. (This is the order of the linear model supplied to us by the Automation and Robotics Lab.)

In Fig. 6.1, the magnitudes and phases of the reduced order two-stage and SK models are compared to the linear model and the nonlinear model supplied to us by the Automation and Robotics Lab. Note that the linear models are all 13th order and the nonlinear model order is greater than 13. The model mismatch data presented in Fig. 6.2 shows that the model produced by the two-stage and SK algorithms do a better job capturing the natural frequencies of the structure as well as the damping ratios when compared to both the linear and nonlinear simulation data. The reason for the larger model error can be seen by examining Fig. 6.1a and Fig. 6.1b. For example, restricting our attention to the error in modeling the first mode (20 – 40 rad/sec), the natural frequency of the nonlinear model is slightly greater than the true natural frequency, and the damping ratio of the linear model is much larger than the true damping ratio. Similar observations can be made to explain the modeling error at the other modes. Note that the SK algorithm seems to outperform the two-stage algorithm when the model mismatches are compared in Fig. 6.2. However, the SK model is unstable, while the system is known to be stable!! This demonstrates one of the problems with parameter optimization algorithms such as the SK algorithm: the identified model of a

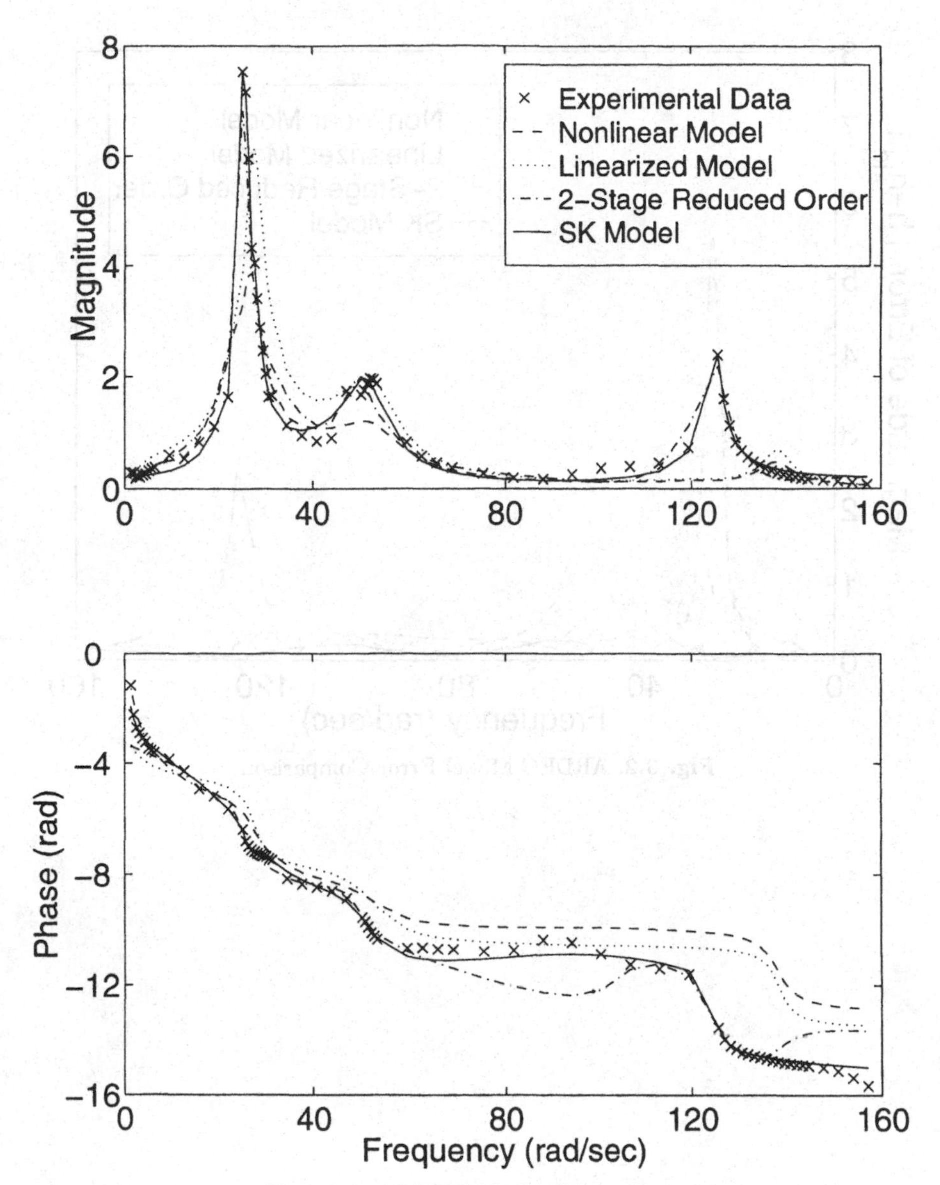

Fig. 6.1. ARDEC Model Comparison

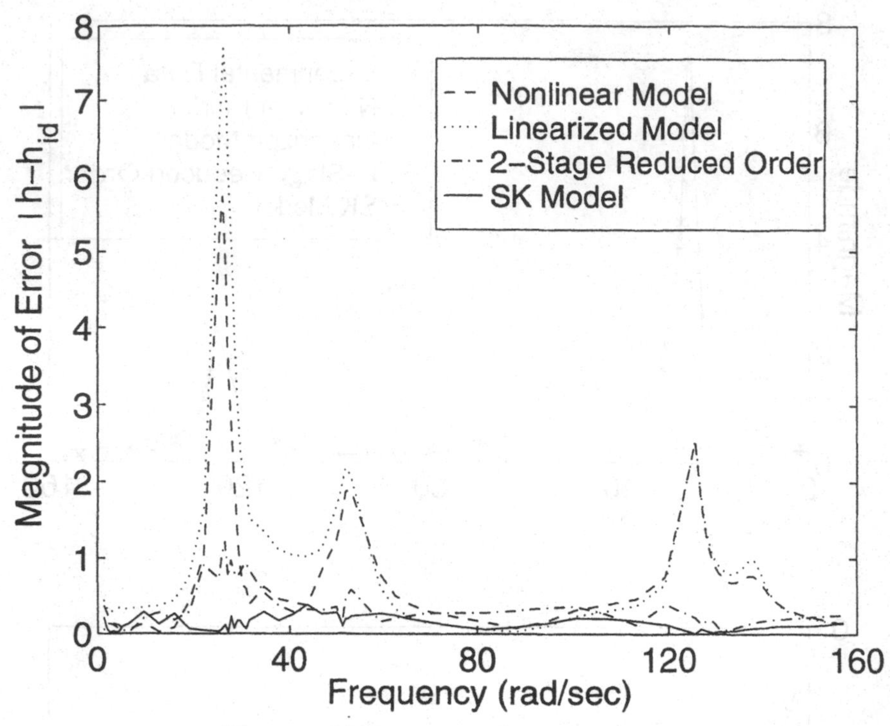

Fig. 6.2. ARDEC Model Error Comparison

stable system may be unstable. On the other hand, the two-stage algorithm will always produce stable models.

Unlike the ATB1000, the ARC structure has very lightly damped, closely spaced flexible modes (see the frequency response data in Fig. 6.3). By examining the inverse Fourier coefficients we determined that the best approach is to use the linear algorithm with $n = N$ and $w_{n,k} = 1$ for all k. This of course, allows us to exactly match the experimental data. Note that this choice is reasonable in light of the fact that very little noise appears in the experimental data. As with the ARDEC model, this model is too large for most engineering applications. Thus, since the model is FIR, we can use FIR balanced truncation model reduction [12] to arrive at a $30th$ order model, shown in Fig. 6.3 and Fig. 6.4. Notice in Fig. 6.4 that, despite our initial misgivings, the worst-case model errors of equal order SK and reduced two-stage models are similar. In fact, as might be expected, the two-stage model has a smaller worst-case model mismatch, while the SK model has a smaller average model mismatch.

The results in [9, 10] have turned out to be quite satisfactory. We have shown that our algorithms produce very good models with very standard computational tools such as the FFT and SVD algorithms. Based on this experience, we feel that the work in identification in $\mathcal{H}_\infty$ has reached a certain level of maturity, and can deal with challenging applications problems.

Acknowledgment

This work was supported in part by Airforce Office of Scientific Research under contract no. F-49620-93-1-0246DEF, F49620-94-1-0415DEPSCoR, the Army Research Office under grant no. DAAH04-93-G-0012, and a National Science Foundation Fellowship.

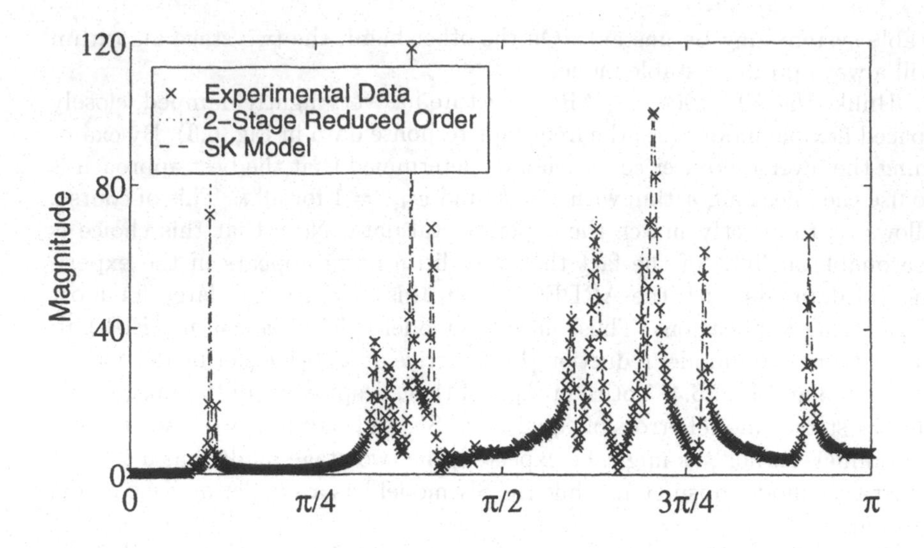

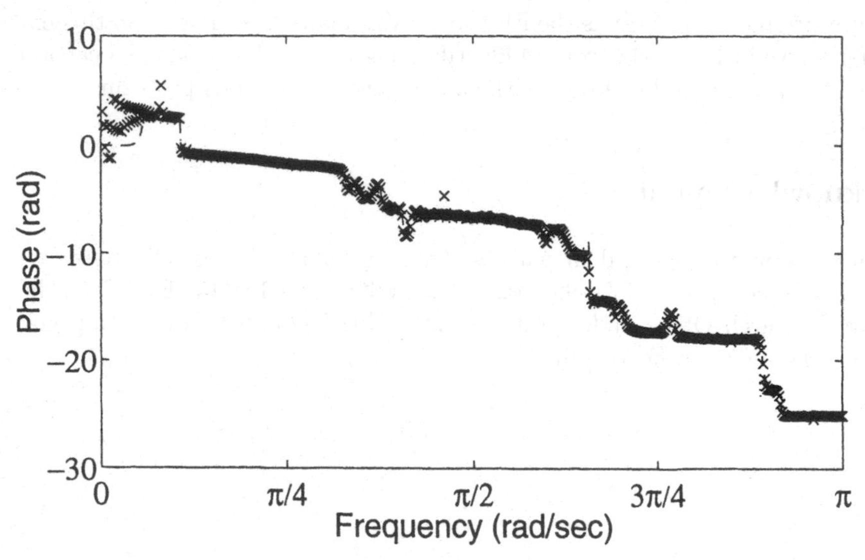

Fig. 6.3. JPL Model Comparison

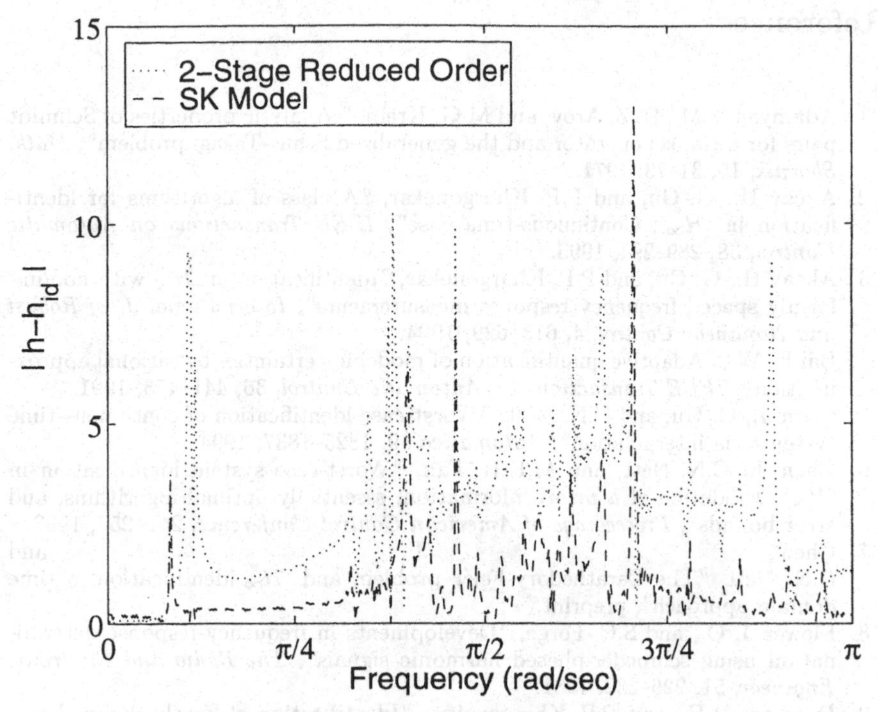

Fig. 6.4. JPL Model Error Comparison

References

1. Adamyan V.M., D. Z. Arov, and M.G. Krein, " Analytic properties of Schmidt pairs for a Hankel operator and the generalized Schur-Takagi problem", *Math. Sbornik*, 15, 31–73, 1971.
2. Akçay H., G. Gu, and P.P. Khargonekar, "A class of algorithms for identification in $\mathcal{H}_\infty$: Continuous-time case", *IEEE Transactions on Automatic Control*, 38, 289–294, 1993.
3. Akçay H., G. Gu, and P.P. Khargonekar, "Identification in $\mathcal{H}_\infty$ with nonuniformly spaced frequency response measurements", *International J. of Robust and Nonlinear Control*, 4, 613–629, 1994.
4. Bai E.-W., "Adaptive quantification of model uncertainties by rational approximation", *IEEE Transactions on Automatic Control*, 36, 441–453, 1991.
5. Chen J., G. Gu, and C.N. Nett, "Worst-case identification of continuous-time systems via interpolation", *Automatica* 39, 1825–1837, 1994.
6. Chen J., C.N. Nett, and M.K.H. Fan, "Worst-case system identification in $\mathcal{H}_\infty$: validation of *a priori* information, essentially optimal algorithms, and error bounds", *Proceedings of American Control Conference*, 251–257, 1992.
7. Chen J., and C.N. Nett, "The Carathédory-Fejér problem and $\mathcal{H}_\infty$ identification: a time domain approach", preprint.
8. Flower J. O., and S.C. Forge, "Developments in frequency-response determination using schroeder-phased harmonic signals", *The Radio and Electronic Engineer*, 51, 226–232, 1981.
9. Friedman J.H., and P.P. Khargonekar, "Identification of lightly damped systems using frequency domain techniques", *Proc. IFAC Conference on System Identification*, 201–206, 1994.
10. Friedman J.H., and P.P. Khargonekar, "Application of Identification in $\mathcal{H}_\infty$ to Lightly Damped Systems: Two Case Studies", *IEEE Transactions on Control Systems Technology*, Accepted for publication.
11. Goodwin G.C., M. Gevers, and B. Ninness, "Quantifying the error in estimated transfer functions with application to model selection", *IEEE Transactions on Automatic Control*, 37, 913–928, 1992.
12. Gu G., P.P. Khargonekar, and E.B. Lee, "Approximation of infinite dimensional systems", *IEEE Transactions on Automatic Control*, AC-34, 610–618, 1989.
13. Gu G., and P.P. Khargonekar, "Linear and nonlinear algorithms for identification in $\mathcal{H}_\infty$ with error bounds", *IEEE Transactions on Automatic Control*, 37, 953-963, 1992.
14. Gu G., and P.P. Khargonekar, "A class of algorithms for identification in $\mathcal{H}_\infty$ ", *Automatica*, 28, 199–312, 1992.
15. Gu G., and P.P. Khargonekar, "Frequency domain identification of lightly damped systems: The JPL example", *Proc. of American Control Conference*, 3052–3057, 1993.
16. Gu G., P.P. Khargonekar, and Y. Li, "Robust convergence of two-stage nonlinear algorithms for system identification in $\mathcal{H}_\infty$ ", *Systems and Control Letters*, 18, 253–263, 1992.
17. Gu G., D. Xiong, and K. Zhou, "Identification in $\mathcal{H}_\infty$ using Pick's interpolation" *Systems and Control Letters*, 20, 263–272, 1993.
18. Hakvoort R.G., "Worst-case system identification in $\mathcal{H}_\infty$: Error bounds, interpolation, and optimal models", Internal Report, Delft University of Technology, The Netherlands. (1992)

19. Helmicki A.J., C.A. Jacobson, and C.N. Nett, "Control oriented system identification: a worst-case/deterministic approach in $\mathcal{H}_\infty$ ", *IEEE Transactions on Automatic Control*, 36, 1163–1176, 1991.
20. Helmicki A.J., C.A. Jacobson, and C.N. Nett, "Worst-case/deterministic identification in $\mathcal{H}_\infty$: The continuous-time case", *IEEE Transactions on Automatic Control*, 37, 604–610, 1992.
21. Helmicki A.J., C.A. Jacobson, and C.N. Nett, "Identification in $\mathcal{H}_\infty$: linear algorithms", *Proc. American Control Conference*, 2418–2423,1990.
22. Kosut R., and H. Ailing, "Worst case control design from batch least squares", *Proc. American Control Conference*, 318–322,1992.
23. Krause J.M., and P.P. Khargonekar, "Parameter identification in the presence of non-parametric dynamic uncertainty", *Automatica*, 26, 113–124, 1990.
24. Lau M., R.L. Kosut, and S. Boyd, "Parameter set identification of systems with uncertain nonparametric dynamics and disturbances", *Proc. 29th IEEE Conference on Decision and Control*, 3162–3167, 1990.
25. Lee W., B.D.O. Anderson, and R. Kosut, "On adaptive robust control and control relevant identification", *Proc. American Control Conference*, 1992.
26. Lin L., L. Wang, and G. Zames, "Uncertainty principles and identification n-widths for LTI and slowly varying systems". *Proc. American Control Conference*, 296–300, 1992.
27. Ljung L., *System Identification, Theory for the User*, Englewood Cliffs, NJ: Prentice-Hall, 1987.
28. Mäkilä P.M., and J.R. Partington, "Robust identification of strongly stabilizable systems", *IEEE Transactions on Automatic Control*, 37, 1709–1716, 1992.
29. Mo S.H., and J.P. Norton, "Parameter-bounding identification algorithms for bounded-noise records", *IEE Proceedings Part D, Control Theory and Applications*, 135, 127–132, 1988.
30. Nehari Z., "On bounded linear forms" *Annals of Mathematics*, 65, 153–162, 1957.
31. Ninness B.M., "Stochastic and deterministic estimation in $\mathcal{H}_\infty$ ", *Proc. 33rd IEEE Conference on Decision and Control*, 62–67, 1993.
32. Parker P.J., and R.R. Bitmead, "Adaptive frequency response identification", *Proc. 28th IEEE Conference on Decision and Control*, 348–353, 1987.
33. Partington J.R., "Robust identification in $\mathcal{H}_\infty$ ", *J. Mathematical Analysis and Applications*, 166, 428–441, 1992.
34. Partington J.R., "Robust identification and interpolation in $\mathcal{H}_\infty$ ", *International Journal of Control*, 54, 1281–1290, 1991.
35. Partington J.R., "Algorithms for identification in $\mathcal{H}_\infty$ with unequally spaced function measurements", *International Journal of Control*, 58, 21–32, 1993.
36. Partington J.R., "Interpolation in normed spaces from the values of linear functionals", *Bulletin of the London Mathematical Society*, 26, 165–170, 1994.
37. Pinkus A., *n-Widths in Approximation Theory*, Berlin: Springer 1985.
38. Poolla K., P.P. Khargonekar, A. Tikku, J. Krause, and K. Nagpal, "A time-domain approach to model validation", *IEEE Transactions on Automatic Control*, 39, 951–959, 1994.
39. Poolla K., and A. Tikku, "Time complexity of worst-case identification" *IEEE Transactions on Automatic Control*, 39, 944–950, 1994.
40. Raman S., and E.W. Bai, "A linear, robust and convergent interpolatory algorithm for quantifying midel uncertainties", *Systems and Control Letters*, 18, 173–178, 1992.
41. Sanathanan C.K., and J. Koerner, "Transfer function synthesis as a ratio of two complex polynomials", *IEEE Transactions on Automatic Control*, AC-8, 56–58, 1963.

42. Schrama R.J.P., "Accurate models for control design: the necessity of an iterative scheme", *IEEE Transactions on Automatic Control*, 37, 991–993, 1992.
43. Smith R.S., and J.C. Doyle, "Towards a methodology for robust parameter identification", *IEEE Transactions on Automatic Control*, AC-37, 942–952, 1992.
44. Tempo R., "IBC: a working tool for robust parameter estimation", *Proc. American Control Conference*, 237–240, 1992.
45. Tse D.N.C., M.A. Dahleh, and J.N. Tsitsiklis, "Optimal Asymptotic Identification under Bounded Disturbances", *Proc. American Control Conference*, 1786–1787, 1991.
46. van den Boom T., "MIMO-systems identification for $\mathcal{H}_\infty$ robust control: A frequency domain approach with minimum error bounds", Ph. D. Thesis: Eindhoven University, The Netherlands.
47. Willems J.C., "From time series to linear system — Parts 1, 2, 3", *Automatica*, 22–23, 1986–87.
48. Younce R.C., and C.E. Rohrs, "Identification with non–parametric uncertainty", *IEEE Transactions on Automatic Control*, 37, 715–128, 1992.
49. Zames G., "On the metric complexity of causal linear systems: ϵ-entropy and ϵ-dimension for continuous time", *IEEE Transactions on Automatic Control*, AC-24, 222–230, 1979.
50. Zang Z., R. Bitmead, and M. Gevers, " $\mathcal{H}_\epsilon$ Iterative model refinement and control robustness enhancement", *Proc. 30th IEEE Conference on Decision and Control*, 279-284, 1991.
51. Zhou T., and H. Kimura, "Identification for robust control in time domain", *Systems and Control Letters*, 20, 167–178, 1993.
52. Zygmund A., *Trigonometric Series*, Cambridge, England: Cambridge University Press 1959.

System Identification with Information Theoretic Criteria

A.A. Stoorvogel[1] * and J.H. van Schuppen[2]

[1] Department of Mathematics and Computing Science
Eindhoven University of Technology
P.O. Box 513
5600 MB Eindhoven
The Netherlands.

[2] CWI
P.O. Box 94079
1090 GB Amsterdam
The Netherlands.

1. Introduction

The purpose of this paper is to clarify and to explore the relationship between several approximation criteria in system identification.

The original motivation for the research reported on in this paper was to clarify the relation between LEQG optimal stochastic control and H_∞ optimal control with an entropy criterion, and to explore the use of this relation in system identification. The investigation lead to a study of information theoretic criteria. The unified framework that appeared seems quite useful for the theory of system identification.

A motivating problem of this paper is the choice of an approximation criterion for system identification. Up to recently the approximation criteria used mainly in system identification of stochastic processes were:

(*i*) The likelihood function (maximum likelihood method).
(*ii*) A quadratic cost function (least squares method).

For stationary Gaussian processes these criteria are identical. The resulting algorithms for parameter estimation with these criteria are well known.

The approximation criteria considered in this paper are:

(*i*) Mutual information rate.
(*ii*) A weighted likelihood function and divergence rate between the measure associated with a simple model and that associated with an element in the model class.

In the following sections these criteria are discussed in detail.

---

* The research of dr. A.A. Stoorvogel has been made possible by a fellowship of the Royal Netherlands Academy of Sciences and Arts. His cooperation with the second author is supported in part by CWI.

It will be shown that the criterion of mutual information rate is related to that of the LEQG optimal stochastic control problem and to that of an H_∞ optimal control problem with an entropy criterion. Mutual information is related to the information theoretic concept of divergence. The divergence concept is the same as the Kullback-Leibler pseudo-distance, which in turn is related to the likelihood function.

The discussion of the approximation criteria is rather general. Thus the discussion is formulated in terms of stationary Gaussian processes but can easily be extended to other classes of processes.

In summary, in this paper several approximation criteria for system identification of stationary Gaussian processes will be introduced and their relationship will be discussed. Subsequently several parameter estimation problems are posed and solved.

The character of this paper is expository. The paper is primarily addressed to doctoral students in engineering, mathematics, and econometrics. The reader is assumed to be familiar with system and control theory, and with probability at a first year graduate level. The paper contains many definitions and elementary results that are known although not available in a concise and compact publication. The body of the paper contains the main results while the appendices contain details on concepts and theorems. A brief conference version of this paper appeared as [68].

A description of the contents by section follows. Section 2. contains a brief problem formulation. Parameter estimation by the approximation criterion of mutual information is the subject of Section 3. and by the approximation criterion of the likelihood function and divergence is the subject of Section 4. Section 5. presents concluding remarks. Appendix A. introduces concepts from probability theory and stochastic processes and Appendix B. concepts from system theory. Appendix C. contains definitions from information theory, Appendix D. formulas of information measures for Gaussian random variables, and Appendix E. formulas of information measures for stationary Gaussian processes. Appendix F. summarizes results for the LEQG optimal stochastic control problem and Appendix G. the H_∞ optimal control problem with an entropy criterion.

2. Problem Formulation

System identification is a topic of the research area of systems and control. The problem of system identification is to obtain from data a mathematical model in the form of a dynamic system for a phenomenon. Examples of applications are system identification of a wind turbine, of a gas boiler, and of the flow of benzo-a-pyreen through the human body.

System identification of a phenomenon often proceeds by the following procedure:

(*i*) Physical modeling and selection of a model class of dynamic systems.
(*ii*) Experiment design, data collection, and preprocessing of the data.
(*iii*) Check on the identifiability of the parametrization of the model class.
(*iv*) Selection of an element in the model class that is an approximation of the data.
(*v*) Evaluation of the model determined and, possibly, redoing one or more of the previous steps.

In this paper attention is restricted to step (iv) of the above procedure, the selection of an element in the model class that is an approximation of the data.

In this paper attention is restricted to phenomena that, after preliminary physical or domain modeling, can be modeled by a stationary Gaussian process. See the appendices for concepts and terminology used in the body of the paper. It is well known from stochastic realization theory for stationary Gaussian processes that such a process can under certain conditions be represented as the output of a Gaussian system in state space form

$$\begin{aligned} x(t+1) &= Ax(t) + Kv(t), \\ y(t) &= Cx(t) + v(t), \end{aligned}$$

where $v : \Omega \times T \to \mathbb{R}^p$ is a Gaussian white noise process. Such a process can also be represented by an ARMA (Auto-Regressive-Moving-Average) representation of the form

$$\sum_{i=0}^{n} a_i y(t-i) = \sum_{j=0}^{m} b_j v(t-j).$$

Below attention is sometimes restricted to an AR (Auto-Regressive) representation of the form

$$y(t) + \sum_{i=1}^{n} a_i y(t-i) = v(t). \tag{2.1}$$

With the notation

$$\theta = (-a_1, \ldots, -a_n)^{\mathrm{T}}, \quad \phi(t) = (y(t-1), \ldots, y(t-n))^{\mathrm{T}},$$

this representation may be rewritten as

$$y(t) = \phi(t)^{\mathrm{T}}\theta + v(t). \tag{2.2}$$

Problem 2.1. *Parameter estimation problem for time-invariant Gaussian systems.* Consider observations of a stationary Gaussian process with values in $\mathbb{R}^p$. Consider the model class of time-invariant Gaussian systems

$$\begin{aligned} x(t+1) &= Ax(t) + Kv(t), \\ y(t) &= Cx(t) + v(t), \end{aligned} \tag{2.3}$$

or, properly restricted, in AR representation,

$$y(t) = \phi(t)^{\mathrm{T}}\theta + v(t). \tag{2.4}$$

Select an element in the model class that approximates the observations by one or more of the approximation criteria mentioned in the following sections.

3. Approximation with Mutual Information

Successively this section presents the topics of mutual information, a parameter estimation problem, the relation between mutual information, LEQG, and H_∞ entropy, parameter estimation by LEQG optimal stochastic control, and parameter estimation by the H_∞ method.

3.1 Mutual Information

There follows an introduction to the concept of mutual information. For a detailed treatment and references see the Appendices C., D., and E..

The mutual information of two discrete valued random variables, say $x : \Omega \to X = \{a_1, \ldots, a_n\}$ and $y : \Omega \to Y = \{b_1, \ldots, b_m\}$, is defined by the formula

$$J(x,y) = \sum_{i=1}^{n}\sum_{j=1}^{m} p_{xy}(a_i,b_j) \ln\left(\frac{p_{xy}(a_i,b_j)}{p_x(a_i)p_y(b_j)}\right), \tag{3.1}$$

where

$$\begin{aligned} p_{xy}(a_i,b_j) &= P(\{x = a_i, y = b_j\}), \\ p_x(a_i) &= P(\{x = a_i\}), \qquad p_y(b_j) = P(\{y = b_j\}). \end{aligned}$$

The mutual information of two random variables describes the dependence between the associated probability measures. In general $0 \le J(x,y) \le +\infty$ and $J(x,y) = 0$ if and only if x and y are independent.

For random variables with continuous values the mutual information is defined as the limit of the mutual information of discrete valued random variables, in which the discrete random variables are required to approximate the continuous valued random variables. In case the random variables x, y are jointly Gaussian random variables with a strictly positive definite variance, or $(x,y) \in G(0,Q)$ with $Q = Q^{\mathrm{T}} > 0$ where

$$Q = \begin{pmatrix} Q_x & Q_{xy} \\ Q_{xy}^{\mathrm{T}} & Q_y \end{pmatrix},$$

then

$$J(x,y) = -\frac{1}{2}\ln\left(\frac{\det Q}{\det Q_x \det Q_y}\right). \tag{3.2}$$

For stationary processes the corresponding concept is the mutual information rate. Let $x : \Omega \times T \to \mathbb{R}^m$, $y : \Omega \times T \to \mathbb{R}^p$, be stationary stochastic processes on $T = \mathbb{Z}$. The *mutual information rate* of x and y is defined by the formula

$$J(x,y) = \limsup_{n\to\infty} \frac{1}{2n+1} J\left(x|_{[-n,n]}, y|_{[-n,n]}\right), \tag{3.3}$$

where $x|_{[-n,n]}$ denotes the random variable defined by the restriction of the process x to the interval $\{-n,\ldots,-1,0,1,\ldots,n\}$.

Let for $T = \mathbb{Z}$, $x : \Omega \times T \to \mathbb{R}^m$ and $y : \Omega \times T \to \mathbb{R}^p$ be two jointly stationary Gaussian processes that admit a spectral density. Denote their joint spectral density function by

$$\hat{W} : \mathbb{C} \to \mathbb{C}^{(m+p)\times(m+p)},\ \hat{W} = \begin{pmatrix} \hat{W}_x & \hat{W}_{xy} \\ \hat{W}_{xy} & \hat{W}_y \end{pmatrix}.$$

It follows from [24] that the mutual information of the processes x, y is invariant under scaling by nonsingular linear systems. Let $S_x : \mathbb{C} \to \mathbb{C}^{m\times m}$ and $S_y : \mathbb{C} \to \mathbb{C}^{p\times p}$ be minimal and square spectral factors of respectively $\hat{W}_x$ and $\hat{W}_y$, hence

$$\hat{W}_x(z) = S_x(z)S_x(z^{-1})^{\mathrm{T}},\ \ \hat{W}_y(z) = S_y(z)S_y(z^{-1})^{\mathrm{T}}.$$

Define

$$S(z) = S_x(z)^{-1}\hat{W}_{xy}(z)S_y(z^{-1})^{-T}.$$

The mutual information rate of these processes is given by the formula

$$J(x,y) = \frac{-1}{4\pi i}\int_{\mathbb{D}} \frac{1}{z}\ln\det[I - S(z^{-1})^{\mathrm{T}}S(z)]dz. \tag{3.4}$$

Assume that the function S admits a realization as a finite-dimensional linear system of the form

$$S(z) = C(zI-A)^{-1}B + D,\ \ sp(A) \subset \mathbb{C}^-, \tag{3.5}$$

and that the following set of relations admits a unique solution $Q \in \mathbb{R}^{n\times n}$

$$\begin{aligned} Q &= Q^{\mathrm{T}} \geq 0, && (3.6)\\ Q &= A^{\mathrm{T}}QA + C^{\mathrm{T}}C + (A^{\mathrm{T}}QB + C^{\mathrm{T}}D)N^{-1}(B^{\mathrm{T}}QA + D^{\mathrm{T}}C), && (3.7)\\ N &= I - D^{\mathrm{T}}D - B^{\mathrm{T}}QB > 0, && (3.8)\\ & sp(A + BN^{-1}(B^{\mathrm{T}}QA + D^{\mathrm{T}}C)) \subset \mathbb{C}^-. && (3.9) \end{aligned}$$

Then the mutual information rate of the processes x, y is given by the formula

$$J(x,y) = -\frac{1}{2}\ln\det(I - B^{\mathrm{T}}QB - D^{\mathrm{T}}D). \tag{3.10}$$

3.2 A Parameter Estimation Problem

Consider the model class of Gaussian systems in AR representation as stated in Section 2.,

$$\begin{aligned} \theta(t+1) &= \theta(t) + v(t), \;\; \theta(0) = \theta_0, \\ y(t) &= \phi(t)^{\mathrm{T}}\theta(t) + Nw(t), \end{aligned} \tag{3.11}$$

with $\theta_0 \in G(m_0, Q_0)$, $v(t) \in G(0, I)$, $w(t) \in G(0, I)$. The problem is to determine a linear recursive parameter estimator, say,

$$\hat{\theta}(t+1) = \hat{\theta}(t) + K(t)[y(t) - \phi(t)^{\mathrm{T}}\hat{\theta}(t)], \;\; \hat{\theta}(0) = m_0, \tag{3.12}$$

such that the estimation error

$$e : \Omega \times T \to \mathbb{R}^n, \;\; e(t) = \theta(t) - \hat{\theta}(t),$$

is small according to a criterion specified below. The recursion for the estimation error is then

$$e(t+1) = [I - K(t)\phi(t)^{\mathrm{T}}]e(t) + v(t) - K(t)Nw(t), \;\; e(0) = \theta_0 - \hat{\theta}_0.$$

Assume that e is such that there exists a Gaussian random process r, which is independent of e, such that $z := e + r$ is a standard white noise process, i.e. $z \in G(0, I)$.

Our objective is to determine a parameter estimator such that the mutual information rate $J(e, z)$ between e and z is minimized. This implies that the error signal e should be as small as possible. Clearly this measure is not very intuitive but it turns out to be closely related to LEQG and H_∞ parameter estimators.

In particular, assume e is asymptotically stationary with asymptotic density function R. The assumption regarding the existence of r and z implies

$$0 \leq R(z) \leq I.$$

It is then easy to show that the mutual information rate of the processes z and e is given by

$$\begin{aligned} J(z, e) &= \frac{-1}{4\pi i} \int_{\mathbb{D}} \frac{1}{z} \ln \det[I - R(z)] dz \\ &= \frac{-1}{4\pi i} \int_{\mathbb{D}} \frac{1}{z} \ln \det[I - S(z^{-1})^{\mathrm{T}} S(z)] dz \end{aligned}$$

where S is a spectral factor of R. Expressing the mutual information in terms of the spectral factor S will be useful later on.

Problem 3.1. Consider the parameter estimation setting defined above, with the observation equation and the parameter model as given in (3.11). Determine a parameter estimator of the form (3.12) that minimizes the mutual information rate (3.13) between the processes z and e.

Problem 3.1 will not be solved directly but via LEQG optimal stochastic control and via H_∞ optimal control theory.

3.3 Relation of Mutual Information, H_∞ Entropy, and LEQG Cost

An investigation of the authors has established that the criterion of mutual information of two stationary Gaussian processes is identical to the H_∞ entropy criterion and to the limit of the cost of a stochastic control problem with an exponential-of-quadratic cost function. This result is first stated as a theorem and then explained.

Theorem 3.1. *Consider a finite-dimensional linear system*

$$\begin{aligned} x(t+1) &= Ax(t) + Bu(t), \\ y(t) &= Cx(t) + Du(t), \end{aligned} \tag{3.13}$$

that is a minimal realization of its impulse response function and such that $sp(A) \subset \mathbb{C}^-$. *Denote the transfer function of this system by* S,

$$S(z) = C(zI - A)^{-1}B + D. \tag{3.14}$$

Assume that the set of relations (3.6)-(3.9) admits a unique solution $Q \in \mathbb{R}^{n \times n}$. *Consider the expression*

$$-\frac{1}{2}\ln(\det(I - B^{\mathrm{T}}QB - D^{\mathrm{T}}D)). \tag{3.15}$$

a. *If* $e : \Omega \times T \to \mathbb{R}^m$ *and* $z : \Omega \times T \to \mathbb{R}^p$ *are stationary Gaussian processes with spectral density*

$$\begin{pmatrix} I & S(z) \\ S(z^{-1})^{\mathrm{T}} & I \end{pmatrix}, \tag{3.16}$$

then the mutual information of (e, z) *is given by the expression (3.15).*

b. *The* H_∞ *entropy of the system (3.13), defined by:*

$$J = \frac{-1}{4\pi i}\int_{\mathbb{D}} \frac{1}{z}\ln\det\left[I - S(z^{-1})^{\mathrm{T}}S(z)\right]dz, \tag{3.17}$$

is equal to the expression (3.15).

c. *Let* $u : \Omega \times T \to \mathbb{R}^m$ *be a stationary Gaussian process in (3.13) with* $u(t) \in G(0, I)$. *Then*

$$\lim_{t\to\infty}\frac{1}{t}\ln E\left[\exp\left(\frac{1}{2}\sum_{s=0}^{t} y(s)^{\mathrm{T}}y(s)\right)\right]$$

is equal to the expression (3.15).

Proof a. See Theorem E.1. b. See Theorem G.1. c. See Proposition F.2. □

The discovery of the relation between the criterion of LEQG and the H_∞ entropy has been published by K. Glover and J.C. Doyle [26]. That the H_∞ entropy criterion is identical to mutual information is new as far as the authors know. The H_∞ entropy criterion for continuous-time systems was introduced by K. Glover and D. Mustafa in [27] and has been used extensively since then in H_∞ optimal control theory. In that paper the concept of H_∞ entropy is referred to the paper [8] by D.Z. Arov and M.G. Krein. In the latter paper the formula (3.17) is presented with the explanation that this formula 'has a definite entropy sense' under reference to a publication by Kolmogorov, Gelfand, and Yaglom. In fact, the formula (3.17) is the mutual information of two stationary Gaussian processes that was apparently first analyzed in [24]. The expression of mutual information provided in [24] is in a geometric formulation from which (3.17) may be derived. The formula for the mutual information in the case of scalar processes is stated in [57]. The authors therefore conclude that H_∞ entropy is actually mutual information.

3.4 Parameter Estimation with an Exponential-of-Quadratic Cost

A short introduction to the LEQG optimal stochastic control problem follows. A detailed treatment may be found in Appendix F.. Consider the Gaussian stochastic control system

$$\begin{aligned} x(t+1) &= Ax(t) + Bu(t) + Mv(t), \\ y(t) &= C_1x(t) + D_1u(t) + Nv(t), \\ z(t) &= C_2x(t) + D_2u(t), \end{aligned} \tag{3.18}$$

where v is Gaussian white noise with $v(t) \in G(0, V_1)$. A *control law* is a collection of maps $g = \{g_t, t \in T\}$, $g_t : Y^t \times U^t \to U$ such that the input $u(t)$ is given by

$$u(t) = g_t(y(0), \ldots, y(t-1), u(0), \ldots, u(t-1)).$$

Let G be the set of control laws. Consider the cost function on a finite horizon $J_{t_1} : G \to \mathbb{R}$

$$J_{t_1}(g) = E\left[c\exp\left(\frac{1}{2}c\sum_{t=0}^{t_1} z(t)^{\mathrm{T}}z(t)\right)\right], \tag{3.19}$$

for $c \in \mathbb{R}$, $c \neq 0$. The LEQG optimal stochastic control problem is then to determine a control law $g^* \in G$ such that

$$J_{t_1}(g^*) = \inf_{g\in G} J_{t_1}(g). \tag{3.20}$$

Assume that we apply a controller g to the stochastic system (3.18) such that the closed loop system is linear and time-invariant. We can then determine the limit of the cost function

$$\frac{1}{t}E\left[c\exp\left(\frac{1}{2}c\sum_{s=0}^{t}z(s)^{\mathrm{T}}z(s)\right)\right], \tag{3.21}$$

as $t \to \infty$. In Appendix F, it is proven that

$$\lim_{t\to\infty}\frac{1}{t}\ln E\left[c\exp\left(\frac{1}{2}c\sum_{s=0}^{t}z(s)^{\mathrm{T}}z(s)\right)\right]$$
$$= -\frac{1}{2}\ln\det(V_1^{-1} - cM^{\mathrm{T}}QM - cN^{\mathrm{T}}N) + \frac{1}{2}\ln\det V_1^{-1}. \tag{3.22}$$

In (3.22) Q is the solution of a set of relations similar to (3.6)-(3.9).

Next the parameter estimation problem is posed in terms of the LEQG problem.

Problem 3.2. Consider the Gaussian system

$$\begin{aligned} \theta(t+1) &= \theta(t) + v(t),\ \theta(0) = \theta_0, \\ y(t) &= \phi(t)^{\mathrm{T}}\theta(t) + w(t), \end{aligned} \tag{3.23}$$

where $T = \{0, 1, \ldots, t_1\}$, $\theta_0 : \Omega \to \mathbb{R}^n$, $\theta_0 \in G(m_0, Q_0)$, $v : \Omega \to \mathbb{R}^n$ and $w : \Omega \to \mathbb{R}^p$ are Gaussian white noise processes with $v(t) \in G(0, Q_v)$, $w(t) \in G(0, Q_w)$ with $Q_w > 0$, θ_0, v, w are independent, and $\phi : \Omega \times T \to \mathbb{R}^{n\times p}$ is as specified in (2.2).

Determine the process $\hat{\theta} : \Omega \times T \to \mathbb{R}^n$ and a recursion for it such that $\hat{\theta}(t)$ is F^y_{t-1} measurable for all $t \in T$, and such that the following expression is minimized

$$E\left[c\exp\left(\frac{1}{2}c\sum_{s=0}^{t_1}z(s)^{\mathrm{T}}z(s)\right)\right], \tag{3.24}$$

where $z(t) = C[\theta(t) - \hat{\theta}(t)]$, $c \in \mathbb{R}$, $c \neq 0$.

Theorem 3.2. *Consider Problem 3.2. Define the functions $K : T \to \mathbb{R}^{n\times p}$, $Q : T \to \mathbb{R}^{n\times n}$ by the recursions:*

$$K(t) = Q(t)\begin{pmatrix}\phi(t) & C^{\mathrm{T}}\end{pmatrix} S(t)^{-1}\begin{pmatrix} I \\ 0\end{pmatrix}, \tag{3.25}$$

$$Q(t+1) = Q(t) + Q_v - Q(t)\begin{pmatrix}\phi(t) & C^{\mathrm{T}}\end{pmatrix} S(t)^{-1}\begin{pmatrix}\phi(t)^{\mathrm{T}} \\ C\end{pmatrix} Q(t), \tag{3.26}$$

$$Q(0) = Q_0, \tag{3.27}$$

$$S(t) = \begin{pmatrix}\phi(t)^{\mathrm{T}} \\ C\end{pmatrix} Q(t)\begin{pmatrix}\phi(t) & C^{\mathrm{T}}\end{pmatrix} + \begin{pmatrix} Q_w & 0 \\ 0 & \frac{-1}{c}I\end{pmatrix}. \tag{3.28}$$

Assume that for all $t \in T$ we have $Q(t) > 0$. The optimal LEQG parameter estimator is then specified by the recursion

$$\hat{\theta}(t+1) = \hat{\theta}(t) + K(t)[y(t) - \phi(t)^{\mathrm{T}}\hat{\theta}(t)]. \tag{3.29}$$

If in addition $Q_v = 0$ *then*

$$\begin{aligned} Q(t+1)^{-1} &= Q(t)^{-1} + \phi(t)Q_w^{-1}\phi(t)^{\mathrm{T}} - cC^{\mathrm{T}}C. & (3.30)\\ K(t) &= Q(t)\phi(t)\left(\phi(t)^{\mathrm{T}}Q(t)\phi(t) + Q_w\right)^{-1} & (3.31) \end{aligned}$$

The details of the proof of Theorem 3.2 will not be provided in this paper because of space limitations. The problem is related to Problem F.1. It differs from that problem in that Problem 3.2 has output matrix, $\phi(t)$, that depends on the past observations. Therefore a conditional version of Problem F.1 is obtained. The result of Problem F.1 may be found in [32]. The second author of this paper is preparing a publication that contains the solution to the conditional version of Problem F.1. As pointed out in Appendix F., P. Whittle first solved the discrete time case of Problem F.1 but for a system representation slightly different from that used in this paper.

3.5 Parameter Estimation with H_∞ Entropy

A short introduction to the H_∞ optimal control problem follows. A detailed treatment may be found in Appendix G. Consider the linear control system

$$\begin{aligned} x(t+1) &= Ax(t) + Bu(t) + Mv(t),\\ y(t) &= C_1x(t) + D_1u(t) + Nv(t),\\ z(t) &= C_2x(t) + D_2u(t), \end{aligned} \tag{3.32}$$

where v is this time an unknown *deterministic* signal. As before, a control law is a collection of maps $g = \{g_t, t \in T\}$, $g_t : Y^t \times U^t \to U$ such that the input $u(t)$ is given by

$$u(t) = g_t(y(0), \ldots, y(t-1), u(0), \ldots, u(t-1)).$$

Let G be the set of control laws. Consider the cost function $J : G \to \mathbb{R}$

$$J(g) = \sup_{v, x_0} \frac{\|z\|_{2,t_1}^2}{\|v\|_{2,t_1}^2 + x_0^{\mathrm{T}}R^{-1}x_0} \tag{3.33}$$

where

$$\|v\|_{2,t_1}^2 := \sum_{s=0}^{t_1} v(t)^{\mathrm{T}}v(t). \tag{3.34}$$

Note that R plays a role similar to the covariance of the initial condition for the LEQG problem. The larger R, the more uncertainty we have for the initial condition. If $t_1 = \infty$ then we must constrain v to ℓ_2, the class of signals for which the infinite sum (3.34) converges. The H_∞ optimal control problem is then to determine a control law $g^* \in G$ such that

$$J(g^*) = \inf_{g \in G} J(g). \tag{3.35}$$

This problem turns out to be very hard. Hence instead we will look for a control law g such that:

$$J(g) < c^{-1}.$$

This is equivalent to minimizing the following criterion

$$J_c(g) = \sup_{v, x_0} c\|z\|^2_{2,t_1} - \|v\|^2_{2,t_1} - x_0^{\mathrm{T}} R^{-1} x_0.$$

As a matter of fact $J(g) < \infty$ if $J_c(g) < c^{-1}$. We will actually solve the problem to find g^* such that

$$J_c(g^*) = \inf_{g \in G} J_c(g). \tag{3.36}$$

If $t_1 = \infty$, $x(0) = 0$ ($R = 0$), and both the system and the controller are time-invariant, then this problem has another interpretation. Let us define

$$J_e(g) = \frac{-1}{4\pi i} \int_{\mathbb{D}} \frac{1}{z} \ln \det \left[I - cS^{\mathrm{T}}(z^{-1}) S(z) \right] dz$$

where S is the closed loop transfer matrix from v to z. Then g^* satisfies (3.36) if and only if

$$J_e(g^*) = \inf_{g \in G} J_e(g).$$

Next the parameter estimation problem is posed in terms of an H_∞ control problem.

Problem 3.3. Consider the Gaussian system

$$\begin{aligned} \theta(t+1) &= \quad \theta(t) + v(t), \quad \theta(0) = \theta_0, \\ y(t) &= \phi(t)^{\mathrm{T}} \theta(t) + w(t), \end{aligned} \tag{3.37}$$

where $T = \{0, 1, \ldots, t_1\}$ and ϕ is as specified in (2.2).

Determine the process $\hat{\theta} : T \to \mathbb{R}^n$ and a recursion for it such that for all $t \in T$ we have that $\hat{\theta}(t)$ is a function of past measurements $y(0), \ldots, y(t-1)$ and such that the following expression is minimized

$$\sup_{w, v, x_0} \left(-x_0^{\mathrm{T}} Q_0^{-1} x_0 + \sum_{t=0}^{t_1} c z(t)^{\mathrm{T}} z(t) - v(t)^{\mathrm{T}} Q_v^{-1} v(t) - w(t)^{\mathrm{T}} Q_w^{-1} w(t) \right) \tag{3.38}$$

where $z(t) = C[\theta(t) - \hat{\theta}(t)]$, $c \in \mathbb{R}$, $c \neq 0$, $Q_v = Q_v^{\mathrm{T}} > 0$, $Q_w = Q_w^{\mathrm{T}} > 0$.

Proposition 3.1. *Consider Problem 3.3. There exists a $\hat{\theta}$ such that the supremum (3.38) is finite if and only if there exists a matrix Q satisfying (3.26) and (3.27). Moreover, in that case the optimal estimator is given by (3.29).*

4. Approximation with Likelihood and Divergence

In this section the reader is presented successively with text on the approximation with likelihood, divergence, the relation of likelihood and divergence, approximation with divergence, and parameter estimation by divergence minimization.

4.1 Approximation with the Likelihood Function

In this subsection a particular recursive weighted maximum likelihood problem will be formulated and solved.

The maximum likelihood method for parameter estimation has been proposed and analyzed by Sir Ronald Fisher, see [21, 22]. In most lectures and most textbooks the likelihood function is defined by example only. In general the likelihood function may be defined as the Radon-Nikodym derivative of the measure of the observations with respect to the measure with respect to which the observations are independent in discrete-time or to the measure with respect to which the observation process is an independent increment process in continuous-time. The likelihood function for a Gaussian or ARMA representation is presented in [33, 63].

Problem 4.1. Consider the parameter estimation model of the first paragraph of Section 3., with the representation

$$\begin{aligned} \theta(t+1) &= \theta(t), \qquad \theta(0) = \theta, \\ y(t) &= \phi(t)^{\mathrm{T}}\theta(t) + w(t). \end{aligned} \tag{4.1}$$

Let $c \in \mathbb{R}$, $c \neq 0$, $\theta \in G(0, Q_0)$, $Q_0 = Q_0^{\mathrm{T}} > 0$. Suppose that at $t \in T$ the parameter estimates $\{\hat{\theta}(s), s = 0, \ldots, t\}$ have been determined. Choose the next parameter estimate $\hat{\theta}(t+1) \in \mathbb{R}^n$ as the solution of the optimization problem

$$\sup_{\theta \in \mathbb{R}^n} \exp\left(\frac{1}{2}c\sum_{s=0}^{t} \|\theta - \hat{\theta}(s)\|^2\right) f(\bar{y}(0), \ldots, \bar{y}(t), \theta), \tag{4.2}$$

where f is the joint density function of the observation process $\{y(s), s = 1, \ldots, t\}$ and $\{\bar{y}(s), s = 1, \ldots, t\}$ are the numerical values of the observations. The density function f depends on the parameter vector θ.

Theorem 4.1. *Consider Problem 4.1 with $Q_w = I$. Assume that the model is such that $Q(t)$ is well-defined by*

$$Q(t+1)^{-1} = Q(t)^{-1} + \phi(t)Q_w^{-1}\phi(t)^{\mathrm{T}} - cI. \tag{4.3}$$

with $Q(0) = Q_0$ and $Q(t) > 0$ for all $t \in T$. Then the optimal estimate at $t \in T$ is given by the recursion

$$\hat{\theta}(t+1) = \hat{\theta}(t) + K(t)[\bar{y}(t) - \phi(t)^{\mathrm{T}}\hat{\theta}(t)], \; \hat{\theta}(0) = 0, \tag{4.4}$$

where

$$K(t) = Q(t)\phi(t)[\phi(t)^{\mathrm{T}}Q(t)\phi(t) + 1]^{-1}. \tag{4.5}$$

Note that the parameter estimator (4.4) is identical to that of Theorem 3.2.

Proof: The logarithm of the criterion is, up to a term that does not depend on θ, equal to

$$2h(\theta) = -\theta^{\mathrm{T}}Q_0^{-1}\theta - \sum_{s=0}^{t} \|\bar{y}(s) - \phi(s)^{\mathrm{T}}\theta\|^2 + c\sum_{s=0}^{t-1} \|\theta - \hat{\theta}(s)\|^2. \tag{4.6}$$

By assumption

$$2d^2h(\theta)/d\theta^2 = ctI - Q_0^{-1} - \sum_{s=0}^{t} \phi(s)\phi(s)^{\mathrm{T}} = -Q(t)^{-1} - cI < 0.$$

The first order conditions then yield:

$$Q(t)^{-1}\theta = \sum_{s=0}^{t} \bar{y}(s)\phi(s) - c\sum_{s=0}^{t-1} \hat{\theta}(s),$$

and after simple calculations one gets (4.4). □

4.2 Divergence

The Kullback-Leibler pseudo-distance or divergence is a function that measures the relation between probability measures. For system identification it is an important approximation criterion. There follows a brief introduction to divergence, the full details may be found in the Appendices C., D., and E.

Consider the measurable space (Ω, F) and two probability measures P_1, P_2 defined on it. Let Q be a σ-finite measure on (Ω, F) such that P_1 and P_2 are absolutely continuous with respect to Q, say with $r_1 = dP_1/dQ$ and $r_2 = dP_2/dQ$. Such a measure always exists. The divergence of P_1, P_2 is defined by the formula

$$D(P_1\|P_2) = E_Q\left[r_1 \ln(\frac{r_1}{r_2})I_{(r_2>0)}\right]. \tag{4.7}$$

It may be shown that the definition of $D(P_1\|P_2)$ does not depend on the measure Q chosen, that for all P_1, P_2 we have $D(P_1\|P_2) \geq 0$, and $D(P_1\|P_2) = 0$ iff $P_1 = P_2$. The triangle inequality is not satisfied by D hence it is not a distance but a pseudo-distance. S. Kullback and R.A. Leibler in [46] introduced this concept which is therefore also known as the Kullback-Leibler pseudo-distance. The term divergence is used in information theory.

Divergence and mutual information for measures induced by two real valued random variables are related by

$$J(x,y) = D(P_{xy} \| P_x \times P_y).$$

Let $G(m_1, Q_1)$ and $G(m_2, Q_2)$ be two Gaussian measures on $\mathbb{R}^n$. Assume that $Q_1, Q_2 > 0$. The divergence between these measures is given by the formula (see D.5),

$$\begin{aligned} &2D\left(G(m_1, Q_1) \| G(m_2, Q_2)\right) \\ &= -\ln\left(\frac{\det Q_1}{\det Q_2}\right) + \mathrm{tr}([Q_2^{-1} - Q_1^{-1}]Q_1) + (m_2 - m_1)^{\mathrm{T}} Q_2^{-1}(m_2 - m_1). \end{aligned} \tag{4.8}$$

Consider next two jointly stationary Gaussian processes $x_1, x_2 : \Omega \times T \to \mathbb{R}^n$ on $T = \mathbb{Z}$. Assume that they have a mean value function that is zero and that the joint processes admit a spectral density. Under additional conditions, see Theorem E.2, the divergence between the measures associated with these processes is given by the formula

$$D(P_1 \| P_2) = -\frac{1}{2}\ln\det(B^{\mathrm{T}}QB + D^{\mathrm{T}}D) + \frac{1}{2}\mathrm{tr}(B^{\mathrm{T}}RB + D^{\mathrm{T}}D - I), \tag{4.9}$$

where $Q \in \mathbb{R}^{n\times n}$ is the solution of an algebraic Riccati equation and $R \in \mathbb{R}^{n\times n}$ is the solution of a Lyapunov equation.

4.3 Relation of Likelihood Function and Divergence

The likelihood function as an approximation criterion is related to the approximation criterion of divergence discussed in the previous subsection. The relation between these criteria will be described for Gaussian random variables.

Consider n real valued observations. The model class of the variable considered is the class of Gaussian random variables with values in $\mathbb{R}$, say $G(m, q)$, $m \in \mathbb{R}$, $q \in \mathbb{R}$, $q > 0$. The observations are assumed to be independent. The measure of these observations and their representation are given by $y : \Omega \times T \to \mathbb{R}^n$, $y \in G(me, Q)$, where

$$e = (1, \ldots, 1)^{\mathrm{T}} \in \mathbb{R}^n, \quad Q = qI.$$

The density function of y with respect to Lebesgue measure is

$$p(v; me, Q) = \frac{1}{\sqrt{(2\pi)^n \det Q}} \exp\left(-\frac{1}{2}(v - me)^{\mathrm{T}} Q^{-1}(v - me)\right).$$

The likelihood function $L : \mathbb{R} \times \mathbb{R}_+ \to \mathbb{R}_+$ is therefore equal to

$$L(m, q; \bar{y}) = \frac{1}{\sqrt{(2\pi)^n \det Q}} \exp\left(-\frac{1}{2}(\bar{y} - me)^{\mathrm{T}} Q^{-1}(\bar{y} - me)\right),$$

where $\bar{y} \in \mathbb{R}^n$ is the vector with the numerical values of the observations. As is well known, the maximum likelihood estimate of the parameters m, q are given by the formulas

$$\hat{m} = \frac{1}{n}\sum_{k=1}^{n} \bar{y}_k, \quad \hat{q} = \frac{1}{n}\sum_{k=1}^{n}(\bar{y}_k - \hat{m})^2.$$

The density function of a Gaussian measure with $\hat{m}, \hat{q}$ as parameters is denoted by $p(., \hat{m}, \hat{q})$.

Formula (4.8) for the divergence of two Gaussian measures on $\mathbb{R}$ reduces in this case to (assuming $q_1 > 0$ and $q_2 > 0$)

$$2D\left(p(.; m_1, q_1) \| p(., m_2, q_2)\right) = -\ln\left(\frac{q_1}{q_2}\right) - 1 + \frac{q_1}{q_2} + \frac{(m_1 - m_2)^2}{q_2}.$$

Proposition 4.1. *Consider the likelihood function and the divergence for the Gaussian random variables defined above. Then*

$$L(m, q; \bar{y}) = \frac{1}{\sqrt{(2\pi)^n}} \exp\left(\frac{-n}{2}\left[2D\left(G(\hat{m}, \hat{q}) \| G(m, q)\right) + 1 + \ln \hat{q}\right]\right).$$

Proof: From the discussion above follows that

$$\begin{aligned} \ln L(m, q; \bar{y}) &= -\frac{n}{2}\ln(2\pi) - \frac{1}{2}n[\ln(q) + \frac{q_1}{q}], \\ 2D(G(\hat{m}, \hat{q}) \| G(m, q)) &= -\ln(\frac{\hat{q}}{q}) - 1 + \frac{\hat{q}}{q} + \frac{(\hat{m} - m)^2}{q}, \end{aligned}$$

where

$$q_1 = \frac{1}{n}\sum_{k=1}^{n}(\bar{y}_k - m)^2.$$

Note that

$$q_1 = \frac{1}{n}\sum_{k=1}^{n}(\bar{y}_k - m)^2 = \frac{1}{n}\sum_{k=1}^{n}(\hat{y}_k - \hat{m} + \hat{m} - m)^2 = \hat{q} + (\hat{m} - m)^2,$$

hence the result. □

The conclusion from Proposition 4.1 is that maximizing the likelihood function is equivalent to minimizing the divergence between a measure in the model class and the measure associated with the maximum likelihood estimates. However, note that this fact cannot be used to find the maximum likelihood estimates since the maximum likelihood estimates $\hat{m}$ and $\hat{q}$ appear in the expression. Rather, it shows that the likelihood function is equivalent to the divergence criterion. In the example above both measures appearing in the divergence, $G(m, q)$ associated with the model class and $G(\hat{m}, \hat{q})$ associated with the maximum, are members of the model class. Another way to consider the relation between the likelihood function and divergence has been proposed by H. Akaike in [1].

Does the relation between the likelihood function and divergence extend to other classes of stochastic systems? In the maximum likelihood approach one maximizes the likelihood function with respect to the parameters of the model class. The divergence between the measure associated with the observations and the measure associated with the model class is a pseudo-distance. Therefore the maximum likelihood approach and the minimum divergence approach are related. The exponential transformation between both criteria as in Proposition 4.1 for the Gaussian case may not hold in general, although this relation does hold for a larger class of distributions than the Gaussian one.

S.-I. Amari has investigated the geometric structure of exponential families of probability distributions and explored the use of the likelihood function and of divergence for this family, see [5, 4, 6].

The minimum divergence method differs from the maximum likelihood method in that for the first method a measure must be chosen that represents the observations. In this point the maximum likelihood approach requires less.

4.4 Approximation with Divergence

For system identification an approximation problem with the divergence criterion may be considered. The relationship between the likelihood function and divergence, see Proposition 4.1, suggests such an approach.

In system identification one is given observations. According to the procedure outlined in Section 2. one then selects a model class. In the case considered in this paper this is the class of stationary Gaussian processes which can be represented as a Gaussian system. With the observations one can associate a measure of a stationary Gaussian process. For example, one can take the measure associated with the mean value function of this process to be the zero function and the covariance function to be an estimate, say

$$\hat{W}(t) = \begin{cases} \frac{1}{t_1 - t} \sum_{s=0}^{t_1} y(s+t)y(s)^{\mathrm{T}}, & t = 0, \ldots, t_2, \\ 0, & \text{else}. \end{cases} \tag{4.10}$$

Another choice is a measure associated with an AR representation of high order.

Problem 4.2. *Approximation of a stationary Gaussian process by a time-invariant Gaussian system according to the divergence criterion.* Consider given a probability measure P_1 on the space of Gaussian processes with values in $\mathbb{R}^p$ and zero mean value function. Consider the model class $GS\Sigma(n)$ of Gaussian systems up to order n and the associated set of probability measures $\{P(\theta), \theta \in GS\Sigma(n)\}$ on the output processes of such systems. Solve the approximation problem

$$\inf_{\theta \in GS\Sigma(n)} D(P_1 \| P(\theta)). \tag{4.11}$$

A solution to Problem 4.2, if it exists, is a Gaussian system of which the probability measure on the output process minimizes the divergence to the probabiblity measure associated with the observations. Note that the probability measure associated with the observations may not be in the model class, for example because the covariance estimate is not a rational function. Problem 4.2 is therefore equivalent to the determination of an element in the model class that minimizes the divergence to the measure associated with the observations.

The principle of approximation by the divergence criterion is used in the literature. It has apparently first been proposed by H. Akaike in [1] who explored its use in [2, 3] and who derived the Akaike Information Criterion (AIC) from it. Other publications on this approach are those of I. Csiszár, [14, 15, 17] and, in the case of the ARMA model class, [20, 53].

The following model reduction problem for Gaussian random variables is of interest to system identification. In system identification a technique is used called *subspace methods* that is closely related to the following problem. It is expected that solution of this problem will be useful to subspace methods. The problem was first formulated in [69, Subsec. 3.8].

Problem 4.3. *Model reduction for a pair of Gaussian random variables.* Let $y_1 : \Omega \to \mathbb{R}^{p_1}$, $y_2 : \Omega \to \mathbb{R}^{p_2}$ with $(y_1, y_2) \in G(0, S)$,

$$S = \begin{pmatrix} S_{11} & S_{12} \\ S_{12}^T & S_{22} \end{pmatrix}, \quad \text{rank}(S_{12}) = n_1.$$

Consider the model class, for $n_2 \in Z_+$, $n_2 < n_1$,

$$\mathbf{G}(n_2) = \{G(0,Q) \text{ on } \mathbb{R}^{p_1+p_2} | \text{rank}(Q_{12}) = n_2, \}, \tag{4.12}$$

$$Q = \begin{pmatrix} Q_{11} & Q_{12} \\ Q_{12}^T & Q_{22} \end{pmatrix}, \tag{4.13}$$

in which the components of Q are compatible with the decomposition (y_1, y_2). Solve for fixed $n_2 \in Z_+$, $n_2 < n_1$,

$$\inf_{G(0,Q) \in \mathbf{G}(n_2)} D\left(G(0,S) \| G(0,Q)\right). \tag{4.14}$$

An interpretation of this problem follows. In stochastic realization theory of stochastic processes one uses the framework of past, future, and present or state. If one restricts attention to the case of Gaussian random variables, thus not of processes, then the past and the future are represented by Gaussian random variables, say y_1, y_2 as above. From observations one can estimate the measure of the past and future observations, say $G(0, S)$. The state space associated with the past and the future then has dimension $n_1 = \text{rank}(S_{12})$. In the model reduction problem one seeks a model or measure of the observations in which the dimension of the state space is less than that initially fixed, thus $\text{rank}(Q_{12}) = n_2 < n_1$. The approximation criterion of divergence seems quite appropriate, also because of its relation with the likelihood function.

For initial results on this problem see [69, Subsec. 3.8]. The problem has a rich convex structure that remains to be explored.

4.5 Parameter Estimation by Divergence Minimization

Consider Problem 2.1 of parameter estimation. Consider the time moment $t \in T$. Suppose that the observations $\{\bar{y}(s), s = 0, \ldots, t\}$ are available and that the estimates $\{\hat{\theta}(s), s = 0, \ldots, t-1\}$ have been chosen. The question is then how to choose $\hat{\theta}(t)$.

With the observations $\{\bar{y}(s), s = 0, \ldots, t\}$ we associate the probability measure $G(\bar{y}, I)$ and with the model class the probability measure $G(\Phi\theta, I)$ where

$$\Phi = \begin{pmatrix} \phi(0)^{\mathrm{T}} \\ \phi(1)^{\mathrm{T}} \\ \vdots \\ \phi(t)^{\mathrm{T}} \end{pmatrix}, \quad \bar{y} = \begin{pmatrix} \bar{y}(0) \\ \bar{y}(1) \\ \vdots \\ \bar{y}(t) \end{pmatrix}.$$

The unknown parameter θ has to be estimated. In this setting $\phi : T \to \mathbb{R}^{1\times n}$ is a deterministic function. Associate with the past parameter estimates the probability measure $G(\bar{\theta}, \bar{W})$ and with the model class the probability measure $G(\theta, c^{-1}I)$, where $c \in \mathbb{R}$, $c \neq 0$,

$$\bar{\theta} = \frac{1}{t}\sum_{s=0}^{t-1} \hat{\theta}(s), \quad \bar{W} = \frac{1}{t}\sum_{s=0}^{t-1} (\hat{\theta}(s) - \bar{\theta})(\hat{\theta}(s) - \bar{\theta})^{\mathrm{T}}.$$

Problem 4.4. Consider Problem 2.1 and the notation introduced above. For $t \in T$ choose the parameter estimate $\hat{\theta}(t)$ as the solution of the minimization problem

$$\inf_{\theta \in \mathbb{R}^n} \left[\frac{1}{t} D\left(G(\bar{y}, I) \| G(\Phi\theta, I)\right) - D\left(G(\bar{\theta}, \bar{W}) \| G(\theta, c^{-1}I)\right) \right]. \tag{4.15}$$

The problem formulated above is related to but different from that of the papers [34, 35].

Theorem 4.2. *Consider Problem 4.4.*

a. The divergence criterion of that problem is equivalent to the criterion

$$\inf_{\theta \in \mathbb{R}^n} \frac{1}{t} \left[\|\bar{y} - \Phi\theta\|^2 - c\|\theta - \bar{\theta}\|^2 \right]. \tag{4.16}$$

b. The parameter estimator that solves this problem is given by the formula

$$\hat{\theta}(t+1) = \hat{\theta}(t) + K(t)[\bar{y}(t) - \phi(t)^{\mathrm{T}}\hat{\theta}(t)], \quad \hat{\theta}(0) = 0, \tag{4.17}$$

where the formula for the gain and the Riccati recursion are as stated in Theorem 4.1.

Note that the parameter estimator (4.17) is identical to the estimators of Theorem 3.2 and Theorem 4.1.

Proof a. A simple calculation yields that

$$\begin{aligned} 2D\left(G(\bar{y}, I) \| G(\Phi\theta, I)\right) &= -\ln\left(\frac{\det I}{\det I}\right) - t + \mathrm{tr}(I^{-1}I) + \|\bar{y} - \Phi\theta\|^2 \\ &= \|\bar{y} - \Phi\theta\|^2, \\ 2D\left(G(\bar{\theta}, \bar{W}) \| G(\theta, c^{-1}I)\right) &= -\ln\left(\frac{\det \bar{W}}{\det(c^{-1}I)}\right) - t + \mathrm{tr}(c\bar{W}) + c\|\theta - \bar{\theta}\|^2, \end{aligned}$$

$$\begin{aligned} &\frac{2}{t}D\left(G(\bar{y}, I) \| G(\Phi\theta, I)\right) - 2D(G(\bar{\theta}, \bar{W}) \| G(\theta, c^{-1}I)) \\ &\quad = \frac{1}{t}\|\bar{y} - \Phi\theta\|^2 - c\|\theta - \bar{\theta}\|^2 + \ln\det\bar{W} + t - c\,\mathrm{tr}\bar{W} + t\ln c. \end{aligned}$$

Note that the last four terms depend on the observations and not on the parameter θ. Hence they can be neglected in the criterion.
b. It will be shown that the criterion of part a of the theorem is equivalent to the criterion

$$\|\bar{y} - \Phi\theta\|^2 - c\|\theta - \hat{\theta}\|^2.$$

This will be established by showing that infimization of the functions

$$f(\theta) = \|\theta - \bar{\theta}\|^2, \quad g(\theta) = \frac{1}{t}\|\theta - \hat{\theta}\|^2 = \frac{1}{t}\sum_{s=0}^{t-1}(\theta - \hat{\theta}(s))^{\mathrm{T}}(\theta - \hat{\theta}(s)),$$

is equivalent. Both functions are quadratic forms in θ. The derivatives are

$$\begin{aligned} \frac{df(\theta)}{d\theta} &= 2(\theta - \bar{\theta})^{\mathrm{T}}, \\ \frac{dg(\theta)}{d\theta} &= \frac{2}{t}\sum_{s=0}^{t-1}(\theta - \hat{\theta}(s))^{\mathrm{T}} = 2\theta - 2\frac{1}{t}\sum_{s=0}^{t-1}\hat{\theta}(s) = 2(\theta - \bar{\theta})^{\mathrm{T}}. \end{aligned}$$

Hence the first order conditions are identical and therefore the optimal θ is the same for both optimization problems. This optimization problem is then given by:

$$\inf_{\theta \in \mathbb{R}^n} \frac{1}{t}\left[\|\bar{y} - \Phi\theta\|^2 - c\|\theta - \hat{\theta}\|^2\right].$$

The result then follows from Theorem 4.1. □

5. Concluding Remarks

What is the Contribution of this Paper to the Theory of System Identification?

The main contribution of the paper is the relation between several information theoretic criteria and their use in system identification. The concept of mutual information rate for stationary Gaussian processes is identical to H_∞ entropy and to the exponential-of-quadratic cost in optimal stochastic control. The likelihood function is identical to divergence for Gaussian random variables. As a consequence of this relation a parameter estimator is presented that may be derived by four different approximation criteria. The problem concerns the approximation of observations from a stationary Gaussian process by the output of a Gaussian system of a specified order. Another product of the paper are formulas for information theoretic criteria of stationary Gaussian processes in case these processes are outputs of time-invariant Gaussian systems.

The relationship between the H_∞ entropy for a finite-dimensional linear system and the exponential-of-quadratic cost for a Gaussian stochastic control system is understood at the level of the formulas. It is a question as to whether a deeper understanding is possible. By rephrasing the relation as a relation between mutual information and the exponential-of-quadratic cost, the interpretation becomes different but it is still not enlightening.

The parameter estimator derived in this paper should be tested on data. Its robustness properties require further study.

Which Problems of System Identification Theory Require Further Attention?

The framework for system identification with information theoretic criteria should be explored further. For stationary Gaussian processes also the case of ARMA representations may be considered. Besides Gaussian processes also finite valued processes, with the hidden Markov model as stochastic system, and counting processes should be considered.

Theoretical problems of system identification and of realization may be considered using the framework of this paper. The relation between the likelihood function and divergence rate may be explored for model reduction of Gaussian systems. Possibly minimization of the divergence can be solved partly analytically. The approximation problem as such also requires further study.

System identification theory in general may benefit from studying other classes of dynamic systems in relation with practical problems. Examples of such classes are positive linear systems, particular classes of nonlinear systems, and the class of errors-in-variables systems. Approximation techniques for nonlinear systems, such as artificial neural nets, wavelets, and fuzzy modeling, may be explored.

Acknowledgements

The authors thank S. Bittanti and G. Picci for the organization of the NATO Advanced Study Institute From Identification to Learning and for the invitation to the second named author to lecture on the topic of this paper. The meeting and its proceedings are very useful ways to transfer knowledge in the area of system identification.

The second named author of this paper also thanks L. Finesso of LADSEB, Padova, Italy, for a discussion on the subject of this paper and for useful references.

The research of this paper has been sponsored in part by the Commission of the European Communities through the SCIENCE Program by the project SC1*-CT92-0779.

A. Concepts from Probability and the Theory of Stochastic Processes

In this appendix concepts and notation of probability and of the theory of stochastic processes are introduced.

General mathematics notation follows. The set of integers is denoted by $\mathbb{Z}$ and the set of positive integers by $\mathbb{Z}_+$. The set of real numbers is denoted by $\mathbb{R}$ and the set of the positive real numbers by $\mathbb{R}_+ = [0, \infty)$. The n-dimensional vector space over $\mathbb{R}$ is denoted by $\mathbb{R}^n$ and the set of $n \times m$ matrices over this vector space by $\mathbb{R}^{n\times m}$. The transpose of a matrix $A \in \mathbb{R}^{n\times n}$ is denoted by A^{T}. The matrix $A \in \mathbb{R}^{n\times n}$ is said to be positive definite if $x^{\mathrm{T}}Ax \geq 0$ for all $x \in \mathbb{R}^n$ and strictly positive definite if $x^{\mathrm{T}}Ax > 0$ for all $x \in \mathbb{R}^n, x \neq 0$. The set of complex numbers is denoted by $\mathbb{C}$ and

$$\begin{aligned} \mathbb{C}^- &:= \{c \in \mathbb{C} \mid |c| < 1\}, \\ \mathbb{D} &:= \{c \in \mathbb{C} \mid |c| = 1\}. \end{aligned}$$

For $A \in \mathbb{R}^{n\times n}$, $sp(A) \subset \mathbb{C}$ denotes the spectrum of the matrix; in other words the set of eigenvalues.

A.1 Probability Concepts

A measurable space, denoted by (Ω, F), is defined as a set Ω and a σ-algebra F. A probability space is defined as a triple (Ω, F, P) where (Ω, F) is a measurable space and $P : F \to \mathbb{R}$ is a probability measure.

Let $I_A : \Omega \to \mathbb{R}$ be the indicator function of the event $A \in F$ defined by $I_A(\omega) = 1$, if $\omega \in A$ and $I_A(\omega) = 0$ otherwise. For any random variable x let F^x denote the smallest σ-algebra on which the random variable x is measurable.

Let P, Q be two probability measures on a measurable space (Ω, F). Then P is said to be absolutely continuous with respect to Q, denoted by $P \ll Q$, if $P(A) = 0$ is implied by $Q(A) = 0$. If $P \ll Q$ then it follows from the Radon-Nikodym theorem that there exists a random variable $r : \Omega \to \mathbb{R}_+$ such that

$$P(A) = E_Q[rI_A] = \int r(\omega)I_A(\omega)dQ.$$

A.2 Gaussian Random Variables

The Gaussian probability distribution function with parameters $m \in \mathbb{R}^n$ and $Q \in \mathbb{R}^{n\times n}$, with Q strictly positive definite, is defined by the probability density function

$$p(v) = \frac{1}{\sqrt{(2\pi)^n \det Q}} \exp\left(-\frac{1}{2}(v-m)^{\mathrm{T}}Q^{-1}(v-m)\right). \tag{A.1}$$

A random variable $x : \Omega \to \mathbb{R}^n$ is said to be a Gaussian random variable with parameters $m \in \mathbb{R}^n$ and $Q \in \mathbb{R}^{n \times n}$, Q positive definite, if for all $u \in \mathbb{R}^n$

$$E[\exp(iu^{\mathrm{T}}x)] = \exp\left(iu^{\mathrm{T}}m - \frac{1}{2}u^{\mathrm{T}}Qu\right). \tag{A.2}$$

The notation $x \in G(m, Q)$ will be used in this case. Moreover, $(x_1, \ldots, x_n) \in G(m, Q)$ denotes that, with $x = (x_1, \ldots, x_n)^{\mathrm{T}}$, $x \in G(m, Q)$. In this case $x_1, \ldots, x_n$ are said to be jointly Gaussian random variables.

A random variable with Gaussian probability distribution function is a Gaussian random variable. A Gaussian random variable does not necessarily have a Gaussian probability density function, only so in the case its variance is strictly positive definite.

In the following the geometric approach to Gaussian random variables is introduced. In this approach one considers the space that the random variables generate rather than the random variables themselves.

Proposition A.1. *Let $x : \Omega \to \mathbb{R}^n$ $x \in G(m, Q)$, $S \in \mathbb{R}^{n_1 \times n}$.*

a. Then $F^{Sx} \subset F^x$.
b. $F^{Sx} = F^x$ iff $\ker Q = \ker SQ$.

The above result motivates the following definition.

Definition A.1. *Let $x : \Omega \to \mathbb{R}^n$, $x \in G$ and consider the σ-algebra F^x.*

a. A basis for F^x is a triple $(n_1, m_1, Q_1) \in \mathbb{N} \times \mathbb{R}^{n_1} \times \mathbb{R}^{n_1 \times n_1}$ such that there exists a $x_1 : \Omega \to \mathbb{R}^{n_1}$ satisfying $x_1 \in G(m_1, Q_1)$, $Q_1 = Q_1^{\mathrm{T}} \geq 0$, and $F^x = F^{x_1}$.
b. A minimal basis for F^x is a basis (n_1, m_1, Q_1) such that $\mathrm{rank}(Q_1) = n_1$.
c. A basis transformation of F^x is a map $x \mapsto Sx$, with $S \in \mathbb{R}^{n_1 \times n}$ such that $F^{Sx} = F^x$.

By use of linear algebra one can, given a basis (n_1, m_1, Q_1), always construct a minimal basis for F^x.

Proposition A.2. *Let $x_1 : \Omega \to \mathbb{R}^{n_1}$, $x_1 \in G(0, Q_1)$, and $Q_1 = Q_1^{\mathrm{T}} \geq 0$. Then there exists a $n_2 \in \mathbb{N}$ and a basis transformation $S \in \mathbb{R}^{n_2 \times n_1}$ such that, if $x_2 : \Omega \to \mathbb{R}^{n_2}$ $x_2 = Sx_1$, then $x_2 \in G(0, Q_2)$, $Q_2 = Q_2^{\mathrm{T}} > 0$, and $F^{x_2} = F^{x_1}$.*

Problem A.1. Let $y_1 : \Omega \to \mathbb{R}^{k_1}$ and $y_2 : \Omega \to \mathbb{R}^{k_2}$ be jointly Gaussian random variables with $(y_1, y_2) \in G(0, Q)$. Determine a canonical form for the spaces F^{y_1}, F^{y_2}.

Note that a basis transformation of the form $S = block - diag(S_1, S_2)$ with S_1, S_2 nonsingular, leaves the spaces F^{y_1}, F^{y_2} invariant. Therefore this operation introduces an equivalence relation on the spaces F^{y_1}, F^{y_2}, hence one can speak of a canonical form.

Definition A.2. *Let $y_1 : \Omega \to \mathbb{R}^{k_1}$ and $y_2 : \Omega \to \mathbb{R}^{k_2}$ be jointly Gaussian random variables with $(y_1, y_2) \in G(0, Q)$. Then (y_1, y_2) are said to be in* canonical variable form *if*

$$Q = \begin{pmatrix} I & 0 & \Lambda & 0 \\ 0 & I & 0 & 0 \\ \Lambda & 0 & I & 0 \\ 0 & 0 & 0 & I \end{pmatrix} \in \mathbb{R}^{(k_1+k_2)\times(k_1+k_2)}$$

where $\Lambda \in \mathbb{R}^{k_{12}\times k_{12}}$, $k_{12} \in \mathbb{N}_+$, $\Lambda = diag(\lambda_1, \dots, \lambda_{k_{12}})$, $1 \geq \lambda_1 \geq \dots \geq \lambda_{k_{12}} > 0$. One then says that $(y_{11}, \dots, y_{1k_1})$, $(y_{21}, \dots, y_{2k_2})$ are the canonical variables *and $(\lambda_1, \dots, \lambda_{k_{12}})$ the* canonical correlation coefficients.

Theorem A.1. *Let $y_1 : \Omega \to \mathbb{R}^{k_1}$ and $y_2 : \Omega \to \mathbb{R}^{k_2}$ be jointly Gaussian random variables with $(y_1, y_2) \in G(0, Q)$. Then there exists a basis transformation $S = block - diag(S_1, S_2)$ such that with respect to the new basis $(S_1 y_1, S_2 y_2) \in G(0, Q_1)$ has the canonical variable form.*

The transformation to canonical variable form is not unique in general. The remaining invariance of the canonical variable form will not be stated here because of space limitation. For additional results on canonical variables the reader is referred to [7, 25].

A.3 Concepts from the Theory of Stochastic Processes

A real valued *stochastic process* $x : \Omega \times T \to \mathbb{R}$ on a probability space (Ω, F, P) and a time index set T is defined to be a function such that for all $t \in T$ the map $x(., t) : \Omega \to \mathbb{R}$ is a random variable.

A stochastic process $x : \Omega \times T \to \mathbb{R}^n$ is said to be a *Gaussian process* if every finite dimensional distribution is Gaussian. Thus, if for all $m \in \mathbb{Z}_+$ and $(t_1, \dots, t_m) \in T$ the collection of random variables $x(t_1), \dots, x(t_m)$ is jointly Gaussian. Such a process is completely specified by its mean value function $m : T \to \mathbb{R}^n$, $m(t) = E[x(t)]$ and its covariance function $W : T \times T \to \mathbb{R}^{n\times n}$, $W(t,s) = E[(x(t) - m(t))(x(s) - m(s))^{\mathrm{T}}]$. A (discrete time) *Gaussian white noise* process with values in $\mathbb{R}^k$ and intensity $V : T \to \mathbb{R}^{k\times k}$, $V(t) = V(t)^{\mathrm{T}} \geq 0$, is a stochastic process such that (1) $\{v(t), t \in T\}$ is a collection of independent Gaussian random variables; (2) v is a Gaussian process with $v(t) \in G(0, V(t))$.

A stochastic process $x : \Omega \times T \to \mathbb{R}^n$ on $T = \mathbb{Z}$ or $T = \mathbb{R}$ is said to be *stationary* if for all $m \in \mathbb{Z}_+$, $s \in T$, and $t_1, \dots, t_m \in T$ the joint distribution of $(x(t_1), \dots, x(t_m))$ equals the joint distribution of $(x(t_1 + s), \dots, x(t_m + s))$. If for all $m \in \mathbb{Z}_+$ and $t_1, \dots, t_m \in T$ the distribution function of $(x(t_1 + s), \dots, x(t_m + s))$ converges to a limit as $s \to \infty$ then we call the process asymptotically stationary. A Gaussian process with mean value function $m : T \to \mathbb{R}^n$ and covariance function W is stationary iff $m(t) = m(0)$ $\forall t \in T$ and

$W(t,s) = W(t+u, s+u)$, $\forall t, s, u \in T$. In this case the function $W_1 : T \to \mathbb{R}^{n \times n}$, $W_1(t) = W(t, 0)$ is also called the covariance function.

The concept of canonical correlation process has been defined for stationary Gaussian processes in analogy with the canonical variable form and canonical correlation coefficients for Gaussian random variables. That this may be done follows from the fact that many properties of a stationary Gaussian process depend only on the spaces generated by such a process. For references on this topic see [28, 42, 43, 54].

B. Concepts from System Theory

This appendix contains several definitions of concepts from system theory that are needed at several places in the paper.

Definition B.1. *A* discrete-time finite-dimensional linear dynamic system *or, by way of abbreviation, a* linear system, *is a dynamic system*

$$\sigma = \{T, \mathbb{R}^p, \mathbf{Y}, \mathbb{R}^n, \mathbb{R}^m, \mathbf{U}, \phi, r\}$$

in which the state transition map ϕ and the read-out map r are specified by

$$\begin{aligned} \phi : x(t+1) &= A(t)x(t) + B(t)u(t), \; x(0) = x_0, \\ r : y(t) \quad &= C(t)x(t) + D(t)u(t). \end{aligned} \tag{B.1}$$

Here $T \subset \mathbb{Z}, m, n, p \in \mathbb{Z}_+$, $u : T \to \mathbb{R}^m, u \in \mathbf{U}$, $x_0 \in \mathbb{R}^n$, $A : T \to \mathbb{R}^{n \times n}$, $B : T \to \mathbb{R}^{n \times m}$, $C : T \to \mathbb{R}^{p \times n}$, $D : T \to \mathbb{R}^{p \times m}$, and $x : T \to \mathbb{R}^n, y : T \to \mathbb{R}^p$ are determined by the above recursions and are called respectively the state function *and the* output function. *Such a system is called* time-invariant *if for all $t \in T$ $A(t) = A(0), B(t) = B(0), C(t) = C(0), D(t) = D(0)$. The parameters of such a system are denoted by $\{n, m, p, A, B, C, D\}$.*

Definition B.2. *A Gaussian stochastic control system is defined by the representation*

$$\begin{aligned} x(t+1) &= A(t)x(t) + B(t)u(t) + M(t)v(t), \quad x(0) = x_0, \\ y(t) \quad &= C(t)x(t) + D(t)u(t) + N(t)v(t), \end{aligned} \tag{B.2}$$

where $T = \{0, 1, \ldots, t_1\}$, $t_1 \in \mathbb{Z}_+$, $X = \mathbb{R}^n$, $U = \mathbb{R}^m$, $Y = \mathbb{R}^p$, $x_0 : \Omega \to X$, $x_0 \in G(m_0, Q_0)$, $v : \Omega \times T \to \mathbb{R}^r$ is a Gaussian white noise process with for all $t \in T$, $v(t) \in G(0, V(t))$, $V : T \to \mathbb{R}^{r \times r}$, $V(t) = V(t)^{\mathrm{T}} \geq 0$, the σ-algebras F^{x_0} and $F^v_{t_1}$ are independent, $A : T \to \mathbb{R}^{n \times n}$, $B : T \to \mathbb{R}^{n \times m}$, $C : T \to \mathbb{R}^{p \times n}$, $D : T \to \mathbb{R}^{p \times m}$, $M : T \to \mathbb{R}^{n \times r}$, $N : T \to \mathbb{R}^{p \times r}$, $x : \Omega \times T \to X$, and $y : \Omega \times T \to Y$.

Stochastic Realization

The problem of obtaining a representation of a stochatic process as the output of a stochastic system is called the *stochastic realization problem.* Below the weak stochastic realization problem for stationary Gaussian processes is mentioned. For references on this problem see [19] and the paper by G. Picci elsewhere in this volume.

Consider a stationary Gaussian process with mean value function equal to zero and covariance function $W : T \to \mathbb{R}^{p\times p}$. The problem is whether there exists a time-invariant Gaussian system, say of the form

$$\begin{aligned} x(t+1) &= Ax(t) + Mv(t), \\ y(t) &= Cx(t) + Nv(t), \end{aligned} \tag{B.3}$$

with $sp(A) \subset \mathbb{C}^-$ such that the covariance function of the output process y equals the given covariance function. If so, then this system is said to be a *stochastic realization* of the given process. There is a necessary and sufficient condition for the existence of such a realization. When a realization exists there may be many realizations. Attention is then restricted to minimal realizations for which the dimension of the state space is as small as possible. There is a classification of all minimal stochastic realizations. In addition, there is an algorithm to construct a minimal realization.

C. Concepts from Information Theory

The purpose of this appendix is to describe for the reader the main concepts of information theory. In Appendix D. formulas are presented for information measures of Gaussian random variables and in Appendix E. formulas for information measures of stationary Gaussian processes.

Information theory originated with the work of C.E. Shannon [61, 62]. The Russian mathematicians A.N. Kolmogorov, I.M. Gelfand, and A.M. Yaglom partly developed the measure theoretic formulation of information theory, see [24]. The book [57] by M.S. Pinsker partly summarizes information theory as developed by the Russian school. The publications of A. Pérez, [55, 56], offer another measure theoretic formulation of information theory in which martingale theory is used as a technique to establish convergence results. Of more recent contributions to information theory we mention the work of I. Csiszár [16].

A recent textbook on information theory is the one by T.M. Cover and J.A. Thomas [13]. Information theory of continuous stochastic processes is treated in the book by S. Ihara [37]. Other books are [30, 45].

Definition C.1. *Let $x : \Omega \to X$ be a finite valued random variable with $X = \{a_1, a_2, \ldots, a_n\}$ and let $p_x : X \to \mathbb{R}$ be its frequency function, $p_x(a_i) = P(\{x = a_i\})$. Define the* entropy of the finite valued random variable *x as*

$$H_C(x) = \sum_{i=1}^{n} p_x(a_i) \ln\left(\frac{1}{p_x(a_i)}\right). \tag{C.1}$$

The concept of entropy of a finite valued random variable should be regarded as entropy with respect to a counting measure. Entropy is a property of a probability measure, not of a random variable. Nevertheless, the terminology of information theory is followed in referring to entropy of a random variable.

Definition C.2. *Let $x : \Omega \to \mathbb{R}^n$ be a random variable whose probability distribution function is absolutely continuous with respect to Lebesgue measure with density $p_x : \mathbb{R}^n \to \mathbb{R}_+$. Define the* entropy *with respect to Lebesgue measure of such a random variable by*

$$H_L(x) = \int_{\mathbb{R}^n \cap S} p_x(v) \ln\left(\frac{1}{p_x(v)}\right) dv, \tag{C.2}$$

where S is the support of p_x.

The entropy defined above depends on Lebesgue measure, or in general on the measure with respect to which it is defined.

Definition C.3. *[57, p. 19] Let (Ω, F) be a measurable space consisting of a set Ω and a σ-algebra F. Let P_1, P_2 be two probability measures defined on (Ω, F). Define the* entropy *of P_1 with respect to P_2 by the formula*

$$H(P_1, P_2) = \sup_{\{E_1,\ldots,E_n\}} \sum_{i=1}^{n} P_1(E_i) \ln\left(\frac{P_1(E_i)}{P_2(E_i)}\right), \tag{C.3}$$

where the supremum is taken over all finite measurable partitions of Ω.

Besides entropy there is the concept of mutual information.

Definition C.4. *[24, p. 200] Let $x : \Omega \to X$, $y : \Omega \to Y$ be random variables taking values in the finite sets $X = \{x_1, \ldots, x_n\}$, $Y = \{y_1, \ldots, y_m\}$ respectively. The* mutual information of x and y *is defined by the formula*

$$J(x, y) = \sum_{i=1}^{n} \sum_{j=1}^{m} p_{xy}(i, j) \ln\left(\frac{p_{xy}(i, j)}{p_x(i) p_y(j)}\right), \tag{C.4}$$

where

$$\begin{aligned} p_{xy}(i, j) &= P(\{x = x_i, y = y_j\}), \\ p_x(i) &= P(\{x = x_i\}), \qquad p_y(j) = P(\{y = y_j\}). \end{aligned}$$

Mutual information of arbitrary real valued random variables may be defined by a limiting argument whereby the entropy is defined as the limit of the entropy of a discrete approximation of the real valued random variables, see [24].

Proposition C.1. *[24, pp. 209-210]. Let $x : \Omega \to \mathbb{R}^n$, $y : \Omega \to \mathbb{R}^p$ be random variables. Assume that the probability distribution functions associated with the probability measures of P_{xy}, P_x, P_y are absolutely continuous with respect to Lebesgue measure with densities denoted by p_{xy}, p_x, p_y. Then the mutual information, if it is finite, is given by the formula*

$$J(x,y) = \int\int p_{xy}(u,v) \ln\left(\frac{p_{xy}(u,v)}{p_x(u)p_y(v)}\right) dudv, \tag{C.5}$$

where the integral is a Lebesgue integral.

Mutual information of two random variables satisfies $0 \leq J(x,y) \leq +\infty$. Moreover, $J(x,y) < \infty$ if the probability measure P_{xy} is absolutely continuous with respect to the product measure $P_x \times P_y$. Also, $J(x,y) = 0$ iff x, y are independent random variables.

The third concept from information theory introduced here is divergence.

Definition C.5. *Given a measurable space (Ω, F). Let*

$$\begin{aligned} F_{2s} &= \{f : \mathbb{R}_+ \to \mathbb{R} \mid; f \, \textit{strictly convex and } f(1) = 0\}, \\ \mathbf{P} &= \{P : F \to \mathbb{R}_+ \mid P \textit{ is probability measure } \}. \end{aligned}$$

For $f \in F_{2s}$ define the pseudo-distance *d_f on $\mathbf{P}$ as $d_f : \mathbf{P} \times \mathbf{P} \to \mathbb{R}$*

$$d_f(P_1, P_2) = E_Q[f(\frac{r_1}{r_2})r_2] = E_{P_2}[f(\frac{r_1}{r_2})], \tag{C.6}$$

where Q is a σ-finite measure on (Ω, F) such that

$$P_1 \ll Q, \; \frac{dP_1}{dQ} = r_1, \qquad P_2 \ll Q, \; \frac{dP_2}{dQ} = r_2.$$

A σ-finite measure Q such as used above always exists, $Q = P_1 + P_2$ will do. The definition of d_f does not depend on the choice of Q. It can then be shown that for all P_1, P_2, we have $d_f(P_1, P_2) \geq 0$ and $d_f(P_1, P_2) = 0$ if and only if $P_1 = P_2$. In general d_f does not satisfy the triangle inequality hence it is not a distance.

Definition C.6. *The* divergence *or the* Kullback-Leibler pseudo-distance *on $\mathbf{P}$ is defined as a special case of the pseudo-distance defined above with*

$$f : \mathbb{R}_+ \to \mathbb{R}, \; f(x) = \begin{cases} x \ln x, & \textit{if } x > 0, \\ 0, & x = 0, \end{cases} \tag{C.7}$$

$$\begin{aligned} D(P_1 \| P_2) &= d_f(P_1, P_2) \\ &= E_Q\left[f(\frac{r_1}{r_2})r_2\right] = E_{P_2}\left[f(\frac{r_1}{r_2})\right] = E_Q\left[r_1 \ln(\frac{r_1}{r_2}) I_{(r_2>0)}\right]. \end{aligned} \tag{C.8}$$

In general $D(P_1\|P_2) \neq D(P_2\|P_1)$.

Divergence is a special case of a pseudo-distance. The choice of the function f specified by (C.7) seems arbitrary. Information and communication theory provide ample evidence that the choice of this function and the concepts of entropy, mutual information, and divergence derived from it are useful for engineering.

Mutual information of real-valued random variables is related to divergence of the probability measures associated with the same variables by the formula

$$J(x,y) = D(P_{xy}\|P_x \times P_y). \tag{C.9}$$

Theorem C.1. *[57, Th. 2.4.2., p. 20] Let (Ω, F) be a measurable space with two probability measures P_1, P_2 defined on it. If the entropy $H(P_1, P_2)$ is finite then P_1 is absolutely continuous with respect to P_2 and*

$$H(P_1, P_2) = D(P_1\|P_2). \tag{C.10}$$

The last stated theorem and the remark above it illustrate the relationship between entropy, mutual information, and divergence. Divergence of probability measures is the basic concept, mutual information may be derived from it, and so may entropy.

D. Information Measures of Gaussian Random Variables

Proposition D.1. *[13, Th. 9.4.1] Let $y : \Omega \to \mathbb{R}^k$, $y \in G(m,Q)$. Assume that $Q = Q^{\mathrm{T}} > 0$. The entropy of the random variable y, or of the measure $G(m,Q)$, with respect to Lebesgue measure is given by the formula*

$$H_L(G(m,Q)) = \frac{1}{2}\ln\left((2\pi e)^k \det Q\right). \tag{D.1}$$

Note that the entropy does not depend on the mean of the Gaussian distribution.

Proposition D.2. *[24, Th. 1.2., p.209] Let $x : \Omega \to \mathbb{R}^n$, $y : \Omega \to \mathbb{R}^k$ be Gaussian random variables and $S \in \mathbb{R}^{n \times n}$. Then $J(Sx,y) \leq J(x,y)$. If in addition the matrix S is nonsingular then $J(Sx,y) = J(x,y)$.*

Proposition D.3. *[24, Th. 2.1] Let $x : \Omega \to \mathbb{R}^n$, $y : \Omega \to \mathbb{R}^k$, $(x,y) \in G(0,Q)$, $Q = Q^{\mathrm{T}} > 0$, and*

$$Q = \begin{pmatrix} Q_x & Q_{xy} \\ Q_{xy}^{\mathrm{T}} & Q_y \end{pmatrix}.$$

a. *The mutual information of the random variables x, y is given by the formula*

$$J(x,y) = -\frac{1}{2}\ln\left(\frac{\det Q}{\det Q_x \det Q_y}\right). \tag{D.2}$$

b. Let $\lambda_1, \ldots, \lambda_r \in (0,1)$ be the nonzero canonical correlations of the random variables x, y. Then

$$J(x,y) = -\frac{1}{2}\sum_{i=1}^{r} \ln(1-\lambda_i^2). \tag{D.3}$$

If in Proposition D.3

$$Q = \begin{pmatrix} I & Q_{xy} \\ Q_{xy}^{\mathrm{T}} & I \end{pmatrix},$$

then the mutual information is given by the formula

$$J(x,y) = -\frac{1}{2} \ln \det(I - Q_{xy}^{\mathrm{T}} Q_{xy}). \tag{D.4}$$

Because of Proposition D.2 mutual information is invariant under scaling of the individual random variables. Hence mutual information of Gaussian random variables can be expressed in terms of canonical correlations as in Proposition D.3. Such a representation was first derived in [24].

Proposition D.4. *Let $G(m_1, Q_1)$ and $G(m_2, Q_2)$ be two Gaussian measures on $\mathbb{R}^n$. Assume that $Q_1 > 0$ and $Q_2 > 0$. The divergence between these measures is given by the formula*

$$\begin{aligned} &2D\left(G(m_1,Q_1)\|G(m_2,Q_2)\right) \\ &= -\ln\left(\frac{\det Q_1}{\det Q_2}\right) + \mathrm{tr}([Q_2^{-1} - Q_1^{-1}]Q_1) + (m_1 - m_2)^{\mathrm{T}} Q_2^{-1}(m_1 - m_2) \end{aligned} \tag{D.5}$$

$$= \sum_{i=1}^{n} [\lambda_i(Q_1,Q_2) - \ln \lambda_i(Q_1,Q_2) - 1] + (m_1 - m_2)^{\mathrm{T}} Q_2^{-1}(m_1 - m_2), \tag{D.6}$$

where $\{\lambda_i(Q_1,Q_2), i \in 1,\ldots,n\}$ are the generalized eigenvalues of Q_1 with respect to Q_2, or the solutions of the polynomial equation in λ

$$\det(Q_2\lambda - Q_1) = 0. \tag{D.7}$$

The authors have not found this result in the literature although it is an elementary calculation. Note the well known convex function $f(x) = x - \ln x - 1$ in (D.6).

Proof of D.4 Both measures are absolutely continuous with respect to Lebesgue measure. Thus

$$D(G(m_1,Q_1)\|G(m_2,Q_2)) = E_L\left[r_1 \ln\left(\frac{r_1}{r_2}\right)\right],$$

where

$$r_1(v) = \frac{1}{\sqrt{(2\pi)^n \det Q_1}} \exp\left(-\frac{1}{2}(v-m_1)^{\mathrm{T}} Q_1^{-1}(v-m_1)\right),$$

$$2\ln\left(\frac{r_1(v)}{r_2(v)}\right) = -\ln\left(\frac{\det Q_1}{\det Q_2}\right) - (v-m_1)^{\mathrm{T}} Q_1^{-1}(v-m_1) + (v-m_2)^{\mathrm{T}} Q_2^{-1}(v-m_2).$$

$$2D(G(m_1,Q_1)\|G(m_2,Q_2)) = 2\int r_1(v)\ln\left(\frac{r_1(v)}{r_2(v)}\right) dv$$

$$= -\ln\left(\frac{\det Q_1}{\det Q_2}\right) - \mathrm{tr}(Q_1^{-1}Q_1) + \mathrm{tr}(Q_2^{-1}Q_1) + (m_1-m_2)^{\mathrm{T}} Q_2^{-1}(m_1-m_2).$$

The expression in terms of generalized eigenvalues is then easy to derive. □

Proposition D.5. *Let* $x : \Omega \to \mathbb{R}^n$, $z : \Omega \to \mathbb{R}^p$, $x \in G(m_1, V_1)$, $m_1 \in \mathbb{R}^n$, $V_1 \in \mathbb{R}^{n\times n}$, $V_1 = V_1^{\mathrm{T}} > 0$, $u \in \mathbb{R}^m$, $C \in \mathbb{R}^{p\times n}$, $D \in \mathbb{R}^{p\times m}$, $c \in \mathbb{R}$,

$$z = Cx + Du. \tag{D.8}$$

Assume that

$$V_2^{-1} := V_1^{-1} - cC^{\mathrm{T}}C > 0. \tag{D.9}$$

Then

$$E[c\exp(\frac{1}{2}cz^{\mathrm{T}}z)] = c\left(\frac{\det V_2}{\det V_1}\right)^{1/2} \exp\left[\frac{1}{2}c\begin{pmatrix} m_1 \\ u\end{pmatrix}^{\mathrm{T}} M \begin{pmatrix} m_1 \\ u\end{pmatrix}\right], \tag{D.10}$$

where

$$M = \begin{pmatrix} \frac{1}{c}[V_1^{-1}V_2V_1^{-1} - V_1^{-1}] & V_1^{-1}V_2C^{\mathrm{T}}D \\ D^{\mathrm{T}}CV_2V_1^{-1} & D^{\mathrm{T}}D + cD^{\mathrm{T}}CV_2C^{\mathrm{T}}D \end{pmatrix}. \tag{D.11}$$

Proof: The proof is a tedious calculation. Let

$$r = -\left[V_1^{-1} - cC^{\mathrm{T}}C\right]^{-1}\begin{pmatrix} -V_1^{-1} & -cC^{\mathrm{T}}D \end{pmatrix}\begin{pmatrix} m_1 \\ u \end{pmatrix}. \tag{D.12}$$

Then

$$E\left[\exp\left(\frac{1}{2}cz^{\mathrm{T}}z\right)\right]$$

$$= \int \frac{1}{\sqrt{(2\pi)^n \det V_1}} \exp\left(\frac{1}{2}cz^{\mathrm{T}}z - \frac{1}{2}(v-m_1)^{\mathrm{T}}V_1^{-1}(v-m_1)\right) dv.$$

The exponent can be computed as

$$\begin{aligned}
&-cz^{\mathrm{T}}z + (v - m_1)^{\mathrm{T}}V_1^{-1}(v - m_1) \\
&= \begin{pmatrix} v \\ u \end{pmatrix}^{\mathrm{T}} \begin{pmatrix} -cC^{\mathrm{T}}C & -cC^{\mathrm{T}}D \\ -cD^{\mathrm{T}}C & -cD^{\mathrm{T}}D \end{pmatrix} \begin{pmatrix} v \\ u \end{pmatrix} \\
&\quad + \begin{pmatrix} v \\ m_1 \end{pmatrix}^{\mathrm{T}} \begin{pmatrix} V_1^{-1} & -V_1^{-1} \\ -V_1^{-1} & V_1^{-1} \end{pmatrix} \begin{pmatrix} v \\ m_1 \end{pmatrix} \\
&= (v - r)^{\mathrm{T}}(V_1^{-1} - cC^{\mathrm{T}}C)(v - r) - c \begin{pmatrix} m_1 \\ u \end{pmatrix}^{\mathrm{T}} M \begin{pmatrix} m_1 \\ u \end{pmatrix}.
\end{aligned}$$

Then:

$$\begin{aligned}
&E\left[\exp\left(\frac{1}{2}cz^{\mathrm{T}}z\right)\right] \\
&= \int \frac{1}{\sqrt{(2\pi)^n \det V_1}} \exp\left(-\frac{1}{2}(v - r)^{\mathrm{T}}(V_1^{-1} - cC^{\mathrm{T}}C)(v - r)\right) \\
&\quad \times \exp\left(\frac{1}{2}c \begin{pmatrix} m_1 \\ u \end{pmatrix}^{\mathrm{T}} M \begin{pmatrix} m_1 \\ u \end{pmatrix}\right) dv \\
&= \left(\frac{\det(V_1^{-1} - cC^{\mathrm{T}}C)^{-1}}{\det V_1}\right)^{\frac{1}{2}} \exp\left(\frac{1}{2}c \begin{pmatrix} m_1 \\ u \end{pmatrix}^{\mathrm{T}} M \begin{pmatrix} m_1 \\ u \end{pmatrix}\right).
\end{aligned}$$

□

Suppose that in Proposition D.5 there holds $m_1 = 0$, $u = 0$, $c > 0$, and $V_1 = I$. Condition (D.9) is then equivalent to $I - cC^TC > 0$ and

$$\begin{aligned}
\ln E[c^{1/2}\exp(\frac{1}{2}cz^Tz)] &= -\frac{1}{2}\ln\det(I - cC^TC) + \frac{1}{2}\ln c \\
&= -\frac{1}{2}\ln\det(\frac{1}{c}I - C^TC). \qquad \text{(D.13)}
\end{aligned}$$

Note the analogy of equation (D.13) with (D.4).

E. Information Measures of Stationary Gaussian Processes

Definition E.1. *[13, p. 63] Let $x : \Omega \to X$ be a stochastic process on $T = \mathbb{Z}$. The* entropy rate *of this process is defined by the formula*

$$H(x) = \lim_{n\to\infty} \frac{1}{n} H(x_1, x_2, \ldots, x_n), \qquad \text{(E.1)}$$

if the limit exists.

Let $x : \Omega \times \mathbb{Z} \to \mathbb{R}$ be a stationary Gaussian process with spectral density S. Then the entropy rate of the process x is given by

$$h(x) = \frac{1}{2}\ln(2\pi e) + \frac{1}{4\pi}\int_{-\pi}^{\pi} \ln S(e^{i\lambda})d\lambda. \tag{E.2}$$

This result is due to A.N. Kolmogorov [44].

Definition E.2. *[29, p. 86; p.135], [30], [37, 2.1.5]. Let $x : \Omega \times T \to X$, $y : \Omega \times T \to Y$, be stochastic processes on $T = \mathbb{Z}$. The* mutual information rate *of x and y is defined by the formula*

$$J(x,y) = \limsup_{n\to\infty} \frac{1}{2n+1} J(x|_{[-n,n]}, y|_{[-n,n]}), \tag{E.3}$$

where $J(x|_{[-n,n]}, y|_{[-n,n]})$ is the mutual information of the processes x, y restricted to the interval $\{-n, -n+1, \ldots, -1, 0, 1, \ldots, n\}$.

Let for $T = \mathbb{Z}$, $x : \Omega \times T \to \mathbb{R}^m$ and $y : \Omega \times T \to \mathbb{R}^p$ be two jointly stationary and Gaussian processes that admit a joint spectral density. Denote their joint spectral density function by

$$\hat{W} : \mathbb{C} \to \mathbb{C}^{(m+p)\times(m+p)}, \quad \hat{W} = \begin{pmatrix} \hat{W}_x & \hat{W}_{xy} \\ \hat{W}_{xy}^{\mathrm{T}} & \hat{W}_y \end{pmatrix}.$$

Assume that the spectral density is nonsingular. It follows from [24] that the mutual information of two stationary Gaussian processes is invariant under scaling by nonsingular finite-dimensional linear systems. Therefore transformation to the following canonical form is useful. Let $S_x : \mathbb{C} \to \mathbb{C}^{m\times m}$ and $S_y : \mathbb{C} \to \mathbb{C}^{p\times p}$ be minimal square spectral factors of respectively $\hat{W}_x$ and $\hat{W}_y$, or

$$\hat{W}_x(z) = S_x(z)S_x(z^{-1})^{\mathrm{T}}, \quad \hat{W}_y(z) = S_y(z)S_y(z^{-1})^{\mathrm{T}}.$$

Define

$$S(z) = S_x(z)^{-1}\hat{W}_{xy}(z)S_y(z^{-1})^{-T}.$$

The spectral density of the transformed process is then

$$\begin{pmatrix} I & S(z) \\ S(z^{-1})^{\mathrm{T}} & I \end{pmatrix}.$$

From the fact that $\hat{W}$ is a spectral density follows that $S(z)S(z^{-1})^{\mathrm{T}} \le I$.

Theorem E.1. *a. The mutual information rate of the stationary Gaussian processes defined above is given by the formula*

$$J(x,y) = \frac{-1}{4\pi i}\int_{\mathbb{D}} \frac{1}{z} \ln\det[I - S(z^{-1})^{\mathrm{T}}S(z)]dz. \tag{E.4}$$

b. *Assume that the function S admits a realization as a finite-dimensional linear system with*

$$S(z) = C(zI - A)^{-1}B + D, \quad sp(A) \subset \mathbb{C}^{-}. \tag{E.5}$$

Assume that the following set of relations admits an unique solution $Q \in \mathbb{R}^{n \times n}$

$$\begin{aligned} Q &= Q^{\mathrm{T}} \geq 0, && \text{(E.6)} \\ Q &= A^{\mathrm{T}}QA + C^{\mathrm{T}}C \\ &\qquad +(A^{\mathrm{T}}QB + C^{\mathrm{T}}D)N^{-1}(B^{\mathrm{T}}QA + D^{\mathrm{T}}C), && \text{(E.7)} \\ N &= I - B^{\mathrm{T}}QB - D^{\mathrm{T}}D > 0, && \text{(E.8)} \\ & sp(A + BN^{-1}(B^{\mathrm{T}}QA + D^{\mathrm{T}}C)) \subset \mathbb{C}^{-}. && \text{(E.9)} \end{aligned}$$

Then

$$J(x,y) = -\frac{1}{2}\ln\det(I - B^{\mathrm{T}}QB - D^{\mathrm{T}}D). \tag{E.10}$$

I.M. Gelfand and A.M. Yaglom in [24] derived a formula for the mutual information rate in a geometric formulation. In [57] a formula analogous to (E.4) is presented for the scalar case. See also [37, Ch. 5]. In the case the stationary Gaussian processes are generated by finite dimensional Gaussian systems the result of Theorem E.1 seems new.

Proof: a. A proof of part a. can be given along the same line as the proof of theorem E.2. However, it also automatically follows from theorem E.2 by using the relation between divergence and mutual information as given by (C.9).

b. If $\|S\|_\infty = 1$ then $J(x,y) = +\infty$. If $\|S\|_\infty < 1$ then it follows from the bounded real lemma that there exists a $Q \in \mathbb{R}^{n \times n}$ that satisfies the relations of part b. of the theorem. A realization $\tilde{\Sigma}$ of the transfer function $I - S^{\mathrm{T}}(z^{-1})S(z)$ is given by

$$\begin{aligned} \sigma^{-1}x &= A^{\mathrm{T}}x + C^{\mathrm{T}}Cp + C^{\mathrm{T}}Du, \\ \sigma p &= Ap + Bu, \\ z &= -B^{\mathrm{T}}x - D^{\mathrm{T}}Cp + (I - D^{\mathrm{T}}D)u. \end{aligned}$$

Let

$$A_x = B^{\mathrm{T}}QA + D^{\mathrm{T}}C, \quad x_n = x - QAp - QBu.$$

Then a realization of $\tilde{\Sigma}$ is given by

$$\begin{aligned} \sigma^{-1}x_n &= A^{\mathrm{T}}x_n - A_x^{\mathrm{T}}N^{-1}A_xp + A_x^{\mathrm{T}}u, \\ \sigma p &= Ap + Bu, \\ z &= -B^{\mathrm{T}}x_n - A_xp + Nu. \end{aligned}$$

It is then easy to check that the transfer matrix of $\tilde{\Sigma}$ is equal to $H^{\mathrm{T}}(z^{-1})H(z)$ where

$$H(z) = -N^{-1/2}A_x(zI - A)^{-1}B + N^{1/2}.$$

Note that $H(z)$ and $H(z)^{-1}$ are, as a consequence of (E.5) and (E.9), both analytic outside the unit disc. Then

$$\begin{aligned}
J(x,y) &= \frac{-1}{4\pi i}\int_{\mathbb{D}} \frac{1}{z} \ln\det\left[I - S^{\mathrm{T}}(z^{-1})S(z)\right] dz \\
&= \frac{-1}{4\pi i}\int_{\mathbb{D}} \frac{1}{z} \ln\det\left[H^{\mathrm{T}}(z^{-1})H(z)\right] dz \\
&= \frac{-1}{4\pi i}\int_{\mathbb{D}} \frac{1}{z} \ln|\det H^{\mathrm{T}}(z^{-1})|^2 dz \\
&= \frac{-1}{4\pi}\int_{-\pi}^{\pi} \ln|\det H^{\mathrm{T}}(e^{-i\theta})|^2 d\theta \\
&= Re\frac{-1}{4\pi}\int_{-\pi}^{\pi} 2\ln\det H^{\mathrm{T}}(e^{-i\theta}) d\theta \\
&= Re\frac{-1}{2\pi i}\int_{\mathbb{D}} \frac{1}{z} \ln\det H^{\mathrm{T}}(z^{-1}) dz \\
&= Re\frac{-1}{2\pi i} 2\pi i \ln\det H^{\mathrm{T}}(\infty) \\
&= -\frac{1}{2}\ln\det(I - B^{\mathrm{T}}QB - D^{\mathrm{T}}D).
\end{aligned}$$

□

Definition E.3. *[37, 2.1.6] Let $x_1, x_2 : \Omega \times T \to \mathbb{R}^p$ be two stationary processes on $T = \mathbb{Z}$. Denote by P_1, P_2 the measures induced by x_1, x_2 respectively on $(\mathbb{R}^n)^{\mathrm{T}}$. The* divergence rate *between P_1, P_2 is defined by the formula*

$$D(P_1\|P_2) = \lim_{n\to\infty} \frac{1}{2n+1} D(P_1|_{[-n,n]}\|P_2|_{[-n,n]}), \tag{E.11}$$

if the limit exists, where $P_1|_{[-n,n]}, P_2|_{[-n,n]}$ denote the restrictions of P_1, P_2 respectively to the time index set $\{-n, \ldots, -1, 0, 1, \ldots, n\}$.

Theorem E.2. *Let $y_1, y_2 : \Omega \times T \to \mathbb{R}^p$ be two jointly stationary Gaussian processes on $T = \mathbb{Z}$ with values in $\mathbb{R}^p$. Assume that they have a mean value function that is zero, that their covariance functions are denoted by $W_1, W_2 : T \to \mathbb{R}^{p\times p}$, and that they admit spectral densities, say $\hat{W}_1, \hat{W}_2 : \mathbb{C} \to \mathbb{C}^{p\times p}$ respectively.*

a. The divergence rate between the measures induced by these processes exists and is given by the formula

$$\begin{aligned}
&D(P_1\|P_2) \\
&= \frac{1}{4\pi}\int_{-\pi}^{\pi} \ln\left(\frac{\det \hat{W}_2(e^{i\lambda})}{\det \hat{W}_1(e^{i\lambda})}\right)
\end{aligned}$$

$$+\mathrm{tr}\left(\hat{W}_2(e^{i\lambda})^{-1}[\hat{W}_1(e^{i\lambda}) - \hat{W}_2(e^{i\lambda})]\right) d\lambda, \tag{E.12}$$

$$= \frac{1}{4\pi}\int_{-\pi}^{\pi} \mathrm{tr}\left(S^{\mathrm{T}}(e^{-i\lambda})S(e^{i\lambda}) - I\right) - \ln[\det S(e^{i\lambda})]^2 d\lambda, \tag{E.13}$$

where

$$\hat{W}_1(z) = G_1^{\mathrm{T}}(z^{-1})G_1(z), \quad \hat{W}_2(z) = G_2^{\mathrm{T}}(z^{-1})G_2(z), \tag{E.14}$$

$$S(z) = G_1(z)G_2(z)^{-1}. \tag{E.15}$$

b. Assume in addition that the transfer function S admits a realization as a finite-dimensional linear system with a minimal realization parametrized by

$$\begin{aligned} x(t+1) &= Ax(t) + Bu(t), \\ y(t) &= Cx(t) + Du(t), \end{aligned} \tag{E.16}$$

where $(A, B, C, D) \in \mathbb{R}^{n\times n} \times \mathbb{R}^{n\times p} \times \mathbb{R}^{p\times n} \times \mathbb{R}^{p\times p}$ and $sp(A) \subset \mathbb{C}^-$. Assume further that there exists a matrix $Q \in \mathbb{R}^{n\times n}$ such that the following relations are satisfied

$$Q = Q^{\mathrm{T}} \geq 0, \tag{E.17}$$

$$Q = A^{\mathrm{T}}QA + C^{\mathrm{T}}C - [A^{\mathrm{T}}QB + C^{\mathrm{T}}D][D^{\mathrm{T}}D + B^{\mathrm{T}}QB]^{-1} \times [A^{\mathrm{T}}QB + C^{\mathrm{T}}D]^{\mathrm{T}}, \tag{E.18}$$

$$N = D^{\mathrm{T}}D + B^{\mathrm{T}}QB > 0, \tag{E.19}$$

$$sp(A - BN^{-1}(B^{\mathrm{T}}QA + D^{\mathrm{T}}C)) \subset \mathbb{C}^-, \tag{E.20}$$

and that this set of relations admits an unique solution. Let $R \in \mathbb{R}^{n\times n}$ be the unique solution to the Lyapunov equation

$$R = A^{\mathrm{T}}RA + C^{\mathrm{T}}C. \tag{E.21}$$

Then the divergence rate between the measures induced by the two processes is given by the formula

$$D(P_1\|P_2) = -\frac{1}{2}\ln\det(B^{\mathrm{T}}QB + D^{\mathrm{T}}D) + \frac{1}{2}\mathrm{tr}(B^{\mathrm{T}}RB + D^{\mathrm{T}}D - I). \tag{E.22}$$

Proof: a. Consider the processes restricted to the finite interval $\{-m, \ldots, -1, 0, 1, \ldots, m\}$. Denote the measures associated with the processes y_1, y_2 by P_1, P_2 respectively and their restrictions to the finite interval by $P_1|_{[-m,m]}, P_2|_{[-m,m]}$. Define for $i = 1, 2$ the block-Toeplitz matrix

$$R_i^m = [W_i(j-k)]_{j,k=-m,\ldots,0,\ldots,m}$$

From Proposition D.4 follows that

$$D(P_1|_{[-m,m]}\|P_2|_{[-m,m]}) = -\frac{1}{2}\ln\left(\frac{\det R_1^m}{\det R_2^m}\right)+\frac{1}{2}\mathrm{tr}\left([(R_2^m)^{-1}-(R_1^m)^{-1}]R_1^m\right).$$

From Szegö's limit theorem for block-Toeplitz matrices (see e.g. [58]) follows that

$$\lim_{m\to\infty}[\det R_i^m]^{1/2m} = \exp\frac{1}{2\pi}\int_{-\pi}^{\pi}\ln\det\hat{W}_i(e^{i\lambda})d\lambda.$$

Hence

$$\lim_{m\to\infty}\frac{1}{2m}\ln\left(\frac{\det R_2^m}{\det R_1^m}\right)=\frac{1}{2\pi}\int_{-\pi}^{\pi}\ln\left(\frac{\det\hat{W}_2(e^{i\lambda})}{\det\hat{W}_1(e^{i\lambda})}\right)d\lambda.$$

Let

$$S^m(t)=[(1-t)(R_1^m)^{-1}+t(R_2^m)^{-1}]^{-1}[(R_2^m)^{-1}-(R_1^m)^{-1}].$$

As a consequence of Szegö's limit theorem, we know that the spectrum of R_i^m is contained in a region $[\bar{m}_i,\bar{M}_i]$ with $\bar{m}_i>0$ where $\bar{m}_i,\bar{M}_i\in\mathbb{R}$ are independent of m. This implies that the spectrum of $S^m(t)$ is also contained in some region $[\bar{m},\bar{M}]$ where $\bar{m},\bar{M}\in\mathbb{R}$ are independent of m,t. Hence for a neighborhood of $t=0$, $\frac{1}{m}\mathrm{tr}S^m(t)$ is a uniformly continuous function in m and t. Therefore, making use of the formula

$$\frac{d}{dt}\log\det K(t)=\mathrm{tr}\left(K^{-1}(t)\frac{d}{dt}K(t)\right),$$

we get

$$\begin{aligned}
&\lim_{m\to\infty}\frac{1}{2m}\mathrm{tr}(R_1^m[(R_2^m)^{-1}-(R_1^m)^{-1}])\\
&= \lim\frac{1}{2m}\mathrm{tr}S^m(0)\\
&= \lim\frac{d}{dt}\frac{1}{2m}\ln\det((1-t)(R_1^m)^{-1}+t(R_2^m)^{-1})\Big|_{t=0}\\
&= \frac{d}{dt}\left(\lim\frac{1}{2m}\ln\det((1-t)(R_1^m)^{-1}+t(R_2^m)^{-1})\right)\Big|_{t=0}\\
&= \frac{d}{dt}\Big[\lim\Big(-\ln(\det R_1^m)^{1/2m}-\ln(\det R_2^m)^{1/2m}\\
&\qquad +\ln\det((1-t)R_2^m+tR_1^m)^{1/2m}\Big)\Big]\Big|_{t=0}\\
&= \frac{d}{dt}\Big[\frac{1}{2\pi}\int_{-\pi}^{\pi}-\ln\det\hat{W}_1(e^{i\lambda})-\ln\det\hat{W}_2(e^{i\lambda})\\
&\qquad +\ln\det[(1-t)\hat{W}_2(e^{i\lambda})+t\hat{W}_1(e^{i\lambda})]d\lambda\Big]\Big|_{t=0}\\
&= \frac{1}{2\pi}\int_{-\pi}^{\pi}\mathrm{tr}\left(\hat{W}_2^{-1}[\hat{W}_1(e^{i\lambda})-\hat{W}_2(e^{i\lambda})]\right)d\lambda.
\end{aligned}$$

Thus

$$\begin{aligned}
D(P_1\|P_2) &= \lim_{m\to\infty}\frac{1}{2m+1}D(P_1|_{[-m,m]}\|P_2|_{[-m,m]}) \\
&= \frac{1}{4\pi}\int_{-\pi}^{\pi}\ln\left(\frac{\det \hat{W}_2(e^{i\lambda})}{\det \hat{W}_1(e^{i\lambda})}\right)d\lambda \\
&\quad +\frac{1}{4\pi}\int_{-\pi}^{\pi}\mathrm{tr}\left(\hat{W}_2^{-1}[\hat{W}_1(e^{i\lambda})-\hat{W}_2(e^{i\lambda})]\right)d\lambda \\
&= \frac{1}{4\pi}\int_{-\pi}^{\pi}-\ln\det(S(e^{i\lambda})^2)d\lambda \\
&\quad +\frac{1}{4\pi}\int_{-\pi}^{\pi}\mathrm{tr}\left(S^{\mathrm{T}}(e^{-i\lambda})S(e^{i\lambda})-I\right)d\lambda.
\end{aligned}$$

b. A realization $\tilde{\Sigma}$ of $S^{\mathrm{T}}(z^{-1})S(z)$ is provided by

$$\begin{aligned}
\sigma^{-1}x &= A^{\mathrm{T}}x + C^{\mathrm{T}}Cp + C^{\mathrm{T}}Du, \\
\sigma p &= Ap + Bu, \\
z &= B^{\mathrm{T}}x + D^{\mathrm{T}}Cp + D^{\mathrm{T}}Du.
\end{aligned}$$

Let

$$A_x = B^{\mathrm{T}}QA + D^{\mathrm{T}}C, \quad x_n = x - QAp - QBu.$$

Then the realization of $\tilde{\Sigma}$ may be written as

$$\begin{aligned}
\sigma^{-1}x_n &= A^{\mathrm{T}}x_n + A_x^{\mathrm{T}}N^{-1}A_xp + A_x^{\mathrm{T}}u, \\
\sigma p &= Ap + Bu, \\
z &= B^{\mathrm{T}}x_n + A_xp + Nu,
\end{aligned}$$

and it is then easy to check that the transfer matrix of $\tilde{\Sigma}$ is equal to $H(z^{-1})^{\mathrm{T}}H(z)$ where

$$H(z) = N^{-1/2}A_x(zI-A)^{-1}B + N^{1/2}.$$

Note that $H(z)$ and $H(z)^{-1}$ are both analytic outside the unit disc. Then, analogously to the proof of Theorem E.1,

$$\begin{aligned}
&D(P_1\|P_2) \\
&= \frac{1}{4\pi i}\int_{\mathbb{D}}\frac{1}{z}\left(\mathrm{tr}\left(S^{\mathrm{T}}(z^{-1})S(z)-I\right)-\ln|\det(H^{\mathrm{T}}(z^{-1})|^2\right)dz \\
&= -\mathrm{Re}\ \ln\det H(\infty) + \frac{1}{2}\mathrm{tr}(B^{\mathrm{T}}RB + D^{\mathrm{T}}D - I) \\
&= -\frac{1}{2}\ln\det(D^{\mathrm{T}}D + B^{\mathrm{T}}QB) + \frac{1}{2}\mathrm{tr}(B^{\mathrm{T}}RB + D^{\mathrm{T}}D - I).
\end{aligned}$$

□

F. LEQG Optimal Stochastic Control

The purpose of this appendix is to introduce the reader to the discrete-time optimal stochastic control problem with a Gaussian stochastic control problem and an exponential-of-quadratic cost function (LEQG). In addition an expression is derived for the cost in case of an infinite horizon.

The problem formulation consists of a discrete-time partially observed stochastic control system driven by Gaussian noise processes. The cost function over a finite-horizon is the expected value of the exponent of an additive form that is quadratic in the state and in the input process. The class of control laws consists of all measurable nonlinear functions of the past observations and inputs.

The history of the LEQG optimal stochastic control problem is summarized below. D.H. Jacobson formulated and solved the discrete-time and the continuous-time complete observations case of the problem in [38]. Complete observations refers to the case in which the input may depend on the current and the past state. Various special cases of the partial observations case were solved [64, 65, 47, 48] but the general case was considered as not to admit a solution in the form of a finite-dimensional control law. P. Whittle [70] solved the discrete-time partial observations LEQG optimal stochastic control problem and established the existence of a finite-dimensional control law. Additional publications on this and on related problems by Whittle are [71, Ch. 19] and [76, 72, 73, 74, 75]. The solution to the continuous-time partial observations LEQG optimal stochastic control problem was presented in [11]. Related publications on the LEQG problem are [32, 39, 59, 60]. Recently a rigorous and elegant approach to the LEQG optimal stochastic control problem has been proposed by M.R. James, see [12, 40, 41].

The LEQG optimal stochastic control problem is of interest for several reasons. Of interest to the development of stochastic control theory is the fact that a partially observed optimal stochastic control problem admits a control law with a finite dimensional representation. It may point the way to other optimal stochastic control problems with the same property. The solution of the LEQG problem has been shown to be equivalent to the solution of a H_∞-optimal control problem with the entropy criterion. Therefore the solution to the LEQG problem has certain robustness properties that are of interest in control engineering.

Problem F.1. *The discrete-time linear-exponential-quadratic-Gaussian (LEQG) optimal stochastic control problem with partial observations.* Consider the Gaussian stochastic control system

$$\begin{aligned} x(t+1) &= A_1(t)x(t) + B_1(t)u(t) + M(t)v(t), \quad x(0) = x_0, \\ y(t) &= C_1(t)x(t) + D_1(t)u(t) + N(t)v(t), \\ z(t) &= C_2(t)x(t) + D_2(t)u(t), \end{aligned} \tag{F.1}$$

where $T = \{0, 1, \ldots, t_1\}$, $T_1 = \{0, 1, \ldots, t_1 - 1\}$, $t_1 \in \mathbb{Z}_+$. See Appendix B. for the full details of the definition of a Gaussian stochastic control system. Assume that for all $t \in T$, $N(t)V(t)N(t)^{\mathrm{T}} > 0$.

Define the class G of control laws by $g \in G$, $g = \{g_0, g_1, \cdots, g_{t_1 - 1}\}$, $g_0 \in U$, and for $t = 1, \ldots, t_1 - 1$, $g_t : Y^t \times U^t \to U$, where g_t are measurable maps.

For the definition of the cost function let $c \in \mathbb{R}$, $c \neq 0$, Assume that for all $t \in T_1$ $D_2(t)^{\mathrm{T}} D_2(t) > 0$ and that $L_{1e} = L_{1e}^{\mathrm{T}} \geq 0$. Define the cost function $J : G \to \mathbb{R}$

$$J(g) \quad = \quad E[c \exp\left(\frac{1}{2}c\left[x^g(t_1)^{\mathrm{T}} L_{1e} x^g(t_1) + \sum_{s=0}^{t_1-1} z(s)^{\mathrm{T}} z(s)\right]\right)]. \quad \text{(F.2)}$$

The problem is to determine

$$\inf_{g \in G} J(g), \tag{F.3}$$

and to determine a control law $g^* \in G$ such that

$$J(g^*) = \inf_{g \in G} J(g). \tag{F.4}$$

The representation of the stochastic control system (F.1) differs from that considered by P. Whittle in [70] given by

$$\begin{aligned} x(t+1) &= A(t)x(t) + B(t)u(t) + v(t), \\ y(t+1) &= C(t)x(t) \qquad\qquad\qquad + w(t). \end{aligned} \tag{F.5}$$

It is well known in system theory for discrete-time systems that the solutions to the control problems for the representations (F.1) and (F.5) differ and these solutions cannot be directly transformed into each other.

Consider the Gaussian stochastic control system (F.1). Define the class G of control laws by $g \in G$, $g = \{g_0, g_1, \cdots, g_{t_1-1}\}$, $g_0 \in U$, and for $t = 1, \ldots, t_1 - 1$, $g_t : Y^t \times U^t \to U$, where g_t are measurable maps. For any $g \in G$ the closed-loop control system is specified by the relations

$$\begin{aligned} x^g(t+1) &= A_1(t)x^g(t) + B_1(t)g_t + M(t)v(t), \quad x^g(0) = x_0, \\ y^g(t) &= C_1(t)x^g(t) + D_1(t)g_t + N(t)v(t). \end{aligned} \tag{F.6}$$

where $g_t = g_t(y^g(0), \ldots, y^g(t-1), u^g(0), \ldots, u^g(t-1))$. In the sequel the superindex g on x^g, y^g will be omitted.

The formulation of a closed-loop stochastic control system includes the case of a dynamic compensator. The solution to Problem F.1 above will not be presented here. Instead an expression will be presented for the cost of an LEQG problem. This expression is needed in the body of the paper.

Proposition F.1. *Consider a Gaussian system*

$$\begin{aligned} x(t+1) &= A(t)x(t) + M(t)v(t), \quad x(t_0) = x_0, \\ z(t) &= C(t)x(t) + N(t)v(t), \end{aligned} \tag{F.7}$$

with $x_0 \in G(m_0, Q_0)$ and $v : \Omega \times T \to R^r$ Gaussian white noise with $v(t) \in G(0, V_1)$. Let $c \in \mathbb{R}$. Assume that

(i) $V_1 = V_1^{\mathrm{T}} > 0$;
(ii) $Q_0 = Q_0^{\mathrm{T}} > 0$;
(iii) *for all $t \in T$,* $V_1^{-1} - c[N^{\mathrm{T}}(t)N(t) + M^{\mathrm{T}}(t)Q_1(t+1)M(t)] > 0$;
(iv) $Q_0^{-1} - cQ_1(0) > 0$.

Define $Q_1 : T \to \mathbb{R}^{n\times n}$, $V_2 : T \to \mathbb{R}^{r\times r}$, $r : T \to \mathbb{R}$ recursively by

$$Q_1(t_1+1) = 0, \tag{F.8}$$

$$r(t_1+1) = c, \tag{F.9}$$

$$V_2(t)^{-1} = V_1^{-1} - c[N(t)^{\mathrm{T}}N(t) + M(t)^{\mathrm{T}}Q_1(t+1)M(t)], \tag{F.10}$$

$$\begin{aligned} Q_1(t) = {} & C(t)^{\mathrm{T}}C(t) + A(t)^{\mathrm{T}}Q_1(t+1)A(t) \\ & + c[C(t)^{\mathrm{T}}N(t) + A(t)^{\mathrm{T}}Q_1(t+1)M(t)]V_2(t) \\ & \times [C(t)^{\mathrm{T}}N(t) + A(t)^{\mathrm{T}}Q_1(t+1)M(t)]^{\mathrm{T}}, \end{aligned} \tag{F.11}$$

$$r(t) = r(t+1)\left(\frac{\det V_2(t)}{\det V_1}\right)^{1/2}, \tag{F.12}$$

$$Q_2 = Q_0^{-1}[Q_0^{-1} - cQ_1(0)]^{-1}Q_0^{-1} - Q_0^{-1}, \tag{F.13}$$

$$Q_3^{-1} = Q_0^{-1} - cQ_1(0). \tag{F.14}$$

Then

$$E\left[c\exp\left(\frac{1}{2}c\sum_{s=0}^{t_1} z(s)^{\mathrm{T}}z(s)\right)\right] = r(0)\left(\frac{\det Q_3}{\det Q_0}\right)^{1/2} \exp\left(\frac{1}{2}m_0^{\mathrm{T}}Q_2m_0\right).$$

Proof 1. Let $W : T \times X \to \mathbb{R}$

$$W(t, x(t)) = E\left[c\exp\left(\frac{1}{2}c\sum_{s=t}^{t_1} z(s)^{\mathrm{T}}z(s)\right)\middle| F_t^x \vee F_{t-1}^v\right]. \tag{F.15}$$

This function is well defined because x is a Markov process and v is Gaussian white noise. It will be shown that W satisfies the recursion

$$W(t, x(t)) = E\left[\exp\left(\frac{1}{2}cz(t)^{\mathrm{T}}z(t)\right)W(t+1, x(t+1))\middle| F_t^x \vee F_{t-1}^v\right],$$

with $W(t_1+1, x(t_1+1)) = c$. By definition, the formula at the terminal time holds. Let $t \in T$ $t < t_1$ and assume that the recursion formula holds for $s = t+1, \ldots, t_1$. Then

$$\begin{aligned}
&W(t,x(t)) \\
&= E\left[c\exp\left(\frac{1}{2}c\sum_{s=t}^{t_1} z(s)^{\mathrm{T}}z(s)\right)\middle| F_t^x \vee F_{t-1}^v\right] \\
&= E\left[\exp\left(\frac{1}{2}cz(t)^{\mathrm{T}}z(t)\right)\times\right. \\
&\qquad \left.\times E\left\{c\exp\left(\frac{1}{2}\sum_{s=t+1}^{t_1} z(s)^{\mathrm{T}}z(s)\right)\middle| F_{t+1}^x \vee F_t^v\right\}\middle| F_t^x \vee F_{t-1}^v\right] \\
&= E\left[\exp\left(\frac{1}{2}cz(t)^{\mathrm{T}}z(t)\right)W(t+1,x(t+1))\middle| F_t^x \vee F_{t-1}^v\right].
\end{aligned}$$

where the second equality follows by reconditioning and because $z(t)$ is measurable on $F_{t+1}^x \vee F_t^v$ The result then follows by induction.
2. The proof proceeds as in dynamic programming, except that there is no optimization in each step. Below use is made of Proposition D.5. Thus

$$\begin{aligned}
&W(t_1,x(t_1)) \\
&= E\left[c\exp\left(\frac{1}{2}cz(t_1)^{\mathrm{T}}z(t_1)\right)\middle| F_{t_1}^x \vee F_{t_1-1}^v\right] \\
&= E\left[c\exp\left(\frac{1}{2}c\begin{pmatrix}v(t_1)\\x(t_1)\end{pmatrix}^{\mathrm{T}}\begin{pmatrix}N^{\mathrm{T}}N & N^{\mathrm{T}}C\\C^{\mathrm{T}}N & C^{\mathrm{T}}C\end{pmatrix}\begin{pmatrix}v(t_1)\\x(t_1)\end{pmatrix}\right)\right. \\
&\qquad \left.\middle| F_{t_1}^x \vee F_{t_1-1}^v\right] \\
&= c\exp\left(\frac{1}{2}cx(t_1)^{\mathrm{T}}Q_1(t_1)x(t_1)\right)\left(\frac{\det V_2(t_1)}{\det V_1}\right)^{1/2} \\
&= r(t_1)\exp\left(\frac{1}{2}cx(t_1)^{\mathrm{T}}Q_1(t_1)x(t_1)\right).
\end{aligned}$$

Suppose that for $s = t+1, t+2, \ldots, t_1$

$$W(s,x(s)) = r(s)\exp\left(\frac{1}{2}cx(s)^{\mathrm{T}}Q_1(s)x(s)\right).$$

It will be shown that this formula holds for $s = t$.

$$\begin{aligned}
&W(t,x(t)) \\
&= E\left[\exp\left(\frac{1}{2}cz(t)^{\mathrm{T}}z(t)\right)W(t+1,x(t+1))\middle| F_t^x \vee F_{t-1}^v\right] \\
&= E\left[r(t+1)\exp\left(\frac{1}{2}c[z(t)^{\mathrm{T}}z(t) + x(t+1)^{\mathrm{T}}Q_1(t+1)x(t+1)]\right)\right. \\
&\qquad \left.\middle| F_t^x \vee F_{t-1}^v\right]
\end{aligned}$$

$$
\begin{aligned}
= \quad & E\left[r(t+1)\exp\left(\frac{1}{2}c \begin{pmatrix} v(t) \\ x(t) \end{pmatrix}^{\mathrm{T}} \right. \right. \\
& \times \begin{pmatrix} N^{\mathrm{T}}N + M^{\mathrm{T}}Q_1(t+1)M & N^{\mathrm{T}}C + M^{\mathrm{T}}Q_1(t+1)A \\ C^{\mathrm{T}}N + A^{\mathrm{T}}Q_1(t+1)M & C^{\mathrm{T}}C + A^{\mathrm{T}}Q_1(t+1)A \end{pmatrix} \\
& \left. \left. \times \begin{pmatrix} v(t) \\ x(t) \end{pmatrix} \right) \middle| F_t^x \vee F_{t-1}^v \right] \\
= \quad & r(t)\exp\left(\frac{1}{2}cx(t)^{\mathrm{T}}Q_1(t)x(t) \right)
\end{aligned}
$$

where the first equality follows from step 1 of the proof. The second equality is a consequence of the induction step. For the last steps we used Proposition D.5 and the formulas for $V_2(t)^{-1}$, $Q_1(t)$, and $r(t)$. Finally

$$
\begin{aligned}
& E\left[c\exp\left(\frac{1}{2}c\sum_{s=0}^{t_1} z(s)^{\mathrm{T}}z(s) \right) \right] \\
& = E\left[E\left[c\exp\left(\frac{1}{2}c\sum_{s=0}^{t_1} z(s)^{\mathrm{T}}z(s) \right) \middle| F_0^x \vee F_{-1}^v \right] \right] \\
& = E\left[W(0, x(0)) \right] = E\left[r(0)\exp\left(\frac{1}{2}cx(0)^{\mathrm{T}}Q_1(0)x(0) \right) \right] \\
& = r(0)\left(\frac{\det Q_3}{\det Q_0} \right)^{1/2} \exp\left(\frac{1}{2}m_0^{\mathrm{T}}Q_2m_0 \right).
\end{aligned}
$$

□

Proposition F.2. *Consider the time-invariant Gaussian system*

$$
\begin{aligned}
x(t+1) &= Ax(t) + Mv(t), \quad x(t_0) = x_0, \\
z(t) &= Cx(t) + Nv(t),
\end{aligned} \tag{F.16}
$$

with $x_0 \in G(m_0, Q_0)$ and $v : \Omega \times T \to \mathbb{R}^r$ Gaussian white noise with $v(t) \in G(0, V_1)$. Let $c \in \mathbb{R}$, $c > 0$. Define $Q_1 : T \to \mathbb{R}^{n \times n}$ as in Proposition F.1. Assume that:

(i) $V_1 = V_1^{\mathrm{T}} > 0$;
(ii) $Q_0 = Q_0^{\mathrm{T}} > 0$;
(iii) for all $t \in T$, $V_1^{-1} - c[N^{\mathrm{T}}N + M^{\mathrm{T}}Q_1(t+1)M] > 0$;
(iv) $Q_0^{-1} - cQ_1(0) > 0$;
(v) the following set of relations admits an unique solution $Q \in \mathbb{R}^{n \times n}$

$$
\begin{aligned}
Q \quad &= \quad A^{\mathrm{T}}QA + C^{\mathrm{T}}C \\
& \qquad + c[A^{\mathrm{T}}QM + C^{\mathrm{T}}N]V_2[A^{\mathrm{T}}QM + C^{\mathrm{T}}N]^{\mathrm{T}}, \qquad \text{(F.17)} \\
V_2^{-1} \quad &= \quad V_1^{-1} - cM^{\mathrm{T}}QM - cN^{\mathrm{T}}N > 0, \qquad \text{(F.18)} \\
& \qquad Q_0^{-1} - cQ > 0; \qquad \text{(F.19)}
\end{aligned}
$$

(vi) *For all* $t \in T$, $\lim_{T \to \infty} Q_{1,T}(t) = Q$ *where* $Q_{1,T}(t)$ *is the solution of (F.11) with terminal condition* $Q_{1,T}(T+1) = 0$.

Then

$$\begin{aligned} &\lim_{t\to\infty} \frac{1}{t} \ln E\left[c\exp\left(\frac{1}{2}c\sum_{s=0}^{t} z(s)^{\mathrm{T}}z(s)\right)\right] \\ &= -\frac{1}{2}\ln\det(V_1^{-1} - cM^{\mathrm{T}}QM - cN^{\mathrm{T}}N) + \frac{1}{2}\ln\det V_1^{-1}. \qquad \text{(F.20)} \end{aligned}$$

Proof Consider the horizon $\{0,1,\ldots,T\}$ with $x_0 \in G(0,Q_0)$. By (F.10) and assumption (vi)

$$\begin{aligned} \lim_{T\to\infty} V_2(t)^{-1} &= \lim_{T\to\infty} [V_1^{-1} - cM^{\mathrm{T}}Q_{1,T}(t)M - cN^{\mathrm{T}}N] \\ &= V_1^{-1} - cM^{\mathrm{T}}QM - cN^{\mathrm{T}}N, \\ \lim_{T\to\infty} Q_3^{-1} &= \lim_{T\to\infty} [Q_0^{-1} - cQ_{1,T}(0)] \\ &= Q_0^{-1} - cQ > 0, \quad \text{and finite}, \\ \lim_{T\to\infty} Q_2 &< \infty. \end{aligned}$$

Then

$$\begin{aligned} &\lim_{t\to\infty} \frac{1}{t} \ln E\left[c\exp\left(\frac{1}{2}c\sum_{s=0}^{t} z(s)^{\mathrm{T}}z(s)\right)\right] \\ &= \lim \frac{1}{t}\ln\left(c\left(\frac{\det Q_3}{\det Q_0}\right)^{1/2} \prod_{s=0}^{t}\left(\frac{\det V_2(s)}{\det V_1}\right)^{1/2}\right) \\ &\quad \text{by Proposition F.1,} \\ &= \lim\left[\frac{1}{t}\ln c + \frac{1}{2t}\ln\left(\frac{\det Q_3}{\det Q_0}\right) + \frac{1}{2t}\sum_{s=0}^{t}\ln\left(\frac{\det V_2(s)}{\det V_1}\right)\right] \\ &= \frac{1}{2}\ln\left(\frac{\det V_2}{\det V_1}\right) \\ &= -\frac{1}{2}\ln\det(V_1^{-1} - cM^{\mathrm{T}}QM - cN^{\mathrm{T}}N) + \frac{1}{2}\ln\det(V_1^{-1}). \end{aligned}$$

□

G. H-Infinity Control with an Entropy Criterion

The purpose of this section is to give the reader a very brief introduction to the discrete time H_∞ control problem. The H_∞ control problem was formulated by G. Zames in [77]. However its roots go back much further to work on differential games. One of the first publications directly related to H_∞ was

[51]. A first solution of the H_∞ control problem was based on frequency domain techniques and for an overview we refer to [23]. A breakthrough was the paper [18] which presented a complete solution of the H_∞ control problem based on time-domain techniques. This approach was based on Riccati equations and hence could be implemented easily. A first solution of the discrete time H_∞ problem can be found in [66, 49, 9]. Currently several good books are available [10, 67, 50, 31].

The main reason for studying the H_∞ control problem is model uncertainty. Using H_∞, it is possible to suppress the effect of model uncertainty on the behaviour of our plant. Also dependence on our knowledge of noise characteristics can be handled via H_∞ control.

We consider the following control system:

$$\Sigma: \begin{array}{ll} x(t+1) &= A(t)x(t) + B_1(t)u(t) + M(t)v(t), \quad x(0) = x_0, \\ y(t) &= C_1(t)x(t) + D_1(t)u(t) + N(t)v(t), \\ z(t) &= C_2(t)x(t) + D_2(t)u(t). \end{array} \tag{G.1}$$

Here $x \in \mathbb{R}^n$ is a state vector in which we are interested. $y \in \mathbb{R}^p$ is a measurement vector and $v \in \mathbb{R}^r$ is a disturbance affecting both the state evolution as well as our measurement. Note that v might consist of two independent components; one affecting the state evolution and one the measurement. In contrast with LEQG we do not assume that v is white noise but an arbitrary ℓ_2 signal. Finally, u denotes our control input and z is an output we want to make small.

Define the class G of control laws by $g \in G$, $g = \{g_0, g_1, \cdots, g_{t_1-1}\}$, $g_0 \in U$, and for $t = 1, \ldots, t_1 - 1$, $g_t : Y^t \times U^t \to U$, where g_t are measurable maps. We define the following cost function:

$$J(g) = \sup_{v, x_0} \frac{\|z\|^2_{2,t_1}}{\|v\|^2_{2,t_1} + x_0^{\mathrm{T}} R^{-1} x_0} \tag{G.2}$$

where

$$\|v\|^2_{2,t_1} := \sum_{s=0}^{t_1} v(t)^{\mathrm{T}} v(t). \tag{G.3}$$

If $t_1 = \infty$ then we must constrain v to ℓ_2, the class of signals for which the infinite sum (G.3) converges. The problem is then to determine

$$J(g^*) = \inf_{g \in G} J(g).$$

This problem turns out to be very hard but the related suboptimal problem does have an elegant solution. For each $c \in \mathbb{R}$ we are looking for a controller $g \in G$ such that $J(g) < c^{-1}$. In the literature, necessary and sufficient conditions for the solvability of this suboptimal problem have been derived. Moreover, if it exists, one particular suboptimal controller is given.

This controller is suboptimal for the original optimization problem. However, this controller minimizes the following criterion:

$$J_c(g) = \sup_{v,x_0} c\|z\|_{2,t_1}^2 - \|v\|_{2,t_1}^2 - x_0^{\mathrm{T}} R^{-1} x_0.$$

If $t_1 = \infty$, $x(0) = 0$, and both the system and the controller are time-invariant, then this particular controller has another interpretation.

First note that in this case

$$J(g) = \|S\|_\infty^2 := \sup_{\lambda \in [-\pi,\pi]} \|S(e^{i\lambda})\|^2$$

where S denotes the closed loop transfer matrix from v to z. As shown in [52] for the continuous time and later in [36] for the discrete time, the particular controller mentioned above is optimal for the following optimization problem:

$$J_e(g^*) = \inf_{g \in G} J_e(g)$$

with

$$J_e(g) = \frac{-1}{4\pi i} \int_{\mathbb{D}} \frac{1}{z} \ln\det\left(I - cS^{\mathrm{T}}(z^{-1})S(z)\right) dz. \tag{G.4}$$

Clearly this interpretation is only valid for a time-invariant system. Note that this expression is the same as the expression for the mutual information given in Theorem E.1. Hence given a controller and hence the closed loop transfer matrix then (G.4) is equal to an expression of the form (E.10). For the sake of completeness we formulate this in the following theorem.

Theorem G.1. *Assume that the transfer matrix S admits a realization as a finite-dimensional linear system with*

$$S(z) = C(zI - A)^{-1}B + D, \quad sp(A) \subset \mathbb{C}^-. \tag{G.5}$$

The following set of relations admits an unique solution $Q \in \mathbb{R}^{n\times n}$

$$\begin{aligned} Q &= Q^{\mathrm{T}} \geq 0, && \text{(G.6)}\\ Q &= A^{\mathrm{T}}QA + C^{\mathrm{T}}C \\ &\quad +(A^{\mathrm{T}}QB + C^{\mathrm{T}}D)N^{-1}(B^{\mathrm{T}}QA + D^{\mathrm{T}}C), && \text{(G.7)}\\ N &= c^{-1}I - B^{\mathrm{T}}QB - D^{\mathrm{T}}D > 0, && \text{(G.8)}\\ & sp(A + BN^{-1}(B^{\mathrm{T}}QA + D^{\mathrm{T}}C)) \subset \mathbb{C}^-, && \text{(G.9)} \end{aligned}$$

if and only if $\|S\|_\infty < c^{-1/2}$. Moreover, in that case

$$\frac{-1}{4\pi i} \int_{\mathbb{D}} \frac{1}{z} \ln\det\left(I - cS^{\mathrm{T}}(z^{-1})S(z)\right) dz$$

$$= -\frac{1}{2} \ln\det(I - cB^{\mathrm{T}}QB - cD^{\mathrm{T}}D). \tag{G.10}$$

References

1. Akaike H., "Information theory and an extension of the maximum likelihood principle", in B.N. Petrov and F. Csaki, editors, *Proceedings 2nd International Symposium Information Theory*, pages 267–281. Akademia Kiado, Budapest, 1973.
2. Akaike H., "Canonical correlation analysis of time series and the use of an information criterion", in R.K. Mehra and D.G. Lainiotis, editors, *System identification - Advances and case studies*, pages 27–96. Academic Press, New York, 1976.
3. Akaike H., "On entropy maximization principle", in P.R. Krishnaiah, editor, *Applications of statistics*, pages 27–41. North-Holland, Amsterdam, 1977.
4. Amari S.I., "Differential-geometrical methods in statistics", Lecture Notes in Statistics, volume 28. Springer, Berlin, 1985.
5. Amari S.I., "Differential geometry of curved exponential families – curvatures and information loss", *Ann. Statist.*, 10:357–385, 1982.
6. Amari S.I., "Differential geometry of a parametric family of invertible linear systems - Riemannian metric, dual affine connections, and divergence", *Math. Systems Th.*, 20:53–82, 1987.
7. Anderson T.W., *An introduction to multivariate statistical analysis*. John Wiley & Sons, New York, 1958.
8. Arov D.Z., and M.G. Krein, "Problem of search of the minimum of entropy in indeterminate extension problem", *Functional Analysis and its Applications*, 15:123–126, 1981.
9. Başar T., "A dynamic games approach to controller design: disturbance rejection in discrete time", in *Proc. CDC*, pages 407–414, Tampa, 1989.
10. Başar T., and P. Bernhard, *H_∞ -optimal control and related minimax design problems: a dynamic game approach*. Birkhäuser, Boston, 1991.
11. Bensoussan A., and J.H. van Schuppen, "Optimal control of partially observable stochastic systems with an exponential-of-integral performance index", *SIAM J. Control Optim.*, 23:599–613, 1985.
12. Campi M.C., and M.R. James, "Nonlinear discrete-time risk-sensitive optimal control", Preprint X, Department of Systems Engineering, Australian National University, Canberra, 1993.
13. Cover T.M., and J.A. Thomas, *Elements of information theory*. John Wiley & Sons, New York, 1991.
14. Csiszár I., "Information type measures of difference of probability distributions and indirect observations", *Studia Sci. Math. Hungar.*, 2:229–318, 1967.
15. Csiszár I., "I-divergence geometry of probability distributions and minimization problems", *Ann. Probab*, 3:146–158, 1975.
16. Csiszár I., and J. Körner, *Information Theory*. Academic Press, New York, 1981.
17. Csiszár I., and G. Tusnady, "Information geometry and alternating minimization procedures", in X, editor, *Statistics and Decisions, Supplement issue no.1*, pages 205–237. R. Oldenbourg Verlag, München, 1984.
18. Doyle J.C., K. Glover, P.P. Khargonekar, and B.A. Francis, "State space solutions to standard H_2 and H_∞ control problems", *IEEE Trans. Aut. Contr.*, 34(8):831–847, 1989.
19. Faurre P., M. Clerget, and F. Germain, *Opérateurs rationnels positifs*. Dunod, Paris, 1979.
20. Findley D.F., "On the unbiasedness property of AIC for exact or approximating linear stochastic time series models", *J. Time Series Analysis*, 6:229–252, 1985.

21. Fisher R.A., "On an absolute criterion for fitting frequency curves", *Mess. of Math.*, 41:155–160, 1912.
22. Fisher R.A., "Theory of statistical estimation", *Proc. Cambridge Philos. Soc.*, 22:700–725, 1925.
23. Francis B.A., *A course in H_∞ control theory*, Lecture Notes in Control and Information Sciences, volume 88. Springer, 1987.
24. Gelfand I.M., and A.M. Yaglom, "Calculation of the amount of information about a random function contained in another such function", *American Mathematical Society Translations, Series 2*, 12:199–246, 1959.
25. Gittens R.D., *Canonical analysis - A review with applications in ecology.* Springer, Berlin, 1985.
26. Glover K., and J.C. Doyle, "State-space formulae for all stabilizing controllers that satisfy an H_∞-norm bound and relations with risk sensitivity", *Systems & Control Lett.*, 11:167–172, 1988.
27. Glover K., and D. Mustafa, "Derivation of the maximum entropy H_∞-controller and a state-space formula for its entropy", *Int. J. Control*, 50:899–916, 1989.
28. Gombani A., and M. Pavon, "On the Hankel-norm approximation of linear stochastic systems", *Systems & Control Lett.*, 5:283–288, 1985.
29. Gray R.M., *Probability, random processes, and ergodic properties.* Springer, Berlin, 1988.
30. Gray R.M., and L.D. Davisson, *Ergodic and information theory.* Dowden, Hutchinson, Ross, 1977.
31. Green M., and D.J.N. Limebeer, *Linear robust control.* Information and System Sciences Series. Prentice Hall, 1994.
32. Fan C.H., J.L. Speyer, and C.R. Jaensch, "Centralized and decentralized solutions of the linear-exponential-gaussian problem", *IEEE Trans. Automatic Control*, 39:1986–2003, 1994.
33. Hannan E.J., and M. Deistler, *The statistical theory of linear systems.* John Wiley & Sons, New York, 1988.
34. Hassibi B., A.H. Sayed, and T. Kailath, "Recursive linear estimation in Krein spaces - Part I Theory", in *Proc. 32nd IEEE Conference on Decision and Control*, pages 3489–3494, New York, 1993. IEEE.
35. Hassibi B., A.H. Sayed, and T. Kailath, "Recursive linear estimation in Krein spaces - Part II Applications", in *Proc. 32nd IEEE Conference on Decision and Control*, pages 3495–3500, New York, 1993. IEEE.
36. Iglesias P.A., D. Mustafa, and K. Glover, "Discrete time H_∞ controllers satisfying a minimum entropy criterion", *Syst. & Contr. Letters*, 14:275–286, 1990.
37. Ihara S., *Information theory for continuous systems.* World Scientific, Singapore, 1993.
38. Jacobson D.H., "Optimal stochastic linear systems with exponential performance criteria and their relation to deterministic differential games", *IEEE Trans. Automatic Control*, 18:124–131, 1973.
39. Jaensch C.R., and J.L. Speyer, "Centralized and decentralized stochastic control problems with an exponential cost criterion", in *Proceedings of the 27th Conference on Decision and Control*, pages 286–291, New York, 1988. IEEE.
40. James M.R., and J.S. Baras, "Robust output feedback control for discrete-time nonlinear systems", Preprint X, Department of Systems Engineering, Australian National University, Canberra, 1993.
41. James M.R., J.S. Baras, and R.J. Elliott, "Risk-sensitive control and dynamic games for partially observed discrete-time nonlinear systems", Preprint X, Department of Systems Engineering, Australian National University, Canberra, 1993.

42. Jewell N.P., and P. Bloomfield, "Canonical correlations of past and future for time series: Definitions and theory", *Ann. Statist.*, 11:837–847, 1983.
43. Jewell N.P., P. Bloomfield, and F.C. Bartmann, "Canonical correlations of past and future for time series: Bounds and computation", *Ann. Statist.*, 11:848–855, 1983.
44. Kolmogorov A.N., "On the Shannon theory of information in the case of continuous signals", *IRE. Trans. Inform. Theory*, 2:102–108, 1956.
45. Kullback S., J.C. Keegel, and J.H. Kullback, "Topics in statistical information theory", Lecture Notes in Statistics, volume 42. Springer-Verlag, Berlin, 1987.
46. Kullback S., and R.A. Leibler, "On information and sufficiency", *Ann, Math. Statist.*, 22:79–86, 1951.
47. Kumar P.R., "Stochastic optimal control and stochastic differential games", PhD thesis, Department of Systems Science and Mathematics, Washington University, St. Louis, MO, USA, 1977.
48. Kumar P.R., and J.H. van Schuppen, "On the optimal control of stochastic systems with an exponential-of-integral performance index", *J. Math. Anal. Appl.*, 80:312–332, 1981.
49. Limebeer D.J.N, M. Green, and D. Walker, "Discrete time H_∞ control", in *Proc. CDC*, pages 392–396, Tampa, 1989.
50. Maciejowski J.M., *Multivariable feedback design.* Addison-Wesley, Reading, MA, 1989.
51. Medanic J., "Bounds on the performance index and the Riccati equation in differential games", *IEEE Trans. Aut. Contr.*, 34:613–614, 1967.
52. Mustafa D., and K. Glover. *Minimum entropy H_∞ control*, Lecture Notes in Control and Information Sciences, volume 146. Springer, Berlin, 1990.
53. Nishii R., "Maximum likelihood principle and model selection when the true model is unspecified", *J. Multivar. Anal.*, 27:392–403, 1988.
54. Pavon M., "Canonical correlations of past inputs and future outputs for linear stochastic systems", *Syst. Control Lett.*, 4:209–215, 1984.
55. Perez A., "Notions generalisees d' incertitude d' entropie et d' information du point de vue de la theorie de martingales", in *Transactions of the First Prague Conference on Information Theory, Statistical Decision Functions, and Random Processes*, pages 183–208, Prague, 1957. Publishing House of the Czechoslovak Academy of Sciences.
56. Perez A., "Sur la theorie de l' information dans le cas d'un alphabet abstrait", in *Transactions of the First Prague Conference on Information Theory, Statistical Decision Functions, and Random Processes*, pages 209–243, Prague, 1957. Publishing House of the Czechoslovak Academy of Sciences.
57. Pinsker M.S., *Information and information stability of random variables.* Holden-Day, San Francisco, 1964.
58. Rosenblatt M., Asymptotic distribution of eigenvalues of block Toeplitz matrices. *Bull. Am. Math. Soc.*, 66:320–321, 1960.
59. Runolfsson T., "Stationary risk-sensitive LQG control and its relation to LQG and H-infinity control", in *Proceedings 29th IEEE Conference on Decision and Control*, pages 1018–1023, New York, 1990. IEEE Press.
60. Runolfsson T., "On the stationary control of a controlled diffusion with an exponential-of-integral performance criterion", in *Proceedings of the 30th Conference on Decision and Control*, pages 935–936, New York, 1991. IEEE Press.
61. Shannon C.E., "A mathematical theory of communication", *Bell System Techn. Journal*, 27:379–423, 623–656, 1948.
62. Shannon C.E., and W.W. Weaver, *The mathematical theory of communication.* University of Illinois, Urbana, 1949.

63. Solo V., "The exact likelihood for a multivariate ARMA model", *J. Multivar. Anal.*, 15:164–173, 1984.
64. Speyer J., J. Deyst, and D.H. Jacobson, "Optimization of stochastic linear systems with additive measurement and process noise using exponential performance criteria", *IEEE Trans. Automatic Control*, 19:358–366, 1974.
65. Speyer J.L., "An adaptive terminal guidance scheme based on an exponential cost criterion with application to homing missile guidance", *IEEE Trans. Automatic Control*, 21:371–375, 1976.
66. Stoorvogel A.A., "The discrete time H_∞ control problem with measurement feedback", *SIAM J. Contr. & Opt.*, 30(1):182–202, 1992.
67. Stoorvogel A.A., *The H_∞ control problem: a state space approach.* Prentice-Hall, Englewood Cliffs, NJ, 1992.
68. Stoorvogel A.A., and J.H. van Schuppen, "An H_∞-parameter estimator and its interpretation", in *Proc. 10th IFAC Symposium System Identification*, volume 3, pages 267–270. Pergamon Press, 1994.
69. van Schuppen J.H., "Stochastic realization problems", in J.M. Schumacher H. Nijmeijer, editor, *Three decades of mathematical system theory*, Lecture Notes in Control and Information Sciences, volume 135, pages 480–523. Springer, Berlin, 1989.
70. Whittle P., "Risk-sensitive Linear/Quadratic/Gaussian control", *Adv. Appl. Prob.*, 13:764–777, 1981.
71. Whittle P., *Optimization over time.* John Wiley & Sons, New York, 1982.
72. Whittle P., "Scheduling and characterization problems for stochastic networks", *J.R. Statist. Soc. B.*, 47:407–415, 1985.
73. Whittle P., "A risk-sensitive maximum principle", *Systems & Control Lett.*, 15:183–192, 1990.
74. Whittle P., *Risk-sensitive optimal control.* John Wiley Sons, Chichester, UK, 1990.
75. Whittle P., "A risk-sensitive maximum principle: The case of imperfect state observations", *IEEE Trans. Automatic Control*, 36:793–801, 1991.
76. Whittle P., and J. Kuhn, "A Hamiltonian formulation of risk-sensitive Linear/Quadratic/Gaussian control", *Int. J. Control*, 43:1–12, 1986.
77. Zames G., "Feedback and optimal sensitivity: model reference transformations, multiplicative seminorms, and approximate inverses", *IEEE Trans. Aut. Contr.*, 26:301–320, 1981.

Least Squares Based Self-Tuning Control Systems

Convergence, Stability and Optimality

Sergio Bittanti[1] and Marco Campi[2]

[1] Dipartimento di Elettronica e Informazione
Politecnico di Milano
Piazza Leonardo da Vinci 32
20133 Milano
Italy
Fax: ++39.2.23993587,
E-mail: bittanti@elet.polimi.it

[2] Dipartimento di Elettronica per l'Automazione
Università di Brescia
Via Branze 38
25123 Brescia
Italy
Fax: ++39.30.380014,
E-mail: Campi@ipmel1.polimi.it

1. Introduction

The challenge of adaptive control is to regulate plants of unknown dynamics, possibly subject to time variations. In this realm, the so called self-tuning rationale plays a central role. A self-tuning control system is a feedback control system in which the controller is synthesized on the basis of a model of the unknown plant identified on line, see e.g. [2, 3, 4, 12, 18, 21]. In the controller design, the estimated parameters are typically used as if they were "the true ones". This "plug in" rationale is known as *certainty equivalence principle*.

The theoretical analysis of self-tuning control systems is an old question, dating back to the seventies, or even earlier. The difficulty arises from the fact that, even when the plant is governed by linear equations, the interplay between the identifier and the controller leads to an overall system with an inherently complex nonlinear dynamics. This is why in the literature attention has been mainly focused on adaptive control systems with specific characteristics. One of the most extensively studied cases is the self-tuning implementation of the minimum variance control law applied to minimum phase plants; the literature on this paradigm of self-tuning control covers almost two decades, see e.g. [5, 6, 7, 8, 13, 14, 24].

In recent years, some new self-tuning theories of wide applicability have been developed. In particular, in [15] it is proven that a wide class of self-tuning control systems based on the Recursive Least Squares (RLS) estimator exhibits good convergence properties under the assumption that the plant is minimum phase. We notice in passing that the minimum phase hypothesis is a recurrent assumptions in adaptive control theory. The main reason behind this requirement is that, with this condition, the boundedness of the plant input signal follows from the boundedness of the output. This makes the analysis easier when trying to secure the stability properties of the control system.

Another relevant reference is [23], where a fairly general notion of *tuning* has been introduced. Precisely, assuming parameter convergence, a control system is said to be tuning if its performance in the long run is the same as the one of the *ideal system*, namely the feedback system which would be designed if the true parameters were known. By resorting to a stochastic realization approach, it is shown in [23] that tuning holds in the case of minimum variance and pole placement control. However, many other control systems are not tuning in the above sense. In the sequel, this notion of tuning will be referred to as *ideal tuning.*

In the present contribution, we provide a general theory of convergence, stability and self-optimality of self-tuning control systems, without resorting to the minimum phase assumption. For, we will introduce a new notion of tuning, dealing with the possibility that the asymptotic behaviour of the adaptive control system resembles that of the *imaginary system* rather than the one of the *ideal system.* By *imaginary system* we mean the feedback system in which the plant is replaced by the identified model, and the controller – no more adaptive – is designed on the basis of the parameters of such a model. Such a notion of tuning is not surprising if one considers that resorting to the certainty equivalence principle means to see the identified model as if it were the true plant. Therefore, any control law based on certainty equivalence is synthesized so as to achieve a desired behavior for the *imaginary system.* Of course, should the true parameter be known, a better performance could be achieved in general (*ideal tuning*).

The key in the development of the theory is the concept of *excitation subspace* introduced in [12]. Roughly, this subspace is the hyperplane in the parameter space where the information conveyed by data diverges with time. As we will show, this concept allows one to put sharply into focus the properties of the parameter estimate which are relevant to control, so gaining a deeper comprehension in the overall behaviour of the control system.

In the analysis, we will take advantage of the fact that, whether the plant is noise-free or it is affected by a white and Gaussian disturbance, RLS converges *irrespective of the excitation characteristics of the involved signals* (this is fundamental in a closed loop setup, where the signals may be only partially exciting). Thanks to such a convergence result, one comes to the conclusion that, whenever the identifier relies on RLS, the estimated model parameters converge; as a consequence, the "plug in" controller parameters converge, too. Thus, one can expect that LS self-tuning control systems asymptotically behave as linear and time invariant systems.

Our paper is tutorial in nature. As such, after the introduction of many preliminary basic notions (Section 2), we will bring into light the many facets of the self-tuning control problem (Section 3) and explain the least squares identification technique (Section 4). Many young researchers complain that most papers on adaptive control are so technical that their reading and comprehension is difficult. In the hope to make the theory accessible to a large audience, we will thoroughly discuss the noise-free case, so as to present the whole theory in its simplest fully deterministic version (Section 5). Then, we will outline the convergence, stability and self-optimality results for noisy plants (Section 6).

2. Self-Tuning Adaptive Control

2.1 Basic Self-Tuning Concepts

Consider a plant $\boldsymbol{P}$ with a scalar input $u(t)$ and a scalar output $y(t)$, $t \in Z$. The control problem consists in interconnecting $\boldsymbol{P}$ with another system, the controller, in such a way that the control system exhibits a desired performance. For the selection of the controller, many techniques are available, the majority of which are based on a mathematical model of the plant (model based control design). When such a mathematical description is not *a priori* known, one can resort to an identification algorithm for the estimation of a black box model. In this framework, let $\{\boldsymbol{M}(\theta)\}$ be the adopted family of models each of which is defined by a parameter vector θ. A controller associated with model $\boldsymbol{M}(\theta)$ will be denoted by $\boldsymbol{C}(\theta)$. Letting $\hat{\theta}(t)$ be the estimate at time t of the parameter vector θ obtained by the recursive identification algorithm, a straightforward control rationale consists in taking $\boldsymbol{C}(\hat{\theta}(t))$ as controller for the unknown plant $\boldsymbol{P}$. As already pointed out in the introduction, this approach is known as *certainty equivalence principle* and leads to the adaptive control system represented in Fig. 2.1.

The overall control system is characterized by two loops; the first one is closed through the controller $\boldsymbol{C}(\hat{\theta}(t))$ for the fast control action; the second one performs a slow adaptation of the controller via the identifier $\boldsymbol{I}$.

In this paper, we will consider various properties of the self tuning control system, under the assumptions that

- the plant $\boldsymbol{P}$ is linear and belongs to the family of models $\{\boldsymbol{M}(\theta)\}$
- the identification is performed by the Recursive Least Squares (RLS) algorithm.

As far as the control law is concerned, we will not confine ourselves to a specific case; we will only require that

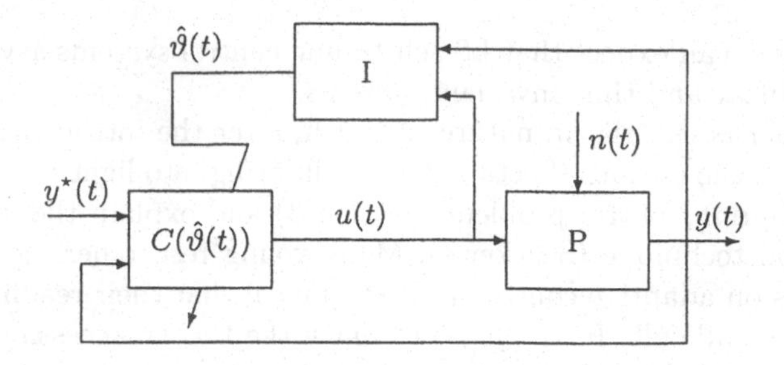

Fig. 2.1. The adaptive control system (*real system*)

– the controller is linear.

No minimum phase assumption on the plant is made.

2.2 Mathematical Framework

2.2.1 The Plant. Consider the SISO plant

$$\begin{aligned} \boldsymbol{P}: y(t) &= a_1^o y(t-1) + a_2^o y(t-1) + \ldots + a_n^o y(t-n) \\ &\quad + b_0^o u(t-d) + \ldots + b_m^o u(t-d-m) + n(t), \end{aligned}$$

$$b_0^o \neq 0, \; d \geq 1. \tag{2.1}$$

In (2.1), it is possible to recognize two regressions, one on the past values of $y(\cdot)$ (the *AutoRegression* part, characterized by parameters a_i^o) and one on the past values of the control variable $u(\cdot)$ (the *eXogeneous* part, with parameters b_i^o). This is why eq. (2.1) is known under the acronym ARX. Integers n and m are referred to as the *orders* of the regressions. The positive integer d is the input-output *delay*, namely the interval of time which elapses before an input action may cause an effect on the output.

We will assume that system (2.1) is initialized at time $t = 1$ with deterministic values $y(0), y(-1), \ldots, y(-n+1), u(0), u(-1), \ldots, u(-d-m+1)$. Signal $n(t)$ represents the effect of disturbances on the given plant. Below, we will first consider the noise-free case ($n(t) = 0$) and develop the theory for the associated deterministic adaptive control problem. Then, we will turn to the disturbed situation, where $n(\cdot)$ is treated as a stochastic process. In this last case $n(\cdot)$ will be characterized as indicated in the following

Assumption 1. *$\{n(t)\}$ is an independent sequence of normally distributed random variables with $E[n(t)] = 0$, $E[n(t)^2] = \sigma^2 > 0$.*

The plant parameters are grouped in the vector

$$\theta^o = [a_1^o\, a_2^o \ldots\, a_n^o\, b_0^o\, b_1^o \ldots\, b_m^o]^T.$$

By introducing the polynomials

$$A(\theta^o; q^{-1}) = 1 - \sum_{i=1}^{n} a_i^o\, q^{-i}, \qquad B(\theta^o; q^{-1}) = \sum_{i=0}^{m} b_i^o\, q^{-i},$$

in the delay operator q^{-1} ($q^{-1}v(t) = v(t-1)$), eq. (2.1) can be written as

$$\boldsymbol{P}: \quad A(\theta^o; q^{-1})y(t) = B(\theta^o; q^{-1})u(t-d) + n(t),$$

Throughout the paper, we assume minimality in the plant representation, i.e.:

Assumption 2. *$q^n A(\theta^o; q^{-1})$ and $q^m B(\theta^o; q^{-1})$ are coprime.*

2.2.2 The Model. The main motivation for adaptive control is the lack of knowledge on some characteristics of the plant. Herein, we will conform to the usual paradigm that the structure of the ARX equation is known, i.e. integers d, m and n are *a priori* given, but the true parameter vector θ^o is not available. Correspondingly, one can resort to an identification algorithm to get a model of the plant.

To perform identification a family of model is needed. For, we consider the family of (deterministic) ARX models parameterized in the vector $\theta = [a_1\ a_2\ \ldots\ a_n\ b_0\ b_1\ \ldots\ b_m]^T$:

$$\begin{aligned} \boldsymbol{M}(\theta): y(t) = \; & a_1 y(t-1) + a_2 y(t-2) + \ldots + a_n y(t-n) \\ & + b_0 u(t-d) + \ldots + b_m u(t-d-m), \end{aligned}$$

whose equation is a mimic of that of $\boldsymbol{P}$.

By introducing the polynomials

$$A(\theta; q^{-1}) = 1 - \sum_{i=1}^{n} a_i\, q^{-i}, \qquad B(\theta; q^{-i}) = \sum_{i=0}^{m} b_i\, q^{-i},$$

this model can also be written as:

$$\boldsymbol{M}(\theta): \quad A(\theta; q^{-1})y(t) = B(\theta; q^{-1})u(t-d).$$

Then, denoting by $\hat{\theta}(t)$ the parameter estimate at time t, the estimated model is given by $\boldsymbol{M}(\hat{\theta}(t))$.

2.2.3 The Controller. We finally introduce a general linear control law for model $M(\theta)$. This is defined by the equation:

$$\begin{aligned} C(\theta) : u(t) = \quad & r_1(\theta)u(t-1) + \ldots + r_\alpha(\theta)u(t-\alpha) \\ & + s_0(\theta)y(t) + \ldots + s_\beta(y)y(t-\beta) \\ & + t_{-\gamma}(\theta)y^*(t+\gamma) + \ldots + t_\delta(\theta)y^*(t-\delta), \end{aligned}$$

where $y^*(\cdot)$ is a reference signal subject to

Assumption 3. *$y^*(\cdot)$ is bounded and deterministic.*

Again, one can resort to a polynomial form of the above difference equation, as follows:

$$C(\theta) : u(t) = R(\theta; q^{-1})u(t) + S(\theta; q^{-1})y(t) + T(\theta; q^{-1})y^*(t),$$

with

$$R(\theta; q^{-1}) = \sum_{i=1}^{\alpha} r_i(\theta)q^{-i};$$

$$S(\theta; q^{-1}) = \sum_{i=0}^{\beta} s_i(\theta)q^{-i};$$

$$T(\theta; q^{-1}) = \sum_{i=-\gamma}^{\delta} t_i(\theta)q^{-i}.$$

The above control law is general enough to encompas as special cases all the most popular techniques which have been proposed in the literature.

Usually, functions $r_i(\cdot)$, $s_i(\cdot)$ and $t_i(\cdot)$ are rational. Then, there are values of θ for which these functions are not continuous. For subsequent use, we are well advised to denote the continuity set of $r_i(\cdot)$, $s_i(\cdot)$ and $t_i(\cdot)$ by Ξ_1:

$$\Xi_1 = \left\{\theta \in \mathbb{R}^{n+m+1} \mid r_i(\theta), s_i(\theta), t_i(\theta) \text{ are continuous}\right\}.$$

2.2.4 The Self-Tuning Control System (Real System). The certainty equivalence principle amounts to take $C(\hat{\theta}(t))$ as present control law for the true plant. Therefore, the *real system*, that is the true plant together with the actual control law (see Fig. 2.1) is given by the equations:

$$\Sigma(\theta^o, \hat{\theta}(t)) : \begin{cases} A(\theta^o; q^{-1})y(t) = B(\theta^o; q^{-1})u(t-d) \\ \qquad\qquad\qquad + n(t) \\ u(t) = R(\hat{\theta}(t); q^{-1})u(t) + S(\hat{\theta}(t); q^{-1})y(t) + \\ \qquad + T(\hat{\theta}(t); q^{-1})y^*(t). \end{cases}$$

In the literature of adaptive control, many expedients have been proposed to ensure that $\hat{\theta}(t) \in \Xi_1 \ \forall\, t$, so that the control law is well defined at any time point. Moreover, this question has also challenging theoretical aspects, which have attracted the attention of various investigators. For instance, in [17], it is shown that, in a rather general stochastic setting, the probability of the occurrence of the nasty event $\hat{\theta}(t) \notin \Xi_1$ is fortunately 0. We will not dwell upon this issue in the sequel.

3. What Makes a ST Control System Nice?

3.1 Imaginary and Asymptotic Imaginary Systems

In analyzing the desired properties of a self tuning control system, one has to bear in mind that, if one conforms to the certainty equivalence principle, $\boldsymbol{M}(\hat{\theta}(t))$ is regarded as if it were the actual plant. Hence, $\boldsymbol{M}(\hat{\theta}(t))$ is so to say the *imaginary plant.* Correspondingly model $\boldsymbol{M}(\hat{\theta}(t))$ controlled through $\boldsymbol{C}(\hat{\theta}(t))$ (Fig. 3.1) can be named *imaginary control system*, or *imaginary system* by short:

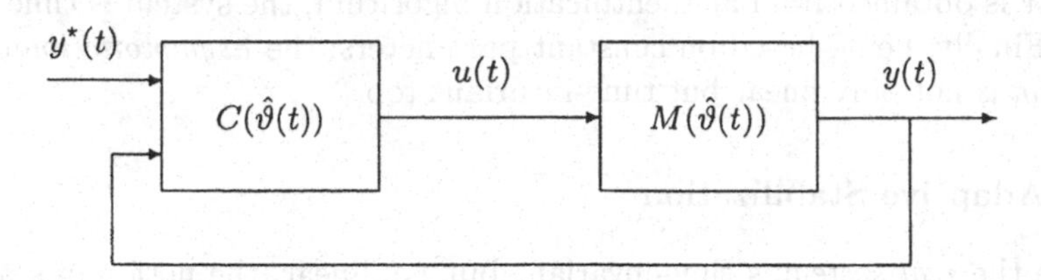

Fig. 3.1. The *imaginary system*

$$\Sigma(\hat{\theta}(t),\hat{\theta}(t)) \begin{cases} A(\hat{\theta}(t);q^{-1})y(t) = B(\hat{\theta}(t);q^{-1})u(t-d) \\ u(t) = R(\hat{\theta}(t);q^{-1})u(t) + S(\hat{\theta}(t);q^{-1})y(t) + \\ \quad + T(\hat{\theta}(t);q^{-1})y^*(t). \end{cases}$$

The adopted terminology is reminiscent of the one introduced in [23]. However, the reader is warned that the notion of imaginary system given in [23] is slightly different from the one introduced here.

If we push our luck and assume for a moment that the estimate converges, say $\hat{\theta}(t) \rightarrow \hat{\theta}(\infty)$, then the *imaginary system* would tend to the following time-invariant system (*asymptotic imaginary system*, Fig. 3.1):

$$\boldsymbol{\Sigma}(\hat{\theta}(\infty),\hat{\theta}(\infty)) \begin{cases} A(\hat{\theta}(\infty);q^{-1})y(t) = B(\hat{\theta}(\infty);q^{-1})u(t-d) \\ u(t) = R(\hat{\theta}(\infty);q^{-1})u(t) + S(\hat{\theta}(\infty);q^{-1})y(t) + \\ \quad + T(\hat{\theta}(\infty);q^{-1})y^*(t). \end{cases}$$

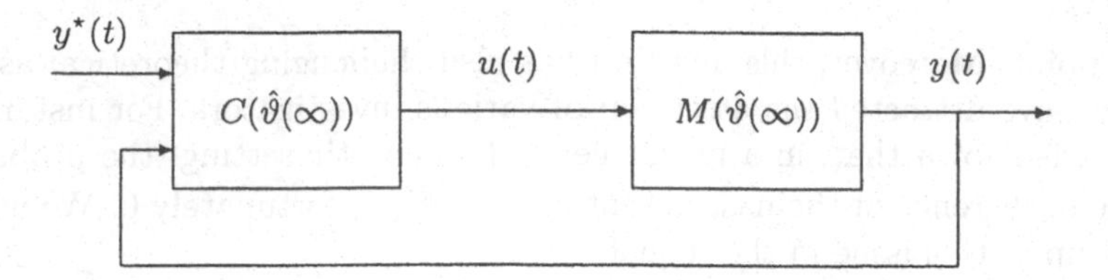

Fig. 3.2. The *asymptotic imaginary system*

Remark 3.1. So far, we have introduced three control systems, the real system $\Sigma(\theta^o, \hat{\theta}(t)) = \boldsymbol{P} + \boldsymbol{C}(\hat{\theta}(t)) +$ identifier $\boldsymbol{I}$ (Fig. 2.1), the imaginary system $\Sigma(\hat{\theta}(t), \hat{\theta}(t)) = \boldsymbol{M}(\hat{\theta}(t)) + \boldsymbol{C}(\hat{\theta}(t))$ (Fig. 3.1), and the aymptotic imaginary system $\Sigma(\hat{\theta}(\infty), \hat{\theta}(\infty)) = \boldsymbol{M}(\hat{\theta}(\infty)) + \boldsymbol{C}(\hat{\theta}(\infty))$ (Fig. 3.1). Note that the plant is time-invariant; however, the self-tuning controller of Fig. 2.1 is nonlinear due to the presence of the identifier. Hence, the real system is time-invariant and nonlinear. On the opposite, since the model class and the controller are linear, the *imaginary system* is linear as well. However, as the parameter vector is obtained from an identification algorithm, the system is time varying. Finally, being based on constant parameters, the *asymptotic imaginary system* is not only linear but time-invariant too.

3.2 Adaptive Stabilization

Since the *real system* is time-invariant but nonlinear, the notion of stability is introduced by making reference directly to the boundedness properties of the signals. Precisely, in the noise-free case we say that the plant is stabilized by the adaptive controller if the input and the output of the plant remain deterministically bounded:

$$| u(t) | < c \text{ and } | y(t) | < c, \quad \forall t.$$

In the noisy case, the boundedness is required to hold in the average:

$$\limsup_{N \to \infty} \frac{1}{N} \sum_{t=1}^{N} [| u(t) | + | y(t) |] < \infty \quad \text{w.p. 1.}$$

To achieve the objective of adaptive stabilization, the design of the controller cannot be made starting from plant $\boldsymbol{P}$ (whose parameters are unknown!). Rather, according to the certainty equivalence principle, one can try to design a stabilizing controller based on the surrogate of $\boldsymbol{P}$ provided by the model $\boldsymbol{M}(\hat{\theta}(t))$. On the other hand, at time t the future evolution of the estimated parameter vector is unknown too. Therefore, one is led to simply consider the *frozen imaginary system*, given by the imaginary system with fixed parameters $\Sigma(\theta, \theta) = \boldsymbol{M}(\theta) + \boldsymbol{C}(\theta)$. The set of parameters for which such a frozen imaginary system is stable will be denoted by Ξ_2 :

$$\Xi_2 = \{\theta \in \mathbb{R}^{n+m+1} \mid \boldsymbol{M}(\theta) + \boldsymbol{C}(\theta) \text{ is asymptotically stable}\}.$$

Of course, even if $\hat{\theta}(t) \in \Xi_2$ for any t, there is no guarantee that the time-varying *imaginary system* $\Sigma(\hat{\theta}(t), \hat{\theta}(t)) = \boldsymbol{M}(\hat{\theta}(t)) + \boldsymbol{P}(\hat{\theta}(t))$ will be stable.

3.3 Self-Optimality

For the discussion of self-optimality, we are well advised to introduce two further control systems. The first one is the *noisy imaginary system*, which is simply the *imaginary system*, with the model of the plant incorporating the disturbance $n(\cdot)$. Precisely, it is given by the equations:

$$\boldsymbol{\Sigma}_n(\hat{\theta}(t), \hat{\theta}(t)) \begin{cases} A(\hat{\theta}(t); q^{-1})y(t) = B(\hat{\theta}(t); q^{-1})u(t-d) \\ \qquad + n(t) \\ u(t) = R(\hat{\theta}(t); q^{-1})u(t) + S(\hat{\theta}(t); q^{-1})y(t) \\ \qquad + T(\hat{\theta}(t); q^{-1})y^*(t). \end{cases}$$

and it is depicted in Fig. 3.3.

Again, if $\hat{\theta}(t) \to \hat{\theta}(\infty)$, the noisy imaginary system is asymptotically governed by a time invariant dynamics. The limit time invariant and linear system will be called *asymptotic noisy imaginary system* (Fig. 3.3):

$$\boldsymbol{\Sigma}_n(\hat{\theta}(\infty), \hat{\theta}(\infty)) \begin{cases} A(\hat{\theta}(\infty); q^{-1})y(t) = B(\hat{\theta}(\infty); q^{-1})u(t-d) \\ \qquad + n(t) \\ u(t) = R(\hat{\theta}(\infty); q^{-1})u(t) + S(\hat{\theta}(\infty); q^{-1})y(t) \\ \qquad + T(\hat{\theta}(\infty); q^{-1})y^*(t). \end{cases}$$

The output $y(t)$ of this last system can be equivalently written in transfer function terms as follows

$$y(t) = W_1(q^{-1})y^*(l) + W_2(q^{-1})n(t),$$

where

$$W_1(q^{-1}) = \frac{q^{-d}B(\hat{\theta}(\infty); q^{-1})\, T(\hat{\theta}(\infty); q^{-1})}{\Delta(\hat{\theta}(\infty); q^{-1})} \tag{3.1}$$

$$W_2(q^{-1}) = \frac{1 - R(\hat{\theta}(\infty); q^{-1})}{\Delta(\hat{\theta}(\infty); q_{-1}} \tag{3.2}$$

$$\begin{aligned} \Delta(\hat{\theta}(\infty); q^{-1}) &= A(\hat{\theta}(\infty); q^{-1})[1 - R(\hat{\theta}(\infty); q^{-1})] \\ &\quad - q^{-d}B(\hat{\theta}(\infty); q^{-1})S(\hat{\theta}(\infty); q^{-1}). \end{aligned}$$

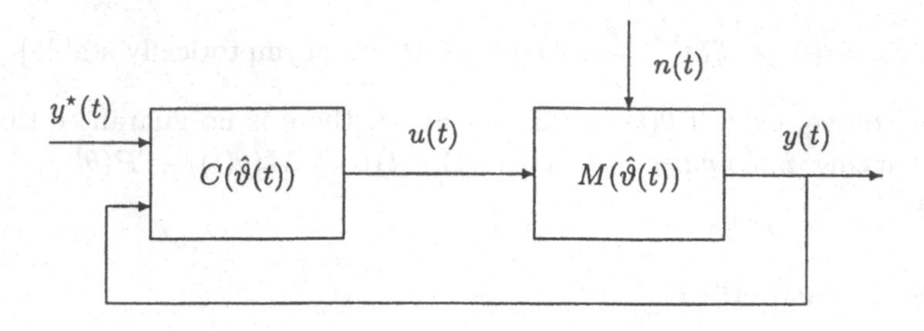

Fig. 3.3. The *noisy imaginary system*

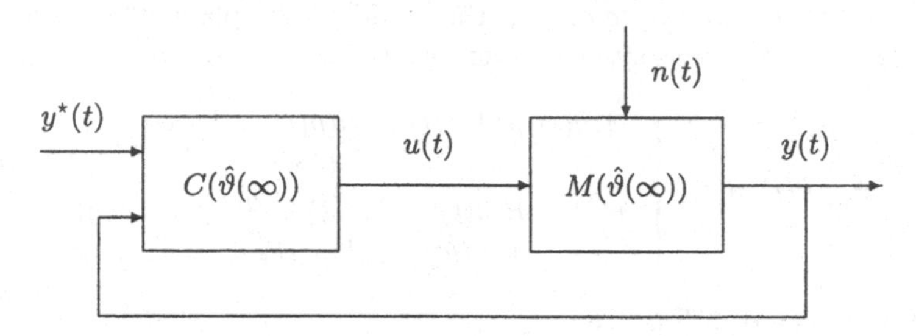

Fig. 3.4. The *asymptotic noisy imaginary system*

Remark 3.2. As is well known, in adaptive control, one should not expect that $\hat{\theta}(\infty)$ coincides with the true parameterization θ^o. In general, $\hat{\theta}(\infty)$ will depend upon the specific control law adopted by the designer and upon the excitation characteristics of the reference signal $y^*(\cdot)$. Consequently, both the control law and signal $y^*(\cdot)$ will influence transfer functions $W_1(q^{-1})$ and $W_2(q^{-1})$.

To the controller's eyes, the plant is not accessible and the control is designed by relying on the knowledge of the identified model only. Since in determining the control policy the identified model is considered as if it were the true plant, transfer functions $W_1(q^{-1})$ and $W_2(q^{-1})$ represent the *desired behavior* in the long run. Therefore, we say that self-optimality holds if the output of the *real system* resembles the one of the *asymptotic noisy imaginary system* for each bounded and deterministic reference signal. Precisely, in the noise-free case, $(n(t) = 0)$ we require:

$$\lim_{t\to\infty} (y(t) - W_1(q^{-1})y^*(t)) = 0, \tag{3.3}$$

whereas, in the noisy case:

$$\lim_{t\to\infty} \frac{1}{N} \sum_{t=1}^{N} | y(t) - W_1(q^{-1})y^*(t) - W_2(q^{-1})n(t) | = 0, \text{ w.p.1} \tag{3.4}$$

for self-optimality.

Remark 3.3. It is important to point out that conditions (3.3) and (3.4) do not imply that the transfer functions of the *asymptotic real system* (that is $\boldsymbol{P} + \boldsymbol{C}(\hat{\theta}(\infty))$) coincide with the corresponding transfer functions of the *asymptotic imaginary system.* Indeed, the above limits refer to the signals only and the possibility of inferring the above conclusions on the transfer functions obviously depends on the excitation characteristics of $y^*(\cdot)$.

3.4 List of Symbols

For easy reference, we summarize in this section the introduced symbols:

$\boldsymbol{P}$	real plant;
$\boldsymbol{M}(\theta)$	model of the plant associated with parameter θ;
$\boldsymbol{C}(\theta)$	controller designed for model $\boldsymbol{M}(\theta)$;
$\hat{\theta}(t)$	present parameter estimate;
$\hat{\theta}(\infty)$	asymptotic parameter estimate;
$\sum(\theta^o, \hat{\theta}(t))$	real system = real plant $\boldsymbol{P}$ + identifier $\boldsymbol{I}$ + controller designed for $\boldsymbol{M}(\hat{\theta}(t))$;
$\sum(\theta^o; \hat{\theta}(\infty))$	asymptotic real system = ideal plant $\boldsymbol{P}$ + controller designed for $\boldsymbol{M}(\hat{\theta}(\infty))$;
$\sum(\theta^o; \theta^o)$	ideal system = real plant $\boldsymbol{P} = \boldsymbol{M}(\theta^o)$ + controller designed for $\boldsymbol{M}(\theta^o)$;
$\sum(\theta; \theta)$	frozen system = model $\boldsymbol{M}(\theta)$ + controller designed for $\boldsymbol{M}(\theta)$;
$\sum(\hat{\theta}(t); \hat{\theta}(t))$	imaginary system = model of the plant $\boldsymbol{M}(\hat{\theta}(t))$ + controller designed for $\boldsymbol{M}(\hat{\theta}(t))$;
$\sum(\hat{\theta}(\infty); \hat{\theta}(\infty))$	asymptotic imaginary system = asymptotic model of the plant $\boldsymbol{M}(\hat{\theta}(\infty))$ + controller designed for $\boldsymbol{M}(\hat{\theta}(\infty))$;
$\sum_n(\hat{\theta}(t); \hat{\theta}(t))$	noisy imaginary system = model of the plant $\boldsymbol{M}(\hat{\theta}(t))$ with the noise $n(t)$ added + controller designed for $\boldsymbol{M}(\hat{\theta}(t))$;
$\sum_n(\hat{\theta}(\infty); \hat{\theta}(\infty))$:	noisy asymptotic imaginary system = asymptotic model of the plant $\boldsymbol{M}(\hat{\theta}(\infty))$ with the noise $n(t)$ added + controller designed for $\boldsymbol{M}(\hat{\theta}(\infty))$.

4. The Least Squares Identification Algorithm

This section is devoted to the introduction of the identification algorithm. For, we start by writing the adopted ARX model is in the regression form:

$$\boldsymbol{M}(\theta) : y(t) = \varphi(t-1)^T\,\theta,$$

where the observation vector $\varphi(t-1)$ is given by

$$\varphi(t-1) = [y(t-1)\dots y(t-n)u(t-d)\dots u(t-d-m)]^T.$$

Parameter θ can be estimated from the measurements of the input and the output of plant $\boldsymbol{P}$ in various ways. One of the most popular is least squares, which amounts to finding θ so as to minimize the criterion:

$$J(\theta) = \sum_{\tau=1}^{t} \left[y(\tau) - \varphi(\tau-1)^T\theta\right]^2.$$

This is a quadratic function of θ; all minima are solutions of the linear system (*normal equations*):

$$\left[\sum_{\tau=1}^{t} \varphi(\tau-1)\varphi(\tau-1)^T\right]\theta = \sum_{\tau=1}^{t} \varphi(\tau-1)y(\tau).$$

Here, matrix

$$M(t) = \sum_{\tau=1}^{t} \varphi(\tau-1)\varphi(\tau-1)^T. \tag{4.1}$$

appears. Note that $M(t)$ is positive semidefinite. Moreover, $M(t+1) \geq M(t)$ is positive semidefinite too. Therefore, if for some $\bar{t}$ $M(\bar{t})$ is non singular, then $M(t)$ is non singular for any $t > \bar{t}$, and the normal equations admit a unique solution:

$$\hat{\theta}(t) = M(t)^{-1} \sum_{\tau=1}^{t} \varphi(\tau-1)y(\tau).$$

The least squares estimate can be given a recursive form (Recursive Least Squares - RLS), enabling the computation of $\hat{\theta}(t)$ from $\hat{\theta}(t-1)$, see e.g. [16, 25]. Precisely,

$$\hat{\theta}(t) = \hat{\theta}(t-1) + P(t)\varphi(t-1)[y(t) - \varphi(t-1)^T\,\hat{\theta}(t-1)], \tag{4.2}$$

$$P(t) = P(t-1) - \frac{P(t-1)\varphi(t-1)\varphi(t-1)^T P(t-1)}{1 + \varphi(t-1)^T\,P(t-1)\,\varphi(t-1)}. \tag{4.3}$$

Note that, if one sets:

$$P(\bar{t}) = M(\bar{t})^{-1}$$

for some $\bar{t}$ such that the inversion can be performed, and initiate the above algorithm at $\bar{t}$ with $P(\bar{t})$ and

$$\hat{\theta}(\bar{t}) = P(\bar{t}) \sum_{\tau=1}^{\bar{t}} \varphi(\tau - 1) y(\tau),$$

then the estimate provided by the recursive equations coincides with the unique solution of the normal equations.

More frequently, however, the equations are initialized at time $t = 0$ with arbitrary initial conditions $\hat{\theta}(0)$. As for $P(O)$, it is obvious that it must meet the condition of positive definiteness, $P(O) > 0$. Then, as it is easy to see, $P(t) > O$, for any t. It goes without saying that, in general, with such an arbitrary initialization, $P(t)$ does not coincide with $M(t)^{-1}$ any more. Precisely, note that eq. (4.3) can can be equivalently written as a recursion in $M(t) = P(t)^{-1}$ as follows:

$$P(t)^{-1} = P(t-1)^{-1} + \varphi(t-1)\varphi(t-1)^T, \tag{4.4}$$

so that $P(t)^{-1} = P(0)^{-1} + M(t)$. Therefore, in general, $P(t)^{-1} \geq M(t)$.

5. Deterministic Analysis of Self-Tuning Control Systems

5.1 Excitation Subspace and Convergence of RLS

We now address the long standing issue of convergence of RLS. In this connection, note that eqs. (4.1), (4.2) can be seen as a discrete-time dynamical system with $\{\hat{\theta}(t), P(t)\}$ as state. It is easy to see that the component $P(t)$ of such a state is convergent. Indeed, from (4.2) it is apparent that, whatever the initialization of RLS be, $P(t)$ is a monotonically decreasing sequence of positive definite matrices. As such, it must converge to some (positive semidefinite) matrix $P(\infty)$. Thus, the real question of convergence is whether $\hat{\theta}(t)$ does converge or not.

We start by showing that, in the noise-free case, the estimate is asymptotically convergent too, $\hat{\theta}(t) \rightarrow \hat{\theta}(\infty)$. With this objective in mind, introduce the parameter error

$$\tilde{\theta}(t) = \hat{\theta}(t) - \theta^o.$$

From eq. (4.1), the following recursion for $\tilde{\theta}(t)$ is easily derived

$$\tilde{\theta}(t) = \tilde{\theta}(t-1) - P(t)\varphi(t-1)\varphi(t-1)^T\tilde{\theta}(t-1). \tag{5.1}$$

Then, using eq. (4.3), we get

$$\begin{aligned} P(t)^{-1}\tilde{\theta}(t) &= P(t)^{-1}\tilde{\theta}(t-1) - \varphi(t-1)\varphi(t-1)^T \tilde{\theta}(t-1) \\ &= P(t-1)^{-1}\tilde{\theta}(t-1), \end{aligned}$$

which corresponds to

$$P(t)^{-1}[\hat{\theta}(t) - \theta^o] = \text{ const.}$$

Since $P(t)$ converges, this obviously implies that $\hat{\theta}(t)$ is convergent to some limit too, a limit which will be denoted by $\hat{\theta}(\infty)$. Of course in general $\hat{\theta}(\infty) \neq \theta^o$.

In order to say more about the asymptotics of $\hat{\theta}(t)$, consider again the positive semidefinite matrix $M(t)$ defined in eq. (6)

Since the information contained in the data is conveyed to the identification algorithm via the observation vector $\varphi(\cdot)$, matrix $M(t)$ can be interpreted as the overall information over the time interval $[0, t-1]$. In general, this information may be nonuniformly distributed. If $x \in \mathbb{R}^{n+m+1}$ is a unit norm vector in the parameter space, the nonnegative scalar $x^T M(t) x$ can be seen as the quantity of information in the x direction of the parameter space. Obviously, such a quantity is monotonically increasing with time t. There are then two possibilities, depending on the specific direction x. Either the quantity of information keeps bounded or it diverges. Correspondingly, we define the notions of *excitation* and *unexcitation* subspaces, originally introduced in [12].

Definition 5.1 (Excitation and unexcitation subspaces). *The subspace* $\bar{\mathcal{E}} = \{x \in \mathbb{R}^{n+m+1} \mid x^T \sum_{\tau=1}^{\infty} \varphi(\tau-1)\varphi(\tau-1)^T x < \infty\}$ *is named* unexcitation subspace.
Its orthogonal complement $\mathcal{E} = \bar{\mathcal{E}}^{\perp}$ *is named* excitation subspace.

For the asymptotic analysis, introduce the Lyapunov-like function

$$V(t) = \tilde{\theta}(t)^T P(t)^{-1}\, \tilde{\theta}(t). \tag{5.2}$$

From (4.2), (4.3) and (5.1), the recursive expression for $V(t)$ can be worked out:

$$V(t) = V(t-1) - \frac{[\varphi(t-1)^T\, \tilde{\theta}(t-1)]^2}{1 + \varphi(t-1)^T\, P(t-1)\varphi(t-1)}.$$

This implies that function $V(t)$ is monotonically decreasing, so that

$$V(0) \geq V(t).$$

As already observed, $P(t)^{-1} \geq M(t)$, so that

$$V(0) \geq \tilde{\theta}(t)^T M(t) \tilde{\theta}(t),$$

from which, considering that $M(t)$ is monotonically increasing,

$$\lim_{t\to\infty} \tilde{\theta}(\infty)^T M(t) \tilde{\theta}(\infty) < \infty, \tag{5.3}$$

where

$$\tilde{\theta}(\infty) = \hat{\theta}(\infty) - \theta^o.$$

Partition now $\tilde{\theta}(\infty)$ in the excitation and unexcitation components: $\tilde{\theta}(\infty) = \tilde{\theta}_E(\infty) + \tilde{\theta}_U(\infty)$ (where $\tilde{\theta}_E(\infty) \in \mathcal{E}$ and $\tilde{\theta}_U(\infty) \in \bar{\mathcal{E}}$). Analogously, partition the observation vector as $\varphi(t) = \varphi_E(t) + \varphi_U(t)$, with $\varphi_E(t) \in \mathcal{E}$ and $\varphi_U(t) \in \bar{\mathcal{E}}$, and let

$$M_E(t) = \sum_{\tau=1}^{t} \varphi_E(\tau-1)\, \varphi_E(\tau-1)^T$$

$$M_U(t) = \sum_{\tau=1}^{t} \varphi_U(\tau-1)\, \varphi_U(\tau-1)^T.$$

Then, from (6) we obtain

$$\begin{aligned}
&\tilde{\theta}(\infty)^T M(t) \tilde{\theta}(\infty) \\
&= \sum_{\tau=1}^{t} \left(\varphi(\tau-1)^T\, \tilde{\theta}(\infty)\right)^2 \\
&= \sum_{\tau=1}^{t} \left(\varphi_E(\tau-1)^T\, \tilde{\theta}_E(\infty) + \varphi_U(\tau-1)^T\, \tilde{\theta}_U(\infty)\right)^2 \\
&\geq \Big\{\tilde{\theta}_E(\infty)^T\, M_E(t)\, \tilde{\theta}_E(\infty) + \tilde{\theta}_U(\infty)^T\, M_U(t)\, \tilde{\theta}_U(\infty) \\
&\quad -2 \left[\tilde{\theta}_U(\infty)^T M_E(t)\, \tilde{\theta}_E(\infty)\right]^{1/2} \left[\tilde{\theta}_U(\infty)^T\, M_U(t)\, \tilde{\theta}_U(\infty)\right]^{1/2}\Big\}.
\end{aligned}$$

In view of the very definition of excitation subspace, $x'\, M_E(t)\, x$ tends to infinity $\forall\, x \in \mathcal{E}$, so that it is necessary that $\tilde{\theta}_E(\infty) = 0$ in order to inequality (12) be true. On the opposite, since $M_U(t)$ does not diverge, $\tilde{\theta}_U(\infty)$ may be different than 0.

For the sake of exposition clarity, the results derived in this subsection are summarized in the following theorem.

Theorem 5.1 (RLS properties; noise-free case). *The RLS estimate $\hat{\theta}(t)$ is asymptotically convergent:*

$$\lim_{t\to\infty} \hat{\theta}(t) = \hat{\theta}(\infty).$$

Moreover, denoting by $\hat{\theta}_E(\infty)$ and θ^o_E the projections of $\hat{\theta}(\infty)$ and θ^o onto the excitation subspace $\mathcal{E}$, it turns out that

$$\hat{\theta}_E(\infty) = \theta^o_E.$$

Remark 5.1. In system identification, one often encounters the condition of *persistent excitation*. In the deterministic case, this condition amounts to requiring the existence of a parameter α and an integer r such that

$$\sum_{\tau=t+1}^{t+r} \varphi(\tau-1)\varphi(\tau-1)^T \geq \alpha I, \ \forall\, t.$$

This implies that the eigenvalues of the information matrix $M(t)$ go to infinity (at least linearly with t). In this way, the excitation subspace coincides with the entire parameter space and the estimate is consistent: $\lim_{t\to\infty}\hat{\theta}(t) = \theta^o$. In control problems, however, vector $\varphi(t)$ is formed by closed-loop observations and, generally, one cannot expect persistency in excitation. Then, only some eigenvalues of the information matrix will tend to infinity (possibly at a lower rate than linearly) and only partial consistency can be claimed: $\lim_{t\to\infty}\hat{\theta}_E(t) = \theta^o_E$. In this connection, note that the excitation subspace is in fact dependent on the specific control law and on reference signal $y^*(\cdot)$. Thus, the same dependence holds for the "directions" of partial consistency.

5.2 Adaptive Stabilization and Self-Optimality

Due to the interplay between identification and control, a self-tuning control system is described by complex nonlinear equations. On the contrary, the *asymptotic imaginary system* $\sum(\hat{\theta}(\infty);\hat{\theta}(\infty))$ is linear and time-invariant. The achievement of this section can be concisely stated as follows. If the adopted control law is continuous in the asymptotic estimate $\hat{\theta}(\infty)$ (i.e. if $(\hat{\theta}(\infty) \in \ \Xi_1)$ and the *aymptotic imaginary system* is asymptotically stable $(\hat{\theta}(\infty) \in \ \Xi_2)$, then the *real system* is adaptively stabilized and meets the self-optimality property. Basically, this result reduces the though analysis of the nonlinear self-tuning control system to the stability analysis of the linear time-invariant system $\sum(\hat{\theta}(\infty);\hat{\theta}(\infty))$.

For the theoretical study, we first represent the *real system* $\sum(\theta^o;\hat{\theta}(t))$ as a variation system with respect to the *imaginary system* $\sum(\hat{\theta}(t);\ \hat{\theta}(t))$.

Set

$$e(t) = \varphi(t-1)^T(\theta^o - \hat{\theta}(t)).$$

Then, the time evolution of vector $z(t) = [y(t)\ u(t)]^T$ generated by $\sum(\theta^o;\hat{\theta}(t))$ is given by the equation

$$z(t) = D(\hat{\theta}(t);q^{-1})z(t-1) + \begin{bmatrix} e(t) \\ T(\hat{\theta}(t);q^{-1})y^*(t) \end{bmatrix}, \tag{5.4}$$

where matrix

$$D(\hat{\theta}(t);q^{-1}) = \begin{bmatrix} [1 - A(\hat{\theta}(t);q^{-1}]q & B(\hat{\theta}(t);q^{-1})q^{-d+1} \\ S(\hat{\theta}(t);q^{-1})q & R(\hat{\theta}(t);q^{-1})q \end{bmatrix}$$

describes the dynamics of $\sum(\hat{\theta}(t);\hat{\theta}(t))$.

Note that $\sum(\theta^o;\hat{\theta}(t))$ and $\sum(\hat{\theta}(t);\hat{\theta}(t))$ are different from each other only due to the presence of the input $e(t) = \varphi(t-1)^T(\theta^o - \hat{\theta}(t))$ of $\sum(\theta^o;\hat{\theta}(t))$. Therefore, in order for the adaptive control scheme to behave closely to $\sum(\hat{\theta}(t);\hat{\theta}(t))$, and consequently to asymptotically resemble the desired behaviour expressed by equations (4.1) and (4.2) (with $n(t)=0$), the quantity $\varphi(t-1)^T(\theta^o - \hat{\theta}(t))$ should be small and eventually vanish. In Fig. 5.2, such a quantity is seen as the output of a *perturbation system* which, in turn, is fed by vector $z(t)$. This generates a perturbation loop whose stability plays a crucial role in determining the self-optimality of the entire scheme. Indeed, thanks to the properties enjoyed by the RLS estimate, the stability of such a loop is sufficient to prove the asymptotic vanishing (in mean) of $\varphi(t-1)^T(\theta^o - \hat{\theta}(t))$. Then, the key is the following theorem which states that the stability of the perturbed system (i.e. the *real system*) is ensured under the mild assumption that $\sum(\hat{\theta}(\infty);\hat{\theta}(\infty))$ is stable.

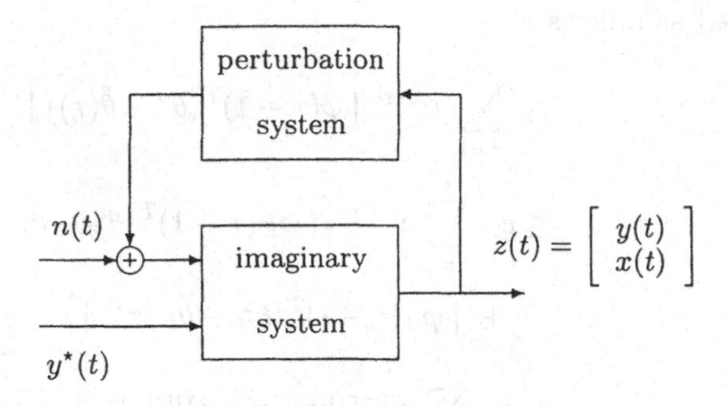

Fig. 5.1. The perturbation system

Theorem 5.2 (Stability; noise-free case). *If* $\hat{\theta}(\infty) \in \Xi_1 \cap \Xi_2$, *then* $|\, u(t)\,| < c$ *and* $|\, y(t)\,| < c, \;\forall\, t$.

Proof. Since $\hat{\theta}(\infty) \in \Xi_2$, the movement of the autonomous system $z(t) = D(\hat{\theta}(\infty);\, q^{-1})z(t-1)$ initialized at time t_0 tends exponentially to zero: $\| \bar{z}(t) \| \leq \alpha\rho^{\Delta t} \| \bar{z}(t_o) \|$, $\rho < 1$, $\Delta t = t - t_o$, where $\bar{z}(t) = [y(t) \quad y(t-1) \;\ldots\; y(t+1-\max\{n,\beta\})\; u(t)u(t-1) \;\ldots\; u(t+1-\max\{d+m,\alpha\})]^T$. Taking into account that $\hat{\theta}(t) \to \hat{\theta}(\infty)$ and $D(\theta;q^{-1})$ is continuous in $\hat{\theta}(\infty)$, for the autonomous *imaginary system* ($z(t) = D(\hat{\theta}(t);q^{-1})z(t-1)$) we then have $\| \bar{z}(t) \| \leq \alpha\rho^{\Delta t} \| \bar{z}(t_0) \| + \gamma(t_0,\Delta t) \| \bar{z}(t_0) \|$, where $\gamma(\cdot,\cdot)$ is such that $\gamma(t_0,\Delta t) \to 0$, $t_0 \uparrow \infty$, Δt fixed. Choose $\Delta\bar{t}$ such that $\alpha\rho^{\Delta\bar{t}} = \beta < 1$ and $\bar{t}_0$ such that $|\,\gamma(t_0,\Delta\bar{t})\,| \leq \delta < 1-\beta$, $\forall\, t_0 \geq \bar{t}_0$. Then, $\forall t_0 \geq \bar{t}_0$, one has $\| \bar{z}(t_0+\Delta\bar{t}) \| \leq (\beta+\delta) \| \bar{z}(t_0) \|$, from which the uniform exponential stability of the autonomous *imaginary system* follows. Turn now to consider the *real system*. In view of the stability of the autonomous *imaginary system* and representation (5.5) of $\sum(\theta^o;\hat{\theta}(t))$, $z(t)$ generated by the *real system* can be bounded as follows:

$$\begin{aligned} ||z(t)|| &\leq c_1 + c_2 \sum_{\tau=1}^{t} \nu^{t-\tau} \left|\left|\left[\begin{array}{c} e(\tau) \\ T(\hat{\theta}(\tau);q^{-1})y^*(\tau) \end{array} \right]\right|\right| \\ &\leq c_1 + c_2 \sum_{\tau=1}^{t} \nu^{t-\tau} \mid \varphi(\tau-1)^T(\theta^o - \hat{\theta}(\tau)) \mid \\ &\quad + c_2 \sum_{\tau=1}^{t} \nu^{t-\tau} \mid T(\hat{\theta}(\tau);q^{-1})y^*(\tau) \mid, \end{aligned} \tag{5.5}$$

c_1, c_2 and $\nu < 1$ being suitable constants. By virtue of the boundedness of the reference output, the term $c_2 \sum_{\tau=1}^{t} \nu^{t-\tau} \mid T(\hat{\theta}(\tau);q^{-1})y^*(\tau) \mid$ turns out to be itself bounded. As for the second term at the right hand side it can be handled as follows:

$$\begin{aligned} & c_2 \sum_{\tau=1}^{t} \nu^{t-\tau} \mid \varphi(\tau-1)^T(\theta^o - \hat{\theta}(\tau)) \mid \\ & \leq c_2 \sum_{\tau=1}^{t} \nu^{t-\tau} \Big\{ \mid \varphi_E(\tau-1)^T(\theta^o_E - \hat{\theta}_E(\tau)) \mid \\ & \quad + \mid \varphi_U(\tau-1)^T(\theta^o_U - \hat{\theta}_U(\tau)) \mid \Big\} \\ & \leq c_2 \sum_{\tau=1}^{t} \nu^{t-\tau} ||\varphi_E(\tau-1)||\ ||\theta^o_E - \hat{\theta}_E(\tau)|| \\ & \quad + c_2 \sum_{\tau=1}^{t} \nu^{t-\tau} ||\varphi_U(\tau-1)||\ ||\theta^o_U - \hat{\theta}_U(\tau)||. \end{aligned}$$

The second term in this last expression tends to zero because of the boundedness of $\| \theta^o_U - \hat{\theta}_u(\tau) \|$ and the fact that $\varphi_U(\tau-1) \to 0$ as $\tau \to \infty$ (see the definition of unexcitation subspace). Therefore, bearing in mind the definition of observation vector, from (5.6) we finally get

$$\begin{aligned} \| \varphi(t) \| &\leq c_3 + c_4 \sum_{\tau=1}^{t} \nu^{t-\tau} \| \varphi_E(\tau-1) \| \ \| \theta^o_E - \hat{\theta}_E(\tau) \| \\ &\leq c_3 + c_4 \sum_{\tau=1}^{t} \nu^{t-\tau} \| \varphi(\tau-1) \| \ \| \theta^o_E - \hat{\theta}_E(\tau) \|, \end{aligned}$$

c_3 and c_4 suitable constants.

Since $\theta^o_E - \hat{\theta}_E(\tau) \to 0$, this inequality implies that $\| \varphi(t) \|$ remains bounded, from which the thesis immediately follows.

By exploiting the stability result just proven in the light of the properties of the RLS estimate stated in Theorem 1, it is easy to conclude that the perturbation term $e(t) = \varphi(t-1)^T(\theta^o - \hat{\theta}(t))$ asymptotically vanishes. Indeed, write

$$\begin{aligned} \mid \varphi(t-1)^T(\theta^o - \hat{\theta}(t)) \mid \leq \quad & \parallel \varphi_E(t-1) \parallel \parallel \theta^o_E - \hat{\theta}_E(t) \parallel \\ & + \parallel \varphi_U(t-1) \parallel \parallel \theta^o_U - \hat{\theta}_U(t) \parallel . \end{aligned}$$

The second term vanishes since $\parallel \theta^o_U - \hat{\theta}_U(t) \parallel$ is bounded and $\parallel \varphi_U(t-1) \parallel$ tends to zero in view of the very definition of unexcitation subspace. As for the first term, taking into account that $\parallel \theta^o_E - \hat{\theta}_E(t) \parallel$ tends to zero (Theorem 1) and $\parallel \varphi_E(t-1) \parallel \leq \parallel \varphi(t-1) \parallel$ remains bounded (Theorem 2), it follows that $\parallel \varphi_E(t-1) \parallel \parallel \theta^o_E - \hat{\theta}_E(t) \parallel$ goes to zero too.

We are now in a position to prove the self-optimality theorem.

Theorem 5.3 (Self-optimality; noise-free case). *If* $\hat{\theta}(\infty) \in \Xi_1 \cap \Xi_2$, *then* $\lim_{t\to\infty} \left(y(t) - W_1(q^{-1})y^*(t)\right) = 0$, *where* $W_1(q^{-1})$ *was defined in eq. (3.1).*

Proof. By virtue of the assumption $\hat{\theta}(\infty) \in \Xi_2$, polynomial $\Delta(\hat{\theta}(\infty); q^{-1}) = A(\hat{\theta}(\infty); q^{-1})\, [1 - R(\hat{\theta}(\infty); q^{-1})]\, q^{-d} - B(\hat{\theta}(\infty); q^{-1})S(\hat{\theta}(\infty); q^{-1})$ is Hurwitz, so that from the fact that $\varphi(t-1)^T(\theta^o - \hat{\theta}(t)) \to 0$, we have

$$\begin{aligned} & \frac{1 - R(\hat{\theta}(\infty); q^{-1})}{\Delta(\hat{\theta}(\infty); q^{-1})}\varphi(t-1)^T(\theta^o - \hat{\theta}(t)) \\ = & \frac{1 - R(\hat{\theta}(\infty); q^{-1})}{\Delta(\hat{\theta}(\infty); q^{-1})}\left\{y(t) - [1 - A(\hat{\theta}(t); q^{-1})]y(t) - B(\hat{\theta}(t); q^{-1})q^{-d}u(t)\right\} \\ & \to 0. \end{aligned}$$

Since $\hat{\theta}(t) \to \hat{\theta}(\infty)$ (Theorem 1) and $\mid y(t) \mid$ is bounded (Theorem 2), we can replace $A(\hat{\theta}(t); q^{-1})y(t)$ in the previous expression with $A(\hat{\theta}(\infty); q^{-1})y(t)$. For an analogous reason, in place of $[1 - R(\hat{\theta}(\infty); q^{-1})]B(\hat{\theta}(t); q^{-1})q^{-d}u(t)$, we can write $B(\hat{\theta}(t); q^{-1})[1 - R(\hat{\theta}(t-d); q^{-1})]q^{-d}u(t)$, which equals $B(\hat{\theta}(t); q^{-1})$ $[S(\hat{\theta}(t-d); q^{-1})y(t-d) + T(\hat{\theta}(t-d); q^{-1})y^*(t-d)]$ and this last expression can be finally replaced by $B(\hat{\theta}(\infty); q^{-1})[S(\hat{\theta}(\infty); q^{-1})y(t-d) + T(\hat{\theta}(\infty); q^{-1})y^*(t-d)]$. In conclusion, we have

$$y(t) - \frac{q^{-d}B(\hat{\theta}(\infty); q^{-1})T(\hat{\theta}(\infty); q^{-1})}{\Delta(\hat{\theta}(\infty); q^{-1})}\, y^*(t) \to 0,$$

which is the self-optimality condition.

Remark 5.2. – The assumption of Theorems 2 and 3 ($\hat{\theta}(\infty) \in \Xi_1 \cap \Xi_2$) requires the control law to be continuous in $\hat{\theta}(\infty)$ and that it is selected so as to stabilize the asymptotic model of the plant. Both these conditions are easily fulfilled provided that the asymptotically estimated polynomials $q^n A(\hat{\theta}(\infty); q^{-1})$ and $q^m B(\hat{\theta}(\infty); q^{-1})$ are coprime. This asymptotic coprimeness issue is further elaborated in the next section.

- Theorem 4 guarantees that the desired behavior is asymptotically achieved in mean. In this regard, note that the notion of desired behavior is based on the asymptotically identified model (see the discussion at the end of Section 3.4). Consequently, the theorem says that the adaptively controlled true system (*real system*) behaves closely to the nonadaptively controlled asymptotically identified system (*asymptotic imaginary system*).
- The previous conclusion does not entail that the performance of the adaptive scheme (*real system*) is close to the one which would be obtained if the real plant were actually known (*ideal system.* As a matter of fact, it is easy to work out examples where this is not the case.

5.3 Specific Control Laws

In the previous subsection, we have shown that the self-tuning control system attains stability and optimality under the basic assumption that the *asymptotic imaginary system* is asymptotically stable. On the other hand, the controller is synthetized on the knowledge of the identified model and, therefore, asymptotic stability of the *asymptotic imaginary system* seems to be a quite natural requirement.

In the literature, many well established techniques can be found which ensure asymptotic stability when applied to a known time-invariant plant under various technical conditions. When one of these techniques is applied in a certainty equivalent fashion, stability and optimality is guaranteed provided that these conditions are met by the asymptotically estimated model. Here, we introduce as an example a couple of control laws which lead to stable control systems under the coprimeness condition of the plant transfer function.

Infinite-horizon LQ control
Consider the plant

$$A(\bar{\theta}; q^{-1})y(t) = B(\bar{\theta}; q^{-1})u(t-d), \tag{5.6}$$

where

$$A(\bar{\theta}; q^{-1}) = 1 - \sum_{i=1}^{n} \bar{a}_i q^{-i}, \; B(\bar{\theta}; q^{-1}) = \sum_{i=0}^{m} \bar{b}_i q^{-i}, \; \bar{b}_0 \neq 0. \tag{5.7}$$

Given a bounded reference signal $y^*(\cdot)$, the linear-quadratic (LQ) infinite-horizon control law is obtained by minimizing the cost function

$$J = \sum_{\tau=1}^{\infty} q(y(\tau) - y^*(\tau))^2 + \sum_{\tau=1}^{\infty} ru(\tau)^2,$$

with $q \geq 0$ and $r > 0$. The first term is the penalty for the discrepancy between the reference signal and the actual output, whereas the second term is the penalty for the required energy of the control action. The relative importance of the two terms can be modulated by a suitable choice of parameters q and r.

As for the stability of LQ control system, a very important result is the following one, [1]:

Proposition 5.1. *If $q^n A(\bar{\theta}; q^{-1})$ and $q^m B(\bar{\theta}; q^{-1})$ have no unstable common factors, then the closed-loop system is asymptotically stable and the control law is a continuous function of the system parameter in a neighborhood of the actual parameterization $\bar{\theta}$.*

Receding-horizon control

For the same plant introduced above ((5.7) and (5.8)), compute the input sequence $\bar{u}(t), \bar{u}(t+1), \ldots, \bar{u}(t+N-1)$ so as to minimize the cost function

$$J = \sum_{\tau=t+d+1}^{t+d+N-1} q(y(\tau) - y^*(\tau))^2 + \sum_{\tau=t}^{t+N-1} ru(\tau)^2,$$

where $q \geq 0$, $r > 0$, $N \geq \max\{n, m+d\}$ (n and m being the degrees of polinomyals $A(\cdot)$ and $B(\cdot)$, respectively), subject to the final constraints

$$u(t+N) = u(t+N+1) = \cdots = u(t+N+n-2) = 0,$$

and

$$\begin{aligned} y(t+d+N) &= y(t+d+N+1) = \cdots = y(t+d+N+n-1) \\ &= y^*(t+d+N). \end{aligned}$$

Then, apply only the first sample of the input sequence at time t (namely, set $u(t) = \bar{u}(t)$) and repeat the whole procedure at time $t+1$ to select $u(t+1)$.

The cost function is again quadratic, but it involves only a finite horizon. Integer N can be used to modulate the amplitude of the "forward window" considered for the design of $u(\cdot)$ at time t.

In this case, the statement of Proposition 1 is still valid, [18, 19, 11]. □

Infinite-horizon LQ control and receding-horizon control are just two examples of a large class of control techniques with stabilizing properties under the coprimeness condition of the (known) plant trasfer function. Therefore, in view of the discussion at the beginning of this subsection, the coprimeness of $q^n A(\hat{\theta}(\infty); q^{-1})$ and $q^m B(\hat{\theta}(\infty); q^{-1})$ becomes a fundamental issue in self-tuning control. To the best knowledge of the authors, no complete theory is to date available in the literature concerning this issue and an effort should be made in an attempt to work out satisfying results in this direction.

6. Stochastic Analysis of Self-Tuning Control Systems

6.1 Stochastic Excitation Subspace and Convergence of RLS

We turn now to the noisy case. The convergence analysis of the RLS algorithm is thoroughly presented in [26, 22, 10, 15].

Following the lead originally given in [26] the convergence analysis can be performed in a *Bayesian embedding context*. This amounts to viewing the true parameterization as a random vector rather than an unknown constant. In this way, two are the sources of stochasticity in the analysis, namely randomness in noise and randomness in true parameter. This leads to an enlarged probability space given by the product of the probability space of the noise times the auxiliary probability space of the parameter vector.

The main advantage of using the Bayesian embedding approach is that, if θ^o is assumed to be Gaussian, the RLS equations can be regarded as a Kalman filter applied to the following dynamic system:

$$\begin{cases} \theta^o(t+1) = \theta^o(t) \\ y(t) = \varphi(t-1)^T \theta^o(t) + n(t). \end{cases}$$

Since the Kalman filter recursively produces the conditional expectation of θ^o given the observations, it is then possible to study the convergengence of the RLS algorithm via standard Martingale theory. Using this approach, in [26, 22] it was proven that the RLS estimate generally converges. Further refinements on the stochastic Kalman filter interpretation, necessary for the rigorous use of the theory in control problems, were given in [10, 15].

The notion of excitation subspace can be given in the noisy case as well, without any modification, i.e. Definition 1 still holds. Note however that this subspace is now stochastic, due to randomness in observation. By making reference to such a notion, we can reformulate the results of the quoted papers which are relevant to the study of self-tuning control schemes in the following way.

Theorem 6.1 (RLS properties; noisy case). *There exists a set $\mathcal{N} \subset \mathbb{R}^{n+m+1}$ with $\mathcal{L}(\mathcal{N}) = 0$ ($\mathcal{L}(\cdot)$ denotes Lebesgue measure in $\mathbb{R}^{n+m+1}$) such that, for any deterministic $\theta^o \notin \mathcal{N}$,*

$$\lim_{t \to \infty} \hat{\theta}(t) = \hat{\theta}(\infty) \quad w.\ p.\ 1,$$

where $\hat{\theta}(\infty)$ is an almost surely bounded random variable. Moreover, denoting by $\hat{\theta}_E(\infty)$ and θ^o_E the projections of $\hat{\theta}(\infty)$ and θ^o onto the excitation subspace $\mathcal{E}$, it turns out that

$$\hat{\theta}_E(\infty) = \theta^o_E \quad w.\ p.\ 1.$$

Remark 6.1. – The condition $\theta^o \notin \mathcal{N}$ of Theorem 1 cannot be dropped. As a matter of fact, there are situations in which, if the system parameter vector belongs to a certain singular set with zero Lebesgue measure, then the RLS estimate drifts out of any bounded set, see [20]. However, this fact should not be of too much concern: in much the same way as almost all the stochastic results hold true with probability one, and still provide a powerful tool in the comprehension of random phenomena, the Bayesian approach is of help to gain insight in the behaviour of the RLS algorithm even though the corresponding results may fail to hold in a zero Lebesgue measure set.

- According to the above statement, set $\mathcal{N}$ contains the true parametrizations for which convergence of the RLS estimate is not guaranteed. This is why we can identify $\mathcal{N}$ with the set of (possibly) "pathological systems".
- The Bayesian embeding viewpoint is just instrumental for the proof of Theorem 4. However, one should bear in mind that in the final statement of this result, reference is made to a deterministic value of θ^o. To better comprehend the reason of this, we make a digression on a wise way to interprete the Bayesian approach.

As already said, in the Bayesian embedding approach one artificially enlarges the original probability space so as to encompass an auxiliary probability space for the true parameterization θ^o viewed as a stochastic gadget too.

For such an enlarged probability space, let $\tilde{\Omega}$ be the product space and $\tilde{p}$ the probability product, see Fig. 6.1.

In the proof of the above theorem, one actually shows that $\hat{\theta}(t)$ is convergent with probability 1 with respect to $\tilde{p}$. This means that there is a set – say $\tilde{\mathcal{N}}$ – in the product space $\tilde{\Omega}$, with $\tilde{p}(\tilde{\mathcal{N}}) = 0$, such that convergence holds everywhere in the complement of $\tilde{\mathcal{N}}$. In order to recover the deterministic viewpoint on θ^o, one should simply consider what happens on a horizontal line in the rectangle $\tilde{\Omega}$. In Fig. 6.1, three possibilities are displayed, corresponding to the values θ^o_1, θ^o_2 and θ^o_3. Since the θ^o_1 line does not intersect $\tilde{\mathcal{N}}$, it is plain that θ^o_1 does not correspond to a "pathological system" ($\theta^o_1 \notin \mathcal{N}$). Obviously, a system may be "nonpathological" even in the case of intersection between the horizontal line associated with its parameterization and set $\tilde{\mathcal{N}}$. This is the case of parameter θ^o_2 granted that the intersection set A_2 has zero probability (in the original probability space Ω). On the contrary, the intersection set for θ^o_3 is given by A_3 which, as

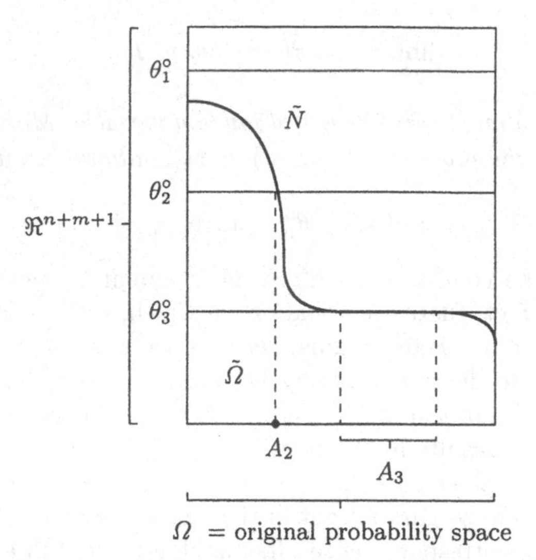

Fig. 6.1. The enlarged probability space used in the Bayesian embedding context

suggested by the drawing, has a nonzero probability measure. This means that the system associated with θ_3^o is a "pathological system".

6.2 Stochastic Adaptive Stabilization and Self-Optimality

The analysis of noisy self-tuning control systems can be sketched as follows, see [9, 6] for the proofs and more details. In analogy with the noise-free case, the *real system* can be represented as a variation system with respect to the *noisy imaginary system.* Under the assumption that the *aymptotic noisy imaginary system* is stable ($\hat{\theta}(\infty) \in \Xi_2$) and no discontinuity in the control law computation takes place ($\hat{\theta}(\infty) \in \Xi_1$), one can prove that the effect of this variation system is asymptotically negligeable. This is a consequence of the fact that the parameter estimation error vanishes along the excitation directions (Theorem 4). Thus, one comes to the following main conclusion.

Theorem 6.2 (Stability and self-optimality; noisy case). *If* $\hat{\theta}(\infty) \in \Xi_1 \cap \Xi_2$, *then*

a) $\limsup_{N\to\infty} \frac{1}{N} \sum_{t=1}^{N} [\mid u(t) \mid + \mid y(t) \mid] < \infty$, *w.p. 1;*

b) $\lim_{N\to\infty} \frac{1}{N} \sum_{t=1}^{N} \mid y(t) - W_1(q^{-1})y^*(t) - W_2(q^{-1})n(t) \mid = 0$, *w.p. 1,*

where $W_1(q^{-1})$ *and* $W_2(q^{-1})$ *were defined in eqs. (3.1) and (3.2).*

7. Concluding Remarks – Towards a Theory of Tuning

In this contribution, we have considered a fairly general class of linear self-tuning control systems which does not assume the plant to be minimum phase. Stability and optimality results have been worked out in full for the deterministic case, and the extension to the stochastic framework has been sketched as well.

In our theory, generality is achieved thanks to a new analysis approach which outdoes previous techniques based on Lyapunov functions and/or the direct use of the RLS normal equations. The basic idea consists in first proving partial consistency of RLS. Then, uncertainty in the system parameter is described by means of a perturbation system whose output acts as an extra disturbance. In this setting, the partial consistency property enables one to show that the effect of the perturbation system is in fact asymptotically vanishing. Among other things, this line of reasoining has the advantage to put sharply into focus the properties of the identification algorithm which are essential for a good performance of the overall control system.

The main conclusion of the paper is that we are now close to a *theory of tuning.* Adaptation in control was originally introduced in the hope to achieve, at least asymptotically, the same performance of the control system one would desing if the parameters were known. But this *ideal tuning* turns out to be utopian in most cases. In this paper, however, we have rather shown that in fairly general situations one should expect that the asymptotic behaviour exhibits the performance of the control system one could obtain if the control design were based on the asymptotycally identified model.

Acknowledgment

Paper supported by the European HCM (Human Capital and Mobility) project "*Nonlinear and Adaptive Control*", the Italian MURST (Ministry of University and Technical and Scientific Research) project "*Model Identification, Systems Control and Signal Processing*", and the CNR *Centro di Teoria dei Sistemi* of Milano.

Reference

1. Anderson B.D.O., and J.B. Moore, *Optimal control - linear quadratic methods*, Prentice-Hall, 1989.
2. Åström K.J., *Introduction to stochastic control*, Academic Press, New York, 1970.
3. Åström K.J., and B. Wittenmark, "On self-tuning regulators", *Automatica*, 9, 185–189, 1973.
4. Åström K.J., and B. Wittenmark, *Adaptive Control*, Addison-Wesley, Reading, Mass., 1989.
5. Bittanti S., P. Bolzern, and M. Campi, "Recursive least squares identification algorithms with incomplete excitation: convergence analysis and application to adaptive control", *IEEE Trans. on Automatic Control*, **AC-35 (12)**, 1371–1373, 1990.
6. Bittanti S., and M. Campi, "Self-optimality of adaptive control systems based on the certainty equivalence principle", *5th IFAC Symp. on Adaptive Systems in Control and Signal Processing*, Budapest, 1995.
7. Bittanti S., M. Campi, and F. Lorito, "Effective identification algorithms for adaptive control", *Int. Journal of Adaptive Control and Signal Processing, Special Issue on "Adaptive and Expert Control: A Russian Perspective"*, (S. Bittanti and A. Fradkov eds.) **6**, 221–235, 1992.
8. Campi M., "On the convergence of minimum-variance directional-forgetting adaptive control schemes", *Automatica*, **28**, 221–225, 1992.
9. Campi M., "Adaptive control of nonminimum phase systems", *Int. J. Adaptive Control and Signal Processing*, **9**, 137–149, 1995.
10. Chen H.F., P.R. Kumar, and J.H. van Schuppen, "On Kalman filtering for conditionally gaussian systems with random matrices", *Systems & Control Letters* **13**, 397–404, 1989.
11. Chisci L., and E. Mosca, "Zero terminal state receding horizon regulation: the singular state transition matrix case", Università di Firenze, Int. report (private communication), 1994.
12. Clarke D.W., and P.J. Gawthrop, "Self-tuning control", *Proc. IEE*, 126, D, 633–640, 1979.
13. Goodwin G.C., P.J. Ramadge, and P.E. Caines. "Discrete-time multivariable adaptive control", *IEEE Trans. on Automatic Control*, **AC-25 (3)**, 449–453, 1980.
14. Goodwin G.C., P.J. Ramadge, and P.E. Caines, "Discrete-time stochastic adaptive control", *SIAM J. Control and Optimization*, **19**, 829–853, 1981.
15. Kumar P.R., "Convergence of adaptive control schemes using least-squares parameter estimates", *IEEE Trans. on Automatic Control*, **AC-35 (4)**, 416–424, 1990.
16. Ljung L., and T. Söderström, *Theory and practice of recursive identification*, MIT Press, Cambridge, Mass., 1983.
17. Meyn S.P., P.E. Caines, "The zero divisor problem of multivariable stochastic adaptive control", *Systems & Control Letters* **6**, 235–238, 1985.
18. Mosca E., *Optimal, predictive, and adaptive control*, Prentice-Hall, 1995.
19. Mosca E., and J. Zhang, "Stable redesign of predictive control", *Automatica*, **28 (6)**, 1229–1233, 1992.
20. Nassiri-Toussi K., and W. Ren, "On the convergence of least squares estimates", *Proc. 31st Conference on Decision and Control*, Tucson, 2233-2238, 1992.

21. Peterka, V., "On steady state minimum variance control strategy", *Kybernetica*, 8, 219–232, 1972.
22. Rootzen H., and J. Sternby, "Consistency in least-squares estimation: a Bayesian approach", *Automatica*, **20 (4)**, 471–475, 1984.
23. van Schuppen J.H., "Tuning of Gaussian Stochastic Control Systems", *IEEE Trans. on Automatic Control*, **AC-39 (11)**, 2178–2190, 1994.
24. Sin K.S., and G.C. Goodwin, "Stochastic adaptive control using a modified least squares algorithm", *Automatica* **18**, 315–321, 1982.
25. Söderstöm T., and P. Stoica, *System identification*, Prentice-Hall, Englewood Cliffs, N.J., 1989.
26. Sternby J., "On consistency for the method of least squares using martingale theory", *IEEE Trans. on Automatic Control*, **AC-22 (3)**, 346–352, 1977.

On Neural Network Model Structures in System Identification

L. Ljung, J. Sjöberg, and H. Hjalmarsson

Department of Electrical Engineering
Linköping University
S-581 83 Linköping
Sweden
E-mail: ljung@isy.liu.se, sjoberg@isy.liu.se, hakan@isy.liu.se

1. Introduction and Summary

1.1 What is the Problem?

The identification problem is to infer relationships between past input-output data and future outputs. Collect a finite number of past inputs $u(k)$ and outputs $y(k)$ into the vector $\varphi(t)$

$$\varphi(t) = [y(t-1) \dots y(t-n_a)\ u(t-1) \dots u(t-n_b)]^T \tag{1.1}$$

For simplicity we let $y(t)$ be scalar. Let $d = n_a + n_b$. Then $\varphi(t) \in \mathbb{R}^d$. The problem then is to understand the relationship between the next output $y(t)$ and $\varphi(t)$:

$$y(t) \leftrightarrow \varphi(t) \tag{1.2}$$

To obtain this understanding we have available a set of observed data (sometimes called the "training set")

$$Z^N = \{[y(t), \varphi(t)]|\ t = 1, \dots N\} \tag{1.3}$$

From these data we infer a relationship

$$\hat{y}(t) = \hat{g}_N(\varphi(t)) \tag{1.4}$$

We index the function g with a "hat" and N to emphasize that it has been inferred from (1.3). We also place a "hat" on $y(t)$ to stress that (1.4) will in practice not be an exact relationship between $\varphi(t)$ and the observed $y(t)$. Rather $\hat{y}(t)$ is the "best guess" of $y(t)$ given the information $\varphi(t)$.

1.2 Black Boxes

How to infer the function $\hat{g}_N$? Basically we search for it in a parameterized family of functions

$$\mathcal{G} = \{g(\varphi(t), \theta) | \theta \in D_{\mathcal{M}}\} \tag{1.5}$$

How to choose this parameterization? A good, but demanding, choice of parameterization is to base it on physical insight. Perhaps we know the relationship between $y(t)$ and $\varphi(t)$ on physical grounds, up to a handful of physical parameters (heat transfer coefficients, resistances,...). Then parameterize (1.5) accordingly.

This paper only deals with the situation when physical insight is *not* used; i.e. when (1.5) is chosen as a flexible set of functions capable of describing almost any true relationship between y and φ. This is the *black-box approach.*

Typically, function expansions of the type

$$g(\varphi, \theta) = \sum_k \theta(k) g_k(\varphi) \tag{1.6}$$

are used, where

$$g_k(\varphi) : \mathbb{R}^d \to \mathbb{R}$$

and $\theta(k)$ are the components of the vector θ. For example, let

$$g_k(\varphi) = \varphi_k \quad (k\text{:th component of } \varphi) \; k = 1, \ldots, d.$$

Then, with (1.1)

$$y(t) = g(\varphi(t), \theta)$$

reads

$$y(t) + a_1 y(t-1) + \ldots + a_{n_a} y(t - n_a) =$$
$$b_1 u(t-1) + \ldots + b_{n_b} u(t - n_b)$$

if

$$a_i = -\theta(i) \qquad b_i = \theta(n_a + i)$$

so the familiar ARX-structure is a special case of (1.6), with a linear relationship between y and φ.

1.3 Nonlinear Black Box Models

The challenge now is the non-linear case: to describe general, non-linear, dynamics. How to select $\{g_k(\varphi)\}$ in this general case? We should thus be prepared to describe a "true" relationship

$$\hat{y}(t) = g_0(\varphi(t))$$

for any reasonable function $g_0 : \mathbb{R}^d \to \mathbb{R}$. The first requirement should be that $\{g_k(\varphi)\}$ is **a basis** for such functions, *i.e.* that we can write

$$[R1]: \quad g_0(\varphi) = \sum_{k=1}^{\infty} \theta(k) g_k(\varphi) \tag{1.7}$$

for any reasonable function g_0 using suitable coefficients $\theta(k)$. There is of course an infinite number of choices of $\{g_k\}$ that satisfy this requirement, the classical perhaps being the basis of polynomials. For $d = 1$ we would then have

$$g_k(\varphi) = \varphi^k$$

and (1.7) becomes **Taylor** or **Volterra** expansion. In practice we cannot work with infinite expansions like (1.7). A second requirement on $\{g_k\}$ is therefore to produce "good" approximations for finite sums: In loose notation:

$$[R2]: \quad \| g_0(\varphi) - \sum_{k=1}^{n} \theta(k) g_k(\varphi) \|$$

$$\text{"decreases quickly as } n \text{ increases"} \tag{1.8}$$

There is clearly no uniformly good choice of $\{g_k\}$ from this respect: It will all depend on the class of functions g_0 that are to be approximated.

1.4 Estimating $\hat{g}_N$

Suppose now that a basis $\{g_k\}$ has been chosen, and we try to approximate the true relationship by a finite number of the basis functions:

$$\hat{y}(t|\theta) = g(\varphi(t), \theta) = \sum_{k=1}^{n} \theta(k) g_k(\varphi(t)) \tag{1.9}$$

where we introduce the notation $\hat{y}(t|\theta)$ to stress that $g(\varphi(t), \theta)$ is a "guess" for $y(t)$ given the information in $\varphi(t)$ and given a particular parameter value θ. The "best" value of θ is then determined from the data set Z^N in (1.9) by

$$\hat{\theta}_N = \arg\min \sum_{k=1}^{N} |y(t) - \hat{y}(t|\theta)|^2 \tag{1.10}$$

The model will be

$$\hat{y}(t) = \hat{y}(t|\hat{\theta}_N) = \hat{g}_N(\varphi(t)) = g(\varphi(t), \hat{\theta}_N) \tag{1.11}$$

1.5 Properties of the Estimated Model

Suppose that the actual data have been generated by

$$y(t) = g_0(\varphi(t)) + e(t) \tag{1.12}$$

where $\{e(t)\}$ is white noise with variance λ. The estimated model (1.11) (*i.e.* the estimated parameter vector $\hat{\theta}_N$) will then be a random variable that depends on the realizations of both $e(t), t = 1, \ldots, N$ and $\varphi(t), t = 1, \ldots, N$. Denote its expected value by

$$\mathrm{E}\hat{g}_N = g_n^* = \sum_{k=1}^{n} \theta^*(k) g_k \tag{1.13}$$

where we used subscript n to emphasize the number of terms used in the function approximation.

Then under quite general conditions

$$\mathrm{E}|\hat{g}_N(\varphi(t)) - g_n^*(\varphi(t))|^2 = \lambda \cdot \frac{m}{N} \tag{1.14}$$

where E denotes expectation both with respect to $\varphi(t)$ and $\hat{\theta}_N$. Moreover, m is the number of estimated parameters, *i.e.*, dim θ. The total error thus becomes

$$\mathrm{E}|\hat{g}_N(\varphi(t)) - g_0(\varphi(t))|^2 =$$
$$\| g_0(\varphi(t)) - g_n^*(\varphi(t)) \|^2 + \lambda \cdot \frac{m}{N} \tag{1.15}$$

The first term here is an approximation error of the type (1.8). It follows from (1.15) that there is a trade-off in the choice of how many basis functions to use. Each included basis function increases the variance error by λ/N, while it decreases the bias error by an amount that could be less than so. A third requirement on the choice of $\{g_k\}$ is thus to

[R3] Have a scheme that allows the exclusion of spurious basis functions from the expansion.

Such a scheme could be based on a priori knowledge as well as on information in Z^N.

1.6 Basis Functions

Out of the many possible choice of basis functions, a large family of special ones have received most of the current interest. They are all based on just one fundamental function $\sigma(\varphi)$, which is scaled in various ways, and centered at different points, *i.e.*

$$g_k(\varphi) = \sigma(\beta_k^T(\varphi + \tilde{\gamma}_k)) = \sigma(\beta_k^T \varphi + \gamma_k) = \sigma(\varphi, \eta_k) \tag{1.16}$$

where $\gamma_k = \beta_k^T \tilde{\gamma}_k$ and η_k is the $d+1$-vector

$$\eta_k = [\beta_k, \gamma_k] \tag{1.17}$$

Such a choice is not at all strange. A very simplistic approach would be to take $\sigma(\varphi)$ to be the indicator function (in the case $d = 1$) for the interval $[0, 1]$:

$$\sigma(\varphi) = \begin{cases} 1 & \varphi \in [0,1] \\ 0 & \varphi \notin [0,1] \end{cases}$$

For a countable collection of η_k (e.g. assuming all rational numbers) the functions $g_k(\varphi)$ would then contain indicator functions for any interval, arbitrarily small and placed anywhere along the real axis. Not surprisingly, these $\{g_k\}$ will be a basis for all continuous functions. Equivalently, it could be threshold function

$$\sigma(\varphi) = \begin{cases} 1 & \varphi > 0 \\ 0 & \varphi \leq 0 \end{cases} \tag{1.18}$$

since the basic indicator function is just the difference between two threshold functions.

1.7 What Is the Neural Network Identification Approach?

The basic Neural Network (NN) used for System Identification (one hidden layer feedforward net) is indeed the choice (1.16) with a smooth approximation for (1.18), often

$$\sigma(x) = \frac{1}{1+e^{-x}}$$

Include the parameter η in (1.16)-(1.17) among the parameters to be estimated, θ, and insert into (1.9). This gives the Neural Network model structure

$$\hat{y}(t|\theta) = \sum_{k=1}^{n} \alpha_k \sigma(\beta_k \varphi + \gamma_k)$$

$$\theta = [\alpha_k, \beta_k, \gamma_k], \ k = 1, \ldots, n \tag{1.19}$$

The $n \cdot (d+2)$-dimensional parameter vector θ is then estimated by (1.10).

1.8 Why Have Neural Networks Attracted So Much Interest?

This tutorial points at two main facts.

1. The NN function expansion has good properties regarding requirement [R2] for nonlinear functions g_0 that are "localized"; *i.e.* there is not much nonlinear effects going to the infinity. This is a reasonable property for most real life physical functions. More precisely, see (7.1) in Section 7..
2. There is a good way to handle requirement [R3] by *implicit or explicit regularization* (See Section 3.)

1.9 Related Approaches

Actually, the general family of basis functions (1.16), is behind both *Wavelet Transform Networks* and estimation of *Fuzzy Models*. The paper [2] explains these connections in an excellent manner.

2. The Problem

2.1 Inferring Relationships from Data

A wide class of problems in disciplines such as classification, pattern recognition and system identification can be fit into the following framework.

A set of observations (data)

$$Z^N = \{y(t),\ \Phi(t)\}_{t=1}^N$$

of two physical quantities $y \in \mathbb{R}^p$ and $\Phi \in \mathbb{R}^r$ is given. It may or may not be known which variables in Φ influence y. There may also be other, non-measured, variables v that influence y. Based on the observations Z^N, infer how the variables in Φ influence y.

Let φ be the variables in Φ that influence y, then we could represent the relation between φ, v and y by a function g_0

$$y = g_0(\varphi, v) \tag{2.1}$$

The problem is thus two-fold:

1. Find which variables in Φ that should be used in φ.
2. Determine g_0.

In identification of dynamical systems, finding the right φ is the model order selection problem. Then t represents the time index and $\Phi(t)$ would be the collection of all past inputs and outputs.

There are two issues that have to be dealt with when determining g_0:

1. Only finite observations in the φ-space are available.
2. The observations are perturbed by the non-measurable variable $\{v(t)\}$.

1) represents the function approximation problem, *i.e.* how to do interpolation and extrapolation, which in itself is an interesting problem. Notice that there would be no problem at all if y was given for all values of φ (if we neglect the non-measurable input) since the function then in fact would be defined by the data. 2) increases the difficulty further since then we cannot infer exactly how φ influences y even at the points of observations. Blended together, these two problems are very challenging. Below we will try to disclose the essential ingredients. For further insight in this problem see also [2].

2.2 Prior Assumptions

Notice that as stated, the problem is ill-posed. There will be far too many unfalsified models, i.e., models satisfying (2.1), if any function g and any non-measurable sequence $\{v(t)\}$ is allowed. Thus, it is necessary to include some a priori information in order to limit the number of possible candidates. However, often it is difficult to provide a priori knowledge that is so precise that the problem becomes well-defined. To ease the burden it is common to resort to some general principles:

1) Non-measurable inputs are additive. This means that g_0 is additive in its second argument, i.e.,

$$g_0(\varphi, v) = g_0(\varphi) + v$$

This is, for example, a relevant assumption when $\{v(t)\}$ is due mainly to measurement errors. Therefore v is often called disturbance or noise.

2) Try simple things first (Occam's razor). There is no reason to choose a complicated model unless needed. Thus, among all unfalsified models, select the simplest one. Typically, the simplest means the one that in some sense has the smoothest surface. An example is spline smoothing. Among the class C^2 of all twice differentiable functions on an interval I, the solution to

$$\min_{g \in C^2} \sum (y(t) - g(\varphi(t))^2 + \lambda \int_I (g''(\varphi))^2 d\varphi$$

is given by the cubic spline, [38]. Other ways to penalize the complexity of a function are information based criteria, such as AIC, BIC and MDL, regularization (or ridge penalty) and cross-validation.

2.3 Function Classes

Thus, g_0 is assumed to belong to some quite general family $\mathcal{G}$ of functions. The function estimate $\hat{g}_N^n$ however, is restricted to belong to a possibly more limited class of functions, $\mathcal{G}_n$ say. This family $\mathcal{G}_n$, where n represents the complexity of the class[1], is a member of a sequence of families $\{\mathcal{G}_n\}$ that satisfy $\mathcal{G}_n \to \mathcal{G}$. As explained above, the complexity of $\hat{g}_N^n$ is allowed to depend on Z^N, i.e., n is a function of Z^N. We will indicate this by writing $n(N)$.

In this perspective, an identification method can be seen as a rule to choose the family $\{\mathcal{G}_n\}$ together with a rule to choose $n(N)$ and an estimator that given these provides an estimate $\hat{g}_N^{n(N)}$. Notice that both the selection of $\{\mathcal{G}_n\}$ and $n(N)$ can be driven by data. This possibility is, as we shall see in Section 7., very important.

Typical choices of $\mathcal{G}$ are Hölder Balls which consist of Lipschitz continuous functions:

[1] Typically n is the number of basis functions in the class.

$$\Lambda^{\alpha}(C) = \{f : |f(x) - f(y)| \leq C \cdot |x - y|^{\alpha}\} \tag{2.2}$$

and L_p Sobolev Balls which have derivatives of a certain degree which belongs to L_p:

$$W_p^m(C) = \{f : \int |f^{(m)}(t)|^p dt \leq C^p\} \tag{2.3}$$

Recently, Besov classes and Triebel classes, [35] have been employed in wavelet analysis. The advantage with these classes are that they allow for spatial inhomogenity. Functions in these classes can be locally spiky and jumpy.

3. Some General Estimation Results

The basic estimation set-up is what is called *non-linear regression* in statistics. The problem is as follows. We would like to estimate the relationship between a scalar y and $\varphi \in \mathbb{R}^d$. For a particular value $\varphi(t)$ the corresponding $y(t)$ is assumed to be

$$y(t) = g_0(\varphi(t)) + e(t) \tag{3.1}$$

where $\{e(t)\}$ is supposed to be a sequence of independent random vectors, with zero mean values and variance

$$\mathrm{E}\, e(t)e^T(t) = \lambda \tag{3.2}$$

To find the function g_0 in (3.1) we have the following information available:

1. A parameterized family of functions

$$\mathcal{G}_m = \{g(\varphi(t), \theta) | \theta \in D_{\mathcal{M}} \subset \mathbb{R}^m\} \tag{3.3}$$

2. A collection of observed y, φ-pairs:

$$Z^N = \{[y(t), \varphi(t)], t = 1, ..., N\} \tag{3.4}$$

The typical way to estimate g_0 is then to form the scalar valued function

$$V_N(\theta) = \frac{1}{N} \sum_{t=1}^{N} |y(t) - g(\varphi(t), \theta)|^2 \tag{3.5}$$

and determine the parameter estimate $\hat{\theta}_N$ as its minimizing argument:

$$\hat{\theta}_N = \arg\min V_N(\theta) \tag{3.6}$$

The estimate of g_0 will then be

$$\hat{g}_N(\varphi) = g(\varphi, \hat{\theta}_N) \tag{3.7}$$

Sometimes a general, non-quadratic, norm is used in (3.4)

$$V_N(\theta) = \frac{1}{N}\sum_{t=1}^{N} \ell(\varepsilon(t,\theta)) \tag{3.8}$$

$$\varepsilon(t,\theta) = y(t) - g(\varphi(t),\theta)$$

Another modification of (3.4) is to add a *regularization term*,

$$W_N(\theta) = V_N(\theta) + \delta|\theta - \theta^{\#}|^2 \tag{3.9}$$

(and minimize W rather than V) either to reflect some prior knowledge that a good θ is close to $\theta^{\#}$ or just to improve numerical and statistical properties of the estimate $\hat{\theta}_N$. Again, the quadratic term in (3.9) could be replaced by a non-quadratic norm.

Now, what are the properties of the estimated relationship $\hat{g}_N$? How close will it be to g_0? Following some quite standard results (see, e.g., [20, 31]), we have the following properties. We will not state the precise assumptions under which the results hold. Generally it is assumed that $\{\varphi(t)\}$ is (quasi)-stationary and has some mixing property (i.e., that $\varphi(t)$ and $\varphi(t+s)$ become less and less dependent as s increases). The estimate $\hat{\theta}_N$ is a random variable that depends on Z^N. Let E denote expectation with respect to both $e(t)$ and $\varphi, t = 1, ..., N$. Let

$$\theta^* = \mathrm{E}\hat{\theta}_N$$

and

$$g^*(\varphi) = g(\varphi, \theta^*)$$

Then $g^*(\varphi)$ will be as close as possible to $g_0(\varphi)$ in the following sense:

$$\arg\min_{g\in\mathcal{G}_m} \mathrm{E}|g(\varphi) - g_0(\varphi)|^2 = g^*(\varphi) \tag{3.10}$$

where expectation E is over the distribution of φ that governed the observed sample Z^N. We shall call

$$g^*(\varphi) - g_0(\varphi)$$

the *bias error*. Moreover, if the bias error is small enough, the variance will be given approximately by

$$\mathrm{E}|\hat{g}_N(\varphi) - g^*(\varphi)|^2 \approx \frac{m}{N}\lambda \tag{3.11}$$

Here m is the dimension of θ (number of estimated parameters), N is the number of observed data pairs and λ is the noise variance. Moreover, expectation both over $\hat{\theta}_N$ and over φ, assuming, the same distribution for φ as in the sample Z^N. The total integrated mean square error (IMSE) will thus be

$$\mathrm{E}|\hat{g}_N(\varphi) - g_0(\varphi)|^2 = ||g^*(\varphi) - g_0(\varphi)||^2 + \frac{m}{N}\lambda \tag{3.12}$$

Here the double bar norm denotes the functional norm, integrating over φ with respect to its distribution function when the data were collected. Now, what happens if we minimize the regularized criterion W_N in (3.9)?

1. The value $g^*(\varphi)$ will change to the function that minimizes

$$\mathrm{E}|g(\varphi, \theta) - g_0(\varphi)|^2 + \delta|\theta - \theta^\#|^2 \tag{3.13}$$

2. The variance (3.11) will change to

$$\mathrm{E}|\hat{g}_N(\varphi) - g^*(\varphi)|^2 \approx \frac{r(m, \delta)}{N} \cdot \lambda \tag{3.14}$$

   where

$$r(m, \delta) = \sum_{k=1}^{m} \frac{\sigma_i^2}{(\sigma_i + \delta)^2} \tag{3.15}$$

   where σ_i are the eigenvalues (singular values) of $\mathrm{E}V_N''(\theta)$, the second derivative matrix (the Hessian) of the criterion.

How to interpret (3.15)? A redundant parameter will lead to a zero eigenvalue of the Hessian. A small eigenvalue of V'' can thus be interpreted as corresponding to a parameter (combination) that is not so essential: "A spurious parameter". The regularization parameter δ is thus a threshold for spurious parameters. Since the eigenvalues σ_i often are widely spread we have

$$r(m, \delta) \simeq m^\# = \# \text{ of eigenvalues of } V'' \text{ that are larger than } \delta$$

We can think of $m^\#$ as "the efficient number of parameters in the parameterization". Regularization thus decreases the variance, but typically increases the bias contribution to the total error.

4. The Bias/Variance Trade-Off

Consider now a sequence of parameterized function families

$$\mathcal{G}_n = \{g_n(\varphi(t), \theta) | \theta \in D_{\mathcal{M}} \subset \mathbb{R}^m\}$$

$$n = 1, 2, 3 \ldots \tag{4.1}$$

where n denotes the number of basis function (1.9).

In the previous section we saw that the integrated mean square error is typically split into two terms the *variance* term and the *bias* term

$$V_2(\hat{g}_N^n, g_0) = V_2(\hat{g}_N^n, g_n^*) + V_2(g_n^*, g_0) \tag{4.2}$$

where, according to (3.11),

$$V_2(\hat{g}_N^n, g_n^*) \sim \frac{m}{N}. \tag{4.3}$$

The bias term, which is entirely deterministic, decreases with n. Thus, for a given family $\{\mathcal{G}_n\}$ there will be an optimal $n = n^*(N)$ that balances the variance and bias terms.

Notice that (4.3) is a very general expression that holds almost regardless of how the sequence $\{\mathcal{G}_n\}$ is chosen. Thus, it is in principle only possible to influence the bias error. In order to have a small integrated mean square error it is therefore of profound importance to choose $\{\mathcal{G}_n\}$ such that the bias is minimized. An interesting possibility is to let the choice of $\{\mathcal{G}_n\}$ be data driven. This may not seem like an easy task but here wavelets have proven to be useful, see Section 7.

When the bias and the variance can be exactly quantified, the integrated mean square error can be minimized with respect to n. This gives the optimal model complexity $n^*(N)$ as a function of N. However, often it is only possible to give the rate with which the bias decreases as a function of n and the rate with which the variance increases with n. Then it is only possible to obtain the rate with which $n^*(N)$ increases with N. Another problem is that if g_0 in reality belongs not to $\mathcal{G}$ but to some other class of functions, the rate will not be optimal. These considerations has lead to the development of methods where the choice of n is based on the observations Z^N. Basically, n is chosen so large that there is no evidence in the data that g_0 is more complex than the estimated model, but not larger than that. Then, as is shown in [14], the bias and the variance are matched. These adaptive methods are discussed in Section 7.

5. Neural Nets

What is meant by the term neural nets depends on the author. Lately neural net has become a word of fashion and today almost all kinds of models can be found by the names neural network somewhere in the literature. Old types of models, known for decades by other names, have been converted to, or reinvented as neural nets. This makes it impossible to cover all types of neural networks and only what is called feedforward and recurrent will be considered, which are the networks most commonly used in system identification. Information about other neural network models can be found in any introductory book in this field, e.g., [18, 29, 15].

In [16, 25, 34] alternative overviews of neural networks in system identification and control can be found. Also the books [41, 42] contain many interesting articles on this topic.

5.1 Feedforward Neural Nets

The step from the general function expansion (1.9) to what is called neural nets is not big. With the choice $g_k(\varphi) = \alpha_k \sigma(\beta_k \varphi + \gamma_k)$ where β_k is a parameter vector of size dimφ, and γ_k and α_k are scalar parameters we obtain

$$g(\varphi) = \sum_{k=1}^{n} \alpha_k \sigma(\beta_k \varphi + \gamma_k) + \alpha_0 \tag{5.1}$$

where a mean level parameter α_0 has been added. This model is referred to as a *feedforward network* with one hidden layer and one output unit in the NN literature. In Fig. 5.1 it is displayed in the common NN way. The basis functions, called *hidden units*, *nodes*, or *neurons*, are univariate which makes the NN to an expansion in simple functions. The specific choice of $\sigma(\cdot)$ is the *activation function* of the units which is usually chosen identically for all units.

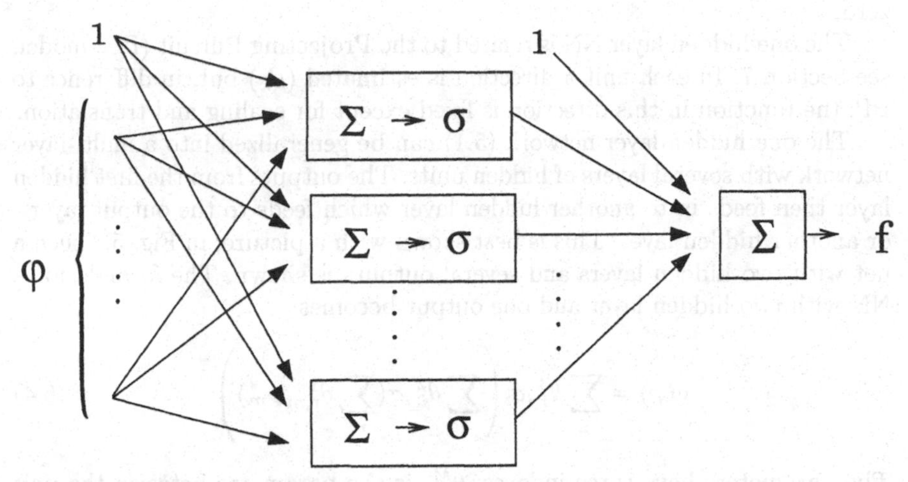

Fig. 5.1. Feedforward network with one hidden layer and one output unit. The arrows symbolize the parameters of the model.

The name *feedforward* is explained by the figure; there is a specific direction of flow in the computations when the output g is computed. First the weighted sums calculated at the input at each unit, then these sums pass the activation function and form the outputs of the hidden units. To form g, a weighted sum of the results from the hidden units is formed. This sum at the output is called the *output unit*. If g is vector function there are several output units forming an *output layer*. The input, φ, is sometimes called the *input layer*. The weights at the different sums are the parameters of the network.

In [6] it was shown that condition [R1], (1.7), holds if the activation function is chosen to be *sigmoidal* which is defined as

Definition 5.1. *Let $\sigma(x)$ be continuous. Then $\sigma(x)$ is called a sigmoid function if it has the following properties*

$$\sigma(x) \to \begin{cases} a & as\ x \to +\infty \\ b & as\ x \to -\infty \end{cases} \tag{5.2}$$

where $a, b,\ b < a$ are any real values.

The most common choice is

$$\sigma(x) = \frac{1}{1 + e^{-x}} \tag{5.3}$$

which gives a smooth, differentiable, model with the advantage that gradient based parameter estimate methods can be used, see Section 6.. However, in [19] it is shown that (5.1) is a universal approximator, i.e., [R1] holds for all non-polynomial $\sigma(\cdot)$ which are continuous except at most in a set of measure zero.

The one hidden layer NN is related to the Projecting Pursuit (PP) model, see Section 7. In each unit a direction is estimated (β_k) but, in difference to PP, the function in this direction is fixed except for scaling and translation.

The one hidden layer network (5.1) can be generalized into a multi-layer network with several layers of hidden units. The outputs from the first hidden layer then feeds in to another hidden layer which feeds to the output layer - or another hidden layer. This is best shown with a picture; in Fig. 5.2 such a net with two hidden layers and several outputs is shown. The formula for a NN with two hidden layer and one output becomes

$$g(\varphi) = \sum_i \theta^3_{1,i} \sigma \left(\sum_j \theta^2_{i,j} \sigma (\sum_m \theta^1_{r,m} \varphi_m) \right) \tag{5.4}$$

The parameters have three indexes. $\theta^M_{j,i}$, is the parameter between the unit i in one layer and unit j in the following layer. M denotes which layer the parameter belongs to. The translation parameters corresponding to γ_k in (5.1) has not been written out.

At first, because of the general approximation ability of the NN with one hidden layer, there seems to be no reason to add more hidden layers. However, the rate of convergence might be very slow for some functions and it might be possible with a much faster convergence with two hidden layers (i.e., condition [R2], (1.8) might favor two layers). Also, in [33] it is shown that in certain control applications a two hidden layer NN can stabilize systems which cannot possibly be stabilized by NN with only one hidden layer.

5.2 Recurrent Neural Nets

If some of the inputs of a feedforward network consist of delayed outputs from the network, or some delayed internal state, then the network is called a *recurrent network*, or sometimes a dynamic network. In Fig. 5.3 an example of a recurrent net with two past outputs fed back into the network.

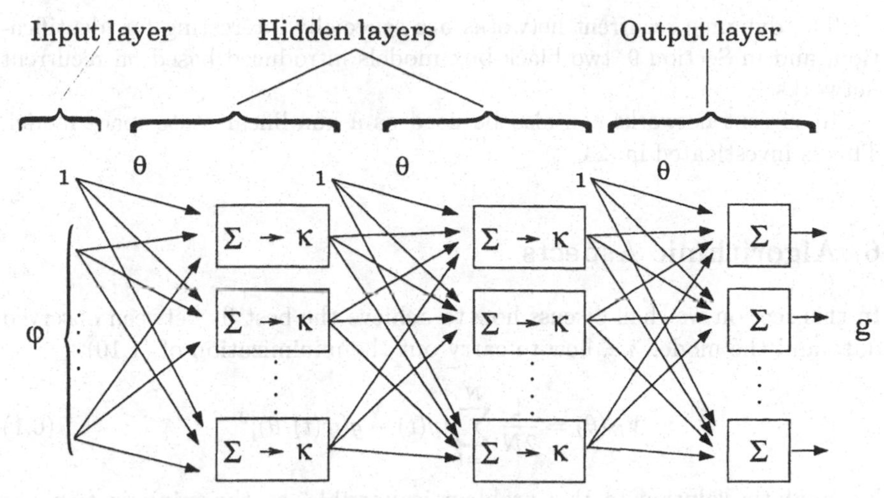

Fig. 5.2. Feedforward network with two hidden layers.

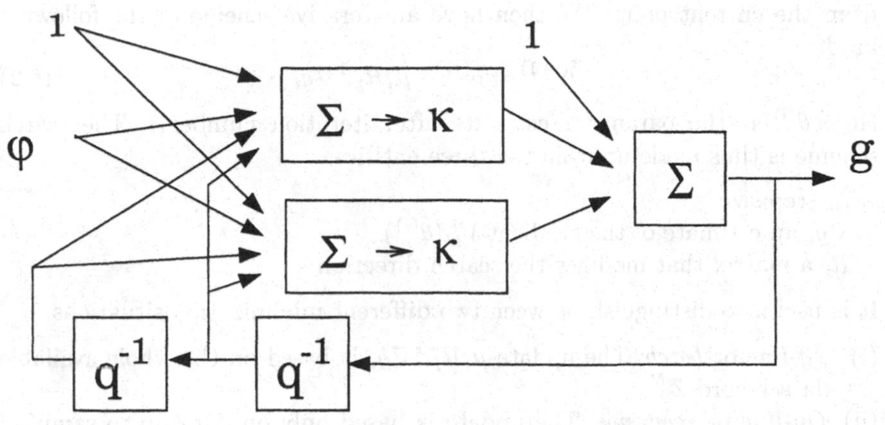

Fig. 5.3. Recurrent network. q^{-1} delays the signal one time sample.

The dynamic recurrent networks are especially interesting for identification, and in Section 9. two black-box models introduced based on recurrent networks.

Recurrent networks can also be used as a non-linear state-space model. This is investigated in [23].

6. Algorithmic Aspects

In this section we shall discuss how to achieve the best fit between observed data and the model, *i.e.* how to carry out the minimization of (1.10).

$$V_N(\theta) = \frac{1}{2N}\sum_{t=1}^{N} |y(t) - g(\varphi(t), \theta)|^2 \tag{6.1}$$

No analytic solution to this problem is possible, so the minimization has to be done by some numerical search procedure. A classical treatment of the problem of how to minimize sum of squares is given in [7]. A survey of methods for the NN application is given in [18] and in [36]. Most efficient search routines are based on iterative local search in a "downhill" direction from the current point. We then have an iterative scheme of the following kind

$$\hat{\theta}^{(i+1)} = \hat{\theta}^{(i)} - \mu_i R_i^{-1} \nabla \hat{g}_i \tag{6.2}$$

Here $\hat{\theta}^{(i)}$ is the parameter estimate after iteration number i. The search scheme is thus made up from the three entities

- μ_i step size
- $\nabla \hat{g}_i$ an estimate of the gradient $V_N'(\hat{\theta}^{(i)})$
- R_i a matrix that modifies the search direction

It is useful to distinguish between two different minimization situations

(i) *Off-line* or *batch*: The update $\mu_i R_i^{-1} \nabla \hat{g}_i$ is based on the whole available data record Z^N.

(ii) *On-line* or *recursive*: The update is based only on data up to sample i (Z^i), (typically done so that the gradient estimate $\nabla \hat{g}_i$ is based only on data just before sample i.)

We shall discuss these two modes separately below. First some general aspects will be treated.

6.1 Search Directions

The basis for the local search is the gradient

$$V_N'(\theta) = -\frac{1}{N}\sum_{t=1}^{N} (y(t) - g(\varphi(t), \theta))\psi(\varphi(t), \theta) \tag{6.3}$$

where

$$\psi(\varphi(t),\theta) = \frac{\partial}{\partial\theta} g(\varphi(t),\theta) \quad (d|1 - \text{vector}) \tag{6.4}$$

It is well known that gradient search for the minimum is inefficient, especially close to the minimum. Then it is optimal to use the *Newton search direction*

$$R^{-1}(\theta)V_N'(\theta) \tag{6.5}$$

where

$$R(\theta) = V_N''(\theta) = \frac{1}{N}\sum_{t=1}^{N} \psi(\varphi(t),\theta)\psi^T(\varphi(t),\theta) + \frac{1}{N}\sum_{t=1}^{N}(y(t) - g(\varphi(t),\theta))\frac{\partial^2}{\partial\theta^2} g(\varphi(t),\theta) \tag{6.6}$$

The true Newton direction will thus require that the second derivative

$$\frac{\partial^2}{\partial\theta^2} g(\varphi(t),\theta)$$

be computed. Also, far from the minimum, $R(\theta)$ need not be positive semidefinite. Therefore alternative search directions are more common in practice:

- *Gradient direction.* Simply take

$$R_i = I \tag{6.7}$$

- *Gauss-Newton direction.* Use

$$R_i = H_i = \frac{1}{N}\sum_{t=1}^{N} \psi(\varphi(t),\hat{\theta}^{(i)})\psi^T(\varphi(t),\hat{\theta}^{(i)}) \tag{6.8}$$

- *Levenberg-Maquard direction.* Use

$$R_i = H_i + \delta I \tag{6.9}$$

  where H_i is defined by (6.8).
- *Conjugate gradient direction.* Construct the Newton direction from a sequence of gradient estimates. Loosely, think of V_N'' as constructed by difference approximation of d gradients. The direction (6.5) is however constructed directly, without explicitly forming and inverting V''.

It is generally considered, [7], that the Gauss-Newton search direction is to be preferred. For ill-conditioned problems the Levenberg-Maquard modification is recommended. However, good results with conjugate gradient methods have also been reported in NN applications ([36]).

6.2 Back-Propagation: Calculation of the Gradient

The only model-structure dependent quantity in the general scheme (6.2) is the gradient of the model structure (6.4). For a one-hidden-layer structure (5.1) this is entirely straightforward, since

$$\begin{aligned}\frac{d}{d\alpha}\alpha\sigma(\beta\varphi+\gamma) &= \sigma(\beta\varphi+\gamma)\\ \frac{d}{d\gamma}\alpha\sigma(\beta\varphi+\gamma) &= \alpha\sigma'(\beta\varphi+\gamma)\\ \frac{d}{d\beta}\alpha\sigma(\beta\varphi+\gamma) &= \alpha\sigma'(\beta\varphi+\gamma)\varphi\end{aligned}$$

For multi-layer NNs the gradient is calculated by the well known *Back-Propagation* (BP) method which can be described as the chain rule for differentiation applied to the expression (5.4). It also makes sure to re-use intermediate results which are needed at several places in the algorithm. Actually, the only complicated with the algorithm is to keep track of all indexes.

Backpropagation has been "rediscovered" several times (see, e.g., [40, 28]). Here the algorithm will be derived for the case where the network has two hidden layers and one output unit. For multi output models and with less- or more hidden layers only minor changes have to be done.

Consider the NN model (5.4). Denote by x_b^M and f_b^M the result at unit b in layer M before and after the activation function, respectively. That is

$$f_b^M = \sigma(x_b^M)$$

We can then write $g(\varphi) = x_1^3 = \sum_i \theta_{1,i}^3 f_i^2$ and the derivative with respect to one of the parameters in the output layer becomes

$$\psi(\varphi)_{3,1,b} = \frac{\partial g}{\partial \theta_{1,b}^3} = f_b^2$$

In the same way $x_a^2 = \sum_m \theta_{a,m}^2 f_m^1$ and the derivative of a parameter in the middle layer becomes

$$\psi(\varphi)_{2,a,b} = \frac{\partial g}{\partial \theta_{a,b}^2} = \delta_a^2 f_b^1$$

where

$$\delta_a^2 = \theta_{1,a}^3 \sigma'(x_a^2)$$

For the first layer we can write $x_a^1 = \sum_m \theta_{a,m}^1 \varphi_m$ and the derivative of a parameter in this layer becomes

$$\psi(\varphi)_{1,a,b} = \frac{\partial g}{\partial \theta_{a,b}^1} = \delta_a^1 \varphi_b$$

where

$$\delta_a^1 = \sum_j \delta_j^2 \theta_{j,a}^2 \sigma'(x_a^1)$$

The nice feature is that $\{f_b^M\}$ and $\{x_b^M\}$ are obtained as intermediate results when $g(\varphi)$ is calculated (forward propagation in the network). Calculating $\{\delta_a^M\}$ can be viewed as propagating $g(\varphi)$ backwards through the net, and this is the origin of the name of the algorithm.

The calculations are further simplified by the relation of the derivative of the sigmoid which follows from (5.3).

$$\sigma'(\cdot) = \sigma(\cdot)\,(1 - \sigma(\cdot)) \tag{6.10}$$

6.3 Implicit Regularization

Recall the discussion about regularization in Section 3.. We pointed out that the parameter δ in (3.9) acts like a knob that affects the "efficient number of parameters used". It thus plays a similar role as the model size:

- Large δ: small model structure, small variance, large bias
- Small δ: large model structure large variance, small bias

It is quite important for NN applications to realize that there is a direct link between the iterative process (6.2) and regularization in the sense that *aborting the iterations before the minimum has been found, has a quite similar effect as regularization.* This was noted in [37] and pointed out as the cause of "overtraining" in [30]. More precisely, the link is as follows (when quadratic approximations are applicable)

$$(I - \mu R^{-1} V'')^i \sim \delta(\delta I + V'')^{-1}$$

so, as the iteration number increases, this corresponds to a regularization parameter that decreases to zero as

$$\log \delta \sim -i \tag{6.11}$$

How to know when to stop the iterations? As $i \to \infty$ the value of the criterion V_N will of course continue to decrease, but as a certain point the corresponding regularization parameter becomes so small that increased variance starts to dominate over decreased bias. This should be visible when the model is tested on a fresh set – the *Validation data* (often called *generalization data* in the NN context). We thus evaluate the criterion function on this fresh data set, and plot the fit as a function of the iteration number. A typical such plot is shown in Fig. 9.3. The point where the fit starts to be worse for the validation data is the iteration number (the degree of regularization or the effective model flexibility) where we are likely to strike the optimal balance between bias and variance error. Experience with NN applications has shown that this often is a very good way of limiting the actual model flexibility by effectively eliminating spurious parameters, i.e., dealing with requirement [R3], mentioned in Section 1.

6.4 Off-line and On-line Algorithms

The expressions (6.3) and (6.6) for the Gauss-Newton search clearly assume that the whole data set Z^N is available during the iterations. If the application is of an off-line character, i.e., the model $\hat{g}_N$ is not required during the data acquisition, this is also the most natural approach.

However, in the NN context there has been a considerable interest in on-line (or recursive) algorithms, where the data are processed as they are measured. Such algorithms are in NN contexts often also used in off-line situations. Then the measured data record is concatenated with itself several times to create a (very) long record that is fed into the on-line algorithm. We may refer to [21] as a general reference for recursive parameter estimation algorithm. In [32] the use of such algorithms is the off-line case is discussed.

It is natural to consider the following algorithm as the basic one:

$$\hat{\theta}(t) = \hat{\theta}(t-1) + \mu_t R_t^{-1} \psi(\varphi(t), \hat{\theta}(t-1)) \varepsilon(t, \hat{\theta}(t-1)) \tag{6.12}$$

$$\varepsilon(t, \theta) = y(t) - g(\varphi(t), \theta) \tag{6.13}$$

$$R_t = R_{t-1} + \mu_t [\psi(\varphi(t), \hat{\theta}(t-1)) \psi^T(\varphi(t), \hat{\theta}(t-1)) - R_{t-1}] \tag{6.14}$$

The reason is that if $g(\varphi(t), \theta)$ is linear in θ, then (6.12) - (6.14), with $\mu_t = 1/t$, provides the analytical solution to the minimization problem (6.1). This also means that this is a natural algorithm close to the minimum, where a second order expansion of the criterion is a good approximation. In fact, it is shown in [21], that (6.12)–(6.14) in general gives an estimate $\hat{\theta}(t)$ with the same ("optimal") statistical, asymptotic properties as the true minimum to (6.1).

In the NN literature, often some averaged variants of (6.12)–(6.14) are discussed:

$$\hat{\hat{\theta}}(t) = \hat{\hat{\theta}}(t-1) + \mu_t R_t^{-1} \nabla \hat{g}_t \tag{6.15}$$

$$\nabla \hat{g}_t = \nabla \hat{g}_{t-1} + \gamma_t [\psi(\varphi(t), \hat{\theta}(t-1)) \varepsilon(t, \hat{\theta}(t-1)) - \nabla \hat{g}_{t-1}] \tag{6.16}$$

$$\hat{\theta}(t) = \hat{\theta}(t-1) + \rho_t [\hat{\hat{\theta}}(t) - \hat{\theta}(t-1)] \tag{6.17}$$

The basic algorithm (6.12)–(6.14) then corresponds to $\gamma_t = \rho_t = 1$. Now, when do the different averages accomplish?

Let us first discuss (6.17) (and take $\gamma_t \equiv 1$). This is what has been called "accelerated convergence". It was introduced by [27] and has been extensively discussed by Kushner and others. The remarkable thing with this averaging is that we achieve the same asymptotic statistical properties of $\hat{\theta}(t)$ by (6.15)–(6.17) with $R_t = I$ (gradient search) as by (6.12)–(6.14) if

$$\gamma_1 = 1$$

$$\rho_t = 1/t$$

$$\mu_t >> \rho_t \qquad \mu_t \to 0$$

It is thus an interesting alternative to (6.12)–(6.14), in particular if $\dim \theta$ is large so R_t is a big matrix.

We now turn to the averaging in (6.16).For $\gamma < 1$ this gives what is known as a *momentum* term. Despite its frequent use in NN applications, it is more debatable. An immediate argument for (6.16) is that the averaging makes the gradient $\nabla \hat{g}_t$ more reliable (less noisy) so that we can take larger steps in (6.15). It is, however, immediate to verify that exactly the same averaging takes place in (6.15) if smaller steps are taken. A second argument is that (6.16) lends a "momentum" effect to the gradient estimate $\nabla \hat{g}_t$. That is, due to the low pass filter, $\nabla \hat{g}_t$ will reflect not only the gradient at $\hat{\theta}(t-1)$, but also at several previous values of $\hat{\theta}(k)$. This means that the update push in (6.15) will not stop immediately at value θ where the gradient is zero. This could of course help to push away from a non-global, local minimum, which is claimed to be a useful feature. However, there seems to be no systematic investigation of whether this possible advantage is counter balanced by the fact that more iterations will be necessary for convergence.

6.5 Local Minima

A fundamental problem with minimization tasks like (6.1) is that $V_N(\theta)$ may have several or many local (non-global) minima, where local search algorithms may get caught. There is no easy solution to this problem. It is usually well used effort to spend some time to come up with a good initial value $\theta^{(0)}$ where to start the iterations. Other than that, only various global search strategies are left, such as random search, random restarts, simulated annealing, the genetic algorithm and whathaveyou.

7. Adaptive Methods

The use of data to select the basis functions characterize adaptive methods. The adaptation can be more or less sophisticated. In its simplest form, only the number of basis functions is selected. The merits and limitations of this procedure are explained in the first subsection while the second subsection deals with more advanced methods where also the basis functions themselves are adapted to the data.

7.1 Adaptive Basis Function Expansion

Suppose that we have a set of basis functions $\{b_k\}$ that span $\mathcal{G}$. Each set of n basis functions would generate a function class $\mathcal{G}_n$ and a good idea would be to select these n basis function such that the approximation error is minimized among all possible choices of these sets of n basis functions. The problem of finding an n-dimensional subspace that minimizes the worst approximation error is is known as Kolmogorovs n-width problem, [26]. Depending on $\mathcal{G}$, the problem can be more or less complicated. For example, for the functions $\sum_{k=0}^{\infty} a_k z^k$that are analytic inside the disc of radius $r \leq 1$ satisfying $\sum_{k=0}^{\infty} |a_k|^2 r^{2k} < 1$, the optimal subspace is given by span$\{1, z, \ldots, z^{n-1}\}$.

Wavelets. For orthonormal basis functions, the basis functions that correspond to the largest coefficients in the expansion of g_0 give the best approximation. Thus, an idea is to estimate a large number of coefficients and to select the n largest ones. It is interesting to note that with this procedure one get adaptation of the $\{\mathcal{G}_n\}$ to the smoothness of $\mathcal{G}$ for free; if the basis functions span several (or a scale of) spaces of functions, the approach will be optimal for all these spaces.

This approach has been exploited in the wavelet theory. Wavelet theory is based on orthonormal bases of L_2 that also span a wide scale of function spaces with a varying degree of smoothness, Besov and Triebel spaces, [35].

The basic problem with such a method is to determine which parameters are small and which are large, respectively. [10] has shown that the use of *shrinkage* gives (near) minimax rates in these spaces. Shrinkage essentially means that a threshold is determined that depends on the number of data. Parameter estimates less than this threshold are set to zero. Often, for technical reasons, a soft threshold is used instead. In that case, every wavelet coefficient is "pulled" towards zero by a certain non-linear function. This is conceptually closely related to the *regularization* procedure outlined in Section 3. Then, parameters are attracted towards the nominal value $\theta^{\#}$. However, so far explicit regularization does not seem to have been exploited in wavelet theory.

Neural Networks. Neural networks is an example of a structure where the basis functions appear more implicit. Consider the expression (5.1). This is an expansion with $\{\sigma(\varphi, \eta_k)\}$ as basis functions. The fact that the $\eta_k s$ are estimated from data means that the basis functions are chosen adaptively. In other words, the basis functions are selected from data. Below we shall see that they have an important property when it comes to high-dimensional systems.

7.2 The "Curse" of Dimensionality

Almost all useful approximation theorems are asymptotic, i.e., they require the number of data to approach infinity, $N \to \infty$. In practical situations this cannot be done and it is of crucial importance how fast the convergence is. A general estimation of a function $\mathbb{R}^d \to \mathbb{R}$ becomes slower in N when d is larger and in most practical situations it becomes impossible to do general estimation of functions for d larger than, say, 3 or 4. For higher dimensions the number of data required becomes so large that it is in most cases not realistic. This is the curse of dimensionality. This can be shown with the following example

Example 7.1. Approximate a function $\mathbb{R}^d \rightarrow \mathbb{R}$ within the unit cube with the resolution 0.1. This requires that the distance between data is not larger than 0.1 in every direction, requiring $N = 10^d$ data. This is hardly realistic for $d > 4$. When there are noisy measurements the demand of data increase further.

7.3 Methods to Avoid the "Curse"

From the discussion in the preceding subsection it should be clear that general nonlinear estimation is not possible. Nevertheless, a number of methods have been developed to deal with the problems occurring for high-dimensional functions. The idea is to be able to estimate functions that in some sense have a low-dimensional character. *Projection pursuit regression*, [11], uses an approximation of the form

$$\hat{g}(\varphi) = \sum_{k=1}^{n} g_k \left(\varphi^T \theta(k)\right)$$

where the g_ks are smooth univariate functions. The method thus expands the function in n different directions. These directions are selected to be the most important ones and, for each of these, the functions g_k are optimized. Thus, it is a joint optimization over the directions $\{\theta(k)\}$ and the functions g_k. The claim is that for small n a wide class of functions can be well approximated by this expansion, [9]. The claim is supported by the fact that any smooth function in d variables can be written in this way, [8]. It is supposed to be useful for moderate dimensions, $d < 20$.

Projection pursuit regression is closely related to *neural networks* where the same function, any sigmoid function σ satisfying Definition 5.1, is used in all directions. The effectiveness of such methods has been illustrated in [1]: Consider the class of functions $\{g\}$ on $\mathbb{R}^d$ for which there is a Fourier representation $\tilde{g}$ which satisfies

$$C_f = \int |\omega||\tilde{g}(\omega)|d\omega < \infty.$$

Then there is a linear combination of sigmoidal functions such that

$$\int_{B_r} |g(x) - g_n(x)|^2 dx \leq \frac{(2rC_f)^2}{n} \tag{7.1}$$

where B_r is a ball with radius r. The important thing to notice here is that the degree of approximation as a function of n does not depend on the dimension d.

This work originated with the result in [17] where sinusoidal functions where used to prove a similar result. The above result is not limited to sinusoidal and sigmoidal functions and the same idea has been applied to projection pursuit regression, [43], *hinging hyperplanes*, [3], and *radial basis functions*, [12].

Notice, however, that the result is only an approximation result and a stochastic counterpart still awaits its proof.

[1] also showed that $1/n^{(2/d)}$ is a lower bound for the minimiax rate for linear methods. For large d, this rate is exceedingly slow compared with $1/n$. Thus, this is a serious disadvantage for the methods described in the previous section. In higher dimensional spaces, the convergence rate of linear method is much slower compared with certain non-linear methods.

8. Specific Properties of NN Structures

So, what are the special features of the Neural Net structure that motivate the strong interest? Based on the discussion so far, we may point to the following list of properties:

- The NN expansion is a basis, even for just one hidden layer, i.e., Requirement [R1] is satisfied.
- The NN structure does extrapolation in certain, adaptively chosen, directions and is localized across these directions. Like Projection Pursuit it can thus handle larger regression vectors, if the data pattern $[y(t), \varphi(t)]$ cluster along subspaces.
- The NN structure uses adaptive bases functions, whose shape and location are adjusted by the observed data.
- The approximation capability (Requirement [R2]) is good as manifested in (7.1).
- Regularization, implicit (stopped iterations) or explicit (penalty for parameter deviations, usually from zero) is a useful tool to effectively include only those basis functions that are essential for the approximation, without increasing the variance. (Requirement [R3]).
- In addition, NNs have certain advantages in implementation, both in hardware and software, due to the repetitive structure. The basis functions are built up from only one core function, σ. This also means that the structure is resilient to failures, since any node can play any other node's role, by adjusting its weights.

9. Models of Dynamical Systems Based on Neural Networks

We are now ready to take the step from general "curve fitting" to system identification. The choice of a model structure for dynamical systems contains two questions

1. What variables, constructed from observed past data, should be chosen as regressors, *i.e.*, as components of $\varphi(t)$?
2. What non-linear mapping should $\varphi(t)$ be subjected to, i.e., How many hidden layers in (5.1) should be used, and how many nodes should each layer have?

The second question is related to more general NN considerations, as discussed in Section 5.. The first one is more specific for identification applications. To get some guidance about the choice of regressors φ, let us first review the linear case.

9.1 A Review of Linear Black Box Models

The simplest dynamical model is the Finite Impulse Response model (FIR):

$$y(t) = B(q)u(t) + e(t) = b_1 u(t-1) + \ldots + b_n u(t-n) + e(t) \tag{9.1}$$

Here we have used q to denote the shift operator, so $B(q)$ is a polynomial in q^{-1}. The corresponding predictor is $\hat{y}(t|\theta) = B(q)u(t)$ is thus based on a regression vector

$$\varphi(t) = [u(t-1), u(t-2), \ldots, u(t-n)]$$

As n tends to infinity we may describe the dynamics of all ("nice") linear systems. However, the character of the noise term $e(t)$ will not be modeled in this way.

A variant of the FIR model is the Output Error model (OE):

$$y(t) = \frac{B(q)}{F(q)}u(t) + e(t) \tag{9.2}$$

where

$$F(q) = 1 + f_1 q^{-1} + \ldots + f_{n_f} q^{-n_f}$$

The predictor is

$$\hat{y}(t|\theta) = \frac{B(q)}{F(q)}u(t) \tag{9.3}$$

Also this predictor is based on past inputs only. It can be rewritten

$$\hat{y}(t|\theta) = b_1 u(t-1) + \ldots + b_{n_b} u(t-n_b) -$$

$$- f_1\hat{y}(t-1|\theta) - \ldots - f_{n_f}\hat{y}(t-n_f|\theta) \tag{9.4}$$

It is thus based on the regression variables

$$[u(t-1),\ldots,u(t-n_b),\hat{y}(t-1|\theta),\ldots,\hat{y}(t-n_f|\theta)] \tag{9.5}$$

Note that these regressors are partly constructed from the data, using a current model. As n_b and n_f tend to infinity, also this model is capable of describing all reasonable linear dynamic systems, but not the character of the additive noise $e(t)$. The advantage of (9.2) over (9.1) is that fewer regressors are normally required to get a good approximation. The disadvantage is that the minimization over θ becomes more complicated. Also, the stability of the predictor (9.3) depends on $F(q)$, and thus has to be monitored during the minimization.

A very common variant is the ARX model

$$A(q)y(t) = B(q)u(t) + e(t) \tag{9.6}$$

with the predictor

$$\begin{aligned} \hat{y}(t|\theta) &= -a_1 y(t-1) - \ldots - a_{n_a} y(t-n_a) \\ &\quad +b_1 u(t-1) + \ldots + b_{n_b} u(t-n_b) \end{aligned} \tag{9.7}$$

thus using the regressors

$$[y(t-1),\ldots,y(t-n_a),u(t-1),\ldots,u(t-n_b)] \tag{9.8}$$

As shown, e.g., in [22] this structure is capable of describing all (reasonable) linear systems, including their noise characteristics, as n_a and n_b tend to infinity. The ARX model is thus a "complete" linear model from the black box perspective. The only disadvantage is that n_a and n_b may have to be chosen larger than the dynamics require, in order to accomodate the noise description. Therefore, a number of variants of (9.6) have been suggested, where the noise model is given "parameters of its own". The best known of these is probably the ARMAX model

$$A(q)y(t) = B(q)u(t) + C(q)e(t) \tag{9.9}$$

Its predictor is given by

$$\begin{aligned} \hat{y}(t|\theta) &= (c_1 - a_1)y(t-1) + \ldots \\ &+ (c_n - a_n)y(t-n) \\ &+ b_1 u(t-1) + \ldots + b_n u(t-n) \\ &+ c_1\hat{y}(t-1|\theta) + \ldots + c_n\hat{y}(t-n|\theta) \end{aligned} \tag{9.10}$$

It thus uses the regression vector (9.8) complemented with past predictors, just as in (9.5) (although the predictors are caclulated in a different way).

A large family of black box linear models is treated, e.g., in [21]. It has the form

$$A(q)y(t) = \frac{B(q)}{F(q)}u(t) + \frac{C(q)}{D(q)}e(t) \tag{9.11}$$

The special case $A(q) = 1$ gives the well known Box-Jenkins (BJ) model. The regressors used for the corresponding predictor are given, e.g., by equation (3.114) in [21]. These regressors are based on $y(t-k)$, $u(t-k)$, the predicted outputs $\hat{y}(t-k|\theta)$ using the current model, as well as the simulated outputs $\hat{y}_u(t-k|\theta)$, which are predicted outputs based on an output error model (9.2).

Let us repeat that from a black box perspective, most variants of (9.11) are equivalent, in the sense that they can be transformed into each other, at the expense of changing the orders of the polynomials. The ARX model (C=D=F=1) covers it all. The rationale for the other variants is that we may come closer to the true system using fewer regressors.

9.2 Choice of Regressors for Neural Network Models

The discussion on linear systems clearly points to the possible regressors:

- Past Inputs $u(t-k)$
- Past Measured Outputs $y(t-k)$
- Past Predicted Outputs, using current model, $\hat{y}(t-k|\theta)$
- Past Simulated Outputs, using past inputs only and current model, $\hat{y}_u(t-k|\theta)$

A rational question to ask would be: Given that I am prepared to use m regressors (the size of the input layer is m), how should I distribute these over the four possible choices? There is no easy and quantitative answer to this question, but we may point to the following general aspects:

- Including $u(t-k)$ only, requires that the whole dynamic response time is covered by past inputs. That is, if the maximum response time to any change in the input is Υ, and the sampling time is T, then the number of regressors should be Υ/T. This could be a large number. On the other hand, models based on a finite number of past inputs cannot be unstable in simulation, which often is an advantage.
  A variant of this approach is to form other regressors from u^t, e.g., by Laguerre filtering, (e.g., [39]). This retains the advantages of the FIR-approach, at the same time as making it possible to use fewer regressors. It does not seem to have been discussed in the NN-context yet.
- Adding $y(t-k)$ to the list of regressors makes it possible to cover slow responses with fewer regressors. A disadvantage is that past outputs bring in past disturbances into the model. The model is thus given an additional task to also sort out noise properties. A model based on past outputs may also be unstable in simulation from input only. This is caused by the fact that the past measured outputs are then replaced by past model outputs.

- Bringing in past predicted or simulated outputs $\hat{y}(t-k|\theta)$ typically increases the model flexibility, but also leads to non-trival difficulties. For neural networks, using past outputs at the input layer gives *recurrent* networks. See Section 5. Two problems must be handled:
  - It may lead to instability of the network, and since it is a non-linear model, this problem is not easy to monitor.
  - The simulated/predicted output depends on θ. In order to do updates in (6.2) in the true gradient direction, this dependence must be taken into account, which is not straightforward. If the dependence is neglected, convergence to local minima of the criterion function cannot be guaranteed.

The balance of this discussion is probably that the regressors (9.8) should be the first ones to test.

9.3 Neural Network Dynamic Models

Following the nomencalture for linear models it is natural to coin similar names for Neural Network models. This is well in line with, e.g., [5, 4]. We could thus distinguish between

- *NNFIR*-models, which use only $u(t-k)$ as regressors
- *NNARX*-models, which use $u(t-k)$ and $y(t-k)$ as regressors
- *NNOE*-models, which use $u(t-k)$ and $\hat{y}_u(t-k|\theta)$
- *NNARMAX*-models, which use $u(t-k), y(t-k)$ and $\hat{y}(t-k|\theta)$
- *NNBJ*-models, which use all the four regressor types.

In [25] another notation is used for the same models. The NNARX model is called Series-Parallel model and the NNOE is called Parallel model.

From a structural point of view, these black-box models are just slightly more troublesome to handle than their linear counterparts. When the regressor has been decided upon, it only remains to decide how many hidden units which should be used. The linear ARX model is entirely specified by three structural parameters $[n_a\ n_b\ n_k]$. [n_k is here the delay, which we have taken as 1 so far. In general we would work with the input regressors $u(t-n_k), \ldots, u(t-n_k-n_b+1)$.] The NNARX model has just one index more, $[n_a\ n_b\ n_k\ n_h]$, where n_h is the number of units in the hidden layer which in some way corresponds to "how non-linear" the system is. The notation for NNOE and NNARMAX models follow the same simple rule.

If more then one hidden layer is used there will be one additional structural parameter for each layer.

It follows from Section 5.2 that NNOE, NNBJ, and NNARMAX correspond to recurrent neural nets because parts of the input to the net (the regressor) consist of past outputs from the net. As pointed out before, it is in general harder to work with recurrent nets. Among other things, it becomes difficult to check under what conditions the obtained model is stable, and it takes an extra effort to calculate the correct gradients for the iterative search.

9.4 Some Other Structural Questions

The actual way that the regressors are combined clearly reflect structural assumptions about the system. Let us, for example, consider the assumption that the system disturbances are additive, but not necessarily white noise:

$$y(t) = g(u^t) + v(t) \tag{9.12}$$

Here u^t denotes all past inputs, and $v(t)$ is a disturbance, for which we only need a spectral description. It can thus be described by

$$v(t) = H(q)e(t)$$

for some white sequence $\{e(t)\}$. The predictor for (9.12) then is

$$\hat{y}(t) = (1 - H^{-1}(q))y(t) + H^{-1}(q)g(u^t) \tag{9.13}$$

In the last term, the filter H^{-1} can equally well be subsumed in the general mapping $g(u^t)$. The structure (9.12) thus leads to a NNFIR or NNOE structure, complemented by a *linear* term containing past y.

In [25] a related Neural Network based model is suggested. It can be described by

$$\hat{y}(t) = f(\theta_1, \varphi_1(t)) + g(\theta_2, \varphi_2(t)) \tag{9.14}$$

where $\varphi_1(t)$ consists of delayed outputs and $\varphi_2(t)$ of delayed inputs. The parameterized functions f and g can be chosen to be linear or non-linear by a neural net. A further motivation for this model is that it becomes easier to develop controllers from (9.14) than from the models discussed earlier.

In [24], it is suggested first to build a linear model for the system. The residuals from this model will then contain all unmodelled non-linear effects. The Neural Net model could then be applied to the residuals (treating inputs and residuals as input and output), to pick up the non-linearities. This is attractive, since the first step to obtain a linear model is robust and often leads to reasonable models. By the second Neural Net step, we are then assured to obtain at least as good a model as the linear one.

The question of how many layers to use is not easy. The paper [34] contains many useful and interesting insights into the importance of second hidden layers in the NN structure. See also the comments on this in Section 5.1.

9.5 The Identification Procedure

A main principle in identification is the rule *try simple things first*. The idea is to start with the simplest model which has a possibility to describe the system and only to continue to more complex ones if the simple model does not pass validation tests.

When a new more complex model is investigated the results with the simpler model give some guidelines how the structural parameters should be chosen in the new model. For example, it is common to start with an ARX model. The delay and number of delayed inputs and outputs give a good initial guess how the structure parameters should be chosen for the more complex ARMAX model. In this way less combinations of structural parameters have to be tested and computer time is saved.

Many non-linear systems can be described fairly well by linear models and for such systems it is a good idea to use insights from the best linear model how to select the regressors for the NN model. To begin with, only the number of hidden units the needs to be varied. Also, there might be more problems with local minima for the non-linear than for the linear models which makes it necessary to do several parameter estimates with different initial guesses. This further limits the number of candidate models which can be tested.

In the following example a hydraulic actuator is identified. First a linear model is proposed which does not capture all the fundamental dynamical behavior and then a NNARX model is tried. The same problem is considered in [2] using wavelets as model structure.

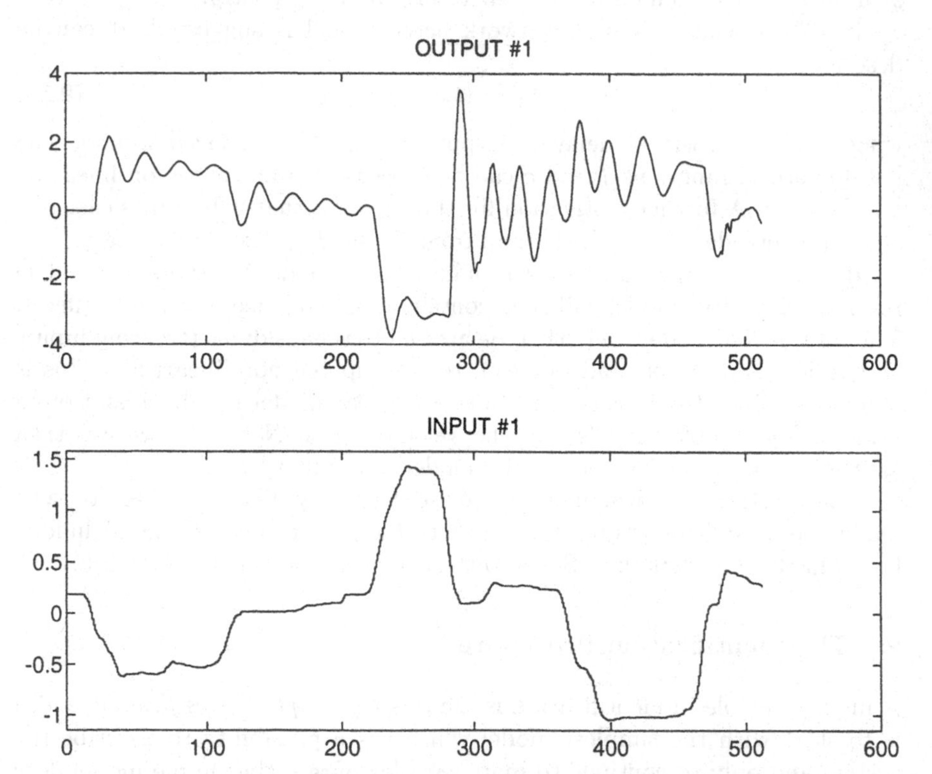

Fig. 9.1. Measured values of oil pressure (top) and valve position (bottom).

Example 9.1. Modeling a Hydraulic Actuator. The position of a robot arm is controlled by a hydraulic actuator. The oil pressure in the actuator is controlled by the size of the valve opening through which the oil flows into the actuator. The position of the robot arm is then a function of the oil pressure. In [13] a thorough description of this particular hydraulic system is given. Fig. 9.1 shows measured values of the valve size and the oil pressure, which are input- and output signals, respectively. As seen in the oil pressure, we have a very oscillative settling period after a step change of the valve size. These oscillations are caused by mechanical resonances in the robot arm.

Following the principle "try simple things first" gives an ARX model with structural parameters $[n_a \; n_b \; n_k] = [3 \; 2 \; 1]$. In Fig. 9.2 the result of a simulation with the obtained linear model on validation data is shown. The result is not very impressive.

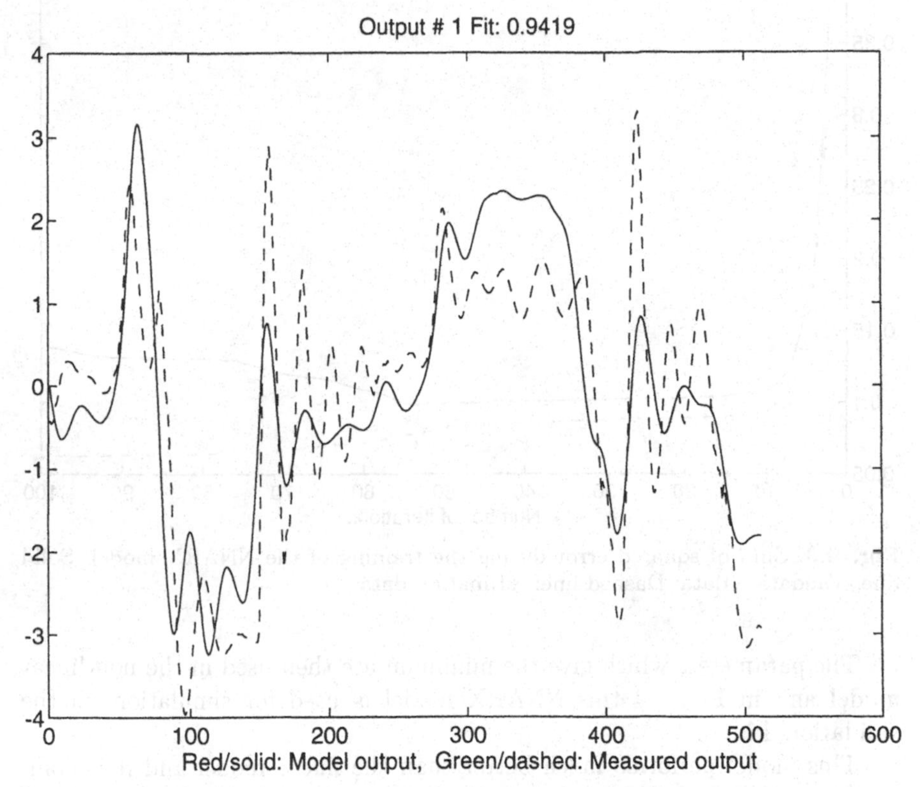

Fig. 9.2. Simulation of the linear model on validation data. Solid line: simulated signal. Dashed line: true oil pressure.

Instead a NNARX model is considered with the same regressor as the linear model, i.e., with the same first three structural indexes, and with 10 hidden units, $n_h = 10$. In Fig. 9.3 it is shown how the quadratic criterion develops during the estimation for estimation and validation data, respectively. For the validation data the criterion first decrease and then it starts to increase again. This is the overtraining which was described in Section 6.3. The best model is obtained at the minimum and this means that not all parameters in the non-linear model have converged and, hence, the "efficient number of parameters" is smaller than the dimension of θ.

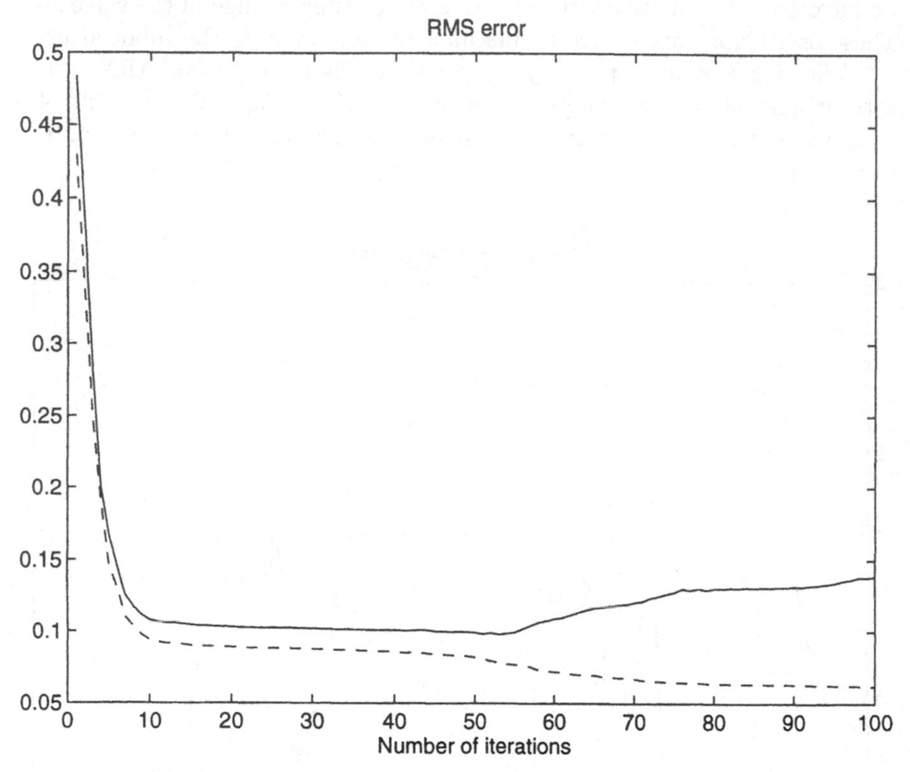

Fig. 9.3. Sum of squared error during the training of the NNARX model. Solid line: Validation data. Dashed line: estimation data.

The parameters which give the minimum are then used in the non-linear model and in Fig. 9.4 this NNARX model is used for simulation on the validation data.

This model performs much better than the linear model and it is compatible to the result obtained with a wavelet model in [2].

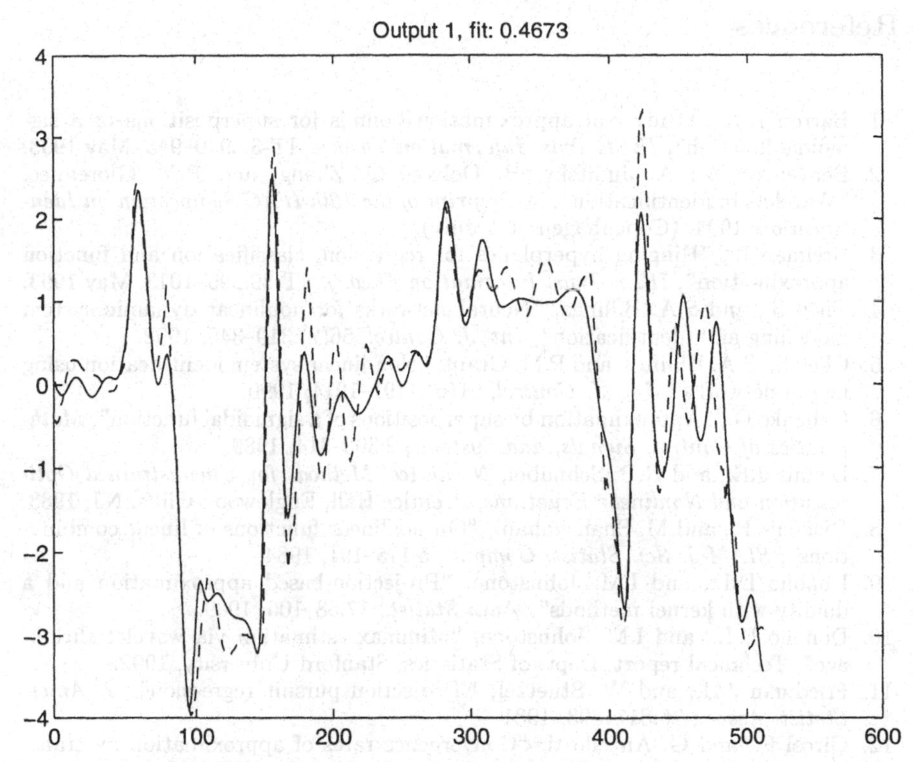

Fig. 9.4. Simulation of the non-linear model on validation data. Solid line: simulated signal. Dashed line: true oil pressure.

References

1. Barron A.R., "Universal approximation bounds for superpositions of a sigmoidal function", *IEEE Trns. Information Theory.*, IT-39:930–945, May 1993.
2. Benveniste A., A. Juditsky, B. Delyon, Q. Zhang, and P-Y. Glorennec, "Wavelets in identification", in *Preprint of the 10th IFAC Symposium on Identification*, 1994. (Copenhagen, 4–6 July).
3. Breiman L., "Hinging hyperplanes for regression, classification and function approximation", *IEEE Trns. Information Theory.*, IT-39:999–1013, May 1993.
4. Chen S., and S.A. Billings, "Neural networks for nonlinear dynamic system modelling and identification", *Int. J. Control*, 56(2):319–346, 1992.
5. Chen S., S.A. Billings, and P.M. Grant, "Non-linear system identification using neural networks", *Int. J. Control*, 51(6):1191–1214, 1990.
6. Cybenko G., "Approximation by superpositions of a sigmoidal function", *Mathematics of Control, Signals, and Systems*, 2:303–314, 1989.
7. Dennis J.E., and R.B. Schnabel, *Numerical Methods for Unconstrained Optimization and Nonlinear Equations*, Prentice-Hall, Englewood Cliffs, NJ, 1983.
8. Diaconis P., and M. Shahshahani, "On nonlinear functions of linear combinations", *SIAM J. Sci. Statist. Comput.*, 5:175–191, 1984.
9. Donoho D.L., and I.M. Johnstone, "Projection-based approximation and a duality with kernel methods", *Ann. Statist.*, 17:58–106, 1989.
10. Donoho D.L., and I.M. Johnstone, "Minimax estimation via wavelet shrinkage", Technical report, Dept. of Statistics, Stanford University, 1992.
11. Friedman J.H., and W. Stuetzel, "'Projection pursuit regression", *J. Amer. Statist. Assoc.*, 76:817–823, 1981.
12. Girosi F., and G. Anzellotti, "Convergence rates of approximation by translates", Technical report, Dept. Math. Art. Intell. Lab, MIT, 1992.
13. Gunnarsson S. and P. Krus, "Modelling of a flexible mechanical system containing hydralic actuators", Technical report, Dep. of Electrical Engineering, Linköping University, S-581 83 Linköping, Sweden, April 1990.
14. Guo L., and L. Ljung, "The role of model validation for assessing the size of the unmodelled dynamics", Technical report, Report LiTH-ISY, Department of Electrical Engineering, Linköping University, Sweden, 1994.
15. Hertz J., A. Krogh, and R.G. Palmer, *Introduction to the Theory of Neural Computation.* Addison-Wesley, Redwood City, CA, 1991.
16. Hunt K.J., D. Sbarbaro, R. Żbikowski, and P.J. Gawthrop, "Neural networks for control systems - a survey", *Automatica*, 28(6):1083–1112, Nov. 1992.
17. Jones L.K., "A simple lemma on greedy approximations in Hilbert space and convergence rates for projection pursuit regression and neural network training", *The Annals of Statistics*, 20:608–613, 1992.
18. Kung S.Y., *Digital Neural Networks.* Prentice-Hall, Englewood Cliffs, NJ, 1993.
19. Leshno M., V.Y. Lin, A. Pinkus, and S. Schocken, "Multilayer feedforward netwoks with a non-polynomial activation function can approximate any function", *Neural Networks*, 6:861–867, 1993.
20. Ljung L., *System Identification: Theory for the User.* Prentice-Hall, Englewood Cliffs, NJ, 1987.
21. Ljung L., and T. Söderström, *Theory and Practice of Recursive Identification.* MIT Press, Cambridge, MA, 1983.
22. Ljung L., and B. Wahlberg, "Asymptotic properties of the least-squares method for estimating transferfunctions and disturbance spectra", *Adv. Appl. Prob.*, 24:412–440, 1992.

23. Matthews M.B., "On the Uniform Approximation of Nonlinear Discrete-Time Fading-Memory Systems using Neural Network Models", PhD thesis, ETH, Zürich, Switzerland, 1992.
24. McAvoy T.J., 1992. Personal communication.
25. Narendra K.S., and K. Parthasarathy, "Identification and control of dynamical systems using neural networks", *IEEE Trans. Neural Networks*, 1:4–27, 1990.
26. Pinkus A., *n-Widths in Approximation Theory.* Springer, Berlin, 1985.
27. Polyak B.T., and A. Juditsky, "Acceleration of stochastic approximation by averaging", *SIAM J. Control*, 30(4):838–855, 1992.
28. Rumelhart D.E., G.E. Hinton, and R.J. Williams, "Learning representations by back-propagating errors", *Nature*, 323(9):533–536, October 1986.
29. Rumelhart D.E., and J.L. McClelland, *Parallel Distributed Processing: Explorations in the Microstructure of Cognition*, volume 1. MIT Press, Cambridge MA, 1986.
30. Sjöberg J., and L. Ljung, "Overtraining, regularization, and searching for minimum in neural networks", in *Preprint IFAC Symposium on Adaptive Systems in Control and Signal Processing*, pages 669–674, Grenoble, France, 1992.
31. Söderström T., and P. Stoica, *System Identification.* Prentice-Hall International, Hemel Hempstead, UK, 1989.
32. Solbrand G., A. Ahlén, and L. Ljung, "Recursive methods for off-line identification", *Int. J. Control*, 41:177–191, 1985.
33. Sontag E.D., "Feedback stabilization using two-hidden-layer nets", technical report, Report SYCON-90-11, Rutgers Center for Systems and Control, Dept. of Mathematics, Rutgers University, New Brunswick, NJ, October 1990.
34. Sontag E.D., "Neural networks for control", in H.L. Trentelman and J.C. Willems, editors, *Essays on Control: Perspectives in the Theory and its Applications*, volume 14 of *Progress in Systems and Control Theory*, pages 339–380. Birkhäuser, Basel, 1993.
35. Triebel H., *Theory of Function Spaces.* Birkhäuser, Basel, 1983.
36. van der Smagt P.P., "Minimisation methods for training feedforward neural networks", *Neural Networks*, 7(1):1–11, 1994.
37. Wahba G., "Three topics in ill-posed problems", in H.W. Engl and C.W. Groetsch, editors, *Inverse and Ill-posed Problems.* Academic Press, 1987.
38. Wahba G., *Spline Models for Observational Data.* SIAM, University City Science Center, Philadelphia, PA, 1990.
39. Wahlberg B., "System identification using laguerre models", *IEEE Tabx. AC*, 36(5):551–562, May 1991.
40. Werbos P., *Beyond Regression: New Tools for Prediction and Analysis in the Behavioral Science.* Ph.D. thesis, Harvard University, 1974.
41. White D.A., and D.A. Sofge, editors, *Handbook of Intelligent Control, Neural, Fuzzy, and Adaptive Approaches.* Van Nostrand Reinhold, New York, 1992.
42. Miller III W.T., R.S. Sutton, and P.J. Werbos, editors, *Neural Networks for Control.* Neural Network Modeling and Connectionism. MIT Press, 1992. Series editor: J.L. Elman.
43. Zhao Y., *On projection pursuit learning.* Ph.D. thesis, Dept. Math., AI Lab. MIT, 1992.

An Overview of Computational Learning Theory and Its Applications to Neural Network Training

M. Vidyasagar

Centre for Artificial Intelligence and Robotics,
Raj Bhavan Circle,
High Grounds
Bangalore 560 001,
India
E-mail: sagar@cair.ernet.in

1. Introduction

Recently there has been a great deal of interest in non-parametric classification/estimation according to a probably-approximately correct (PAC) criterion. Much early and fundamental work related to this area was done in the probability and statistics literature — e.g., by Vapnik and Chervonenkis [15, 16, 17], Dudley [6], and Pollard [13]. The paper of Valiant [14] spurred recent work in this area in the computer science community as well. Haussler [9] has recently refined and consolidated some of the work in the statistics and computer science communities. ¿From the standpoint of the neural networks community, PAC learning theory is attractive because it permits one to turn a vague statement such as "neural networks can generalize" into a precise statement such as: "After adequate training, neural networks can correctly classify previously untrained inputs with high accuracy, most of the time." Moreover, PAC learning theory permits one to make *quantitative estimates* about the number of training samples required to achieve a desired level of accuracy and confidence in the ability of the neural network to generalize. Though these bounds are often quite pessimistic, nevertheless they are a welcome replacement for the "hand-waving" that abounds in most neural networks papers. In the present paper, the PAC learning problem formulation is introduced and its significance in the context of neural network training is explained. Then some known results are summarized. This part of the paper is more like a survey and there is no original contribution in this part of the paper.

In the PAC learning community, almost all authors have studied one of two extreme cases, namely: (i) learnability under a *fixed* probability measure, or (ii) *distribution-free* learning, in which learning has to take place under all possible probability measures. In the former case (pp. 149-151 of [17], [1]), a set of concepts is PAC-learnable if and only if it has a finite ϵ-cover for each positive ϵ. In the latter case [17, 2], a set of concepts is PAC-learnable if and only if it has finite Vapnik-Chervonenkis (VC)-dimension. Thus, the two extreme situations are well-understood, but very little is known about learnability with respect to other families of probability measures. Clearly a *necessary* condition for PAC-learnability with respect to an arbitrary family of probabilities is that the concept class should have uniformly bounded metric entropy. However, it has been recently shown that the condition is *not* sufficient [8].

Unless some structure is imposed on the family of probabilities under which learning is to take place, it is evident that the problem is hopelessly intractable. In the latter part of the paper, some new results are presented regarding the problem of PAC learning under an arbitrary family of probability measures. In particular, it is shown that uniform boundedness of the metric entropy of the class of concepts is both necessary and sufficient for learnability under each of two conditions: (i) the family of probability measures is *totally bounded*, with respect to the total variation metric, and (ii) the family of probability measures *contains an interior point*, when equipped with the same metric. These two results establish that in any counterexample showing that uniform boundedness of the metric entropy is *not* sufficient for learnability, the family of probability measures cannot be totally bounded and yet must have an empty interior. Then two sufficient conditions for learnability are presented. Specifically, it is shown that learnability with respect to each of a finite collection of families of probability measures implies learnability with respect to their union; also, learnability with respect to each of a finite number of measures implies learnability with respect to the convex hull of these measures.

2. Problem Formulation

The basic ingredients of PAC-learning (in the case of binary-valued functions) are:

- A set X,
- A σ-algebra $\mathcal{S}$ of subsets of X,
- A subset $\mathcal{C} \subset \mathcal{S}$, called the **concept class**, and
- A family of probability measures $\mathcal{P}$ on the pair $(X, \mathcal{S})$.

For the present purposes, an **algorithm** is a family of maps $A_m : [X \times \{0,1\}]^m \rightarrow \mathcal{C}$, for each integer $m \geq 1$. "Learning" takes place as follows: A **target** concept $T \in \mathcal{C}$ is fixed, as well as a probability measure $P \in \mathcal{P}$. A sequence of independent identically distributed random samples $x_1, x_2, \ldots, x_m$ is selected from X, in accordance with the fixed but unknown probability $P \in \mathcal{P}$. For each integer $k \leq m$, an "oracle" informs the learner whether or not x_k belongs to the unknown target concept T; in other words, one is given the value of the indicator function $I_T(x_k)$. After m runs of the experiment, one generates a hypothesis

$$H_m = A_m[(x_1, I_T(x_1)), \ldots, (x_m, I_T(x_m))].$$

The hope is that H_m is a "probably approximately correct" representation of the target concept T, as made precise next.

Given two sets $A, B \subset X$, define $A \Delta B$ to be the symmetric difference of the sets, i.e.,

$$A \Delta B = (A \cap B^c) \cup (A^c \cap B) = (A \cup B) - (A \cap B).$$

In other words, $A \Delta B$ is the set of elements in X that belong to *exactly one* of the two sets A and B. For $A, B \in \mathcal{S}$, define

$$d_P(A, B) = P(A \Delta B). \tag{2.1}$$

Then d_P is a *pseudometric* on $\mathcal{S}$; thus d_P satisfies all the axioms of a metric, except that $d_P(A, B) = 0$ does not necessarily imply that $A = B$. In fact $d_P(A, B) = 0$ if and only if $A \Delta B$ is a set of measure zero.

Definition 2.1. *An algorithm*$\{A_m\}$ *is said to be* **probably approximately correct (PAC)** *if, for each* $\epsilon, \delta > 0$, *there exists an integer* $m_0 = m_0(\epsilon, \delta)$ *such that*

$$r(m, \epsilon) := P^m\{(x_1, \ldots, x_m) \in X^m : d_P(H_m, T) > \epsilon\} \leq \delta, \ \forall m \geq m_0, \tag{2.2}$$

for all target concepts $T \in \mathcal{C}$, *and all probability measures* $P \in \mathcal{P}$, *where* $P^m(\cdot)$ *denotes the m-fold product probability on* X^m. *The concept class* $\mathcal{C}$ *is said to be* **PAC-learnable** *with respect to the family* $\mathcal{P}$ *if there exists a PAC algorithm for* $\mathcal{C}$.

In the context of neural networks, one can interpret the PAC learning problem as follows: Suppose one *fixes* a particular neural network architecture, but permits the parameters of the neural network to vary. For example, one could think of the class of perceptrons that accepts a k-dimensional vector $\mathbf{x}$ as input and produces an output of either 0 or 1 in accordance with the familiar formula

$$y = \begin{cases} 1 & \text{if } \sum_{i=1}^k w_i x_i - \theta \geq 0, \text{ and} \\ 0 & \text{if } \sum_{i=1}^k w_i x_i - \theta < 0. \end{cases}$$

For each fixed choice of weights w_i and threshold θ, the set of inputs that get mapped into an output of 1 consists of a closed hyperplane in $\Re^k$. Now, as we vary the weight vector and the threshold over $\Re^k$ and $\Re$ respectively, we generate all possible perceptrons. Clearly there is an onto (but not one-to-one!) association between the set of all perceptrons and the collection of all possible closed half-planes in $\Re^k$. For neural networks with one or more hidden layers, there is a similar association between the neural network architecture and an appropriate collection of subsets of $\Re^k$. In other words, one can associate, with each neural network architecture, a corresponding family $\mathcal{C}$ of subsets of $\Re^k$. Now, in the context of *generalization* by neural networks, one can ask the following question: Suppose a network architecture is fixed, and that one is given a set of training inputs, selected at random in accordance with some unknown probability measure P chosen from some known family $\mathcal{P}$; let these training inputs be labelled as $x_1, \ldots, x_m$. Suppose further that, for each training input x_i, one is given a corresponding desired output $y_i \in \{0, 1\}$, and that there is a "true" neural network that is capable of producing all these desired outputs. Now let us "train" a candidate neural network, belonging to the class of networks under consideration, by adjusting all of its parameters so that it faithfully produces the correct output on all training inputs. What then is the probability that, on the *next randomly chosen input* $x \in X$, the "trained" neural network H_m gives the same output as the unknown "true" neural network T? To put this question in the format of PAC learning, observe that an input $x \in X$ is incorrectly classified if and only if it belongs to the symmetric difference $H_m \Delta T$, and that $d_P(H_m, T)$ is precisely the probability that a randomly chosen element of X is misclassified by H_m. Thus the PAC inequality (2.2) can be intrepreted as "generalization with high accuracy $1-\epsilon$ and high confidence $1-\delta$." The above discussion pertains to so-called "classification" neural networks, i.e., neural networks whose output is binary. However, using the results of Haussler [9], it is also possible to intercorporate neural networks with real-valued outputs (i.e., neural networks employing sigmoidal neurons) into the PAC framework.

3. Summary of Known Results

We now introduce some standard definitions and summarize the known conditions for PAC-learnability in the cases of fixed distribution and distribution-free learnability.

Definition 3.1. *A set $S = \{x_1, \ldots, x_n\} \subset X$ is said to be* **shattered** *by the concept class $\mathcal{C}$ if, for every subset A of S, there exists a concept $C \in \mathcal{C}$ such that $S \cap C = A$. The* **Vapnik-Chervonenkis (VC)-dimension** *of the concept class $\mathcal{C}$ is the largest integer d such that there exists a set of cardinality d that is shattered by $\mathcal{C}$. If there exist sets of arbitrarily large (integer) cardinality that are shattered by $\mathcal{C}$, then $\mathcal{C}$ is said to have infinite VC-dimension.*

The following result characterizes distribution-free learnability. Certain additional, but mild, measurability conditions are actually required (e.g., see [17] or [2]), but we ignore these conditions throughout.

Theorem 3.1. *[2] Let $\mathcal{P}^*$ denote the set of all probability measures on S. Then a concept class $\mathcal{C}$ is PAC-learnable with respect to $\mathcal{P}^*$ if and only if the VC-dimension of $\mathcal{C}$ is finite.*

The results in [2] also provide a quantitative upper bound for the number $r(m, \epsilon)$ defined in (2.2), provided one uses a so-called "consistent" algorithm. For the purposes of the present discussion, an algorithm is said to be consistent if the hypothesis produced by the algorithm always correctly classifies the data. More precisely, an algorithm $\{A_m\}$ is **consistent** if, for every $T \in \mathcal{C}$ and every set of samples $x_1, \ldots, x_m \in X$, the hypothesis H_m satisfies

$$I_{H_m}(x_i) = I_T(x_i),\ i = 1, \ldots, m.$$

Now it is shown in [2] that, if a concept class has finite VC-dimension, say d, and if the hypothesis H_m is produced using a consistent algorithm, then

$$r(m, \epsilon) \leq 2 \left(\frac{2em}{d} \right)^d 2^{-m\epsilon/2},$$

where as usual e denotes the base of the natural logarithm. This result is directly applicable to neural networks, as follows: Suppose a neural network uses only a single hidden layer, and consists only of perceptron-type neurons. Then such a network has finite VC-dimension, and the VC-dimension is linearly related to the number of adjustable parameters in the network. Further details can be found in the paper by Baum and Haussler [3]. Analyzing various neural network architectures with a view towards estimating their VC-dimension is an active area of research, and new results are constantly being discovered. The reader is referred to an excellent survey by Sontag [12] for a description of currently available results.

In learning with respect to a fixed distribution, the main idea is that since we only need to approximate the unknown target concept, if we can replace $\mathcal{C}$ with a finite number of concepts (one of which is close to the target concept), then we can simplify the problem to one of deciding between a finite number of alternatives. This motivates the following definition.

Definition 3.2. *Suppose ρ is a pseudometric on a set $\mathcal{S}$. Suppose $\mathcal{C} \subset \mathcal{S}$, and let $\epsilon > 0$. A finite set $\{B_1, \ldots, B_n\}$ where each $B_i \in \mathcal{C}$ is said to be an* **ϵ-cover** *of $\mathcal{C}$ with respect to ρ if, for each $C \in \mathcal{C}$, there exists an index i such that $\rho(B_i, C) < \epsilon$. The smallest integer $N = N(\epsilon, \mathcal{C}, \rho)$ such that $\mathcal{C}$ has an ϵ-cover of cardinality N is called the* **ϵ-covering number** *of $\mathcal{C}$ with respect to ρ. If $\mathcal{C}$ does not have a finite ϵ-cover, then N is taken to be infinite.*

Theorem 3.2. *[1] Suppose P_0 is a probability measure on $\mathcal{S}$, and define the associated pseudometric d_{P_0} in analogy with (2.1). Suppose $\mathcal{C} \subset \mathcal{S}$, and let $\mathcal{P}$ equal the singleton set $\{P_0\}$. Then $\mathcal{C}$ is PAC-learnable with respect to $\{P_0\}$ if and only if $\mathcal{C}$ has a finite ϵ-cover with respect to d_{P_0} for each $\epsilon > 0$.*

The proof of the above theorem as given in [1] also leads to a learning algorithm, as described next. Suppose $\mathcal{C}$ has a finite $\epsilon/2$-cover with respect to d_{P_0}. Let $N := N(\epsilon/2, \mathcal{C}, d_{P_0})$, and choose an $\epsilon/2$-cover $B_1, \ldots, B_N$ for $\mathcal{C}$. Select i.i.d. samples $x_1, \ldots, x_m \in X$, and choose the hypothesis H_m as the concept B_k that misclassifies the smallest number of samples. This algorithm is known as the "minimum risk" algorithm. It is shown in [1] that the algorithm is PAC provided one chooses at least

$$m(\epsilon, \delta) \geq \frac{32}{\epsilon} \ln \frac{N}{\delta}.$$

samples.

The two results above naturally suggest that there may be connections between the covering numbers of a concept class with respect to various distributions and the VC-dimension of a concept class. The known results typically provide upper and lower bounds on $\sup_{P \in \mathcal{P}^*} N(\epsilon, \mathcal{C}, d_P)$ in terms of the VC-dimension of $\mathcal{C}$ (e.g., some known results are summarized in [11]). The upper bounds are the deeper results and the fundamental result along this line was obtained by Dudley [6], which was subsequently refined by Pollard [13] and more recently by Haussler [9]. These bounds imply the following result which is sufficient for the present paper. Again, the result requires some weak measurability conditions which we ignore here.

Theorem 3.3. *For any concept class $\mathcal{C}$, $\sup_{P \in \mathcal{P}^*} N(\epsilon, \mathcal{C}, d_P) < \infty$ for all $\epsilon > 0$ iff the VC-dimension of $\mathcal{C}$ is finite. If the VC-dimension of $\mathcal{C}$ is finite, say d, then*

$$\sup_{P \in \mathcal{P}^*} N(\epsilon, \mathcal{C}, d_P) \leq 2 \left(\frac{2e}{\epsilon} \ln \frac{2e}{\epsilon} \right)^d.$$

For future convenience, one more concept is defined.

Definition 3.3. *Suppose $\mathcal{C}$ is a class of concepts, and $\mathcal{P}$ is a family of probability measures. Then the class $\mathcal{C}$ is said to have* **uniformly bounded metric entropy** *with respect to the family $\mathcal{P}$, or to satisfy the* **UBME condition** *with respect to $\mathcal{P}$, if*

$$\bar{N}(\epsilon, \mathcal{C}, \mathcal{P}) := \sup_{P \in \mathcal{P}} N(\epsilon, \mathcal{C}, d_P) < \infty \text{ for each } \epsilon > 0. \tag{3.1}$$

Now suppose $\mathcal{P}$ is an *arbitrary* family of probability measures. For each $P \in \mathcal{P}$, define d_P as in (2.1). Then, by a slight extension of the results in pp. 149-151 of [17] and [1], one can prove the following:

Lemma 3.1. *Suppose $\mathcal{C}$ is a class of concepts and $\mathcal{P}$ is a family of probability measures. Suppose $\mathcal{C}$ is learnable with respect to $\mathcal{P}$. Then $\mathcal{C}$ satisfies the UBME condition with respect to $\mathcal{P}$.*

Lemma 3.1 shows that a *necessary* condition for $\mathcal{C}$ to be learnable with respect to $\mathcal{P}$ is that $\mathcal{C}$ satisfies the UBME condition with respect to $\mathcal{P}$. In other words, $\mathcal{C}$ has a finite ϵ-cover for each $\epsilon > 0$, and each probability $P \in \mathcal{P}$; moreover, the ϵ-covering number is uniformly bounded with respect to P for each ϵ. Furthermore, the preceding results show that the condition (3.1) is also *sufficient* in the two "extreme" cases, namely when $\mathcal{P}$ is a singleton set, and when $\mathcal{P} = \mathcal{P}^*$. Thus it is natural to ask whether the condition (3.1) is also sufficient for $\mathcal{C}$ to be PAC-learnable for *all* families of probabilities $\mathcal{P}$. The answer is "No" as shown in [8]. Thus the problem of deriving necessary and sufficient conditions for PAC-learnability under a general family of probability measures is still open. The present paper derives a few results in this direction.

While Theorem 3.2 is for learnability with respect to a *single* measure, a careful examination of the proof of this result shows that it is actually possible to prove a rather stringent *sufficient* condition for learnability.

Corollary 3.1. *Suppose $\mathcal{C} \subseteq \mathcal{S}$, $\mathcal{P} \subseteq \mathcal{P}^*$, and define the pseudometric $\bar{d}_{\mathcal{P}}$ as follows: For each $A, B \in \mathcal{S}$, let*

$$\bar{d}_{\mathcal{P}} = \sup_{P \in \mathcal{P}} d_P(A, B). \tag{3.2}$$

Then $\mathcal{C}$ is learnable to an accuracy ϵ with respect to $\mathcal{P}$ if $\mathcal{C}$ has an $\epsilon/2$-cover with respect to the pseudometric $\bar{d}_{\mathcal{P}}$. In particular, the minimum risk algorithm described above is uniformly correct to accuracy ϵ.

The reason why (3.1) is *not* sufficient in general for learnability, whereas the condition of Corollary 3.1 *is* sufficient, lies in the order of the quantifiers. In Corollary 3.1, the *same* sets $B_1, \ldots, B_N$ can serve as the "centers" of the N "balls" that provide a $\epsilon/2$-cover of $\mathcal{C}$, for *every* distribution $P \in \mathcal{P}$. In contrast, if (3.1) holds, then the *number* of balls in an ϵ-cover of $\mathcal{C}$ is independent of $P \in \mathcal{P}$, but the "centers" of these balls may depend on P (as it does in the counterexample of [8]).

Let us now examine the question: How close is the condition of Corollary 3.1 to being necessary for learnability? To shed some light on this question, let us study the case where $\mathcal{P} = \mathcal{P}^*$, the set of *all* probability measures on $\mathcal{S}$, and compare the sufficient condition of Corollary 3.1 with the known necessary condition provided by Theorem 3.1.

Suppose $\mathcal{P}$ is the set of all probability measures on $\mathcal{S}$, and define $\bar{d} := \bar{d}_{\mathcal{P}^*}$ as in (3.2). Of course, if $A = B$ then $\bar{d}(A, B) = 0$. However, if $A \neq B$ then $A \Delta B$ is nonempty. In this case, if P is taken to be the purely atomic measure concentrated at a point $x \in A \Delta B$ then $d_P(A, B) = 1$. It follows that $\bar{d}(A, B) = 1$ whenever $A \neq B$. Hence, if $\mathcal{P}$ is the set of all distributions on $\mathcal{S}$, then any two distinct concepts in $\mathcal{C}$ are a distance 1 apart with respect to $\bar{d}$. Therefore, $\mathcal{C}$ has a finite ϵ-cover with respect to $\bar{d}$ for each $\epsilon > 0$ if and only if $\mathcal{C}$ is finite.

Thus, in the extreme case of distribution-free learning, the sufficient condition of Corollary 3.1 reduces to the requirement that the class $\mathcal{C}$ be finite, which is much more stringent than the requirement that the VC-dimension of $\mathcal{C}$ be finite.

4. Families of Measures with a Nonempty Interior

The objective of this section is to show that the UBME condition is sufficient as well as necessary in the case where the family of probability measures $\mathcal{P}$ has a nonempty interior. Actually, we will see that as long as there is some positive amount of nonparametric uncertainty (measured in various ways) around some nominal P_0, then a concept class is learnable with respect to $\mathcal{P}$ if and only if it is distribution-free learnable. Thus, if one ignores the issue of the number of samples needed for learning, the learning problem with nonparametric uncertainty is just as difficult as distribution-free learning.

Let ρ denote the total variation metric on $\mathcal{P}^*$. That is, given $P, Q \in \mathcal{P}^*$, define

$$\rho(P, Q) = \sup_{A \in \mathcal{S}} |P(A) - Q(A)|. \tag{4.1}$$

Let $\mathcal{B}(P, \lambda)$ denote the open sphere of radius λ centered at P, and let $\bar{\mathcal{B}}(P, \lambda)$ denote the closed sphere of radius λ centered at P, with respect to ρ. Then P_0 is an interior point of $\mathcal{P}$ if (and only if) there exists a $\lambda > 0$ such that $\mathcal{B}(P_0, \lambda) \subseteq \mathcal{P}$.

Suppose P_1, P_2 are probability measures on $\mathcal{S}$; then their **convex combination** $\lambda P_1 + (1 - \lambda) P_2$ is also a probability measure on $\mathcal{S}$ for each $\lambda \in [0, 1]$. Now, for $P_0 \in \mathcal{P}^*$ and $\lambda \in [0, 1]$, define

$$\mathcal{P}_c(P_0, \lambda) = \{(1 - \eta) P_0 + \eta P : \eta \in [0, \lambda], P \in \mathcal{P}^*\}. \tag{4.2}$$

The collection $\mathcal{P}_c(P_0, \lambda)$ can be thought of as those distributions nominally equal to P_0, but with some nonparametric uncertainty with respect to mixtures up to λ in extent. Note that

$$\mathcal{P}_c(P_0, \lambda) \subseteq \bar{\mathcal{B}}(P_0, \lambda).$$

To see this, suppose $Q = (1 - \eta) P_0 + \eta P$ for some $P \in \mathcal{P}^*, \eta \in [0, \lambda]$. Then, for each $A \in \mathcal{S}$, we have

$$\begin{aligned} |Q(A) - P(A)| &= |(1 - \eta) P_0(A) + \eta P(A) - P_0(A)| \\ &= \eta\, |P(A) - P_0(A)| \leq \eta \leq \lambda. \end{aligned}$$

Also, observe that

$$\mathcal{P}_c(P_0, 0) = \bar{\mathcal{B}}(P_0, 0) = \{P_0\}, \quad \mathcal{P}_c(P_0, 1) = \bar{\mathcal{B}}(P_0, 1) = \mathcal{P}^*.$$

Theorem 4.1. *Let C be a class of concepts, P_0 a fixed probability measure, and $0 < \lambda \leq 1$. Then the following are equivalent:*

1. *C is learnable with respect to $\mathcal{P}_c(P_0, \lambda)$.*
2. *C is learnable with respect to $\bar{\mathcal{B}}(P_0, \lambda)$.*
3. *C is learnable with respect to $\mathcal{P}^*$.*
4. *C has finite VC-dimension.*

Proof. **"(2) $\Rightarrow$ (1)"** Obvious, because $P_c(P_0, \lambda) \subseteq \bar{\mathcal{B}}(P_0, \lambda)$.

"(3) $\Rightarrow$ (2)" Obvious, because $\bar{\mathcal{B}}(P_0, \lambda) \subseteq \mathcal{P}^*$.

"(4) $\Rightarrow$ (3)" This is a consequence of Theorem 3.1.

"(1) $\Rightarrow$ (4)" If $\mathcal{C}$ is learnable with respect to $\mathcal{P}_c(P_0, \lambda)$, then $\mathcal{C}$ satisfies the UBME condition with respect to $P_c(P_0, \lambda)$. Now choose $P \in \mathcal{P}^*$ arbitrarily, and define $Q = (1 - \lambda)P_0 + \lambda P \in \mathcal{P}_c(P_0, \lambda)$. If $A, B \in \mathcal{S}$ are any measurable sets, then

$$\begin{aligned} d_Q(A, B) &= Q(A \Delta B) = (1 - \lambda)P_0(A \Delta B) + \lambda P(A \Delta B) \\ &\geq \lambda P(A \Delta B) = \lambda d_P(A \Delta B) \end{aligned}$$

Therefore, $N(\lambda\epsilon, \mathcal{C}, d_Q) \geq N(\epsilon, \mathcal{C}, d_P)$ and so

$$\sup_{P \in \mathcal{P}^*} N(\epsilon, \mathcal{C}, d_P) \leq \sup_{P \in \mathcal{P}^*} N(\lambda\epsilon, \mathcal{C}, d_{(1-\lambda)P_0 + \lambda P}) = \sup_{Q \in \mathcal{P}_c(P_0, \lambda)} N(\lambda\epsilon, \mathcal{C}, d_Q) < \infty$$

Hence, from Theorem 3.3, C has finite VC-dimension.

Corollary 4.1. *Let $\mathcal{C}$ be a class of concepts, and suppose the family $\mathcal{P}$ has a nonempty interior. Then $\mathcal{C}$ is learnable with respect to $\mathcal{P}$ if and only if $\mathcal{C}$ has finite VC-dimension.*

Proof. **"If"** This is a consequence of Theorem 3.1 and the observation after Definition 2.1.

"Only if" Since $\mathcal{P}$ has a nonempty interior, it is possible to select $P_0 \in \mathcal{P}, \lambda > 0$ such that $\bar{\mathcal{B}}(P_0, \lambda) \subseteq \mathcal{P}$. By hypothesis, $\mathcal{C}$ is learnable with respect to $\bar{\mathcal{B}}(P_0, \lambda)$ as well. The desired conclusion now follows from Theorem 4.1.

Note that Theorem 4.1 can actually be improved by considering the class $\mathcal{P}_a^*$ of all purely atomic probability measures. We can define $\mathcal{P}_{ac}(P_0, \lambda)$ in analogy with equation (4.2) where convex combinations are taken only with atomic measures. It turns out that Theorem 3.3 can be improved by taking the supremum only over all $P \in \mathcal{P}_a^*$. Then using this result, it is easy to see that learnability with respect to $\mathcal{P}_{ac}(P_0, \lambda)$ is equivalent to the other four conditions in Theorem 4.1. Another interesting equivalent condition involves all measures close to some measure P_0 in relative entropy. Namely, let

$$\bar{\mathcal{B}}_I(P_0, \lambda) = \{Q : I(P \parallel Q) \leq \lambda\} \tag{4.3}$$

where

$$I(P \parallel Q) = \int \log \frac{P(x)}{Q(x)} \, dP(x)$$

Then if $Q = (1 - \eta)P_0 + \eta P$ we have $Q(x) \geq (1 - \eta)P(x)$ so that $I(P \parallel Q) \leq -\log(1 - \eta)$. Thus, for any $\lambda > 0$, we have $\bar{\mathcal{B}}_I(P_0, \lambda) \supset \mathcal{P}_c(P_0, 1 - e^{-\lambda})$, so that learnability with respect to $\bar{\mathcal{B}}_I(P_0, \lambda)$ for any $\lambda > 0$ is also equivalent to all four conditions of Theorem 4.1. This result may be useful in analyzing the robustness of statistical procedures that assume a certain P_0, and may also be useful in analyzing the tradeoff between labeled and unlabeled examples (e.g., see [4]).

5. Totally Bounded Families of Measures

We begin with a preliminary result that serves as a counterpoint to Theorem 4.1.

Lemma 5.1. *Suppose $\mathcal{C}$ is a class of concepts, P_0 is a given probability measure, and that $N(\epsilon/4, \mathcal{C}, d_{P_0})$ is finite. Then $\mathcal{C}$ is learnable with respect to $\bar{\mathcal{B}}(P_0, \epsilon/4)$ to accuracy ϵ.*

Proof. Let $N := N(\epsilon/4, \mathcal{C}, d_{P_0})$, and let $B_1, \ldots, B_N$ be an $\epsilon/4$-cover for $\mathcal{C}$ with respect to the pseudometric d_{P_0}. Let $P \in \bar{\mathcal{B}}(P_0, \epsilon/4)$ be arbitrary. Then, for each $A \in \mathcal{C}$, we have

$$d_P(A, B_i) = P(A \Delta B_i) \leq P_0(A \Delta B_i) + \epsilon/4 \leq \epsilon/2 \text{ for some } i.$$

Hence the collection $\{B_1, \ldots, B_N\}$ is also an $\epsilon/2$-cover for $\mathcal{C}$ with respect to the pseudometric $\bar{d}_{\bar{\mathcal{B}}(P_0, \epsilon/4)}$. The result now follows from Corollary 3.1.

In fact, an algorithm for learning $\mathcal{C}$ to accuracy ϵ with respect to $\bar{\mathcal{B}}(P_0, \epsilon)$ can be readily constructed by adapting Corollary 3.1. Given $\delta > 0$, choose

$$m = \frac{32}{\epsilon} \ln \frac{N}{\delta}$$

i.i.d. samples, and choose H_m to be the "minimum risk" hypothesis from $B_1, \ldots, B_N$.

The seemingly contradictory results of Theorem 4.1 and Lemma 5.1 can be reconciled as follows: In both cases, the family $\mathcal{P}$ is a sphere of probability measures, representing some nonparametric uncertainty about a nominal distribution P_0. Theorem 4.1 states that, if it is desired to learn a class of concepts to *arbitrarily small* accuracy in spite of the uncertainty regarding the probability measure, then the problem reduces to one of distribution-free learning. In contrast, Lemma 5.1 states that, if we are content to learn a class of concepts to a *fixed finite* accuracy, then a correspondingly small amount of nonparametric uncertainty in the distribution can be tolerated, so long as the class is learnable with respect to the nominal distribution. Together, Theorem 4.1 and Lemma 5.1 suggest that there is a trade-off between the amount of nonparametric uncertainty under which learning is to take place, and the specified accuracy of learning.

The focus of attention in this section is the case where the set $\mathcal{P}$ is totally bounded. The study is commenced with a few relevant definitions and properties. Recall that the metric ρ on $\mathcal{P}$ is the total variation metric defined by equation 4.1.

Definition 5.1. *A family $\mathcal{P} \subseteq \mathcal{P}^*$ is said to be* **totally bounded** *if for each $\epsilon > 0$ the ϵ-covering number of $\mathcal{P}$ (with respect to ρ) is finite — that is, if for every $\epsilon > 0$ there exists a finite set $\{P_1, \ldots, P_M\}$ with $P_i \in \mathcal{P}$ for all i, such that for all $P \in \mathcal{P}$ there exists an index i such that $\rho(P, P_i) \leq \epsilon$.*

The next result shows that the UBME condition is sufficient as well as necessary for learnability, when the family of probability measures is a totally bounded set.

Theorem 5.1. *Suppose $\mathcal{C}$ is a class of concepts and $\mathcal{P}$ is a totally bounded family of probability measures. Then $\mathcal{C}$ is learnable with respect to $\mathcal{P}$ if and only if $\mathcal{C}$ satisfies the UBME condition with respect to $\mathcal{P}$.*

Proof. **"Only if"** This is a consequence of Lemma 3.1.

"If" Given ϵ, δ, choose $\{P_1, \ldots, P_M\}$ such that each $P \in \mathcal{P}$ is within $\epsilon/8$ of some P_i. For each i, choose an $\epsilon/8$-cover $\{B_1^i, \ldots, B_{N_i}^i\}$ for $\mathcal{C}$ with respect to the pseudometric d_{P_i}. Draw

$$m_0 = \max_{1 \le i \le M} \frac{32}{\epsilon/2} \ln \frac{N_i}{\delta/2}$$

i.i.d. samples according to the unknown probability measure P, and for each $i = 1, \ldots, M$ generate a "minimum risk" hypothesis $H_i \in \{B_1^i, \ldots, B_{N_i}^i\}$ such that the corresponding indicator function misclassifies the fewest of the m_0 samples. Now run off another

$$m_{extra} = \frac{32}{\epsilon} \ln \frac{M}{\delta}$$

samples, and let H be a hypothesis among $H_1, \ldots, H_M$ that misclassifies the fewest of these last samples. It is now shown that H satisfies (2.2), establishing that the algorithm is PAC.

By Lemma 5.1, it follows that if $P \in \bar{\mathcal{B}}(P_i, \epsilon/8)$ for some i, then the corresponding hypothesis H_i satisfies $d_P(H_i, T) \le \epsilon/2$ with probability at least $1 - \delta/2$. Thus, among $H_1, \ldots, H_M$, at least one H_i satisfies $d_P(H_i, T) \le \epsilon/2$ with probability at least $1 - \delta/2$. Now, one can repeat the line of reasoning in [17, 1] used to establish Theorem 3.2 to show that the probability that the minimum risk hypothesis H satisfies $d_P(H, T) \le \epsilon$ is at least $1 - \delta$.

Taken together, Theorems 4.1 and 5.1 show that the UBME condition is both sufficient as well as necessary under each of two conditions: (i) the family $\mathcal{P}$ has an interior point, and (ii) the family $\mathcal{P}$ is totally bounded. To put it another way, in any counterexample along the lines of [8] showing that the UBME condition is not sufficient for learnability, the family of probability measures cannot be totally bounded, and yet, at the same time, has to have an empty interior.

The proof of Theorem 5.1 also gives an upper bound on the sample complexity of the above learning algorithm. In analogy with Definition 3.2, let $M(\epsilon)$ denote the ϵ-covering number of the set $\mathcal{P}$ with respect to the metric ρ, and observe that each of the integers N_i in the proof above is at most equal to $\bar{N}(\epsilon/8) := \bar{N}(\epsilon/8, \mathcal{C}, \mathcal{P})$ as defined in Definition 3.3. Therefore it suffices to take

$$m(\epsilon,\delta) \leq \frac{64}{\epsilon}\ln\frac{2\bar{N}(\epsilon/8)}{\delta} + \frac{32}{\epsilon}\ln\frac{M(\epsilon/8)}{\delta}$$

The next result shows that, in some sense, Theorem 5.1 is just Corollary 3.1 in disguise.

Theorem 5.2. *Suppose $\mathcal{C}$ is a class of concepts, $\mathcal{P}$ is a totally bounded family of probability measures, and define the pseudometric $\bar{d}_{\mathcal{P}}$ as in (3.2). Then $\mathcal{C}$ satisfies the UBME condition with respect to $\mathcal{P}$ if and only if $\mathcal{C}$ has a finite ϵ-cover with respect to the pseudometric $\bar{d}_{\mathcal{P}}$ for each $\epsilon > 0$.*

Proof. **"If"** Obvious.

"Only If" Given $\epsilon > 0$, select an $\epsilon/2$-cover $\{P_1, \ldots, P_M\}$ for $\mathcal{P}$ with respect to the metric ρ. For each index i, select an $\epsilon/4$-cover $\{B_1^i, \ldots, B_{\bar{N}}^i\}$ for $\mathcal{C}$ with respect to the pseudometric d_{P_i}, where $\bar{N} = \bar{N}(\epsilon/4)$. It is shown next that $\mathcal{C}$ has an $\epsilon/2$-cover of cardinality at most $N_0 = \bar{N}^M$ with respect to the pseudometric $\bar{d}_\epsilon$ defined by

$$\bar{d}_\epsilon(A,B) = \max_{1\leq i\leq M} d_{P_i}(A,B).$$

Assume for a moment that the above claim has been established, and let $\{A_1, \ldots, A_{N_0}\}$ denote an $\epsilon/2$-cover for $\mathcal{C}$ with respect to $\bar{d}_\epsilon$. Then $\{A_1, \ldots, A_{N_0}\}$ is also an ϵ-cover for $\mathcal{C}$ with respect to $\bar{d}_{\mathcal{P}}$. To see this, let $A \in \mathcal{C}, P \in \mathcal{P}$ be arbitrary, and select indices i, j such that $\rho(P, P_i) \leq \epsilon/2$, and $\bar{d}_\epsilon(A, A_j) \leq \epsilon/2$. Then certainly $d_{P_i}(A, A_j) \leq \epsilon/2$. Therefore

$$d_P(A, A_j) = P(A\Delta A_j) \leq P_i(A\Delta A_j) + \rho(P, P_i) \leq \frac{\epsilon}{2} + \frac{\epsilon}{2} = \epsilon.$$

Thus, in order to complete the proof, it remains only to establish the existence of an $\epsilon/2$-cover for $\mathcal{C}$ with respect to the pseudometric $\bar{d}_\epsilon$. Such a cover can be constructed as follows: For each $i \in \{1, \ldots, M\}, j \in \{1, \ldots, \bar{N}\}$, define

$$\mathcal{C}_{ij} = \{D \in \mathcal{C} : d_{P_i}(D, B_j^i) \leq \epsilon/2\}.$$

Next, for each vector $\mathbf{k} = [k_1 \ldots k_M] \in \{1, \ldots, \bar{N}\}^M$, define

$$\mathcal{A}_{\mathbf{k}} = \bigcap_{i=1}^{M} \mathcal{C}_{i,k_i}.$$

Then each (nonempty) set $\mathcal{A}_{\mathbf{k}}$ has "diameter" at most $\epsilon/2$ in each of the pseudometrics d_{P_i}, and hence in the pseudometric $\bar{d}_\epsilon$. Also, it is easy to see that the collection $\{\mathcal{A}_{\mathbf{k}}\}$ covers $\mathcal{C}$. Now choose some $A_{\mathbf{k}} \in \mathcal{A}_{\mathbf{k}}$ for each $\mathbf{k}$, provided of course that $\mathcal{A}_{\mathbf{k}}$ is nonempty. This is the desired $\epsilon/2$-cover for $\mathcal{C}$.

The proof of Theorem 5.2 enables us to obtain a different upper bound for the sample complexity $m(\epsilon, \delta)$ than the proof of Theorem 5.1. The proof of Theorem 5.2 shows that, for each $\epsilon > 0$, the class $\mathcal{C}$ has an ϵ-cover of cardinality at most $[N(\epsilon/4)]^{M(\epsilon/2)}$ with respect to the pseudometric $\bar{d}_{\mathcal{P}}$. Hence $\mathcal{C}$ has an $\epsilon/2$-cover of cardinality at most $N = [\bar{N}(\epsilon/8)]^{M(\epsilon/4)}$. Therefore the "minimum risk" algorithm described in Corollary 3.1 is PAC, provided we take at least

$$m(\epsilon, \delta) = \frac{32}{\epsilon} \ln \frac{N}{\delta} = \frac{32}{\epsilon}[M(\epsilon/4) \ln \bar{N}(\epsilon/8) + \ln(1/\delta)].$$

The next set of results is aimed at showing that, if a class of concepts is learnable with respect to a totally bounded family of probability measures, then the class is also learnable with respect to the *closed convex hull* of the family.

Definition 5.2. *Suppose $\mathcal{P} \subseteq \mathcal{P}^*$. Then the* **convex hull** *of $\mathcal{P}$ is defined as the set of all convex combinations*

$$P = \sum_{i=1}^{n} \lambda_i P_i, \quad \text{where } n \geq 1, P_i \in \mathcal{P}\ \forall i,\ \lambda_i \geq 0\ \forall i,\ \sum_{i=1}^{n} \lambda_i = 1,$$

and is denoted by $\mathbf{C}(\mathcal{P})$. The **closed convex hull** *of $\mathcal{P}$ is defined as the closure of $\mathbf{C}(\mathcal{P})$ and is denoted by $\bar{\mathbf{C}}(\mathcal{P})$.*

Observe that $\mathbf{C}(\mathcal{P})$ and $\bar{\mathbf{C}}(\mathcal{P})$ are also families of probability measures and are thus subsets of $\mathcal{P}^*$. The following lemma is absolutely standard, and holds in any topological linear vector space (e.g. see Theorem 3, p. 70 of [10], or p. 644 of the 1964 edition). This lemma will be used in the subsequent theorem which is the result of interest for the present purposes.

Lemma 5.2. *Suppose $\mathcal{P} \subseteq \mathcal{P}^*$ is totally bounded. Then $\bar{\mathbf{C}}(\mathcal{P})$ is totally bounded.*

Theorem 5.3. *Suppose $\mathcal{C}$ is a class of concepts, $\mathcal{P}$ is a totally bounded family of probability measures, and that $\mathcal{C}$ satisfies the UBME condition with respect to $\mathcal{P}$. Then $\mathcal{C}$ satisfies the UBME condition with respect to $\bar{\mathbf{C}}(\mathcal{P})$.*

Proof. By Lemma 5.2, it is enough to establish that $\mathcal{C}$ satisfies the UBME condition with respect to $\mathbf{C}(\mathcal{P})$.

Accordingly, suppose $\mathcal{C}$ satisfies the UBME condition (3.1). From Theorem 5.2, it follows that $\mathcal{C}$ also satisfies the finite metric entropy condition with respect to the pseudometric $\bar{d}_{\mathcal{P}}$. Specifically, given any $\epsilon > 0$, there exists a set $\{A_1, \ldots, A_N\}$ where $N \leq [\bar{N}(\epsilon/4)]^{M(\epsilon/2)}$, that forms an ϵ-cover for $\mathcal{C}$ with respect to $\bar{d}_{\mathcal{P}}$. It is shown now that the *same* set is also an ϵ-cover for $\mathcal{C}$ with respect to the pseudometric $\bar{d}_{\mathbf{C}(\mathcal{P})}$. To see this, suppose $P \in \mathbf{C}(\mathcal{P})$, and suppose to be specific that

$$P = \sum_{i=1}^{n} \lambda_i P_i, \; P_i \in \mathcal{P}, \; \lambda_i \geq 0, \; \sum_{i=1}^{n} \lambda_i = 1.$$

Suppose $B \in \mathcal{C}$. By assumption, there exists an index j such that

$$\bar{d}_{\mathcal{P}}(A_j, B) \leq \epsilon.$$

or in other words,

$$d_Q(A_j, B) \leq \epsilon, \; \forall Q \in \mathcal{P}.$$

Therefore

$$d_P(A_j, B) = P(A_j \Delta B) = \sum_{i=1}^{n} \lambda_i \, d_{P_i}(A_j, B) \leq \epsilon \sum_{i=1}^{n} \lambda_i = \epsilon.$$

As this inequality holds for *each* $P \in \mathbf{C}(\mathcal{P})$, it follows that $\bar{d}_{\mathbf{C}(\mathcal{P})}(A_j, B) \leq \epsilon$. Therefore $\{A_1, \ldots, A_n\}$ is an ϵ-cover for $\mathcal{C}$ with respect to the pseudometric $\bar{d}_{\mathbf{C}(\mathcal{P})}$.

Corollary 5.1. *Suppose $\mathcal{C}$ is a class of concepts, $\mathcal{P}$ is a totally bounded family of probability measures, and that $\mathcal{C}$ satisfies the UBME condition with respect to $\mathcal{P}$. Then, for each $\epsilon > 0$,*

$$N(\epsilon, \mathcal{C}, \bar{d}_{\mathcal{P}}) = N(\epsilon, \mathcal{C}, \bar{d}_{\mathbf{C}(\mathcal{P})}).$$

Note that Theorem 5.3 is *false* in general if $\mathcal{P}$ is not totally bounded. In fact, the counterexample of [8] is also a counterexample to Theorem 5.3 in case $\mathcal{P}$ is not totally bounded. To avoid disrupting the flow of the paper, the counterexample is given in an appendix.

We end this section with a brief discussion of a specific example of a class of concepts and family of distributions. Let $\mathcal{C}$ be the set of all convex subsets of some bounded domain in $\mathbf{R}^n$. It is easy to show that this class of concepts has infinite VC-dimension (e.g., any finite set of points on the surface of a sphere can be shattered) and so is not distribution-free learnable. However, if P_0 is the uniform distribution on the bounded domain, then Dudley [7] has shown that $N(\epsilon, \mathcal{C}, P_0)$ is finite for all $\epsilon > 0$ and in fact he provides explicit bounds on $N(\epsilon, \mathcal{C}, P_0)$. Hence, $\mathcal{C}$ is learnable with respect to the uniform distribution. From Theorems 5.1 and 5.2 it is easy to see that for any $b < \infty$, $\mathcal{C}$ is also learnable with respect to the family of all distributions that have a density bounded by b. Actually, this result is also a special case of Theorem 6.2 in the next section.

6. Two Sufficient Conditions

Thus far the emphasis has been on finding families of probability measures for which the UBME condition is both sufficient and necessary for learnability. In this section, two *sufficient* conditions for learnability are presented.

The first result shows that learnability of a class of concepts is retained under finite unions of families of distributions.

Theorem 6.1. *Let C be a class of concepts, and let $\mathcal{P}_1, \ldots, \mathcal{P}_n$ be n families of probability measures. If C is learnable with respect to $\mathcal{P}_i$ for $i = 1, \ldots, n$ then C is learnable with respect to $\cup_{i=1}^n \mathcal{P}_i$.*

Proof. Let f_i be an algorithm that learns C with respect to $\mathcal{P}_i$, and let $m_i(\epsilon, \delta)$ be the number of samples required by f_i to learn with accuracy ϵ and confidence δ. Define an algorithm f as follows. Choose

$$m(\epsilon, \delta) = \max_{1 \leq i \leq n} m_i(\frac{\epsilon}{2}, \frac{\delta}{2}) + \frac{32}{\epsilon} \ln \frac{n}{\delta/2}$$

samples. Using the first $\max_i m_i(\epsilon/2, \delta/2)$ samples, form hypotheses $H_1, \ldots, H_n$ using algorithms $f_1, \ldots, f_n$ respectively. Then, using the last $(32/\epsilon) \ln \frac{n}{\delta/2}$ samples, let f output the hypothesis H_i that is inconsistent with the fewest number of this second group of samples, and call it H. It is claimed that f is a learning algorithm for C with respect to $\cup_{i=1}^n \mathcal{P}_i$.

Let T denote the target concept, and suppose $P \in \cup_{i=1}^n \mathcal{P}_i$. Then $P \in \mathcal{P}_k$ for some k. Since the f_i are learning algorithms with respect to the $\mathcal{P}_i$, at least one H_i satisfies $d_P(H_i, T) \leq \epsilon/2$ with probability (with respect to product measures of P) greater than $1 - \delta/2$. Now the arguments from [17, 1] used in establishing Theorem 3.2 show that the "minimum risk" algorithm above will return a hypothesis H that satisfies $d_P(H, T) \leq \epsilon$ with a probability of at least $1 - \delta/2$. Thus the probability that $d_P(H, T) \leq \epsilon$ is at least $1 - \delta$.

Note that the above result is not true in general for an infinite number of families of distributions since the sample complexity of the corresponding algorithms may be unbounded (i.e. we may have $\sup_i N(\epsilon, C, \mathcal{P}_i) = \infty$). However, even if $N(\epsilon, C, \mathcal{P}_i)$ is uniformly bounded the result is not true. In fact, this is the case for the counterexample provided in [8].

The second set of results pertains to learnability with respect to a family consisting of probability measures that are "commensurate" with respect to a given nominal distribution. In Section 4., the uncertainty regarding the probability measure P under which learning is to take place was modelled by representing $\mathcal{P}$ as a sphere centered at a nominal measure P_0. However, there was no restriction about the nature of the "perturbation" about P_0. In particular, even if P_0 is a nonatomic measure, a sphere centered about P_0 may contain measures with an atomic part. As shown in Theorem 4.1, learning to arbitrarily small accuracy under such a family of distributions is equivalent (in terms of feasibility) to distribution-free learning. The notion of "commensurate" measures is intended to permit the family $\mathcal{P}$ to contain a large variety of distributions, while at the same time excluding "malicious" choices of probability measures. To motivate the definition, recall the proof of the fact that if a class $\mathcal{C}$ does not have finite VC-dimension, then it is not learnable if $\mathcal{P}$ is the class of *all* probability measures on $\mathcal{S}$ (e.g., see [2]). Given an arbitrarily large integer m, one selects a set of cardinality $2m$ that is shattered by $\mathcal{C}$ (call it Γ), and chooses P to be a purely atomic measure concentrated on Γ. With this choice of P, the inequality (2.2) is shown to be violated for suitable ϵ, δ. Choosing P to be purely atomic, and choosing the support set of P to be a set that will cause difficulties, may be thought of as "malicious." The notion of a commensurate family is intended specifically to exclude such extreme choices of P.

Let P_0 be a given fixed probability measure, referred to as the nominal distribution, and suppose $b \geq 1$. We define $\mathcal{M}(b, P_0)$ to be the set of all probability measures P on $\mathcal{S}$ such that

$$P(S) \leq b\, P_0(S),\ \forall\, S \in \mathcal{S}.$$

Thus $\mathcal{M}(b, P_0)$ consists of all distributions that are absolutely continuous with respect to the nominal distribution P_0, whose Radon-Nikodym derivatives are (essentially) bounded by b. The family $\mathcal{M}(b, P_0)$ is said to be **commensurate** with P_0. A distribution P in $\mathcal{M}(b, P_0)$ is "nonmalicious" in the sense that it will never assign a positive measure to a set that has zero measure with respect to P_0. For example, if X is a subset of $\Re^l$ for some integer l and P_0 is the Lebesgue measure on X (suitably normalized such that $P_0(X) = 1$), then every purely atomic measure is "malicious" and does not belong to $\mathcal{M}(b, P_0)$ for any b; we may say therefore that such a measure is "noncommensurate" with P_0. Similarly, the family of distributions studied in [8] to show that (3.1) is not sufficient for learnability is also *not* of the form $\mathcal{M}(b, P_0)$ for any b and P_0. The reason is that, given any number $\epsilon > 0$, one can find measures P_a and P_b in this family and sets A and B such that

$$\frac{P_a(A)}{P_b(A)} > \epsilon,\ \frac{P_b(B)}{P_a(B)} > \epsilon.$$

Theorem 6.2. *Suppose $\mathcal{C}$ is a class of concepts, P_0 is a probability measure, and $b \geq 1$. Then the following statements are equivalent:*

1. *The class $\mathcal{C}$ is learnable with respect to $\mathcal{M}(b, P_0)$.*
2. *The class $\mathcal{C}$ is learnable with respect to P_0.*
3. *The class $\mathcal{C}$ has a finite ϵ-cover with respect to the pseudometric d_{P_0} on $\mathcal{S}$ induced by P_0, for all $\epsilon > 0$.*

Proof. "(1)⇒(2)" Obvious, because $P_0 \in \mathcal{M}(b, P_0)$.

"(2)⇒(3)" This follows from Theorem 2.

"(3)⇒(1)" Define $\bar{d}_{\mathcal{M}} := \bar{d}_{\mathcal{M}(b,P_0)}$ as in (3.2). Then it is easy to see that

$$\bar{d}_{\mathcal{M}}(A,B) \leq b\, d_{P_0}(A,B),\ \forall\, A,B \in \mathcal{S}.$$

Hence an ϵ/b-cover of $\mathcal{C}$ with respect to d_{P_0} is also an ϵ-cover of $\mathcal{C}$ with respect to $\bar{d}_{\mathcal{M}}$. Now Statement 1 follows from Corollary 3.1.

Note that, in general, the family $\mathcal{M}(b, P_0)$ is not totally bounded (though it can be, for specific choices of P_0). Also, $\mathcal{M}(b, P_0)$ has an empty interior unless the measurable space $(X, \mathcal{S})$ is rather trivial. Thus, in spite of its simplicity, Theorem 6.2 cannot be derived as a consequence of earlier results.

By combining Theorem 6.2 and the method of proof used in Section 5., it is possible to prove the following result; the proof is omitted as it is simple.

Theorem 6.3. *Suppose $P_1, \ldots, P_l$ are probability measures on $\mathcal{S}$, and $b_1, \ldots, b_l$ are constants such that $b_i \geq 1$ for all i. Suppose $\mathcal{C} \subseteq \mathcal{S}$ is a class of concepts. Then $\mathcal{C}$ is learnable with respect to the family*

$$\mathcal{P} = \bar{\mathbf{C}}(\cup_{i=1}^{l} \mathcal{M}(b_i, P_i)) \tag{6.1}$$

if and only if $\mathcal{C}$ is learnable with respect to each P_i, for $i = 1, \ldots, l$.

7. Conclusions

In the present paper, the PAC learning problem formulation has been introduced, and has been interpreted as "generalization with high accuracy and confidence" in the context of neural networks. Some known results on PAC learning have been summarized, both for learning with respect to a fixed distribution and for learning with respect to the family of all possible probability measures. Then some new results have been stated and proven for learning under intermediate families of probability measures. A metric is defined on the set of all probability measures on a given measurable space, whereby the distance between two probability measures is defined as the maximum variation between the two. The uniform boundedness of the metric entropy of the class of concepts, which is known to be a *necessary* condition for learnability always, has been shown to be *sufficient* as well, under each of two conditions: (i) the family of probability measures is *totally bounded*, or else (ii) the family of probability measures contains an *interior point*. These two results establish that, in any counterexample along the lines of [8] (showing that the uniform boundedness of the metric entropy is *not* sufficient for learnability), the family of probability measures cannot be totally bounded, and yet at the same time must have an empty interior. Second, two sufficient conditions for learnability are presented. (i) It is shown that learnability with respect to each of a finite collection of families of probability measures implies learnability with respect to the *union* of these families. (ii) It is shown that learnability with respect to each of a finite number of probability measures implies learnability with respect to the convex hull of the families of all "commensurate" probability measures.

There are a number of interesting questions that are left open by the present paper. First, one can conjecture that the uniform boundedness of the metric entropy of the class of concepts is sufficient for learnability whenever the family of distributions under which learning is to take place is *closed and convex.* The counterexample given in [8] to the Benedek-Itai conjecture (namely, that uniform boundedness of the metric entropy is sufficient for learnability for an *arbitrary* family of probability measures) is not a counterexample to this more restricted conjecture, as shown in the Appendix. Second, one can ask whether the result of Lemma 5.1 holds for an *arbitrary* family of distributions. That is: is it true that, whenever a class $\mathcal{C}$ can be learned to an accuracy ϵ with respect to a given family of probability measures $\mathcal{P}$, it can also be learned to some degraded accuracy $\nu_1(\epsilon)$ for all P that lie within some distance $\nu_2(\epsilon)$ of $\mathcal{P}$? In case $\mathcal{P}$ is a singleton set, Lemma 5.1 states (roughly) that the above statement is indeed true with $\nu_1(\epsilon) = 2\epsilon$ and $\nu_2(\epsilon) = \epsilon/2$; but it would be interesting to determine whether an analogous statement is true for more general families $\mathcal{P}$. If such a statement were to be true, then it would provide a trade-off between the accuracy of learning, and the accuracy to which the family $\mathcal{P}$ (under which learning is to take place) needs to be known.

Appendix: A Counterexample

In this appendix, it is shown by example that Theorem 5.3 is *not* true in general if the family of distributions $\mathcal{P}$ is not totally bounded.

Our example is the same as in [8]. The set X equals $\{0,1\}^{\mathcal{N}}$, the set of all binary sequences indexed over the set of natural numbers $\mathcal{N}$ (beginning with 1). $\mathcal{S}$ equals the Borel σ-field over X. Define the sequence

$$p_i = \frac{1}{\lg(i+1)},$$

where lg denotes the logarithm to the base 2. A product measure P_I can be induced on X by identifying $p_i = P(x_i = 1)$. Let $\sigma : \mathcal{N} \to \mathcal{N}$ denote a permutation (possibly infinite) of the integers; thus σ is a one-to-one and onto map on $\mathcal{N}$. Let Σ denote the set of all such permutations. Let P_σ denote the probability measure on X defined by $P_\sigma(x_{\sigma(i)} = 1) = p_i$. Now let $\mathcal{P} = \{P_\sigma, \sigma \in \Sigma\}$. This specifies the family of probability measures. Next, let $C_i = \{x \in X : x_i = 1\}$, and define $\mathcal{C} = \{C_i, i \in N\}$. Since any C_i with $p_{\sigma^{-1}(i)} < \epsilon$ satisfies $d_{P_\sigma}(C_i, \emptyset) < \epsilon$, it is easy to see that the sets

$\{C_{\sigma(1)}, \ldots, C_{\sigma(n)}, \emptyset\}$ form an ϵ-cover for $\mathcal{C}$ with respect to the pseudometric d_{P_σ} provided $n \geq 2^{1/\epsilon}$. It follows therefore that the class $\mathcal{C}$ satisfies the UBME condition with respect to the family $\mathcal{P}$, and that

$$N(\epsilon, \mathcal{C}, \mathcal{P}) \leq 2^{1/\epsilon}.$$

It is shown in [8] that the class $\mathcal{C}$ is *not* learnable with respect to $\mathcal{P}$. Our objective here is somewhat different: It is shown here that $\mathcal{C}$ *does not* satisfy the UBME condition with respect to the convex hull of $\mathcal{P}$. In order to do this, a bit of terminology is introduced. Suppose S is a set and ρ is a pseudometric on S. A subset $M \subseteq S$ is said to be **ϵ-separated** with respect to the pseudometric ρ if the distance between every pair of nonidentical points in M equals or exceeds ϵ. It is clear that the cardinality of such a set M is a lower bound on the $\epsilon/2$-covering number of the set S.

Lemma 7.1. *Define*

$$\alpha = 1 - \frac{1}{\lg 3} \approx 0.36907, \quad d = 2^\alpha \approx 1.2915.$$

For each sufficiently small $\epsilon < \alpha$ and each integer n, there exists a probability measure $P \in \mathbf{C}(\mathcal{P})$ such that $\mathcal{C}$ contains a set of cardinality $nd^{1/\epsilon}$ that is ϵ-separated with respect to d_P. Therefore, for each sufficiently small $\epsilon < \alpha$,

$$\sup_{P \in \mathbf{C}(\mathcal{P})} N(\epsilon, \mathcal{C}, d_P) = \infty.$$

The proof of the lemma makes use of the following preliminary result.

Lemma 7.2. *For each sufficiently small $\delta > 0$ and each sufficiently large integer n, there exists another integer $M = 2^{c(\delta,n)/\delta}$, where $c(\delta, n) \to 1$ as $n \to \infty, \delta \to 0$, such that*

$$\frac{1}{n} \sum_{i=1}^{n} \frac{1}{\lg iM} \geq \delta.$$

Proof. **(Lemma 7.2)** Let $x = \lg M$. Then the above summation can be written as

$$\frac{1}{n} \sum_{i=1}^{n} \frac{1}{\lg iM} = \frac{1}{n} \sum_{i=1}^{n} \frac{1}{x + \lg i} = \frac{N(x)}{D(x)},$$

where $N(x)$ and $D(x)$ are polynomials in x. Specifically,

$$D(x) = n \prod_{i=1}^{n} (x + \lg i) = nx^n + n \left(\sum_{i=1}^{n} \lg i \right) x^{n-1} + \ldots + \left(\prod_{i=2}^{n} \lg i \right) x$$

after observing that $\lg 1 = 0$. Note that there is no constant term (x^0) in $D(x)$. Similarly,

$$N(x) = \sum_{i=1}^{n} \prod_{j \neq i} (x + \lg j) = nx^{n-1} + \left(\sum_{i=1}^{n} \sum_{j \neq i} \lg j \right) x^{n-2} + \ldots + \prod_{i=2}^{n} \lg i.$$

Now note that $\sum_{i=1}^{n} \lg i = \lg n!$. If we define

$$\beta_n = \sum_{i=1}^{n} \sum_{j \neq i} \lg j,$$

then $\beta_n < n \lg n!$, because $\sum_{j \neq i} \lg j < \lg n!$ for all $i > 1$. Now rewrite the desired inequality as

$$N(x) \geq \delta D(x).$$

Observe that $D(0) = 0$, while $N(0) > 0$. Hence the polynomial

$$\phi(x) := \delta D(x) - N(x)$$

satisfies $\phi(0) < 0$, and $\phi(x) \to \infty$ as $x \to \infty$ (because the degree of $D(x)$ is higher than that of $N(x)$). Let $r(\delta, n)$ denote the smallest positive root of the equation $\phi(x) = 0$. It is claimed that

$$r(\delta, n) \approx 1/\delta$$

for sufficiently large n and sufficiently small δ. To show this, we proceed by establishing that (i) there is a root of the form

$$x_0 = \frac{c(\delta, n)}{\delta}$$

where $c(\delta, n) \to 1$ as $\delta \to 0, n \to \infty$, and (ii) $\phi(x) < 0 \; \forall x < x_0$. To prove (i), substitute $x = c/\delta$ into $\phi(x)$. This gives

$$\phi(c/\delta) = n \frac{c^n}{\delta^{n-1}} + n \lg n! \frac{c^{n-1}}{\delta^{n-2}} + \ldots - n \frac{c^{n-1}}{\delta^{n-1}} - \beta_n \frac{c^{n-2}}{\delta^{n-2}} \ldots$$

For $\delta \to 0$, these are the dominant terms. Observe first that

$$\phi(1/\delta) = \frac{n \lg n! - \beta_n}{\delta^{n-2}} + \ldots > 0$$

because $\beta_n < n \lg n!$. So the root x_0 is less than $1/\delta$ as $\delta \to 0$. However, as $\delta \to 0$,

$$\phi(c/\delta) = \frac{1}{\delta^{n-1}} n(c^n - c^{n-1}) + \ldots$$

equals zero when $c \approx 1$. The same argument shows that if $c < 1$, then $c^n < c^{n-1}$, so that $\phi(c/\delta) < 0$ when $\delta \to 0, n \to \infty$. So the *smallest* positive root of the equation $\phi(x) = 0$ roughly equals $1/\delta$.

Proof. **(Lemma 7.1)** Observe that if $C_i, C_j \in \mathcal{C}$, then

$$C_i \Delta C_j = \{x \in X : x_i = 1 \text{ and } x_j = 0, \text{ or } x_i = 0 \text{ and } x_j = 1\}.$$

Therefore

$$P(C_i \Delta C_j) = \phi_i(1 - \phi_j) + (1 - \phi_i)\phi_j,$$

where

$$\phi_i = P(x_i = 1).$$

Let us use d_σ as an abbreviation for d_{P_σ}. If $P = P_I$, then

$$d_I(C_i, C_j) = p_i(1 - p_j) + (1 - p_i)p_j.$$

If $P = P_\sigma$, then

$$d_\sigma(C_i, C_j) = p_{\sigma(i)}(1 - p_{\sigma(j)}) + (1 - p_{\sigma(i)})p_{\sigma(j)}.$$

Given ϵ, n, choose $M \approx 2^{\alpha/\epsilon} = d^{1/\epsilon}$ such that

$$\frac{1}{n} \sum_{i=1}^{n} \frac{1}{\lg i(M+1)} \geq \frac{\epsilon}{\alpha}.$$

This is possible by Lemma 7.2. Now define a permutation σ on the natural numbers as follows:

$$\sigma(i) = i + M \text{ for } 1 \leq i \leq (n-1)M,$$

$$\sigma(i) = i - (n-1)M \text{ for } (n-1)M < i \leq nM,$$

$$\sigma(i) = i \text{ for } i > nM.$$

In other words, σ is a "block"-cyclic permutation, and $\sigma^n = I$. Now define

$$P = \frac{1}{n} \sum_{l=0}^{n-1} P_{\sigma^l} \in \mathbf{C}(\mathcal{P}),$$

where σ^0 is taken as the identity permutation. It is now shown that the set $\{C_1, \ldots, C_{nM}\}$ is ϵ-separated with respect to the metric d_P. For this purpose, let us compute $d_P(C_i, C_j)$. There are two cases to consider, namely: (i) i and j belong to the same "block" of length M, that is, $(k-1)M + 1 \leq i, j \leq kM$ for some integer k, and (ii) i and j belong to different "blocks."

Case (i) Suppose i, j belong to the same block. In this case, it is easy to see that $\sigma^l(i), \sigma^l(j)$ also belong to the same block for each l. Also, as l varies from 0 to $n-1$, $\sigma^l(i)$ and $\sigma^l(j)$ will visit each of the n blocks $\{1, \ldots, M\}, \ldots, \{(n-1)M + 1, \ldots, nM\}$ exactly once. So it may be assumed without loss of generality that $i, j \in \{1, \ldots, M\}$. In this case, we have

$$d_I(C_i, C_j) = p_i(1 - p_j) + (1 - p_i)p_j.$$

If $i = 1, j > 1$, then

$$d_I(C_i, C_j) = 1 - p_j = 1 - \frac{1}{\lg(j+1)} \geq 1 - \frac{1}{\lg(M+1)} \geq \frac{2\alpha}{\lg(M+1)}$$

whenever $M \geq 2^{2\alpha+1} = 2d^2 \approx 3.3360$. If $i, j > 1$, then

$$1 - p_i, 1 - p_j \geq 1 - p_2 = \alpha,$$

$$p_i, p_j \geq \frac{1}{\lg(M+1)},$$

$$d_I(C_i, C_j) \geq \frac{2\alpha}{\lg(M+1)}.$$

Next, $\sigma(i), \sigma(j) \in \{M+1, \ldots, 2M\}$. So

$$d_\sigma(C_i, C_j) = p_{\sigma(i)}(1 - p_{\sigma(j)}) + (1 - p_{\sigma(i)})p_{\sigma(j)} \geq \frac{2\alpha}{\lg(2M+1)}.$$

Similarly,

$$d_{\sigma^l}(C_i, C_j) \geq \frac{2\alpha}{\lg[(l+1)M+1]}, \; l = 0, 1, \ldots, n-1.$$

Hence

$$d_P(C_i, C_j) = \frac{1}{n} \sum_{l=0}^{n-1} d_{\sigma^l}(C_i, C_j) \geq \frac{1}{n} \sum_{i=1}^{n} \frac{2\alpha}{\lg(iM+1)} \geq 2\epsilon.$$

Case (ii) Suppose i, j belong to different blocks. By the same logic as in Case (i), it can be assumed without loss of generality that $i \in \{1, \ldots, M\}$ and $j \in \{M+1, \ldots, nM\}$. In this case

$$d_I(C_i, C_j) \geq p_i(1 - p_j) \geq p_M(1 - p_2) = \frac{\alpha}{\lg(M+1)},$$

because $p_i \geq p_M$ and $1 - p_j \geq 1 - p_2$. Similarly

$$d_{\sigma^l}(C_i, C_j) \geq \frac{\alpha}{\lg[(l+1)M+1]},$$

and as a consequence,

$$d_P(C_i, C_j) \geq \frac{1}{n} \sum_{i=1}^{n} \frac{\alpha}{\lg(iM+1)} \geq \epsilon.$$

This shows that the set $\{C_1, \ldots, C_{nM}\}$ is ϵ-separated with respect to the pseudometric d_P.

References

1. G.M. Benedek and A. Itai, "Learnability by fixed distributions," *Proc. First Workshop on Computational Learning Theory*, pp. 80-90, 1988.
2. Blumer A., A. Ehrenfeucht, D. Haussler, and M. Warmuth, "Learnability and the Vapnik-Chervonenkis dimension," *J. ACM*, 36(4), pp. 929-965, 1989.
3. Baum E., and D. Haussler, "What size net gives valid generalization?", *Neural Computation*, 1(1), pp. 151-160, 1989.
4. Castelli V., and T.M. Cover, "Classification rules in the unknown mixture parameter case: relative value of labeled and unlabeled samples," *Proc. 1994 IEEE Int. Symposium on Information Theory*, p. 111, 1994.
5. Duda R.O., and P.E. Hart, *Pattern Classification and Scene Analysis*, Wiley, 1973.
6. Dudley R.M., "Central limit theorems for empirical measures" *Ann. Probability* 6(6), pp. 899-929, 1978.
7. Dudley R.M., "Metric entropy of some classes of sets with differentiable boundaries," *J. Approximation Theory*, Vol. 10, No. 3, pp. 227-236, 1974.
8. Dudley R.M., S.R. Kulkarni, T.J. Richardson, and O. Zeitouni, "A metric entropy bound is not sufficient for learnability," *IEEE Trans. Information Theory*, Vol. 40, pp. 883-885, 1994.
9. Haussler D., "Decision theoretic generalizations of the PAC model for neural net and other learning applications," *Information and Computation*, vol. 100, pp. 78-150, 1992.
10. Kantorovich L.V., and G.P. Akilov, *Functional Analysis*, Second Edition, Pergamon Press, New York, 1982.
11. Kulkarni S.R., S.K. Mitter, and J.N. Tsitsiklis, "Active learning using arbitrary binary valued queries," *Machine Learning*, 11, pp. 23-35, 1993.
12. Sontag E.D., "Some Topics in Neural Networks and Control," *Siements Corporate Research Inc.*, Report No. LS93-02, 1993.
13. Pollard D., *Convergence of Stochastic Processes*, Springer-Verlag, 1984.
14. Valiant L.G., "A theory of the learnable," *Comm. ACM*, 27(11), pp. 1134-1142, 1984.
15. Vapnik V.N., and A.Ya. Chervonenkis, "On the uniform convergence of relative frequencies to their probabilities," *Theory of Prob. and its Appl.* 16(2), pp. 264-280, 1971.
16. Vapnik V.N., and A.Ya. Chervonenkis, "Necessary and sufficient conditions for the uniform convergence of means to their expectations," *Theory of Prob. and its Appl.*, Vol. 26, No. 3, pp. 532-553, 1981.
17. Vapnik V.N., *Estimation of Dependences Based on Empirical Data*, Springer-Verlag, 1982.

Just-in-Time Learning and Estimation*

George Cybenko

Thayer School of Engineering
Dartmouth College
Hanover, NH 03755 USA
E-mail: gvc@dartmouth.edu

1. Introduction

Machine learning, nonlinear regression and system identification share a very general common framework. A *sample set* of data $S = \{(x_i, y_i)\}$ for $1 \leq i \leq N$ is observed. Here $x_i \in \mathbf{R}^d$ and $y_i \in \mathbf{R}^k$ and the pairs (x, y) are drawn according to a joint distribution on $\mathbf{R}^{d+k}$, say $p(x, y)$. Our goal is to estimate the marginal distribution

$$f(x) = y = \int_{\mathbf{R}^k} yp(x, y)dy.$$

In simpler terms, y is determined by an unknown function of x but subject to noisy observations. Given such noisy observations of $f(x) = y$, we need to estimate f at points which are not in the sample set. That is, given a new operating point, x, we seek to estimate $f(x)$ using some $\hat{f}_S(x)$ which is derived from S and any prior knowledge of the application. Introductions to such problems from the perspectives of various approaches can be found in [13, 16] for example.

This problem, assuming that sufficient data is collected to model the underlying *state*, is the canonical challenge of nonlinear regression and machine learning. It has received considerable research attention in the past decade. In the literature dealing with such problems, two main opposing paradigms have emerged – *local* versus *global* methods. To be sure, the distinction between these paradigms is objective. Nonetheless, the extreme cases are easy to articulate.

- *GLOBAL MODELS* - Global modeling builds a single monolithic functional model of the dataset, S. This has traditionally been the approach taken in neural network modeling, ridge regression and other forms of nonlinear statistical regression. Given S, an algorithm produces a model of f after which the dataset is essentially discarded and only the model, $\hat{f}_S$, is kept.

* Partially supported by AFOSR contract F49620-93-1-0266, AFOSR/DARPA 89-0536, NSF MIP-89-11025 and The Hewlett-Packard Corporation

– *LOCAL MODELS* - The classical nearest neighbor method is the prototypical approach for local modeling. The dataset is always kept and the estimate for a new operating point is derived from a vote or other interpolation based on a neighborhood of the new point in the dataset. Not only is no global model of the dataset built, but no sophisticated model of the local neighborhood is typically built.

Table 1.1.

Property vs. Model	GLOBAL	LOCAL	JIT
Data Requirements	small	large	medium to large
Data Fate	discarded after model built	kept throughout	kept or edited down
Key Algorithms	large-scale optimization	large dataset searching	small-scale optimization large dataset searching
Compute/Memory Ratio	large	small	balanced
Examples	neural networks, splines, Fourier, regression	nearest neighbors, case-based reasoning, memory-based learning, radial basies	time-frequency analysis, JIT models
Appropriate Applications	physical systems with global laws	pattern analysis and cognition	general nonlinear systems
Theory	mature	middle aged	adolescent

Just-in-time models are a hybrid approach, leaning more in the direction of local modeling but using the power of global modeling in the local neighborhood. We emphasize at this point that the major applications of interest are those that involve very large datasets in very high dimensional spaces. Applications involving low dimensional spaces can be handled by a variety of methods and do not present a major challenge. In Sections 2 and 3, we will discuss the advantages and disadvantages of global and local modeling approaches respectively, from both theoretical and practical points of view. Our hope is that the hybrid, just-in-time modeling methodology will inherit the advantages and minimize the disadvantages of these two approaches. In Section 4 we outline some theoretical results justifying JIT models. Section 5 is a summary with ideas for future work.

2. Global Models

Global models are typically parametrically defined. That is, suppose that the model class consists of functions of the form

$$f(x, \Theta) = \sum_{i=1}^{m} g_i(x, \Theta).$$

This formulation allows for the parameters Θ to enter linearly according to $g_i(x, \Theta) = \theta_i h_i(x, \Theta)$ so that the dependence of f on some of the components of Θ can be linear and while on others it will be nonlinear. For instance, this is the situation for the following special cases:

Table 2.1.

Modeling Approach	$h_i(x, \Theta)$
Fourier	$\sin(x \cdot \theta + \phi)$
Feedforward neural networks	$\sigma(x \cdot \theta + \phi)$ $\sigma(t) = 1/(1 + e^{-t})$
Linear regression	$x \cdot \theta + \phi$
Projection Pursuit	$r(x \cdot \theta + \phi)$, r general scalar

Global models are characterized by the fact that the regressor functions h_i are not locally supported. That is, $h_i(x, \Theta) \neq 0$ for $x \in \Omega$, an unbounded set. In such a case, the value of $f(x, \Theta)$ for a fixed x will depend on values drawn from the data sample set which are not local to x. In general, such a property makes sense from the modeling point of view only if there is a compelling reason to believe that the application is driven by a physical-like law. Such a law would be succinctly expressed by a mathematical equation which would essentially determine the functional form of the relationship between the inputs, x, and the output values, y, over the whole space of applicable values.

This global modeling property is a great advantage when the model and its parameter values can have a simple interpretation. A linear relationship as computed by classical linear regression is attractive and simple when it works. Generally speaking, global models involve a relatively small number of parameter values because there is often a simple, succinct meaningful model behind it.

Global models are obtained from the sample data by an optimization process in most cases. The global model estimate of the true relationship, $f_S(\cdot, \Theta_0)$ is obtained by solving a minimization problem of the form:

$$\Theta_0 = \text{argmin}_{\Theta} \sum_j | \sum_{i=1}^{m} g_i(x_j, \Theta) - y_j|^2$$

where we have used least squares but other distance measures can of course be used.

This optimization problem is in general nonlinear and will have numerous local minima which makes its solution difficult. The convergence rates of optimization methods for these problems have been observed to be slow [2]. In most cases, efficient gradient calculation can be based on *automatic differentiation* ideas [1]. In the neural networks case, this is known as the backpropogation algorithm. This gradient evaluation technique makes the inner loop of an optimization method using derivative information efficient but does not allay the basic problems of slow convergence due to ill-conditioning of the Jacobian and Hessians underlying the problem. See [2] for more details. More specific negative results have also been obtained in the neural network case [11, 12, 6]. The upshot is that finding a local minima is difficult and finding the global minimum can be NP-complete in this case.

The benefit of performing a hard optimization is that the memory required to store a model is small once the parameters are computed. This approach has been undoubtedly influenced by the fact that for many years most computing systems imposed severe memory limitations on users. The data set may occupy several megabytes of memory but the model is measured in kilobytes or less quite often. Moreover, the evaluation of the function estimate $f_S(\cdot, \Theta)$ will in general require a short program that can be efficiently run.

Over time, the practical limitations imposed by available computer memory in previous hardware generations has become a mindset. Practitioners are often reluctant to use models that require large memories fearing problems with implementation when such problems no longer exist. On the other hand, some nontrivial form of data compression is *required* for learning or inference [7, 15]. Loosely speaking, the sample data *must* be compressed by the model in order to eliminate noise and uncertainty about the true function. Conversely, if a modeling approach yields a function approximation that can be represented by a much smaller memory than the original data and still interpolates the data, then some degree of extrapolation/generalization can be guaranteed.

To be more specific, global models generally result in a high degree of data compression in the process of going from a data set to a model. A precise relationship between sample set size and model class complexity can be captured by the so-called Vapnik-Chervonenkis (V-C) dimension of a model class. The V-C dimension of a function class describes the effective "degrees of freedom" in that model class. In most cases, the V-C dimension is closely related to the nontrivial parameters that comprise Θ. It is closely related to the notion of metric entropy as introduced by Kolmogorov for applications in data compression [20].

There are many results about V-C dimension and machine learning in the recent literature [4, 16]. The basic relationship between V-C dimension and modeling is a simple description of the relationship between data sample size (randomly sampled) and quality of the resulting model performance. These results are typically *distribution independent*, meaning they hold for all distributions and so are typically worst case. Much work has been devoted to computing the V-C dimension of various model classes. For instance, take the class of polynomials of bounded degree in a scalar x as our model class. If our data sample size matches the polynomial order, then it is classically known that this is a notoriously poor estimator of the true f because the interpolation is very sensitive to both sampling bias and noise in the data. So we need to have more data than there are polynomial coefficients. How much more? V-C dimension machine learning theory tells us in a distribution free way.

Important and fundamental refinements of the V-C theory for sigmoidal neural networks have recently been made by Barron and Jones [5, 11] wherein it is shown that naturally described function classes can be very well approximated by networks with a small number of ncdes (m in the summations above).

In summary, the advantages of global models are that they enjoy a mature mathematical and computational theory and that the resulting models are easy to interpret in the framework of the underlying application (as physical laws for example). The drawbacks are that the optimizations are difficult to perform and, with the data essentially replaced by the model, there are no good methods to update models should new data become available.

3. Local Models

Local models typically keep the whole data sample set intact, as opposed to global methods which discard the data after use. Given x, local methods extract a neighborhood, $\mathcal{N}_S(x)$, from S of points which are close to x in some natural metric. For example, the k-nearest neighbor method uses precisely the k points, $x_{i_j}, 1 \leq j \leq k$ which are nearest to x in the metric. Once those points are determined, an estimate

$$\hat{y}(x) = \sum_{j=1}^{k} \alpha_j y_{i_j}$$

is used. Here the α_j are weights that depend on the closeness of x_{i_j} to x and other factors perhaps, closer points being weighed more heavily. In the case of a discrete output variable, y, one can use majority voting. The nearest neighbor method is perhaps the cannonical local method.

Another important example of a local method is the radial basis function approach. Radial basis function methods use an approximation of the form

$$\hat{y}(x) = \sum_{i=1}^{N} \alpha_i r(||x - x_i||/\sigma_i)$$

where α_i and σ_i are determined from a datafitting optimization. That is, the choices of α_i and σ_i should yield good approximations to y_j if the dataset points x_j are substituted for x. The function r is positive and decreasing with $r(t) = e^{-t^2}$ (Gaussian) being very popular.

The radial basis function approach is strictly speaking *not* a local method but since the radial function r decreases so rapidly with respect to its argument, only points local to an x contribute significantly to the summation above. In fact, an implementation of this method would typically search the dataset for x_j sufficiently close to x so that $r(||x - x_j||/\sigma_j)$ is above a threshold. This implicitly constructs a neighborhood around x which is based on absolute distance to x as opposed to nearest neighbor methods which use distance relative to the k closest points.

There are two main difficulties with local models – searching and sample size. Both have to do with the inherent geometry of multidimensional space. Searching is a problem because there are no multidimensional data structures that allow nearest neighbors to be computed efficiently. The best known methods currently require a search that is exponential in the dimension of the space. Such methods effectively reduce to brute force, exhaustive searches for dimensions greater than $d = 20$ on databases involving up to several million samples. Recent research that explores approximate neighborhood searching is only beginning now [18, 3].

The other difficulty with local methods involves the density of sample data points. For general data distributions, the worst case demands that an exponential amount of data be collected for a given estimation accuracy. Such pessimistic estimates do not take into account the fact that for many machine learning problems, the data is highly clustered around a small number of operating points or discrete classes. In the next section, we present some analysis for JIT models that indicates small data samples may be acceptable in cases where the distribution of data has a special relationship with the function being estimated, f.

4. Just-In-Time Models

Global models require computing a single global model which in practice is extremely compute intensive and rarely yields good approximations on large problems. Local models have traditionally been avoided because of memory contraints and search time limitations. However, memory is no longer a major resource limitation because of the remarkable growth in memory technology, both solid state and magnetic. Just-in-time modeling is a compromise – the idea is to use all or most available data but to build models dynamically as the need arises. Whan a model is needed in the neighborhood of an operating point, JIT modeling proposes to retrieve a subset of the data closest to the operating point desired and performs a modeling operation on that subset of data. Recent interest in such approaches has been growing [19, 9, 8]. In the following section, we discuss some new analysis that shows learning and estimation using local data is feasible for problems in which there is a close interplay between the data distribution and the function being estimated.

4.1 Analysis of Just-In-Time Models

We explore a new approach to analysing JIT modeling methods in terms of the interaction between the underlying probability distribution and the target function. Let $D \subset \mathbf{R}^d$ be the support of a probability distribution μ. If μ is continuous with respect to Lesbegue measure then $d\mu(x) = g(x)dx$ for $x \in D$ and $g(x) > 0,\ x \in D$.

A basic measure of the variation of a target function, f, with respect to μ is

$$\int |\nabla f(x)| g(x) dx = \int |\nabla f(x)| d\mu(x)$$

Later we also uses the slightly modified measure

$$V(f, g) = \int |\nabla f(x)| g(x)^{\frac{n-1}{n}} dx$$

when μ is continuous with respect to Lesbegue measure.

For $\rho > 0$, let $B(x, \rho)$ be the ball centered at x of sufficient radius, ϵ, so that

$$\int_{B(x,\rho)} d\mu(x) = \rho.$$

Note that when g exists as above and is continuous, asymptotically ϵ is related to ρ via the relationship

$$g(x) C_d \epsilon^d \approx \rho$$

where $C_d = \pi^{d/2}/\Gamma(\frac{n}{2}+1)$ is the volume of the ball of radius 1 in $\mathbf{R}^d$. Then $\epsilon \approx \rho^{\frac{1}{n}} g(x)^{\frac{-1}{n}} C_d^{\frac{-1}{n}} \approx \rho^{\frac{1}{n}} g(x)^{\frac{-1}{n}} \pi^{\frac{-1}{2}} (n/2e)^{\frac{1}{2}}$ by Stirling's formula.

Introduce the average variation in f over balls of probability ρ with respect to μ as

$$V(f, \mu, \rho) = \int_D \frac{1}{\rho} \int_{B(x,\rho)} |f(x) - f(y)| d\mu(y) d\mu(x).$$

Compare this with *uniformly Lipshitz on average* functions introduced by Haussler [10]. Also define

$$\begin{aligned} W(f, \mu, \rho) &= \int_D \frac{1}{\rho} \int_{B(x,\rho)} |f(y) \\ &- \frac{1}{\rho} \int_{B(x,\rho)} f(z) d\mu(z)|^2 d\mu(y) d\mu(x). \end{aligned}$$

as the variance of f over balls of probability ρ averaged over D. For smooth f we know that $V(f, \mu, \rho) \to 0$ and $W(f, \mu, \rho) \to 0$ as $\rho \to 0$ (by dominated convergence for example).

To get a feeling for these measures of variation, it is useful to apply them to the previously mentioned extreme cases than can arise. In the case of constant f, the measures are 0 for all ρ. In the case of an arbitrary f but with a distribution that is concentrated at a finite number of point masses, the measures are 0 when ρ is smaller than the smallest point mass weight.

Theorem – Let $\alpha, \delta, k > 0$. Pick ρ so that $V(f, \mu, \rho) < \alpha\delta/8$ and $W(f, \mu, \rho) < \sqrt{k}\alpha\delta^{3/2}/16$. Then for a sample of size N for which $N\rho - 2\sqrt{(N/\delta)}\sqrt{\rho(1-\rho)} > k$ we will have

$$|f(x) - \frac{1}{k}\sum_{j=1}^{k} f(x_j)| < \alpha$$

with probability at least $1-\delta$. Here x_j are the k nearest neighbors of x from the sample of size N.

Outline of Proof – The basic idea is to break the problem down into four events, each one of whose probability can be made arbitrarily close to 1. Three of the events have to do with the local variations in f and ultimately measure the rate at which a Monte Carlo quadrature method should work for estimating f locally. The fourth event arises from purely sampling considerations, namely, how many samples are needed to guarantee enough local values on which to base a Monte Carlo estimate with high enough probability. The basic tool used is a Tchebyshev type inequality which arises repeatedly.

Proof – By the above definitions, we have

$$\int_D |f(x) - \frac{1}{\rho}\int_{B(x,\rho)} f(y)d\mu(y)|d\mu(x) \le V(f, \mu, \rho).$$

Now,

$$\begin{aligned}\text{Prob}\{\ x\ \ &\text{such that } |f(x) \\ &- \frac{1}{\rho}\int_{B(x,\rho)} f(y)d\mu(y)| \ge \alpha/2\} \\ &\le 2V(f, \mu, \rho)/\alpha \\ &\le 2\rho^{\frac{1}{n}}\pi^{\frac{-1}{2}}(n/2e)^{\frac{1}{2}}V(f, g)/\alpha\end{aligned}$$

so that

$$\begin{aligned}\text{Prob}\{ \quad x & \quad \text{such that } |f(x) \\ & - \frac{1}{\rho}\int_{B(x,\rho)} f(y)d\mu(y)| \leq \alpha/2\} \\ & \geq 1 - 2\rho^{\frac{1}{n}}\pi^{\frac{-1}{2}}(n/2e)^{\frac{1}{2}}V(f,g)/\alpha \\ & \geq 1 - 2V(f,\mu,\rho)/\alpha \geq 1 - \delta/4\end{aligned}$$

by the choice of ρ as stated in the theorem. Similarly,

$$\begin{aligned}\text{Prob}\{ \quad x & \quad \text{such that } \frac{1}{\rho}\int_{B(x,\rho)} |f(y) \\ & - \frac{1}{\rho}\int_{B(x,\rho)} f(z)d\mu(z)|^2 d\mu(y) \leq \sqrt{\delta k}\alpha/4\} \\ & \geq 1 - 4W(f,\mu,\rho)/(\alpha\sqrt{\delta k}) \\ & \geq 1 - \delta/4\end{aligned}$$

by the choice of ρ again.

Thus the set of x for which both

$$\frac{1}{\rho}\int_{B(x,\rho)} |f(y) - \frac{1}{\rho}\int_{B(x,\rho)} f(z)d\mu(z)|^2 d\mu(y) \leq \alpha\sqrt{\delta k}/4$$

and

$$|f(x) - \frac{1}{\rho}\int_{B(x,\rho)} f(y)d\mu(y)| \leq \alpha/2$$

has probability at least $1 - \delta/2$.

For a sample of size N where

$$N\rho - 2\sqrt{(N/\delta)}\sqrt{\rho(1-\rho)} > k,$$

the number of samples in the ball $B(x,\rho)$ is at least k with probability at least $1 - \delta/4$. To see this, we use Tchebyshev's inequality. Let $X_i = 1$ if the ith sample among the N drawn is in $B(x,\rho)$ and $X_i = 0$ otherwise. Then the sequence X_i is Bernoulli with probabilities ρ and $1-\rho$ of being 1 and 0 respectively. We have

$$\text{Prob}\{|\frac{1}{N}\sum X_i - \rho| < S\sigma'\} \geq 1 - \frac{1}{S^2} \geq 1 - \delta/4$$

for $S = 2/\sqrt{\delta}$ where $\sigma' = \sqrt{\rho(1-\rho)/N}$ is the variance of $\frac{1}{N}\sum X_i$. Thus, with probability at least $1 - \delta/4$, we have

$$\sum_i X_i \geq N(\rho - S\sigma') = N(\rho - 2\sqrt{\rho(1-\rho)}/\sqrt{N\delta} > k.$$

These k samples, say $x_j, j = 1, ..., k$ can be used for a Monte Carlo estimate of $\int_{B(x,\rho)} f(y)d\mu(y)$ according to

$$\int_{B(x,\rho)} f(y)d\mu(y) \approx \frac{1}{k}\sum_j f(x_j)$$

which has variance

$$\begin{aligned}\sigma_x &= \frac{1}{\rho\sqrt{k}}\int_{B(x,\rho)} |f(y) - \frac{1}{\rho}\int_{B(x,\rho)} f(z)d\mu(z)|^2 d\mu(y)\\ &\leq \sqrt{\delta}\alpha/4\end{aligned}$$

when x is in the previously specified set.

By Tchebyshev's inequality again,

$$\begin{aligned}\text{Prob}\{|\frac{1}{k}\sum_j f(x_j) &- \frac{1}{\rho}\int_{B(x\rho)} f(y)d\mu(y)| < R\sigma\}\\ &\geq 1 - 1/R^2\end{aligned}$$

With $R = 2/\sqrt{\delta}$ and $\sigma \leq \sqrt{\delta}\alpha/4$, we have

$$\begin{aligned}\text{Prob}\{|\frac{1}{k'}\sum_j f(x_j) &- \frac{1}{\rho}\int_{B(x\rho)} f(y)d\mu(y)| < \alpha/2\}\\ &\geq 1 - \delta/4.\end{aligned}$$

Combining all of the above, we have with probability at least $1 - \delta$, that both

$$|\frac{1}{k}\sum_j f(x_j) - \frac{1}{\rho}\int_{B(x\rho)} f(y)d\mu(y)| < \alpha/2$$

and

$$|f(x) - \frac{1}{\rho}\int_{B(x,\rho)} f(y)d\mu(y)| \leq \alpha/2$$

from which

$$|\frac{1}{k}\sum_j f(x_j) - f(x)| < \alpha$$

follows by the triangle inequality. □

4.2 Discussion

The main result of the previous section does not, nor cannot, defeat the curse of dimensionality in all cases. To get a sense of this note that

$$
\begin{aligned}
V(f,\mu,\rho) &\leq \int_D \frac{1}{\rho}\int_{B(x,\rho)} |\nabla f(x)|\cdot|x-y|d\mu(y)d\mu(x) \\
&\leq \int_D |\nabla f(x)|\frac{1}{\rho}\int_{B(x,\rho)} \epsilon d\mu(y)d\mu(x) \\
&\leq \int_D |\nabla f(x)|\rho^{\frac{1}{n}} g(x)^{\frac{-1}{n}} \pi^{\frac{-1}{2}} (n/2e)^{\frac{1}{2}} d\mu(x) \\
&\leq \rho^{\frac{1}{n}} \pi^{\frac{-1}{2}} (n/2e)^{\frac{1}{2}} \int_D |\nabla f(x)| g(x)^{\frac{n-1}{n}} d(x) \\
&= \rho^{\frac{1}{n}} \pi^{\frac{-1}{2}} (n/2e)^{\frac{1}{2}} V(f,g)
\end{aligned}
$$

to the first order in $\rho^{1/n}$. The same bound can be derived for $W(f,\mu,\rho)$ (this is left to the reader). This suggests that the convergence of $V(f,\mu,\rho)$ to zero is going to be slow in most cases. It is governed by both $\rho^{1/d}$ and $V(f,g)$ when g exists. Now $\rho^{1/d}$ approaches 0 very slowly for large d but $V(f,g)$ can be small for a problem and herein lies at least one explanation for the good observed performance of many memory-based learning methods and by extension JIT methods.

As previously noted, this analysis can deal with both extreme cases: that of a trivial function and uniform probability distribution; and that of a complex function with a simple point mass distribution. We know of no other analysis demonstrating that both cases are "learnable."

It would be interesting to see whether the proof technique we use can be extended to other memory-based methods. We suspect that it can and this should form the basis for further work. The question of efficiently estimating $V(f,\mu,\rho)$, $W(f,\mu,\rho)$ and $V(f,g)$ in a specific case is interesting of course and should be attempted for some learning problems where memory-based methods are both successful and a failure.

The Monte Carlo interpretation of memory-based methods suggests that approximate nearest neighbor searches should be acceptable for some problems but with improved efficiency. This has been observed by Saarinen [18].

References

1. "Neural networks, backpropagation and automatic differentiation", in *Automatic Differentiation*, edited by A. Greiwank, SIAM Publications, 1991.
2. "The conditioning of neural network training problems", *SIAM J. on Scient. and Stat. Computing*, 13, 1992.
3. "An approximate k-nearest neighbor method", in *Adaptive and Learning Systems II,* Firooz A. Sadjadi, Editor, Proc. SPIE 1962, 1993.
4. Proceedings of Annual Workshops on Computational Learning Theory.
5. Barron A.R., "Universal approximation bounds for superpositions of a sigmoidal function", *IEEE Transactions on Information Theory*, 39:930–946, 1993.
6. Blum A., and R. Rivest, "Training a 3-node neural network is NP-complete" in *Proceedings First Workshop on Computational Learning Theory MIT*, pages 8–14, 1988.
7. Blumer A., A. Ehrenfeucht, D. Haussler, and M.K. Warmuth. "Classifying learnable geometric concepts with the Vapnik-Chervonenkis dimension", in *Proceedings 18th ACM Symposium on Theory of Computation*, pages 273–282, 1986.
8. Cybenko G. et al. "on the unreasonable effectiveness of memory-based learning", in *Proceedings of IEEE CMP'94*, pages 1–8, Prague, CZ, 1994.
9. Friedman J., "Flexible metric nearest neighbor classification", technical report, Department of Statistics, Stanford University, Stanford, CA, 1994.
10. Haussler D., "Generalizing the PAC model for neural net and other applications", technical report UCSC-CRL-89-30, Computer Research Laboratory, UC-Santa Cruz, 1989.
11. Jones L. K., "Constructive approximations for neural networks by sigmoidal functions", preprint, 1988.
12. Judd J.S., "The intractibility of learning in connectionist networks", technical report, University of Massachusetts, March 1988.
13. Olshen R.A., L. Breiman, J.H. Friedman, and C.J. Stone, *Classification and Regression Trees*, Wadsworth International Group, Belmont, CA, 1984.
14. Lee Y., "Handwritten digit recognition using k-nearest neighbor, radial-basis function, and backpropagation neural networks" *Neural Computation*, 3:440–449, 1991.
15. Pitt L., M. Kearns, M. Li, and L. Valiant, "Recent results on boolean concept learning", *Proc. 4th Int. Workshop on Machine Learning*, pages 337–352, 1987.
16. Natarajan B.K., *Machine Learning*, Morgan Kaufmann, San Mateo, CA, 1991.
17. Ripley B.D., "Statistical aspects of neural networks", technical report, Oxford University, Department of Statistics, 1992.
18. Saarinen S., Ph.D. Thesis, Department of Computer Science, University of Illinois at Urbana, 1994.
19. Hastie T., and R. Tibshirani, "Discriminant adaptive nearest neighbor classification" technical report, Department of Statistics, Stanford University, Stanford, CA, 1994.
20. Vitushkin A.G., *Theory of the Transmission and Processing of Information*, Pergamon Press, New York, 1961.

Wavelets in Identification*

A. Benveniste[1], A. Juditsky[1], B. Delyon[1], Q. Zhang[1], and P-Y. Glorennec[2]

[1] IRISA-INRIA
Campus Universitaire de Beaulieu
35042 Rennes Cedex
France
E-mail: name@irisa.fr

[2] IRISA-INSA
Campus Universitaire de Beaulieu
35042 Rennes Cedex
France
E-mail: name@irisa.fr

1. Introduction, Motivations, Basic Problems

In his inspiring tutorial [36], L. Ljung noted the following:

> An engineer, who is faced with [characterizing, or predicting, the behaviour of his plant based on recorded data] has the following perspective:
> - How can I best use the information in the observed data to calculate a model of the system's properties?
> - How can I know if the model is any good, and how can I trust it for simulation and design purposes?
> - How shall I manipulate the input signals to obtain as much information as possible about the system?
> - What kind of software support is available for doing the tasks?

Later on in the same article, L. Ljung discusses the question of model nature and structure. By model nature, we have in mind the following classification:
- physical models,
- semi-physical models, also called "grey-box" models,
- black-box models.

This paper mainly concentrates on the last category, namely black-box models. And, within black-box models, we shall concentrate on the less popular ones in control community, namely those that are *nonlinear and nonparametric* in nature. Here "nonlinear" means that our model class will not be restricted to linear input-output maps. And "nonparametric" means that our models do have parameters, but in a quantity that is not a priori fixed, but fully depends on data; consequently, convergence issues and quality of fit cannot be assessed in terms of the involved parameters, but rather more globally in terms of the global behaviour. A typical form of the kind of model class

* This work has been supported in part by Alcatel-Alsthom-Recherche and European Gas Turbine SA under several contracts, C. de Maindreville, P. Durand, T. Pourchot, F. Costa, and D. Cavalerie are gratefully acknowledged.

that we shall consider is the popular single hidden layer neural network for static systems:

$$f_n(x) = \sum_{i=1}^{n} c_i \sigma(a_i^T x + t_i) + c_0 , \tag{1.1}$$

where σ is the sigmoid function, $x \in \mathbf{R}^d$ is the input, n is the number of neurons, and the (c_i, a_i, t_i)'s are the adjustable parameters. This is clearly nonlinear in x, and the size n of the network is to be tuned on the data. In addition, in this case, the model is also nonlinear in the parameters.

Such models have gained increasing interest, as reflected for instance in the articles [45, 27, 42]. This is due to their ability to encompass truly nonlinear behaviours, including those involved in classification and, more generally, decision procedures. Referring to Ljung's practical problem setting above, the following practical questions must be investigated when using nonlinear nonparametric models such as (1.1):

- *How good nonlinear nonparametric models can extrapolate or predict behaviours outside the range of data used for their identification, fitting, tuning, or training* [1] *?*
- *How nonlinear nonparametric models can be used for system monitoring and diagnostics ?*
- *How can one take advantage of any kind of prior knowledge for some partial or pre-tuning of the model ?*
- *What kind of software support is available for doing the tasks ?*

Moving one step further toward mathematical formulation of our problems, we may translate some of the above questions into the more technical following ones:

- *How to assess the quality of approximation ?* Given a true system f, and an approximation $\hat{f}$ of it, how to measure the quality of approximation ?
- *How to measure the quality of fit from noisy data ?*
- *What plays the role of "Cramer-Rao bounds", and what means for an estimator to be "optimal" ?*

These are some of the issues that we shall discuss throughout this tutorial.

1.1 Two Application Examples

Modelling a gas turbine system, an example of identification of a static nonlinear system. We briefly present the case study of a gas turbine system, as an example of identification of a static nonlinear system. Results and experiments will be reported in Section 10.. One of the purposes of our joint study

[1] These are more or less equivalent words used by different communities, we shall use anyone of these indifferently.

with European Gas Turbine SA, Belfort, and Alcatel-Alsthom-Recherche, Marcoussis, was to develop a monitoring and diagnostics system for the joint system {combustion chambers, turbine}. Monitoring is based on the measured pressure in the compressor, the rotation velocity of the turbine and measurements from the thermocouples available at the exhaust of the turbine. Thus no direct observation is available on the status of the combustion chambers, see Fig. 1.1. Hence a coarse semi-physical model has been developed that predicts the profile of temperature at the exhaust of the turbine using the pressure in the compressor, the mean temperature at the exhaust of the turbine, and the rotation velocity of the turbine, see [56, 60]. This model is static but strongly nonlinear. This semi-physical modelling was for the purpose of monitoring the turbine system. Despite its inaccurate nature, the model has been successfully used for developing a monitoring system of the combustion chambers, see [60]. Unfortunately, this model is not entirely satisfactory for some other purposes, such as the monitoring of the thermocouples installed at the exhaust of the turbine. The purpose of this discussion is to compare results from this semi-physical model with some alternative nonparametric identification method based on wavelets, and discuss the two questions of respective accuracy of fit and explicative power of these two styles of models.

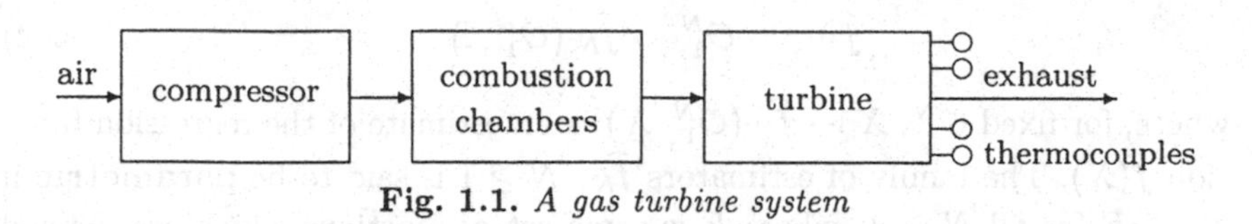

Fig. 1.1. *A gas turbine system*

Modelling the hydraulic actuator of a robot arm, an example of identification of a dynamic nonlinear system. Now let us consider the modelling of the actuator of a robot arm[2]. It is a hydraulically driven arm. By controlling the position of a valve, the oil pressure in the transmission circuit is regulated. The oil pressure drives the motion of the arm. What we want to model is the relationship between the position of the valve and the oil pressure, both quantities being measured. In fact the valve directly regulates the oil streams injected in the transmission circuit. Hence variation of the oil pressure depends not only on the position of the valve, but also on the quantity of the oil accumulated in the transmission circuit, which in turn is reflected by the oil pressure. Clearly this is a dynamic system: variation of its output (oil pressure) depends on both its input (the position of the valve) and its state (reflected by the oil pressure). We tried to model this dynamic system with linear ARX models, but the results were not satisfactory. Therefore,

[2] This application has been borrowed from Linköping University, while Q. Zhang was visitor at the Automatic Control Group.

we decided to apply some nonlinear nonparametric model and see if we can improve the performance of the modelling.

2. Basic Mathematical Problems

Here we establish the general framework of *nonparametric regression* we shall use throughout this article, and we justify the use of particular distance measures between true system and its estimate we shall deal with in the sequel.

Problem 2.1 (nonparametric regression). Let (X,Y) be a pair of random variables with values in $\mathcal{X} = \mathbf{R}^d$ and $\mathcal{Y} = \mathbf{R}$ respectively. A function $f : \mathcal{X} \mapsto \mathcal{Y}$ is said to be the **regression function of** Y **on** X if

$$\mathbf{E}(Y|X) = f(X) . \tag{2.1}$$

A typical case is $Y = f(X) + e$, where e is zero mean and independent of X. For $N \geq 1$, $\widehat{f}_N$ shall denote an estimator of f based on the random sample $\mathcal{O}_1^N = \{(X_1,Y_1),\ldots,(X_N,Y_N)\}$ of size N from the distribution of (X,Y), i.e., a map

$$\widehat{f}_N \quad : \quad \mathcal{O}_1^N \mapsto \widehat{f}_N\left(\mathcal{O}_1^N,.\right) \tag{2.2}$$

where, for fixed $\mathcal{O}_1^N$, $X \mapsto \widehat{f}_N\left(\mathcal{O}_1^N, X\right)$ is an estimate of the regression function $f(X)$. The family of estimators $\widehat{f}_N$, $N \geq 1$ is said to be **parametric** if $\widehat{f}_N \in F$ for all $N \geq 1$, where F is some set of functions which are defined in terms of a *fixed* number of unknown parameters. Otherwise the family of estimators $\widehat{f}_N$, $N \geq 1$ is said to be **nonparametric.**

For the sake of convenience, we shall often refer to X and Y as the *input* and *output* respectively (although they do not need to be such in actual applications). Our objective in this section is to give a short overview of some basic instances of nonparametric regression.

Nonparametric Regression for Static Systems. The considered system has the form

$$Y_i = f(X_i) + e_i, \qquad i = 1,\ldots,N , \tag{2.3}$$

where $f(x) : \mathbf{R}^d \mapsto \mathbf{R}$, and, for the sake of simplicity, we assume that e_i are independent Gaussian random variables with $\mathbf{E}e_i = 0$ and $\mathbf{E}e_i^2 = \sigma_e^2$.

Adaptive classification and density estimation. The problem of classification (discriminant analysis, or statistical pattern recognition) is usually formulated as follows. Let X be a random variable with values in $\mathbf{R}^d$, and let the label Z denote a random variable which takes values in some finite set $\mathcal{Z} = \{z_1, \ldots, z_M\}$; the symbol z shall denote a generic element of this finite set. We want to guess the value of Z when X is observed. We consider the case in which the random vector X has probability density $f(x)$ and conditional density $f(x|z)$ given that $Z = z$, the general case is handled similarly. Bayesian or Maximum A Posteriori (MAP) classifiers minimize the error probability, and are thus optimal. In *adaptive classification,* density $f(x)$ and conditional density $f(x|z)$ are unknown. Efficient adaptive classifiers are obtained by 1/ estimating the above densities, and 2/ applying MAP classification using these estimates; i.e., we apply the certainty equivalence principle. Setting $\mathbf{p}(z) = \mathbf{P}(Z = z)$, Devroye and Györfi [17] have shown that the loss in error probability due to performing *adaptive* classification is bounded from above by the quantity $\sum_{z \in \mathcal{Z}} \int |\mathbf{p}(z)f(x|z) - \widehat{\mathbf{p}}(z)\widehat{f}(x|z)|\, dx$, where $\widehat{\mathbf{p}}, \widehat{f}$ denote estimates. Thus we have related the problem of adaptive classification to that of estimating the density of a random variable in L_1-norm.

Nonparametric Regression with Dynamics. Consider the following dynamical system:

$$Y_i = f(\Phi_i) + e_i, \qquad i = 1, \ldots, N ,$$

where $Y_i \in \mathbf{R}$ and $\Phi_i \in \mathbf{R}^d$ are observed, and e_i is a white noise as above. We assume that

$$\Phi_i = (Y_{i-1}, \ldots, Y_{i-m}; U_i, \ldots, U_{i-p}) , \tag{2.4}$$

where $U_i \in \mathbf{R}$ denote the inputs ($m + p = d$). For example, if $\Phi_i = (Y_{i-1}, \ldots, Y_{i-d})$, then

$$Y_i = f(Y_{i-1}, \ldots, Y_{i-d}) + e_i . \tag{2.5}$$

In analogy with the corresponding parametric model we call this system a *nonparametric autoregression* or a *functional autoregression* of dimension d (FAR(d)). As an interesting application, we can consider a simple controlled FAR model for adaptive control:

$$Y_i = f(\Phi_i) + U_i + e_i , \tag{2.6}$$

where $\Phi_i = (Y_{i-1}, \ldots, Y_{i-m})$, and U_i is the control. The following question can be considered: how to choose the control (U_i) for the system (2.6) to track some reference trajectory $y = (y_i)$, or, at least, how to choose U_i in order to minimize $\mathbf{E}Y_i^2$, or, simply, to stabilize the system (2.6)? If the function $f(\Phi)$ was known, we could use control

$$U_i = -f(\Phi_i)$$

to obtain $Y_i = e_i$. Clearly, this is a "minimum variance" control, since $\mathbf{E}Y_i^2 \geq \sigma_e^2 = \mathbf{E}e_i^2$. If f is unknown, a possible solution consists in performing nonparametric "certainty equivalence control": compute an estimate $\widehat{f}_N$ of the regression function f based on the observations of the input/output pair $(\Phi_i, Y_i - U_i)$, and then take

$$U_i = -\widehat{f}_i(\Phi_i) \ . \tag{2.7}$$

To analyse the certainty equivalence control (2.7), let us consider the control cost

$$Q_N = \frac{1}{t}\sum_{i=1}^{N} Y_i^2 = \frac{1}{t}\sum_{i=1}^{N}\left[(f(\Phi_i) - \widehat{f}_i(\Phi_i))^2 + e_i^2\right]$$

It is easily checked that

$$\mathbf{E}(\widehat{f}_i(\Phi_i) - f(\Phi_i))^2 \to 0 \text{ when } i \to \infty \tag{2.8}$$

implies $\mathbf{E}Q_N \to \sigma_e^2$, and $\widehat{f}_i(\Phi_i) - f(\Phi_i) \to 0$ a.e. implies $Q_N \to \sigma_e^2$ a.e. Thus condition (2.8) is instrumental in analysing this problem. It can be checked that this condition can be ensured provided that bounds are obtained for estimation error norms of the form $\mathbf{E}\|\widehat{f}_{i-1} - f\|_\infty^2$ or $C_p\|\widehat{f}_{i-1} - f\|_2^2$. Thus, as a conclusion, in any case, the crux in analysing this adaptive minimum variance nonlinear control consists in getting bounds for the error in estimating the unknown function f.

Discussion. This section about basic mathematical issues can be summarized as follows:

1. Nonparametric estimation of regression functions is instrumental in various problems such as adaptive identification, classification, and control.
2. Averaged L_p-norms of estimation error for various p's are natural candidates as a figure of merit. We shall see later that error measures involving also derivatives of f and $\widehat{f}$ will be useful, so that smoothness of estimates can also be guaranteed.
3. Having bounds for the estimation error is of paramount importance. This has been illustrated on the adaptive control example.

3. Classical Methods of Nonlinear System Identification: Linear Nonparametric Estimators

Throughout this section, Problem 2.1 is considered. We first discuss some estimators that are *linear*, i.e., that satisfy $\widehat{f+g} = \widehat{f} + \widehat{g}$; note that functions f, g, and their estimates, are generally nonlinear as functions of their input

x. For the sake of clarity, *and unless otherwise stated,* algorithms will be presented for the one-dimensional case only, i.e., in the framework of Problem 2.1, the input space $\mathcal{X}$ is of dimension one. However convergence theorems will be given for the general case of higher dimension.

All estimators presented in this section are linear ones, i.e. they have a common general form

$$\begin{aligned} \widehat{f}_N(x) &= \sum_{i=1}^{N} Y_i \, W_{N,i}(x) \\ W_{N,i}(x) &= W_{N,i}(x, X_1, \ldots, X_N) \end{aligned} \tag{3.1}$$

where we recall that $\mathcal{O}_1^N = \{(X_1, Y_1), \ldots, (X_N, Y_N)\}$ is the given random sample observation, and the weights $W_{N,i}(x)$ only may differ.

3.1 Projection Estimates as an Example of Linear Nonparametric Estimators

Kernel estimators for regression functions, Parzen-Rosenblatt estimators for densities, piecewise polynomial and spline estimators for regression functions, are also discussed in the extended version [30] of this paper.

Projection estimates. Another class of function estimates was introduced by Cencov [9], who called them *projection estimates.* The idea consists in expanding the unknown function into its "empirical" Fourier series. We consider again the one-dimensional case. Consider the set $\mathcal{W}_2^s(L)$ of functions $f(x)$, $x \in [0,1]^d$, defined as follows. Each f can be represented by its Fourier series

$$f(x) = \sum_{j=1}^{\infty} c_j \varphi_j(x), \tag{3.2}$$

where $\varphi_1 \equiv 1$, $\varphi_{2k}(x) = \sqrt{2}\sin(2\pi kx)$ and $\varphi_{2k+1}(x) = \sqrt{2}\cos(2\pi kx)$, $k = 1, \ldots$ and the following condition is satisfied: $\sum_{j=1}^{\infty} |c_j|^2(1 + |j|^{2s}) < L^2$. Note that this is again a smoothness prior. We construct the estimate $\widehat{f}_N$ as follows:

$$\widehat{f}_N(x) = \sum_{j=1}^{m} \widehat{c}_j^N \varphi_j(x), \tag{3.3}$$

where m is the "model order", and the empirical estimates $\widehat{c}_j^N$ of Fourier coefficients

$$\widehat{c}_j^N = \frac{1}{N} \sum_{i=1}^{N} Y_i \varphi_j(X_i) \tag{3.4}$$

are substituted for the true ones c_j, $j = 1, \ldots, m$. Note that the estimate (3.3)–(3.4) is linear (cf. (3.1)) with weights given by

$$W_{N,i}(x) = \sum_{j=1}^{m} \frac{1}{N} \varphi_j(x)\, \varphi_j(X_i).$$

Cencov, [9] has proved the following result. Assume that the observations X_i are independent and identically distributed on $[0,1]^d$ with density $g(x) \geq c > 0$, $g \in \mathcal{W}_2^s(L)$, and that the noise satisfies $\mathbf{E}e_i = 0$ and $\mathbf{E}e_i^2 \leq \sigma_e^2 < \infty$. Then, uniformly over $f \in \mathcal{W}_2^s(L)$ and $x \in [0,1]^d$,

$$\mathbf{E}\|\widehat{f}_N(x) - f(x)\|_2^2 \leq C\left(L^2 m^{-2s} + \frac{\sigma_e^2 m^d}{N}\right) \tag{3.5}$$

The optimal order m of the model is

$$m = \lfloor \left(\frac{L^2 N}{\sigma_e^2}\right)^{1/(2s+d)} \rfloor, \tag{3.6}$$

it balances bias and variance error estimates, and yields the bound

$$\mathbf{E}\|\widehat{f}_N(x) - f(x)\|_2^2 \leq C L^{2/(d+2s)} \left(\frac{\sigma_e^2}{N}\right)^{2s/(2s+d)}. \tag{3.7}$$

The following result, due to Ibragimov and Khas'minskij [28], provides a uniform bound. Take

$$m = \lfloor (\frac{N}{\ln N})^{1/(2s+d)} \rfloor$$

for the model order (note that this is slightly different from (3.6)). Then, uniformly over $f \in \mathcal{C}^s(L)$, it holds that

$$\mathbf{E}\|\widehat{f}_N - f\|_\infty^2 \leq CO\left(\frac{\ln N}{N}\right)^{2s/(2s+d)}. \tag{3.8}$$

The bound for the estimation error variance given on the right hand side of (3.5) is decomposed into *bias* + *variance* terms. And, as expected, the optimal choice of h_N in (3.6) exactly balances these two terms.

3.2 Choice of Model Order, Bandwidth, or Binwidth: the Generalized Cross Validation – GCV – Method

As we have seen, the convergence of the estimates strongly depends on the choice of the bandwidth h_N for kernel estimator, the model order m for the projection estimator, and the binwidth δ_N (or, equivalently, the "model order" $M = \delta^{-1}$) for piecewise polynomial estimator. *These design parameters* *depend on the parameters of the smoothness class $\mathcal{W}_2^s(L)$, which are a priori unknown.* Instead of trying to estimate them, we shall proceed differently.

The model order (or bandwith, or binwidth, depending on the different estimates) shall be estimated from data using a procedure usually referred to as the *Generalized Cross Validation* (GCV) test. GCV procedures were studied for kernel (see, for instance, [48], [25]), spline ([34], [12]), and projection estimates (c.f. [46], [35]). Let us consider, for instance, the procedure for the projection estimates. To make the model order explicit in formula (3.3) we shall write $\widehat{f}_{m,N}$ instead of $\widehat{f}_N$. Set $S^2_{m,N} = N^{-1}\sum_{i=1}^N \|Y_i - \widehat{f}_{m,N}(X_i)\|^2$. As for the prediction error variance estimate in parametric prediction error methods, $S^2_{m,N}$ is a *biased* estimate of the error. Thus one cannot minimize $S^2_{m,N}$ with respect to m directly: the result of such a bruteforce procedure would give a function $\widehat{f}_{m_N,N}(x)$ which perfectly fits the noisy data, this is known as "overfitting" in the neural network litterature. The solution rather consists in introducing a penalty which is proportional to the model order m, i.e., we search for m_N such that

$$m_N = \arg\min_{m \leq N} \left(S^2_{mN} + \frac{2\sigma_e^2 m}{N} \right). \tag{3.9}$$

This technique is clearly equivalent to the celebrated Mallows-Akaike criterion [39], [1]. The following result, due to Polyak and Tsybakov [46], shows the consistency of this procedure. Assume that the Fourier coefficients of f in expansion (3.2) satisfy $|c_j| \leq \varepsilon_j$, $\sum_{j=1}^\infty \varepsilon_j < \infty$, $(j\,\varepsilon_j)$ is non-increasing, and σ_e^2 is known. Set $V_{m,N} = \|\widehat{f}_{m,N} - f\|_2^2$. Then for the estimate (3.3), (3.4), and (3.9), it holds that

$$\frac{V_{m_N,N}}{\min_m V_{m,N}} \to 1 \text{ a.e. as } N \to \infty\,.$$

4. Performance Analysis of the Nonparametric Estimators

The performance analysis of nonparametric estimation algorithms and/or identification procedures is much more difficult than for parametric estimation. In fact, the following specific issues are important:

1. What plays the role of Cramer-Rao bound and Fisher Information Matrix in our case? Recall that the Cramer-Rao bound reveals the best performance one can expect in identifying the unknown parameter θ from sample data arising from some parametrized distribution $\mathbf{p}_\theta, \theta \in \Theta$, where Θ is the domain over which the unknown parameter θ ranges. In the nonparametric case, lower bounds for the best achievable performance are provided by *minimax risk functions.* We shall introduce these lower bounds and discuss associated notions of optimality.

2. For lower bounds, what is the class of systems on which best achievable performance is considered, is another important issue. For nonparametric representations of linear systems, L_2, L_∞, H_2, H_∞ with their associated norm are typical spaces to work with. For (even static) nonlinear systems, however, the choice is much wider. How wide should be the class $\mathcal{F}$ of systems in consideration, what kind of smoothness should be required?

4.1 Lower Bounds for Best Achievable Performance

In order to compare different nonparametric estimators it is necessary to introduce suitable figures of merit. It seems first reasonable to build on the mean square deviation (or mean absolute deviation) of some measures of the error, we denote it by $\|\widehat{f}_N - f\|$. Typical instances are L_p-norm, L_∞-norm, or the absolute value at a fixed point x_0. Then we consider the *risk function*

$$R_{a_N}(\widehat{f}_N, f) = \mathbf{E}\left[a_N^{-1}\|\widehat{f}_N - f\|\right]^2, \tag{4.1}$$

where a_N is a normalizing positive sequence. Letting a_N decrease as fast as possible so that the risk still remains bounded yields a notion of a convergence rate. Let $\mathcal{F}$ be a set of functions which contains the "true" regression function f, then the maximal risk $r_{a_N}(\widehat{f}_N)$ of estimator $\widehat{f}_N$ on $\mathcal{F}$ is defined as follows:

$$r_{a_N}(\widehat{f}_N) = \sup_{f\in\mathcal{F}} R_{a_N}(\widehat{f}_N, f)\ .$$

If the maximal risk is used as a figure of merit, the optimal estimator $\widehat{f}_N^*$ is the one for which the maximal risk is minimized, i.e., such that [3]

$$r_{a_N}(\widehat{f}_N^*) = \min_{\widehat{f}_N} \sup_{f\in\mathcal{F}} R_{a_N}\left(\widehat{f}_N, f\right)$$

We call $\widehat{f}_N^*$ *the minimax estimator* and the value

$$\min_{\widehat{f}_N} \sup_{f\in\mathcal{F}} R_{a_N}\left(\widehat{f}_N, f\right)$$

the minimax risk on $\mathcal{F}$. The construction of minimax nonparametric regression estimators for different sets $\mathcal{F}$ is a hard problem. More precisely, we state the following definition:

[3] To properly understand the statement to follow, the reader should pay attention to definition (2.2) of an estimator.

Definition 4.1.

1. *The positive sequence a_N is a* **lower rate of convergence for the set $\mathcal{F}$ in the semi-norm $\|\cdot\|$** *if*

$$\liminf_{N\to\infty} \min_{\widehat{f}_N} \sup_{f\in\mathcal{F}} R_{a_N}\left(\widehat{f}_N, f\right) \geq C_0 \tag{4.2}$$

for some positive C_0. The inequality (4.2) is a kind of negative statement that says that no estimator of function f can converge to f faster than a_N. This notion can be refined as follows.

2. *The positive sequence a_N is called* **minimax rate of convergence** *for the set $\mathcal{F}$ in semi-norm $\|\cdot\|$, if it is lower rate of convergence, and if, in addition, there exists an estimator $\widehat{f}^*_N$ achieving this rate, i.e., such that*

$$\limsup_{N\to\infty} \min_{\widehat{f}_N} \sup_{f\in\mathcal{F}} R_{a_N}\left(\widehat{f}_N, f\right) < \infty$$

Thus, a coarser, but easier approach consists in assessing the estimators by their convergence rates. In this setting, by definition, optimal estimators reach the lower bound as defined in (4.2) (recall that the minimax rate is not unique: it is defined to within a constant). In [30], some negative results are stated, which show that, *in order to obtain any interesting rate of convergence, smoothness conditions should be imposed* on the considered class of functions — boundedness or even continuity are not sufficient. We now state some positive results.

We concentrate on the case of deterministic uniform design, i.e., the input data X are uniformly sampled in the considered interval. The following result can be acknowledged to [28] (for the random design case, see [51], [32]). Consider the Hölder class $\mathcal{C}^s(L)$ on $[0,1]^d$, defined as follows: for s and L positive, let $\mathcal{C}^s(L)$ be the family of functions $f(x)$, $x \in [0,1]^d$ defined by [4]

$$\mathcal{C}^s(L) = \left\{f : |f^{(k)}(x) - f^{(k)}(x')| \leq L|x-x'|^{s-k}\right\} \tag{4.3}$$

for any $x, x' \in [0,1]^d, k = \lfloor s \rfloor$. If $s \geq 1$ is integer, then $\mathcal{C}^s(L)$ contains continuous functions having Lipschitz $(s-1)$-th derivative.

Theorem 4.1. *Consider*

$$\|g\| = \left(\int |g(x)|^p dx\right)^{1/p}, \quad 0 < p \leq \infty$$

Then $N^{-\frac{s}{2s+d}}$ is a lower rate of convergence for the class $\mathcal{C}^s(L)$ in the semi-norm $\|\cdot\|$.

---

[4] $\lfloor s \rfloor$ denotes the maximal integer $k < s$.

The following result is reported in [30] : for the class $C^s(L)$ used in theorem 4.1, kernel and piecewise polynomial estimators have the same rate of convergence (up to a constant) as given in (3.7) for projection estimators, namely $N^{-\frac{s}{2s+d}}$. Thus theorem 4.1 shows that these estimators are optimal in $C^s(L)$ i.e., they reach the minimax rate of convergence [5].

4.2 Discussion

Criticizing the minimax paradigm. Despite many impressive technical achievements in the above work, the general reaction within the statistics community has not been really enthusiastic. For example, David Donoho notes that "*... a large number of computer packages appeared over last fifteen years, but the work on the minimax paradigm* [(as reported before)] *had relatively little impact on software*" [20]. One of the arguments supporting this skepticism about methods based on the minimax paradigm — kernel estimators, spline methods or orthogonal series — is that they are spatially nonadaptive, while real functions exhibit a variety of shapes and spatial inhomogeneities. Thus the authors of statistic software argue that one should construct methods (heuristically, if necessary) which address the "real problem", namely *spatial adaptation.* Interesting spatially adaptive methods include all sorts of neural networks, projection pursuit [23], and others.

This short discussion reveals the crux in the route to both practical efficiency and mathematical support of the methods. It consists in *finding a parametrized family of functional classes which*

1. *fits our prior knowledge about the smoothness of the function to be estimated, (in particular, that f is smooth everywhere, except at a sparse set of points), and*
2. *has associated with it estimation technique which is minimax within these classes.*

It was the merit of David Donoho, Iain Johnstone [18] and Dominique Picard to recognize that the *Besov spaces,* which play a central role in Yves Meyer's mathematical theory of wavelets [40], provide an adequate answer. And wavelet based nonparametric estimators shall provide at the same time practical efficiency and mathematically proved optimality. This material will be the topic of Sections 6. and 7.

However, before discussing wavelets and their use in identification, we briefly scan some popular nonlinear estimates. They all provide the kind of "spatial adaptation" that we advocated before. Some of them are supported by efficient software.

[5] Projection estimates are also minimax on suitable spaces.

5. Nonlinear Estimates

Starting from early 1980s a variety of techniques have been proposed in the statistics literature, which exhibit this desirable feature of "spatial adaptivity". Among them *Projection Pursuit Algorithm* developed in [23] (very good review of these results can be found in [26]), *Recursive Partitioning* [41], [8] and related methods (see for instance [22] with discussion). These methods are derived from some mixture of statistic and heuristic arguments and give impressive results in simulations. Their drawback lies in nearly total absence of any theoretical results on their convergence.

Surprisingly enough, the A.I. literature has proposed independently and at the same time different techniques with the same feature of "spatial adaptivity". These include various forms of neural networks, see the other tutorial [37] by Lennart Ljung, and fuzzy models. We shall briefly describe these. In addition we shall sketch a recent technique due to Leo Breiman [7], which practically combines some advantages of neural networks (in particular the ability to handle very large dimensional inputs) and of constructive wavelet based estimators (availability of very fast training algorithms).

A relationship with neural networks; A. Barron's result. The following result was recently published in [2], it is the most accurate theoretical result about the function approximation using neural networks available today. Let $\sigma(x)$ be a sigmoidal function (i.e. a bounded measurable function on the real line for which $\sigma(x) \to 1$ as $x \to \infty$ and $\sigma(x) \to 0$ as $x \to -\infty$). Consider a compactly supported function f with supp $(f) \subseteq [0,1]^d$, and assume that

$$C_f = \int_{\mathbf{R}^d} |\omega| \, |\widehat{f}(\omega)| \, d\omega < \infty \, . \tag{5.1}$$

The main result of [2] can be roughly stated as follows: there exists an approximation f_n of the compactly supported function f, of the form

$$f_n(x) = \sum_{i=1}^{n} c_i \sigma(a_i^T x + t_i) + c_0 \tag{5.2}$$

(note that f_n is *not* compactly supported), such that

$$\|(f_n - f) \, 1_{[0,1]^d}\|_2 \, \leq \, 2\sqrt{d} \, C_f \, n^{-1/2} \, . \tag{5.3}$$

This result provides a bound of the minimum distance (in L_2-norm) between any f satisfying condition (5.1) and the class of all neural networks of size not larger than n. In the same article, this upper bound is compared with the best achievable convergence rate for any linear estimator in class (5.1). It is shown that a lower rate for linear estimators is $n^{-1/d}$, compare with the much better rate $n^{-1/2}$ for neural networks, especially for large dimension d. No result is available which takes advantage of possible improved smoothness of the unknown system f. An iterative algorithm for the construction of the approximation (5.2) is also proposed. The true problem of system identification, i.e., that of neural network training based on noisy input/output data, is not addressed in this paper. Also, neural networks need the backpropagation procedure for their training, which is known to be of prohibitive cost. In turn, neural network training works even for very large dimensional input data.

Breiman's Hinging Hyperplanes. We now briefly discuss a recent technique due to Leo Breiman [7], which practically combines some advantages of neural networks (in particular the ability to handle very large dimensional inputs) and of constructive wavelet based estimators (availability of very fast training algorithms). Breiman's technique is a very elegant and efficient way of identifying piecewise linear models based on data collected from an unknown nonlinear system, see [50] for the use of such models in control. Following [7], we call *hinge function* a function $y = h(x)$, $x \in \mathbf{R}^d$ which consists of two hyperplanes continuously joined together, i.e., an open book. If the two hyperplanes are given as

$$y = \langle \beta^+, x\rangle + \beta_0^+, \quad y = \langle \beta^-, x\rangle + \beta_0^-,$$

where $\langle .,.\rangle$ denotes scalar product in Euclidian spaces, then an explicit form for the hinge function is either

$$\begin{aligned} h(x) &= \max(\ \langle \beta^+, x\rangle + \beta_0^+\ ,\ \langle \beta^-, x\rangle + \beta_0^-\ )\,, \\ \text{or} \quad h(x) &= \min(\ \langle \beta^+, x\rangle + \beta_0^+\ ,\ \langle \beta^-, x\rangle + \beta_0^-\ )\,. \end{aligned}$$

It is proved in [7], using the methods by Barron [2] that there is a constant C such that for any n there are hinge functions $h_1, \ldots, h_n$ such that

$$\|f - \sum_{i=1}^{n} h_i 1_{[0,1]^d}\|_2 \leq C n^{-1/2} \tag{5.4}$$

for any f such that

$$\int_{\mathbf{R}^d} |\omega|^2 |\widehat{f}(\omega)| d\omega < \infty\,,$$

i.e., Breiman's hinge model is as efficient as neural networks for the L_2-norm. An iterative projection algorithm is proposed to compute the approximation. The interesting point about this iterative approximation technique is that *it converges with an order of magnitude faster than backpropagation does.* Reported experimental results show the efficiency of this technique. These experiments show that practically the approximation obtained is much more accurate than it is suggested by the estimate in (5.4). On the other hand, note that a superposition of hinge functions is not smooth, since it is piecewise linear. Also the use of superposition of hinge functions is especially advocated in (5.4) for large dimensional x's. However, as indicated at the beginning of this section, no convergence rate is given for models identified from noisy data (the bound (5.4) is not a convergence rate for identification, but only a rate of approximation of a given function by some finitely parametrized class of approximants).

6. Wavelets: What They Are, and Their Use in Approximating Functions

Warning:. throughout this section, the notation $\widehat{\varphi}(\omega)$ denotes the Fourier transform of function $\varphi(x)$, and *not* the estimator of φ.

6.1 The Continuous Wavelet Transform

The continuous wavelet transform and inverse transform of a function f are respectively given by equations (6.2) below. These transforms use two functions $\psi(x)$ and $\varphi(x) \in L_2(\mathbf{R}^d)$, both radial (i.e., depending only on $|x|$), known as the *analysis and synthesis wavelets*:

Theorem 6.1. *Let ψ and φ be radial functions satisfying*

$$\forall \omega \in \mathbf{R}^d \ : \ \int_0^\infty a^{-1}\widehat{\varphi}(a\omega)\widehat{\psi}(a\omega)da = 1 \ , \tag{6.1}$$

where we recall that $\widehat{\varphi}(\omega)$ denotes the Fourier transform of function $\varphi(x)$. Then for any function $f \in L_2(\mathbf{R}^d)$, the following formulae define an isometry between $L_2(\mathbf{R}^d)$ and a subspace of $L_2(\mathbf{R}^d \times \mathbf{R}_+)$ [13] :

$$\begin{aligned} u(a,t) &= a^{d-1/2} \int f(x) \ \psi(a(x-t)) \ dx \\ f(x) &= \int u(a,t) \ \varphi(a(x-t)) \ a^{d-1/2} \ da \ dt \end{aligned} \tag{6.2}$$

Here $a \in \mathbf{R}^+$ and $t \in \mathbf{R}^d$ are respectively the dilation and translation factors. Note that the integral (6.1) does not depend on $\omega \neq 0$ since the functions ψ and φ are radial. In order for this integral to be properly defined, it is sufficient that, for example $\widehat{\varphi}(\omega)\widehat{\psi}(\omega) = O(|\omega|)$ at zero; this happens if $\varphi(x)$ and $(1+|x|)\psi(x)$ are in $L_1(\mathbf{R}^d) \cap L_2(\mathbf{R}^d)$ and ψ has zero integral. Once the integral (6.1) is well defined and finite, a simple normalization leads to a pair (φ, ψ) which satisfies the assumption.

Examples:. it can be verified that the following pairs ψ, φ satisfy the assumption :

$$\begin{aligned} \psi(x) &= \sqrt{2}(d - |x|^2)e^{-\frac{|x|^2}{2}} \ , \ \varphi(x) \ = \ \sqrt{2}e^{-\frac{|x|^2}{2}} \\ \psi(x) &= \varphi(x) = \frac{1}{\sqrt{2}}(d - |x|^2)e^{-\frac{|x|^2}{2}} \ . \end{aligned}$$

The choice of possible pairs ψ, φ is very large. In particular, pairs (ψ, φ), with ψ nonsmooth but φ smooth, are allowed.

6.2 The Discrete Wavelet Transform: Orthonormal Bases of Wavelets and Extensions

Multiresolution analysis introduced by Stephane Mallat and further developed by Ingrid Daubechies provides orthonormal bases of $L_2(\mathbf{R})$ of the form $\psi_{j,k}(x) = \{2^{j/2}\psi(2^j x - k) \ : \ j, k \in \mathbf{Z}\}$, i.e., each element of the basis is a translated and dilated version of a single *wavelet* ψ. For a function $f \in L_2(\mathbf{R})$, the inner product $\langle f, \psi_{j,k} \rangle$ performs zooming on f over a $O(2^{-j})$ width interval centered at point $2^{-j}k$. Thus, *large* j *corresponds to checking function* f *at fine scales.* This implies that a local singularity of a function f will affect only a small part of its coefficients in this wavelet basis. This is the main difference with the Fourier basis : a local singularity of f would affect the whole Fourier representation.

Definition and construction of orthogonal wavelet bases. To begin we first discuss the scalar case, i.e., that of functions defined on $\mathbf{R}$. Otherwise explicitly stated, all results in this subsection are borrowed from monograph [13].

Definition 6.1 (Multiresolution Analysis (MA)). *A multiresolution analysis consists of a function* φ, $\|\varphi\|_2 = 1$, *and a sequence* $(V_j)_{j \in \mathbf{Z}}$ *of spaces defined by*

$$\begin{aligned} \varphi_{jk} &= 2^{j/2}\varphi(2^j x - k) \ \ , \ j, k \in \mathbf{Z} \\ V_j &= \text{Span} \ \{\varphi_{jk}, k \in \mathbf{Z}\} \end{aligned}$$

with the properties :

(MA0) : $(\varphi_{0k})_{k \in \mathbf{Z}}$ *is an orthonormal family*
(MA1) : $\bigcap_{j \in \mathbf{Z}} V_j = \{0\}$
(MA2) : $\overline{\bigcup_{j \in \mathbf{Z}} V_j} = L_2(\mathbf{R})$
(MA3) : $V_j \subset V_{j+1}$

Property (MA3) is equivalent to the existence of a square integrable sequence (h_k) such that

$$\varphi(x) = \sqrt{2} \sum h_k \ \varphi(2x - k) \ . \tag{6.3}$$

We call such a function the *scale function* (also known as the *father wavelet* [40]). Theorem 6.2 to follow is the basis of the theory; it shows how, starting from a multiresolution analysis and its scale function φ, we can construct very simply an orthonormal basis of $L_2(\mathbf{R})$.

Theorem 6.2. *Assume that conditions (MA0–3) are satisfied. Set* [6]

$$\begin{aligned} \psi(x) &= \sqrt{2}\sum g_k\, \varphi(2x-k)\ ,\ g_k = (-1)^{k+1}\, \overline{h}_{1-k} \\ \psi_{jk} &= 2^{j/2}\, \psi(2^j x - k) \\ W_j &= \mathrm{Span}(\psi_{jk}, k \in \mathbf{Z}) \end{aligned} \tag{6.4}$$

then

1. $V_{j+1} = V_j \bigoplus W_j$, *and* $\{\psi_{jk} : j,k \in \mathbf{Z}\}$ *is an orthonormal basis of* $L_2(\mathbf{R})$;
2. $L_2(\mathbf{R}) = V_0 \bigoplus W_0 \bigoplus W_1 \bigoplus ...$, *and* $\{\varphi_{0k}, \psi_{jk}\ :\ j \geq 0, k \in \mathbf{Z}\}$ *is an orthonormal basis of* $L_2(\mathbf{R})$.

The function $\psi(x)$ defined in (6.4) is often referred to as the "mother wavelet". Then theorem 6.3 gives the basic tool for building scale functions.

Theorem 6.3. *Let* $m_0(\omega)$ *be a trigonometric polynomial*

$$m_0(\omega) = \frac{1}{\sqrt{2}} \sum_{k=K}^{L} h_k e^{-ik\omega}$$

such that

(QMF1): $m_0(0) = 1$,
(QMF2): $m_0(\omega) \neq 0$ *if* $\omega \in [-\pi/2, \pi/2]$,
(QMF3): $|m_0(\omega)|^2 + |m_0(\omega+\pi)|^2 = 1$.

Then the function φ, *with Fourier transform given by*

$$\widehat{\varphi}(\omega) = \prod_{j=1}^{\infty} m_0(2^{-j}\omega)$$

satisfies assumptions (MA0–3) and $supp(\varphi) \subset [K, L]$.

Examples of polynomials satisfying assumption (QMF1–3) are given in [13] and smoothness properties of φ and ψ are studied.

We now move on discussing the multidimensional case. There exist two main types of constructions of the wavelet basis with dilation factor 2 in $\mathbf{R}^d$ ([13], 10.1). A first guess simply consists in taking tensor product functions generated by d one-dimensional bases:

$$\psi_{j_1,k_1,\ldots,j_d,k_d}(x) = \psi_{j_1,k_1}(x_1) \times \ldots \times \psi_{j_d,k_d}(x_d) \tag{6.5}$$

[6] $\overline{h}$ denotes the complex conjugate of h.

This construction has the drawback of mixing different resolution levels j_i. Alternatively, if such a mixing is not desired, we proceed as follows. Introduce the scale function

$$\varphi(x) = \varphi(x_1) \times \ldots \times \varphi(x_d) \tag{6.6}$$

and the $2^d - 1$ mother wavelets $\psi^{(i)}(x), i = 1, \ldots, 2^d - 1$ obtained by substituting in (6.6) some $\varphi(x_j)$'s by $\psi(x_j)$'s. Then the following family is an orthonormal basis of $L_2(\mathbf{R}^d)$:

$$\left\{\varphi_{0k}(x), \psi_{jk}^{(1)}(x), \ldots, \psi_{jk}^{(2^d-1)}(x)\right\}$$
$$j \in \mathbf{N}_0,\ k = (k_1, \ldots, k_d) \in \mathbf{Z}^d \tag{6.7}$$

where $\mathbf{N}_0 = \mathbf{N} \cup 0$, and

$$\begin{aligned} \varphi_{jk}(x) &= 2^{jd/2}\, \varphi(2^j x_1 - k_1, \ldots, 2^j x_d - k_d) \\ \psi_{jk}^{(i)}(x) &= 2^{jd/2}\, \psi^{(i)}(2^j x_1 - k_1, \ldots, 2^j x_d - k_d)\,. \end{aligned}$$

NOTA :. as formula (6.7) shows, constructing and storing orthonormal wavelet bases become of prohibitive cost for large dimension d. This is the main limitation for using the otherwise very efficient techniques which rely on orthonormal wavelet bases (and their generalizations).

Orthogonal wavelet bases and Quadrature Mirror Filters (QMF). Equations (6.3) and (6.4) imply that [7], for $f \in L_2(\mathbf{R}^d)$,

$$\alpha_{jk} = \langle f, \varphi_{jk}\rangle, \quad \beta_{jk} = \langle f, \psi_{jk}\rangle \tag{6.8}$$

satisfy [8]

$$\alpha_{jk} = \sum_l \overline{h}_{l-2k}\, \alpha_{j+1,l} \tag{6.9}$$

$$\beta_{jk} = \sum_l \overline{g}_{l-2k}\, \alpha_{j+1,l}. \tag{6.10}$$

Introduce the polynomial filters $H(z) = \sum_k h_k z^{-k}$, $G(z) = \sum_k g_k z^{-k}$ where coefficients h_k, g_k are as in (6.3) (6.4). Property (QMF3) expresses that the pair (H, G) is QMF [54] [4]. Recursions (6.9) and (6.10) are used to compute recursively from fine scales to coarse scales the orthonormal wavelet decomposition, with α_{j_0k} as initial condition (index "j_0" denotes the finest scale in these recursions). Assume that, in addition, scale function φ is selected so that the computation of inner product $\langle f, \varphi_{j_0k}\rangle$ in (6.8) is performed efficiently. Then, *formulas (6.8), (6.9), and (6.10) together build a highly efficient procedure for computing the wavelet decomposition of f*, see [30] for efficient computation of inner product $\langle f, \varphi_{jk}\rangle$.

[7] recall that $\langle .,. \rangle$ denotes the inner product in L_2.

[8] recall that $\overline{h}$ denotes the complex conjugate of h.

Equations (6.9) and (6.10) can be inverted to yield the synthesis equation

$$\alpha_{jk} = \sum_l h_{k-2l}\, \alpha_{j-1,l} + g_{k-2l}\, \beta_{j-1,l} \tag{6.11}$$

For $f \in V_{j_0}$, we have, by definition of this space,

$$f = \sum_k \alpha_{j_0 k}\, \varphi_{j_0 k}\ , \tag{6.12}$$

and, since $V_{j_0} = V_0 \bigoplus W_0 \bigoplus W_1 .... \bigoplus W_{j_0}$,

$$f = \sum_k \alpha_{0k}\, \varphi_{0k} + \sum_{j,k} \beta_{jk}\, \psi_{jk}\ . \tag{6.13}$$

Formulas (6.9) and (6.10) allow us to switch from representation (6.12) to representation (6.13). The latter one is generally much more compact since, when f is smooth, most β_{jk} are negligible.

6.3 Wavelets and Besov Spaces

Besov spaces as spaces of smooth functions with localized singularities. Smooth functions with localized singularities are typically encountered in nonlinear systems, e.g., in mechanical and chemical systems. As we shall see, Besov spaces are spaces

- of smooth functions with possibly localized singularities,
- in which norms are easily evaluated using wavelet coefficients.

For the sake of clarity we consider only compactly supported functions f: $\text{supp}\, f \subseteq [0,1]^d$, though all the definitions below can be generalized for a noncompact case. For $f \in L_1$ and $M \in \mathbf{N}$ we define the local oscillation of order M (or M-oscillation for short) at the point $x \in [0,1]^d$ by

$$\text{osc}_M f(x,t) \triangleq \inf_P \frac{1}{t^d} \int_{|x-y|<t} |f(y) - P(y)|\, dy \tag{6.14}$$

where the infimum is taken over all polynomials P of degree less than or equal to M. This quantity measures the quality of local fit of f by polynomials on balls of radius t.

Select $p, q > 0$, $s > d(p^{-1} - 1)$, and take $M = \lfloor s \rfloor$. Denote by $\mathcal{B}^s_{pq}$ the set of functions $f \in L_{1\wedge p}$ such that the quantity

$$\|f\|_{\mathcal{B}^s_{pq}} \triangleq \|f\|_p + \left(\sum_{j=1}^{\infty} (2^{js} \|\, \text{osc}_M f(x, 2^{-j}) \|_p)^q \right)^{1/q}$$

is finite. Then the parametrized family $\mathcal{B}^s_{pq}$ of spaces (with the usual modification for p or $q = \infty$) coincides with the *Besov spaces* of functions [6], and it is shown in [53] that $\|\cdot\|_{\mathcal{B}^s_{pq}}$ is equivalent to the classical Besov norm.

The triple parametrization using s, p, and q provides a very accurate characterization of smoothness properties. As usual for Hölder or Sobolev spaces, index s indicates how many derivatives are smooth. Then, for larger p, $\|f\|_{\mathcal{B}^s_{pq}}$ is more sensitive to details. Finally, index q has no useful practical interpretation, but it is a convenient instrument that serves to compare Besov spaces with the more usual Sobolev spaces W^s_p (see [30]). It is interesting to notice that the indicator functions of intervals belong to the spaces $\mathcal{B}^s_{s^{-1}\infty}$ for all $s > 0$, this illustrates our claim in the title of this subsection.

Approximation in Besov spaces via free knots splines and rational fractions is discussed in [44]. It is stated there that both types of approximations are optimal in Besov spaces, but linear methods of Section 3. are not. It is amazing that *wavelet approximations are as good as spline or rational ones, but are much easier to construct.* We discuss this next.

Wavelets and Besov spaces: efficient and effective. Let φ be a piecewise continuous scale function satisfying the following conditions:

$$\exists a > 0 \quad : \quad \text{supp}\ \varphi \in \{|x| \leq a\} \tag{6.15}$$

$$\exists r > s \quad : \quad \varphi \in \mathcal{B}^r_{u\infty} \tag{6.16}$$

We have the following result (c.f. theorem 4 in [49]):

Theorem 6.4 (Besov norms and wavelet decompositions). *Let $s > d(1/p - 1)$ and φ be a scale function satisfying conditions (6.15) and (6.16). For any $f \in \mathcal{B}^s_{pq}$ define*

$$\|f\|_{spq} = \left(\sum_k |\alpha_k|^p\right)^{1/p} + \left(\sum_{j=0}^{\infty}\left[2^{j(s+d/2-d/p)}\|\beta_{j\cdot}\|_p\right]^q\right)^{1/q} \tag{6.17}$$

and $\|\beta_{j\cdot}\|_p = (\sum_k |\beta_{jk}|^p)^{1/p}$, see (6.8) (6.13) for the definition of coefficients $\alpha_k = \alpha_{0k}$ and β_{jk}. Then (6.17) is a equivalent to the norm of Besov space $\mathcal{B}^s_{pq}$, i.e., there exist constants C_1 and C_2, independent of f, such that

$$C_1\ \|f\|_{\mathcal{B}^s_{pq}} \leq \|f\|_{spq} \leq C_2\ \|f\|_{\mathcal{B}^s_{pq}}\ . \tag{6.18}$$

Theorem 6.4 states that norms in Besov spaces are suitably evaluated using orthonormal wavelet decompositions. This explains why truncating wavelet expansions can yield very efficient approximations.

We now indicate how such a wavelet approximation w_n of f can be constructed. Consider the full wavelet decomposition of f:

$$f(x) = \sum_k \alpha_k \varphi_k(x) + \sum_{j=0}^{\infty} \sum_k \beta_{jk} \psi_{jk}(x) . \tag{6.19}$$

1. Keep the projection of f on the subspace V_0, this corresponds to the left most sum in (6.19). When f and φ are both compactly supported this requires computing only a fixed amount of coefficients, say m. And then,
2. Select in the second (double) sum those coefficients β_{jk} with largest absolute value, denote by Λ the set of the $n-m$ so selected wavelet coefficients. Finally,
3. Add $n-m$ detail terms $\beta_{jk}\psi_{jk}$ to the sum taken in step 1.

This procedure yields the approximation

$$w_n(x) = \sum_k \alpha_k \varphi_k(x) + \sum_{\lambda \in \Lambda} \beta_\lambda \psi_\lambda(x) , \tag{6.20}$$

and the following theorem provides corresponding approximation bounds.

Theorem 6.5 (DeVore, Jawerth, Popov, [16]). *Consider $f \in \mathcal{B}^s_{pp}$, $s, p > 0$ and $s - d/p + d/u \geq 0$. Let w_n denote the approximation (6.20) of f. If the scale function satisfies conditions (6.15) and (6.16), then*

$$\|f - w_n\|_u \;\leq\; C(s,p)\, n^{-s}\, \|f\|_{\mathcal{B}^s_{pp}}$$

holds. If, in addition, u satisfies $s - d/p + d/u = 0$, $u < \infty$, and it is a priori known that $f \in L_u$, then the following converse bound holds.

$$\|f\|_{\mathcal{B}^s_{pp}} \;\leq\; C(s,p,q)\, n^s\, \|f - w_n\|_u .$$

Discussion. At this point we have the requested background for understanding how to perform wavelet based estimation. Roughly speaking, the crux is the following. The function $f \in \mathcal{B}^s_{pq}$ to be estimated can be *approximated* using expansion w_n in (6.20) with n terms. This is achieved with a rate of $O(n^{-s})$. Then coefficients α_k and β_λ in (6.20) are estimated via empirical means based on N noisy observations, exactly as for the projection estimates in formula (3.4). The mean square error on the estimate of each coefficient is $O(1/N)$. Thus the total mean square error of the estimate will be, as usual, the sum of the stochastic part and of the bias due to the approximation error: this yields $O(n/N) + O(n^{-2s})$. The optimal choice for n balances these two terms: $n = N^{\frac{1}{2s+1}}$. This choice for n yields a quadratic error of order $N^{-\frac{2s}{2s+1}}$ (independent of p, q). As we shall see, this is the typical minimax rate of convergence on Besov spaces. Thus we might be ready to deduce that wavelet estimators are minimax optimal in Besov spaces. Unfortunately, the set Λ of "important" coefficients in truncation (6.20) is *not* known a priori when noisy data sets are at hand for estimation. Thus some kind of hypothesis testing problem must be solved in order to obtain the optimal approximation. This adds to the estimation problem a nice stochastic flavour. We address this point in the next section.

7. Wavelets: Their Use in Nonparametric Estimation

We consider here some simple results concerning the estimation of a regression function or a density $f : \mathbf{R}^d \to \mathbf{R}$, and we assume f to be compactly supported ($\text{supp} f \subseteq [0,1]^d$). For the sake of simplicity we measure the estimation error in L_2-norm. We discuss the problem of non-parametric regression, density estimation is also discussed in [30].

Assume a N-sample of input/output observations of the following system are available:

$$Y_i = f(X_i) + w_i ,$$

where (X_i) and (w_i) are i.i.d. sequences of random variables, X_i is *uniformly distributed* on $[0,1]^d$ and $Ew_i = 0$, $Ew_i^2 \leq \sigma_w^2$. These assumptions are introduced for the sake of simplicity. They can be weakened, in particular the (unusual) assumption that X is uniformly distributed can easily be relaxed, see [14] [30], this would introduce additional burden to our presentation, however. For $f \in L_2$, recall the wavelet expansion

$$f(x) = \sum_{k \in \mathbf{Z}} \alpha_k \varphi_{0k}(x) + \sum_{j=0}^{\infty} \sum_{k \in \mathbf{Z}} \beta_{jk} \psi_{jk}(x), \tag{7.1}$$

where

$$\begin{aligned} \alpha_k &= \int f(x) \varphi_{0k}(x) dx \\ \beta_{jk} &= \int f(x) \psi_{jk}(x) dx \end{aligned} \tag{7.2}$$

To construct an estimate of f a first idea consists in using the law of large numbers and replacing, in expansion (7.1), the coefficients α_k and β_{jk} by their empirical estimates

$$\begin{aligned} \hat{\alpha}_k &= \frac{1}{N} \sum_{i=1}^{N} Y_i \varphi_{0k}(X_i) \\ \hat{\beta}_{jk} &= \frac{1}{N} \sum_{i=1}^{N} Y_i \psi_{jk}(X_i) \end{aligned} \tag{7.3}$$

Note that the assumption that input X is uniformly distributed has been used at this point.

Obviously, in order to compute the empirical coefficient $\hat{\beta}_{jk}$, we need that at least several observations X_i hit the support of $\psi_{jk}(x)$. Statistical laws of *loglog* type guarantee that this would generically hold for scales that are not too fine. More specifically, for $j \leq j_{\max}$, where

$$\frac{N}{\log N} \leq 2^{dj_{\max}} \leq \frac{2N}{\log N}$$

Thus we bruteforce set $\hat{\beta}_{jk} = 0$ for $j > j_{\max}$. At this point we have built an estimator of the linear projection type, as in the case of Fourier series in section 3.. Since these estimators are linear, as argued before, we cannot expect them to be efficient for Besov spaces [31].

To circumvent the issue of selecting the largest coefficients in appropriate quantity, the idea is to perform "shrinking" as proposed by D. Donoho, I. Johnstone, G. Kerkyacharian and D. Picard, we give a typical instance of this method [18] [15] : set

$$\tilde{\beta}_{jk} \quad = \quad \hat{\beta}_{jk}\, 1_{\{|\hat{\beta}| \geq \lambda_j\}} \tag{7.4}$$

where λ_j is a threshold parameter, and then take

$$\hat{f}_N(x) = \sum_k \hat{\alpha}_k \varphi_{0k}(x) + \sum_{jk} \tilde{\beta}_{jk} \psi_{jk}(x)\ . \tag{7.5}$$

In other words, in expansion (7.1), we keep those empirical estimates of wavelet coefficients which exceed some properly selected threshold. How this threshold should be selected is provided by the following result:

Theorem 7.1 ([19] and [20]). *Let $f \in \mathcal{B}^s_{p\infty}$ with $s \geq d/p$, $\|f\|_\infty < \infty$. Select $\lambda_j = \lambda = \sqrt{\frac{C \log N}{N}}$, with an appropriate $C < \infty$. Then*

$$E\|\hat{f}_N - f\|_2^2 = O\left(\frac{\log N}{N}\right)^{\frac{2s}{2s+d}}\ .$$

The constant C in the expression for the threshold parameter λ is a sort of an "hyperparameter" of the procedure, it can be easily estimated, see [15] and [19] for related discussions. It is heuristically explained in [30] why the above choice for a threshold is good. Compare this result with the lower rate of convergence for this problem obtained in [43], where it is shown that

$$\inf_{\hat{f}_N} \sup_{f \in \mathcal{B}^s_{pq}} E\|\hat{f}_N - f\|_2 \geq C N^{-2s/(2s+d)} \tag{7.6}$$

for any estimator $\hat{f}_N$. We see that the proposed wavelet estimator is (almost) minimax optimal, up to the $\log N$ factor, which is small compared to N.

Finally, in [15] the authors of this paper showed that properly selecting the threshold λ for shrinking provides the optimal rate of convergence (without a logarithmic factor). An adaptive version of this algorithm is developed in [29].

Summary of the wavelet estimation procedure for an N-sample length:.

1. Select $j_{\max}$ scales for the wavelet expansion, where

$$\frac{N}{\log N} \leq 2^{dj_{\max}} \leq \frac{2N}{\log N}$$

2. For $j \leq j_{\max}$ compute the empirical estimates

$$\begin{aligned} \hat{\alpha}_k &= \frac{1}{N}\sum_{i=1}^{N} Y_i \varphi_{0k}(X_i) \\ \hat{\beta}_{jk} &= \frac{1}{N}\sum_{i=1}^{N} Y_i \psi_{jk}(X_i) \end{aligned}$$

3. Skrink these estimates according to

$$\tilde{\beta}_{jk} = \hat{\beta}_{jk}\, 1_{\{|\hat{\beta}| \geq \lambda_j\}}$$

   where λ_j is a properly selected threshold (cf. theorem 7.1).
4. The final estimate is given by

$$\hat{f}_N(x) = \sum_k \hat{\alpha}_k \varphi_{0k}(x) + \sum_{jk} \tilde{\beta}_{jk} \psi_{jk}(x)$$

Steps 1–4 yield an efficient and cost effective estimation procedure for low dimensional inputs. For larger dimensions, the construction and storage of wavelet basis become prohibitive. This motivates the alternative procedure we describe in the next section.

8. A Wavelet Network for Practical System Identification

In this section we present a method for constructing estimators with non orthogonal wavelets, corresponding software is available [59]. This method falls into the category of wavelet estimators with adaptive dilation and translation sampling, and combines regressor selection and backpropagation algorithms. Related works have been reported in [61, 57, 58]. We investigate problem 2.1 of Section 2. in the case of additive noise, i.e., the pair of random variables X, Y satisfies

$$Y = f(X) + e\,, \tag{8.1}$$

where $f(x) : \mathbf{R}^d \mapsto \mathbf{R}$ and e is some noise of zero mean and independent of X. We want to estimate f based on a sample of size N that we shall refer to as the *training data set*: $\mathcal{O}_1^N = \{(X_1, Y_1), \ldots, (X_N, Y_N)\}$. We are particularly interested in training with sparse data sets. Sparse data often occur in classification problems and in the modeling of control systems, where available data can be relatively few as compared to the dimension of input X.

8.1 The Wavelet Network and Its Structure

For a wavelet function $\varphi : \mathbf{R}^d \to \mathbf{R}$, the wavelet network is written as follows

$$f_n(x) = \sum_{i=1}^{n} u_i \varphi(a_i \star (x - t_i)) \tag{8.2}$$

where $u_i \in \mathbf{R}, a_i \in \mathbf{R}^d, t_i \in \mathbf{R}^d$, and "$\star$" means component-wise product of two vectors. Note that we could have used scalar dilation parameters a_i, but we prefer the vector dilation parameters because they considerably increase the flexibility of network (8.2) at a reasonable price. To justify the use of (8.2) for solving regression problems, we can refer to the theoretical results of sections 6. and 7. The purpose of this section is to present an efficient comprehensive method for wavelet network training. The outline of this procedure is the following.

1. Construct a library W of dilated/translated versions of a given wavelet φ. This library W is adapted to the available training data set, by selecting a subset from all dilated/translated versions of φ on a regular grid. This technique makes it feasible to build the library W even for significantly large input dimension when the training data are sparse.
2. Not all wavelets from library W are useful in fitting f from noisy data, however. Thus we are faced with a problem of selecting the best regressors among W. Three methods will be proposed for this. When the regressors are conveniently selected, fitting model (8.2) amounts to identifying the u_i coefficients, which is a standard least squares estimation problem.
3. Steps 1 and 2 above yield a fast training procedure. The result can still further be improved by subsequently applying an iterative backpropagation algorithm with steps 1 and 2 as fast initialization. In fact, since initialization was good, a faster Newton procedure can be used.

Details of steps 1 and 2 are given now.

8.2 Constructing the Wavelet Library W

We have to restrict ourselves to a finite set of regressor candidates. It consists in selecting some subset of the continuously parameterized family

$\{\varphi(a(x-t)) : a \in \mathbf{R}^+, t \in \mathbf{R}^d\}$ generated by dilating and translating a single wavelet function $\varphi(x)$. This is in principle the same problem as discretizing the continuous wavelet reconstruction (6.2) to obtain discretized version of the reconstruction in (6.2). The standard discretization is a regular lattice:

$$\{\varphi(a_0^n x - mt_0)) : n \in \mathbf{Z}, m \in \mathbf{Z}^d\} \tag{8.3}$$

where $a_0, t_0 > 0$ are two scalar constants defining the discretization step sizes for dilation and translation, respectively. Typically we take the dyadic grid for discretization. Now the countable family (8.3) should be truncated into a finite set. Usually we only want to estimate $f(x)$ on a compact domain $D \subset \mathbf{R}^d$ and the wavelet function $\varphi(x)$ is chosen to have compact or rapidly vanishing support, therefore we can replace in (8.3) $m \in \mathbf{Z}^d$ by $m \in S_t$ with a finite set $S_t \subset \mathbf{Z}^d$; on the other hand, $n \in \mathbf{Z}$ should be replaced by $n \in S_a$ with a finite set $S_a \subset \mathbf{Z}$ corresponding to the "desired" resolution levels of the estimation. In practice, 4 or 5 consecutive dilation levels are generally sufficient. After such a truncation being performed, family (8.3) is replaced by

$$\{\varphi(a_0^n x - mt_0)) : n \in S_a, m \in S_t(n)\} \tag{8.4}$$

Note that the cardinality of this wavelet library grows exponentially with the dimension d. The following procedure is used to overcome this for sparse training data set : scan the training data set $\mathcal{O}_1^N$, for each sample point in $\mathcal{O}_1^N$, determine the wavelets in (8.4) whose supports contain this data point. When using this method, the dimension d is not a critical factor of complexity, since family(8.4) does not need to be actually created. In practice the number N of available observations is limited, this method allows to handle problems of relatively large dimension d. In particular, if the supports of the wavelets are approximated by hyper-cubes in $\mathbf{R}^d$, this method is easily implemented. From now on we denote by W the resulting *library of wavelet regressor candidates.* For computational convenience we normalize the wavelets and get the library W composed of the wavelets :

$$\varphi_i(x) \quad = \quad \alpha_i \varphi(a_i(x-t_i)), i = 1, \ldots, L$$

where L is the number of elements in W, a_i, t_i correspond to the dilation and translation parameters a_0^n and $a_0^{-n} m t_0$ of the wavelet φ_i, and α_i is the normalizing factor. The numbering order with i is arbitrary.

8.3 Selecting Best Wavelet Regressors

The problem of regressor selection is then to select a number $M \leq L$ of wavelets, the "best" ones from W for building the regression.

$$f_M(x) = \sum_{i \in I} u_i \varphi_i(x) \tag{8.5}$$

where I is a M-elements subset of the index set $\{1, 2, \ldots, L\}$. This is a classical problem in regression analysis [21]. Let $\mathcal{I}_M$ be the set of all the M-elements subsets of $\{1, 2, \ldots, L\}$. For any $I \in \mathcal{I}_M$ the optimal linear weights u_i of (8.5) are found using the least squares method. Then the question is how to choose $I \in \mathcal{I}_M$ which minimizes the averaged prediction error

$$J(I) = \frac{1}{N} \sum_{k=1}^{N} \left(Y_k - \sum_{i \in I} u_i \varphi_i(X_k) \right)^2$$

Determining the optimal number M should be performed using Generalized Cross Validation, cf. Section 3.. For a chosen M, selecting the M optimal regressors must be performed via exhaustive search. To overcome this, three different heuristics are proposed instead, details are found in [30].

The residual based selection (RBS). The idea of this method is to select for the first stage the wavelet in W that fits best the observations $\mathcal{O}_1^N$, then repeatedly select the wavelet that fits best the residual of the fitting of the previous stage. In the literature of the classical regression analysis, this method is considered to be inefficient, for example in [21] where it is called *stagewise regression procedure*. For classical regressions the number of regressor candidates is usually small, hence alternative more complicated and more efficient procedures are preferred. In our situation the number of regressor candidates may reach several hundreds or even more, the computational efficiency becomes more important and the simple residual based selection should be a first choice. Recently it has also been used in the matching pursuit algorithm of S. Mallat and Z. Zhang [38] and the adaptive signal representation of S. Qian and D. Chen [47].

Stepwise selection by orthogonalization (SSO). The idea of this method is to select for the first stage the wavelet in W that fits best the observations $\mathcal{O}_1^N$, then repeatedly select the wavelet that fits best $\mathcal{O}_1^N$ while working together with the previously selected wavelets. For computational efficiency, later selected wavelets are orthogonalized to earlier selected ones. This explains the name of this method. It has been used in radial basis function (RBF) networks and other nonlinear modeling problems by S. Chen *et al.* [10, 11].

Backward elimination (BE). In contrast to the previous methods, the backward elimination method starts building the regression (8.5) by using all wavelets in W, then eliminates one wavelet per stage, while trying to increase as less as possible the residual at each stage. For each stage a least squares problem must be solved. One iterative scheme is proposed to reduce the computational cost. Experimentally this method gives good results, but computationally it is not efficient for very large wavelet libraries.

9. Fuzzy Models: Expressing Prior Information in Nonlinear Nonparametric Models

Fuzzy models such as typically used in fuzzy control [33] are obtained as follows (we refer the reader to [30] for an extended discussion of this topic).

1. Input variables are scalar and are written $x_1, \ldots, x_d$. Input locations are encoded via fuzzy set membership functions, i.e., functions $\mu_A(x_i)$ with values in $[0,1]$ where symbol A is just a label; fuzzy set membership function μ_A is the mathematical meaning of "fuzzy set A". Thus, for each value of x_i, the statement "$\mathbf{x}_i$ **is A** " has a value equal to $\mu_A(x_i)$, such statements are premisses of so-called "fuzzy rules". Be careful that a typical form of such statements is "$\mathbf{x}_i$ **is large** ", which does not convey as much information as formula $\mu_A(x_i)$ does, since function μ_A is not explicitly specified by this statement.
2. Fuzzy rules are statements of the form "**if x is A then y is B** ". Note that more complex premisses can be used, using **and, or, not**. We now consider the particular case in which both the fact and the conclusion of the rule are *crisp* statements, i.e., have the standard forms "**x is** x" and "**y is** y_o" respectively, where x, y_o are ordinary values. Membership functions of crisp statements are just Dirac, i.e., for "**y is** y_o" we have $\mu_{y_o}(y) = 1$ if $y = y_o$ and $\mu_{y_o}(y) = 0$ otherwise. For such a case, fuzzy rule

   if x is A then y is B

   translates into the following function in the usual sense:

   $$x \mapsto y \quad : \quad y = \mu_A(x) \times y_o \ , \tag{9.1}$$

   where y_o appears as a parameter.
3. A "fuzzy rule basis" is a collection of fuzzy rules of the form, say,

   $$\begin{array}{lll} \textbf{if}(x_1 \ \textbf{is} A_{1,1}) \ \textbf{and}... \ \textbf{and}(x_d \ \textbf{is} A_{1,d}) & \textbf{then} & (y \ \textbf{is} B_1) \\ & \cdots\cdots\cdots & \\ \textbf{if}(x_1 \ \textbf{is} A_{p,1}) \ \textbf{and}... \ \textbf{and}(x_d \ \textbf{is} A_{p,d}) & \textbf{then} & (y \ \textbf{is} B_p) \end{array}$$

   where the $A_{j,i}$ are doubly indexed labels, i is the index of the input coordinate, and j is the index of the rule. The mathematical translation of this rule basis is now given. When the rule basis is such that

   $$\forall i = 1, \ldots, d : \sum_{j=1}^{p} \mu_{A_{j,i}}(x_i) \ \equiv \ 1 \tag{9.2}$$

   i.e., the fuzzy sets form a *"fuzzy partition"* of the space, then combining rules translates into the sum of their associated membership functions. Thus, for crisp input and output statements again, the above fuzzy rule basis translates into the ordinary function

$$
\begin{aligned}
y &= \sum_{j=1}^{p} y_j \left(\prod_{i=1}^{d} \mu_{A_{j,i}}(x_i) \right) \\
&\triangleq \sum_{j=1}^{p} y_j \; w_j(x) \; , \qquad (9.3)
\end{aligned}
$$

where $x = (x_1, \ldots, x_d)$, this defines the weights $w_j(x)$. The mechanism we have used in (9.3) to combine rules is called *"defuzzification"*, it is a kind of weighted average of the fuzzy rules.

Usually, fuzzy set membership functions are parametrized functions of the form

$$\mu_A(x) \quad = \quad \mu(a(x-t)) \qquad (9.4)$$

where $\mu(x)$ is a given function with values in $[0,1]$, a is a dilation factor, and t is a translation factor, and the pair (a, t) encodes the fuzzy set A. Mostly used is the piecewise linear function μ such that $\mu(1) = 1$ and $\mu(x) = 0$ for x outside the interval $[0,2]$, i.e., a spline of order 1. In this case, the defuzzification mechanism (9.3) just performs interpolation. If the fuzzy partition is fixed and not adjustable, then we get a particular case of the Kernel estimate (3.1). Obviously, fuzzy models such as (9.3) are amenable of identification since they have some unknown parameters for tuning, namely the y's, a's, and t's. Identified fuzzy models are often referred to as "neuro-fuzzy models" in the A.I. literature [24], since standard backpropagation (i.e., stochastic gradient) can be used for their training, exactly as for neural networks. It is also proved that fuzzy models are universal approximants [55], which is not surprising.

To summarize, fuzzy models are described by fuzzy rule bases, plus some additional parameters which make vague statements such as "large", "small", etc., to be precise in terms of fuzzy set membership functions. The fuzzy rule basis exhibits the structure of the model, plus some coarse features related to the location of the elementary functions in the decomposition (9.3). Thus *fuzzy models are just particular instances of the kind of nonlinear nonparametric model we consider here, with the advantage of providing the fuzzy rules as a way to describe some possibly available prior knowledge.* In the experiments reported in Section 10., neuro-fuzzy modelling is used in the above sense. In the extended version [30] of this tutorial, we present a proposal for blending the practical advantages of fuzzy models with the mathematical quality of wavelet based identification techniques. This proposal is based on the notion of *hierarchical fuzzy models,* which reflect the multiresolution nature of the approximants in the syntactic structure of the rules.

10. Experimental Results

10.1 Modelling the Gas Turbine System

Using the wavelet network. In the gas turbine system we introduced in Section 1.1, the temperature profile at the exhaust of the turbine is considered as the output. We need a model that predicts this temperature profile from available measurements. For the semi-physical model we mentioned in subsection 1.1, the temperature profile is predicted from the mean temperature in the combustion chambers (T_e), the mean temperature at the exhaust of the turbine (T_s), and the rotation velocity of the turbine (N). N is directly measured, T_s is given by the average of a set of thermocouples installed at the exhaust of the turbine, T_e is computed from T_s and the compression rate π of the compressor [56, 60]. By substituting T_e, the temperature profile at the exhaust of the turbine depends on T_s, π and N. As suggested by this semi-physical model, we assume that the temperature measured by each of the thermocouples installed at the exhaust of the turbine is a function of T_s, π and N, which all are measured. Therefore, we can try to construct, for each of the thermocouples, a wavelet network with T_s, π and N as its input variables, and train it to predict the temperature measured by the thermocouples.

We have experimented this approach on the data taken from a gas turbine of European Gas Turbine SA. The training data were collected during about 48 hours. We have resampled the data and kept only 1000 measurement points. This gas turbine system is equipped with 18 thermocouples at its exhaust. For the sake of briefness, we show only the results concerning the first thermocouple. The resampled data are depicted in Fig. 10.1 where the plots correspond to T_s, π, N and $y = t_1 - T_s$, where t_1 is the measurement of the first thermocouple. These 1000 measurement points, that we refer to as the *training data*, are used for training models whose input vector is $x = (T_s, \pi, N)^T$. The obtained models are tested on another set of measured data, that we refer to as the *test data* set and depict in Fig. 10.2.

We have chosen the wavelet function $\varphi(x) = (d - x^T x)e^{-\frac{1}{2}x^T x}$ with $d = \dim(x)$. The number of wavelets used in the networks is chosen to be 40.

We initialize the wavelet networks with each of the proposed (RBS, SSO, BE) procedures and train them with the Gauss-Newton procedure.

In order to get some idea on the performance of the resulting models, we compare their results with those of the semi-physical model and a third order polynomial model. In Fig. 10.3 and Fig. 10.4 are shown the results obtained with the semi-physical model and the third order polynomial model, respectively. The results obtained with the wavelet networks initialized with algorithm BE, and the results after 10 iterations of the Gauss-Newton procedure are given in Fig. 10.5. In Table 10.1 we listed the mean of square errors (MSE) of all these models on the training data set as well as on the test data set. For each of these networks we give the result of its initialization

Table 10.1. *Performance evaluation of the models*

models	RBS net	SSO net	BE net
train. init. MSE	1.2656	1.0453	1.0381
train. final MSE	0.5395	0.4239	0.4503
test. init. MSE	1.2368	1.1229	1.1576
test. final MSE	1.1886	1.2348	1.0898
init. flops	2.0718×10^7	4.3714×10^8	7.5143×10^7
train. flops	1.5365×10^9	1.5365×10^9	1.5365×10^9

models	semi-physical	polynomial
train. final MSE	3.5268	2.8438
test. final MSE	2.8914	2.1135
train. flops	9.8041×10^8	4.7056×10^5

(init. MSE) and the result after 10 iterations of the Gauss-Newton procedure (final MSE). The time of computation for building these models is also listed in Table 10.1, based on our programs in Matlab 4.1 language run on a Sun Sparc-2 work station, as well as corresponding Matlab's Flop which measures the computational burden of programs.

By examining these results, we can make the following observations:

- The semi-physical model performs quite poorly in predicting the output of the system.
- The system is truly nonlinear, in addition the results obtained with the polynomial model are quite poor.
- The wavelet networks do improve the performance on prediction, but at the price of increasing computational complexity and loss of the physical meaning of the model parameters.

Results using fuzzy modelling are reported in [30].

10.2 Modelling the Hydraulic Actuator of the Robot Arm

Let us denote by $u(t)$ and $p(t)$ the position of the valve and the oil pressure at time t, respectively. A sample of 1024 pairs of $(u(t), p(t))$ was registered[9]. We divide it into two equal parts for training and testing the models. The training data are depicted in Fig. 10.6, and the test data in Fig. 10.7.

We first tried to model this system with linear autoregressive exogenous (ARX) models. More precisely, we tried to use models of the following form:

$$\begin{aligned} & p_t + a_1 p_{t-1} + \cdots + a_n p_{t-n} \\ & = b_1 u_{t-\tau-1} + \cdots + b_m u_{t-\tau-m} + e_t \end{aligned}$$

where the pure time delay τ is assumed to be an integer and e_t is some noise independent of u_t and past values of p_t. After the identification of the model

[9] We gratefully acknowledge Jonas Sjöberg and Svante Gunnarsson from Linköping University for providing the data.

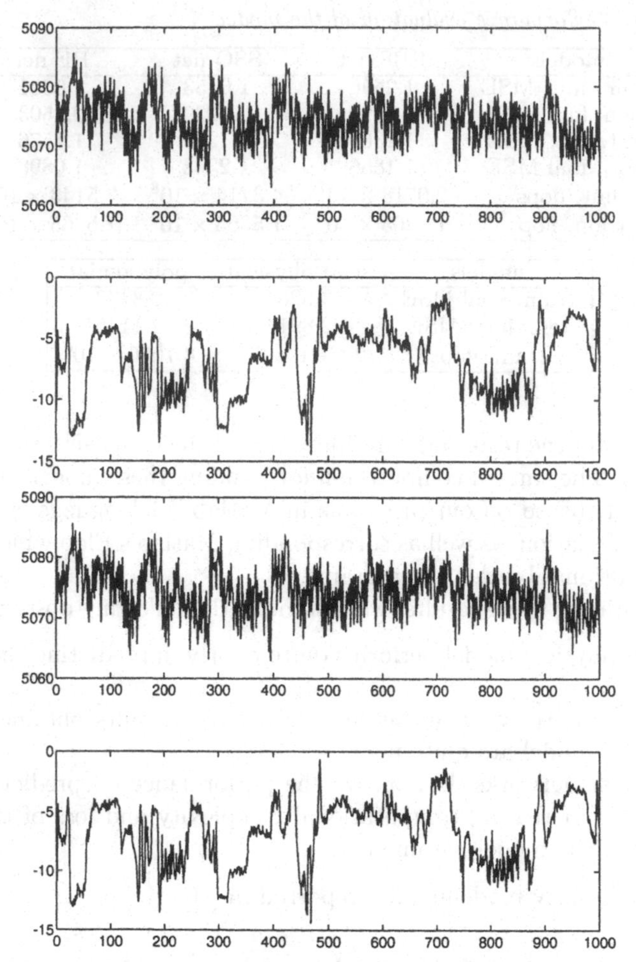

Fig. 10.1. *Training data. The plots correspond to, from top to bottom, T_s, π, N and $y = t_1 - T_s$.*

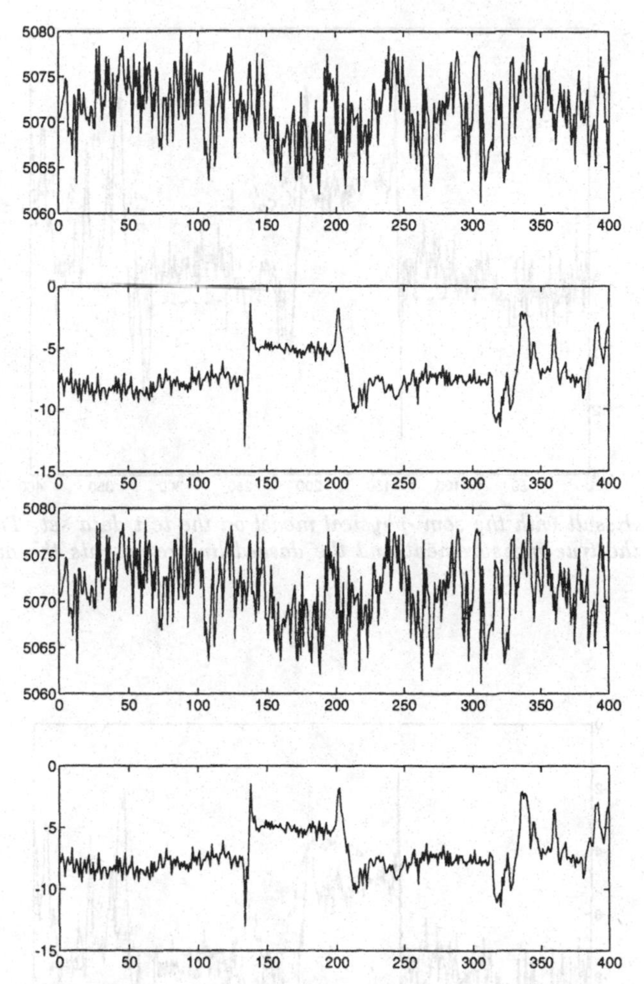

Fig. 10.2. *Test data. The plots correspond to, from top to bottom, Ts, π, N and* $y = t_1 - T_s$.

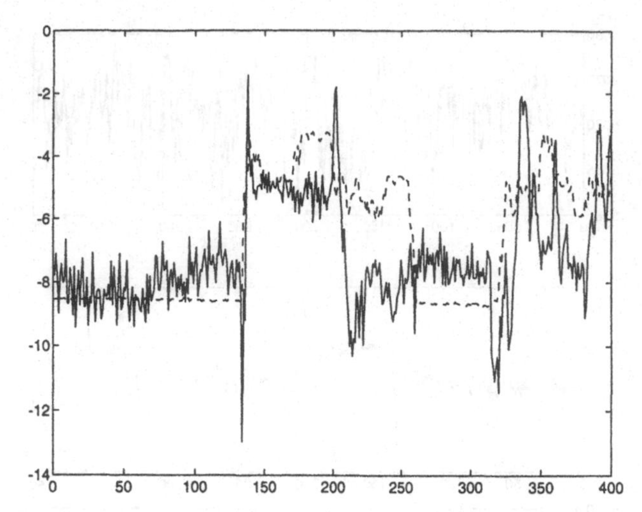

Fig. 10.3. *Result with the semi-physical model on the test data set. The solid line represents the true measurement and the dashed line represents the output of the model.*

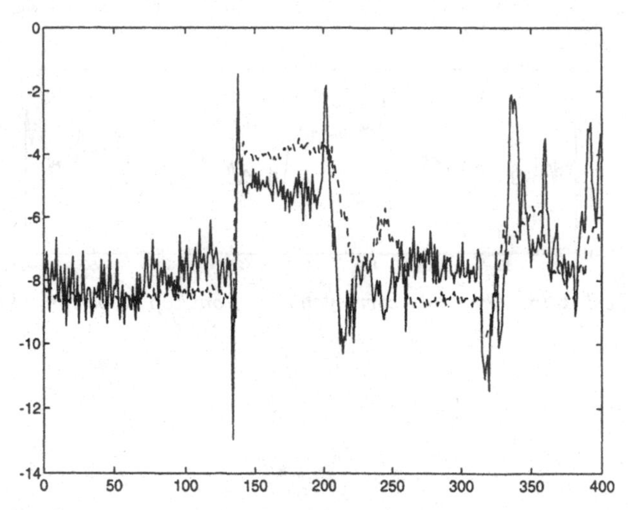

Fig. 10.4. *Result with the third order polynomial model on the test data set. The solid line represents the true measurement and the dashed line represents the output of the model.*

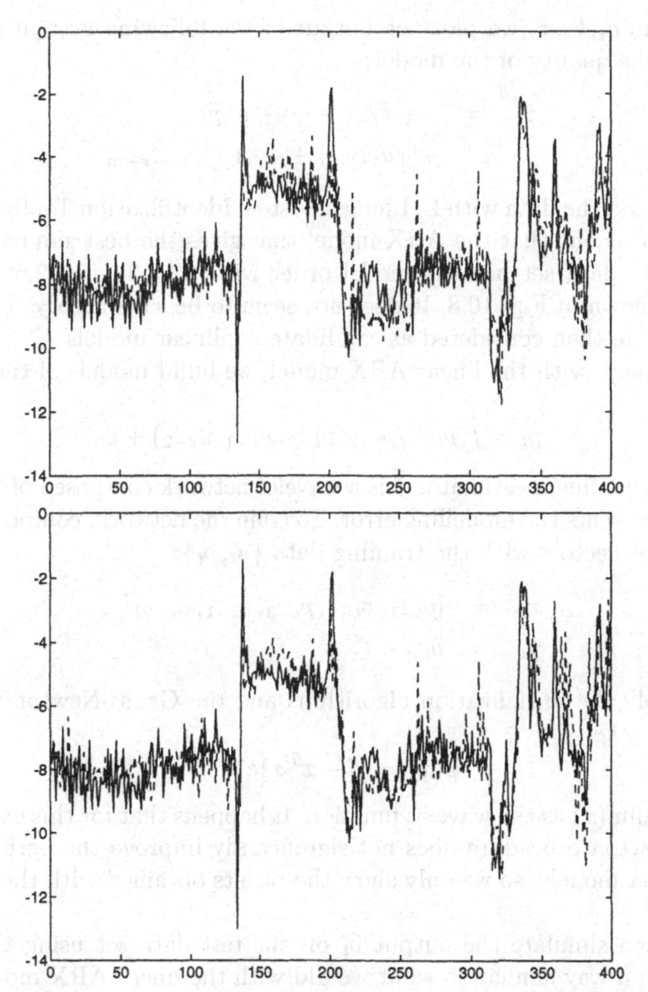

Fig. 10.5. *Results with wavelet network initialized by algorithm BE (top) and after 10 iterations of the Gauss-Newton procedure (bottom). The solid lines represent the true measurement and the dashed lines represent the output of the model.*

parameters a_i, b_j, τ, we plot the output of the following system to visually evaluate the quality of the model:

$$\begin{aligned}\widehat{p}_t &= -a_1\widehat{p}_{t-1} - \cdots - a_n\widehat{p}_{t-n} \\ &\quad + b_1 u_{t-\tau-1} + \cdots + b_m u_{t-\tau-m}\end{aligned}$$

We processed the data with L. Ljung's System Identification Toolbox, Version 3.0a. It turns out that the ARX model that gives the best simulation result on the test data set has the model order with $n = 3, m = 2, \tau = 0$. This result is shown in Fig. 10.8. It does not seem to be satisfactory. The wavelet networks are then considered as candidate nonlinear models.

In analogy with the linear ARX model, we build models of the following form:

$$p_t = \widehat{f}(p_{t-1}, p_{t-2}, p_{t-3}, u_{t-1}, u_{t-2}) + e_t$$

where the nonlinear estimator $\widehat{f}$ is a wavelet network composed of 6 wavelets, and e_t represents the modelling error. To train the network, compose its input and output vectors with the training data $\{u_t, p_t\}$:

$$\begin{aligned}x_t &= [p_{t-1}, p_{t-2}, p_{t-3}, u_{t-1}, u_{t-2}]^T , \\ y_t &= p_t .\end{aligned}$$

Then apply the initialization algorithms and the Gauss-Newton procedure. Again we take

$$\varphi(x) = (d - x^T x)e^{-\frac{1}{2}x^T x}$$

with $d = \dim(x)$ as the wavelet function. It happens that for this example, the Gauss-Newton procedure does not significantly improve the performance of the wavelet models, so we only show the results obtained with the initialized network.

We then simulate the output $\widehat{p}_t$ on the test data set using the wavelet models, in a way similar to what we did with the linear ARX model:

$$\widehat{p}_t = \widehat{f}(\widehat{p}_{t-1}, \widehat{p}_{t-2}, \widehat{p}_{t-3}, u_{t-1}, u_{t-2})$$

The simulation results obtained with the wavelet network initialized with algorithm BE are depicted in Fig. 10.9. Clearly, the wavelet models significantly improve the result of the simulation.

11. Discussion and Conclusions

In this tutorial we have discussed the wide area of nonparametric nonlinear estimation from the point of view of system identification. We have seen that a huge amount of work has been pursued in the statisticians community. We also know from numerous press releases that, in parallel, the A.I. community revitalized the same area by advertizing neural networks, fuzzy models,

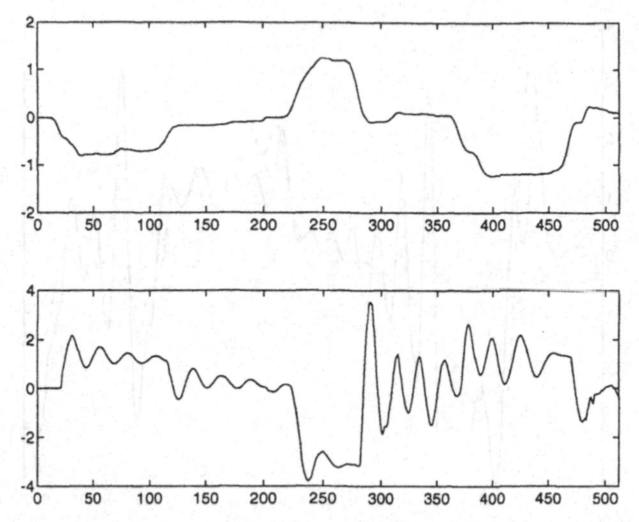

Fig. 10.6. *Training data: the input u_t (top) and the output p_t (bottom).*

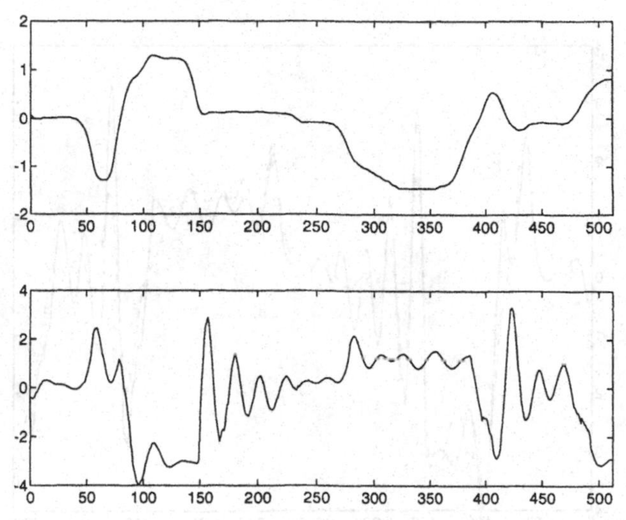

Fig. 10.7. *Test data: the input u_t (top) and the output p_t (bottom).*

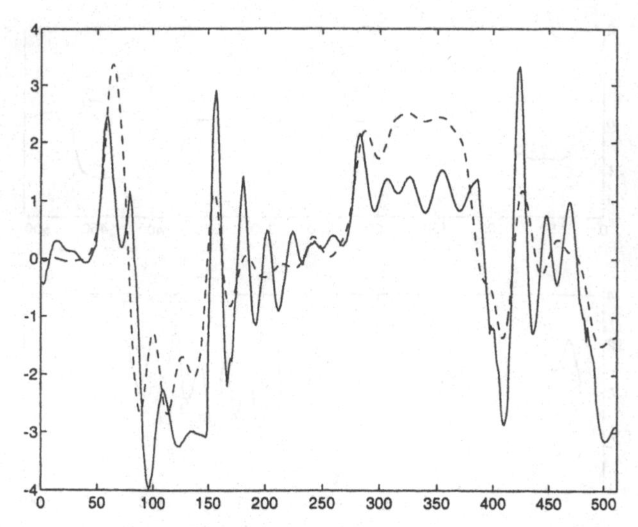

Fig. 10.8. *Result with the linear ARX model on the test data set. The solid line represents the true measurement and the dashed line represents the simulated output.*

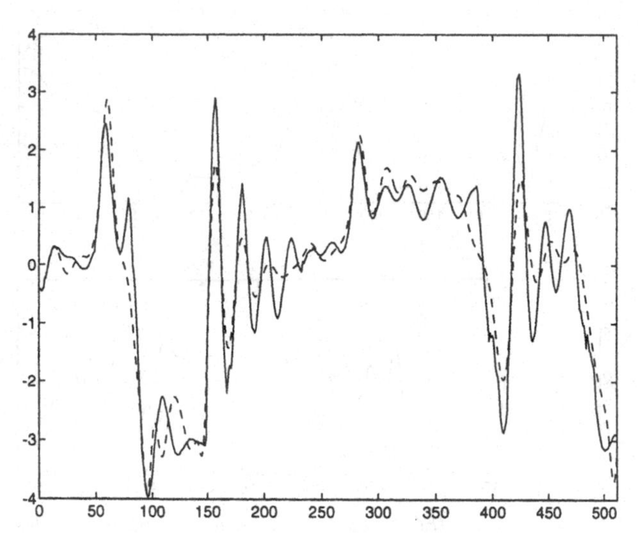

Fig. 10.9. *Result with the wavelet network initialized with algorithm BE. The solid line represents the true measurement and the dashed line represents the simulated output.*

and neuro-fuzzy models. In addition, A.I. scientists and engineers packaged these techniques with user oriented software and even hardware. It is not until recently that the A.I. community went interested in the mathematical developments and algorithms from statistics. At the same time, statisticians became involved in the mathematical study of the methods advertized by the A.I. community and engineering practice. In parallel, the control community recognized those models and estimation algorithms as possible candidates for nonlinear black-box system identification. In this tutorial, we have tried to put together material — both classical and very modern — from different areas, and have discussed both mathematical and practical issues. Here is a summary of tentative conclusions and suggestions for future work.

Practical issues.

- *Models for prediction and simulation.* As reflected by the reported experiments, our experience has been that nonlinear nonparametric models are very good at predicting behaviours, even for sparse training data set, provided that the training data set reflects all actual operating conditions that can occur. The quality of prediction can rapidly vanish outside the range of training data set, however.
- *System monitoring and diagnostics.* The reported experiments on the gas turbine case study show that data fit is much better for our wavelet network (and even for the neuro-fuzzy network) than for our semi-physical model. Accordingly one may expect a better performance in slight change detection by using the wavelet network. Designing a change detection procedure based on the wavelet network can be performed by applying the general asymptotic local approach discussed in [5] [60]. However since the parameters of the network have no useful interpretation, diagnostics would require learning the failure modes from training data sets: this is unrealistic since real data corresponding to failure modes are (fortunately) seldom. Thus diagnostics requires a combination of data *and* prior knowledge, preferably in the form of a model: data are the current data (from safe or failured mode), and the model is used to describe prior knowledge about failure modes. In fact, gas turbine monitoring and diagnostics was successfully performed using our seemingly poor semi-physical model, see [60] [3] for an account of the results.
- *Describing prior knowledge.* Fuzzy models and their associated rules can be used to describe prior knowledge for nonlinear nonparametric models. Now if it is desired to blend the style of fuzzy rules with the mathematical quality of modern nonparametric models. We have discussed a possible proposal toward this objective. This has to be further explored. In addition, it would be interesting to develop statistical methods checking for violation of a particular subset of fuzzy rules, this would blend methods from A.I. and statistics model based diagnostics.
- *Software support.* Our current experience can be summarized as follows. There are three different kinds of needs for nonlinear black-box identifi-

cation: low dimensional input (say, $1,2,3$), medium dimensional input (in the range of tens), and large dimensional input (in the range of hundreds or thousands). The first case typically corresponds to curve fitting and is useful in signal or image processing, and sometimes in control. High performance algorithms based on wavelets are available today, which outperform others in both accuracy and computational cost, see Section 7, and software is available, such as [52].
The second case has its main applications in system identification and control. There, RBF (radial basis function) networks, which provide fast noniterative training procedures, are preferred; theoretical studies and experiments suggest that wavelet networks, such as discussed in this paper [59], are likewise more efficient candidates as well as semi-heuristic methods listed in the beginning of Section 5 [23],[26], [22] and others.
Finally, sigmoid based neural networks with their iterative backpropagation algorithm, both simple and time consuming, are still effective for very large dimensional cases such as encountered in some pattern recognition applications. We have seen that alternative models with much more efficient iterative training procedures can also work well, such as Breiman's hinge functions; Breiman's hinging hyperplane algorithm fits piecewise linear models on nonlinear systems in a very efficient way.

Mathematical issues.

- *Assessing the quality of an approximation.* What is the convenient figure of merit for the estimation error $\|f-\widehat{f}\|$? We have emphasized in this tutorial the central role played by *Besov spaces*: this is a triply parametrized family of spaces of functions, that are generally smooth but may have sparse singularities. Being smooth outside localized singularities is a common feature of most of the nonlinear systems encountered in practice: Besov spaces should be used to assess the quality of an estimator.
- *Quality of fit from noisy data, and "Cramer-Rao bounds".* Maximal risks and lower rates of convergence provide adequate frameworks; they have to be used in combination with Besov spaces. And we have shown that wavelet based estimators are optimal for systems in Besov spaces.
- *How efficient identification algorithms really are in terms of computational cost and quality of conditioning?* When orthonormal wavelet libraries can be efficiently built (this is feasible for low dimensional input, say, up to 4 or slightly more), wavelet estimators from Section 7 are the fastest ones. For very large dimensions, wavelet libraries cannot be built today, and standard sigmoid based neural networks are preferred; Breiman's hinging hyperplane models are very promising alternative candidates. In the medium range situation, wavelet networks using partial wavelet libraries seem to be efficient alternatives to RBF networks.

Research directions. Based on the material of this tutorial, we can suggest the following three major challenges for future research.

- Providing wavelet based identification methods for higher dimensional inputs. The central question here lies in the efficient construction of wavelet libraries in higher dimensions.
- Taking advantage of multiresolution in both *time and space* is a major challenge for dynamical system identification. Functional nonlinear autoregressions of the form
$$Y_k = f(Y_{k-1}, \ldots, Y_{k-p}) + e_k,$$
or their state space counterpart, are naturally used with both neural and wavelet networks. These models do not allow playing with multiresolution for time, however, since discretization is fixed and rigid. Thus a new framework would be needed for this purpose.
- Investigating the interplay between the syntax of fuzzy modelling and modern nonparametric models certainly is a topic of major practical interest. It would provide the user with ways of describing prior knowledge within nonparametric models.

References

1. Akaike H., "Statistical predictor identification", *Ann. Inst. Math. Statist.*, 22:203–217, 1970.
2. Barron A., "Universal approximation bounds for superpositions of a sigmoidal function", *IEEE Trans. on Information Theory*, 39(3), 1993.
3. Basseville M., A. Benveniste, G. Mathis, and Q. Zhang, "Monitoring the combustion set of a gas turbine", in *Proceedings of SAFEPROCESS'94*, 1994.
4. Benveniste A., *Multiscale signal Processing : from QMF to Wavelets*, chapter "Digital Signal Processing Techniques and Applications", Advances in Control and Dynamic Systems. Academic Press, 1993.
5. Benveniste A., M. Basseville, and G. Moustakides, "The asymptotic local approach to change detection and model validation", *IEEE Trans. on Automatic Control*, 32:583–592, 1987.
6. O.V. Besov. On a family of functional spaces: Embedding theorems and applications. *Doklady Acad. Nauk SSSR*, 126, 1163-1165, 1959.
7. Breiman L., "Hinging hyperplanes for regression, classification and function approximation", *IEEE Trans. Information Theory*, 39(3):999–1013, 1993.
8. Breiman L., J.M. Friedman, J.H. Olshen, and C.J. Stone, "Classification and regression trees". Wadsworth, Belmont, CA, 1984.
9. Cencov N.N., "Statistical decision rules and optimal inference", *Amer. Math. Soc. Transl.*, 53, 1982. Providence, RI.
10. Chen S., S.A. Billings, and W. Luo, "Orthogonal least squares methods and their application to non-linear system identification", *Int. J. Control*, 50(5):1873–1896, 1989.
11. Chen S., C.F.N. Cowan, and P.M. Grant, "Orthogonal least squares learning algorithm for radial basis function networks", *IEEE Trans. on Neural Networks*, 2(2):302–309, March 1991.
12. Craven P. and G. Wahba, "Smoothing noisy data with spline functions", *Numer. Math.*, 31:337–403, 1979.
13. Daubechies I., "Ten lectures on wavelets", *CBMS-NSF regional series in applied mathematics*, 1992.
14. Delyon B., and A. Juditsky, "Optimal estimators for functional autoregression", *Tech. rep. IRISA*, in preparation, 1994.
15. Delyon B., and A. Juditsky, "Wavelet estimators, global error measures: Revisited", *Technical report 782*, IRISA, 1993.
16. DeVore R.A., B. Jawerth, and V.A. Popov, "Compression of wavelet decompositions", *Amer. J. Math.*, To appear, 1994.
17. Devroye L., and L. Györfi, *Nonparametric Density Estimation L_1 View*. John Wiley & Sons, New York, 1985.
18. Donoho D., and I. Johnstone, "Minimax risk over l_p-balls", *Technical report*, Department of Statistics, Stanford University, 1992.
19. Donoho D., I. Johnstone, G. Kerkyacharian, and D. Picard, "Density estimation by wavelet thresholding", *Technical report*, Department of Statistics, Stanford University, 1993.
20. Donoho D., I. Johnstone, G. Kerkyacharian, and D. Picard, *Wavelet shrinkage: Asymptotics*, Manuscript, 1993.
21. Draper N., and H. Smith, *Applied regression analysis.* Series in Probability and Mathematical Statistics. John Wiley & Sons, 1981. Second edition.
22. Friedman J.H., "Multivariate adaptive regression splines (with discussion)", *Ann. Statist.*, 19:1–141, 1991.

23. Friedman J.H., and W. Stuetzle, "Projection pursuit regression", *J. Amer. Stat. Assoc.*, 76:817–823, 1981.
24. Glorennec P.Y., "A general class of fuzzy inference systems", in *Proc. of CES2 Conf.*, Prague, 1993.
25. Härdle W., and J.S. Marron, "Optimal bandwidth selection in nonparametric regression function estimation", *Ann. of Statist.*, 13:1465–1481, 1985.
26. Huber P.J., "Projection pursuit (with discussion)", *Ann. Statist.*, 13:435–475, 1985.
27. Hunt K.J., D. Sbarbaro, R. Zbikowski, and P.J. Gawthrop, "Neural networks for control systems — a survey", *Automatica*, 28 n°6:1083–1112, 1992.
28. Ibragimov I., and R. Khasminskij, *Statistical Estimation Asymptotic Theory*. Springer-Verlag, Mosow, Nauka, 1981.
29. Juditsky A., "Wavelet estimators: Adapting to unknown smoothness", *Technical report IRISA*, 815, 1994.
30. Juditsky A., Q. Zhang, B. Delyon, P-Y. Glorennec, and A. Benveniste, "Wavelets in identification", *Technical Report 849*, IRISA, July 1994.
31. Kerkyacharian G., and D. Picard, "Density estimation in besov spaces", *Stat. and Prob. Letters*, 13, 15-24, 1992.
32. Korostelev A., and A. Tsybakov, *Minimax Theory of Image Reconstruction*. Springer-Verlag, Berlin, 1981.
33. Lee C.C., "Fuzzy logic in control systems, parts i and ii", *IEEE Trans. on Systems, Man, and Cybernetics*, 20 n°2, 1990.
34. Li K.C., "Asymptotic optimality of c_L and generalized cross-validation in ridge regression and application to the spline smoothing", *Ann. of Statist.*, 14:1101–1112, 1986.
35. Li K.C., "Asymptotic optimality of c_L and generalized cross-validation : discrete index set", *Ann. of Statist.*, 15:958–975, 1987.
36. Ljung L., "Perspectives on the process of identification", in *Proceedings of the 12th IFAC World Congress*, Sydney, 1993.
37. Ljung L., "Neural networks in identification, a tutorial", in *Proc. of the 10th IFAC Symposium on Identification and System Parameter Estimation*, Kopenhagen, July 4–6, 1994.
38. Mallat S., and Z. Zhang, "Matching pursuit with time-frequency dictionaries", *Technical Report 619*, New-York University, Computer Science Department, August 1993.
39. Mallows C., "Statistical predictor identification", *Technometrics*, 15:661–675, 1973.
40. Meyer Y., *Ondelettes et Opérateurs*. Hermann, 1990.
41. Morgan J.N., and J.A. Sonquist, "Problems in the analysis of survey data, and a proposal", *J. Amer. Stat. Assoc.*, 58:415–434, 1963.
42. Narendra K.S., and K. Parthasarathy, "Identification and control of dynamical systems using neural networks", *IEEE Trans. on Neural Networks*, 1 n°1:4–27, 1990.
43. Nemirovskij A., "Nonparametric estimation of smooth regression functions", *Izv. Acad. Nauk SSSR, Techn. Kibern.*, 3, 50-60, 1985.
44. Petrushev P.P., and V.A. Popov.
45. Poggio T., and F. Girosi, "Networks for approximation and learning", *Proceedings of the IEEE*, 78(9):1481–1497, 1990.
46. Polyak B.T., and A.B. Tsybakov, "Asymptotical optimality of c_p criterion for projection regression estimates", *Theory of Prob. and Appl.*, 35:305–317, 1990.
47. Qian S., and D. Chen, "Signal representation using adaptive normalized gaussian functions", *IEEE Trans. on Signal Processing*, 36(1), January 1994.

48. Rice J., "Bandwidth choice for nonparametric regression", *Ann. Statist.*, 12:1215–1230, 1984.
49. Sickel W., "Spline representations of functions in besov-triebel-lizorkin spaces on ^{n}", *Forum Math.*, 2, 451-476, 1990.
50. Sontag E.D., "Nonlinear regulation: the piecewise linear approach", *IEEE Trans. on Automatic Control*, 26:346–358, 1981.
51. Stone C.J., "Optimal global rates of convergence for nonparametric regression", *The Annals of Statistics*, 10:1040–1053, 1982.
52. Taswell C., Wavbox. Public domain MATLAB toolbox. Anonymous ftp: simplicity.stanford.edu : /pub/taswell, 1993.
53. Triebel H., *Theory of Function Spaces.* Birkhäuser Verlag, Berlin, 1983.
54. Vaidyanathan P.P., "Quadrature mirror filters banks, m-band extensions and perfect reconstruction techniques", *IEEE-ASSP Magazine*, 4(3):4–20, 1987.
55. Wang L.X., "Fuzzy systems are universal approximators", in *Proc. First IEEE Conf. on Fuzzy Systems, 1163-1169*, San Diego, 1992.
56. Zhang Q., *Contribution à la surveillance de procédés industriels.* Thesis, Université de Rennes I, December 1991.
57. Zhang Q., "Wavelet networks : the radial structure and an efficient initialization procedure", *Technical Report LiTH-ISY-I-1423*, Linköping University, October 1992.
58. Zhang Q., "Regressor selection and wavelet network construction", *Technical Report 709*, Inria, April 1993.
59. Zhang Q., Wavenet. Public domain MATLAB toolbox. Anonymous ftp: ftp.irisa.fr : /local/wavenet, 1993.
60. Zhang Q., M. Basseville, and A. Benveniste, "Early warning of slight changes in systems and plants with application to condition based maintenance", *Automatica*, 30(1):95–113, 1994. Special Issue on Statistical Methods in Signal Processing and Control.
61. Zhang Q., and A. Benveniste, "Wavelet networks", *IEEE Trans. on Neural Networks*, 3(6):889–898, November 1992.

Fuzzy Logic Modelling and Control

P. Albertos

Departamento de Ingeniería de sistemas, Computadores y Automática
Universidad Politécnica de Valencia
Spain
e-mail: pedro@aii.upv.es

1. Motivation

In any control problem two main components appear. On one hand, there is a process to be controlled which is totally or partially determined but not always well known. On the other hand, there are some control specifications the controlled system must fulfil. For that purpose, either a new subsystem, the controller, is added or some parameters of the existing components are tuned.

There are many control problems where the knowledge about the process to be controlled, the control specifications, or both, is not complete. This partial knowledge can be expressed in an approximated way or by heuristics. Let us consider the problem of driving a vehicle from a starting point to a target point, avoiding some obstacles and minimising the travelling time, Fig. 1.a. In this case, the available process model can be as accurate as required. The same can be said for the optimisation index, as well as for the operational constraints. Nevertheless, the on-line optimisation solution, under unexpected obstacles, may have a very high computational cost leading to an inapplicable solution. But a skilled driver can react and drive the car just having an approximated information about some relevant variables, such as vehicle speed, position, and orientation, distance from the obstacles and target position. Also, the driving actions such as vehicle acceleration or steering angle are determined in an approximate way. What is important is to improve the knowledge about a variable as far as it becomes more relevant for the problem (the proximity to an obstacle, for instance).

In other applications, mainly in the process industry, the process model is not well known. The number of variables is to high, the process dynamic behaviour requires distributed parameters models or it is partially known. Also the variables to be controlled can be difficult to measure and indirect measurements only provide approximate information about the control achievements. Moreover, the performance indices may be expressed in a qualitative way based on properties impossible to measure at the process level, requiring long time delay, external resources, or experimental conditions only available in a later step.

This is the case, for instance, of the quality control in the ceramic tile production, Fig. 1.b. There are subprocesses to be controlled, such as the

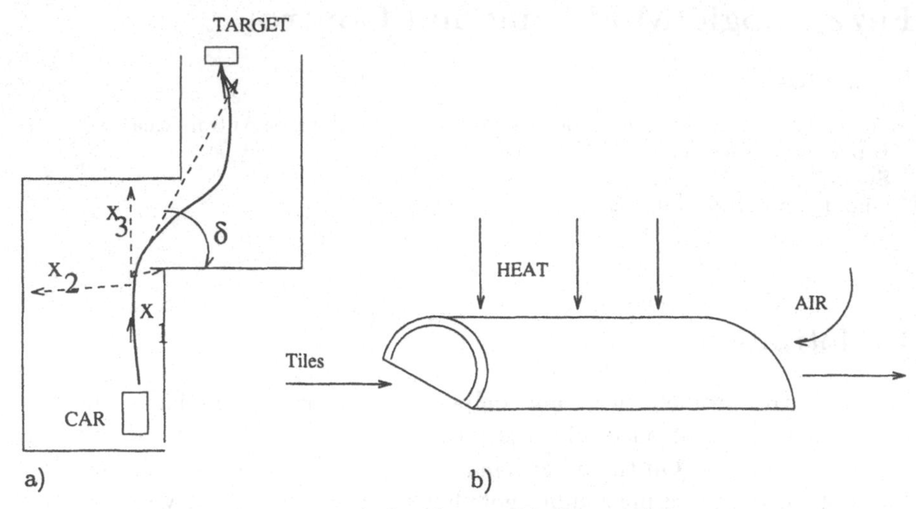

Fig. 1.1. a) Car driving. b) Ceramic kiln.

mill, the press, or the kiln, where some variables can be measured, such as the granulometry, composition, density, or local temperatures. All them influence what is really specified as requirements: tiles within measurement tolerances, appropriated brightness, mechanical resistance and so on. In some cases, there are not devices to directly get these measurements but an expert may have a good estimate based on indirect information. Of course, this information is often a qualitative or heuristic one.

Control specifications can be stated as set of specifications. In many situations (we could say in most human dependant situations) set boundaries are not sharply defined, they are better expressed linguistically and this always implies some vagueness. Probably because in order to (more) properly define the problem we must consider a larger number of variables, operating conditions, or indices.

The information available for control can be logical, qualitative or numeric. If we consider a structured and hierarchical control system, with a number of levels, all these kinds of data are available but, the closer to the process the control is the more computationally efficient it should be. On the other hand, at the upper levels, aggregated information is handled and the control actions deal with decisions. Fuzzy logic concepts are suitable at decision levels, as an improvement of the classical binary logic reasoning, [12]. But our main interest is focused in control subsystems directly connected to the process, both, receiving crisp data from the plant data acquisition system and sending control actions or references to the manipulated variables. The real time operating conditions initially lead to very simple reasoning structures involving only one level of implication. This poor reasoning structure does not allow to deal with complex systems. The number of rules increases and the knowledge appears quite difficult to understand or modify. Thus, selective reasoning as well as deep reasoning are necessary to handle the required control under a variety of uncertainly defined operating conditions.

The purpose of this chapter is to present the use of Fuzzy Logic basic concepts as a powerful tool to deal with approximate knowledge, [11]. These are the cases beforehand discussed. This methodology is an alternative tool to model the behaviour of a system and to implement its control. One of the advantages of this model should be its suitability to represent processes with non-linear behaviour as well as the possibility to be enhanced by user experience. The definition of Fuzzy Systems and some comments about their capability to approximate non linear functions are reported. These systems are the base of the so called Fuzzy Logic Controllers, (**FLC**), [1], that is controllers where the control action is computed by means of fuzzy logic operations. The design of a controller with approximate process model or specifications can not be done following the classical control design methodologies. These approaches are mainly based on a good process model, linear/non-linear, deterministic/stochastic, with bounded uncertainties, and functionally defined control objectives. The different methodologies to design FLC, as well as their structure, main components, and use are also discussed. Finally, some results about an application to the control of a cement kiln are presented.

2. Fuzzy Logic Basic Concepts

Let us start by reviewing the general control problem, its different ways of approaching, and its solution by the use of the Fuzzy Logic theory.

The controller is considered as a system, a dynamic one, with a set of input/output variables, and a mathematical model expressed by the control algorithms or control laws, with some tuneable parameters. The controller may be interpreted as either a subsystem providing appropriated performances to the controlled process or a module to carry out the activities assigned to an already defined subsystem or device (or human expert operator). As later shown, these two different viewpoints will lead to two different FLC design approaches.

The control design methodology largely depends on both, the process model and the goals. And, of course, they both are conditioned to the kind of signals the measurement or processing systems allow to use. In our case, FLC will deal with fuzzy variables. Any other kind of data should be converted into/from fuzzy information. But let us briefly define some basic concepts.

2.1 Fuzzy Logic Variables

In order to illustrate some concepts, we will refer to a physical process, like a heat exchanger. We can consider the water flow and the steam feeding pressure as input variables, and $x(t)$, the outlet temperature as the output. Some times, we have an approximate knowledge of these variables and their relationship and we are only interested in some qualitative properties. For instance, for the temperature, we are interested in its current value if and only if it belongs to some precise interval named **universe of discourse**, $\mathcal{X}$, [i.e. $\mathcal{X} = (20, 95)$].Otherwise, the variable will be out of range and its value may be associated to the corresponding extreme.

In this universe of discourse, some **linguistic variables** are defined, [Cold $= X_1$, OK $= X_2$, Hot $= X_3$]. Each one defines a fuzzy subset. Attached to each fuzzy subset is a **membership function**, (MF), such that $0 \leq \mu_{X_i}(x) \leq 1$, for all $x \in \mathcal{X}$. That is, $\mu_{X_i} : \mathcal{X} \rightarrow [0,1]$.

For any linguistic variable, X_i, some parameters to simply define the fuzzy subset may be considered:

- Support: $\mathcal{X}_i = \{x | \mu_{X_i}(x) > 0 \}$, that is $\mu_{X_i}(x) = 0$, if $x \notin \mathcal{X}_i$. For instance, Cold support may be (20,65), (50,75) for the OK support, and (65,95) for Hot support.
- Crossover: $\eta_i, \rightarrow \mu_{X_i}(\eta_i) = 0.5$.
- Bandwidth: $w_i = \{x | x \in \mathcal{X}$ and $\mu_{X_i}(x) \geq .5 \}$. In general, w_i is a compact subset and its measure is also called *bandwidth.*
- Peakvalue: $p_i, \rightarrow \mu_{X_i}(p_i) = 1$.

Fig. 2.1.a shows these concepts for the OK outlet temperature fuzzy subset.

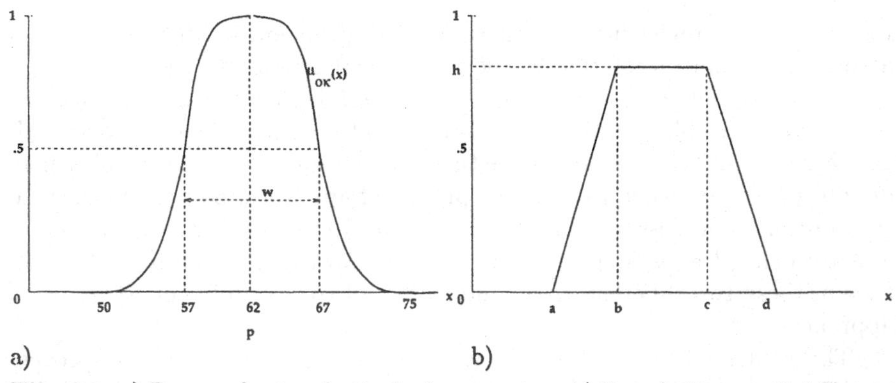

Fig. 2.1. a) Fuzzy subset and attached parameters. b) Pseudo Trapezoidal Shape.

In some cases, for instance if we have a measurement without error and it is expressed as a fuzzy subset, the support may be reduced to a single value s_i, that is, the bandwidth is zero, and the crossover and the peakvalue are the same. This fuzzy subset is call a **singleton**.

In a *Normal Fuzzy Set* X, there is an interval $[\,x_m, x_M]$, called the Normal Subset, $M(X)$, such that: i) $\mu_X(x) < 1$ if $x < x_m \cup x > x_M$, and ii) $\mu_X(x) = 1$ if $x \in [\,x_m, x_M]$. That is $M(x) = \{x | x \in \mathcal{X}$ and $\mu(x) = 1\}$

Different basic MF can be defined. For instance, the *Pseudo Trapezoidal Shape* MF is represented in Fig. 2.1.b. The support is $[a, d]$. It will be *normal,* if $h = 1$. In this case, the normal subset is $[b, c]$. It will become *triangular,* if $b = c$, and *linear,* if $c = b = a$ or $b = c = d$. The *rectangular* one, ($a = b$ and $c = d$), converts the fuzzy subset into a logical variable.

Many other geometrical functions are used but, in order to easily modify the MF, those including a reduced number of parameters are preferable. Functions like the exponential or the sigmoidal one are rather common. The exponential MF is defined by,

$$\mu_X(x) = e^{\frac{-(x-m_X)^2}{2\sigma_X^2}}, \quad \forall x \in \mathcal{X}$$

where m is the peakvalue and σ determines the bandwidth. The sigmoidal is:

$$\begin{aligned} \mu_X(x) = S(x, a, c) &= 0, & x \le a \\ &= 2\left[\frac{x-a}{c-a}\right]^2, & a \le x \le \frac{a+c}{2} \\ &= 1 - 2\left[\frac{c-x}{c-a}\right]^2, & \frac{a+c}{2} \le x \le c \\ &= 1, & x \ge c \end{aligned}$$

or in symmetrical shape,

$$\begin{aligned} \Pi(x, p, w) &= S(x, p-w, p), & x \le p \\ &= 1 - S(x, p, p+w), & x \ge p \end{aligned}$$

where w is directly the bandwidth.

In order to reduce the number of parameters, and to make easier the tuning of the MF, discretized fuzzy subsets may be used. For instance, if the universe of discourse of the variable, x is $\mathcal{X} = (-10, 10)$, and a set of seven linguistic variables is used (NB, NM, NS, Z, PS, PM, PB, where N := negative, P := positive, S := small, M := medium, B := big, and Z := zero), $\mathcal{X}$ may be divided into a number of intervals, say (-10,-7), (-7,-4), (-4,-1), (-1,1), (1,4), (4,7), and (7,10). For any value of x, the range it belongs to is selected and the most appropriated linguistic variables with attached discretized membership function are assigned to it. Table 2.1 is an example of such a discretization.

If the number of intervals is larger than the linguistic variables one, the MF values may be expressed by $[0, \ldots, .3, .7, 1, .7, .3, 0, \ldots]$, the unity value being assigned to the interval characterising the linguistic variable.

Fuzzy Variables Definitions. For a given physical variable x, with universe of discourse $\mathcal{X}$, let us assume a set $\bar{X}$ of fuzzy subsets X_i, $i = 1, 2, \ldots, N$, such that $\forall x, x \in \mathcal{X} \Rightarrow \quad 0 \le X_i(x) \le 1$.

Table 2.1. Discretized MF

Membership functions values								
		Linguistic Variables						
Int.	Range	NB	NM	NS	Z	PS	PM	PB
1	$x < -7$	1.	.7	.3	0	0	0	0
2	$-4 > x \geq -7$	.7	1	.7	.3	0	0	0
3	$-1 > x \geq -4$	.3	.7	1	.7	.3	0	0
4	$1 > x \geq -1$	0	.3	.7	1	.7	.3	0
5	$4 > x \geq 1$	0	0	.3	.7	1	.7	.3
6	$7 > x \geq 4$	0	0	0	.3	.7	1	.7
7	$x \geq 7$	0	0	0	0	.3	.7	1

-$\bar{X}$ is a *Complete Partition* in $\mathcal{X}$ if $\forall x,\ x \in \mathcal{X},\ \exists X_i$, such that $X_i(x) > 0$.

-$\bar{X}$ is a *Consistent Partition* in $\mathcal{X}$ if $\forall x_0$, such that $X_i(x_0) = 1, X_j(x_0) = 0$ if $j \neq i$,

-$\bar{X}$ is an *Ordered Partition* in $\mathcal{X}$ if $\forall x_1 \in M(X_i), x_2 \in M(X_j)$, and $i > j$, $\Rightarrow x_1 > x_2$.

In the discretization example above considered, the consistency property is not held because for a given value, for instance $x = 0$, the MF are: $\mu_Z(0) = 1$, but $\mu_{NS}(0) = \mu_{PS}(0) = .7, \mu_{NM}(0) = \mu_{PM}(0) = .3$. On the other hand, it is a complete and ordered partition.

Uncertainty measurement. It is clear that the membership function of a fuzzy subset gives an idea of the degree of fuzziness in the subset definition. For instance, following the classical set theory, if there is a fuzzy subset X such that $\mu_X(x) = 0$ for $x < x_0$ and $\mu_X(x) = 1$ for $x > x_0$, no uncertainty is attached to such a definition. On the other hand, if $\mu_X(x) = .5$ for all $x \in X$, the membership of any x value is totally uncertain. Generally, there is an increasing zone $[\ x_a, x_b]$, (a-b in the trapezoidal example), where $0 = \mu_X(x_a) < \mu_X(x) < \mu_X(x_b)$, a decreasing zone $[\ x_c, x_d]$ where $\mu_X(x_c) > \mu_X(x) > \mu_X(x_d) = 0$, or both. The amplitude of the transition zone $[x_a, x_b] \cup [x_c, x_d]$, can be considered as a measurement of the uncertainty. The shorter the transition zones are the sharper the membership function is. It can be said that if X_1, X_2 are two fuzzy subsets over the same universe of discourse, $\mathcal{X}$, if $\mu_{X1}(x) \geq \mu_{X2}(x)$, for all $x \in X$, X_1 is *sharper* than X_2 . The X_2 uncertainty measurement should be larger than the X_1 one. If the MF bandwidth is narrow, the MF will be a discriminant one. In an heuristic sense it means a clear meaning is given to the linguistic variable. On the other hand, a flat MF will be non-discriminant, that is, almost any variable value belongs to the fuzzy subset.

In [9], some uncertainty measurements are reviewed. In particular, following Ebans, [5], there are some properties any uncertainty measurement must satisfy. Let us call $U(X)$ a normalised $(0 \leq U(X) \leq 1)$ uncertainty measurement of a fuzzy subset X_1, defined by $\mathcal{X}$ and $\mu_{X_1}(x)$. It must be:

1) $U(X_1) = 0$ iff $\mu_{X_1}(x) = 0$ (or 1) $\forall x \in \mathcal{X}$
2) $U(X_1) = 1$ iff $\mu_{X_1}(x) = .5\ \forall x \in \mathcal{X}$
3) $U(X_1 \geq U(X_2)$ if X_1 is sharper than X_2.
4) $U(X_1) = U(1 - X_1)$, where $\mu_{X_1}(x) = 1 - \mu_{1-X_1}(x)$, $\forall x \in \mathcal{X}$
5) $U(X_1 \cup X_2) + U(X_1 \cap X_2) = U(X_1) + U(X_2)$

A simpler way to look at the uncertainty is to consider the relative size of the union of intervals [a,b] ∪ [c,d], although this measurement does not satisfy condition 3.

Fuzziness Versus Probability. The concept of fuzziness is sometimes misunderstood as equivalent to probability. It is clear that probability measurements take sense in the context of huge amount of data. Fuzziness can represent also this idea, but more properly reflects for an object or element the degree of membership to a given set of them.

In [4], there is an example clearly illustrating this difference. Assume that you have a set of water bottles that may contain poison. If you know that the probability of being poisoned water is 50%, if you are very thirsty and lucky you can try to take one bottle and drink water. It is one over two possibilities to choose a good one. On the other hand, if you know that the bottles belong to the set of poisoned ones with a membership degree of .5, you'd better refrain from drinking and look for some alternative.

But sometimes, when developing a fuzzy model, the frequency of an event can be taken as a degree of membership, in the same way that in expert systems knowledge acquisition steps, this frequency is assumed as a certainty coefficient.

2.2 Fuzzy Logic Operations

Let us define the LV and universe of discourse attached to the other two variables for the heat exchanger example: the water input flow, [Y_1 = Low; Y_2 = OK; Y_3 = High; $\mathcal{Y} = (-10, 10)$], and the steam pressure [Z_1 = Low; Z_2 = OK; Z_3 = High; $\mathcal{Z} = (1.2, 2.8)$]. Membership functions, similar to that in Fig. 2.1 may be defined for any of these fuzzy subsets. Some operations may be defined between fuzzy subsets.

We know that in Fuzzy algebra we can define, with a particular interpretation, some basic operations similar to those of the basic boolean algebra operations: Equality, Inclusion, Complementation, Intersection, and Union. In particular, these operations are valid for fuzzy sets with common universe of discourse. Also, classical properties like Commutativity, Associativity, Idempotency, Distributivity, Absorption, Identity, and the Morgan's law are applicable. Moreover, from the point of view of FLC, very useful operations can be defined when different physical variables are considered. For instance:

Multivariable fuzzy relation. Between n fuzzy subsets, a joint fuzzy relation, F may be defined. The joint membership function is

$$\mu_F = \mu_{\mathcal{X}_1,\mathcal{X}_2,\ldots,\mathcal{X}_n}(x_1, x_2, \ldots x_n).$$

The fuzzy relation, output temperature OK - water flow High, that is, $\mu_F = \mu_{\mathcal{X},\mathcal{Y}}(x, y)$, will express the membership function of a given pair (temperature, flow) to this combined fuzzy subset. A particular case is the *Cartesian Product*, where the joint membership function is either the minimum of the components or their product. That is:

$$\begin{aligned} \mu_{\mathcal{X}_1,\mathcal{X}_2,\ldots,\mathcal{X}_n}(x_1, x_2, \ldots x_n) &= min(\mu_{X_i}(x_i)), \qquad \text{or} \\ &= \prod_{i=1}^{n} \mu_{X_i}(x_i) \end{aligned}$$

Sum-star Composition. If F_1 and F_2 are fuzzy relations in $\mathcal{X} \times \mathcal{Y}$ and $\mathcal{Y} \times \mathcal{Z}$, respectively, it is possible to compose both relations, the result being denoted by $F_1 \circ F_2$. The sum-star composition of F_1 and F_2, is defined by the membership function

$$\mu_F = \mu_{\mathcal{X},\mathcal{Z}}(x, z) = sum(\mu_{F_1}(x, y) * \mu_{F_2}(y, z))$$

where $*$ stands for any operator in the class of triangular norms (minimum, algebraic product,...), and *sum* stands for any operator in the class of triangular conorms (maximum, algebraic sum, bounded sum,...).

Let us assume a fuzzy relation F_1 between Low flow, Y_1, and High steam pressure, Z_3, and a second fuzzy relation F_2 between the same pressure and High temperature, X_3. If we apply the max-min composition operation, that is, the Sum-star composition where for the triangular conorm the maximum is computed and the minimum for the triangular norm, the resulting fuzzy relation will be:

$$\mu_F(y, x) = \mu_{Y_1,X_3}(y, x) = \max_z(min(\mu_{Z_3,Y_1}(z, y), \mu_{X_3,Z_3}(x, z))$$

2.3 Approximated Reasoning

In classical reasoning there are two basic inference rules, the Modus Ponens (MP) and the Modus Tollens (MT), that can be generalised for the case of approximated knowledge. By the **Generalised modus ponens**, (GMP), given the following premises:

```
premise 1:   IF x is A THEN y is B
premise 2:   x is A'
```

we can conclude:

```
conclusion:  y is B'
```

By the **Generalised modus tollens**, (GMT), given the following premises:

```
premise 1:   IF x is A THEN y is B
premise 2:   y is not B'
```

we can conclude:

```
conclusion:  x is not A'
```

If $A' = A$, and $B' = B$, then MP and MT inference rules are obtained. For instance, let us assume the following premise:

```
premise:   IF x1   is  A1            THEN  y2   is  B2
```

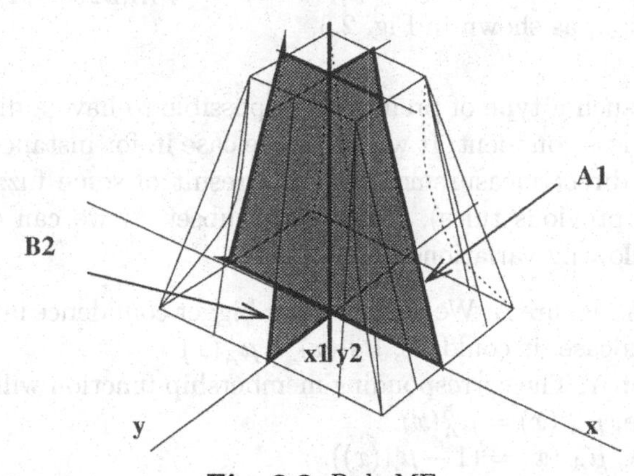

Fig. 2.2. Rule MF

The joint MF expressed by this rule is represented in Fig. 2.3, where to implement the conjunction, two options (product and minimum) are shown.

When dealing with data from physical processes, they may be crisp data. If, for instance, for $x_1 = x_0$ the MF is $\mu_{A_1}(x_0) = .7$, and we apply the cartesian product to this fuzzy relation we obtain, Fig. 2.3:

$$\mu_{B'2}(y_2) = \mu_{A'}(x_1) \circ \mu_{A1,B2}(x_1, y_2) = 0.7\mu_{B2}(y_2)$$

as a conclusion about the fuzzy value of the output y_2.

If such a rule is applied, and we know that x_1 is A_1', the conclusion will be:

$$\mu_{B'}(y_2) = \mu_{A'}(x_0) \circ \mu_{A1,B2}(x_1, y_2)$$

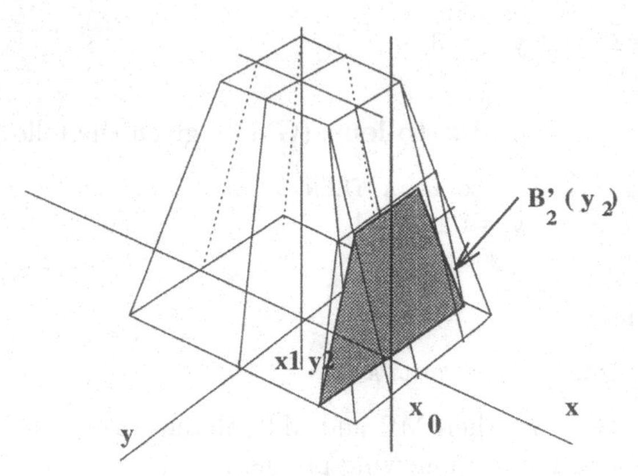

Fig. 2.3. Crisp input.

where $\mu_{A'}(x_0)$ is the MF of the input data, and $\mu_{A1,B2}(x_1, y_2)$ is MF of the rule composition, as shown in Fig. 2.3.

But, with such a type of premise, it is possible to have a different knowledge, more or less confident. It would be the case if, for instance, the variable x_2 was not a direct measurement but the result of some fuzzy implication (the result of previous rules). On a fuzzy subset A, we can define, among others, the following variations:

- A' := very A, or sure A. We point out our bigger confidence in the peakvalue of A. In this case, it could be, $\mu_{A'}(x) = \mu_A^2(x)$.
- A' ;= almost A. The corresponding membership function will be smoother. For instance, $\mu_{A'}(x) = \mu_A^{.5}(x)$
- A' := not A. $\mu_{A'}(x) = (1 - \mu_A(x))$.
- A' := few A. The MF will be almost that of "not A". For instance, $\mu_{A'}(x) = (1 - \mu_A^{.5}(x))$.

Similar approximated fuzzy sets should be considered at the conclusion B'. Possible criteria to relate the linguistic variables A' and B' are intuitive but not unique. Among the many freedoms we have in selecting operations and compositions to express a fuzzy inference, the correspondence between A' and B' fuzzy subsets in GMP or GMT inferences is one of the selections requiring more user experience.

This is not the only way we have to compose rules, [6]. Let us just mention some other options:

-*Fuzzy conjunction.* It is the general case we applied in the example above (Fig. 2.3).

If $A \rightarrow B$, is the rule R, then $\mu_R(x.y) = \mu_A(x) * \mu_B(y)$.

-*Fuzzy disjunction.* Under the same rule of inference,

$$\mu_R(x.y) = \mu_A(x) + \mu_B(y)$$

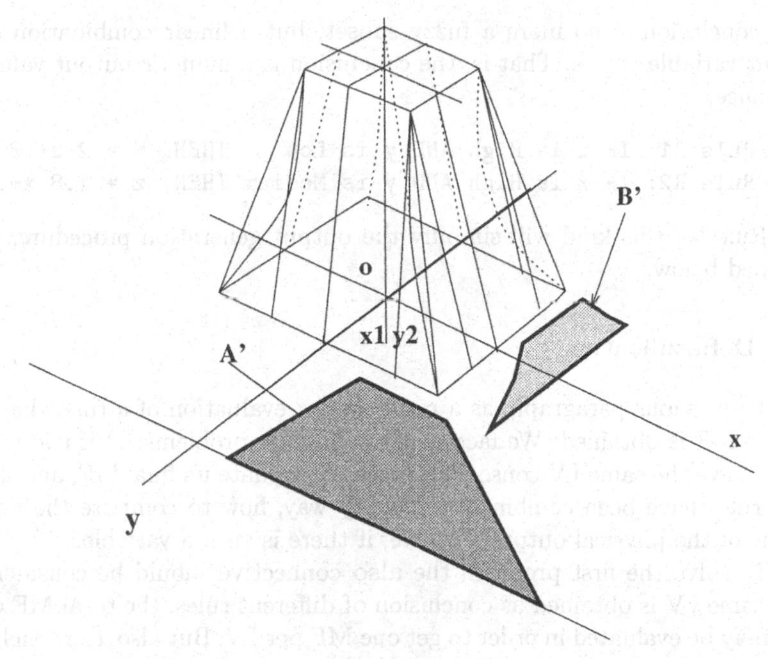

Fig. 2.4. Fuzzy subset input

- *Fuzzy implication.* Many other ways to relate A and B have been proposed in the literature, again using some intuitive relations, such as $(notA) + B, (notA) + (A * B), (notA) \times (notB) + B$, and so on.

2.3.1 Types of Rules. The typical rules, as in the previous examples, are the so called *state evaluation* rules, such as:

```
rule R1 :  IF x is X1 and   y is  Y1  THEN    z is Z1
```

where X1, X2, ..., are the *antecedents*, and Z1 is the *consequent*.

Using the fuzzy composition notation, it means: $z = (x \cap y) \circ R_1$. The conclusion MF will be

$$\mu_{R_1}(z) \doteq sum(\mu_{Z1}(z) * sum(\mu_{X1}(x) * \mu_{Y1}(y)))$$

Also, rules may have as antecedent a performance evaluation. For instance, the rule

```
rule R2 : IF [(u is C1 ---> p is P1) p is P1] THEN u is C1
```

is read as follows: If the control C1 was applied, the performance p would be P1. P1 is the desired performance value. Then, make $u \in C1$.

This kind of rules will provide something like a "predictive fuzzy control".

In FLC, usually, input is a set of physical variables. As described above, these values are converted to LV, loosing any further interest. Another kind of rules are the so called **linear combination conclusion** rules. In them, the conclusion is no more a fuzzy subset, but a linear combination of the input variable values. That is, the conclusion is a numeric output value. For instance,

```
Rule R1: IF x is High AND y is Low    THEN  z = 2 x+.3 y
Rule R2: IF x is High AND y is Medium THEN  z = 1.8 x+.5 y
```

Rules of this kind will simplify the output generation procedure, as described below.

2.4 Defuzzification

In the previous paragraph, as a result of the evaluation of a rule, the consequent MF is obtained. We face now two further problems: 1) if two or more rules have the same LV consequent, how to evaluate its final MF, and 2) once the rules have been combined in a fuzzy way, how to compute the numeric value of the physical output variable, if there is such a variable.

To solve the first problem, the **also** connective should be considered. If the same LV is obtained as conclusion of different rules, the total MF of this LV may be evaluated in order to get one MF per LV. But also, the conclusions may be treated as a set of LV attached to a unique physical variable, and then apply the defuzzifier strategy to the ensemble.

In order to simplify the rule base structure, rules consider neither, the connective "OR" in the antecedent nor the connective "AND" in the consequent. The second case is easily decomposed in as many rules as consequents are, owing to some sort of linearity in the output. All the rules will have the same antecedents.

To deal with the first case, and in order to allow a possible different fuzzy implication strategy for each antecedent AND-connected set, the rule base incorporates the also connective, that is, rules with different antecedents having the same LV in the consequent.

To combine conclusions referred to the same LV any of the triangular conorm or sum operations, $(+)$, may be used: union (maximum value), algebraic sum, bounded sum and so on. The result is bounded between 0 and 1. This will be the case if the result of this reasoning should be used as antecedent for some other rules.

In Fig. 2.4, two rules with disjoint LV consequent, are shown. The max-prod sum-star operation is performed.

Once a set of conclusions is obtained, a numeric variable may be required. The defuzzification step determines this value. For that purpose, different strategies may be, again, used.

- **Most Significant**. It is the simplest approach. If the conclusion MF set is $\{A_i'(x)\}$, the output is the MF peak value having the greatest MF value. The remaining conclusions are not considered.

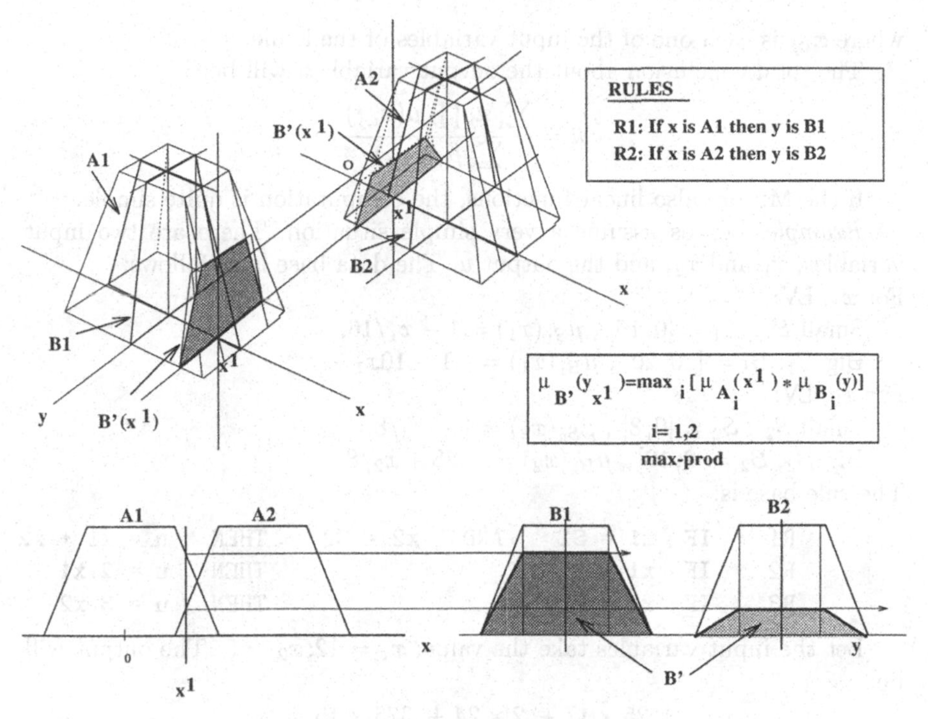

Fig. 2.5. Also connection

$$u = arg \max_x \{\mu_{A'i}(x)\}$$

- Mean of Maxima. Also trying to reduce the computation burden, but taking into account all the conclusions, the final output is computed averaging the MF peak values, p_i.

$$\bar{u} = \frac{\sum_i p_i \times \mu_{A'i}(p_i)}{\sum_i \mu_{A'i}(p_i)}$$

- Mass Centroid. In this case, not only the peak values but the full MF are considered. If we call $\mu_A(x) = sum(\mu_{A'i}(x))$, the triangular conorm of MF for each x, the computed output is:

$$u = \frac{\int_X \sum_i \mu_{Ai}(x) x.dx}{\int_X \sum_i \mu_{Ai} dx}$$

The term mass centroid refers to the weighted average of all the concluded MF.

- Linear Combination If the rules are of the linear combination type, each one will provide an output value:

$$u_i = \sum_j a_{i,j} x_{i,j} + a_{0,j}$$

where $x_{i,j}$ is each one of the input variables of the i-rule.

The total conclusion about the output variable u will be:

$$\bar{u} = \frac{\sum_i u_i \prod_j \mu(x_{i,j})}{\sum_i \mu_j(x_{i,j})}$$

If the MF are also linear functions, the computation is quite simple.

Example. Let us assume a very simple situation. There are two input variables, x_1 and x_2, and the output u. The data base is as follows:
For x_1, LV:

Small S_1 , $\mathcal{S}_1 = [0, 16]$, $\mu_{S_1}(x_1) = 1 - x_1/16$,
Big B_1, $\mathcal{B}_1 = [10, 20]$, $\mu_{B_1}(x_1) = -1 + 10x_1$.

For x_2, LV:

Small S_2 , $\mathcal{S}_2 = [0, 8]$, $\mu_{S_2}(x_2) = 1 - x_2/8$,
Big B_2, $\mathcal{B}_2 = [2, 10]$, $\mu_{B_2}(x_2) = -.25 + x_2/8$

The rule base is:

```
R1    IF  x1 = S1    AND    x2 = S2    THEN    u = x1 + x2
R2    IF  x1 = B1                      THEN    u = 2.x1
R3    IF  x2 = B2                      THEN    u = 3.x2
```

Let the input variables take the value: $x_1 = 12; x_2 = 5$. The output will be:

$$u = \frac{.25 \times 17 + .2 \times 24 + .375 \times 15}{.25 + .2 + .375} = 17.79$$

2.5 Fuzzy Systems

We can consider a fuzzy system as a black-box, where both, inputs and outputs are physical variables. Internally, some operations are carried out. Based on the above concepts, the schema of a fuzzy system can be summarised as follows. It always will have: i) a fuzzification approach, ii) a set of fuzzy rules, iii) a fuzzy inference (or many) operation, and iv) a defuzzification method. As we have seen, there are a lot of options to implement each one of these operations. As previously mentioned, if there are p outputs, the system can be decomposed in p parallel subsystems.

In order to illustrate this concept, let us define a simple fuzzy system with input, $x(t)$, a vector of dimension m, and one output, $y(t)$. For each input variable x_j there is a set of fuzzy subsets, $\bar{A}_j = \{A_{kj}(x_j)\}$, for $x_j \in \mathcal{X}_j$. The same for the output: $\bar{C} = \{C_k(y)\}$, for $y \in \mathcal{Y}$.

- Fuzzifier. We adopt the singleton shape: If the j input is $\bar{x}_j$, the input MF is $A'_j(x_j) = 1$ if $x_j = \bar{x}_j$, $A'_j(x_j) = 0$ otherwise.
- Fuzzy Rule set. For $i = 1, 2, \ldots, N$:

$$R_i: \quad \text{IF} \quad x_1 \; is \; A_{i1} \quad \text{and} \quad x_2 \; \text{is} \; A_{i2} \quad \text{and} \quad \ldots x_m \; \text{is} \; A_{im}$$
$$\text{THEN} \quad y \; \text{is} \; C_i,$$

Let us call $A_i(x)$ the joint input MF.

– Fuzzy Inference. Let us assume, for all the inferences, the sup-product as sum-star operation. That means, for input $A'(x)$, the i-rule will conclude:

$$C_i'(y) = Y_{A' \circ Ri}(y) = \sup_{x \in \mathcal{X}} [A'(x).A_i(x).C_i(y)]$$

– Defuzzifier. The mean of maxima. For each rule we get $C_i'(y)$. So, if the peakvalue of the M output MF is $p_k = arg \max_y [C_k(y)]$, for $k = 1, 2, \ldots, M$, the crisp output value will be:

$$y = \frac{\sum_{i=1}^{N} p_i C_i'(p_i)}{\sum_{i=1}^{N} C_i'(p_i)}$$

– Fuzzy System. The full system can be expressed by $y = f(x)$.

In order to better show this non-linear function, let us further reduce the fuzzy system assuming that all the output fuzzy subsets are normal, that is, $C_k(p_k) = 1$, $k = 1, 2, \ldots, M$. Then:

$$y = \frac{\sum_{i=1}^{N} p_i A_i(\bar{x})}{\sum_{i=1}^{N} A_i(\bar{x})}$$

This can be also expressed as a linear combination of non-linear functions:

$$y = \sum_{i=1}^{N} p_i . a_i(\bar{x})$$

These functions, $a_i(x)$, are called **Fuzzy Basic Functions (FBF)**:

$$a_i(x) = \frac{A_i(x)}{\sum_{i=1}^{N} A_i(x)}$$

They have some interesting properties. Let us consider a single input single output system. For the input variable, x there is a set of fuzzy subsets, $\bar{A} = \{A_i(x)\}$, for $x \in \mathcal{X}$. Assume $\bar{A}$ is a normal, consistent, complete, and ordered set (NCCO).

- *Structured similarity.* $M(a_i) = \{x | A_j(x) = 0; \forall j \neq i\}$; $a_i(x) > 0$ if $A_i(x) > 0$.
- *Compatibility.* $\bar{a} = \{a_i(x)\}$ is NCCO. So, if $x \in \{x | a_i(x) = 1\} \rightarrow a_j(x) = 0$, $\forall j \neq i\}$
- *Complementarity.* If $x \in \{x | 0 < a_i(x) < 1\}, \rightarrow a_j(x) = 1 - a_i(x)$, for either $j = i - 1$ or $j = i + 1$.
- *Less fuzziness.* $\{x | A_i(x) = 1\} \subset \{x | a_i(x) = 1\}$

That means: the set of FBF is similar but better structured than the set of fuzzy subsets it results from.

3. Fuzzy Logic Controller Structure

In this section, the general structure of a FLC is discussed. It is assumed that the FLC is directly connected to the process, that is, the process/controller interface involves A/D and D/A converters. On the other hand, the FLC also interacts with the operator, exchanging both numeric and heuristic data. A linguistic interface should be also available. This scheme is called Direct FLC, [7].

Figure 3.1 shows the general scheme of such a controller. Measurements from the process may be previously filtered or processed, in order to get trends, compensations, integrals and so on. Then, the FLC input from the process is a set of discrete variables, such as the error and functions of it, state variables, and/or references. This information, together with the numeric one introduced by the operator, if any, must be converted into fuzzy data in order to be processed by the inference engine, a simple one.

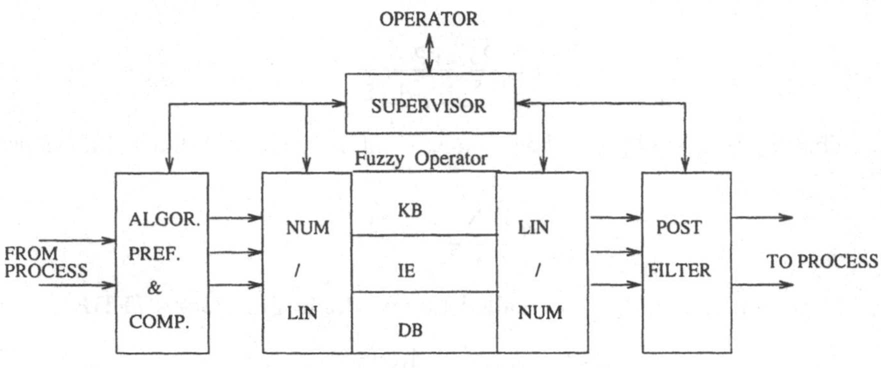

Fig. 3.1. Fuzzy Logic Controller

The output of the FLC is twofold. That to the process must be numeric or logic, asking for some control action such as a manipulated variable variation or a binary actuator switch on/off. The output to the operator should be both, graphical or numeric, to show the process variables evolution, and linguistic or symbolic, to give some explanation about the reasoning process, if required.

As shown in Fig. 3.1, supervision is a natural feature of FLC even when applied in Direct control. The availability of controller and process behaviour data allows some kind of supervision, autotuning or, in the more desirable case, learning.

Other than the digital pre- and post- processor modules, the FLC includes: a fuzzification module, to convert from numeric to linguistic data, a knowledge-based system composed by a data base, a knowledge base and an inference engine, a defuzzifier module converting data from fuzzy to numeric, and a supervisor taking care of some tunable parameters and selective options, also dealing with the operator interface.

Let us go in some detail through the different FLC modules.

3.1 Fuzzifier

The digital preprocessor provides all the physical variables the FLC may deal with. The supervisor module, if there is one, will determine the current used variables, for instance, the error and its increment (PD-like control), the error and the accumulated error (PI-like control) or any other available option. The fuzzifier role is to convert this information into fuzzy data, accessing to the fuzzy variables parameters already defined in the DB.

If the controller input is a physical measurement, the fuzzy subset it generates must be treated by the GMP reasoning to get the conclusion MF. The simplest way to handle physical data, if they are assumed to be deterministic, i.e. without any added random noise, is to consider the conclusion MF as an homotetic function of the predefined one. Another option is to directly handle the measured data, converting them into singletons, attaching a certainty coefficient to them or, using their stochastic properties, if they are random variables, to create a new fuzzy subset. In that case, the KB must be able to add new facts and/or rules from a predefined class of them, or the fuzzy implication will use the generalised reasoning. The block NUM/LIN performs this numeric to linguistic conversion.

3.2 Fuzzy Operator

Deals with the operations between fuzzy subsets. Based on the fuzzified inputs, a fuzzy output should be elaborated. In general, this output will be converted to a numeric value to be applied to the process. To perform these operations some resources are needed. Let us have a look at these components.

Data Base (DB). In the DB, linguistic and physical variables are defined. Attached to each physical variable, the following information must be given: i) The universe of discourse, ii) The linguistic variables and iii) The membership functions parameters.

In general, parametrized fuzzy sets are predefined, and at this level, a set of instantiations is attached to each physical variable. So, the number of LV and their support, and other MF parameters are fixed.

Knowledge Base (KB). As in the Expert Systems structure, the KB includes data (objects) and rules (relation between data objects). Information about the variables and fixed known facts are included in the first block. Parametric and heuristic relations between variables are expressed by rules.

It is clear that the greater the number of rules is the more complex the reasoning process, leading to more difficult to predict needed computer time.

If the FLC handles n physical variables, with m attached linguistic variables each one, the full set of rules will contain n^m rules.

Some kind of hierarchy should be established, in such a way that a set of rules is only fired if a given conclusion has been previously proved as a valid one. On the other hand, this approach may lead to rules nesting and recursion. Again, time constraints must be fulfilled.

The Rule Base must have some properties:

- Completeness. For any input pattern an output must be generated. No-action is considered as one of the possible outputs, of course.
- Consistency. Same input must conclude the same action.
- Interactivity. Any of the possible output values must be the result of, at least, one input pattern.

Inference Engine. Let us consider a simple FLC with two inputs, the error $e(t)$ and the change of the error, $c(t)$, and also to simplify the notation, only three LV attached to each one, i.e., P positive, N negative, and Z zero. For each one of the nine input combination an output will be assigned. Table 3.1 shows this set of rules, where B, M, and S stand for big, medium and small, respectively.

Table 3.1. Rule table for the control, u

		error change, c		
		N	Z	P
	N	NB	NM	PS
error,	Z	NM	Z	PB
e	P	NS	PM	PB

The same table may be used to compute the change in the control, instead of the control itself.

For instance:

```
IF error is Z  AND  change-in-error is P

        THEN  change-in-control PM
```

The fuzzifier provides the fuzzy values of the inputs. Let us see how the inference engine selects the rules to be fired.

Assume a set of MF, $\mu_i(e)$ for the error LV, and $\mu_j(ce)$ for the change-in-error variables. For a given pair of current data, (e, ce) the following matrix, named *inference matrix*, may be formed

$$m_{ij} = [\mu_i(e) * \mu_j(ce)]$$

where, again, * stands for any triangular norm (min, product,...).

The entries of this matrix allow the rule selection.

Some of these entries may be zero. That means inactivation of the corresponding rule. In order to elaborate the output, the rules to be fired, and in which order, must be decided. Usually, a threshold discard the less significant rules. We can dispose to evaluate only a fixed number of rules, those with higher m_{ij} value.

Fuzzy Conclusion. The use of the also connective will allow to merge all the conclusion's fuzzy subsets attached to the same LV. In the expert system context, some confidence improvement is evaluated if the same conclusion is obtained from different rules. In FLC implementation the simplest operation of maximum is usually performed.

3.3 Defuzzification

In the FLC structure, Fig. 3.1, there is a module to convert linguistic to numeric information. The defuzzifier will combine the reasoning process conclusions into a final control action. As we have seen before, the problem is: given a set of LV, X_i, with attached MF, $\mu_{X_i}(x)$, all defined in the universe of discourse $\mathcal{X}$ of a physical variable, x, find the most appropriated value $\bar{x}$ for this variable.

Again, different models may be applied. They have different computational complexity. The more complex they are the more reasoned conclusion try to keep in. But, as we will discuss later, this operation is not the most relevant to determine the FLC effect.

4. FLC Analysis

The FLC, as described in the last section, may be considered as a dynamic system. The preprocessor evaluates increments, accumulations and so on and the fuzzy part may be purely static. The most interesting feature being its user oriented configuration and description, the mathematical analysis is rather complex.

The FLC complexity is also related to the number of inputs and how many LV each input has. In general, simple FLC only consider two inputs, with a number of 7 or 9 LV. But, following the user oriented approach, the FLC can change the set of inputs and outputs it is dealing with. For instance, the input vector may be formed by the error and the change in the error, leading to a PD-like controller, or it may be changed to deal with the error and the accumulated error, leading to a PI-like controller. According to the operating conditions, the same FLC may perform as decided. Another simple option is to change the output variable. Dealing with the error and the change in the error as input, the output may be decided to be the control or its change.

Changing the universe of discourse in both input and output, the same structure may perform as a coarse or a fine controller, without any additional requirement.

As previously discussed, the FLC is linear in the output. This allows us to design a FLC for each control variable, the process input interaction being taken into account in the controller input. On the other hand, the input-output mapping is a non-linear one.

Some techniques to reduce the computation time have been mentioned in the last section. Also, a FLC may be implemented as a **look-up the table** controller. Let us assume a discretized universe of discourse of the input variables, if the output for any possible input combination is precomputed, the output value can be easily computed by a trapezoidal interpolation:

$$u(e,c) = \begin{bmatrix} c_{i+1} - c & c - c_i \end{bmatrix} \begin{bmatrix} u(e_i, c_i) & u(e_{i+1}, c_i) \\ u(e_i, c_{i+1}) & u(e_{i+1}, c_{i+1}) \end{bmatrix} \begin{bmatrix} e_{i+1} - e \\ e - e_i \end{bmatrix}$$

$$c_i < c < c_{i+1}; \qquad e_i < e < e_{i+1}.$$

The control variable will be expressed by the non-linear mapping the Fig. 4.1 shows.

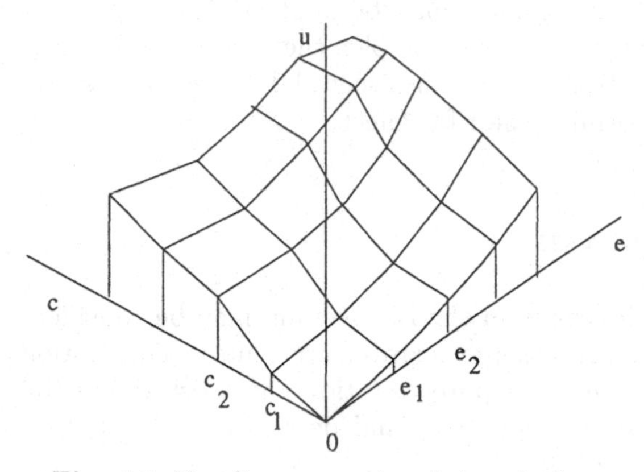

Fig. 4.1. Non-linear mapping. Interpolation

That is, the controller may be considered as an experimentally tuned **non linear controller**, [8]. For instant, in a Proportional-like FLC, the control action is computed from the measurement error. In order to simplify the notation, let us assume a set of rules expressed by Table 4.1, with numerical conclusions.

If the MF are as shown in Fig. 4.2, for any error value, e, the output may be computed by, for instance, addition of the $u_i \times \mu_i(e)$. The controller input/output relation is plotted in the Fig. 4..bb. Note that the gain $\Delta u/\Delta e$

Table 4.1. P - control

Error, e	Control, u
PB	20
PM	10
Z	0
NM	-10
NB	-20

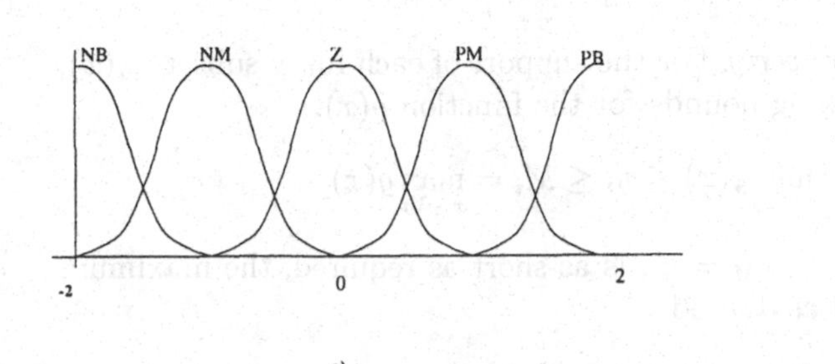

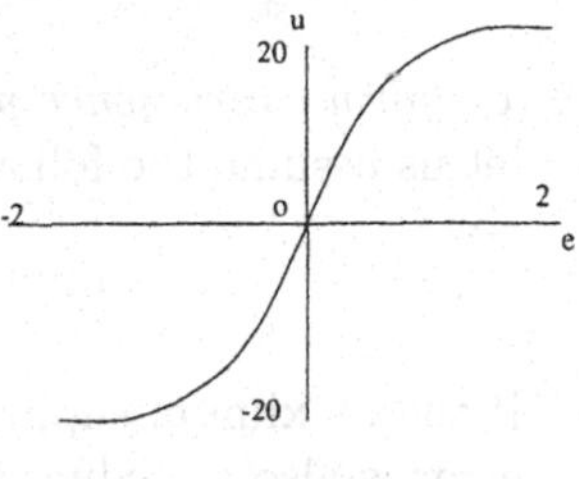

Fig. 4.2. Non-linear P-FLC

is variable, leading to a non-linear controller. The nice property being the user option to change the shape of this non-linear function according to his/her own experience.

4.1 Fuzzy Systems Approximation Properties

The FLC is a fuzzy system, as defined in section 2.5. Let us study the approximation properties of this non-linear input-output mapping, [13]. A S.I.S.O. system is assumed. Reminding the FBF definitions, if for the input variable, x there is a set of N fuzzy subsets, $\bar{A} = \{A_i(x)\}$, for $x \in \mathcal{X}$, which is NCCO, the output may be expressed by:

$$y = \sum_{i=1}^{N} y_i.a_i(x)$$

where $\bar{a} = \{a_i(x)\}$ is the set of FBF, and y_i is the contribution of each one, *a priori* defined.

Given a non-linear continuous function $g(x)$, the following properties are easily derived:

- *Basic approximation property.* The difference between the required function and the fuzzy system output is known or, at least, bounded.

$$\begin{aligned} |g(x) - f(x)| &= |g(x) - y_i| \forall x \in M(a_i) \\ |g(x) - f(x)| &= |g(x) - a_i(x)y_i - a_{i+1}y_{i+1}| \\ &\leq \max\{|g(x) - y_i|, |g(x) - y_{i+1}|\} \forall x, p_i < x < p_{i+1} \end{aligned}$$

– *Uniform approximation property.* The maximum error in this approximation is bounded.

$$\sup_{x \in \mathcal{X}} |g(x) - y(x)| = \epsilon_M = \max_{i,j}\{\epsilon_i, \delta_j\}$$

where

$$\epsilon_i = \sup_x |g(x) - y_i| \quad ; \quad \delta_j = \sup_x |g(x) - a_i(x)y_i - a_{i+1}(x)y_{i+1}|$$

– *Uniform convergence property.* For the support of each fuzzy subset $A_i(x)$, let us assume the following bounds for the function $g(x)$:

$$m_i = \min_{x \in \mathcal{X}i} g(x) \leq y_i \leq M_i = \max_{x \in \mathcal{X}i} g(x)$$

If the maximum support, $\mathcal{X}_M = \Delta$, is as short as required, the maximum error is also as reduced as desired:

$$\lim_{\Delta \to 0} \sup_x |g(x) - f(x)| \to 0$$

The last property points out the possibility to implement any non-linear control function by means of a fuzzy system. Of course, by the use of a number of LV and rules as large as required, ($N \to \infty$). All the difficulties in the design of non-linear control systems appear, but, from an experimental point of view, the controller function may be on-line tuned, as we will see in the next section.

5. Fuzzy Logic Controllers Design

According to the previous discussion, let us review the decisions we must take to define a FLC. Many items are the same than those of the classical controllers design viewpoint. Other items are specific of this control approach.

Structure. At this level, the typical issues usually considered are: the Input/Output variables, the decision/control levels, and the controller flow diagram relating the involved variables.

The controller knowledge includes most of the analysed items: the defined LV, the set of rules, the reasoning period, the way to handle physical data, the fuzzy logic operations, and the defuzzification method. To perform the fuzzy logic operations we must decide about how to evaluate the antecedents conjunction, the also conclusion connective, and the inference rule to be used.

The number of LV is determined by the ability to qualitatively distinguish different variable conditions, that is, the ability to express different operating rules for each LV. They will lead to the set of rules. The reasoning period selection is made in a similar way to the sampling period in sampled data systems. It should be short enough as compared with the dynamics of the controlled process, but large enough to allow the KB system reasoning.

The set of rules express the expert knowledge about how to perform the process control and determine the basic controller structure.

The relevance of the selected defuzzification method as well as the implementation of the fuzzy logic operations is not well defined and strongly depends on the application.

Tuning Parameters. Most of the decisions at the structural level are taken off-line. If a good knowledge of the process and the control goals is provided, we can go further in the off-line controller definition. But, usually, this structure will serve as a first step in the control problem solution and some tuneable parameters shall be decided on line. Let us review these parameters.

The physical variables range (LV universe of discourse) can be tuned according to the practice. Also, MF parameters like shape, support, or gain will allow to establish smoothed or sharper control actions. The rules to carry out this tuning are approximated, as the total approach is.

One of the FLC options is to change the control based on the results obtained. In some sense it could be compared with the multimode classical control. Some changes in the controller structure above outlined can be easily implemented: Removing LV not frequently used, addition/deletion of rules or facts, activation of guard antecedents in the event of new operating conditions, and so on.

As before, the design of a FLC always follows a trial and error procedure. If the plant to be controlled has been running and some data records are available, the FLC design and parameters tuning may be done off-line. If not, an iterative approach will lead to the most convenient controller.

As previously mentioned, one of the most important challenges in FLC design is how to adjust the rule base and the fuzzy set MF to reach some predefined control goals. This has led a lot of research in this area. One approach is to adjust the parameters by trial and error and compare their performance with classical controllers. But another proposal is to construct a FLC behaving similarly to an already existing linear controller, or a controller designed on the basis of an approximate model of the plant. Later on, some methods can be used to refine the FLC and to improve the controlled system performance.

5.1 Experimental FLC Design

In the cement application latter described, an experimental FLC design based on the operator knowledge and trial and error tuning is reported. But, if we have some measurements and a qualitative knowledge of the process to be controlled, something can be done. Let us consider, for instance, an experimental design based on linear combination conclusion rules. From a rough process behaviour knowledge, for each process measured variable, or error function, x_i:

- Define LV, $\mathcal{X}_{ik}$, and Simple MF, $\mu_{x_{ik}}(x_i)$
- Write some control rules, like:

```
Rule R1:  IF  (x1 is  *)  AND  (x2 is  *)  AND  (x3 is  *)
```

$$\text{THEN} \quad u_1 = a_{1,0} + a_{1,1}x_1 + a_{1,2}x_2 + a_{1,3}x_3 + b_{1,1}y_1$$

where y_1 is a variable not appearing in the antecedents. Assume there are m such a rules.

- Form an experimental data table with columns $x_1, x_2, x_3, y_1, \bar{u}_1$ for any of the rules above, where $\bar{u}_1$ is the control action usually applied by a skilled operator. Let N be the number of measurements.

- For any raw, j in the measurement table, express:

$$\bar{u^j} = (a_{i,0} + a_{i,1}x_1^j + a_{i,2}x_2^j + a_{i,3}x_3^j + b_{i,1}y_1^j)$$

- Apply Weighted Least Squares, (WLS), to get $\hat{a}_{i,k}$. For any of the rows above, the weight will be based on the degree of LV matching, β_j:

$$\beta_j = \frac{\mu_*^j(x_1)\mu_*^j(x_2)\mu_*^j(x_3)}{\sum_{j=1}^m \mu_*^j(x_1)\mu_*^j(x_2)\mu_*^j(x_3)}$$

These are the best parameters fitting the experimentally recommended data. Of course, the FLC will operate in conditions which are different to those considered in the design step.

5.2 FLC from Linear Ones

First let us look for a FLC equivalent to a PID one. For this application, some preprocessing to the error signal should be added to the basic structure of a FLC, [7].

Normalised membership functions with a scale factor are used, Fig. 5.1.

The model used for PID fuzzy controller is as depicted in Fig. 5.2. The parameters and design factor used in this model are:

- e_k is the input variable to the controller.
- K_i, scaling factors.
- $e_k^{'}$, $e_\Delta^{'}$, $e_\Sigma^{'}$. Normalised inputs.
- fe is the module transforming the inputs variable to its linguistic variables A_1, A_Δ, A_Σ.
- *RuleBase* Contains the knowledge of the regulator. It is a set of IF-THEN rules.
- u is the controller output.
- A_u are the linguistic variable of the output variable.

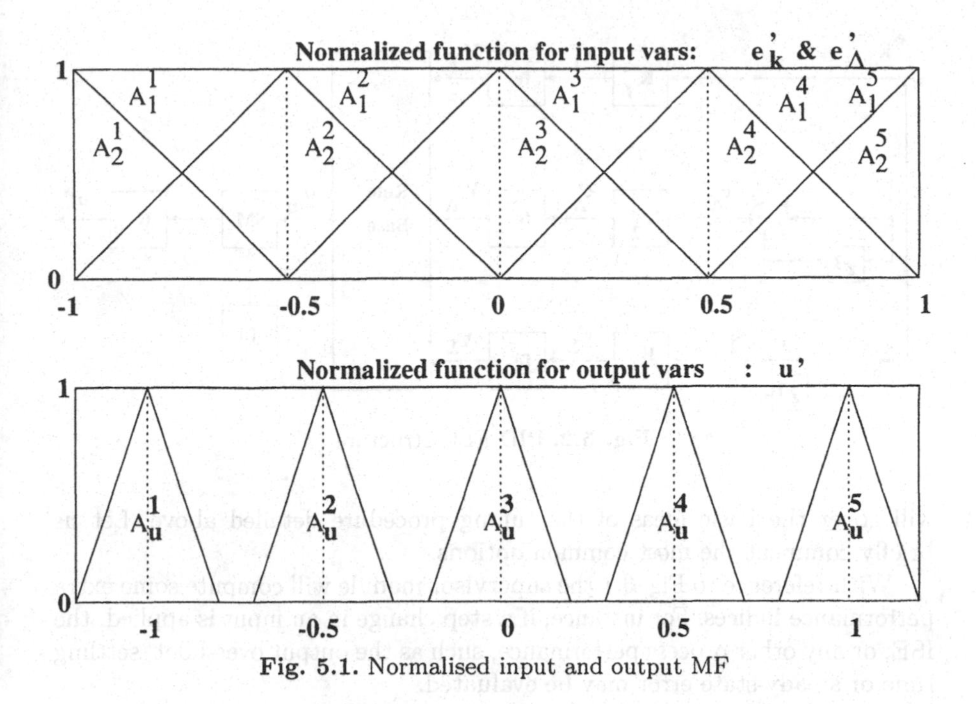

Fig. 5.1. Normalised input and output MF

- NM is the defuzzification method used to obtain a number from the output linguistic variable.
- fu is the module transforming the linguistic outputs values to its normalised numeric value u'.
- K_u, the output scale factor.

The output of a discrete PID controller, based on the proposed fuzzy structure can be expressed by:

$$u_k = K_e e_k + K_\Delta(\Delta e)_k + K_\Sigma \sum_{i=0}^{k-1} e_i$$

This equation is a linear manifold of the input variables to the FLC. By the triangular normalised MF, a linear equivalence can be found. But if we have additional information about the controlled system behaviour, the MF of both, input and output LV can be tuned and the FLC input-output mapping will be conveniently modified. Also, additional antecedents in the rule base, allow an easy treatment of different modes of operation based on range values of the process variables or the error.

5.3 FLC Supervision

As we have seen in the previous sections, the rule-base structure of a FLC allows an easy implementation of its basic supervision. The FLC supervision

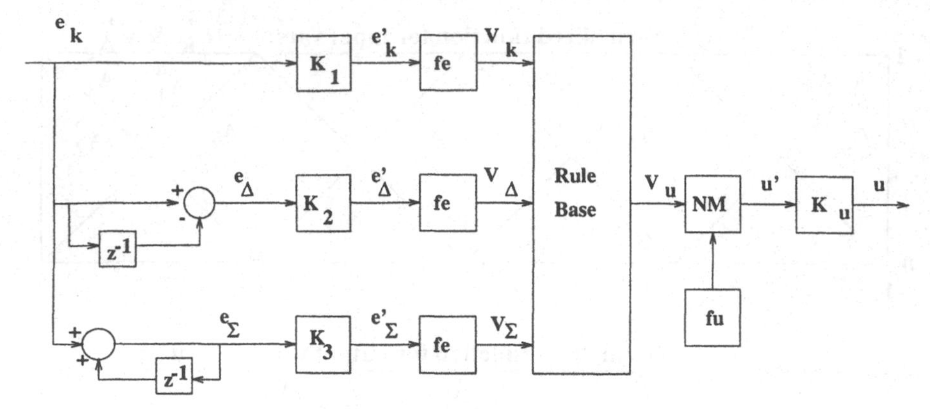

Fig. 5.2. PID FLC structure

will cover the basic ideas of the tuning procedure detailed above. Let us briefly comment the most common options.

With reference to Fig. 3.1 the supervisor module will compute some extra performance indices. For instance, if a step change in an input is applied, the ISE, or any other process performance, such as the output overshoot, settling time or steady-state error may be evaluated.

Based on a long term sampling scheme, it may detect oscillations, slow drifts or a more unusual type of performances, such as unexpected behaviour, unclear response or complex reasoning.

The kind of expected actions from the supervisory module are:

- Changes in parameters.
  In that case, a smoothed or progressive change is recommended.
- Deletion or addition of rules.
  Based on lake of use, rules may be removed. Some indicators may recommend a new rule. For instance, to reduce static errors, rules for narrower universe of discourse of the error may be added, increasing the static gain. In that case, a test of the actions effect will confirm them or recommend to go back to the previous rule set.
- Changes in variables.
  Nondiscriminant MF may lead to neglecting some antecedents in prestablished rules. Unsatisfactory behaviour may ask for the consideration of new variables.
  As before, trial and error procedure will confirm the goodness of the action.

6. Process Fuzzy Modelling

The fuzzy knowledge representation may be also used to model a plant from its qualitative knowledge, [10]. By the term plant we include the proper controller, if there is a previous system controlling the process (human operator, conventional decision based software, or a complex combined system). Following the approach of linear combination conclusion rules, the modelling procedure may be supported by algorithmic computations, enabling a faster model building.

Let us outline the procedure for a process modelling. As previously stated, some *a priori* knowledge of the process behaviour must be available. Otherwise, on plant experiments should provide this information. For instance, we must know which are the input/output physical variables to be considered. Assuming that (if not, we can come back to this assumption and repeat the procedure including/removing some variables), we must decide:

1. Number of LV.
2. Parameters in the MF of the antecedents.
3. Consequent parameters.

The iterative procedure will be:

- **a.-** If 1 and 2 are fixed, compute the optimal consequent parameters as in section 4.1.
  A record of experimental data is required, or the results must be checked by the expert.
- **b.-** If 1 is fixed, assuming linear MF, derive the optimal LV support (which is the only MF parameter), by nonlinear programming techniques. A set of Weighted Least Squares, problems must be solved.
- **c.-** Start with one variable in the antecedent side, assuming two LV attached to it. Write the simplest rules and go to the following sequence, applying **b**.
  - Combine one more input variable.
  - Double the number of LV of the most significant input.
  - .
  -
  - Stop if: i) too much rules, or ii) the WLS index in not longer improved.

The resulting model may be experimentally improved if more flexible MF and/or conclusions are allowed. The advantage of the outlined procedure is its possibility to be automated.

7. Cement Kiln Control

In order to show the practical use of the concepts above defined, an industrial application is presented. This is one of the typical fuzzy logic control applications. In our case, no special tools for FLC developing have been used, and the solution is fully open to be managed by the proper user, [2].

7.1 Process Description

Cement kilns are composed by a large cylinder (about 80 meters long) mounted at a slight inclination to the horizontal, and slowly rotating. Inside the kiln, the inlet flow of raw material (*lime*, *sand*, *pyrite*, *clay*) coming from the mill is heated until the clinker is obtained. The process is formed by three step: the **preheater**, where the material is heated using the residual heat of the combustion fumes; the **kiln**, where the clinker is formed; and the **cooler**, where the clinker interchanges its heat with the combustion air.

Some of the characteristics of this process are:

- Distributed process
- The system can operate in several different situations.
- High number of variables (between 40 and 50): pressures(9), temperatures(14), gas concentration(6), laboratory analysis, etc.
- Large delay in some variables due to the kiln condition.
- Absence of a model due to the process complexity.
- Only few people know how the system works (**expert operators**).

Functionally, two main subprocesses can be distinguished, Fig. 7.1:

- Raw material transport, which is a rather slow process. The raw material feeds the preheater, is then introduced in the kiln, where is heated up to near its fusion point to produce the clinker. The material transport is carried out by the kiln rotation. Finally, the temperature is reduced by a cooler at the end of the kiln.
- Fumes transport, which is a faster process. The combustion air feeds the kiln through the cooler where the clinker residual heat is recovered. Through the kiln and the preheater, the gases, going against the raw material flow, heat it.

The main measured variables are the O_2 and CO gases concentration, their temperature, and the kiln torque. The variables used to control the kiln are: *raw material*, *coal*, and *fumes* flows, and kiln rotation speed. Thus the controller has to conclude a value for each variable to achieve the control goals, [3]. The relationship between these input variables and the controlled variables is complex, with delays and interactions.

The control goals proposed, in a priority order, are: Maximum quality, Maximum production, and Minimum production cost.

The first one is the most important and difficult to achieve because of the delay between the clinker formation and its measurement. There is an unavoidable delay in the sample preparation, transport, and X-Ray analysis (about 40 minutes). Another difficulty is due to the impossibility of the clinker formation temperature measurement, which is directly related to the final quality. This problem forces to estimate the temperature using other variables (kiln torque, kiln temperatures, preheater temperatures, etc.).

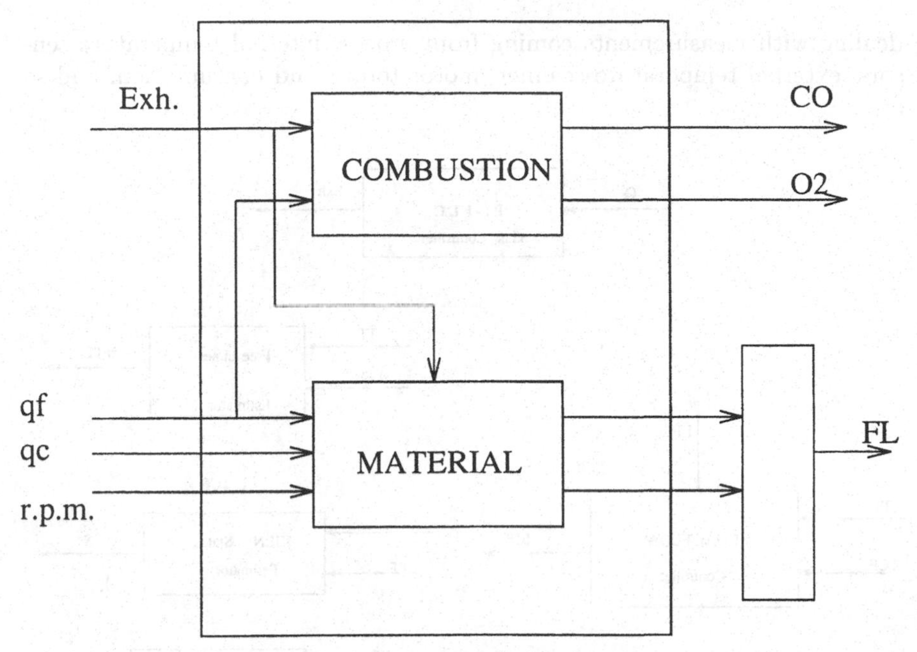

Fig. 7.1. Simplified Kiln Model

Maximum production is achieved maintaining the control variables at their nominal values whenever it is possible.

Minimum production costs are obtained minimising O_2 concentration in the fumes, CO concentration between two limits, and avoiding abnormal situations.

The kiln process is difficult to control because of its time varying non-linear behaviour and the poor quality of available on-line measurements. Moreover, the cement kiln has several functioning modes in such a way that we have to consider different controllers for each mode, with even different control strategies: *normal* situation, the desirable one, *abnormal* situation, if the basic control goals are not fully reached, and *exceptional* situations, in which case the control goal is to return under normal operating conditions. Some of these situations are: *repairing ring*, *inlet coolers obstruction*, *cyclone obstruction*, or *raw cut off*. For each situation a different control policy must be implemented.

7.2 Control Structure

We first define the control domains or modes. A fault detection system, based on a large amount of data, will determine if an emergency situation happens. Under each predefined situation, a FLC will try to recover the kiln situation or will fire the actions to safely stop the kiln. The fault detection system is dealing with measurements coming from process internal temperature sensors, external temperature scanner, motor torque and consumption, and so on.

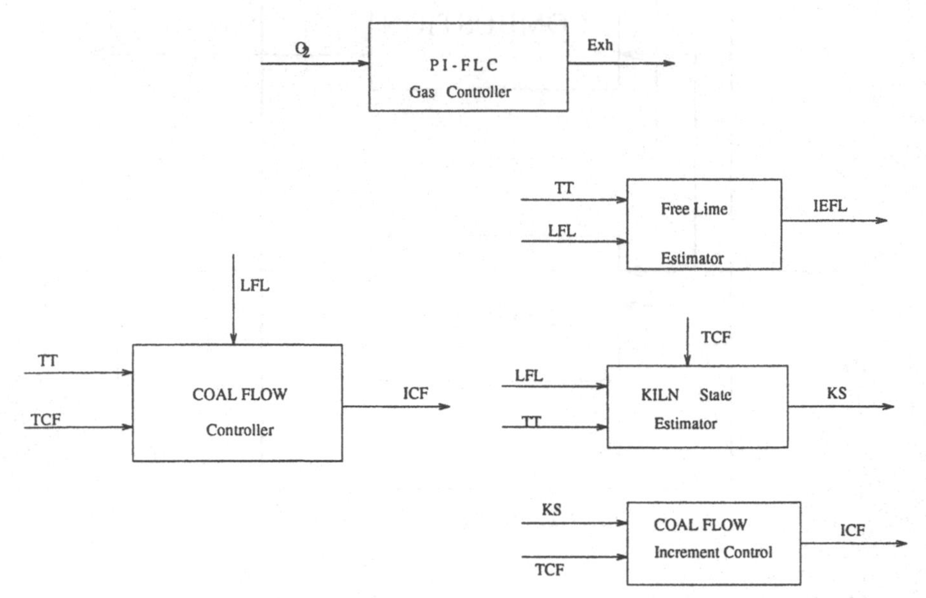

Fig. 7.2. Simplified Kiln control structure

Under normal and abnormal conditions, a unique FLC has been designed. As shown in Fig. 7.2, there are two main control loops: O_2 concentration, by fumes flow, and clinker quality by coal flow. Raw materials flow and motor speed have been devoted to optimise the production.

The fumes concentration controller was implemented using the information of the already available PID controller (out of order most of the time). We included some alarms and data preprocessing to increment the confidence in the closed loop control. The FLC emulates the initial control.

Let us review the clinker quality control. The control structure, as finally decided, includes a first level of measurement assessment, a second level of kiln state estimation, and a final level of coal flow computation. That means a splitting of the KB, per level, easier to understand, to access, and to update.

7.2.1 Free Lime Estimation. Free lime clinker content is very difficult to on-line measure. Laboratory analysis provides a good measurement, but a long delay is required. From a kiln process knowledge we can derive that free lime content in the clinkerization zone is related with the motor torque. Direct torque measurement is not very reliable owing to noise measurement disturbances. Moreover, the absolute torque value is strongly dependent on the kiln operating conditions. For these reasons, a qualitative measurement of the averaged torque trend is considered as the most relevant input to estimate the variations in the free lime.

The fuzzifier converts the crisp measurements and laboratory data into fuzzy variables. The first attempt in the rule base was the following:

-Output variables: Increment in estimated free lime, IEFL.

-Input variable: Torque Trend, TT.

-Linguistic variables. IEFL: Put, Let, Take-off.

TT: Down, Steady, Up.

- *Rule base*: Symmetric, as follows

```
If TT Up then IEFL T-off
If TT Steady then IEFL Let
If TT Down then IEFL Put
```

Refinements. Experiments carried out in the plant pointed out a mismatch in the edges of the considered TT behaviour, the relevance of the information provided by the laboratory, referred to the past value of the free lime, as well as the convenience to modify the initial "linear" relationship.

Under trial and error approach, a new physical variable, the Laboratory Free Lime, LFL, was introduced and the number of LV attached to the TT was increased to five. The rule base structure appears in Table 7.1.

Table 7.1. Rule base: Free Lime Estimation

IEFL		LFL		
		Low	Normal	High
TT	V-Down	Put	Put-M	Put-M
	Down	Let	Put	Put
	Steady	Let	Let	Let
	Up	Let	Let	T-off
	V-Up	T-off	T-off	T-off-M

A graphical interpretation of this set of rules is shown in Fig. 7.3. Observe that coordinate scales are linguistic. The parameters defining the functions (High, Normal, Low) show the LV attached to the LFL.

7.2.2 Kiln State Estimation. This FLC level has the main goal of determining the state of the kiln, in the same way the expert kilner does. This variable is a qualitative one. No physical support can be directly related. For the output variable, a set of 9 LV has been decided. Nevertheless, in order to make a difference on the estimated value, based on the user confidence in his/her conclusion, certainty coefficients are attached to each conclusion. That means proportional MF based on this coefficient.

The first attempt to define the KB was the following:

-Output variables: Kiln State, KS.

-Input variables: Error in Free Lime, EFL, and Torque Trend, TT.

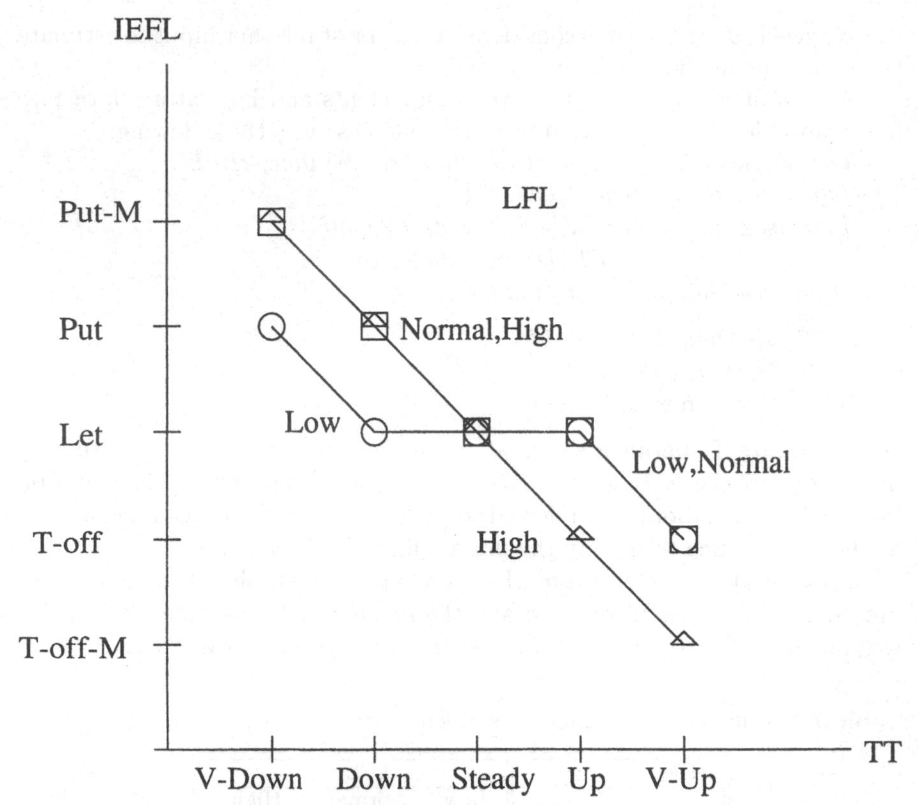

Fig. 7.3. Graphical interpretation of Free Lime Estimation

-Linguistic variables: EFL: Negative, Zero, Positive.
TT: VeryDown, Down, Steady, Up, VeryUp.
KS: Hh, H, Nh, Hc, N, Ch, Nc, C, Cc.
Where H/h stands for Heat/heating, N for Normal, and C/c for Cool/cooling.
- *Rule base*: as shown in Table 7.2:

Refinements. This initial KB settling was shown to be unappropriated in the table edges. This was motivated by the non-linear effect of the heat capacity and the materials melting heat in the current kiln state. For that purpose, a new input variable, the Trend in Coal Flow, TCF, averaged over the past three hours, was considered. This information, in some sense, reflects a pseudo derivative effect on the kiln state, leading to an overheating or undercooling. This new variable only was relevant if the error in the free lime was negative. Thus, the new rule base appears in Table 7.3 where there are only 5 columns instead of 9, the extended table would have if the coal trend was considered under any conditions.

The non-linear relationship between variables can be graphically shown to the operators in a similar way as for the free lime estimation, plotting the Kiln State against Torque Trend, the Error in Free Lime being considered

Table 7.2. Rule Base: Kiln State Estimation

	EFL	Neg	Zero	Pos
TT	V-Down	Hc,1	Nc,1	Cc,1
	Down	Hc,0.7	Nc,0.7	Cc,0.7
	Steady	H,1	N,1	C,1
	Up	Hh,0.7	Nh,0.7	Ch,0.7
	V-Up	Hh,1	Nh,1	Ch,1

Table 7.3. Rule base: Kiln State Estimation Refinement

	EFL	Neg	Neg	Neg	Zero	Pos
	TCF	Steady	Down	V-Down		
TT	V-Down	Hc,1	Nc,1	C,1	Nc,1	Cc,1
	Down	Hc,0.7	Nc,0.7	C,0.7	Nc,0.7	Cc,0.7
	Steady	H,1	Hc,1	Nc,1	N,1	C,1
	Up	Hh,0.7	H,0.7	Hc,0.7	Nh,0.7	Ch,0.7
	V-Up	Hh,1	H,1	Hc,1	Nh,1	Ch,1

as a parameter enlarged, when Negative, by the Trend in Coal Flow. The TCF feedback introduces a dynamical behaviour to the full clinker quality controller.

7.2.3 Coal Flow Computation. The goal of this control level is to determine the action to be taken in the coal flow, that is, to increment or reduce it, based on the kiln state obtained as a conclusion, with an attached certainty coefficient, in the previous level. Although the reasoning is very simple, to get the final control action as a conclusion of this levels allows an easy understanding as well as the possible improvement if the expert would decide that some data, other than the kiln state, should be taken into account to determine any change in the coal flow.

The KB is the following:

-Output variables: Increment in Coal Flow, ICF.

-Input variable: Kiln State, KS

-Linguistic variables. ICF: VP, P, L, T, VT.

Where P, L, and T means as before, and V for very, showing a strong action.

- *Rule base:* As in Table 7.4.

Table 7.4. Rule base: Coal Flow Computation

KS	Cc	C	Nc	Ch	N	Hc	Nh	H	Hh
ICF	VP,1	P,1	P,0.5	L,0.7	L,1	L,0.7	T,0.5	T,1	VT,1

Refinements. According to real tests in the factory, two kinds of refinements have been carried out: 1) A new variable (fumes temperature) was considered because this variable is related to the future evolution of the kiln state without delay, and so, it provides a good information in order to take the appropriate action. 2) It is known that the kiln has a big thermal inertia. So, the consideration of the coal trend seems to be a solution in order to reduce this physical property. In this way, a new variable was considered, the coal trend, as an antecedent in the rules.

Acknowledgements

The author is fully indebted to his colleagues, Professor M. Martínez and Dr. J.L. Navarro for their many suggestions, comments and original and nice figures like those numbered 2.3, 2.4, 2.5, 5.1, 5.2. The application has been partially supported by and developed in the framework of the national project CICYT TAP93-0596-C04-01.

References

1. Albertos P., "Fuzzy Controllers", in Boullart L., A. Krijgsman, R.A. Vingerhoeds (eds.), *Applications of Artificial Intelligence in Process Control*, Pergamon Press, Oxford, 1992.
2. Albertos P., A. Crespo, F. Morant, and J.L. Navarro, "Intelligent Controllers Issues", *IFAC Symp. on Intelligent Components and Instruments for Control Applications*, Malaga, 1992.
3. Albertos P., M. Martínez, J.L. Navarro, and F. Morant, "Fuzzy Controllers Design: a Methodology", *2nd IEEE Conference on Control Applications*, 69-76, Vancouver, Canada, 1993.
4. Bezdek J.C., "Fuzzy models - What are they and Why?", *IEEE Trans. Fuzzy Syst.*, Vol 1, 1-6, (1993).
5. Ebanks B.R., "On measures of fuzziness and their representation", *J. Math. Anal. and Appl.*, Vol 94, 24-37, (1983).
6. Lee C.C., "Fuzzy Logic in Control Systems: Fuzzy Logic Controller", *IEEE Trans on Systems, Man, and Cybernetics*, Vol SMC-20, No 2, 404-435, (1990).
7. Navarro J.L., P. Albertos, and M. Martínez, "Some Experiences in the Design of Industrial Fuzzy Controllers", *International Workshop on Fuzzy Technologies in Automation and Intelligent Systems*, Duisburg, spring 1994.
8. Ollero A., A. Garcia-Cerezo, and J. Aracil, "Design of Rule-based Expert Controllers", *ECC91 European Control Conference*, Grenoble, France, 578-583, 1991.
9. Pal N.R., and J.C. Bezdec, "Measuring Fuzzy Uncertainty", *IEEE Trans. on Fuzzy Systems*, Vol 2 No. 2, 107-118, (1994).
10. Takagi T., and M. Sugeno, "Fuzzy Identification of Systems and its Application to Modelling and Control", *IEEE Trans on Systems, Man, and Cybernetics*, Vol SMC 15 No 1, 116-132, 1985.
11. Zadeh L.A., "Fuzzy sets", *Informat. Control.*, Vol 8, 338-353, 1965.
12. Zadeh L.A., K.S. Fu, K. Tanaka, and M. Shimura (eds.), *Fuzzy sets and their applications to cognitive and decision processes*, Academic, London 1975.
13. Zeng X.J., and M.G. Sing, "Approximation Theory of Fuzzy Systems - SISO Case", *IEEE Trans. on Fuzzy Systems*, Vol. 2 No. 2, 162-176, (1994).

Searching for the Best: Stochastic Approximation, Simulated Annealing and Related Procedures

Georg Pflug

University of Vienna
Universitätsstraße 5
A-1010 Wien/Vienna
Austria.

1. Introduction

Searching for the best has the goal to find a solution x^* of an optimization problem

$$(P) \qquad \left\| \begin{array}{l} \text{Minimize } F(x) \\ x \in S \end{array} \right. \tag{1.1}$$

The function F is called the *objective function*, the set S is the feasible set (*search space*). If F is theoretically known and smooth, then the solution of problem (P) can be reduced by differential calculus to finding all solutions of an equation. But in the typical application, the analytic form of F is unknown, but one can only evaluate F at any desired argument x. This evaluation is costly and – even worse – often corrupted by noise.

The goal of finding an optimal solution x^* should be reached with minimal costs and effort. Therefore, a plan for designing the points for function evaluation is necessary. The plan should be such that the majority of evaluations are made near the candidates for optimizers and only few evaluations are made far away from them.

This principle is the guiding line for practically all solution techniques, like Simulated Annealing, Stochastic Approximation, Response Surface methods and Bandit processes. These methods "learn" to follow the most promising way of searching.

The organization of this presentation is as follows. In section 2, some basic facts about iterations and convergence are collected. It is demonstrated in section 3, how these general facts can be applied to stochastic iterations. Applications include Simulated Annealing and Stochastic Approximation (sections 4 and 5). The dynamical systems aspect of Stochastic Approximation is discussed in section 6. Some results which compare the performance of parallelized stochastic approximation procedures are presented in section 7. Some related topics, namely Response Surface methods and Bandit Processes are treated in sections 8 and 9.

2. Deterministic Iterations, Fixpoints and Convergence

The simplest adaptive plan determines the next search point x_{n+1} only on the basis of the current search point x_n, i.e. is of iterative structure: Starting from some initial guess x_1, functions T_n are applied to the current point x_n to yield a better point x_{n+1}:

$$x_{n+1} = T_n(x_n). \tag{2.1}$$

This iteration should be designed in such a manner that the sequence of generated search points (x_n) converges to an optimal solution of the problem (P).

As an example, take the Newton-Raphson method for unconstrained optimization

$$x_{n+1} = x_n - [\nabla^2 F(x_n)]^{-1} \nabla F(x_n). \tag{2.2}$$

(∇F is the gradient and $\nabla^2 F$ is the Hessian on F). Here the function T_n determining the next search point does not depend on n, since $T_n(x) = x - [\nabla^2 F(x)]^{-1} \nabla F(x)$.

Let us first discuss iterations of the form

$$x_{n+1} = T(x_n). \tag{2.3}$$

If such an iteration converges, then the limit is necessarily a fixpoint of T, i.e. a point x satisfying $T(x) = x$, as is stated in the following theorem.

Theorem 2.1. *Suppose that*

(i) $x \mapsto T(x)$ *is continuous.*
(ii) The recursion $x_{n+1} = T(x_n)$ *converges to* x^*.

Then x^* *is a fixpoint of* T.

Proof. Since $x_n \to x^*$, by continuity $T(x_n) \to T(x^*)$. On the other hand, $T(x_n) = x_{n+1} \to x^*$, whence $T(x^*) = x^*$.

The converse is however not true: Even if T has a unique fixpoint x^*, the iteration (2.3) does not necessarily converge to it. An additional assumption like the "contraction property" stated below is needed.

Theorem 2.2. *Suppose that the following contraction property holds:*

$$\|T(x) - T(y)\| \le \rho \, \|x - y\| \tag{2.4}$$

for some constant $\rho < 1$ *and all* x, y. *Then*

(i) There is a unique fixpoint x^*
(ii) The recursion $x_{n+1} = T(x_n)$ *converges to* x^* *for all starting values* x_1.

Proof. By iterating the contraction property, $\|x_{n+1} - x_n\| \leq \rho^n \|x_2 - x_1\|$ and one may conclude that (x_n) is a Cauchy sequence, which converges to x^* (say). By Theorem 2.1, x^* is a fixpoint of T. Suppose that x_1^* and x_2^* are two different fixpoints. Then

$$\|x_1^* - x_2^*\| = \|T(x_1^*) - T(x_2^*)\| \leq \rho \|x_1^* - x_2^*\| < \|x_1^* - x_2^*\|,$$

a contradiction.

Let us now to turn the more general recusion (2.1), where the iteration function T_n varies from step to step. We will show, that under some regularity assumptions, the generated sequence will track the fixpoints (or more generally quasi-fixpoints) of T_n and converge, if these (quasi-)fixpoints converge.

Theorem 2.3. *Suppose that there are real sequences (ρ_n) and (η_n) such that*

(i) $\|T_n(x) - T_n(y)\| \leq \rho_n \|x - y\| + \eta_n$ *for all* x, y
(ii) $\rho_1 \cdot \rho_2 \cdots \rho_n \to 0$ *as* $n \to \infty$
(iii) $\eta_n \geq 0; \sum \eta_n < \infty$
(iv) x_n^* *are quasi-fixpoints of* T_n*, i.e.*

$$\sum_{n=1}^{\infty} \|T_n(x_n^*) - x_n^*\| < \infty$$

(v) the quasi-fixpoints x_n^* *of* T_n *satisfy* $\sum_{n=1}^{\infty} \|x_{n+1}^* - x_n^*\| < \infty$

then the sequence (x_n^*) *converges to some* x^* *and for all starting values* x_1*, the iteration* $x_{n+1} = T_n(x_n)$ *converges to* x^*.

Proof. Since the quasi-fixpoints form a Cauchy-sequence, they converge to some limit x^*. Let $\epsilon_n = \|T(x_n^*) - x_n^*\| + \|x_{n+1}^* - x_n^*\|$ and $\delta_n = \|x_n - x_n^*\|$. By assumption, $\sum \epsilon_i < \infty$. We have

$$\begin{aligned}
\delta_{n+1} &= \|x_{n+1} - x_{n+1}^*\| \\
&\leq \|T_n(x_n) - T_n(x_n^*)\| + \|T_n(x_n^*) - x_n^*\| + \|x_n^* - x_{n+1}^*\| \\
&\leq \rho_n \delta_n + \eta_n + \epsilon_n
\end{aligned}$$

and therefore

$$\delta_n \leq \delta_1 \prod_{i=1}^{n-1} \rho_i + \sum_{i=1}^{n-1} (\eta_i + \epsilon_i) \prod_{j=i+1}^{n-1} \rho_j.$$

Assumptions (ii) - (v) imply that $\delta_n \to 0$, i.e. $\|x_n - x_n^*\| \to 0$ and since $x_n^* \to x^*$ we have also $x_n \to x^*$.

The just presented result may be used to demonstrate the convergence of the simple gradient algorithm

$$x_{n+1} = x_n - a_n \nabla F(x_n) \tag{2.5}$$

where (a_n) is a sequence of stepsizes.

Theorem 2.4. *Suppose that $F(x)$ satisfies the following condition: There is a constant $c_0 > 0$ such that for all x, y and all sufficiently small a*

$$\|x - a\nabla F(x) - y + a\nabla F(y)\| \leq (1 - c_0 a)\|x - y\|. \tag{2.6}$$

Then there is a unique x^* such that $\nabla F(x^*) = 0$. Suppose further that

$$a_n \geq 0, \quad a_n \to 0, \quad \sum_n a_n = \infty.$$

Then for the sequence defined by (2.5) it follows that

$$x_n \to x^*.$$

Proof. Set $T_n(x) = x - a_n \nabla F(x)$, $\rho_n = (1 - c_0 a_n)$, $\eta_n = 0$ and $x_n^* = x^*$. All assumptions of theorem 2.3 are fulfilled, whence $x_n \to x^*$.

Sometimes, the design rule is based not only on the current guess point x_n, but on some auxiliary values z_n:

$$\begin{aligned} x_{n+1} &= T_n(x_n, z_n) \\ z_{n+1} &= S_n(x_n, z_n) \end{aligned}$$

For instance, the Newton-Raphson iteration is often modified to the numerically easier scheme

$$x_{n+1} = x_n - D_n \nabla F(x_n)$$

where the matrices D_n follow a recursion of their own, namely

- $D_{n+1} = D_n - D_n \cdot u_n^T \cdot u_n \cdot D_n / (u_n^T \cdot D_n \cdot u_n) + v_n \cdot v_n^T / u_n^T \cdot v_n$
  where $\quad u_n = x_{n+1} - x_n$
  and $\quad v_n = \nabla F(x_{n+1}) - \nabla F(x_n)$
  (Broyden/Fletcher/Goldfarb/Shanno (BFGS) method)
  or
- $D_{n+1} = (I - u_n \cdot v_n^T / (u_n^T \cdot v_n)) \cdot D_n \cdot (I - u_n \cdot v_n^T / (u_n^T \cdot v_n)) + v_n \cdot v_n^T / (u_n \cdot v_n)$
  where u_n and v_n have the same meaning as above.
  (Davidon/Fletcher/Powell (DFP) method).

Convergence of these more sophisticated methods can also be established by using "contraction properties".

3. Stochastic Iterations, Stationary Distributions and Convergence

A stochastic iteration is a sequence of random variables (X_n) such that

$$X_{n+1} = T_n(X_n, \xi_n)$$

started at some starting value X_1 (which may be a random variable too). The random sequence (ξ_n) is the *innovation sequence.*

There are two reasons for the need of stochastic iterations:

- The observations may be corrupted by noise, or
- the choice of the next design point may be based on the introduction of some artificial randomness (random search, simulated annealing, etc).

We will assume here that the ξ_i's are independent, an assumption, which guarantees that the process X_n is Markovian. Let $\boldsymbol{P}_n$ be the transition operator of this process, i.e.

$$\boldsymbol{P}_n(u, A) = \mathbb{P}(X_{n+1} \in A | X_n = u) = \mathbb{P}(T_n(u, \xi_n) \in A).$$

Recall that a Markov transition operator $\boldsymbol{P}$ acts by right multiplication on probability measures γ:

$$(\gamma \cdot \boldsymbol{P})(A) := \int \boldsymbol{P}(w, A)\, d\gamma(w).$$

The interpretation is the following: If X_n has distribution γ and $X_{n+1} = T_n(X_n, \xi_n)$, then X_{n+1} has distribution $\gamma \cdot \boldsymbol{P}_n$.

Two transition operators $\boldsymbol{P}_1$ and $\boldsymbol{P}_2$ may be multiplied, yielding another transition operator as their product

$$(\boldsymbol{P}_1 \cdot \boldsymbol{P}_2)(w, A) = \int \boldsymbol{P}_2(v, A)\, \boldsymbol{P}_1(w, dv).$$

This multiplication is not commutative in general. A probability distribution π is called stationary w.r.t. the transition $\boldsymbol{P}$, if

$$\pi \cdot \boldsymbol{P} = \pi.$$

To put it differently, one may say that a stationary distribution is a fixpoint of the mapping $\gamma \mapsto \gamma \cdot \boldsymbol{P}$.

Stochastic algorithms are generalizations of deterministic algorithms: If the randomness is degenerated ($\xi_n \equiv$ constant), then the algorithm is a deterministic recursion of the type discussed in the introduction.

From a higher viewpoint, a stochastic recusion can be seen as a recursion on probability distributions. Let γ_1 be the distribution of X_1 and let γ_n be the distribution of X_n. The recursion for the distributions is

$$\gamma_{n+1} = T_n(\gamma_n) = \gamma_n \cdot \boldsymbol{P}_n \tag{3.1}$$

The formal analogy of (3.1) and (2.1) allows to extend theorem 2.3 for stochastic iterations. A "contraction property" based on the Wasserstein distance for measures is needed here.

Let r be some distance on the search space S. For two probability measures μ and ν on S let $r(\mu, \nu)$ be their Wasserstein-distance:

$$r(\mu, \nu) = \inf\{\mathbb{E}(r(X, Y)) : \quad (X, Y) \text{ is a bivariate r.v. with given marginals } \mu \text{ and } \nu\}.$$

It is evident, that for two point masses δ_x and δ_y, the distance $r(\delta_x, \delta_y)$ is equal to $r(x, y)$ and therefore there is no conflict in notation if we use the same symbol r. The Wasserstein distance has an interpretation as a transportation problem: Mass distribution μ has to be transported such that distribution ν is obtained and the average transport distance is minimal.

For a finite set $S = \{s_1, \ldots, s_N\}$, the distance $r(\mu, \nu)$ of two probabilities $\nu = (\nu_1, \ldots, \nu_N)$ and $\mu = (\mu_1, \ldots, \mu_N)$ is optimal value of the linear minimization problem (LP-problem)

$$\left\| \begin{array}{l} \text{Minimize } \sum_{k,\ell} \kappa_{k\ell} r(s_k, s_\ell) \\ \sum_k \kappa_{k\ell} = \nu_\ell \\ \sum_\ell \kappa_{k\ell} = \mu_k \\ \kappa_{k\ell} \geq 0 \quad \text{for all } k, \ell \end{array} \right. \tag{3.2}$$

The relation of the Wasserstein topology to the weak topolgy is the following:

Let $\mathcal{P}_r$ the set of all probability measures on the Borel sets of the metric space (R, r) with the following property:

$$\mathcal{P}_r = \{\mu \in \mathcal{P} : \int r(u_0, u)\, d\mu(u) < \infty \quad \text{for some} \quad u_0 \in R\}.$$

If $x \mapsto r(x, x_0)$ is uniformly μ_n-integrable, then

$$\mu_n \in \mathcal{P}_r \text{ converges weakly to } \mu \in \mathcal{P}_r$$

if and only if

$$r(\mu_n, \mu) \to 0.$$

([6]).

Introduce the notation

$$\mu_n \Rightarrow \mu$$

for

$$r(\mu_n, \mu) \to 0.$$

Examples 3.1. If μ and ν are two measures on the real line and the distance is the euclidean

$$r_1(x, y) = |x - y|,$$

then

$$r_1(\mu, \nu) = \int_{-\infty}^{\infty} |F(x) - G(x)|\, dx = \int_0^1 |F^{-1}(x) - G^{-1}(x)|\, dx,$$

where F resp. G are the distribution functions of μ resp. ν. ([57]).

If however the distance is the discrete one

$$r_0(x, y) = \left\{ \begin{array}{ll} 1 & if \quad x \neq y \\ 0 & if \quad x = y \end{array} \right. ,$$

then the Wasserstein metric is

$$r_0(\mu,\nu) = 1 - \int \min\left(\frac{d\mu}{d\lambda}, \frac{d\nu}{d\lambda}\right) d\lambda,$$

where λ is some dominating measure (e.g. $\lambda = \frac{1}{2}(\mu+\nu)$).

Definition 3.1. *The* coefficient of ergodicity *of the transition $\boldsymbol{P}$ pertaining to the metric r is*

$$\rho_r(\boldsymbol{P}) = \sup_{\mu \neq \nu} \frac{r(\mu\boldsymbol{P}, \nu\boldsymbol{P})}{r(\mu,\nu)}.$$

If there is a metric r such that $\rho_r(\boldsymbol{P}) < 1$, then the transition $\boldsymbol{P}$ is called geometrically ergodic.

It is easy to see that a geometrically ergodic transition $\boldsymbol{P}$ has a unique fixpoint, since it is a contracting operator in the sense of (2.4).

Definition 3.2. *A infinite sequence of Markov transitions $(\boldsymbol{P}_i)$ is called* weakly ergodic, *if for any $k > 0$ and for any two starting measures $\gamma^{(1)}$ and $\gamma^{(2)}$,*

$$r(\gamma^{(1)} \prod_{i=k}^{n} \boldsymbol{P}_i, \gamma^{(2)} \prod_{i=k}^{n} \boldsymbol{P}_i) \to 0$$

if $n \to \infty$. (Here and in what follows, the product $\prod_i \boldsymbol{P}_i$ is meant in ascending order of the indices).

Technically, weak ergodicity is the following property:
There is an infinite sequence of integers $1 = l_1 < l_2 < \ldots$ such that

$$\prod_{j=1}^{\infty} \rho_r(\boldsymbol{P}_{l_j+1} \cdot \boldsymbol{P}_{l_j+2} \cdots \boldsymbol{P}_{l_{j+1}}) = 0. \tag{3.3}$$

If $(\boldsymbol{P}_i)$ is weakly ergodic, then $\gamma \prod_{i=1}^{n} \boldsymbol{P}_i$ does not converge in general. To ensure convergence, additional assumptions must be fulfilled.

Theorem 3.1. *Let $(\boldsymbol{P}_i)_{i\geq 1}$ be a sequence of geometrically ergodic Markov transitions. Let π_i be the stationary distributions pertaining to P_i. Suppose that*

(i) $(\boldsymbol{P}_i)$ is weakly ergodic;
(ii) $\sum_{i=1}^{\infty} r(\pi_{i+1}, \pi_i) < \infty$.

Then, the measures π_i have a limit π^* and for all starting measures γ_1,

$$\gamma_1 \prod_{i=1}^{n} \boldsymbol{P}_i \Rightarrow \pi^*.$$

Proof. The existence and uniqueness of $\pi^* = \lim_{i\to\infty} \pi_i$ follows from the fact that the π_i's form a Cauchy sequence.

According to (3.3) there are integers l_i such that with $\boldsymbol{Q}_j = \boldsymbol{P}_{l_j+1}\cdot \boldsymbol{P}_{l_j+2}\cdots \boldsymbol{P}_{l_{j+1}}$ we have $\prod_{j=1}^{\infty} \rho_r(\boldsymbol{Q}_j) = 0$. Set $\xi_n = r(\pi_{l_{n+1}}, \pi_{l_n}\boldsymbol{Q}_n)$ and $\eta_n = r(\gamma \prod_{j=1}^{n} \boldsymbol{Q}_j, \pi_{l_{n+1}})$. Notice that

$$\begin{aligned}
\xi_n &= r(\pi_{l_{n+1}}, \pi_{l_n} \prod_{k=l_n+1}^{l_{n+1}} \boldsymbol{P}_k) \\
&\le r(\pi_{l_{n+1}}, \pi_{l_n+1} \prod_{k=l_n+1}^{l_{n+1}} \boldsymbol{P}_k) + r(\pi_{l_n} \prod_{k=l_n+1}^{l_{n+1}} \boldsymbol{P}_k, \pi_{l_n+1} \prod_{k=l_n+1}^{l_{n+1}} \boldsymbol{P}_k) \\
&\le r(\pi_{l_{n+1}}, \pi_{l_n+1} \prod_{k=l_n+2}^{l_{n+1}} \boldsymbol{P}_k) + r(\pi_{l_n}, \pi_{l_n+1}) \\
&\le r(\pi_{l_{n+1}}, \pi_{l_n+2} \prod_{k=l_n+2}^{l_{n+1}} \boldsymbol{P}_k) + r(\pi_{l_n+1} \prod_{k=l_n+2}^{l_{n+1}} \boldsymbol{P}_k, \pi_{l_n+2} \prod_{k=l_n+2}^{l_{n+1}} \boldsymbol{P}_k) \\
&\quad + r(\pi_{l_n}, \pi_{l_n+1}) \\
&\le r(\pi_{l_{n+1}}, \pi_{l_n+2} \prod_{k=l_n+3}^{l_{n+1}} \boldsymbol{P}_k) + r(\pi_{l_n+1}, \pi_{l_n+2}) + r(\pi_{l_n}, \pi_{l_n+1}) \le \cdots \\
&\le \sum_{k=l_n}^{l_{n+1}-1} r(\pi_k, \pi_{k+1})
\end{aligned} \tag{3.4}$$

By (ii) $\sum_{n=1}^{\infty} \xi_n = \sum_{k=1}^{\infty} r(\pi_k, \pi_{k+1}) < \infty$. We have

$$\begin{aligned}
\eta_{n+1} &= r(\gamma \prod_{j=1}^{n+1} \boldsymbol{Q}_j, \pi_{l_{n+2}}) \\
&\le r(\gamma \prod_{j=1}^{n+1} \boldsymbol{Q}_j, \pi_{l_{n+1}} \boldsymbol{Q}_{n+1}) + r(\pi_{l_{n+2}}, \pi_{l_{n+1}} \boldsymbol{Q}_{n+1}) \\
&\le \rho_r(\boldsymbol{Q}_{n+1})\, r(\gamma \prod_{j=1}^{n} \boldsymbol{Q}_j, \pi_{l_{n+1}}) + \xi_{n+1} \\
&= \rho_r(\boldsymbol{Q}_{n+1})\, \eta_n + \xi_{n+1}.
\end{aligned}$$

Consequently

$$\eta_n \le \left[\prod_{j=2}^{n} \rho_r(\boldsymbol{Q}_i) \right] \eta_1 + \sum_{i=1}^{n} \xi_i \prod_{k=i+1}^{n} \rho_r(\boldsymbol{Q}_k) \to 0$$

and

$$r(\gamma \prod_{j=1}^{n} \boldsymbol{Q}_j, \pi^*) \leq \eta_n + r(\pi_{l_{n+1}}, \pi^*) \to 0.$$

It remains to show that $r(\gamma \prod_{i=1}^{N} \boldsymbol{P}_i, \pi^*)$ is arbitrary small, if N is large. Let l_n such large that $r(\gamma \prod_{i=1}^{l_n} \boldsymbol{P}_i, \pi^*) \leq \epsilon$ and also $\sum_{k=l_n}^{\infty} r(\pi_k, \pi_{k+1}) \leq \epsilon$. An argument similar to (3.4) implies that for $N \geq l_n$ we have
$r(\pi^* \prod_{i=l_n}^{N} \boldsymbol{P}_i, \pi^*) \leq 2 \sum_{i=l_n}^{\infty} r(\pi_i, \pi_{i+1}) \leq 2\epsilon$. Then, for $N \geq l_n$,

$$\begin{aligned} r(\gamma \prod_{i=1}^{N} \boldsymbol{P}_i, \pi^*) &\leq r(\gamma \prod_{i=1}^{N} \boldsymbol{P}_i, \prod_{i=l_n}^{N} \boldsymbol{P}_i \pi^*) + r(\pi^* \prod_{i=l_n}^{N} \boldsymbol{P}_i, \pi^*) \\ &\leq r(\gamma \prod_{i=1}^{l_n} \boldsymbol{P}_i, \pi^*) + \sum_{i=l_n}^{\infty} r(\pi_{i+1}, \pi_i) \leq 3\epsilon \end{aligned}$$

which proves the theorem.

It is not necessary, that the operators $\boldsymbol{P}_i$ have all a unique fixpoint (stationary distribution) π_i, quasi-fixpoints in the sense of theorem 2.3 suffice. Here is the reformulation of theorem 2.3 in terms of Markov transitions:

Theorem 3.2. *Suppose that*

(i) $r(\mu \cdot \boldsymbol{P}_n, \nu \cdot \boldsymbol{P}_n) \leq \rho_n \cdot r(\mu, \nu) + \eta_n$ *for all probability measures* μ, ν
(ii) $\rho_1 \cdot \rho_2 \cdots \rho_n \to 0$
(iii) $\eta_n \geq 0; \sum_n \eta_n < \infty$
(iv) $\sum_n r(\pi_n \cdot \boldsymbol{P}_n, \pi_n) < \infty$
(v) $\sum_n r(\pi_n, \pi_{n+1}) < \infty$

Then, the measures π_i *have a limit* π^* *and for all starting measures* γ_1,

$$\gamma_1 \prod_{i=1}^{n} \boldsymbol{P}_i \Rightarrow \pi^*.$$

Proof. The proof of theorem 2.3 may be repeated literally.

4. Simulated Annealing and Global Optimization

If the search space S is finite, then a primitive non-adaptive design for search is *complete enumeration*. However, complete enumeration is possible only for a small cardinality of S.

Large search spaces may be cut into smaller pieces by endowing them with a neighborhood structure. A neighborhood system $\mathcal{N}$ in S is a symmetric subset of $S \times S$ (i.e. $(x, y) \in \mathcal{N}$ implies that $(y, x) \in \mathcal{N}$). The neighborhood $\mathcal{N}_x$ of x is

$$\mathcal{N}_x = \{y : (x, y) \in \mathcal{N}\}.$$

Notice that because of symmetry, $y \in \mathcal{N}_x$ iff $x \in \mathcal{N}_y$.

The *greedy algorithm* for deterministic optimization makes only improvement steps in the neighborhood and never changes the function value to the worse.

$$\begin{array}{|ll} 1. & \text{Choose the starting value } x_1. \\ 2. & \text{For } n = 1, 2, \ldots \text{ set } x_{n+1} = \text{ argmin } \{F(x) : x \in \mathcal{N}_{x_n}\}. \\ 3. & \text{Stop, if } F(x_n) \leq \min\{F(x) : x \in \mathcal{N}_{x_n}\}. \end{array} \quad (4.1)$$

It is clear, that the algorithm is trapped in all *local minima*, i.e. points y which satisfy $F(y) \leq \min\{F(x) : x \in \mathcal{N}_y\}$.

The *simulated annealing* algorithm is a probabilistic improvement of the greedy algorithm. It accepts also worse neighboring points, but with an acceptance probability which tends to 0.

$$\begin{array}{|ll} 1. & \text{Choose the starting value } X_1. \\ 2. & \text{Choose a } y \text{ in } \mathcal{N}_{X_n} \text{ at random with uniform probability}. \\ 3. & \text{Set } X_{n+1} = \left\{ \begin{array}{ll} y & \text{with probability } \exp(-\frac{[F(y)-F(X_n)]^+}{a_n}) \\ X_n & \text{with probability } 1 - \exp(-\frac{[F(y)-F(X_n)]^+}{a_n}) \end{array} \right. . \end{array}$$
$$(4.2)$$

This algorithm was invented by Kirkpatrick et al. [30], the probabilistic acceptance rule goes back to Metropolis et al. [44], for a complete treatment see [1].

Theorem 4.1. *Let x_{j^*} be a global minimizer and M be chosen such that x_{j^*} can be reached from every other x in at most M steps. Set L equal to the Lipschitz constant of F, i.e.*

$$L = \max\{|F(x_1) - F(x_2)| : x_1 \in \mathcal{N}_{x_2}, x_2 \in S\}.$$

If $a_n = \frac{ML}{\log n}$, then X_n defined in (4.2) converges in distribution. The limit distribution sits on all global minimizers of F. If F has a unique minimizer x^*, then X_n converges in probability (but not almost surely) to x^*.

Proof. The proof is an application of Theorem 2.3. The metric on S has to be chosen as the discrete metric

$$r(x, y) = \left\{ \begin{array}{ll} 1 & if \quad x \neq y \\ 0 & if \quad x = y \end{array} \right. .$$

Details are omitted.

The simulated annealing algorithm may be adapted to discrete stochastic optimization i.e. to cases, where the observations of F are corrupted by random noise. However, the algorithm does not work, if simply every exact observation is replaced by a random one. For global convergence, the variance of the random estimates have to decrease to zero. Moreover, the convergence properties depend on the tail behavior of the errors. We assume that by taking multiple observations and their average at each search point x, we obtain at step n an estimate $\hat{F}(x)$, which is normally distributed with mean $F(x)$ and variance σ_n^2.

Here is the algorithm:

1. Let X_1 be a starting value.
2. Choose a y in $\mathcal{N}_{X_n}$ at random with uniform probability.
3. Observe the random function value at y: $\hat{F}(y) \sim N(F(y), \sigma_n^2)$.
4. Set $X_{n+1} = \begin{cases} y & \text{with probability } \exp(-\frac{[\hat{F}(y)-\hat{F}(X_n)]^+}{a_n}) \\ X_n & \text{with probability } 1-\exp(-\frac{[\hat{F}(y)-\hat{F}(X_n)]^+}{a_n}) \end{cases}$.

Theorem 4.2. *Let the assumptions of theorem 4.1 be satisfied, and assume further that $F(x_1) \neq F(x_2)$, if $x_1 \in \mathcal{N}_{x_2}$. If $a_n = \frac{ML}{\log n}$ and $\sigma_n^2 = O(\frac{1}{n})$, then X_n converges in distribution to a limiting distribution sitting only on global minimizers of F.*

For a proof see [23].

5. Stochastic Approximation

If the search space is $\mathbb{R}^d$, then the standard method is gradient search or a variant of it. The most important stochastic approximation procedure, the Robbins-Monro procedure, is the stochastic version of the simple gradient procedure (2.5):

$$X_{n+1} = X_n - a_n \nabla F(X_n) - a_n \xi_n. \tag{5.1}$$

Here ξ_n denotes the noise which corrupts the observation of the gradient $\nabla F(X_n)$. We assume that (ξ_n) is a sequence of independent zero-mean r.v.'s with bounded variances.

Introduce the notation $\nabla F = f$.

The analysis of the algorithm is rather simple, if the regularity assumption (2.6) is met:

There is a constant $c_0 > 0$ such that

$$\|x - af(x) - y + af(y)\| \leq (1 - ac_0)\|x - y\|$$

for all x, y and sufficiently small a.

Under this assumption f has at most one root x^*. We may apply here Theorem 3.4 by taking the metric

$$r_1(x, y) = \|x - y\|$$

here. The pertaining coefficient of ergodicity of the Markov process

$$X_{n+1} = X_n - af(X_n) - a\xi_n$$

is bounded by

$$\rho_1(a) \le \sup_{x \ne y} \frac{\mathbb{E}(x - af(x) - a\xi - (y - af(y) - a\xi))}{\|x - y\|} = 1 - ac_0.$$

Since $\prod_{n=1}^{\infty}(1 - a_n) = 0$ iff $\sum_n a_n = \infty$, we get the following result:

Theorem 5.1. *Under Assumption (2.6), if $a_n \downarrow 0$ and $\sum a_n = \infty$, then for the procedure*

$$X_{n+1} = X_n - a_n \cdot f(X_n) - a_n \xi_n$$

it follows that

$$X_n \to x^* \textit{ in probability.}$$

Notice that the usual condition $\sum a_n^2 < \infty$ is not required for this conclusion. If $\sum a_n^2 < \infty$ then one may show that X_n converges to x^* almost surely.

If the set of equilibria $\mathcal{E} = \{y\colon f(y) = 0\}$ contains more than one point, one may prove that all finite cluster points of (X_n) lie in $\mathcal{E}$. Some of them, namely the unstable points may be excluded as limits by the following result:

Theorem 5.2. *Let*

$$\tilde{\mathcal{E}} = \{x \in \mathcal{E} : \exists \text{ a positive definite matrix } C \text{ such that } f(y)\, C\, (y - x) \ge 0 \text{ for all } y \text{ in some neighborhood of } x\}.$$

Suppose that $Var(\xi_n | X_n) \ge \sigma^2 > 0$. Then all cluster points of the stochastic approximation procedure (5.1) lie in $\mathcal{E} \backslash \tilde{\mathcal{E}}$.

Proof. see [45], Theorem 4.1., chapter 5.

An extension of this result was proved by Pemantle [46]. It turns out, that only local minima and some one-sided minima can appear as limits.

The points in $\tilde{\mathcal{E}}$ are unstable for the dynamical system $dx(t) = -f(x(t))\,dt$. More results about the connection between dynamical systems and the limit behavior of stochastic iterations will be discussed in section 6.

Here, we show how the procedure can be modified such that only global minima can be limits. This modification is in the direction of simulated annealing: The gradient procedure is corrupted by artificial noise ξ_n and this noise is made such large that the process can escape all local minima. The additional noise will be larger by magnitude than the original noise and therefore the original noise will not be considered any more.

For simplicity, we investigate only the univariate situation $S = \mathbb{R}$. Let the stepsize a constant and consider the recursion

$$X_{n+1} = X_n - af(X_n) - \sqrt{a}\xi_n. \tag{5.2}$$

We assume that f satisfies the following assumption:

Assumption 5.3.

(i) There is a $K > 0$ and a $k_1 > 0$ such that sign $(x) \cdot f(x) \geq k_1$ for $|x| \geq K$,
(ii) $|f(x)| \leq k_2 \cdot |f(u)|$ for $K < |x| < |u|$,
(iii) $|f'(x)| \leq k_3$.

This assumption implies that $F(x) \to \infty$ as $|x| \to \infty$ and that

$$\int \exp\left(-\frac{2F(u)}{\sigma^2}\right) du < \infty$$

for all $\sigma^2 > 0$.

Theorem 5.3. *Let ν^a be the stationary distribution of*

$$X_{n+1} = X_n - af(X_n) - \sqrt{a}\,\xi$$

and let ν^0 be the probability measure with density

$$const. \exp\left(-\frac{2F(x)}{\sigma^2}\right),$$

where $\sigma^2 =$ Var (ξ).
Then, under assumption 5.3, ν^a converges weakly to ν^0.

Proof. Define $\mathcal{H}_L = \{\phi : |\phi(\cdot)| \leq L; |\phi'(\cdot)| \leq L, |\phi(\cdot) \cdot f(\cdot)| \leq L\}$ as a class of test functions. We will show that $\int \phi d\nu^a \to \int \phi d\nu^0$ as $a \to 0$ for all $\phi \in \mathcal{H}_L$. Let $\phi \in \mathcal{H}_L$. W.l.o.g. we may assume that $\int \phi \, d\nu^0 = 0$, otherwise we replace ϕ by $\phi - \int \phi \, d\nu^0$. Let

$$\psi'(x) := \int_{-\infty}^{x} \frac{2}{\sigma^2} \cdot \phi(u) \cdot \exp\left(\frac{2}{\sigma^2}(F(x) - F(u))\right) du.$$

ψ' is bounded: Since it is continuous, it suffices to show the boundedness for $|x| > K$. Let $x > K$ (the other case is similar). Then

$$|\psi'(x)| \leq \frac{2}{\sigma^2} \exp(\frac{2F(x)}{\sigma^2}) \int_x^\infty \frac{L}{k_1} f(u) \exp(-\frac{2F(u)}{\sigma^2})\, du = \frac{L}{k_1}.$$

Taking the derivative of ψ', we get

$$\psi''(x) = \frac{2}{\sigma^2}\phi(x) + \frac{2}{\sigma^2}\psi'(x) \cdot f(x). \tag{5.3}$$

Taking another derivative, we obtain

$$\psi'''(x) = \frac{2}{\sigma^2}\phi'(x) + \frac{2}{\sigma^2}\psi'(x)f'(x) + \frac{4}{\sigma^4}\phi(x)f(x) + \frac{4}{\sigma^4}\psi'(x)f^2(x). \quad (5.4)$$

We show that ψ''' is bounded. The first three summands in (5.4) are bounded by assumption. The boundedness of the last summand for $x \geq K$ follows from $f^2(x)|\psi'(x)| \leq$

$$\leq f^2(x)\frac{2}{\sigma^2}\exp(\frac{2F(x)}{\sigma^2})\int_x^\infty |\phi(u)f(u)|f(u)f^{-2}(u)\exp(-\frac{2F(u)}{\sigma^2})\,du$$

$$\leq f^2(x)\frac{2}{\sigma^2}\exp(\frac{2F(x)}{\sigma^2})\int_x^\infty Lk_2^2 f(u)f^{-2}(x)\exp(-\frac{2F(u)}{\sigma^2})\,du = Lk_2^2.$$

The case $x \leq -K$ is analogous. Let $\psi(x) = \int_0^x \psi'(u)\,du$ and (X_n) be stationary. By a Taylor expansion we get

$$\begin{aligned} \psi(X_{n+1}) &= \psi(X_n - af(X_n) - \sqrt{a}\xi) \\ &= \psi(X_n) - (af(X_n) + \sqrt{a}\xi)\psi'(X_n) + \\ &+ \frac{1}{2}(af(X_n) + \sqrt{a}\xi)^2\psi''(X_n) - \\ &- \frac{1}{6}(af(X_n) + \sqrt{a}\xi)^3(\psi'''(\tilde{X}_n)) \end{aligned}$$

for some point $\tilde{X}_n$ lying between X_n and X_{n+1}.
Taking the expectation on both sides and using the stationarity we get

$$0 = -a\,\mathbb{E}(f(X_n)\psi'(X_n)) + \frac{1}{2}a\sigma^2\,\mathbb{E}(\psi''(X_n)) + O(a^{3/2}).$$

Consequently, by (5.3),

$$\int \phi\,d\nu^a = \mathbb{E}(\phi(X_n)) = \mathbb{E}(-f(X)\psi'(X)) + \frac{1}{2}\sigma^2\,\mathbb{E}(\psi''(X)) = O(\sqrt{a}).$$

The class of functions $\mathcal{H}_L$ contains all trigonometric functions that are made zero outside a compact interval in a continuous way. This family is rich enough to determine weak convergence. □

We remark that the limit distribution ν^0 has nonvanishing density and may therefore be viewed as Gibbs distribution:

Definition 5.1. *Let F be such that $F(x) \to \infty$ as $|x| \to \infty$ in such a way that $\int \exp(-F(x))\,dx < \infty$. The family of distributions with densities*

$$c(a)\exp(-\frac{F(x)}{a}) \quad (5.5)$$

is called Gibbs family *with* energy function F.

The following limit result is known for Gibbs families:

Lemma 5.1. *As a tends to 0, the distribution (5.5) converges weakly to the uniform distribution on all global minima of F.*

Proof. see [26].

In order to make use of the limiting property indicated in Lemma 5.6 one should modify the recursion in such a way, that the variance of the stochastic part is of the order $o(\sqrt{a})$ as $a \to 0$.

We consider therefore the modified recursion

$$X_{n+1} = X_n - af(X_n) - \sqrt{a \cdot \delta(a)} \cdot \xi_n$$

for some $\delta(a)$ tending to zero as $a \to 0$. Introduce an additional assumption.
Assumption 5.7.

(i) $F(x) \geq 0$ and $F(x) = 0$ for the global minimizers
(this is no loss of generality, if there are global minimizers at all)
(ii) There are finitely many global minimizers $x_1^*, \ldots, x_k^*$ of F.

Theorem 5.4. *Let γ^a be the stationary distribution of*

$$X_{n+1} = X_n - af(X_n) - \sqrt{a \cdot \delta(a)} \cdot \xi_n. \tag{5.6}$$

Let Assumption 5.7 be fulfilled. If

$$\delta(a) = \frac{1}{\log\log(1/a)}, \tag{5.7}$$

then all weak limits of γ^a as $a \to 0$ are concentrated on the set of global minimizers of F.

Proof. Let $\tilde{\gamma}^a$ be the distribution with density

$$\text{const. } \exp\left(-\frac{2F(x)}{\sigma^2\delta(a)}\right).$$

Fix an $\epsilon > 0$ and a large constant K such that $F(x) > 2\epsilon$ for $|x| > K$. Let

$$\phi_1(x) = \mathbb{1}_{\{F(x) \geq 3\epsilon\} \cap [-K,K]}$$

$$\phi_2(x) = \mathbb{1}_{\{\epsilon \leq F(x) \leq 2\epsilon\}}$$

and

$$\phi_a(x) = \phi_1(x) - \eta(a) \cdot \phi_2(x).$$

Choose $\eta(a)$ such that

$$\int \phi_a(x)\, d\tilde{\gamma}^a(x) = 0.$$

Let $\lambda(\epsilon)$ be the Lebesgue measure of the set $\{\epsilon \leq F(x) \leq 2\epsilon\}$. Since

$$\eta(a) = \frac{\int \phi_1\, d\tilde{\gamma}^a}{\int \phi_2\, d\tilde{\gamma}^a} \leq \frac{2K \exp(-6\epsilon\sigma^{-2}\delta(a)^{-1})}{\lambda(\epsilon)\exp(-2\epsilon\sigma^{-2}\delta(a)^{-1})}$$

and $\delta(a) \to 0$, it follows that

$$\eta(a) \to 0 \text{ as } a \to 0.$$

Consider the function

$$\psi_a'(x) = \int_{-\infty}^{x} \frac{2}{\sigma^2 \cdot \delta(a)} \cdot \phi_a(u) \cdot \exp\left(\frac{2}{\sigma^2 \cdot \delta(a)}(F(x) - F(u))\right) du.$$

If $x \geq K$ then by Assumption 5.3 (i)

$$\begin{aligned} \psi_a'(x) &= \int_x^{\infty} \tfrac{2}{\sigma^2 \cdot \delta(a)} \cdot \phi_a(u) \cdot \exp\left(\tfrac{2}{\sigma^2\cdot\delta(a)}(F(x) - F(u))\right) du \\ &\leq \text{const.} \int_x^{\infty} \tfrac{2f(u)}{\sigma^2\cdot\delta(a)} \exp\left(\tfrac{2}{\sigma^2\cdot\delta(a)}(F(x) - F(u))\right) du \leq \text{const.} \end{aligned}$$

Consequently ψ_a' is bounded for $x \geq K$ and the same is true for $x \leq -K$. Since F is bounded for $|x| < K$ we have

$$\sup_x |\psi_a'(x)| = O(\delta^{-1}\exp(c_1/\delta(a)))$$

for some c_1 as $a \to 0$. Notice that

$$\phi_a(x) = -f(x)\psi_a'(x) + \frac{\delta(a)\cdot\sigma^2}{2}\psi_a''(x) \tag{5.8}$$

and therefore

$$\sup_x |\psi_a''(x)| = O(\delta(a)^{-2}\exp(c_1/\delta(a)))$$

and

$$\sup_x |\psi_a'''(x)| = O(\delta(a)^{-3}\exp(c_1/\delta(a))). \tag{5.9}$$

Let $\psi_a(x) = \int_0^x \psi_a'(u)\, du$ and (X_n) be stationary. By a Taylor expansion we get

$$\begin{aligned} \psi_a(X_{n+1}) &= \psi_a(X_n - af(X_n) - \sqrt{a\cdot\delta(a)}\xi_n) \\ &= \psi_a(X_n) - (af(X_n) + \sqrt{a\cdot\delta(a)}\xi_n)\psi_a'(X_n) + \\ &+ \frac{1}{2}(af(X_n) + \sqrt{a\cdot\delta(a)}\xi_n)^2\psi_a''(X_n) + \\ &+ \frac{1}{6}(af(X_n) - \sqrt{a\cdot\delta(a)}\xi_n)^3(\psi_a'''(\tilde{X}_n)) \end{aligned}$$

for some point $\tilde{X}_n$ lying between X_n and X_{n+1}.
Taking the expectation on both sides and using the stationarity and (5.9) we get

$$\begin{aligned} 0 &= -a\,\mathbb{E}(f(X_n)\psi_a'(X_n)) + \frac{1}{2}a\delta(a)\sigma^2\,\mathbb{E}(\psi_a''(X_n)) \\ &+ O\left(a^{3/2}\delta(a)^{-3/2}\exp(c_1/\delta(a))\right). \end{aligned}$$

Consequently, dividing by a and using (5.7) and (5.8) we get

$$\int \phi_a\,d\gamma^a = \mathbb{E}(\phi_a(X_n)) = \mathbb{E}(-f(X)\psi_a'(X)) + \frac{\delta(a)\sigma^2}{2}\,\mathbb{E}(\psi_a''(X)) = o(1).$$

Since ϕ_a converges uniformly to ϕ_1, one sees that $\gamma^a(\{x : F(x) \geq 3\epsilon\} \cap [-K,K]) \to 0$. Since ϵ and K are arbitrary, all weak limit points of γ^a are concentrated on the global minimizers of F.

Now we consider an algorithm which decreases the stepsize a from step to step such that the whole algorithm approximates only the global minimizers of F.

Suppose that all global minimizers of F are contained in a compact set $[-K,K]$, which is known. We use a "reflected" version of the Robbins–Monro procedure: Let

$$T(x) = \begin{cases} x - 4jK & x \in [(4j-1)K, (4j+1)K] \\ (4j+2)K - x & x \in [(4j+1)K, (4j+3)K] \end{cases}$$

be the reflection function. Define the process

$$X_{n+1} = T(X_n - a_n f(X_n) - \sqrt{a_n \cdot \delta(a_n)} \cdot \xi_n). \tag{5.10}$$

Introduce the discrete metric

$$r_0(x,y) = \begin{cases} 0 & \text{if } x \neq y \\ 1 & \text{if } x = y. \end{cases}$$

and the Wasserstein metric pertaining to it

$$r_0(\mu,\nu) = 1 - \int \min\left(\frac{d\mu}{d\lambda}, \frac{d\nu}{d\lambda}\right) d\lambda,$$

where λ is some measure dominating μ and ν.

Suppose that ξ_n has density $g(\cdot)$. Then the ergodic coefficient ρ_0 pertaining to the discrete metric of (5.10) satisfies

$$\rho_0(a) \leq 1 - \frac{2K}{\sqrt{a\delta(a)}} \inf_{x\in[-2K,2K]} g(x/\sqrt{a\delta(a)}).$$

The appropriate choice of a_n depends on the density g. For the normal density $g(x) = \frac{1}{\sqrt{2\pi}}\exp(-\frac{x^2}{2})$ we have

$$\rho_0(a) \leq 1 - \frac{c_2}{\sqrt{a\delta(a)}} \exp\left(-\frac{c_2^2}{a\delta(a)}\right) \tag{5.11}$$

for some constant c_2, for the Cauchy density $g(x) = \frac{1}{\pi}\frac{1}{1+x^2}$ however

$$\rho_0(a) \leq 1 - \frac{c_3\ a\ \delta(a)}{c_3^2 + a^2\delta^2(a)}$$

for some constant c_3. Using theorem 3.4 we get the following result:

Theorem 5.5. *Let $a_n \downarrow 0$ be chosen such that*

$$\delta(a_n) = \frac{1}{\log\log(1/a_n)}$$

and $a_n\delta(a_n) = \frac{1}{\log(n)}$ *(if W_n is normally distributed)*
$a_n\delta(a_n) = \frac{1}{n}$ *(if W_n is Cauchy-distributed).*
Then all limit points of the recursion

$$X_{n+1} = T(X_n - a_n f(X_n) - \sqrt{a_n\delta(a_n)}W_n)$$

are concentrated on the set of global minimizers of F in $[-K, K]$.

One may also derive a variant of this result by combining several steps of the procedure into one block (compare definition 3.3). Choose $l_j = [e^j]$, where [] denotes the integer part. Then, if $|f'| \leq k_3$ and (ξ_n) are normally distributed, the conditional density of $X_{l_{j+1}}$ given $X_{l_j} = x$ is bounded below by

$$\frac{c_2}{\sqrt{\gamma_j}} \exp(-\frac{c_2}{\gamma_j})$$

uniformly in x, where

$$\gamma_j = \sum_{i=l_j}^{l_{j+1}} a_i\delta(a_i).$$

Choosing

$$a_n = \frac{1}{n}; \qquad \sqrt{a_n\delta(a_n)} = \sqrt{\frac{1}{n\log\log n}} \tag{5.12}$$

it follows that

$$\gamma_j \sim \frac{1}{\log(j+1)}$$

and the same conclusion can be made as in theorem 5.9.

The result just presented uses a reflection and needs the advance knowledge of a compact interval, which contains all global minimizers. A similar assumption was made by Kushner [35]. For practical purposes, the a-priori knowledge of some compact interval which contains the global minimizers is not a big problem. However, it is interesting, that one can get rid of such an assumption by cleverly choosing the metric in the search space, which is used in the "contraction" property.

Assume now that the the set of all critical points ($f(x) = 0$) is contained in some compact set $[-K, K]$, which is unknown. Without loss of generality we may assume that $K = 1/2$. We will consider the unconstrained algorithm

$$X_{n+1} = X_n - a_n f(X_n) - \sqrt{a_n \cdot \delta(a_n)} \cdot \xi_n. \tag{5.13}$$

Introduce the metric

$$r_2(x,y) = \begin{cases} 0 & \text{if } x = y \\ 1 & \text{if } 0 < |x-y| \le 1 \\ |x-y| & \text{if } |x-y| > 1 \end{cases}$$

Let us calculate the pertaining Wasserstein distance ρ_2 of two normal distributions with same variance: Let X be distributed according to a normal $N(0, \sigma^2)$ distribution and Y according to $N(-d, \sigma^2)$. Denote by

$$\phi_\sigma(u) = \frac{1}{\sigma\sqrt{2\pi}} \exp(-\frac{u^2}{2\sigma^2})$$

the density of X. The density of Y is $\phi_\sigma(u + d)$. Recall that calculating the Wasserstein distance is equivalent to finding the minimal way of transporting the distribution of X to the distribution of Y. Here is a possible way:
For $d \ge 1$, let $Y = X + d$.
For $d < 1$, set:
If $X < -1$, then

$$Y = \begin{cases} X & \text{with probability } \frac{\phi_\sigma(X+1)-\phi_\sigma(X+d)}{\phi_\sigma(X+1)-\phi_\sigma(X)} \\ X+1 & \text{with probability } \frac{\phi_\sigma(X+d)-\phi_\sigma(X)}{\phi_\sigma(X+1)-\phi_\sigma(X)} \end{cases}.$$

If $-1 \le X \le 0$, then $Y = X + d$.
If $X \ge 0$, then

$$Y = \begin{cases} X & \text{with probability } \frac{\phi_\sigma(X+d)-\phi_\sigma(X+1)}{\phi_\sigma(X)-\phi_\sigma(X+1)} \\ X+1 & \text{with probability } \frac{\phi_\sigma(X)-\phi_\sigma(X+d)}{\phi_\sigma(X)-\phi_\sigma(X+1)} \end{cases}.$$

It is easy to see that indeed in each case $Y \sim N(-d, \sigma^2)$.

Since the transport over a distance of $d < 1$ and over the distance 1 costs the same, this cost is saved by not transporting (i.e. when $X = Y$). One gets after a short calculation that for $d < 1$

$$\begin{aligned} r_2(N(x,\sigma^2), N(x-d,\sigma^2)) &\le 1 - \int_{-\infty}^{-1} \frac{\phi_\sigma(u+1)-\phi_\sigma(u+d)}{\phi_\sigma(u+1)-\phi_\sigma(u)} \phi_\sigma(u)\, du \\ &\quad - \int_0^{\infty} \frac{\phi_\sigma(u+d)-\phi_\sigma(u+1)}{\phi_\sigma(u)-\phi_\sigma(u+1)} \phi_\sigma(u)\, du \\ &\le 1 - \Phi(\frac{d}{\sigma}) + \Phi(\frac{1}{\sigma}), \end{aligned} \tag{5.14}$$

where $\Phi(x) = \frac{1}{\sqrt{2\pi}} \int_{-\infty}^{x} \exp(-\frac{u^2}{2})\, du$. For $d \geq 1$,

$$r_2(N(x,\sigma^2), N(x-d,\sigma^2)) = d.$$

Theorem 5.6. *Suppose that f' is bounded on $\mathbb{R}$ and that there is a constant $c_0 > 0$ such that for all $x, y \notin [-1/2, 1/2]$*

$$\|x - af(x) - y + af(y)\| \leq (1 - c_0 a)\|x - y\|$$

for sufficiently small a. Then, if ξ_n are normally $N(0,1)$ distributed, and $a_n \downarrow 0$, $a_n\delta(a_n) = \frac{1}{\log n}$, then all limit points of the recursion (5.13) are concentrated on the set of global minimizers of F.

Proof. By the bound (5.14) valid in $[-1/2, 1/2]$ and the strong contraction property outside this interval, one may show that

$$\rho_2(a) \leq 1 - \frac{c_3}{\sqrt{a\delta(a)}} \exp\left(-\frac{c_3}{a\delta(a)}\right)$$

for some constant c_3 (compare (5.11). The theorem is now again an application of theorem 3.4.

By a different approach, Gelfand and Mitter [17] proved a similar result for the procedure

$$X_{n+1} = X_n - \frac{A}{n} f(X_n) - \sqrt{\frac{B}{n \log\log n}}\, W_n,$$

which uses similar stepsize constants as in (5.12)

Recently, Gelfand and Mitter have also introduced an algorithm for global optimization in $\mathbb{R}^d$, which is a "continuous" version of the Metropolis algorithm (4.2) for discrete Simulated Annealing.

Let Q be the transition of a geometrically ergodic Markov process on the search space S, which is a closed subset of $\mathbb{R}^d$. Suppose that the stationary distribution π of Q has support S.

The algorithm is as follows:

1. Let X_1 be a starting point.
2. If X_n is the actual search point, make one transition from X_k to Y (say) according to Q.
3. Set $X_{n+1} = \begin{cases} Y & \text{with probability } \exp(-\frac{[F(Y)-F(X_n)]^+}{a_n}) \\ X_n & \text{with probability } 1 - \exp(-\frac{[F(Y)-F(X_n)]^+}{a_n}) \end{cases}$.

The search sequence (X_n) is an inhomogeneous Markov process. Its transition is

$$P_{k+1}(x,A) = \mathbb{P}\{X_{n+1} \in A | X_n = x\} = \int_A \exp(-\frac{[F(y)-F(x)]^+}{a_k})\, Q(x,dy)$$

$$+\mathbb{1}_A(x)[1 - \int \exp(-\frac{[F(y)-F(x)]^+}{a_k})\, Q(x,dy)].$$

Gelfand and Mitter have shown the convergence in law of the process (X_n) to a limiting law, which is concentrated on the global minimizers of F. They use a Gaussian transition Q with a state dependent variance and stepsizes (a_n), which are also state dependent. ([18]).

6. Robustness of Dynamical Systems

The stochastic approximation algorithm

$$X_{n+1} = X_n - a_n f(X_n) + a_n \xi_n \tag{6.1}$$

can be viewed as a discretized and noise corrupted version of the dynamical system

$$dx(t) = -f(x(t))\, dt. \tag{6.2}$$

It is evident that there is a close relation between the limiting behavior of the stochastic sequence (6.1) and the dynamical system (6.2).

As is well known, if f is locally Lipschitz (i.e. for each bounded set B there is a constant L_B such that $\|f(x) - f(y)\| \le \|x - y\|$ for all $x, y \in B$), then for every starting point x_0, there is a unique solution $x(t, x_0)$ of (6.2)

$$\begin{aligned} \frac{d}{dt}x(t,x_0) &= -f(x(t,x_0)) \\ x(0,x_0) &= x_0 \end{aligned}$$

for all $t \ge 0$. Uniqueness of the solution cannot be guaranteed, if f fails to be locally Lipschitz. The *orbit* $o(x_0)$ is the set of all points y which can be reached from x_0 or from which x_0 can be reached

$$o(x_0) = \{y : \exists t > 0 \quad y = x(x_0,t) \text{ or } x_0 = x(y,t)\}.$$

Let $\omega(x_0)$ be the *omega-limit* set of x_0, i.e.

$$\omega(x_0) = \{y \in \mathbb{R}^d : \exists (t_k) \text{ with } t_k \to \infty \text{ such that } x(t_k,x_0) \text{ converges to } y\}.$$

The *alpha-limit* $\alpha(x_0)$ is the omega-limit of the "inverse" dynamical system

$$\begin{aligned} \frac{d}{dt}x(t,x_0) &= f(x(t,x_0)) \\ x(0,x_0) &= x_0. \end{aligned}$$

A point x_0 is *periodic*, if there is a $T > 0$ such that $x(T, x_0) = x_0$. The orbit of a periodic point is a *cycle*. A set of orbits is a *pseudocycle*, if there are points $x_0, x_1, \ldots x_k = x_0$, such that

$$\omega(x_i) = \alpha(x_{i+1}) \qquad \text{for } 0 \le i \le k-1.$$

If $k = 1$, the pseudocycle is called homocyclic, otherwise heterocyclic. Let further

$$\mathcal{E} = \{x : f(x) = 0\}$$

be the set of *equilibrium points*. Define the set $\mathcal{R}$ of *chain-recurrent* points as

$$\begin{aligned}\mathcal{R} \quad = \quad & \{x : \forall \delta > 0, \forall T > 0, \exists k, (t_i)_{0 \le i \le k}, (y_i)_{0 \le i \le k} \text{ such that } t_i \ge T, \\ & \|x - y_0\| \le \delta, \|x(t_i, y_i) - y_{i+1}\| \le \delta \text{ for } 0 \le i \le k; x = y_k\}.\end{aligned}$$

The characterization of the limiting behavior of the dynamical system is easy, if the existence of a Lyapunov function is guaranteed. This is a differentiable function V, which is bounded from below and which satisfies

$$\frac{dV(x(t, x_0))}{dt} < 0 \qquad \text{if } x(t, x_0) \notin \mathcal{E}.$$

If f is a gradient, i.e. $f = \nabla F$, then $V(x) = F(x)$ is a natural Lyapunov function.

Here are some facts about the omega-limits:

- $\omega(x_0)$ is an invariant set. Each y in $\omega(x_0)$ is nonwandering, i.e. in every neighborhood U of y and every $T > 0$ there is a point z and a time t such that $x(t, z) \in U$.
- If the trajectory $x(t, x_0); t \ge 0$ is bounded, then $\omega(x_0)$ is a nonempty, compact, connected set.
- If, in addition, there is a Lyapunov function V, then $\omega(x_0)$ consists only of equilibria and V is constant on it. In particular, closed orbits and pseudocycles cannot exist.

A closed invariant set A is *attractive* , if there is a neighborhood U of A such that $\omega(x) \in A$ for all $x \in U$. The domain of attraction of A is the set $\{y : \omega(y) \in A\}$. A is called *stable*, if for every ϵ, there is a δ such that $\text{dist}(x_0, A) < \delta$ implies that $\text{dist}(x(t, x_0), A) < \epsilon$ for all $t > 0$.

Let us now turn to the stochastic approximation algorithm (6.1). Suppose for simplicity that (ξ_n) is a sequence of independent, zero mean random variables with bounded variances and (a_n) satisfy $a_n \ge 0$, $\sum_n a_n = \infty$, $\sum_n a_n^2 < \infty$.

If there is one single equilibrium point x^*, which is globally stable, i.e. its domain of attraction is the whole $\mathbb{R}^d$, then

$$X_n \to x^* \qquad \text{a.e. .}$$

The situation is more complicated, if there are more equilibria. Some of them may be asymptotically stable, i.e. attractive and stable. If ∇f exists, then a sufficient condition for asymptotic stability of $x^* \in \mathcal{E}$ is that the eigenvalues of $\nabla f(x^*)$ are all stricty negative.

If f is a gradient ($f = \nabla F$), then the asymptotically stable points are the local minima.

It is unfortunately not true that the stochastic sequence (X_n) given by (6.1) is always convergent to an asymptotically stable point. Without further assumptions, convergence to other equilibrium points or oscillation between several points cannot be excluded. However by the Kushner-Clark Lemma, oscillation between an asymptotically stable point and some other equilibrium is impossible:

Lemma 6.1 (Kushner-Clark). *Let x^* be an equilibrium point, which is a stable attractive set. If X_k visits infinitely often a compact set K contained in the domain of attraction of x^*, then*

$$X_k \to x^* \quad a.e.\,.$$

Proof. (see [33]).

It is also not true, that the sequence (X_n) must converge a.s., if the omega-limits of all points are singletons. Thus there might be the situation, where all limits of the deterministic dynamic system are singletons and X_n has more than one cluster point with positive probability.

There are four main questions about the relation of stochastic iterations and dynamical systems:

1. Can the possible cluster sets of X_n be characterized?
2. Under which circumstances does the sequence (X_n) converge to a random variable, say X^*?
3. Can the possible values of X^* (if existing) be characterized?
4. Is it possible to calculate the distribution of X^* (if existing)?

Some of these questions can be partially answered, but much has to be done yet. To question 4, no answer is known today.

By the mentioned results of Nevelson and Hasminskii [45] and Pemantle [46] some unstable equilibrium points can be excluded as limits of X_n (see section 5).

Benaim [2] has shown that under mild general conditions, all cluster points of (X_n) are contained in $\mathcal{R}$, the set of all chain recurrent points.

Fort and Pages [16] prove that if the dynamical system has no homoclinic or heteroclinic pseudocyles and if the equilibrium set $\mathcal{E}$ is locally finite (finite in every compact set), then $X_n \to X^* \in \mathcal{E}$ a.s.

Delyon [9] deals with the case of non-isolated equilibrium points. He supposes that there is an attractive set L of equilibria and shows that under some regularity assumptions $dist(X_n, L) \to 0$. Under more restrictive assumptions, (X_n) converges to $X^* \in L$.

Kaniovski [28] has shown that if convergence to non-stable points can be excluded, and the set of stable equilibria is locally finite, then convergence to a stable point can be established without the existence of a Lyapunov function.

7. Parallelization

Consider the Robbins-Monro procedure

$$X_{n+1} = X_n - a_n h_n(X_n, \xi_n) \tag{7.1}$$

where $\mathbb{E}(h_n(X_n, \xi_n)) = f(X_n) = \nabla F(X_n)$ and a_n is a sequence of stepsizes. It was demonstrated in section 5, that under some regularity assumptions this sequence recursively approximates the argmin of $F(x)$.

In this section we assume that we have a multi-processor machine with p processors at our disposal and address the problem how to distribute the work among them in an efficient way. We suppose that the most time consuming part of the algorithm is the calculation of $h_n(X_n, \xi_n)$ and that the synchronization, information exchange and calculation of X_{n+1} from X_n and $h_n(X_n, \xi_n)$ is of comparably negligible complexity. We suppose further that the observations $h_n(\cdot, \xi^{(1)}), ..., h_n(\cdot, \xi^{(p)})$ made by different processors are stochastically independent.

Problems of this kind arise in the context of optimizing simulated stationary systems. Here, the calculation of $h_n(X_n, \xi_n)$ requires a time-consuming simulation run (in theory of infinite length, in practice of large and even increasing (in n) length). The crucial problem is the following: At which stage should the processors communicate and interchange their current approximation values? The two extreme positions are

(i) communication with averaging at each step,
(ii) no communication until the very end.

Procedure (i) decreases the variance at each step, but is exposed to the danger that all processors work in the same basin of attraction of a local minimum. Procedure (ii), although subject to larger variances may be superior at least in situations with several local minima, since each approximation sequence may lie in a different basin.

In mathematical notation, we consider the procedures

(i) $\tilde{X}_{n+1} = \tilde{X}_n - a_n \frac{1}{p} \sum_{j=1}^{p} h_n(\tilde{X}_n, \xi_n^{(j)})$
(ii) $X_n^{(j)} = X_n^{(j)} - a_n h_n(X_n^{(j)}, \xi_n^{(j)}) \quad j = 1, ..., p.$

After n steps of procedure (ii) there must be a method of finding a unique approximation value X_n out of the values $X_n^{(1)}, ..., X_n^{(p)}$. Here are two methods:

(ii') Averaging: $\bar{X}_n = \frac{1}{p}\sum_{j=1}^{p} X_n^{(j)}$;

(ii") Making s further observations at each argument value $(X_n^{(1)}, \ldots, X_n^{(p)})$, i.e.

$$H(X_n^{(j)}, \xi_{n+1}^{(j)}), \ldots, H(X_n^{(j)}, \xi_{n+s}^{(j)}) \qquad j = 1, \ldots, p$$

and choose

$$\check{X}_n = X_n^{(j)} \text{ if } \frac{1}{s}\sum_{i=1}^{s} H(X_n^{(j)}, \xi_{n+i}^{(j)}) = \inf\{\frac{1}{s}\sum_{i=1}^{s} H(X_n^{(k)}, \xi_{n+i}^{(k)}) : 1 \le k \le p\}.$$

The results we are going to prove show some stochastic dominance relations either for the arguments X_n or the function values $Y_n = F(X_n)$. The next section introduces therefore the concept of stochastic ordering.

7.1 Stochastic Ordering

Let X and Y be two nonnegative random variables. X is *stochastically smaller* than Y, in notation

$$X \prec Y,$$

if $\mathbb{P}\{X \le t\} \ge \mathbb{P}\{Y \le t\}$ for all t.

If X and Y are real random variables and

$$|X| \prec |Y|, \tag{7.2}$$

we say that X is *more peaked around 0* than Y (cf. [7]). If X and Y are symmetric around 0, then (7.2) is equivalent to

$$\mathbb{P}\{X \le t\} \ge \mathbb{P}\{Y \le t\} \text{ for all } t > 0.$$

Definition 7.1. *Let $D(\cdot)$ be a distribution function on $\mathbb{R}^+$. We say that D belongs to the class*

DAIH (decreasing average integrated hazard) *if* $-x^{-1}\log(1 - D(x))$ *is* decreasing *in x,*

IAIH (increasing average integrated hazard) *if* $-x^{-1}\log(1 - D(x))$ *is* increasing *in x.*

Notice that

$$-\log(1 - D(x)) = \int_0^x \frac{D'(u)}{1 - D(u)}\,du$$

is the integrated hazard function. Consequently, any distribution with decreasing hazard $\frac{D'(u)}{1-D(u)}$ is DAIH and with increasing hazard is IAIH. This implies that the $\Gamma(\delta, b)$ distributions with density

$$\frac{1}{\Gamma(\delta)} b^{-\delta} u^{\delta-1} e^{-u/b}$$

belong to DAIH for $\delta \le 1$ and to IAIH for $\delta \ge 1$. In particular, the $\chi^2(1)$–distribution ($\delta = 1/2$) is DAIH.

The Exponential distibution is the only one lying in the intersection DAIH $\cap$ IAIH.

Lemma 7.1. *Let $Y_1, ..., Y_p$ be nonnegative random variables which are independently and identically distributed with distribution function D. If $D \in$ DAIH then, for each p,*

$$\min(Y_1, \ldots, Y_p) \prec \tfrac{1}{p} Y_1.$$

If $D \in$ IAIH then

$$\min(Y_1, \ldots, Y_p) \succ \tfrac{1}{p} Y_1.$$

Proof. The proof follows from the fact that D belongs to

$$\begin{array}{lll} \text{DAIH} & \text{iff} & 1 - D(px) \geq [1 - D(x)]^p \\ \text{IAIH} & \text{iff} & 1 - D(px) \leq [1 - D(x)]^p \end{array}$$

for all $x \geq 0$ and $p \geq 1$. This can be seen by noticing that D is DAIH iff

$$-\frac{1}{px} \log(1 - D(px)) \leq -\frac{1}{x} \log(1 - D(x))$$

for all $x \geq 0$ and $p \geq 1$.

Lemma 7.2. *The set DAIH is closed under convex combinations.*

Proof. Assume that $1 - D_1(px) \geq [1 - D_1(x)]^p$ and $1 - D_2(px) \geq [1 - D_2(x)]^p$. If $\alpha_1, \alpha_2 \geq 0$; $\alpha_1 + \alpha_2 = 1$, then by the convexity of $u \mapsto u^p$ for $p \geq 1$ one gets

$$\begin{aligned} 1 - \alpha_1 D_1(px) - \alpha_2 D_2(x) &\geq \alpha_1 [1 - D_1(x)]^p + \alpha_2 [1 - D_2(x)]^p \\ &\geq [1 - \alpha_1 D_1(x) - \alpha_2 D_2(x)]^p. \end{aligned}$$

Consider the Robbins-Monro procedure

$$X_{n+1} = X_n - \frac{a}{n} f(X_n) + \frac{a}{n} \xi_n.$$

W.l.o.g. we may and do assume that the unique stationary point is zero, i.e. $f(0) = 0$. The errors are assumed to be i.i.d. with $\mathbb{E}(\xi_n) = 0$, $\text{Var}(X_n) = \sigma^2$.

Assumption 7.4 (A).

(i) $x f(x) \geq c x^2$
(ii) $f(x) = \alpha x + o(x)$, as $x \to 0$

Theorem 7.1. *Let Assumption (A) be satisfied and let $\alpha \cdot a > 1/2$, then*

$$\sqrt{n} X_n \xrightarrow{\mathcal{L}} N\left(0, \frac{a^2 \sigma^2}{2a\alpha - 1}\right).$$

Here $\xrightarrow{\mathcal{L}}$ stands for the convergence in law. Moreover, with $F(x) = \int_0^x f(u)\, du$,

$$n \frac{2(2a\alpha - 1)}{\alpha a^2 \sigma^2} F(X_n) \xrightarrow{\mathcal{L}} \chi^2(1)$$

(see [58]).

Let us now look to the consequences of this result for the distributed algorithm:

Procedure (i) reduces the variance σ^2 to σ^2/p. Hence

$$\sqrt{n}\tilde{X}_n \xrightarrow{\mathcal{L}} N\left(0, \frac{a^2\sigma^2}{p(2a\alpha - 1)}\right). \tag{7.3}$$

Procedure (ii') averages p independent processes $X_n^{(j)}$ at the end and since independently for all j

$$\sqrt{n}X_n^{(j)} \xrightarrow{\mathcal{L}} N\left(0, \frac{a^2\sigma^2}{2a\alpha - 1}\right),$$

we get

$$\sqrt{n}\bar{X}_n \xrightarrow{\mathcal{L}} N\left(0, \frac{a^2\sigma^2}{p(2a\alpha - 1)}\right).$$

We see that there is no distinction between (i) and (ii') what concerns the asymptotic distribution.

However, assuming that the minimizing procedure (ii') finds the index j with mimimal $F(X_n^{(j)})$ without any error, we may infer that procedure (ii") is superior to (i) and (ii'), since the asymptotic distribution of $F(X_n)$ is χ^2 and hence DAIH.

Therefore, under the given assumption,

$$F(\check{X}_n) \quad \text{is asymptotically stochastically smaller than } F(\bar{X}_n) \text{ or } F(\tilde{X}_n).$$

Now we consider an extension of this result to the case of non local quadratic functions F.

Let f_γ the following function:

$$f_\gamma(x) = \text{ sign } (x)\, |x|^{\gamma-1}. \tag{7.4}$$

Assumption 7.6 (A_γ).

(i) $xf(x) \geq cx^2$
(ii) $f(x) = \alpha f_\gamma(x)(1 + o(1))$, as $x \to 0$

Definition 7.2. *The density*

$$g_\gamma(u) = \frac{\gamma^{1-1/\gamma}}{2\Gamma(\frac{1}{\gamma})} \exp(-\frac{1}{\gamma}|u|^\gamma)$$

is called Exponential Power *(γ) density (see [41]), p. 116). Special cases are the Laplace (Double Exponential) distribution for $\gamma = 1$ and the Normal distribution for $\gamma = 2$.*

Remark 7.1. If X has an Exponential Power(γ) density, then $|X|^\gamma$ has a $\Gamma(\frac{1}{\gamma}, \gamma)$ distribution.

Theorem 7.2. *Under Assumption* (A_γ), *for* $1/2 < \gamma < 1$,

$$n^{1/\gamma}\left(\frac{2\alpha}{a\sigma^2}\right)^{1/\gamma} X_n \text{ has an asymptotic Exponential Power } (\gamma) \text{ distribution}$$

and

$$n\frac{2\gamma}{a\sigma^2}F(X_n) \text{ has an asymptotic } \Gamma(\frac{1}{\gamma},\gamma) \text{ distribution .}$$

Proof. (see [27])

Remark 7.2. It may be proved, that under the same assumptions, but for $\gamma > 2$, $(\log n)^{1/(2-\gamma)} X_n$ converges to a constant. (see [40], p.72)

Proposition 7.1. *If* $X^{(1)}$ *and* $X^{(2)}$ *are independent and both distributed according to* g_γ, $\gamma \leq 2$, *then*

$$\frac{|X^{(1)}|}{2^{1/\gamma}} \prec \frac{|X^{(1)} + X^{(2)}|}{2}.$$

Proof. For the proof, we need the following Auxiliary Lemma:

Lemma 7.3. *Let* $f_\gamma(x) = \gamma$ *sign* $(x)|x|^{\gamma-1}$. *Then, if* $\frac{x+y}{2} > 0$, *then*

$$f_\gamma(\frac{x+y}{2}) \leq \frac{1}{2}(f_\gamma(x) + f_\gamma(y)).$$

If $\frac{x+y}{2} < 0$, *then*

$$f_\gamma(\frac{x+y}{2}) \geq \frac{1}{2}(f_\gamma(x) + f_\gamma(y)).$$

(The proof of this Lemma follows easily from geometric considerations.)

Let X^a be distributed according to the stationary distribution of the recursion

$$X^a_{n+1} = X^a_n - af_\gamma(X^a_n) + \sqrt{a}\xi_n,$$

where (ξ_n) is an i.i.d. $N(0,1)$-sequence. By Theorem 3 below, as a tends to 0,

$$X^a \xrightarrow{\mathcal{L}} X,$$

where X has density g_γ.

Let now $(\xi_n^{(1)})$ resp. $(\xi_n^{(2)})$ be two independent sequences and $\bar{\xi}_n = \frac{1}{2}(\xi_n^{(1)} + \xi_n^{(2)})$.

Consider the recursions

$$X^{(i)}_{n+1} = X^{(i)}_n - af_\gamma(X^{(i)}_n) + \sqrt{a}\xi^{(i)}_n; \quad i = 1,2$$

with

$$\bar{X}^a_n = \frac{1}{2}(X^{(1)}_n + X^{(2)}_n)$$

and

$$\tilde{X}^a_{n+1} = \tilde{X}^a_n - af_\gamma(\tilde{X}^a_n) + \sqrt{a}\xi_n.$$

Let $X_0^{(i)} = \tilde{X}_0 = 0$. Then $\bar{X}^a_n$ and $\tilde{X}^a_n$ have symmetric distributions. We show by induction that

$$|\tilde{X}^a_n| \prec |\bar{X}^a_n|$$

for all n. Suppose that

$$|\tilde{X}^a_n| \prec |\bar{X}^a_n|$$

has been established. Let $\hat{X}^a_{n+1} = \bar{X}^a_n - af_\gamma(\bar{X}^a_n) + \sqrt{a}\bar{\xi}_n$. The Auxiliary Lemma implies that $\bar{X}^a_n f_\gamma(\bar{X}^a_n) \geq \bar{X}^a_n \frac{1}{2}(X_n^{(1)} + X_n^{(2)})$ and therefore, the comparison Lemma in [49] yields that $|\hat{X}^a_{n+1}| \prec |\bar{X}^a_{n+1}|$ On the other hand, $|\tilde{X}^a_{n+1}| \prec |\hat{X}^a_{n+1}|$ so that finally $|\tilde{X}^a_{n+1}| \prec |\bar{X}^a_{n+1}|$ is established. This implies that also for the stationary distributions we have $|\tilde{X}^a| \prec |\bar{X}^a|$ and, letting a tend to 0, we have shown the assertion of the theorem.

Remark 7.3. Proschan [54] has shown a kind of "inverse" inequality for random variables with log-concave densities, namely that

$$|X^{(1)}| \succ \frac{|X^{(1)} + X^{(2)}|}{2}.$$

Let us again look at the consequences of this Proposition for the distributed algorithm. Here we consider only the case of two processors, since the just proved proposition is only available for $p = 2$.

Procedure (i) reduces the variance σ^2 to $\sigma^2/2$. Hence

$$2^{1/\gamma} n^{1/\gamma} \left(\frac{2\alpha}{a\sigma^2}\right)^{1/\gamma} \tilde{X}_n \xrightarrow{\mathcal{L}} \text{Exponential Power } (\gamma). \tag{7.5}$$

Procedure (ii') averages two independent processes $X_n^{(j)}$ at the end and we get

$$n^{1/\gamma} \left(\frac{2\alpha}{a\sigma^2}\right)^{1/\gamma} \bar{X}_n \xrightarrow{\mathcal{L}} \text{the mean of two independent}$$
$$\text{Exponential Power}(\gamma) \text{ distributions}.$$

The proposition shows that (i) is superior to (ii'), if $\gamma < 2$,.

Assuming again that the minimizing procedure (ii") finds the index j with mimimal $F(X_n^{(j)})$ without any error, we may infer that procedure (ii") is superior to (i), since the asymptotic distribution of $F(X_n)$ is a $\Gamma(\frac{1}{\gamma})$-distribution, which is DAIH.

Therefore, under the given assumption,

$F(\check{X}_n)$ is asymptotically stochastically smaller than $F(\tilde{X}_n)$,
which, in turn, is asymptotically stochastically smaller than $F(\bar{X}_n)$

8. Response Surface Methods

The stochastic approximation technique is characterized by an interplay between data collection and the calculation of the new design point X_{n+1}.

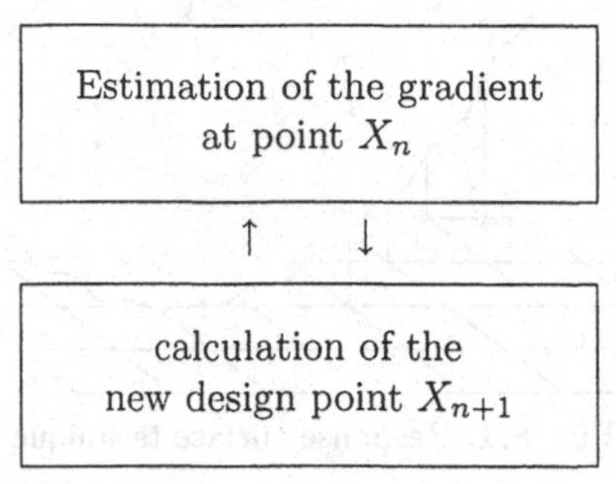

Stochastic Approximation

In contrast, the response surface technique collects the data first, approximates the response function $F(\cdot)$ by some interpolation $\hat{F}(\cdot)$ and optimizes in a second phase this interpolated function.

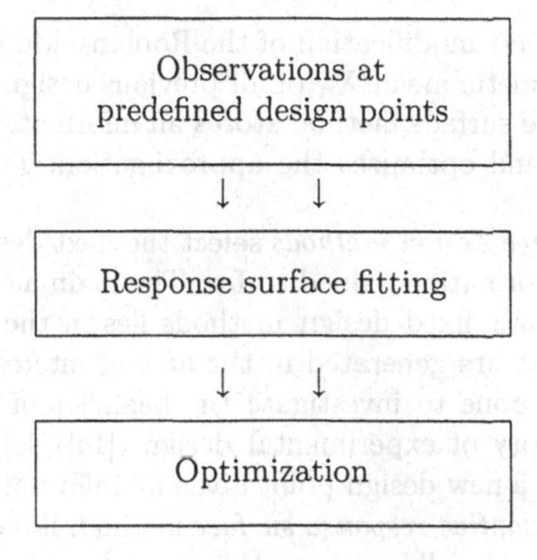

Response surface technique

There is a variety of "mixed" algorithms which combine elements of pure stochastic approximation with those of pure response surface method.

To classify a method, the following two questions are crucial:

1. Which amount of information about F is stored from step to step?
2. How is the function F approximated by a member of some simpler class of functions?

Here are some examples:

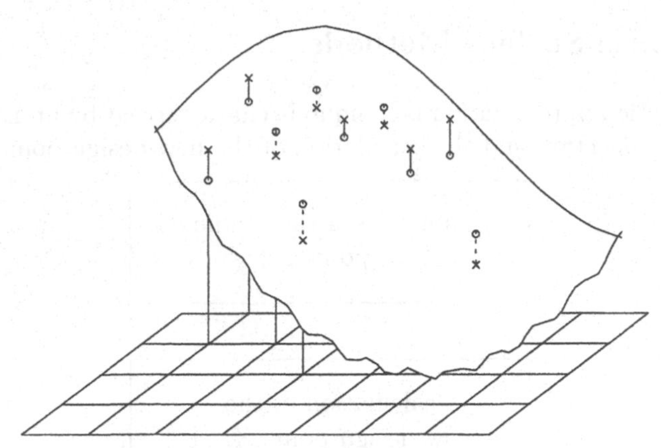

Fig. 8.1. Response surface technique

- The Kiefer–Wolfowitz procedure stores only the current value X_n and approximates F linearly, by taking 2^k additional design points and a linear interpolation.
- The Robbins–Monro procedure stores also only the current approximation value X_n and approximates F locally by a linear function via its gradient f.
- The Polyak–Ruppert modification of the Robbins–Monro procedure stores X_n and the arithmetic mean $\bar{X}_n$ of all previous design points.
- The pure response surface method stores all information about F approximates it finally and optimizes the approximation. This is a *fixed design method.*
- In general, *adaptive design methods* select the next design point X_{n+1} according to the information gained so far. The main advantage of adaptive design methods over fixed design methods lies in the fact that typically more design points are generated in the area of interest near the solution and less effort is done to investigate the behavior of F in regions of no interest. The theory of experimental design ([14]) solves the problem of optimally placing a new design point given all information gathered so far. We speak of an *adaptive response surface* method, if a curve of some parametric class is fitted locally to some design points and the new design point is then found by experimental design technique. Typically *linear*, *quadratic* or *cubic* functions are fitted. See [43] for a review on local response surface methods.

9. Bandit Processes

The search plan in simulated annealing is rather simple: The only information stored from the past is the current search point and the best point seen so far. If the algorithm visits a point it has seen previously, the function value is evaluated again. If the observations are corrupted by noise, the previously gathered information is wasted.

For small search sets $S = \{x_1, \dots, x_m\}$ one can store more information and use a more sophisticated experimental design plan. It is clear, that it does not make sense to invest much time and effort in observations at parameter points x_i, which are far from being optimal. On the other hand, we cannot exclude finally some parameter points in an early stage, since it may happen that due to random fluctuations our observations were much larger than their mean.

Adaptive assignment rules for a finite decision space must exhibit the following properties:

(i) At each stage of the procedure, there is a confidence set available, which contains the argmin of F with a prespecified probability $1 - \alpha$.
(ii) No parameter is excluded from further investigation in a final manner. All of them get an infinite number of observations, if the total number of observations tends to infinity. However the relative frequency of investigation of a particular parameter point can be very low, if it shows itself not a good candidate for the minimizer.
(iii) If it is known that the minimizer is unique, there can be a nice stopping time based on the rules: The search may stop, if the confidence set contains only one point.

Recall that the theory of optimal experimental design used in statistical estimation is based on some measure of the information which is contained in a statistical sample. If we have already collected some data, we may calculate the amount of information we can expect, if we make an additional observation at x. This expected measure of information is often but not necessarily obtained by a Bayesian argument. The next observation is made at the point x, which maximizes the expected information.

We may adopt a similar approach. As a (negative) measure of information about the unknown minimizer, we consider the expected size of a confidence set for this minimizer. This is indeed a good indicator, since to obtain a small confidence set is all that we can expect from random optimization problems. It also makes sure that we never exclude parameters in a final manner, since the sizes of the confidence sets depend on the means *and* variances and therefore the size of the confidence set can only be small, if all variances are small.

The details about this adaptive design plan can be found in [50].

Let us compare this approach to the well known *multi-armed bandit problem.*

The name multi-armed bandit comes from the problem of finding the best slot machine out of a finite number of such machines. The player does not know which machine provides the highest payoff. Therefore, he has to try all machines to find out the one with the highest winning chance. However, it is advisable not to play all machines with the same frequency: The machine with the presumably highest payoff should be played as often as possible.

Multi-armed bandit processes are also successfully applied in biostatistics. Here the problem is to find the best out of a finite number of drugs. The drugs are applied to patients and the experimental plan should be designed such that the best drug is found w.pr.1 and the number of patients who do not get the best drug should be minimized.

The theory of multi-armed bandit processes shows that there is a method of designing the experiment in such a manner that for $N \to \infty$

- The best machine is found almost surely.
- The relative number of experiments conducted with the best machine converges to 1.

In particular, Lai and Robbins [37] show the following: Let m populations be given, each with probability density $f(x, \theta_i)$; $i = 1, \ldots, m$. Let μ_i be the mean of density $f(x, \theta_i)$ and let μ_{i^*} be the best mean $\mu_{i^*} = \max_j \mu_j$. Suppose the sum of outcomes according to some sampling plan after n observations is $S_n = \sum_{i=1}^n X_i$. Call $R_n = n\mu_{i^*} - \mathbb{E}(S_n)$ the *regret.*

If $R_n = o(n^a)$ for some $a > 0$, then necessarily the regret obeys the following lower bound

$$\liminf_{n \to \infty} \frac{R_n}{\log n} \geq \sum_{\mu_j < \mu_{i^*}} \frac{\mu_{i^*} - \mu_j}{I(\theta_j, \theta_{i^*})},$$

where $I(\theta_j, \theta_{i^*})$ is the Kullback-Leibler entropy.

Lai and Robbins construct an allocation rule that achieves the lower bound and call this rule *efficient.*

Notice that efficiency is only an asymptotic property. By changing the rule for finitely many observations, the efficiency is not changed.

If the criterion to be minimized is not the regret, but the discounted sum $\sum \beta^i X_i$, the the best allocation rule must be based on dynamic allocation indices (*Gittins indices*) – see Gittins (1979).

It is important to see that our discrete optimization problem is different in nature from the multi-armed bandit problem. In our problem every observation costs the same and the objective is only to find the minimizer with as few observations as possible. We can make many observations at non-optimal points if we consider them as necessary with no extra costs.

In contrast, the costs of the multi-armed bandit are proportional to the payoff and therefore it is important to play the presumably best machine as often as possible. It is not advisable to use rules from bandit-processes for the optimal design of simulation experiments.

References

1. Aarts E., and J. Korst, *Simulated annealling and Boltzmann machines: A stochastic approach to combinatorial optimization and neural computing*, J. Wiley and Sons, New York, 1989.
2. Benaim M., A dynamical system approach to stochastic approximations, *Mimeo*, Univ. of California at Berkeley, 1993.
3. Benveniste A., M. Métivier M., and P. Priouret, *Adaptive algorithms and stochastic approximation*, Springer, Berlin, 1990.
4. Berry D.A., and B. Fristedt, *Bandit problems*, Chapman and Hall, London, 1985.
5. Bertsekas D.P., and J.N. Tsitsiklis, *Parallel and distributed computing*, Prentice Hall, Englewood Cliffs, 1989.
6. Bickel P.J., and D.A. Freedman, Some asymptotic theory for the bootstrap, *Ann. Statist.*, 9, 1196–1217, 1981.
7. Birnbaum Z.W., On random variables with comparable peakedness, *Ann. Math. Statist.*, 19, 76–81, 1948.
8. Delyon B., and A. Juditsky, Stochastic optimization with averaging of trajectories, *Stochastics and Stochastics Reports*, 39, 107–118 ,1922.
9. Delyon B., A deterministic approach to stochastic approximation, *Publication 789*, IRISA, Rennes, 1994.
10. Duflo M., *Méthodes récursives aléatoires*, Masson, Paris,1990.
11. Dupac V., and U. Herkenrath, Stochastic approximation on a discrete set and the multi-armed bandit problem, *Commun. Statist. Sequential Analysis 1* (1), 1–25,1982.
12. Dupac V., and U. Herkenrath, Stochastic approximation with delayed observations, *Biometrika 72*, 3, 683-685,1985.
13. Fabian V., On asymptotic normality in stochastic approximation, *Ann. Math. Statist. 39*, 1327–1332,1968.
14. Fedorov V., *Theory of optimal experiment*, Academic Press, New York,1972.
15. Fletcher R., *Practical Methods of Optimization*, J. Wiley & Sons, Chichester, 1981.
16. Fort J.C., and G. Pages, Convergence of stochastic algorithms: the Kushner-Clark theorem revisited, *Mimeo*, 1994.
17. Gelfand S.B., and S.K. Mitter, Recursive stochastic algorithms for global optimization in $\mathbb{R}^d$, *SIAM J. Control and Optimization*, 29 (5), 999–1018,1991.
18. Gelfand S. B., and S.K. Mitter, Metropolis-type annealing algorithms for global optimization in $\mathbb{R}^d$, *Siam J. Control and Optimization*, 33 (1) 111–131,1993.
19. Geman S., C.R. Hwang, Diffusions for global optimization, *Siam J. Control and Optimization* 24, 5, 1031–1043,1986.
20. Gibbons J.D., I. Olkin, and M. Sobel, *Selecting and ordering populations*, Wiley, New York, 1977.
21. Gittins J.C., Bandit processes and dynamic allocation indices, *J. Roy. Statist. Soc. 41*, 148–177,1979.
22. Gupta S. S., and S. Panchapakesan, *Multiple decision procedures: theory and methodology of selecting and ranking populations*, Wiley, New York, 1979.
23. Gutjahr W., and G. Pflug, Simulated annealing for noisy cost functions, *Technical Report TR-94-19*, University of Vienna 1994.
24. Herkenrath U., The N-armed bandit with unimodal structure, *Metrika*, 30, 195–210,1983.
25. Ho, Y.C., R.S. Sreenivas, and P. Vakili, Ordinal optimization of DEDS, *J. of Discrete Event Dynamical Systems* 2 (2), 61–88,1992.

26. Hwang C.R., Laplace's method revisited: weak convergence of probability measures, *Ann. Probab.*, Vol 8., 1177–1182,1980.
27. Kaniovski Yu., and G. Pflug, Non-standard limit theorems for urn models and stochastic approximation procedures, to appear in: *Stochastic Models*.
28. Kaniovski Y.M., Strong convergence of stochastic approximation without Lyapunov functions, Working paper 95-19, IIASA, Laxenburg, Austria, 1995.
29. Kersting G., Some results on the asymptotic behavior of the Robbins–Monro process, *Bull. Int. Stat. Inst.*, 47, 327–335
30. Kirkpatrick S., C.D. Gelatt, and M.P. Vecchi, Optimization by simulated annealing, *Science*, 220, 671–680, 1983.
31. Kushner H.J., Stochastic approximation algorithms for the local optimization of functions with nonunique stationary points, *IEEE Trans. Automatic Control*, AC-17, 646–654,1972.
32. Kushner H. J., and E. Sanvincente, Stochastic approximation for constrained systems with observation noise on the system and constraints, *Automatica*, 11, 375–380, 1975.
33. Kushner H.J., and D.S. Clark, Stochastic approximation for constrained and unconstrained systems, *Appl. Math. Sciences*, 26, Springer, 1978.
34. Kushner H.J., Asymptotic behaviour of stochastic approximation and large deviations, *IEEE Trans. Automatic Control*, AC-29, 984–990, 1984.
35. Kushner H.J., Asymptotic global behavior for stochastic approximation and diffusions with slowly decreasing noise effects: global minimization via Monte Carlo, *Siam J. Appl. Math.*, 47, 1, 169–185,1987.
36. Kushner H.J., and G. Yin, Asymptotic properties of distributed and communicating stochastic approximation algorithms, *Siam J. Control and Optimization* 25, 5, 1266–1290, 1987.
37. Lai T.L., and H. Robbins, Asymptotically efficient adaptive allocation rules, *Adv. Appl. Math.* 6, 4–22,1985.
38. Ljung L., Analysis of recursive stochastic algorithms, *IEEE Trans. Automatic Control* AC-22, 551–575, 1977.
39. Ljung L., Strong convergence of a stochastic approximation algorithm, *Ann. Statist.* 6, 680–696, 1978.
40. Ljung L., G. Pflug, and H. Walk, Stochastic approximation and optimization of random systems, *DMV Seminar Band* 17, Birkhäuser Verlag, Basel, 1992.
41. Manoukian E.B. *Modern concepts and theorems of mathematical statistics*, Springer Series in Statistics, Springer, New York, 1986.
42. Marshall A., and I. Olkin, *Inequalities: theory of majorization and its applications*, Academic Press, New York, 1979.
43. Marti K., Semi-stochastic approximation by the response surface methodology, *Optimization* 25, 209–230, 1992.
44. Metropolis N., A. Rosenbluth, M. Rosenbluth, A. Teller, and E. Teller, Equation of state calculations by fast computing machines, *J. Chemical Physics* 21, 1087–1092, 1953.
45. Nevel'son M.B., and R.S. Hasminskij, Stochastic approximation and recurrent estimation, *Nauka*, Moskwa (in Russian). Translated in Amer. Math. Soc. Transl. Monographs, Vol.24, Providence, R.I,1972.
46. Pemantle R., Nonconvergence to unstable points in urn models and stochastic approximations, *Ann. Prob.* 18, 698–712, 1990.
47. Pflug G., Stochastic minimization with constant step-size - asymptotic laws, *SIAM J. of Control*, 14 (4) 655–666, 1985.
48. Pflug G., Non-asymptotic confidence bounds for stochastic approximation algorithms with constant step size, *Monatshefte für Mathematik*, 110, 297–314. Springer, 1990.

49. Pflug G., A note on the comparison of stationary laws of Markov processes, *Statistics and Probability letters* 11, 4, 1991.
50. Pflug G., Adaptive designs in discrete stochastic optimization, *IIASA working paper* WP-94-59, 1994.
51. Pflug G., *Simulation and optimization*, book manuscript, 1995.
52. Polyak B., Novi metod tipa stochasticeskoi approksmacii, *Automatika i Telemechanika* No.7, 98–107 (in Russian), 1991.
53. Polyak B., and A. Juditsky, Acceleration of stochastic approximation by averaging, *SIAM J. Control Optimization* 30, 838–855, 1992.
54. Proschan F., Peakedness of distributions of convex combinations, *Ann. Math. Statist.* 36, 1703–1706, 1965.
55. Robbins H., and S. Monro, A stochastic approximation method, *Ann. Math. Statist.* 22, 400–407, 1951.
56. Ruppert D., Efficient estimators from a slowly convergent Robbins-Monro process, *Technical Report* 781, School of Operations Research and Industrial Engineering, Cornell University, Ithaca, New York. see also: Stochastic Approximation in: Handbook of Sequential Analysis (B.K. Gosh, P.K. Sen, eds.) Marcel Dekker, New York, 1991, 503–529, 1988.
57. Vallender S.S., Calculation of the Wasserstein distance between probability distributions on the line, *Theory Probab. Appl.*, 18, 784–786, 1973.
58. Wasan M.T., *Stochastic approximation.* Cambridge University Press, 1969.
59. Wiggins S., *Introduction to applied nonlinear dynamical systems and chaos.* Springer, 1990.

List of Contributors

P. Albertos
Departamento de Ingeniería de
Sistemas, Computadores y
Automática
Universidad Politécnica de Valencia
Spain.

A. Benveniste
IRISA-INRIA
Campus Universitaire de Beaulieu
35042 Rennes Cedex
France.

S. Bittanti
Dipartimento di Elettronica
e Informazione
Politecnico di Milano
Piazza Leonardo da Vinci 32
20133 Milano
Italy.

M. Campi
Dipartimento di Elettronica
per l'Automazione
Università di Brescia
via Branze 38
25123 Brescia
Italy.

G. Cybenko
Thayer School of Engineering
Dartmouth College
Hanover, NH 03755
USA.

M. Deistler
Institute for Econometrics,
Operations Research and Systems
Theory
Technical University Vienna
Argentinierstraße 8
A-1040 Vienna
Austria.

B. Delyon
IRISA-INRIA
Campus Universitaire de Beaulieu
35042 Rennes Cedex
France.

J. Friedman
Dept. of Aerospace Engineering
The University of Michigan
Ann Arbor, MI 48109-2140
USA.

P-Y. Glorennec
IRISA-INSA
Campus Universitaire de Beaulieu
35042 Rennes Cedex
France.

G. Gu
Dept. of Electrical Engineering
Louisiana State University
Baton Rouge, LA 70803-5901
USA.

H. Hjalmarsson
Department of Electrical
Engineering
Linköping University
S-581 83 Linköping
Sweden.

A. Juditsky
IRISA-INRIA
Campus Universitaire de Beaulieu
35042 Rennes Cedex
France.

P. P. Khargonekar
Dept. of Electrical Engineering
and Computer Science
The University of Michigan
Ann Arbor, MI 48109-2122, USA.

A. Lindquist
Optimization and System Theory
Dept. of Mathematics
Royal Institute of Technology
S-10084
Sweden.

L. Ljung
Department of Electrical
Engineering
Linköping University
S-581 83 Linköping
Sweden.

J. M. Maciejowski
Cambridge University
Engineering Dept.
Cambridge CB2 1PZ, UK.

R. J. Ober
Center for Engineering
Mathematics
University of Texas at Dallas
Richardson, TX75083, USA.

G. Pflug
University of Vienna
Universitätsstraße 5
A-1010 Wien/Vienna
Austria.

G. Picci
Dipartimento di Elettronica
e Informatica
Universita' di Padova
via Gradenigo 6/A
35131 Padua
Italy.

P. Rapisarda
Dipartimento di Ingegneria
Elettronica, Elettrotecnica ed
Informatica
Universita' di Trieste
34127 Trieste
Italy.

W. Scherrer
Institute for Econometrics,
Operations Research
and Systems Theory
Technical University Vienna
Argentinierstraße 8
A-1040 Vienna
Austria.

J. H. van Schuppen
CWI
P.O. Box 94079
1090 GB Amsterdam
The Netherlands.

J. Sjöberg
Department of Electrical
Engineering
Linköping University
S-581 83 Linköping
Sweden.

A. A. Stoorvogel
Department of Mathematics and
Computing Science
Eindhoven University of Technology
P.O. Box 513
5600 MB Eindhoven
The Netherlands.

M. Vidyasagar
Centre for Artificial Intelligence
and Robotics
Raj Bhavan Circle
High Grounds
Bangalore 560 001
India.

J. C. Willems
Institute of Mathematics
and Computing Science,
University of Groningen
9700 AV Groningen
The Netherlands.

Q. Zhang
IRISA-INRIA
Campus Universitaire de Beaulieu
35042 Rennes Cedex
France.

NATO ASI Series F

Including Special Programmes on Sensory Systems for Robotic Control (ROB) and on Advanced Educational Technology (AET)

Vol. 1: Issues in Acoustic Signal - Image Processing and Recognition. Edited by C. H. Chen. VIII, 333 pages. 1983. *(out of print)*

Vol. 2: Image Sequence Processing and Dynamic Scene Analysis. Edited by T. S. Huang. IX, 749 pages. 1983.

Vol. 3: Electronic Systems Effectiveness and Life Cycle Costing. Edited by J. K. Skwirzynski. XVII, 732 pages. 1983. *(out of print)*

Vol. 4: Pictorial Data Analysis. Edited by R. M. Haralick. VIII, 468 pages. 1983.

Vol. 5: International Calibration Study of Traffic Conflict Techniques. Edited by E. Asmussen. VII, 229 pages. 1984.

Vol. 6: Information Technology and the Computer Network. Edited by K. G. Beauchamp. VIII, 271 pages. 1984.

Vol. 7: High-Speed Computation. Edited by J. S. Kowalik. IX, 441 pages. 1984.

Vol. 8: Program Transformation and Programming Environments. Report on a Workshop directed by F. L. Bauer and H. Remus. Edited by P. Pepper. XIV, 378 pages. 1984.

Vol. 9: Computer Aided Analysis and Optimization of Mechanical System Dynamics. Edited by E. J. Haug. XXII, 700 pages. 1984. *(out of print)*

Vol. 10: Simulation and Model-Based Methodologies: An Integrative View. Edited by T. I. Ören, B. P. Zeigler, M. S. Elzas. XIII, 651 pages. 1984.

Vol. 11: Robotics and Artificial Intelligence. Edited by M. Brady, L. A. Gerhardt, H. F. Davidson. XVII, 693 pages. 1984.

Vol. 12: Combinatorial Algorithms on Words. Edited by A. Apostolico, Z. Galil. VIII, 361 pages. 1985.

Vol. 13: Logics and Models of Concurrent Systems. Edited by K. R. Apt. VIII, 498 pages. 1985.

Vol. 14: Control Flow and Data Flow: Concepts of Distributed Programming. Edited by M. Broy. VIII, 525 pages. 1985. *Reprinted as Springer Study Edition 1986.*

Vol. 15: Computational Mathematical Programming. Edited by K. Schittkowski. VIII, 451 pages. 1985.

Vol. 16: New Systems and Architectures for Automatic Speech Recognition and Synthesis. Edited by R. De Mori, C.Y. Suen. XIII, 630 pages. 1985. *(out of print)*

Vol. 17: Fundamental Algorithms for Computer Graphics. Edited by R. A. Earnshaw. XVI, 1042 pages. 1985. *Reprinted as Springer Study Edition 1991.*

Vol. 18: Computer Architectures for Spatially Distributed Data. Edited by H. Freeman and G. G. Pieroni. VIII, 391 pages. 1985. *(out of print)*

Vol. 19: Pictorial Information Systems in Medicine. Edited by K. H. Höhne. XII, 525 pages. 1986.

Vol. 20: Disordered Systems and Biological Organization. Edited by E. Bienenstock, F. Fogelman Soulié, G. Weisbuch. XXI, 405 pages. 1986

Vol. 21: Intelligent Decision Support in Process Environments. Edited by E. Hollnagel, G. Mancini, D. D. Woods. XV, 524 pages. 1986.

Vol. 22: Software System Design Methods. The Challenge of Advanced Computing Technology. Edited by J. K. Skwirzynski. XIII, 747 pages. 1986. *(out of print)*

Vol. 23: Designing Computer-Based Learning Materials. Edited by H. Weinstock and A. Bork. IX, 285 pages. 1986. *(out of print)*

Vol. 24: Database Machines. Modern Trends and Applications. Edited by A. K. Sood and A. H. Qureshi. VIII, 570 pages. 1986.

NATO ASI Series F

Including Special Programmes on Sensory Systems for Robotic Control (ROB) and on Advanced Educational Technology (AET)

Vol. 25: Pyramidal Systems for Computer Vision. Edited by V. Cantoni and S. Levialdi. VIII, 392 pages. 1986. *(ROB)*

Vol. 26: Modelling and Analysis in Arms Control. Edited by R. Avenhaus, R. K. Huber and J. D. Kettelle. VIII, 488 pages. 1986. *(out of print)*

Vol. 27: Computer Aided Optimal Design: Structural and Mechanical Systems. Edited by C. A. Mota Soares. XIII, 1029 pages. 1987.

Vol. 28: Distributed Operating Systems. Theory und Practice. Edited by Y. Paker, J.-P. Banatre and M. Bozyiğit. X, 379 pages. 1987.

Vol. 29: Languages for Sensor-Based Control in Robotics. Edited by U. Rembold and K. Hörmann. IX, 625 pages. 1987. *(ROB)*

Vol. 30: Pattern Recognition Theory and Applications. Edited by P. A. Devijver and J. Kittler. XI, 543 pages. 1987.

Vol. 31: Decision Support Systems: Theory and Application. Edited by C. W. Holsapple and A. B. Whinston. X, 500 pages. 1987.

Vol. 32: Information Systems: Failure Analysis. Edited by J. A. Wise and A. Debons. XV, 338 pages. 1987.

Vol. 33: Machine Intelligence and Knowledge Engineering for Robotic Applications. Edited by A. K. C. Wong and A. Pugh. XIV, 486 pages. 1987. *(ROB)*

Vol. 34: Modelling, Robustness and Sensitivity Reduction in Control Systems. Edited by R.F. Curtain. IX, 492 pages. 1987.

Vol. 35: Expert Judgment and Expert Systems. Edited by J. L. Mumpower, L. D. Phillips, O. Renn and V. R. R. Uppuluri. VIII, 361 pages. 1987.

Vol. 36: Logic of Programming and Calculi of Discrete Design. Edited by M. Broy. VII, 415 pages. 1987.

Vol. 37: Dynamics of Infinite Dimensional Systems. Edited by S.-N. Chow and J. K. Hale. IX. 514 pages. 1987.

Vol. 38: Flow Control of Congested Networks. Edited by A. R. Odoni, L. Bianco and G. Szegö. XII, 355 pages. 1987.

Vol. 39: Mathematics and Computer Science in Medical Imaging. Edited by M. A. Viergever and A. Todd-Pokropek. VIII, 546 pages. 1988.

Vol. 40: Theoretical Foundations of Computer Graphics and CAD. Edited by R. A. Earnshaw. XX, 1246 pages. 1988. *(out of print)*

Vol. 41: Neural Computers. Edited by R. Eckmiller and Ch. v. d. Malsburg. XIII, 566 pages. 1988. *Reprinted as Springer Study Edition 1989, 1990.*

Vol. 42: Real-Time Object Measurement and Classification. Edited by A. K. Jain. VIII, 407 pages. 1988. *(ROB)*

Vol. 43: Sensors and Sensory Systems for Advanced Robots. Edited by P. Dario. XI, 597 pages. 1988. *(ROB)*

Vol. 44: Signal Processing and Pattern Recognition in Nondestructive Evaluation of Materials. Edited by C. H. Chen. VIII, 344 pages. 1988. *(ROB)*

Vol. 45: Syntactic and Structural Pattern Recognition. Edited by G. Ferraté, T. Pavlidis, A. Sanfeliu and H. Bunke. XVI, 467 pages. 1988. *(ROB)*

NATO ASI Series F

Including Special Programmes on Sensory Systems for Robotic Control (ROB) and on Advanced Educational Technology (AET)

Vol. 46: Recent Advances in Speech Understanding and Dialog Systems. Edited by H. Niemann, M. Lang and G. Sagerer. X, 521 pages. 1988.

Vol. 47: Advanced Computing Concepts and Techniques in Control Engineering. Edited by M. J. Denham and A. J. Laub. XI, 518 pages. 1988. *(out of print)*

Vol. 48: Mathematical Models for Decision Support. Edited by G. Mitra. IX, 762 pages. 1988.

Vol. 49: Computer Integrated Manufacturing. Edited by I. B. Turksen. VIII, 568 pages. 1988.

Vol. 50: CAD Based Programming for Sensory Robots. Edited by B. Ravani. IX, 565 pages. 1988. *(ROB)*

Vol. 51: Algorithms and Model Formulations in Mathematical Programming. Edited by S. W. Wallace. IX, 190 pages. 1989.

Vol. 52: Sensor Devices and Systems for Robotics. Edited by A. Casals. IX, 362 pages. 1989. *(ROB)*

Vol. 53: Advanced Information Technologies for Industrial Material Flow Systems. Edited by S. Y. Nof and C. L. Moodie. IX, 710 pages. 1989.

Vol. 54: A Reappraisal of the Efficiency of Financial Markets. Edited by R. M. C. Guimar ães, B. G. Kingsman and S. J. Taylor. X, 804 pages. 1989.

Vol. 55: Constructive Methods in Computing Science. Edited by M. Broy. VII, 478 pages. 1989.

Vol. 56: Multiple Criteria Decision Making and Risk Analysis Using Microcomputers. Edited by B. Karpak and S. Zionts. VII, 399 pages. 1989.

Vol. 57: Kinematics and Dynamic Issues in Sensor Based Control. Edited by G. E. Taylor. XI, 456 pages. 1990. *(ROB)*

Vol. 58: Highly Redundant Sensing in Robotic Systems. Edited by J. T. Tou and J. G. Balchen. X, 322 pages. 1990. *(ROB)*

Vol. 59: Superconducting Electronics. Edited by H. Weinstock and M. Nisenoff. X, 441 pages. 1989.

Vol. 60: 3D Imaging in Medicine. Algorithms, Systems, Applications. Edited by K. H. Höhne, H. Fuchs and S. M. Pizer. IX, 460 pages. 1990. *(out of print)*

Vol. 61: Knowledge, Data and Computer-Assisted Decisions. Edited by M. Schader and W. Gaul. VIII, 421 pages. 1990.

Vol. 62: Supercomputing. Edited by J. S. Kowalik. X, 425 pages. 1990.

Vol. 63: Traditional and Non-Traditional Robotic Sensors. Edited by T. C. Henderson. VIII, 468 pages. 1990. *(ROB)*

Vol. 64: Sensory Robotics for the Handling of Limp Materials. Edited by P. M. Taylor. IX, 343 pages. 1990. *(ROB)*

Vol. 65: Mapping and Spatial Modelling for Navigation. Edited by L. F. Pau. VIII, 357 pages. 1990. *(ROB)*

Vol. 66: Sensor-Based Robots: Algorithms and Architectures. Edited by C. S. G. Lee. X, 285 pages. 1991. *(ROB)*

Vol. 67: Designing Hypermedia for Learning. Edited by D. H. Jonassen and H. Mandl. XXV, 457 pages. 1990. *(AET)*

Vol. 68: Neurocomputing. Algorithms, Architectures and Applications. Edited by F. Fogelman Soulié and J. Hérault. XI, 455 pages. 1990.

Vol. 69: Real-Time Integration Methods for Mechanical System Simulation. Edited by E. J. Haug and R. C. Deyo. VIII, 352 pages. 1991.

NATO ASI Series F

Including Special Programmes on Sensory Systems for Robotic Control (ROB) and on Advanced Educational Technology (AET)

Vol. 70: Numerical Linear Algebra, Digital Signal Processing and Parallel Algorithms. Edited by G. H. Golub and P. Van Dooren. XIII, 729 pages. 1991.

Vol. 71: Expert Systems and Robotics. Edited by T. Jordanides and B.Torby. XII, 744 pages. 1991.

Vol. 72: High-Capacity Local and Metropolitan Area Networks. Architecture and Performance Issues. Edited by G. Pujolle. X, 536 pages. 1991.

Vol. 73: Automation and Systems Issues in Air Traffic Control. Edited by J. A. Wise, V. D. Hopkin and M. L. Smith. XIX, 594 pages. 1991.

Vol. 74: Picture Archiving and Communication Systems (PACS) in Medicine. Edited by H. K. Huang, O. Ratib, A. R. Bakker and G. Witte. XI, 438 pages. 1991.

Vol. 75: Speech Recognition and Understanding. Recent Advances, Trends and Applications. Edited by P. Laface and Renato De Mori. XI, 559 pages. 1991.

Vol. 76: Multimedia Interface Design in Education. Edited by A. D. N. Edwards and S. Holland. XIV, 216 pages. 1992. *(AET)*

Vol. 77: Computer Algorithms for Solving Linear Algebraic Equations. The State of the Art. Edited by E. Spedicato. VIII, 352 pages. 1991.

Vol. 78: Integrating Advanced Technology into Technology Education. Edited by M. Hacker, A. Gordon and M. de Vries. VIII, 185 pages. 1991. *(AET)*

Vol. 79: Logic, Algebra, and Computation. Edited by F. L. Bauer. VII, 485 pages. 1991.

Vol. 80: Intelligent Tutoring Systems for Foreign Language Learning. Edited by M. L. Swartz and M. Yazdani. IX, 347 pages. 1992. *(AET)*

Vol. 81: Cognitive Tools for Learning. Edited by P. A. M. Kommers, D. H. Jonassen, and J. T. Mayes. X, 278 pages. 1992. *(AET)*

Vol. 82: Combinatorial Optimization. New Frontiers in Theory and Practice. Edited by M. Akgül, H. W. Hamacher, and S. Tüfekçi. XI, 334 pages. 1992.

Vol. 83: Active Perception and Robot Vision. Edited by A. K. Sood and H. Wechsler. IX, 756 pages. 1992.

Vol. 84: Computer-Based Learning Environments and Problem Solving. Edited by E. De Corte, M. C. Linn, H. Mandl, and L. Verschaffel. XVI, 488 pages. 1992. *(AET)*

Vol. 85: Adaptive Learning Environments. Foundations and Frontiers. Edited by M. Jones and P. H. Winne. VIII, 408 pages. 1992. *(AET)*

Vol. 86: Intelligent Learning Environments and Knowledge Acquisition in Physics. Edited by A. Tiberghien and H. Mandl. VIII, 285 pages. 1992. *(AET)*

Vol. 87: Cognitive Modelling and Interactive Environments. With demo diskettes (Apple and IBM compatible). Edited by F. L. Engel, D. G. Bouwhuis, T. Bösser, and G. d'Ydewalle. IX, 311 pages. 1992. *(AET)*

Vol. 88: Programming and Mathematical Method. Edited by M. Broy. VIII, 428 pages. 1992.

Vol. 89: Mathematical Problem Solving and New Information Technologies. Edited by J. P. Ponte, J. F. Matos, J. M. Matos, and D. Fernandes. XV, 346 pages. 1992. *(AET)*

Vol. 90: Collaborative Learning Through Computer Conferencing. Edited by A. R. Kaye. X, 260 pages. 1992. *(AET)*

Vol. 91: New Directions for Intelligent Tutoring Systems. Edited by E. Costa. X, 296 pages. 1992. *(AET)*

NATO ASI Series F

Including Special Programmes on Sensory Systems for Robotic Control (ROB) and on Advanced Educational Technology (AET)

Vol. 92: Hypermedia Courseware: Structures of Communication and Intelligent Help. Edited by A. Oliveira. X, 241 pages. 1992. *(AET)*

Vol. 93: Interactive Multimedia Learning Environments. Human Factors and Technical Considerations on Design Issues. Edited by M. Giardina. VIII, 254 pages. 1992. *(AET)*

Vol. 94: Logic and Algebra of Specification. Edited by F. L. Bauer, W. Brauer, and H. Schwichtenberg. VII, 442 pages. 1993.

Vol. 95: Comprehensive Systems Design: A New Educational Technology. Edited by C. M. Reigeluth, B. H. Banathy, and J. R. Olson. IX, 437 pages. 1993. *(AET)*

Vol. 96: New Directions in Educational Technology. Edited by E. Scanlon and T. O'Shea. VIII, 251 pages. 1992. *(AET)*

Vol. 97: Advanced Models of Cognition for Medical Training and Practice. Edited by D. A. Evans and V. L. Patel. XI, 372 pages. 1992. *(AET)*

Vol. 98: Medical Images: Formation, Handling and Evaluation. Edited by A. E. Todd-Pokropek and M. A. Viergever. IX, 700 pages. 1992.

Vol. 99: Multisensor Fusion for Computer Vision. Edited by J. K. Aggarwal. XI, 456 pages. 1993. *(ROB)*

Vol. 100: Communication from an Artificial Intelligence Perspective. Theoretical and Applied Issues. Edited by A. Ortony, J. Slack and O. Stock. XII, 260 pages. 1992.

Vol. 101: Recent Developments in Decision Support Systems. Edited by C. W. Holsapple and A. B. Whinston. XI, 618 pages. 1993.

Vol. 102: Robots and Biological Systems: Towards a New Bionics? Edited by P. Dario, G. Sandini and P. Aebischer. XII, 786 pages. 1993.

Vol. 103: Parallel Computing on Distributed Memory Multiprocessors. Edited by F. Özgüner and F. Erçal. VIII, 332 pages. 1993.

Vol. 104: Instructional Models in Computer-Based Learning Environments. Edited by S. Dijkstra, H. P. M. Krammer and J. J. G. van Merriënboer. X, 510 pages. 1993. *(AET)*

Vol. 105: Designing Environments for Constructive Learning. Edited by T. M. Duffy, J. Lowyck and D. H. Jonassen. VIII, 374 pages. 1993. *(AET)*

Vol. 106: Software for Parallel Computation. Edited by J. S. Kowalik and L. Grandinetti. IX, 363 pages. 1993.

Vol. 107: Advanced Educational Technologies for Mathematics and Science. Edited by D. L. Ferguson. XII, 749 pages. 1993. *(AET)*

Vol. 108: Concurrent Engineering: Tools and Technologies for Mechanical System Design. Edited by E. J. Haug. XIII, 998 pages. 1993.

Vol. 109: Advanced Educational Technology in Technology Education. Edited by A. Gordon, M. Hacker and M. de Vries. VIII, 253 pages. 1993. *(AET)*

Vol. 110: Verification and Validation of Complex Systems: Human Factors Issues. Edited by J. A. Wise, V. D. Hopkin and P. Stager. XIII, 704 pages. 1993.

Vol. 111: Cognitive Models and Intelligent Environments for Learning Programming. Edited by E. Lemut, B. du Boulay and G. Dettori. VIII, 305 pages. 1993. *(AET)*

Vol. 112: Item Banking: Interactive Testing and Self-Assessment. Edited by D. A. Leclercq and J. E. Bruno. VIII, 261 pages. 1993. *(AET)*

Vol. 113: Interactive Learning Technology for the Deaf. Edited by B. A. G. Elsendoorn and F. Coninx. XIII, 285 pages. 1993. *(AET)*

NATO ASI Series F

Including Special Programmes on Sensory Systems for Robotic Control (ROB) and on Advanced Educational Technology (AET)

Vol. 114: Intelligent Systems: Safety, Reliability and Maintainability Issues. Edited by O. Kaynak, G. Honderd and E. Grant. XI, 340 pages. 1993.

Vol. 115: Learning Electricity and Electronics with Advanced Educational Technology. Edited by M. Caillot. VII, 329 pages. 1993. *(AET)*

Vol. 116: Control Technology in Elementary Education. Edited by B. Denis. IX, 311 pages. 1993. *(AET)*

Vol. 117: Intelligent Learning Environments: The Case of Geometry. Edited by J.-M. Laborde. VIII, 267 pages. 1996. *(AET)*

Vol. 118: Program Design Calculi. Edited by M. Broy. VIII, 409 pages. 1993.

Vol. 119: Automating Instructional Design, Development, and Delivery. Edited by. R. D. Tennyson. VIII, 266 pages. 1994. *(AET)*

Vol. 120: Reliability and Safety Assessment of Dynamic Process Systems. Edited by T. Aldemir, N. O. Siu, A. Mosleh, P. C. Cacciabue and B. G. Göktepe. X, 242 pages. 1994.

Vol. 121: Learning from Computers: Mathematics Education and Technology. Edited by C. Keitel and K. Ruthven. XIII, 332 pages. 1993. *(AET)*

Vol. 122: Simulation-Based Experiential Learning. Edited by D. M. Towne, T. de Jong and H. Spada. XIV, 274 pages. 1993. *(AET)*

Vol. 123: User-Centred Requirements for Software Engineering Environments. Edited by D. J. Gilmore, R. L. Winder and F. Détienne. VII, 377 pages. 1994.

Vol. 124: Fundamentals in Handwriting Recognition. Edited by S. Impedovo. IX, 496 pages. 1994.

Vol. 125: Student Modelling: The Key to Individualized Knowledge-Based Instruction. Edited by J. E. Greer and G. I. McCalla. X, 383 pages. 1994. *(AET)*

Vol. 126: Shape in Picture. Mathematical Description of Shape in Grey-level Images. Edited by Y.-L. O, A. Toet, D. Foster, H. J. A. M. Heijmans and P. Meer. XI, 676 pages. 1994.

Vol. 127: Real Time Computing. Edited by W. A. Halang and A. D. Stoyenko. XXII, 762 pages. 1994.

Vol. 128: Computer Supported Collaborative Learning. Edited by C. O'Malley. X, 303 pages. 1994. *(AET)*

Vol. 129: Human-Machine Communication for Educational Systems Design. Edited by M. D. Brouwer-Janse and T. L. Harrington. X, 342 pages. 1994. *(AET)*

Vol. 130: Advances in Object-Oriented Database Systems. Edited by A. Dogac, M. T. Özsu, A. Biliris and T. Sellis. XI, 515 pages. 1994.

Vol. 131: Constraint Programming. Edited by B. Mayoh, E. Tyugu and J. Penjam. VII, 452 pages. 1994.

Vol. 132: Mathematical Modelling Courses for Engineering Education. Edited by Y. Ersoy and A. O. Moscardini. X, 246 pages. 1994. *(AET)*

Vol. 133: Collaborative Dialogue Technologies in Distance Learning. Edited by M. F. Verdejo and S. A. Cerri. XIV, 296 pages. 1994. *(AET)*

Vol. 134: Computer Integrated Production Systems and Organizations. The Human-Centred Approach. Edited by F. Schmid, S. Evans, A. W. S. Ainger and R. J. Grieve. X, 347 pages. 1994.

Vol. 135: Technology Education in School and Industry. Emerging Didactics for Human Resource Development. Edited by D. Blandow and M. J. Dyrenfurth. XI, 367 pages. 1994. *(AET)*

Vol. 136: From Statistics to Neural Networks. Theory and Pattern Recognition Applications. Edited by V. Cherkassky, J. H. Friedman and H. Wechsler. XII, 394 pages. 1994.

NATO ASI Series F

Including Special Programmes on Sensory Systems for Robotic Control (ROB) and on Advanced Educational Technology (AET)

Vol. 137: Technology-Based Learning Environments. Psychological and Educational Foundations. Edited by S. Vosniadou, E. De Corte and H. Mandl. X, 302 pages. 1994. *(AET)*

Vol. 138: Exploiting Mental Imagery with Computers in Mathematics Education. Edited by R. Sutherland and J. Mason. VIII, 326 pages. 1995. *(AET)*

Vol. 139: Proof and Computation. Edited by H. Schwichtenberg. VII, 470 pages. 1995.

Vol. 140: Automating Instructional Design: Computer-Based Development and Delivery Tools. Edited by R. D. Tennyson and A. E. Barron. IX, 618 pages. 1995. *(AET)*

Vol. 141: Organizational Learning and Technological Change. Edited by C. Zucchermaglio, S. Bagnara and S. U. Stucky. X, 368 pages. 1995. *(AET)*

Vol. 142: Dialogue and Instruction. Modeling Interaction in Intelligent Tutoring Systems. Edited by R.-J. Beun, M. Baker and M. Reiner. IX, 368 pages. 1995. *(AET)*

Vol. 144: The Biology and Technology of Intelligent Autonomous Agents. Edited by Luc Steels. VIII, 517 pages. 1995.

Vol. 145: Advanced Educational Technology: Research Issues and Future Potential. Edited by T. T. Liao. VIII, 219 pages. 1996. *(AET)*

Vol. 146: Computers and Exploratory Learning. Edited by A. A. diSessa, C. Hoyles and R. Noss. VIII, 482 pages. 1995. *(AET)*

Vol. 147: Speech Recognition and Coding. New Advances and Trends. Edited by A. J. Rubio Ayuso and J. M. López Soler. XI, 505 pages. 1995.

Vol. 148: Knowledge Acquisition, Organization, and Use in Biology. Edited by K. M. Fisher and M. R. Kibby. X, 246 pages. 1996. *(AET)*

Vol. 149: Emergent Computing Methods in Engineering Design. Applications of Genetic Algorithms and Neural Networks. Edited by D.E. Grierson and P. Hajela. VIII, 350 pages. 1996.

Vol. 152: Deductive Program Design. Edited by M. Broy. IX, 467 pages. 1996.

Vol. 153: Identification, Adaptation, Learning. Edited by S. Bittanti and G. Picci. XIV, 553 pages. 1996.

Vol. 154: Reliability and Maintenance of Complex Systems. Edited by S. Özekici. XI, 589 pages. 1996.

NATO ASI Series F

Springer-Verlag and the Environment

We at Springer-Verlag firmly believe that an international science publisher has a special obligation to the environment, and our corporate policies consistently reflect this conviction.

We also expect our business partners – paper mills, printers, packaging manufacturers, etc. – to commit themselves to using environmentally friendly materials and production processes.

The paper in this book is made from low- or no-chlorine pulp and is acid free, in conformance with international standards for paper permanency.